Food in China

A Cultural and Historical Inquiry

Food in China

A Cultural and Historical Inquiry

Frederick J. Simoons

Maps by
Mary Beth Cunha

CRC Press
Boca Raton   Ann Arbor   Boston

Library of Congress Cataloging-in-Publication Data

Simoons, Frederick J.
Food in China, a cultural and historical inquiry / Frederick J. Simoons : maps by Mary Beth Cunha.
p. cm.
Includes bibliographical references.
ISBN 0-8493-8804-X
1. Food Habits--China--History. 2. Dinners and dining--China--History.
3. Cookery, Chinese--History. 4. China--Social life and customs. I. Title.
Gt2853.C6S56 1991
394.1′2′0951--dc20 89-20356
CIP

Developer: Telford Press

This book represents information obtained from authentic and highly regarded sources. Reprinted material is quoted with permission, and sources are indicated. A wide variety of references are listed. Every reasonable effort has been made to give reliable data and information, but the authors and the publisher cannot assume responsibility for the validity of all materials or for the consequences of their use.

Direct all inquiries to CRC Press, Inc., 2000 Corporate Blvd., N.W., Boca Raton, Florida, 33431.

International Standard Book Number 0-8493-8804-X

Library of Congress Card Number 89-20356

Printed in the United States

This book is dedicated to Andrew H. Clark and William L. Thomas, Jr., who started me on my academic path; Glenn T. Trewartha and Jonathan D. Sauer, whose scholarly example has been a constant inspiration; James J. Parsons and Carl O. Sauer, who led me into the realm of culture history; and Norman Kretchmer, who urged me beyond.

CONTENTS

	List of Illustrations	xv
	List of Tables	xvii
	Map of the Provinces of China	xviii
	Dynasties of China	xix
	Preface	xxi
	Bibliographic Note	xxv
	INTRODUCTION	1
1	FOOD IN CHINESE THOUGHT AND CULTURE	13
	The Central Role of Food in Chinese Life	14
	Chinese Perceptions of Food, Health, and Disease	15
	Food, Eating, and Social Status	26
	Food for Deities and The Ancestors	26
	Vegetarianism in China	31
	Differences among Buddhists, Taoists, and Confucianists in Early China	31
	Buddhism and the Struggle for Vegetarianism	33
	Vegetarianism in Traditional China	34
	Vegetarian Cuisine in China	37
2	MAJOR REGIONAL CUISINES OF CHINA	43
	Northern Cuisine	45
	Eastern Cuisine	48
	Western Cuisine	51
	Southern Cuisine	54
	FOODS UTILIZED	61
3	CEREALS AND PULSES	63
	Importance of Cereals and Pulses	63
	Major Cereals and Pulses of Eurasian Origin	66
	Rice	66
	Soybeans	70
	Common Millet and Foxtail Millet	72
	Wheat and Barley	73
	Broad Bean and Pea	74
	Sorghum: An Early Arrival from Sub-Saharan Africa	75
	Maize: America's Cereal Contribution	76

	Minor Cereals and Pulses	78
	Buckwheat, Rye, and Oats	78
	Barnyard Millet	80
	Job's Tears	81
	Minor Pulses	83
	Preparation and Consumption of Cereals and Pulses	84
4	ROOTS AND TUBERS	99
	Historical Sketch	99
	Root Crops of Eurasian Origin	102
	Yam	102
	Taro	104
	Radish	107
	Carrot and Turnip	108
	Aquatic Roots and Tubers	109
	Old World Arrowhead	110
	Chinese Water Chestnut	111
	Lotus Root	112
	Minor Roots and Tubers: Great Burdock, Kudzu, and Others	116
	Root Crops of American Origin	122
	Sweet Potato	122
	White Potato	124
	Manioc	125
	American Yam Bean	126
	Notes on Other American Root Crops	128
	Sago: Another Starchy Food	128
5	OTHER CULTIVATED VEGETABLES	133
	Cabbages and Related Plants	134
	Mallow: An Early Competitor	137
	Other Green Leafy Vegetables	138
	Spinach	139
	Water Spinach	140
	Amaranth	141
	Malabar Nightshade	143
	Spinach Beet and Swiss Chard	144
	Watercress	145
	Lettuce	146
	Water Shield	148
	Alfalfa	149
	Sow Thistle	149

Some Additional Greens 150
Squashes, Gourds, Cucumbers, and Other Cucurbits 150
Onions and Their Relatives 156
Fungi 159
Lily Bulbs 162
Bamboo Shoots 162
Wild Rice 165
Celery 166
Water Drop-wort 167
Tomato 167
Eggplant 169
Okra 170
Asparagus 171

6 SEAWEEDS AND OTHER ALGAE 179

Their Importance in East Asia 179
Types of Seaweed Eaten 180
A Note on Edible Microscopic Algae in Traditional China 182
Beginnings of Seaweed Cultivation in China 182
Marine Algae in Chinese Diet 183
Marine Algae and Human Nutrition 184
Minerals 184
Iodine 184
Calcium 184
Iron 185
Vitamins 185
Protein 185
Carbohydrates 186
Medicinal Uses 186

7 FRUITS 191

Importance of Fruit in Chinese Agriculture and Diet 192
Kwangtung as a Producer of Fruit 193
Fruits of Chinese or Southern Asian Origin 194
Citrus Fruits 194
History and Variety 194
Sour Oranges, Sweet Oranges, and Mandarin Oranges 195
Pummelo 198
Citron 200
Its Early Cultivation in India, the Near East, and Mediterranean Area 200
Introduction of Citron to China; Its Uses There 201

Lemon and Lime 203
Kumquat, Calamondin, Trifoliate Orange, and Ichang Orange 204
Wampee 205
Litchi 206
Longan 210
Loquat 212
Persimmon and Its Relatives 213
Background 213
The Persimmon in China 213
The Date-Plum in China 216
Peach, Apricot, and Chinese Plums and Cherries 217
Peach 217
Apricot 220
Chinese Plums and Cherries 221
Chinese Jujube 223
Chinese Olive 225
Some Lesser Fruits of Chinese Origin 226
Chinese Quince 226
Chinese Haw 227
Chinese Gooseberry or Kiwifruit 228
"Human Face" 229
Chinese Strawberry 229
Japanese Raisin Tree 230
Chinese Mulberry 230
Additional Fruits from Tropical Asia 232
Banana and Plantain 232
Carambola 235
Myrobalans 237
Other Fruits of Tropical Asian Origin 238
Some Introductions from the Near East, Europe, or Africa 241
Common Pear and Apple 241
Pear 241
Apple 243
Melons 244
Pomegranate 245
Date 247
Fig 248
Olive 250
Grape 251
Fruits from the New World 254

8 EDIBLE NUTS, NUT-LIKE FRUITS, AND SEEDS 267

Nuts and Nut-like Fruits from Species Domesticated in China 267
Chinese Chestnut 267
Chinese Almond 269
Chinese Hazelnuts 270
Ginkgo Nut 271
Ling Nut 274
Southeast Asia's Main Contribution: The Coconut 276
Nuts from Central Asia and the Near East 278
Persian or English Walnut 278
Pistachio 279
The Peanut: A Gift from America 280
Other Edible Nuts and Seeds Consumed by the Chinese 282
From Trees and Shrubs 282
From Other Cultivated Plants 284
Oil Seeds 286
A Few Conclusions 287

9 FLESH AND FOWL 293

Importance in Diet 293
Pork 295
Chicken and Duck 297
Beef and Mutton 301
Minor Forms of Flesh and Fowl 305
Goose 305
Pigeon 306
Muscovy Duck and Turkey 308
Dog 309
Cat 315
Snakes 317
Rats 320
Insects 321
Still Other Animals 326

10 FISH, TURTLES, AND TORTOISES 337

Fish and Fishing 337
Their Importance 337
Fish Consumed 338
Methods of Catching Fish 339
Range of Fishing Methods 339

Two Unusual Methods of Fishing 340
Fishing with Cormorants 340
Fishing with Otters 342
Fish Cultivation in Ponds 343
Oysters and Oyster Farming 346
Fermented Fish Products 348
Fish in Nutrition and Chinese Food Beliefs 349
Turtles and Tortoises 350
Symbolism and Early Role in Divination 350
Turtles as Food 352
Fresh-Water Forms 352
Sea Turtles 354

11 EGGS 361

Eggs of Fowl 361
In Ritual and Belief 361
Types and Amounts of Fowl Eggs Used as Food 362
Food Uses of Fresh Eggs 363
Preserved Eggs: Their Processing and Uses 363
Nutritional Value of Eggs 365
Turtle Eggs 365

12 SPICES AND FLAVORINGS 369

Ginger 370
Garlic and Vinegar 374
Salt 375
Honey, Sugar, and Other Sweeteners 376
Historical Notes 376
Sugar Intake in Modern China 381
Fagara: Chinese or Szechwan Pepper 382
True Pepper 383
Capsicum Peppers 385
Star Anise, Bastard Star Anise, and Anise 386
Cinnamon 390
Cinnamon in China 390
Ancient Sources of Cinnamon Reaching the Mediterranean and Near East 392
Turmeric, Saffron, and Zedoary 396
Cardamom 400
Clove and Nutmeg 402
Coriander 405
Sweet Basil 406

Golden Needles 409
Ginseng 410

13 FOUR CHINESE DELICACIES 427

Bird's Nest 427
Shark's Fin 431
Bear's Paw 433
Bêche-de-mer 434

14 BEVERAGES, INCLUDING DAIRY PRODUCTS 439

Introduction 439
Tea 440
Its History in China 440
Its Use in Traditional Times 443
Alcoholic Beverages 448
Problems with Terminology 448
Alcoholic Beverages in Antiquity 448
Grain Wines 448
Grape Wine, Mead, Toddy, and Koumiss 449
Distilled Liquors 449
Alcoholic Beverages in Traditional China 450
Principal Kinds 450
Use of Alcoholic Beverages 452
Milk and Dairy Products 454
Origin and Spread of Dairying in the Old World 454
Hypotheses Advanced for the Failure of Dairying and Use of Animal Milk to be Accepted Universally 456
History of Dairying in China 458
Dairying and Dairy Products in Modern China 460
Nutritional and Medical Perspectives 462

15 NOTES ON FOOD IN NUTRITION AND HEALTH IN TRADITIONAL CHINA 469

Parasitic and Infectious Diseases 470
Foods, Nutrient Intake, and Nutritional Deficiency 475
Calories: Adequacy and Sources 475
Protein and Protein Deficiency 476
Vitamin A and Deficiency States 478
Thiamine Deficiency and Beriberi 479
Riboflavin Intake and Deficiency 481

Niacin Deficiency and Pellagra 482
Vitamin C Deficiency and Scurvy 483
Calcium, Vitamin D, Rickets, and Osteomalacia 487
Iron and Iron Deficiency 490
Iodine Deficiency and Goiter 491
Selenium Deficiency in Keshan Disease 493
Food, Nutrition, and Care of Those at Greatest Risk: Mothers and Children 494
Pregnancy and Abortion 494
Birth and Birth Weights 496
Disposal of Unwanted Infants 497
Confinement and the Feeding of Infants and Nursing Mothers 500
Weaning 503
Mortality among Infants and Mothers 504
Growth Rates of Chinese Infants and Children 504
Birth Rates, Death Rates, and Population Increase 506
Chinese Customs, Good Health, and Nutritional Balance 507

16 CONCLUSION 513

REFERENCES 517

1. The Yüeh and early China 2
2. Ancient Chinese archeological sites and cultures mentioned in the text 4
3. Southeast Asia: Places mentioned in the text 6
4. The Pearl River Delta and nearby areas 8
5. China: Modern places and peoples mentioned in the text 9
6. Bicyclist in marketplace with beans, eggplant, and stem lettuce 16
7. Spices and other foods in market stalls 16
8. Fish for sale by market vendors 17
9. Fresh eggs and candling device in market 17
10. Marketplace containers for live fish and other aquatic animals 19
11. Live eels for sale 19
12. Live salamanders 20
13. Snails 20
14. Fresh fish for sale 21
15. Newly arrived cages of chickens 21
16. Stall of a poultry vendor in an indoor market 22
17. The Kitchen God 27
18. Offering food to the ancestors at the ancestral tablet 29
19. Main regional cuisines of China 46
20. The two main agricultural regions of China 65
21. Crops providing five percent or more of total calories in each area of China, in order of importance 66
22. China's agricultural areas 67
23. Bundles of yard-long beans (*Vigna unguiculata* subspecies *sesquipedalis*) in marketplace 85
24. Baked wheat bread being removed from oven among the Uighur of Turfan 90
25. Tuber of the yam *Dioscorea alata* 103
26. Taro (*Colocasia antiquorum*) 105
27. Taro tuber 106
28. Old World arrowhead (*Sagittaria sinensis*) 110
29. Flower, seed capsule, and a seed of the lotus (*Nelumbo nucifera*) 113
30. Chinese statuette of an aspiring Buddha with lotus flower 114
31. Lotus rhizome 115
32. Great burdock (*Arctium lappa*) 116
33. Root tuber of kudzu (*Pueraria lobata*) 117
34. Chinese artichoke (*Stachys affinis*) 120
35. *Amorphophallus rivieri* 121
36. Yam bean (*Pachyrrhizus erosus*) 126
37. Yam bean tuber 127
38. Mustard cabbage (Swatow mustard) (*Brassica juncea* var. *rugosa*) 135
39. Chinese kale (*Brassica alboglabra*) 136

40. Celery cabbage (*B. pekinensis* var. *cylindrica*) 137
41. Chinese white cabbage (*Brassica chinensis* var. *chinensis*) 138
42. Chinese spinach (*Amaranthus gangeticus*) 142
43. Ceylon or Malabar spinach (*Basella rubra*) 144
44. Pile of stem or asparagus lettuce (*Lactuca sativa* var. *angustana*) 147
45. Wax gourd (*Benincasa hispida*) 151
46. Bottle gourd (*Lagenaria siceraria*) 152
47. Bitter cucumber (*Momordica charantia*) 153
48. Snake gourd (*Trichosanthes cucumerina*) 154
49. Angled luffa (*Luffa acutangula*) 155
50. Smooth luffa (*Luffa cylindrica*) 156
51. Chinese padi-straw mushrooms (*Volvariella volvacea*) 161
52. Water drop-wort (*Oenanthe stolonifera*) 168
53. Chinese eggplant in market 170
54. Buddha's-hand citron (*Citrus medica* var. *sarcodactylis*) 202
55. Litchi (*Nephelium litchi*) 207
56. Litchi fruit on tray 208
57. Longan (*Euphoria longana*) 211
58. Chinese jujube (*Zizyphus sativa*) 223
59. Japanese raisin tree (*Hovenia dulcis*) 231
60. Fruit of the carambola (*Averrhoa carambola*) 236
61. Jackfruit (*Artocarpus heterophyllus*) 240
62. Leaf, stem, and nut of the ginkgo (*Ginkgo biloba*) 272
63. Ling nut: The nut-like fruit of *Trapa bicornis* 275
64. Men carrying pigs 296
65. Dog carcass in Canton marketplace 311
66. Cages of snakes for sale at a shop in Canton 318
67. Merchant proclaiming the virtues of a snake to potential customers 319
68. Cicada nymph infected with the fungus *Cordyceps sobolifera* 324
69. Bird catchers 328
70. Fisherman with cormorants at Kweilin 341
71. Oyster fishing vessels and oyster shells on beach at Lau Fau Shan, New Territories, Hong Kong 347
72. Carved turtle supporting memorial tablet, Ming dynasty 351
73. Soft-shelled or mud turtle (*Amyda sinensis*) 353
74. Man spearing fresh-water turtles 354
75. Green turtle (*Chelonia mydas*) 356
76. Chinese or Star anise 388
77. Ginseng (*Panax ginseng*) and mandrake (*Mandragora officinarum*) 411
78. Tea sources and trade mentioned in the text 442
79. Calcium intake per adult-male unit per day 488

LIST OF TABLES

1. Sources of calories in rural China and the United States 64
2. Percentage of calories supplied by various important crops 68
3. Animal food calorie intake 294
4. Nutritional value of some insects and insect larvae used as food in East Asia or elsewhere 327
5. Egg output in relation to human population in China 363
6. Comparison of sources and amounts of protein in the average rural Chinese diet with diet in the United States 476
7. Daily intake of protein in China per adult-male unit and percentage of protein supplied by different classes of food 477
8. Loss of thiamine in preparation of rice by traditional means 481
9. Some good plant sources of vitamin C in China 485
10. Loss of vitamin C: Comparison between boiling vegetables in water and cooking in sizzling hot oil 486
11. Effect of duration of sautéing in boiling oil on the vitamin C content of vegetables 486
12. Vitamin C content of four vegetable waters 487
13. Vitamin C content of pickled vegetables 487
14. Foods prescribed and proscribed throughout pregnancy among Chinese women in Australia 495
15. Regional differences in birth weight of Chinese babies 497
16. Infant mortality in various countries 505
17. Average height of Chinese adults in the 1920's 505

Map of the provinces of China

DYNASTIES OF CHINA

Hsia	21st-18th century B.C.
Shang	18th-12th century B.C.
Chou	12th century B.C.-221 B.C.
Ch'in	221-207 B.C.
Han	206 B.C.-A.D. 220
Three Kingdoms	220-265
Chin	265-420
Six Dynasties	420-589
Sui	581-618
T'ang	618-907
Five Dynasties	907-960
Sung	960-1279
Yüan	1271-1368
Ming	1368-1644
Ch'ing	1644-1911

(from K. C. Chang, 1977)

This book is a study of Chinese food from a cultural and historical perspective. Its focus is on traditional China of the nineteenth and early twentieth centuries, before establishment of the People's Republic. It is intended to provide comprehensive information on a broad range of Chinese food plants and animals for general readers as well as for specialists whose interests have led them to questions relating to the food of China. Despite the publication, in the last dozen years, of two excellent works dealing with Chinese food in the context of dynasties or other historical periods, information on individual food plants and animals is often fragmentary and difficult to find. This book attempts to alleviate that problem by dealing, in succession, with those plants and animals, each of which has its own story and role in Chinese culture and diet.

When I began this study of food and China, my primary interest was in identifying, and gaining an understanding of, the broad range of foods used in Kwangtung, the southeast coastal province that lies behind Hong Kong and whose capital is Canton. My secondary interests were in determining how the foods in question were regarded in Chinese views of health and disease, and what their nutritional value might be. As I proceeded, however, I became intrigued by the rich diversity of Chinese food and food habits not just in Kwangtung, but in the entire nation. I was also struck by the fact that little had been done to assemble information on the plants and animals eaten by the Chinese people. One suspects that such a task was not attempted previously in part because relevant information, while abundant, is published in widely-scattered sources and presented from such different points of view as to discourage attempts at synthesis. Despite this, I decided to make such an attempt, and to fit the information into a single perspective, one befitting a geographer of culture historical persuasion. One will find herein a special concern with origins, with how things came to be: with time and place of food-plant, and animal, domestication; with spread of plants and animals to China or from China to the outside world; and with other historical questions about the foods used by the Chinese people.

In approaching the question of China's food, my first step was to assemble the material accessible in translations of Chinese works and in the extensive literature in Western languages. Then I went to China to observe at first hand and to understand better, and, in writing the text that follows, those observations were joined with the library material. The resulting book, though written with Canton and the southeast as points of departure, now embraces all of China, and is directed primarily to persons, whether scholars or others, who are unfamiliar with China. It is intended to fill one gap in the literature: the identity and history of the food plants and animals of China, and the description of how they are used.

This book provides a different view of Chinese food from those presented by other recent scholarly books on the subject. First it deals with foods by kind, not by historical period, and, as a result, provides much more comprehensive information on individual food plants and animals, their ancestry and use. Second it incorporates far more ethno-botanical and ethno-zoological background than normal. Third, it views Chinese food from the vantage point of the nation's southeast, richest of China's culinary regions, and not from the more usual perspective, that of the North Chinese hearth of high civilization. Fourth, it ranges more frequently beyond China's borders, especially to Southeast Asia and India, in seeking origins of plants and animals and influences on China's food scene. Fifth, it makes a greater effort to present information on nutrition and health. And sixth, it draws more fully on and relates to the rich world of English-language Chinese cookbooks, using them not only to cast light on the food and cooking of China, but indicating where one may find recipes for foods being considered. Such cookbooks, though directed toward the culinary needs of Western readers, are invaluable sources, providing insights on China's food scene not to be gained elsewhere in the literature.

I am grateful to Dr. Norman Kretchmer, Director of the Koret Center of Human Nutrition, University of California, San Francisco, for stimulating my interest in the subject of Chinese food. Recognizing the major changes in diet that immigrant groups in San Francisco sometimes undergo, and concerned about their possible adverse nutritional implications for children, Kretchmer organized a series of investigations of immigrant groups in San Francisco. Though he knew that I was not a sinologue, Kretchmer was aware of my long interest in food and food habits in the Old World, and asked that I participate in his study by gathering information on the foods used in the homeland of San Francisco's Chinese, most of whom derive from a single province, Kwangtung. It was through involvement in Kretchmer's study that, for the first time, I came to think seriously about Chinese food, and it was not long before I embarked on this, my own investigation of the food of China as a whole.

I also wish to express my special obligation to the many scholars, both experts on China and others, whose published works provide the data on which this study rests. Those scholars are too numerous to single out individually, but I must acknowledge a special debt to Berthold Laufer, Edward H. Schafer, and Hui-lin Li, whose works are unusually interesting and rich sources on the botany and food history of China. I also want to single out three other works that have been important sources. One is *Land utilization in China*, edited by John Lossing Buck (originally published in 1937), an unparalleled study which presents the results of a survey of farms in twenty-two Chinese provinces in 1929-33. A second is *Food in Chinese culture: Anthropological and historical perspectives* (1977), edited by Kwang-chih Chang, a work outstanding in the insights of its contributors, each of whom focuses on a different historical period, from ancient to modern. One of those contributors, E. N. Anderson, recently authored a book of his own, *The food of China* (1988), a third important source, one written from the perspective of economic, ecological, political, and social history.

I would also single out certain cookbook authors whose works have proved particularly informative and useful: Gloria Bley Miller (1966); Wonona W. and Irving B. Chang and Helene W. and Austin H. Kutscher (1970); Kenneth Lo (1971-79); Buwei Yang Chao (1972); Hsiang Ju Lin and Tsuifeng Lin (1972); Emily Hahn *et al.*, 1973; Grace Zia Chu (1975); Michael Field *et al.* (1975); Florence Lin (1975); Nobuko Sakamoto (1977); Calvin B. T. Lee and Audrey Evans Lee (1980); Ken Hom and Harvey Steiman (1981); Barbara Tropp (1982); Karen Lee and Alaxandra Branyon (1984); and Yan-kit Martin (1984).

I wish to acknowledge the Department of Geography and the Committee on Research of the University of California at Davis for financial support of this research. I wish also to thank Calvin W. Schwabe and Bärbel Leber for editorial suggestions; Amy Lanoux, Jill Earick, and Alan Calhoun for library assistance; the staff of the Department of Geography, including Wendy Rowland, Maria Hope, Sally Akers, and Frances Ramos, for its aid in manuscript preparation and in other ways; and Mary Beth Cunha, Stephen F. Cunha, and Shelby Hall for illustrations. I am also indebted to various publishers for granting copyright permissions, particularly Unwin Hyman for the use of excellent drawings by G. A. C. Herklots from his *Vegetables in South-East Asia*. Last, but not least, I thank my wife Elizabeth for assistance in collecting information in China and in editing and proofing the manuscript, as well as for her patience and encouragement over a several-year period.

Since our vantage point is that of Kwangtung, it is necessary, before launching into a consideration of Chinese food, to provide a sketch, historical and geographical, of the province. This is done in the Introduction which follows, our intent being to provide background to help the reader who is not a China specialist to understand both local conditions and historical ties with North China. In addition, however, we seek to provide for that reader an appreciation of the unusual geographical position of China's southeast, whose provinces of Kwangtung and Fukien dominated China's early sea trade with Southeast Asia. We suspect that it was partly because of such trade contacts that the region came to enjoy a national reputation for its rich culinary diversity.

BIBLIOGRAPHIC NOTE

Because of the enormous amount of English-language material already in the Wade, or Wade-Giles, system of transcribing (romanizing) Chinese words as they sound in the Peking (Mandarin) dialect, we have chosen it as our standard rather than the Pinyin system. For the reader who prefers the Pinyin system, a table of Wade-Giles/Pinyin equivalents may be found in numerous published sources, including *The Times atlas of China* cited below.

For place names outside of China, we have generally followed the spellings given in *The Columbia Lippincott gazetteer of the world* (New York: Columbia University Press, 1962). For place names in China, we follow established English usage for those likely to be familiar to the general English-speaking reader, for example Canton, Peking, Tientsin, Dairen, Soochow, and Amoy. For other place names in China, we have generally followed the Wade-Giles system as given in *The Times atlas of China*, edited by P. J. M. Geelan and D. C. Twitchett (London: Times Books, 1974).

In our Preface, we noted the importance, for our study, of the book *Land utilization in China*, edited by John Lossing Buck. Though certain aspects of that study have been criticized by some later observers, it remains a monumental effort, unmatched by subsequent studies in its scope and sample size. Carried out during the years 1929-33, Buck and his collaborators surveyed a total of 16,786 farms in 168 localities of 22 provinces, and obtained a wealth of information on food and agriculture presented in a book to which we refer commonly as "the 1929-33 village survey," "the 1929-33 survey," or simply "the village survey." In presenting material from Buck's book, we generally use terms as he defines them. Thus our "Wheat Region" and "Rice Region," China's two principal agricultural regions (see our Figure 20), are identical with those delimited by Buck. So are subdivisions of those two regions (see our Figure 22): "Spring Wheat Area," "Winter Wheat-millet Area," "Winter Wheat-kaoliang Area," "Szechwan Rice Area," "Yangtze Rice-wheat Area," "Rice-tea Area," "Southwestern Rice Area," and "Double Cropping Rice Area." Kwangtung, it might be noted, occupies a central geographic position in the Double Cropping Rice Area, and of the villages surveyed there, nearly two-thirds were in Kwangtung.

Introduction

"The Southerners from birth are fond of playing with waves,
They make light of their lives for profits and for gains."

Su Shih (translated by Xu Yuan-zhong)

The hearth of the Shang state, that ancient civilization of China (18th-12th century B.C.), was on loess soils of the north, especially in the middle course of the Yellow River (Figure 1). That region is bounded on the south and southwest by the Tsinling and adjacent mountains, east-west ranges whose peaks rise over 12,000 feet above sea level to separate the arid loess lands of North China and the green, rainy lands of Central and South China. In North China, the early Chinese were subject to recurrent pressures that led them to migrate, commonly in large, organized population movements, to the south. Those pressures were of various sorts, some environmental, some cultural. The North Chinese homeland had modest precipitation, much of it today between 30 to 40 inches annually, an amount similar to that of much of the American Great Plains. There was also considerable rainfall variability, and repeated periods of drought, famine, and pestilence, which provided powerful incentives to migrate. Another pressure was exerted by the northern neighbors of the Chinese, warlike nomads and semi-nomads, such as the Hsiung-nu, who coveted the grasslands of the Chinese and who were tempted, as well, by the prospects for loot. Over the four millennia of Chinese migration to the south, one discerns other major causes as well: floods; political instability and civil war; hostility among neighboring ethnic groups; invasion and defeat by other settled states; prestige and expansion of the empire; and population increase, poverty, and desire for economic betterment. In some cases, migrations were spontaneous, and could involve movement of an entire village or even the people of an entire region. In other cases, migrations were encouraged or decreed by the authorities, and might involve dissidents, criminals, retinues of government appointees, and others.

In their southward migration, the Chinese, a literate people who had the only system of writing in East Asia from the second half of the second millennium B.C. until the second century A.D. (Pulleyblank, 1983: 414-15), faced a similar situation to, but acted very differently from, the Anglo-Americans in their westward expansion across the United States. Whereas the Anglo-Americans largely eliminated or displaced the aboriginal inhabitants, the Chinese commonly settled among, and ultimately absorbed, them. Anglo-American pioneers had a great advantage in arms over American Indians, and even though Indians did obtain guns and ammunition by trade, they remained dependent on the whites for replacements. Such an advantage in arms was not enjoyed by the Chinese, for the aboriginal peoples they faced usually had arms of equal effectiveness or nearly so. The Chinese advantage was mainly in their broader economic base and larger-scale political and military organization. Thus they could

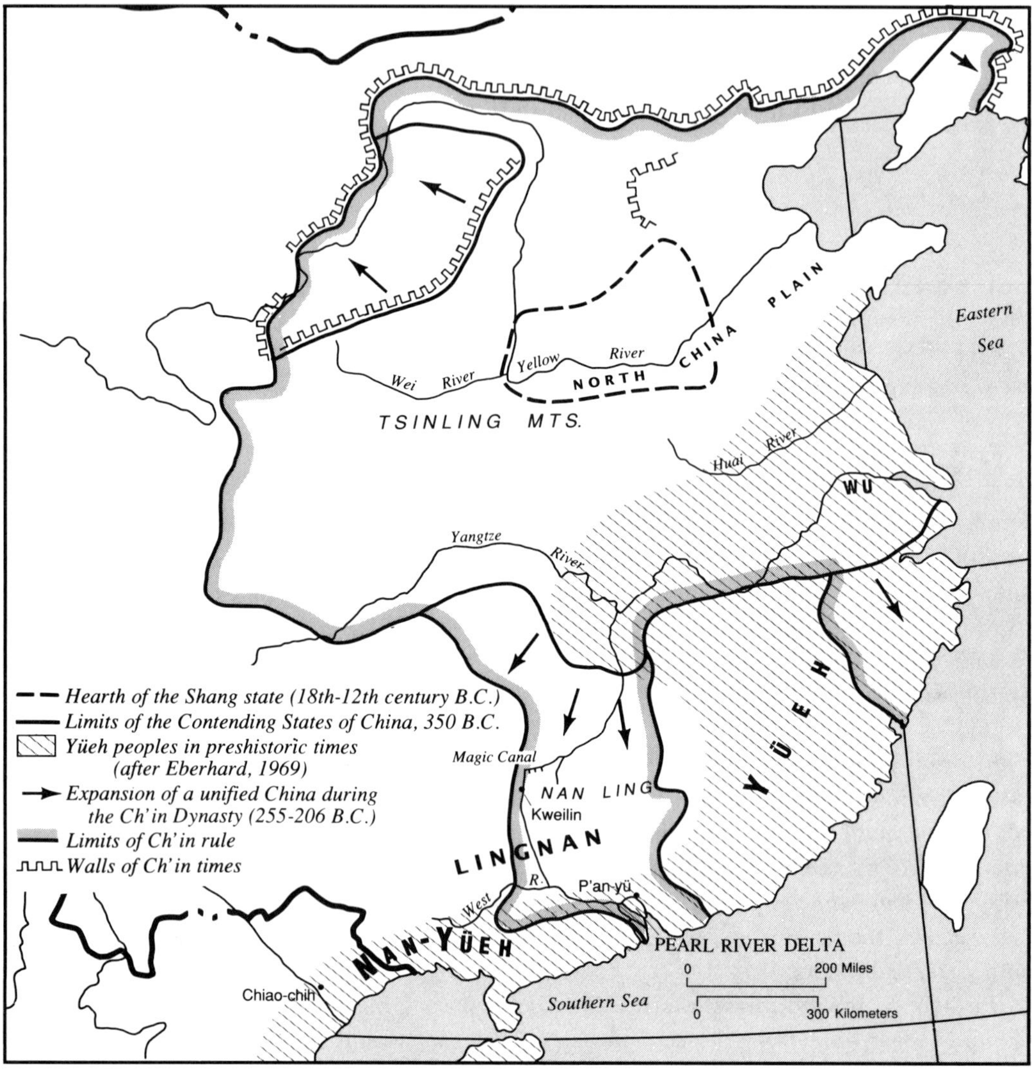

Figure 1. The Yüeh and early China (after Herrmann, 1966, and Eberhard, 1969)

mount an offensive with many more troops, and fight longer campaigns than their aboriginal adversaries. They also possessed the manpower and organization to pacify and rule.

In settling the United States, Anglo-Americans seem to have faced no debilitating diseases new to them. On the contrary, they introduced diseases, such as smallpox, that brought high mortality and undermined the defenses of the American Indians. The opposite was true of the

Chinese in their penetration of the south, for they were a mid-latitude people who entered areas of virulent malaria to which they had little resistance. That disease, moreover, was most severe in river valleys, the very places most favored by the Chinese for settlement (Wiens, 1954: 333-35).

In archeological levels dated 5000 to 3500 B.C. at Ho-mu-tu (Hemudu), a village site south of the Yangtze in coastal Chekiang Province (W. Chen, 1984: 49; B. Lao, 1984; W. Xin, 1984) (Figure 2), there is evidence of a people of aquatic orientation much like that of peoples living along China's southeastern coast today. The inhabitants of Ho-mu-tu were a boat people, fishers, hunters, and farmers, who, with water buffalo and plow, cultivated rice, and perhaps water chestnuts and other aquatic plants. They also kept dogs, as well as pigs and apparently fowl, those mainstays of Chinese animal husbandry today. The rice at Ho-mu-tu is among the earliest found anywhere in the world, and overall, excavations at the site confirm the great antiquity of an aquatic way of life in coastal China.

Already before 500 B.C., Chinese accounts mentioned water-oriented people called "Yüeh," who formed a state centered on what today is Kiangsu, Chekiang, and Kiangsi. In Han Times (206 B.C.-A.D. 220), the name Yüeh was applied to coastal peoples from Chekiang southwards, with "Nan-Yüeh" (Mandarin pronunciation; Vietnamese: "Nam-Viet") or "South-Yüeh" a separate kingdom centered on the Pearl River Delta of Kwangtung but extending southward to northern Vietnam. Peoples with fleets of large war canoes used along inland waterways and the seacoast, the Yüeh, though non-literate, were "civilized" people in Chinese eyes. They were impressive in their subsistence accomplishments and political organization, and noted for their decorated cast bronze drums, symbols of authority used in ceremony and appeals to supernatural beings. In reading of their culture, one discerns essential elements of modern Chinese subsistence in Kwangtung: irrigated rice cultivation; pig, chicken, and dog the main domesticated animals; and a focus on aquatic life and the sea. The Nan-Yüeh lived by farming, fishing, and commerce, with pearls, tortoise shell, rhinoceros horn, fruit such as litchis, and cloth traded to Chinese merchants in exchange for weapons and iron implements (C. W. Bishop, 1922: 35-36; 1938: 421-24; H. L. Lo, 1967b; Eberhard, 1968: 432-33; 1982: 87-88; FitzGerald, 1972: 1; Pirazzoli-t'Serstevens, 1982: 85-86).

The linguistic and ethnic identity of the Yüeh remains a matter of controversy, but the term may have been applied by the Chinese to groups who were not related to one another. Schafer (1967: 13) writes that the Yüeh of Chekiang may have been partly-sinicized Tai who in that area ruled ethnic groups of different affiliations; Eberhard (1968: 432, 436; 1969: 12, 152) proposes that Yüeh culture was formed by a mixing of Tai and Yao cultures; and Pulleyblank (1983: 438-42), who favors a common linguistic and ethnic background among the coastal Yüeh, presents evidence that their language was of Austro-Asiatic form and related to modern Vietnamese. Whatever their identify, some coastal Yüeh, under Chinese pressure, may have migrated southward to Annam from China while others, the "Mountain Yüeh," fled inland (Eberhard, 1968: 432; 1982: 88).

The region that is now Kwangtung remained independent of the Chinese until late in the first millennium B.C. An initial invasion, for reasons of imperial glory and expansion, occurred in 234 B.C. when the Ch'in emperor dispatched a motley army of 500,000 men to subjugate the south. One important Ch'in action was construction of the strategic Lin-ch'ü Canal ("Magic Canal" or "Holy Canal") to connect Yangtze and West River drainage and facilitate supply of the troops. The Han Chinese, successors of the Ch'in, formally annexed the Kwangtung region

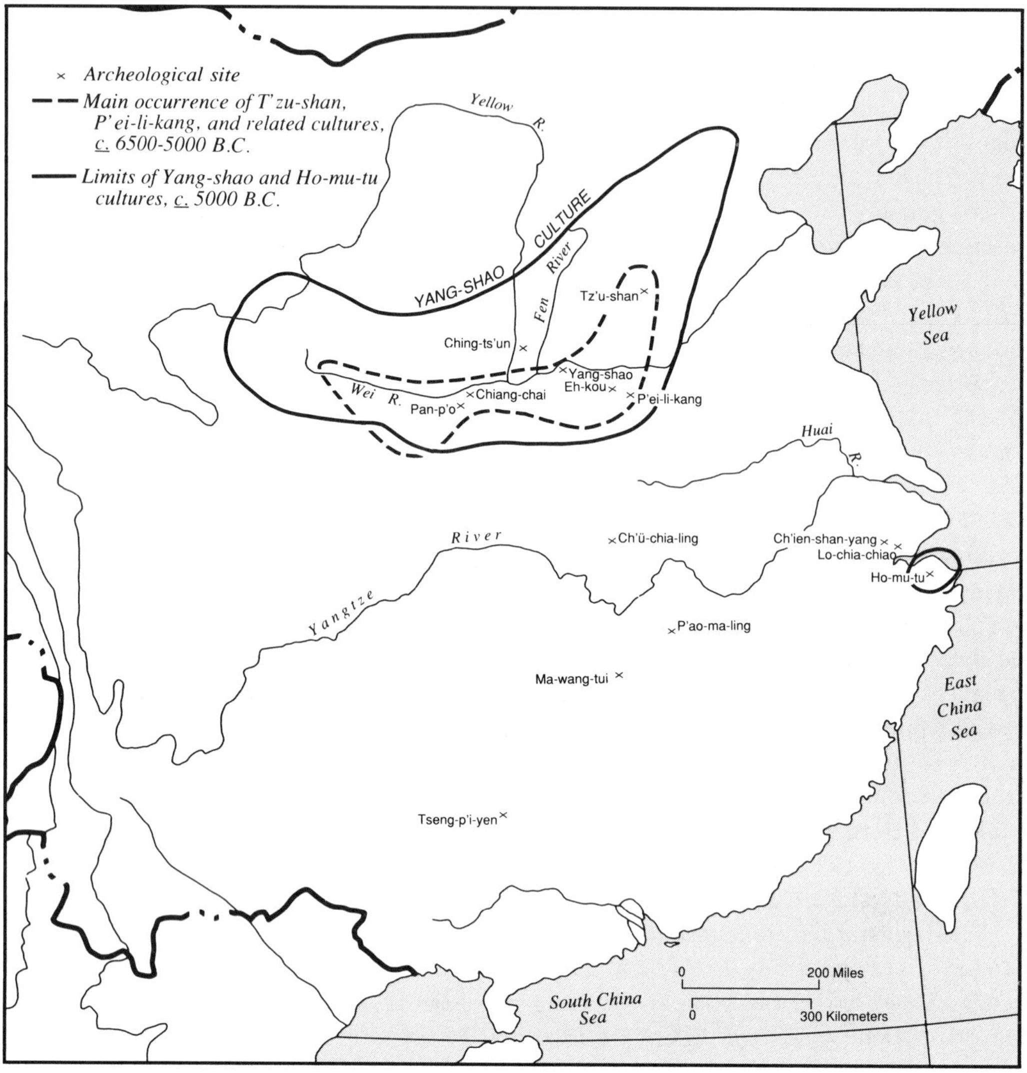

Figure 2. Ancient Chinese archeological sites and cultures mentioned in the text (after K. C. Chang,1986)

in 111 B.C., and in Han times the canal played a major role in gaining access to the deep south (Y. S. Yü, 1967: 29; Cotterell and Cotterell, 1975: 55, 78). It was also important in foreign trade, for it permitted imports from overseas to be shipped by river to the Yangtze Valley from P'an-yü (present-day Canton). P'an-yü in Han times was not only the largest port in coastal South China and a center of the pearl trade, but a capital city and port of entry for goods reaching China by sea from Southeast Asia and beyond (Y. S. Yü, 1967: 29, 32, 178-79). Nevertheless, even after a Chinese presence for many centuries, the region was one of military garrisons thinly spread, of repeated native uprisings requiring large armies to suppress (Wiens, 1954: 132-45, 187-90).[1]

Today Cantonese or Yue (Yüeh) dialects are spoken in Hong Kong, Canton, and the Pearl River Delta westward into southeastern Kwangsi Province. In Kwangtung, the Cantonese compose perhaps two-thirds of the total population of 59.3 million (1982), and in China as a whole they probably account for about 5 percent of the total population. Cantonese is described by Forrest as having developed, over time, in relative independence of influence by Northern Chinese (dialects of the Yangtze Valley and of places to the north), and as being closest among modern Chinese dialects to Ancient Chinese in its basic characteristics. This, he suggests may be due to distance from, and difficulty of contact with, the North Chinese hearth, and possibly to a conservatism deriving partly from Tai linguistic influences (Forrest, 1965: 230-36; Kratochvil, 1968: 16-17; O. K. Y. Hashimoto, 1972: 1-12, 63, Map 4; Y. R. Chao, 1976: 22-25).

Yet in many ways the Cantonese of traditional times were anything but conservative. Indeed they have typically been characterized as progressive, independent, energetic, and enterprising. Cressey (1934: 366-68), writing at a time when environmental determinism was still somewhat in vogue, wondered at the vigor of the Cantonese in view of the enervating climate of their homeland. He attributed that vigor to two probable causes. One was genetic selection during the Chinese migration from the north, which, Cressey suggested, favored capable persons. The second was an unusual exposure to foreign ways, especially through Cantonese emigrants who had returned from abroad with a rich store of knowledge. In this regard, one is reminded of the ancient Egyptians of the Nile Delta who, by stereotype, were clever and sensate, but innovative, in contrast to Upper Egyptians, who were dull, puritanical, and conservative. The similarity between modern Cantonese and ancient delta Egyptians, one might argue, derives from similar geographic situations on the shores of seas that brought, through trade and foreign residents, an unusual exposure to new ideas, products, and technologies from high civilizations abroad. One might argue further that Cantonese innovativeness also derives from association and intermarriage with the Yüeh peoples who preceded them in Kwangtung. The Yüeh, as we have seen, had a long history of seafaring, trade, and willingness to seek new experiences on distant shores, as may have been true of the "Southerners" (people of coastal Chekiang?) referred to by Su Shih the Sung poet and statesman quoted at the beginning of this chapter.

It is true that the Chinese authorities did not in all historical periods look on such sea trade with the same favor as the overland routes that reached the North Chinese heartland from the west; nevertheless there were times, among them those of the Sung and Yüan dynasties, when the merits of the foreign sea trade were appreciated (E. N. Anderson, 1988: 59, 60, 74, 87), and in those times China's southeast was able to play a key role in contacts with Southeast Asia, India, and beyond (Figure 3). One notes, in this regard, Crawfurd's observation (1820, 3: 172) that early in the nineteenth century nearly all Chinese foreign sea trade was carried out from the two maritime provinces of Kwangtung and Fukien.

Kwangtung Province is divided in two by the Tropic of Cancer which passes just north of Canton. Of the states in the continental United States, Florida is perhaps most like Kwangtung in climate, but it has a population density only a fourth that of Kwangtung. Indeed, Kwangtung's total population of 59.3 million (1982) was twice that of all states of the American Deep South together, including Louisiana, Mississippi, Alabama, Georgia, Florida, and South Carolina.

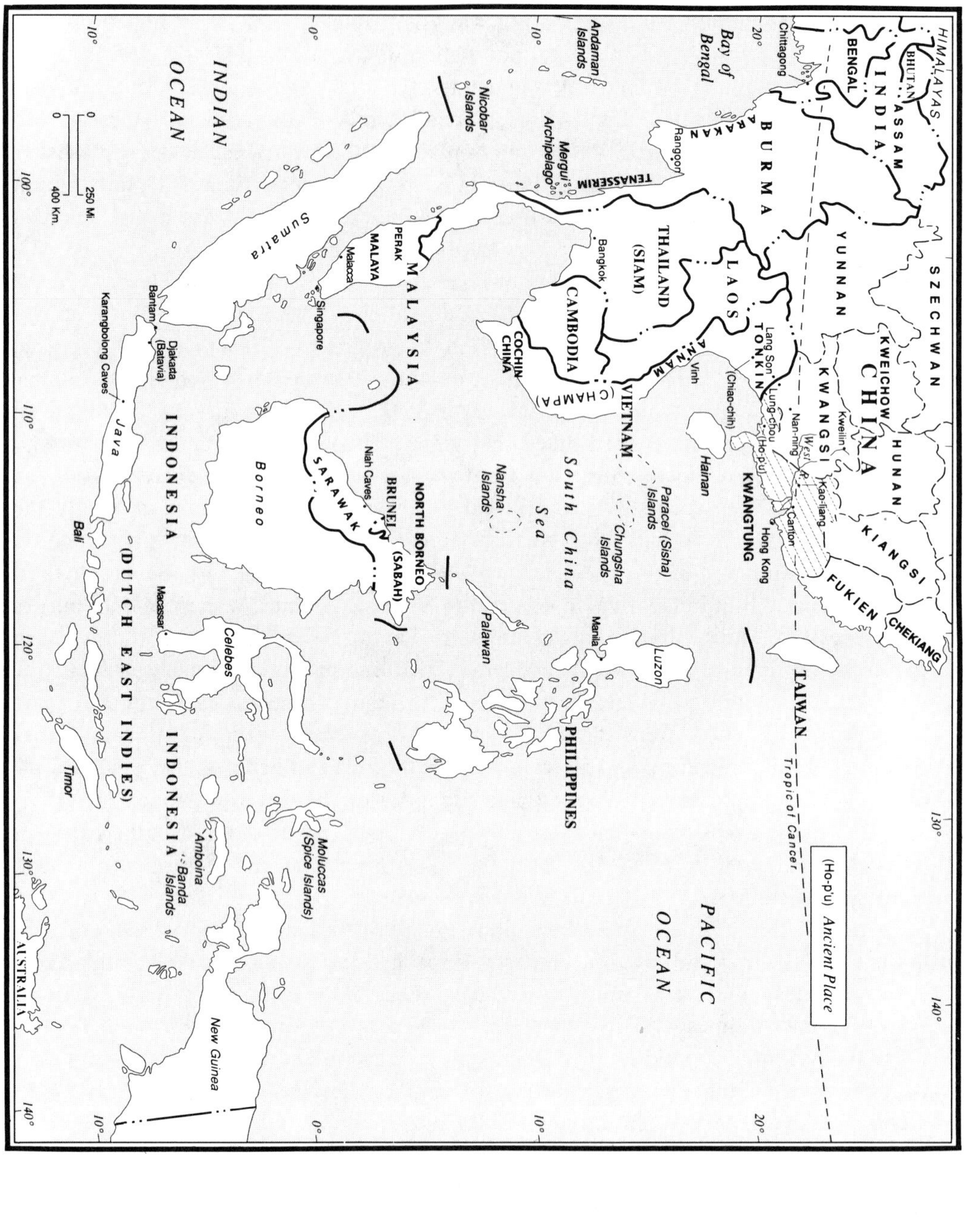

Figure 3. Southeast Asia : Places mentioned in the text

Kwangtung is a coastal province dominated by hills and low mountains. From its northeastern border with Fukien Province, it extends about 500 miles southwestward along the South China Sea. In a usage that dates from T'ang times (A.D. 618-907), it is often grouped with Kwangsi Province, its western neighbor, in a region called "Lingnan" (or "Lingnam"), "South of the Mountain Passes." These passes cross the Nan Ling or Nan Shan, a mountain system which rises almost 6000 feet above sea level. The Nan Ling separates Kwangtung and Kwangsi, on the south, from Hunan and Kweichow, interior provinces to the north.

Nowhere in Kwangtung is one very far from the coast, 50 to 200 miles at most. The coastline, moreover, is one of submergence, in which the sea has advanced on the land to create a rugged shoreline with many sheltered bays. This, together with a shortage of level land suitable for cultivation, has since antiquity led many of Kwangtung's people to focus on the sea. Segmented by hills and mountains, Kwangtung does not have the internal uniformity enjoyed by some places in China, for example those located on the floodplain of a great river such as the Hwang Ho (Yellow River) of North China. Nor is it clearly separated physiographically from nearby areas, as is Szechwan, that mountain-rimmed basin of interior South China. Instead, Kwangtung's hills and mountains merge, sometimes imperceptibly, into those of its neighbors east and west. Only to the south and north are its boundaries set well in physical features, the sea a sharp boundary, the Nan Ling less so. What unity the province enjoys is based on its system of rivers, especially the Hsi Chiang, or West River. The West River is 1250 miles in length, less than half that of China's great rivers, the Yangtze and Yellow. Yet it drains a generally well-watered region in distant areas of Kwangsi, Kweichow, and Yunnan, and is a river of good size and appreciable volume. In most of its course, however, its valley is narrow and limited in the alluvial soils it provides for cultivation. It flows eastward, most of its length within Kwangsi Province, and finally breaks through the mountains east of the city of Wu-chou to enter Kwangtung. Ultimately it reaches central Kwangtung where, with two other important rivers, it forms the Pearl River Delta (Figure 4). The two rivers in question are the Pei Chiang, North River, which drains the south slopes of the Nan Ling, and the Tung Chiang, East River, which drains Kwangtung's eastern hills and mountains.

Because it administered Hainan Island, off its south coast, and disputed islands, sandbars, and reefs much farther to the south (Paracels, Chungshas, Nanshas), Kwangtung's jurisdiction extended well into Southeast Asia. Nevertheless, except for coastal Hainan Island and southern Lei-chou (Luichow) Peninsula (Figure 5) on the mainland, which have a tropical rainy climate, Hainan and Kwangtung have humid subtropical climates, like Florida's, with long, hot, sultry summers, and mild winters with no frost or snow except at higher elevations. The region is subject to hurricanes, an average of seven a year, that sweep in from Southeast Asia during the summer. Almost everywhere rainfall averages over 40 inches annually, and in some places over 80 inches; most of it falls in summer, from June through September, commonly in thunderstorms (M. Y. Nuttonson in C. S. Chen, 1970: 29, 32-36, 39). The natural vegetation is forest of one sort or another, whether evergreen or mixed evergreen tropical or evergreen subtropical, but the area has experienced serious deforestation.

A subtropical and tropical land with a twelve month growing season, Kwangtung, at least in terms of climate, presents rich possibilities for food crop cultivation. Those possibilities, however, are sharply limited by a shortage of level land. Indeed some areas of the province, in the hills and mountains of Hainan and north Kwangtung for example, have 10 percent or less of all land under cultivation, some of it swidden. In contrast, other areas along the coast and in

Figure 4. The Pearl River Delta and nearby areas

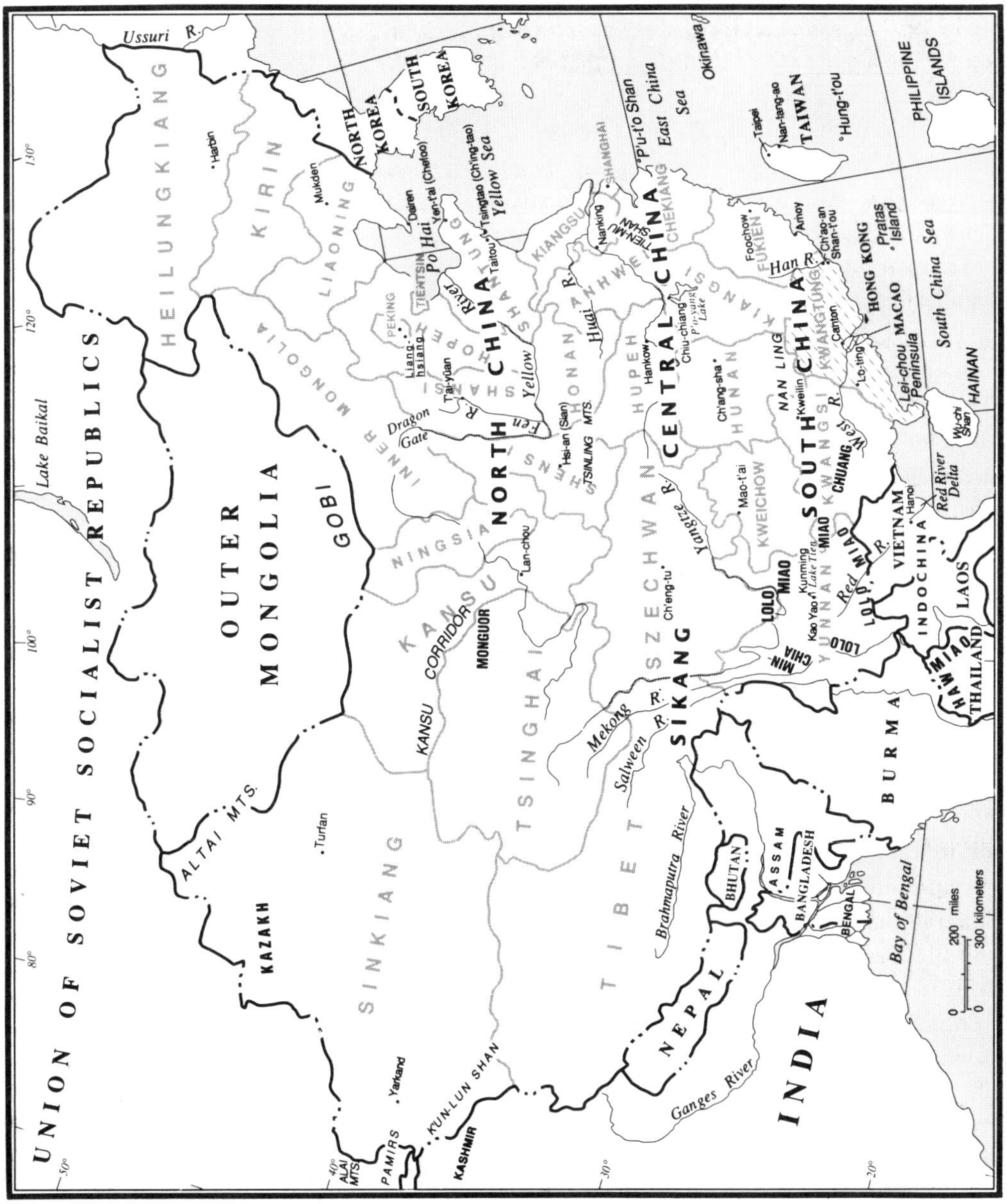

Figure 5. China: Modern places and peoples mentioned in text

deltas such as that of the Han River near Shan-t'ou (Swatow) and the Pearl River near Canton, have quite intensive sedentary agriculture, with 30 percent or more of all land under cultivation (C. M. Hsieh, 1973: 166).

In any sketch of Kwangtung's geography, however brief, the Pearl River Delta (Canton Delta) requires attention, for it is the heart of the province, with only 3 percent of its area but 25 percent or more of its population. An area of unusually intensive agriculture and dense population, it also includes Kwangtung's greatest urban center and port, Canton (1.6 million people in 1953; 3.1 million in 1982). The term "Pearl River," sometimes called "Canton River," is used for one of the many waterways that extend from the latitude of Canton toward the sea. Its lower course is a bay, the Canton Bay, the remains of a large estuary that has been largely filled by alluvial deposition. The bulk of the delta lies from the Pearl River westward to the main course of the West River, and embraces an area about 60 miles long and mostly 20 to 30 miles wide. Like the Mississippi Delta, it consists of an intricate network of tributaries and is subject to floods. Tributaries are connected to a maze of waterways, large and small, which were excavated over the centuries; there are numerous islands, many just a few feet above flood level, and some, formerly rocky islands, hilly. Ocean vessels and ships of the coastal trade reach Canton, a distance of about 80 miles from the sea, by way of the larger channels, but more impressive is the enormous local trade carried on by smaller boats able to use minor channels and canals. Striking, too, are the "boat people," numbering hundreds of thousands of persons, who live on their small boats. The delta is subject to tidal ebb and flow, the maximum tidal range at Canton being about six feet.

The Pearl River Delta is extraordinary for its high densities of rural population, among the highest in China and the world. Its ability to support such dense human populations derives from an impressive reclamation effort, of digging channels and canals and raising levees, carried out over a period of two millennia or more, comparable to the Dutch reclamation of the Zuider Zee. Involved also is an unusual human ecosystem, one of aquatic agriculture in which little is thrown away, in which waste products of one cycle, including human excrement, are used to benefit another. Appropriately termed the "minimax strategy" by E. N. Anderson and M. L. Anderson (1973: 99-107), this system is one of minimum-sized plots of land and achieves maximum food yields and a balanced diet with minimum input from outside the system. This is achieved in the Pearl River Delta not only through a careful choice of crops by farmers, but by their following unusual methods of agriculture, fishing, animal husbandry, and food preparation. It represents a triumph of human ingenuity in feeding dense populations.

The traditional cultural landscape of the delta was one of irrigated rice (90 percent of crop area in paddy, usually providing two crops a year) and fish ponds, of sugar cane, sweet potatoes, and other field crops, and vegetable gardens and fruit trees, especially litchi and mulberry. Silkworms fed on leaves of the mulberry, and the region was a center of silk production and silk textile manufacture. Besides the fish, mainly carp, that were reared in ponds for local consumption, young fish were grown for shipping to other parts of China and abroad. And there was much fishing in streams and canals. Prominent, too, in this cultural landscape, were water buffalo, as draft animal, and ducks and geese, along with the pigs and chickens typical of so much of China. Giving the landscape a special charm were the shrines, pagodas, burial grounds, and groves of trees that dotted the delta's hills (Hubbard, 1929; Cressey, 1934: 361-63; Trewartha, 1939; Shabad, 1972: 184-89; Tregear, 1980: 300-304).

Hainan, since 1988 a province but formerly an administrative district of Kwangtung Province, is a large island about 13,000 square miles in size, separated from the Lei-chou Peninsula by a strait ten to fifteen miles wide. Except for some banks, reefs, and small islands, it is farther south than any part of China, almost all of it at latitudes less than 20°N. This is south of Hanoi and the Red River Delta of Vietnam, and marks the island geographically as part of Southeast Asia. Its name, Hainan, means "South of the Sea," and reflects an early Chinese perspective of the region, as viewed from the north. Early navigators from India, working north to the ports of China, were struck most by the impressive stands of palms along Hainan's coast, which led them to call it the "Isle of Palms."

Hainan is 185 miles long and 100 miles wide, and nearly a third larger than Sicily. Though its northern third is mostly lowland, its southern two-thirds are dominated by hills and mountains, with the highest peaks of the Wu-chi Shan ("Five Finger Mountains") rising more than 6000 feet above sea level. As a tropical rainy area, southern Hainan's natural vegetation is largely rain forest, especially at lower elevations; in the northern lowlands, it is savanna. Rich in tropical birds and exotic mammals, Hainan's winters are mild and frost-free in the lowlands, and its summers are hot and humid, like mainland Kwangtung, but to a greater degree.

Though an early source of valuable trade goods for the Chinese, Hainan's distant location, disease, and the fierce resistance by its aboriginal people have left it, like Yunnan and Kwangsi on the mainland, a southern frontier of Han civilization. In modern times, its population included a sizeable aboriginal minority, which led the People's Republic to create a Hainan Li-Miao Autonomous Prefecture in 1952. Though many tribal hill people of that prefecture remain shifting cultivators, the Chinese and acculturated lowland tribal groups are sedentary farmers able to obtain three crops of rice annually and to raise a broad range of tropical crops, among them sugar cane; coconuts; fruits such as bananas, papayas, and mangoes; and coffee and rubber.

We leave Kwangtung and its peoples now to turn to the broader question of traditional Chinese views relating to food, health, and disease. Then we will identify, describe, and compare China's major regional cuisines, before turning to the material central to our study: the plants and animals consumed in China, their identity, origins, uses, and nutritional value. Following that, and drawing on material already presented as well as on literature on China's customs relating to food, we will sketch China's traditional state of nutrition, nutritional deficiency, and disease. In our conclusion, however, we will look at Cantonese cuisine again, to see whether further light can be cast on the factors responsible for its culinary development.

[1] For an account of routes of migration and successive periods of northern influence on Kwangtung, see H. L. Lo, 1967a.

CHAPTER ONE

Food in Chinese Thought and Culture

"It is [China's] food that stirs the imagination of her thinkers,
sharpens the wits of her scholars,
enhances the talents of those who work by the hand,
and enlivens the spirit of the people."

F. T. Cheng

The Central Role of Food in Chinese Life
Chinese Perceptions of Food, Health, and Disease
Food, Eating, and Social Status
Food for Deities and the Ancestors
Vegetarianism in China
Differences among Buddhists, Taoists, and Confucianists in Early China
Buddhism and the Struggle for Vegetarianism
Vegetarianism in Traditional China
Vegetarian Cuisine in China

It is difficult to imagine a statement like that of F. T. Cheng being made about many Western peoples, and certainly not about the English or Americans. As Lin Yutang has observed, the Chinese consider eating food to be one of the rare joys of living, one to which they are more devoted than to religion or the pursuit of knowledge. In this he sees his people as quite different from the English, whose very words relating to elegance in eating come from the French. The English language, Lin notes (Y. Lin, 1935: 337), has a word for "cooking" but none for *cuisine*; it has one for "cook," but no proper word for *chef*; one for "dishes," but none for *menu*; and as for a *gourmet*, he is likely to be called, at least in English nursery rhymes, "greedy gut." In fact, Lin points out, the only time an English person is likely to speak of his stomach is when it is "sick" or "aching." The same might be said of Americans, of course, though not everyone in the United States (or in England) fits that mold. Nevertheless, the American view all too often is that eating, while necessary for survival, is something that should not interfere excessively with other, more important, things in our lives. Thus we have been pioneers in creating fast food chains of considerable variety, from the automats of Horn and Hardart in New York and Philadelphia early in this century to the present-day drive-up windows of McDonald's and Burger King. Cheng, however, is presenting a quite different perspective, one in which food is seen as playing a central role in the life of a broad cross section of an entire people. It is a perspective so at odds with the standard American view as to deserve further examination.

THE CENTRAL ROLE OF FOOD IN CHINESE LIFE

Though we have already made the point above, it is so critical to understanding the Chinese food scene that we would make it again. Food plays such an important role in Chinese life as to lead many to characterize the Chinese as having a food-centered culture. Not only are they wide-ranging in their choice of foods, but concern with the excellence of food is found in all segments of society and is reflected in the common greeting "Have you eaten already?" Indeed some have noted that food is not only a common topic of conversation, but often the dominant one. In the words of H. J. Lin and T. Lin (1972: 46):

> "From the Chinese point of view, the enjoyment of food is a facet of the art of living. The gourmet has practiced his vocation almost everywhere. These are the scenes through which he moved: the hills and fields, terraced and green, at the side of which stood a shabby farmhouse where a simple meal was being prepared; craggy mountains marked by a steep path, beside which sedan chair-bearers paused, drinking strong bitter tea; a charming courtyard, in which a scholar was ladling rain water for tea; countrywomen at New Year's time, shaping glutinous rice balls between their palms; the crab feasts by moonlight; the temples and the monks and vegetarian food, and the mingled essences of the teeming streets. These were the scenes of life itself, in which eating was one of the more dependable pleasures."

The above quotation emphasizes the centrality of food and cooking in traditional Chinese life, the enjoyment of food, and the universality of efforts aimed at creating outstanding dishes. Cooking was more than a necessity to the Chinese. It was an art whose mastery was sought after, one in which there was constant experimentation in the quest for better dishes, including ones that were visually attractive and sometimes shaped into unusual forms, as of birds and animals.[1] Chinese literature is rich in allusions to food, for it was a matter of pride for scholars to be gourmets. Dishes may bear their names, a situation that would be like the English having "Wordsworth steak" or "Galsworthy cutlet" (Y. Lin, 1935: 338).

Royal households in all parts of the world are, as a rule, well-staffed with persons devoted to food and cooking, but China's royal households seem to have been unusually well endowed in this regard, perhaps a further reflection of the importance of food in China. One royal household in classical times devoted nearly 60 percent of its entire staff, 2271 persons in all, to food and drink. The depth of its culinary talent is impressive: 162 persons responsible for preparing the everyday menus of the king, queen, and crown prince; 128 chefs to meet the needs of the royal family itself and an equal number to provide for guests; 335 persons responsible for grains, vegetables, and fruits; 342 specialists in fish, 62 in pickles and sauces; 340 wine servers;[2] and many more (K. C. Chang, 1977b: 11). In Ming times, a much later period when China had a far larger population, there was a Court of Imperial Entertainments whose role it was to dispense meat and drink, serve meals to officials as required, and provide foodstuffs to the palace staff. In A.D. 1425, that Court employed 6300 cooks who may have fed 10,000 to 15,000 individuals every day (R. Huang, 1969: 90).

More certain as a measure of the importance of food is the impressive range of dishes that might be offered at banquets. At a great dinner in Hong Kong to which the Duke of Connaught was invited around the turn of this century, the menu included not only such delicacies as bird's nest soup, shark's fins with crab, bêche-de-mer, quail, and pigeon eggs, but sixteen other main courses, four side dishes (cold roast suckling pig, cold roast mutton, cold roast duck, cold roast fowl), many "table dishes" (sausages, prawns, preserved eggs, liver, and such), a good selection of fruit (citron, pomegranate, carambola, pineapple, Tientsin pear, etc.), several kinds of sweets (almond custard, sweet lotus soup, and others), champagne, claret, and five other wines, among them rice wine, pear wine, and orange wine (Ball, 1906: 291).

The central role of food is also reflected in the elaborate protocol and ceremony that could accompany dining,[3] not only among royalty but among others as well. Even ordinary people, with only simple ingredients at their disposal, made special efforts at making eating pleasurable, theirs being a perspective reflected in the common saying in China "to the people, food is Heaven" (F. T. Cheng, 1954: 5, 23, 28-29; H. D. R. Baker, 1983).[4]

The focus on food in China has led to an unusually broad knowledge of edible substances, as revealed by the rich diversity of everyday foodstuffs in its lively marketplaces (Figures 6-9).[5] It is also manifested in its oft-crowded restaurants, including some in present-day Canton with large dining halls on several stories of a building, the largest eating establishments, some with lakes and gardens, this writer has seen anywhere. Restaurant names, too, often reflect the primacy of food (for example, "Eating Comes First"), or seek to whet one's appetite ("Pavilion of Collected Delicacies," "Restaurant of Rich and Fine Viands," and "Fragrant Vegetable Root Dining Hall"), or invoke a sense of well being ("Eastern Prosperity House," "Unbounded Virtue and Happiness," "Hall of Happy Old Age," and "House of Abundant Peace"), beauty ("House of Exquisite Beauty" and "Hibiscus Flower Garden"), or both of these ("Prosperous Spring Forest").[6] Whatever restaurants were named, as H. D. R. Baker (1983) has observed, they were "far cries from the appetite-sapping 'Fred's All-Day Diner' or 'Ken's Corner Cafe' of the West." To eat in a Chinese restaurant is a social experience just the opposite from that of breakfast in the typical bed-and-breakfast place in London, where the silence may be so overwhelming as to make one hesitate to bite one's toast. By contrast, Chinese restaurants are noisy places, for the Chinese like noise and dislike quiet, a perspective reflected in the Chinese word for a pleasant time, which, in translation, means "hot and noisy" (H. D. R. Baker, 1983; Tiger and Wolf, 1985: 17).[7]

CHINESE PERCEPTIONS OF FOOD, HEALTH, AND DISEASE

Basic to any consideration of Chinese views of food is the distinction between *fan* and *ts'ai*. In a narrow sense, *fan* means "rice" or "cooked rice," and *ts'ai* originally meant "vegetables." In a broader sense, however, *fan* embraces all cereal and starchy dishes, among them porridge, steamed bread, and noodles, and *ts'ai* today refers to the accompanying dishes—whether made of vegetables, meat, or fish—that make a meal tasty. The Chinese believe that suitable quantities of both *fan* and *ts'ai* must be consumed at a meal or it would lack balance, and the individual diner may play a role, in exercising a choice over dishes and flavorings, to assure that such balance is created. *Fan*, however, is the more critical of the two, the basic, essential element of a meal, so much so that the term is also used in the broader senses of "meal" and "food." *Ts'ai*, in turn, is regarded as an addition, one that accompanies *fan* and makes it palatable. Though a child may prefer *ts'ai*, even in moneyed families he is encouraged to eat

Figure 6. Bicyclist in marketplace with beans, eggplant, and stem lettuce (Ch'eng-tu, Szechwan, 1987)

Figure 7. Spices and other foods in market stalls (Ch'eng-tu, 1987)

Figure 8. Fish for sale by market vendors (Canton, 1987)

Figure 9. Fresh eggs and candling device in market (Canton, 1987)

much *fan* and is complimented when he does so. This situation is usually reversed in restaurants, and also at feasts where, to emphasize the importance of the affair, *ts'ai* dishes are many, prominent, and given precedence in serving. Guests, moreover, are encouraged to eat them, and when the rice and noodles are finally served a guest may have little room for them (B. Y. Chao, 1972: 3, 18, 22; E. N. Anderson and M. L. Anderson, 1977: 327; K. C. Chang, 1977b: 7-8, 10, 41; V. Y. N. Hsu and F. L. K. Hsu, 1977: 300, 303, 307; H. D. R. Baker, 1983; Bray, 1984: 4-5; Tiger and Wolf, 1985: 14; E. N. Anderson, 1988: 156).

Despite such departures from normal, there is an ancient Chinese tradition that one should practice moderation in eating, that a person should eat only to ease hunger and relieve thirst (H. C. Hou, 1938: 347). One notes in the *Analects* that Confucius ate sparingly, and that, however abundant meat was, he consumed it in amounts appropriate to the rice that was available (Legge, 1893: 232-33). This tradition of frugality in eating is related to China's limited food supplies, to the view that cereals are sacred and should not be wasted, and to health concerns (K. C. Chang, 1977b: 10-11). As a result, Chinese people are often painstaking in their quest for edibles wherever they might be found (Ross, 1911: 78-83), and they waste very little food (A. H. Smith, 1894: 19-21). This is well illustrated by the fruit vendors in Canton early in this century who sold oranges without their skins, which they kept to be sold separately as flavoring or for preserving (J. S. Thomson, 1909: 69). It is also revealed by the numerous Chinese recipes that involve leftovers (W. W. Chang *et al.,* 1970: 1). The tradition of frugal eating applied not just to the poor, for whom frugality was a necessity, but to the rich and famous as well. Excessive consumption of food and liquor was considered a sin so serious that it could even bring down a dynasty. To present a righteous image, moreover, ostentatious feasting and the eating of expensive foods might be restricted, as in the 1920's when the Peasants Association, a Communist-led organization, established rules against lavish banquets so as to distinguish themselves from those who abused them (K. C. Chang, 1977b: 10, 15). In the People's Republic, too, such views continue in official circles, as evidenced by recent calls for austerity in dining and official regulations limiting the length of banquets to ninety minutes with no more than four dishes and one soup per diner (Gargan, 1988).

Frugality in eating should not be interpreted as indifference to the quality and freshness of ingredients. Indeed the Chinese may go to great lengths to obtain the freshest ingredients possible (Figures 10-16), with shoppers closely inspecting foodstuffs before buying and with some urban dwellers shopping for perishables twice a day, with purchase of small quantities quite acceptable. The importance of freshness is apparent in Canton's Ch'ing-p'ing (Qingping) market with its rich assortment of vegetables, fruit, and flesh, and its varied assemblage of live animals, including mammals, fowl, fish, reptiles, and other creatures, in water or not. It is not unusual for Chinese housewives to pay two or even three times the regular price for live fish or other sea creatures (Koo, 1973: 128). The concern with freshness is also shown in the Ch'ing-p'ing market by the display for sale of fish that have been carefully eviscerated so as to reveal their still-beating hearts as testimony that they were alive not long before. Also conspicuous in that market are electrical candling devices vendors provide so that buyers may select their own fresh eggs.

Permeating Chinese philosophy, religion, and medical thinking, whether Confucian, Taoist, or Buddhist, is the view that the human body is a microcosm of the universe and that, like the latter, it possesses *ch'i,* a body's "vital energy" or "strength." Internal *ch'i* is seen as pervading a person's body and enabling him to meet the needs of life and maintain good health (E. N.

Figure 10. Marketplace containers for live fish and other aquatic animals (Canton, 1987)

Figure 11. Live eels for sale (Canton, 1987)

Figure 12. Live salamanders (Canton, 1987)

Figure 13. Snails (Canton, 1987)

Figure 14. Fresh fish for sale (Canton, 1987)

Figure 15. Newly arrived cages of chickens (Canton, 1987)

Figure 16. Stall of a poultry vendor in an indoor market (Hong Kong, 1987)

Anderson and M. L. Anderson, 1976: 145; 1977: 369; E. N. Anderson, 1988: 188, 192-94). Internal *ch'i* has exchanges with external ones (cold, warm, hot, windy, dry, wet) found in the atmosphere. Though the term for external, or "six *ch'i*," means literally "air" in Chinese medical thinking, it encompasses a broad range of environmental influences on humans. One's internal *ch'i* may be enhanced or adversely affected by exchanges with external *ch'i*, and sometimes, too, develops a malfunction on its own (Needham and Lu, 1962: 430-31; Pálos, 1971: 24, 45-46, 92-93, 128 ff.; Bridgman, 1974: 12-14; Wei, 1976: 369-71, 372). If a malfunction occurs, a person must take action to set things right, but if possible that condition should be prevented by constantly maintaining one's body in a state of balance among the forces, both internal and external, to which it is subjected.

Essential in maintaining one's balance, and hence a critical matter in food selection in China as well as among overseas Chinese, are the opposing forces of *yin* and *yang*.[8] By the concept of yin and yang, which has also been central to the Chinese view of the world since antiquity, an imbalance between these two can bring on ill-health and disease. Among their other qualities, the forces of yin are cooling, dark, and feminine, those of yang heating, light, and masculine. Women naturally tend be imbued with somewhat more yin and men with more yang, but there are individual differences, with some persons by nature inclining too much in one direction or the other. Individuals with too much yang are commonly of ruddy complexion, and are characterized by quickness of breath, frequent nosebleeds, lively dreams, and constipation. Persons with too much yin, on the other hand, are cold to the touch, lacking in energy, and phlegmatic. Should one's body be imbalanced in a yin direction, one may experience chills, and one risks getting ailments or diseases that are cold, thus consumption of hot (*jeh*) foods is one way of regaining equilibrium. Imbalance in a yang direction leads to fever and the risk of hot ailments or diseases, and cold (*liang*) foods are in order (W. R. Morse, 1926-29: 87-90; 1934: 6-15; K. C. Wong and L. T. Wu, 1936: 15-19; Veith, 1949: 13-24, 49-53; Wallnöfer and von Rottauscher, 1965: 6-12, 95-99; Crozier, 1968: 17; E. N. Anderson, 1970a: 208-10; Shiang and Li, 1971; Pálos, 1971: 26-28, 34-37; Koo, 1973: 133-36; Yeung *et al.*, 1973: 197-99; Bridgman, 1974: 12-14; B. Chang, 1974; Porkert, 1974: 65-66; Topley, 1976: 247-48; Wei, 1976: 356-60; E. N. Anderson and M. L. Anderson, 1977: 368-70; K. C. Chang, 1977b: 10; Ahern, 1978: 17-25; Gould-Martin, 1978: 39; Tan and Wheeler, 1983: 39-42; Unschuld, 1985: 55-58). Eating proper foods is thus one way to help a person maintain such balance, and can return him to a state of balance. Food may be eaten to prevent illness and disease in the first place, or, once those conditions have developed, to cure them. This importance of food in prevention and cure has since antiquity led to the Chinese preoccupation with tonics and health foods (J. Y. P. Chen, 1973: 225), and to a materia medica rich in items that also serve as foods, drinks, or spices.

The classification of foods as "hot" or "cold"—more appropriately "heating" and "cooling"[9]—is based not on their temperature, but on other qualities they possess. Classification of foods, drugs, and diseases as hot or cold are one aspect of humoral systems widespread in the Old and New Worlds,[10] and it is unclear whether such classifications originated in a single place or more than one. European tradition looks to an origin of the hot-cold concept in ancient Greece, where Hippocrates (*c.* 460-*c.* 370 B.C.) was a noted advocate, though not originator of, the concept. From Greece, in the traditional view, the hot-cold concept was incorporated into Islamic medical tradition and introduced by the Spanish to the New World (Manderson, 1987: 329; Messer, 1987: 339); it also spread from Greece by way of Persia to India, perhaps only in the fourteenth century A.D. (B. Walker, 1968, 1: 460-61; Peterson, 1973: 3, 14). Various scholars have questioned the European tradition, suggesting that hot-cold classifications may have existed among American Indians before discovery of the New World (Messer, 1987: 339-40), and that humoral medicine was as early in India and China as in Greece (Manderson, 1987: 329). Some have suggested that the hot-cold concept may have originated not in Greece, but in India (Jelliffe and Bennett, 1960: 251) or the Near East, or that it had independent origins in more than one of these regions or elsewhere (Laderman, 1987: 357-58; E. N. Anderson, 1987: 331).

As for China, the use of yin and yang in philosophy began in about the fourth century B.C., and the concepts themselves are likely no earlier than the sixth century B.C. (Needham and Wang, 1956: 273; Needham and Lu, 1962: 430; Unschuld, 1985: 55-56). There is nothing to link the Chinese dualism with that of ancient Greece (Needham and Wang, 1956: 278), yet the possibility remains that the two may have come from a common source "somewhere between Greece and China" (Unschuld, 1985: 54-55). Similar uncertainty surrounds the classification of foods as hot or cold, which is said to be found first in the writings of T'ao Hung-ching (A.D. 452-536) (E. N. Anderson, 1987: 331; 1988: 49), a famous Taoist naturalist and medical writer who was influenced by Buddhism and drew liberally from much earlier works. Sivin (1980) writes of the Greek humoral concept reaching China by way of India by about A.D. 500. E. N. Anderson, in turn, suggests that the hot-cold concept of foods was Graeco-Indian in origin and that it was likely introduced to China by Buddhists, to be accepted in part because it fit well with the yin-yang system that had long been in force. He notes that to some degree foods had been considered to be yin or yang in earlier China, but that the major force behind Chinese acceptance of the hot-cold system was Buddhism (E. N. Anderson, 1987: 331; 1988: 49, 189, 191-92). The view that Buddhists brought the concept to China from India doesn't fit with the European tradition that it arrived in India only in the fourteenth century A.D.; thus the matter is one that requires further investigation.[11] For our purposes, it is sufficient that the yin-yang and hot-cold concepts in China have continued to be important in both classical and folk medicine, and that today yin-yang enjoys official sanction in the People's Republic, whose authorities perceive in it a parallel to the dialectical materialism of Marx (Crozier, 1968: 204; Pálos, 1971: 28).

As in other parts of the world, there is no unanimity in China as to which foods are hot and which are cold (Whyte, 1972: 29; Peterson, 1973: 19; Yeung *et al.*, 1973: 200; E. N. Anderson and M. L. Anderson, 1977: 369; E. N. Anderson, 1988: 190, 192). Commonly classified as hot are (1) fatty flesh, such as mutton and dog, and oily nuts, such as peanuts; (2) spices, among them chili peppers and ginger; and (3) strong alcoholic drinks. Commonly classified as cold are most fruits as well as bland, low-calorie vegetables, among them water plants such as water cress and seaweed (because water is cold). Most foods, however, are neutral or only slightly hot or cold, with rice most conspicuous among these (Bradshaw, 1934: 172; W. R. Morse, 1934: 15-16; E. N. Anderson and M. L. Anderson, 1968: 24; 1976: 146-48; 1977: 368-69; E. N. Anderson, 1970a: 209-10; S. S. Chan, 1973: 149; Koo, 1973: 134-36, 141; Yeung *et al.*, 1973: 198, 200; K. C. Chang, 1977b: 10; Gould-Martin, 1978: 39; Pillsbury, 1978: 13; Tan and Wheeler, 1983: 41-42; E. N. Anderson, 1988: 190-92). Ingredients used in preparing foods contribute their hotness or coldness, as, for example, making a dish hot by adding chili to it. Methods of preparation also influence hotness or coldness. Steeping in cold water is coldest, and boiling is cold; thus the Cantonese view boiled rice as neutral, but congee (watery porridge),[12] because of the water used in its preparation, is cool or cold. Stir-frying is somewhat less cold, and steaming and stewing are neutral. On the other hand, grilling, roasting, and baking are hot, and deep-frying hottest of all. The longer that baking and deep-frying continue, moreover, the more the end product is heating, regardless of its temperature when eaten. From this perspective, certain Chinese in Hong Kong look on the diet of Westerners, in which long-cooked rich foods are prominent, as leading to persistent overheating, and an increased likelihood of such hot diseases as influenza and dysentery (E. N. Anderson and M. L. Anderson, 1968: 24; 1976: 146-48; 1977: 368-69; Yeung *et al.*, 1973: 198-99; Tan and Wheeler, 1983: 41-42).

The Chinese are constantly alert to their state of yin and yang, sensitive to symptoms of one or the other, and ever ready to take action to maintain a harmonious balance. A mother who deems her children over-heated when they return from school may provide them with cooling drinks and prepare them a boiled dinner of cooling foods. Women are naturally inclined toward the cold side, and are considered at special risk when their femininity is most pronounced, as during menstruation, early pregnancy, and when nursing, times when heating foods are prescribed and cooling ones, such as most fruit and vegetables, proscribed out of fear of illness and disease. It is believed, too, that with increasing age one's body becomes colder and thus many old people think that they cannot include an excess of "cold" foods in their diets (H. C. Hou, 1938: 348; Topley, 1970: 426; B. Chang, 1974: 437-38; Mathews and Manderson, 1981: 10-14; Tan and Wheeler, 1983: 39-40).

Another major concern in food selection is with enhancing one's *ch'i* by eating things that are *pu*, meaning "strengthening" and "restoring." Such foods are intended to correct any bodily imbalance, to create resistance to illness, and to aid recovery (Tan and Wheeler, 1983: 43). Various strengthening foods are also believed to stimulate a person's sexual appetite,[13] but this role is an extension of their general role as *pu* substances (E. N. Anderson and M. L. Anderson, 1977: 369; E. N. Anderson, 1988: 192-93).

Strengthening foods commonly are high in both protein and calories.[14] Some such foods, such as chicken cooked in broth, are readily available. Others are unusual and expensive, and may be used only at times of special need or at feasts or other special occasions. Among the more prominent strengthening foods are bird's nest, shark's fin, and bêche-de-mer, whose strengthening qualities parallel those of ginseng, deer antler, and rhinoceros horn.[15] Sympathetic magic may be involved in determining the classification of foods as strengthening. Walnuts are thought to invigorate the brain[16] because, when unshelled, they resemble that organ; and red jujubes, like all jujubes, are not only heating and strengthening, but beneficial to the blood largely because they are red (E. N. Anderson, 1988: 136, 192).[17] In the same vein are the views of London Chinese mothers: that eating pig's tail toughens the human spine, and consuming heart, kidney, legs and feet strengthens their human counterparts (Tan and Wheeler, 1983: 43). Similar beliefs are responsible for the presence, among foods that are strengthening, of certain animals, such as pangolin, raccoon dog (*Nyctereutes procyonoides*),[18] tiger,[19] large fish, and poisonous snakes, that are unusual in appearance or behavior, impressive, or, for creatures of their size, having great strength or ability to harm (E. N. Anderson and M. L. Anderson, 1968: 25; 1977: 369; E. N. Anderson, 1988: 147, 192-94).

Among the other dichotomies that influence choice of food in China, though less important than hot and cold, is wet versus dry, with shellfish, which are moist and juicy, considered "wet," and dry-roasted peanuts, which bring on a parched, scratchy throat, "dry." This dichotomy, however, was always less important than hot and cold in China, and in South China was seldom cited in matters of food selection (E. N. Anderson and M. L. Anderson, 1968: 24; 1977: 369; E. N. Anderson, 1988: 189, 190). Still another dichotomy is "clean" versus "poisonous/dirty." The "poisonous" category embraces substances of known toxicity, such as the liver of puffer-fish, and others that are not toxic as such but which can nevertheless be injurious to a person eating them. Some poisonous items are combinations, such as persimmon and crab, honey and garlic, and green beans and dog flesh. Some may be poisonous to a sick person, but not to persons who are well. Shrimp, for one, are believed to retard healing "of an open wound or to aggravate a patient's case of venereal disease," but are harmless for those who are uninjured and well (H. C. Hou, 1938: 350; Topley, 1970: 425-26; Pálos, 1971: 91, 92; Koo, 1973: 137, 141; Ahern, 1978: 17, 22; Gould-Martin, 1978: 40-42; Tan and Wheeler, 1983: 44; E. N. Anderson, 1988: 194-96).

Some food items may be avoided because of physical or psychological qualities they possess.[20] Squid, which lacks red blood, tears, and emotion, is considered lacking in food value, and may therefore be passed over as food. Persons with smallpox or measles may avoid sesame seeds, mung beans, and nuts, because of their resemblance to the sufferer's pustules (Koo, 1973: 137; Wheeler and Tan, 1983: 56). San Francisco pregnant women may refuse iron supplements because of fear that these would "harden the bones and contribute to a difficult delivery." They may also reduce the amount of soy sauce they consume "so that the baby will not be too dark" (B. Chang, 1974: 438),[21] or of squid and octopus lest the child "be born with multiple appendages" (Ling *et al.*, 1975: 131). Of particular interest in this regard is the avoidance of lamb or mutton by many pregnant Chinese women today out of fear that the child will be an epileptic; this is based on similarities in Chinese words for "lamb or mutton" and "epilepsy," for the latter translates literally as "mad lamb" or "lamb in convulsions" (Ling *et al.*, 1975: 131; Tan and Wheeler, 1983: 44; Wheeler and Tan, 1983: 53, 54). Positive similarities, whether in sounds of words or of other sorts, may also be recognized and used to advantage. According to Cormack (1935: 32, 35), whose observations dealt largely with the Peking area, the words for "onion" and "clever" sound alike, and an onion is used in an attempt to assure the cleverness of a child. This is made when, in the infant's initial bath, an onion is dipped in the bath water, a line drawn down the child's body, and the onion then cast into the street by a tall man who is considered clever. The hope is that the infant would, as a result, become both tall and clever. On the hundredth day after the child's birth, a time when presents are brought to the house, a chicken is cooked and its tongue removed and rubbed over the lips of the child, to assure that in the future it will be eloquent.

FOOD, EATING, AND SOCIAL STATUS

Wolf (1974: 176-77), from his experiences in Taiwan, has written of Chinese food behavior as part of a broader social system, and has noted that a family's social standing may be discerned in terms of the people with whom the family will eat. A beggar, for example, might be given food to eat outside the family home, but, because of his much lower social status, would not be asked to come in to consume it. In a similar vein, a person of high social status would not accept an invitation to eat with individuals of much lower social status. Though a farm family may try to ingratiate itself with an official, Wolf notes, it would do so by presenting food to him through an intermediary, rather than inviting him to dinner. As a result, the persons invited to eat with a family at home normally include kin, associates, and others of roughly the same social standing.

Social gatherings in China invariably involve a meal, which may differ in elaborateness and formality depending on the occasion and status and wealth of those who sponsor them. Even persons of modest income manage to sponsor feasts at times of birth, marriage, and death, with persons of greater means holding them more frequently, as in celebrating birthdays and festivals.[22] A man of wealth, in turn, may sponsor a feast, providing a dozen main courses for ten guests, as frequently as twice weekly (Koo, 1973: 117). A person of equal social standing attending such a banquet has an obligation to reciprocate, with the result that such people may eat out several times a week. Thus in traditional China the frequency of eating out was a fairly good indication of a person's rank in society (E. N. Anderson and M. L. Anderson, 1977: 372).[23]

FOOD FOR DEITIES AND THE ANCESTORS

Further evidence of the social role of food in traditional China is found in its being offered both to deities and ancestral spirits. One such deity, who had roots in antiquity and is among the most prestigeful in China, is the Kitchen God, guardian of the family, protector of the hearth, and patron deity of professional cooks. The Kitchen God, who resides in the kitchen, is depicted in various ways, from a simple piece of red paper, properly inscribed, to an elaborate portrait (Figure 17).[24] In Chinese homes, offerings, among them food, are made to the Kitchen God on various occasions. In earlier times, the Kitchen God was associated with Confucianism

Figure 17. The Kitchen God (from Burkhardt, 1953-58, 1: 76)

and/or Taoism, and included among food offerings presented to him were meats such as pork, chicken, and lamb. When Buddhism arrived in China, the Kitchen God was included in the Buddhist pantheon as a *deva* (god), and persons influenced by Buddhist thinking substituted, in their offerings, vegetarian dishes for ones of meat (Welch, 1967: 401-2; C. S. Wong, 1967: 168; Reichelt, 1968: 194-95). One period when food is offered to the Kitchen God is that of the Chinese New Year. Several days before that date, the Kitchen God is given a farewell dinner, but only of sweets, such as sweet rice, cake, and candied fruit. In addition, his lips are sealed with sticky sweets, such as honey, molasses, or sugar candy, in the further hope that he will say only sweet things when he visits Yü Huang (Shang Ti), the "Jade Emperor" or supreme Taoist god, to give his annual accounting of family affairs.[25] Then the representation of the Kitchen God is removed and burned, along with spirit-money and other things of use on his journey, and he is sent on his way, accompanied by the sound of firecrackers. A new representation of the Kitchen God is set up several days later, on New Year's Eve, when his return is celebrated with firecrackers and he shares food and other offerings with the family ancestral spirits and other deities, and once again presides over the family hearth (Feng, n.d.: 182-84; Giles, 1876: 105-6; Cormack, 1935: 143-48; Eberhard, 1952: 17-26, 39-42; Burkhardt, 1953-58, 1: 1-2, 68, 71, 147; 2: 51; 3: 161; Tun, 1965: 98; C. S. Wong, 1967: 162-69; Latsch, 1984: 82-96).

Offerings to deities were gifts intended to show respect, but upon which the deities, as powerful beings, were not dependent. Ancestral spirits, on the other hand, were considered to have the same need as the living for food and other necessities, and thus were dependent on the offerings provided by their living descendants (Latourette, 1964: 537; Ahern, 1973: 149; Wolf, 1974: 177-78). The custom of offering food to the ancestors has persisted since antiquity, as indicated in the case of Confucius who, according to the *Analects*, sacrificed to his dead forefathers as if they were actually there (Legge, 1893: 159). Not to have sons and grandsons to assume that obligation was therefore a thing to be feared, as revealed in the observation by Mencius, "There are three ways of being a bad son. The most serious is to have no heir" (H. D. R. Baker, 1979: 46).

Judging from Ahern's research (1973: 149-62) in Taiwan, the obligation for making such offerings was not a simple one borne equally by living descendants, but was tied to various kinds of indebtedness and obligations that were subject to interpretation by the living and dead. The general rule was that the living person who had the greatest obligation toward the dead was responsible for the offerings. The obligation decreased from one generation to another, so that a man, for example, would provide for his father and grandfather, persons he had known, and his great-grandfather, of whom he had heard, but perhaps not earlier ancestors (H. D. R. Baker, 1979: 82-91).

An example of food offerings to the recently deceased is provided by Wieger and Davrout (1981: 531-33, 538-39) in writing of popular practices in Chihli (Peking area, North China) around the turn of this century. Food was prepared as a person was about to die. A table was set near the person's head, adjacent to the bed, and a food-jar was placed on the table and later in the coffin so that the deceased would have food to eat.[26,27] At the funeral, there was a ceremony, "the sending of food for the journey," in which a bag of flour cakes, along with representations of other objects made of bamboo and paper, was placed on a paper cart prepared for the deceased. The cart and its contents were then burned amid cries of lamentation, with mourners vying to obtain and eat what remained of the cakes, for this was believed to give one courage and prevent illness. Ahern (1974: 297-98) reports a similar offering in contemporary Taiwan as part of the final ritual before the coffin of the deceased was removed for burial. Several tables were set up in front of the coffin and covered with prepared foods of various kinds, and then, as part of the ceremony that followed, a relative or acquaintance knelt and read a statement detailing for the dead person the amount of food and paper money that were being provided for his use. This part of the ritual was intended to set up communication with the dead person so as to let him know what his descendants had done for him, and to assure him that he had been provided with everything coming to him. Another part of the ceremony was intended to exorcise the spirit of the deceased so that it would not return to injure the living. Ahern did not mention food being buried with the dead, but, in the past, foods and foodstuffs were commonly placed in graves, and when the deceased was a person of importance the quantities might be quite large indeed (J. J. M. de Groot, 1964, 2: 382-89).[28] Where such graves have been

Figure 18. Offering food to the ancestors at the ancestral tablet (from Doolittle, 1865, 1: 223)

uncovered by modern archaeologists, they have, of course, been quite useful in reconstructing China's food history. A touching example of such an offering, found in some parts of China, is that of placing a cup of mother's milk on the grave of a dead infant for three days after it was buried (Cormack, 1935: 89).

The Chinese continued to honor their dead ancestors, whether at the wooden tablets in which the ancestral spirits were believed to dwell (Figure 18),[29,30] or at the grave site. It is so critical that someone carry out the ancestral ceremonies after a person dies, according to H. C. Hu (1948: 31), that a person who has no male offspring is referred to as "ended," an unfortunate position. Though one brother may perform the necessary rites for another, over time his descendants cannot be expected to follow them properly, if at all. In the more involved sacrificial rites, offerings of food and drink were the focus of attention, with well-off families preparing tempting foods, and the poor, too, making whatever efforts they could to supply food suitable for such occasions. This was done in recognition of the influence ancestors may have on the living and their well-being, and because food was basic to human life and well being. The ancestors were believed to eat the essence of the food; what was left the living would consume after the ceremony.[31]

This offering of food to ancestors reflected the wish that a link between living and dead family members be maintained, and it reaffirmed in a sacred context the loyalty and unity of the family. As a people heavily dependent on agriculture for their livelihood, sacrifices in spring and fall were most critical, those in spring marking hopes for life and growth, and those in fall, realization and plenty (C. K. Yang, 1961: 39-40). One reads of the men of a Chinese family visiting the family tombs in the spring[32] with their offerings of food and with paper money, and representations in paper (or paper and bamboo) of clothing, servants, and other valuables that were burned to enable them to reach the dead ancestors. Such visits were made, as well, to tombs of lineage ("clan")[33] founders, sometimes followed by dining on the ground nearby or by a great banquet at the lineage ancestral hall. Or sacrifices might be made at the ancestral hall itself, the ceremony a carefully-regulated affair commonly followed by the feasting befitting an important time when group ties are restored and strengthened. The participants, when about to leave the hall, divided up the sacrificial meat and took it with them, to be eaten by all families of the lineage, who, as a result, enjoyed the blessing of the ancestors (Gray, 1878, 1: 320-21; S. W. Williams, 1895, 2: 250-54; Ball, 1906: 31-32; Couling, 1917: 27; H. C. Hu, 1948: 31-40, 116-19; C. K. Yang, 1961: 38-43; J. J. M. de Groot, 1964, 2: 385ff.; M. M. Chiu, 1984: 343-50).[34,35]

Ahern (1973: 166-74) noted striking differences in her Taiwan village between the food offered at ancestral tablets and that offered at graves. Arrangements at ancestral tablets were like those for a family meal. Bowls and chopsticks were supplied, and the food prepared was the everyday fare of the village, though perhaps more plentiful in flesh and delicacies. After the ceremony, the assembled family and guests ate the food, typically chicken, pork liver, stir-fried eggs, soup of various kinds, and rice. At the grave, in contrast, raw foodstuffs were provided, most of them dry, tasteless items which had not been prepared for eating. Ahern explains this as representing a change in status of the deceased; at the ancestral tablet, whether at home or in an ancestral hall, a dead person was recognized as an individual with a clear identity: a familiar, accessible being who is given food just like others of the family. At the grave, however, the descendants are dealing with a being residing apart from the everyday activities of the family. Even more striking, Ahern noted, is the fact that the two types of food offerings "cannot be combined or substituted for each other."

From Sung times onwards, offerings of food also figured prominently in the most important and magnificent of national rituals, the annual sacrifice by the emperor, at the time of the winter solstice, to the supreme deity, Shang Ti. A bullock was killed and offered to Shang Ti, and one reads of an altar containing the sacred tablets of Shang Ti and his subsidiary deities, where there were young oxen on tables and foods in front of their shrines. This was a time of thanks both to deities and imperial ancestors, an affair considered essential to the continued well-being of the Chinese people and empire (Rattenbury, 1944: 151-53; D. H. Smith, 1968: 140-45).

Since the People's Republic was established, religious practices of all sorts, including ancestor worship, have been under attack, and with the Cultural Revolution there was far more violent criticism and disruption. The government also acted to deny people items needed for worship, for example by banning the making of ceremonial candles, joss sticks, and paper money for the deceased (H. D. R. Baker, 1979: 206-7). One newspaper article in 1970 put the official attitude in a way reminiscent of Karl Marx's famous dictum that "religion is the opiate of the masses." The article reads: "We materialists just do not believe in any of this stuff. In this world there are no gods and there are no ghosts. So-called gods and ghosts and geomancy and so on were just spiritual shackles employed by the exploiting classes to fool, control and enslave the people." Religious structures of all sorts, including ancestral temples, were converted to secular uses; ancestor worship was also banned and an attempt made to replace ancestral tablets at home with pictures of Mao Tse-tung. Despite this, H. D. R. Baker was able to write in 1979 that ancestor worship at the level of the family (not of the lineage, which would be public) still survives in China, though carried out largely in private (H. D. R. Baker, 1979: 206-11).

VEGETARIANISM IN CHINA

Certain foreign observers have written of the contemporary Chinese or South Chinese as lacking in food taboos that affect an entire people or region. Whyte (1972: 29), for example, writes that the food restrictions of the major religions are of no significance in present-day China, and probably never affected many lay Chinese except Chinese Moslems, who rejected pork. E. N. Anderson and M. L. Anderson (1977: 363), on their part, say flatly that in South China food taboos were eliminated long ago by famines. We shall consider this matter again later, but for the moment would note that Gould-Martin (1978: 40), using field data along with evidence from both contemporary and older accounts, writes of another pair of Chinese opposites applied to an entire diet: *su* and *hun*. A *su* diet is vegetarian, without meat, fish, or dairy products, and also without onions, garlic, and leeks or alcoholic drinks. A *hun* diet, by contrast, is one that contains such food and drink. The *su* diet is the diet of Buddhists, based on their reluctance to take life, but also, according to Gould-Martin, it derives from Taoist concerns with gaining immortality. This raises interesting questions about Chinese vegetarianism, its nature and roots, and the degree to which the vegetarian perspective in China derives from India, that great hearth of vegetarianism in Asia.

Differences among Buddhists, Taoists, and Confucianists in Early China

We have found no work on Chinese vegetarianism comparable to Ludwig Alsdorf's excellent study of Indian vegetarianism.[36] In India, the indigenous religions of Hinduism, Jainism, and Buddhism have, since early times, been committed to respect for life, and active in encouraging adherents to give up killing animals and eating their flesh. The first Indian Buddhist monks to arrive in China (around the beginning of the first century A.D.) were pious men who brought with them a commitment to vegetarianism, but who found widespread meat eating and animal sacrifice. The practice of sacrificing animals (and, especially prior to Chou times, humans as well) was an ancient and important one in China, carried out in honor of ancestral spirits and other supernatural beings in the interest of the lineage and the state; at An-yang in Shang times, for example, as many as a thousand cattle were sacrificed in a single ritual. There are also records of fasting and abstinence from meat in early China, but it was only with the introduction of Buddhism that vegetarianism became commonplace (C. K. Yang, 1961: 253-54; C. S. Wong, 1967: 156-57; D. H. Smith, 1968: 21-25, 83-84; C. Li, 1977: 106-7, 248-49, 252-54; K. C. Chang, 1980: 143, 227-30 and *passim*), as was true of Japan as well (Oiso, 1976: 38-40).

When Buddhism was introduced to China, Taoism and Confucianism dominated the scene. Taoism had developed from several sources, philosophical, religious, and otherwise,[37] that converged from the fourth century B.C. onwards. According to Taoist views in Han times, life is one's most prized possession, and an individual should seek immortality for his body, or at least longevity. This is just the opposite of the Buddhist conception of the impermanence of the human body, and denial of the existence of personality. Taoists viewed life as representing a harmonious balance of material elements belonging to the yin and the yang. If those elements could be maintained in absolute harmony, they believed, life would be extended indefinitely. Such harmony, however, was regularly upset by wrong behavior, emotional excitement, craving for food and drink, and ingestion of elements that upset the body and lead to illness and death. An individual could remove elements disruptive to harmony by various means, among them meditation and respiratory exercises, the use of alchemy to create elixirs,[38] dietary practices, fasting, and acts of virtue. The acts of virtue among Taoists of Han times are given as deliberate poverty; providing for the poor, orphans, and old and sick people; keeping up roads; and repairing bridges. If refraining from taking animal life was then virtuous in Taoist eyes, it cannot have been prominent because it is not alluded to in our sources. In a similar vein, Taoist dietary practices were not based on respect for animal life, but on the view that consumption of certain foods might interfere with bodily harmony. Most important was abstinence from the Five Grains,[39] for these were believed to nourish the Three Worms, malignant beings within one's body capable of bringing on sickness, frailty, and reduced life expectancy. Meat was also among items from which Taoists should abstain, but the Taoist basis for abstaining from meat was quite different from the Buddhist one, with Taoists holding that the many gods sustaining one's body dislike the odor of flesh, and that if one eats it they may escape, thereby shortening one's life (Welch, 1958: 95-105, 108; D. H. Smith, 1968: 69-77, 98-103; L. G. Thompson, 1975: 95-103; Maspero, 1981: 38, 269, 273-74, 321-24, 335, 427-30).

Turning to Confucianism, a moral philosophy favored in certain historical periods as the court religion of China, there seems to have been no general opposition to animal slaughter or meat eating before Buddhism arrived. The ethical teaching of Confucius (551-479 B.C.) focussed on perfection of the individual and on his contributing to general well-being through respect for the great sages of the past and conformity with the ways of heaven and customary practices of family and society. Among those practices, continued from earlier times, were rites involving animal sacrifice. Confucius supported such sacrifices as being in the public interest, though he emphasized their ceremonial aspects and moral and social contributions (Legge, 1893: 161; C. K. Yang, 1961: 253-55; D. H. Smith, 1968: 21, 34-41).

Modern Confucianists suggest that compassion for animal life is a normal thing for humans to feel; they point out that such compassion is revealed in their ancient works; and they conclude that such feelings were known in China before the time of Buddhism. As an example, they cite a story, perhaps apocryphal, about Mencius (*c.* 374-289 B.C.), renowned successor of Confucius. Mencius' mother moved their residence from a neighborhood where there was a butcher shop so that her young son would not become cruel through observing killing on a daily basis.[40] They also cite the adult Mencius as speaking approvingly of a prince who, upset over the fear displayed by an ox being led to sacrifice, ordered that a sheep be substituted for it (Edkins, 1893: 202-3; Legge, 1895: 139-40). Such compassion for animals, especially ones with whom a person has lived or worked, is common among humans in many parts of the world, but, it should be emphasized, this does not normally lead a person to reject all animal sacrifice or turn him to vegetarianism. Nor, one should emphasize, was this the case with Confucius, Mencius, or their early followers. One judges from the *Analects*, the best single source on Confucius, that he carried out sacrifices to spirits and the dead, attended sacrificial rites, and received flesh of the animal victim.[41] In addition, he ate meat and fish, though he was careful about its quality and preparation (Legge, 1893: 158-59, 161, 232-33; 1895: 434-35). When, in 195 B.C., Kao Tsu, the Han emperor, chose to honor Confucius, he sacrificed an ox, a sheep, and a pig to the sage's spirit (Shryock, 1966: 97-99), something unthinkable had such an act been abhorrent to Confucius. Mencius, in turn, spoke favorably of sacrifice, including animal sacrifice, if it were carried out properly, as by a sovereign in the public interest. He was given pork to eat by his mother when he was a boy. That he liked meat is indicated in the quote:

> "Kung-sun Ch'âu asked, saying, 'Which is best,—minced meat and broiled meat, or sheep-dates?' Mencius said, 'Mince and broiled meat, to be sure'" (Legge, 1895: 497).

Mencius was fond of fish and, even more, of bear's paw. His lists of proper behavior mention kindness toward parents and other people, but say nothing about abstaining from killing animals or eating meat. On the contrary, he observed that at age fifty silks were required to keep a person warm, and that at age seventy meat was needed to satisfy hunger (Legge, 1895: 17, 44, 236-37, 266-67, 337, 399, 411, 437-38, 460-62, 483-84, 497-98).

Buddhism and the Struggle for Vegetarianism

In the period immediately after it reached China, Buddhism was mentioned along with a Taoist cult; in another context, its name appeared in connection with sacrifice. Sacrifice was important among Taoists in Han times, and one emperor performed animal sacrifice to the Buddha along with Lao Tzu, the philosopher of early Taoism. It is difficult to imagine the devout Indian Buddhist monks who had come to China condoning such sacrifice, and indeed such acts were likely little more than a modification of existing Taoist practice "slightly tinged with Buddhism" (Zürcher, 1959: 26-27, 36-37). It has been suggested that if contact with India had been cut off after Han times, Chinese Buddhism may well have become just another Taoist cult. If so, one wonders if it might even have come to tolerate such animal sacrifice.

That contact, however, was not severed, and the two movements became competitors, with Taoism waging a struggle over many centuries to retain adherents in the face of inroads by the new religion (D. H. Smith, 1968: 108-9, 119-21). Weinstein's recent history of Buddhism in T'ang times (1987: 43, 49, 83, 88, 102, 114, 123, 139, 145, 146, 201) is especially revealing in demonstrating how shifts in the fortunes of Buddhism and Taoism affected the practices of animal slaughter and vegetarianism. Also important is the suggestion of Buddhist influence on Taoists, who also favored bans on slaughter on their sacred days. The T'ang period was a time when one emperor might favor Buddhism and the next act against it. In 619 A.D., shortly after the T'ang dynasty had been formed, Kao Tsu, the emperor, broadened existing bans on the slaughter of animals and fish (and the execution of criminals) to include the initial nine days of the new year as well as ten Buddhist taboo days every month. In 691, the Empress Wu, a pious Buddhist, placed Buddhism above Taoism, and instituted various laws in its favor, including a ban on all animal slaughter and fish catching. One can only guess at how such a ban may have been enforced. Regardless, it angered Confucianists who argued that it blocked their sacrifices, and the ban was ultimately repealed in 700. Limited bans, however, continued in force, being modified by certain later emperors according to their Taoist or Buddhist leanings, with a pro-Taoist emperor in one case restricting slaughter only on Taoist sacred days.

Great vegetarian feasts for Buddhist clergy, and sometimes Taoist clergy and others, were also held by various T'ang emperors and merchants. At some royal feasts, only distinguished Buddhist monks were invited; other royal feasts were attended by all the monks of a monastery, who might number in the thousands, even as many as ten thousand. Wen Tsung (A.D. 826-840), an emperor who tried to curb the power of the Buddhist church, warned his people not to hold vegetarian banquets for clergy in honor of his birthday as had been the custom with previous emperors. Though he did ban the slaughter of animals on his birthday, he issued a decree to emphasize that this was based on his esteem for life, not that he had any faith in Buddhism or that he was seeking its support. To drive home his point, he permitted the serving of preserved meat at the birthday banquets, an act that would prevent Buddhist clergy from participating. His successor, Wu Tsung (840-846), continued the suppression of Buddhism, and eliminated the ban on slaughtering animals during the three taboo months of Buddhism, but Hsüen Tsung (846-859) reinstituted that ban. In addition, on his first birthday after becoming emperor, Hsüen Tsung held a great vegetarian feast at the palace, one that Buddhist monks, along with Taoist priests, were invited to attend. The Taoist cause was championed by a renowned priest who came to the court to plead for a ban on slaughter on eight days of one of their holy months. Hsüen Tsung's son Li Ts'ui (I Tsung) was even more committed than his father to Buddhism, holding great feasts for thousands of clergy on the six Buddhist taboo days of each month. He also organized a procession for a relic, reputed to be the finger bone of the Buddha, and along the way wealthy families constructed fine pavilions at which vegetarian food was available for clergy and lay people.

Vegetarianism in Traditional China

Though Confucianist and Taoist thinking has been influenced by Buddhism, since T'ang times Buddhism has remained China's greatest pro-vegetarian force, its devout adherents expressing sentiments that echo those of ancient India. The writer of one modern Chinese Buddhist tract, for example, observes that one way to happiness is by loving living creatures and refusing to kill them or eat meat. He singles out the killing of dogs and oxen as being particularly heinous, for, though their owners give them little sustenance, they faithfully serve in the human interest. In another place, the writer says that meat eating is permitted at weddings and funerals, though for regular meals, he argues, vegetables are sufficient, for they not only help one gain happiness but they save money as well. Elsewhere, he attempts to counter the traditional Chinese view that animal sacrifice benefits the individual in whose honor it is carried out. Killing of animals at a funeral, he argued, brings sin on the deceased, and such killing in celebration of the birth of a son may even cause the child to lose its life (Wieger and Davrout, 1981: 197-205).

The writer of the Buddhist tract observes that Taoists and Confucianists also advocate compassion for living creatures and have feelings against killing. This, however, passes over the fact that in the cult of Confucius, which developed in Han times and has persisted since, animals are sacrificed in honor of the sage, especially oxen, sheep or goats, and pigs (Doolittle, 1865, 1: 359-68; Soothill, 1923: 239; C. K. Yang, 1961: 253-55; Shryock, 1966: 97, 99-100, 106, 107, 168-76, 203-4; D. H. Smith, 1968: 140-47). More than a century ago it was estimated that, apart from oxen sacrificed, over 62,000 sheep, pigs, rabbits, and deer were offered at Confucian temples each year. The writer of the Buddhist tract may not be giving an altogether balanced impression of modern Taoism either, though he is right to the degree that it has drawn on Buddhist precepts. Modern Taoists, however, are divided into many sects, which differ greatly in beliefs and practices. Among those beliefs are compassion toward animals and vegetarianism, with the latter found especially in monasteries (D. H. Smith, 1968: 110). One reads, in what has been called the most popular Taoist work in nineteenth century China,[42] a statement in favor of compassion to animals, an admonition against eating flesh on an everyday basis, and an observation that it is sinful to kill an animal just to satisfy one's taste. At the same time, the work recognized exceptions: people can consume flesh on special occasions such as New Year feasts, offerings, and the receiving of guests (Wieger and Davrout, 1981: 246, 252, 283).

The beliefs and practices of Confucianism, Taoism, and Buddhism relating to animal killing and vegetarianism have become so mixed as to present difficulties in gaining an understanding of their differences. Comprehending the everyday conduct of ordinary Chinese is even more difficult, for the Chinese speak of those three faiths as though they embodied the idea of religion, and they do not find it inconsistent to participate in observances and embrace elements of more than one, or even all three. (S. W. Williams, 1895, 2: 193-94; Rattenbury, 1944: 140; Burkhardt, 1953-58, 3: 10-11; L. G. Thompson, 1975: 103-5). This has led to notable differences from one person to another in their views of, and commitment to, vegetarianism. It is widely-held, nevertheless, that flesh eating is sensual and sinful and does not represent appropriate behavior for a person devoted to religion. Indeed the Chinese so associated vegetarian practice with religious persons that initially, until they learned better, they regarded and referred to Christian missionaries as "eaters of vegetables" (Doolittle, 1865, 2: 183).

Turning to other manifestations of respect for animal life and vegetarianism in traditional China and its neighbors, one notes the Buddhist practices of liberating animals and of feeding birds, fish, and other creatures at their temples. Sometimes animals were kept at Buddhist temples or monasteries; fish and turtles, for example, were kept in ponds where people could gain merit by feeding them (Latourette, 1964: 544).[43] Also sometimes found among temple animals was a pig, sheep, duck, or chicken with a malformed foot resembling a human foot, which marked the animal as having been a human being in some previous incarnation (Rattenbury, 1944: 137-38). People in Buddhist Tibet and Mongolia, though they may eat meat, seek to avoid the guilt of killing animals for food. One way is to leave that task in the hands of butchers; those butchers may be Chinese Moslems who are not troubled by taking life, but, whatever their ethnic affiliation, such persons are looked down on because of their occupation. In China proper during periods of drought, when officials and certain members of the public would go to the temples and pray for rain, orders might be issued for butchers to cease their activities. Such orders derived from the view that it smacked of insincerity for meat eaters to pray for rain, and that this might cause supernatural beings to ignore their appeals (Doolittle, 1865, 2: 183-84).

Apart from vegetarianism imposed at times of drought, many traditional Chinese, at one time or another in their lives, made vows not to eat meat. Such vows might be made to a deity out of conviction that one gained merit by avoiding meat, with one's pleas—for longer life, good health, prosperity, male children, and such—more likely to be heeded.[44] Such vows might also be made as part of a religious initiation, or in connection with the vegetarian associations found among both Buddhists and Taoists (Doolittle, 1865, 1: 184-86; Wieger and Davrout, 1913: 205; Reichelt, 1968: 165). Among the Min Chia, a non-Chinese tribe of Yunnan who have been influenced by Buddhism, old women would join vegetarian associations and make promises of one sort or another to avoid eating meat (FitzGerald, 1940: 179-80). In Singapore, a study (Topley, 1954) of Chinese women's vegetarian houses, places where unattached or single women may live, found all but one to be Buddhist.[45]

Vows to avoid meat were usually of limited duration, observed only at stated times or occasions, as at breakfast every day or all day on the third day of each month. A vow might specify the type of vegetable food that was to be consumed, for example nothing but rice. Vows for a lifetime of abstinence from flesh were far less common, made only by clergy and devout lay people, the latter more often women, especially widows.

Of special interest here is the question of what specific "meats" were banned. We use quotes here because some foods, though they were grouped with the prohibited meats, were in fact vegetables. We refer to onions and garlic, which, as we shall see later, were subject to such prohibitions in India as well. The definition of "vegetables" also seems to have been flexible, for Doolittle (1865, 2: 183) writes of three salt-water marine animals or their derivatives (not identified) that were permissible food, which led to the Chinese saying "among the vegetables are three kinds of meats."[46] Writing of Chinese Buddhist monks in the first half of this century, Welch (1967: 112) provides an even longer list of items banned or usually avoided, including flesh, fish, eggs, milk products, onions and their relatives, and all alcoholic drinks. We will consider certain of these bans again in later chapters, and here we would observe only that lists of banned foods can go well beyond meat itself.

Looking at the bans from a local perspective, one finds striking differences in their prevalence from place to place. In a Chinese village in Yunnan in the late 1930's, only Buddhist priests were vegetarian, and apart from them only women of the Buddhist Women's Club observed any restriction on flesh: on two days a month they ate only vegetables and rice (Osgood, 1963: 322-23). In a village in the Yangtze Valley, true vegetarianism was unusual, practiced only by a few widows. Other women, though believing that killing animals is not favored by the deities, took vegetarian meals only once every two weeks (Fei, 1946: 128). This was quite different from the situation on the beautiful, templed island of P'u-t'o Shan (Pootu) off the coast of Chekiang, a destination for Buddhist pilgrims from throughout China; on P'u-t'o Shan, where Buddhist monks comprised more than half the population in 1935, no meat or eggs could be purchased, and there was a ban on importing meat products (E. L. Liu, 1937: 134-35; Rattenbury, 1944: 144, 157).

The case of P'u-t'o Shan calls attention to the special role of Buddhist temples and monasteries in Chinese vegetarianism. Pilgrims travelling to such places would normally eat only vegetarian fare, and one commonly found collections of vegetarian restaurants near temples. Within temples and monasteries, Buddhist monks and nuns would eat only vegetarian dishes, and visitors, too, would partake of such fare.[47] The Buddha himself had permitted monks to eat meat and fish as long as they had no knowledge that they were killed specifically for them. In some Buddhist countries flesh food may be included among the edibles obtained by monks through begging, yet, in accord with the Buddha's ruling, they could eat it with impunity. In China, however, Buddhist monks were not mendicants, and they usually prepared their food in the monastery from foodstuffs they grew, purchased, or obtained as rent (L. G. Thompson, 1975: 88). Even had they wished to do so, monks could not have claimed ignorance of the origin of their foods, and were therefore morally obliged to eat vegetarian fare.

Vegetarian Cuisine in China

Vegetarian fare can be quite dull, and often is in the West. In China, however, monks and nuns, not having lost interest in tasty food, came to develop delicious vegetarian dishes. Some Buddhist temples and monasteries had culinary traditions going back to the medieval period or before, and their dishes were many and varied, though prepared in much larger quantities than at home. Where food had to be prepared for many hundreds of monks, stoves and cooking vessels could be huge, and a basket of chopped vegetables ready to be fried might measure as much as three feet across, so large as to require two men to stir-fry the vegetables with rakes made of bamboo (Holmes, 1948: 136). Visitors commonly added substantially to the numbers of persons at a monastery meal. In the Peking area, for example, lay persons would go to Buddhist temples or monasteries for vegetarian dishes, though most city restaurants in Peking, when requested to do so, would also prepare them (K. Lo, 1971: 25-26; 1979c: 106-7). In some places, vegetarian cooking might also be obtained in specialty restaurants, as in Sung times in Hangchow where there were restaurants dealing in "temple food" (Freeman, 1977: 161). Temple cooking also found its way into the royal kitchen, as shown by the Buddhist dishes served at the Ch'ing court (Davis, 1966: 126-36).

The many vegetarian dishes inspired by temple and monastery cooking sometimes have names indicating an association with Buddhism, as in "Buddha's Delight," made with vegetarian steak, black mushrooms, cloud ears, ginger, lily flowers, soy sauce, and other ingredients (W. W. Chang *et al.*, 1970: 425) and "Lo-Han Grand Vegetarian Ensemble" (Lo-han are the minor gods of Buddhist temples) (Lo, 1979c: 110). People may also combine Buddhist vegetable preparations with meat, as in "Duck with Buddhist Vegetables" (C. B. T. Lee and A. E. Lee, 1980: 44). Or, without using meat or fish, they may consume tasty Buddhist analogues, commonly made of cereals and bean curd, intended to resemble beef, pork, chicken, abalone, and even shark's fin (C. S. Wong, 1967: 157; Jenner, 1968: 41; E. N. Anderson and M. L. Anderson, 1977: 326, 339).[48] Such products could be so much like meat in appearance as to require convincing a pious vegetarian that they were in fact acceptable foods (Mote, 1977: 224).

[1] For a photograph of chefs' assistants carving vegetables, and some of their finished products (rose, bird, and God of Longevity, all made from carrots; flowers, from cucumber; chrysanthemums of yam), see Em. Hahn *et al.*, 1973: 66, 67. For photographs of other modern creations, see Tiger and Wolf, 1985: 30-37, 203. Among the latter are a red and white dessert in the form of the yin-yang symbol; a colorful phoenix carefully crafted from beet, cucumber, egg white, pigeon eggs, and duck skin; and a lotus flower table decoration whose leaves and stems appear to be made from cucumber peels, and flowers of sliced sausage, peas, and egg whites.

[2] Certainly most of the "wine" in this ancient account must have been fermented from grain, for it dates from the Chou dynasty, before distillation was known and before grape wines were commonplace. Though the term "wine," as used in modern China, may embrace all sorts of alcoholic beverages, distilled as well as fermented ones, most Chinese fermented alcoholic beverages, in ancient times as today, are made from grain, and might better be called "beers" or "ales." For a discussion of the confusion surrounding the term "wine" in China, see "Problems with Terminology" in our section on alcoholic beverages in the "Beverages" chapter.

[3] On this in modern times, see "Dining as an art," pp. 130-46 in F. T. Cheng, 1954.

[4] The Chinese food scene is so varied as to make it difficult to describe in words alone; it requires abundant, carefully-chosen illustrations. To this writer, two of the most successful efforts at presenting the food scene visually are the beautiful book by Lionel Tiger and Reinhart Wolf (1985), *China's food: A photographic journey*, and that by Emily Hahn *et al.* (1973), *The cooking of China*.

[5] The percentage of family income spent on food in traditional China was also very high, on average about 60 percent of household expenditures (Koo, 1973: 117). In large part, this was a reflection of the general poverty of China's people, and such percentages dropped significantly with increased family income. In a study of 283 Peking families in 1926-27, for example, Gamble (1933: 47, 290-92) found the share of family budgets spent for food to range from about 14 to 83 percent. All families that spent 70 percent or more of their budgets on food had incomes of below $20 (Chinese silver dollars) a month, whereas all but one of the families that spent less than 25 percent of their budgets on food had incomes of more than $50 a month. We have found no data comparing percentages of income spent for food among Chinese and Americans or Europeans with similar incomes. Whatever such comparisons might show, K. C. Chang (1976: 118-19; 1977b: 12) has argued that the very need for the Chinese to devote such a large portion of their income to food would make them more food-conscious than Americans.

[6] Several of these are the names of well-known restaurants in Peking in the 1930's (Arlington and Lewisohn, 1935: 269-71).

[7] One of the best statements of the Chinese tolerance for noise that we have found is in the account of Peking by Arlington and Lewisohn (1935: 267):

> "The rickshaw man can ring his bell, the chauffeur toot his horn, the hawker cry his wares on the street at all hours of the day and night. Private theatricals raise a perfect pandemonium until the early hours of the morning, whilst the racket kept up at Chinese dinner-parties by the 'morra' players, the high shrill voices of the sing-song girls, and the discordant notes of the various musical instruments is something that must be heard to believed. No one minds it in the least...no one ever dreams of making the slightest complaint. It is all taken as part of the day's work, on the principle of letting the other fellow enjoy himself as he likes. Indeed, nothing less than a revolution, riot, or an attack of bandits is considered here the equivalent of our 'disturbance of the peace.'"

[8] The yin-yang symbol, described by one scholar as calling to mind "two tadpoles interlocked," was placed above the entry doors of homes in traditional China to protect against misfortune (Gray, 1878, 2: 44). It was also a common motif on Chinese pottery and in other forms of art, and it is even represented in certain desserts or main dishes such as "Yin-Yang Rice" (Y. K. Martin, 1984: 174-75; Tiger and Wolf, 1985: 37).

[9] E. N. Anderson (1988: 190-91) writes of a group of foods considered by the Chinese to be "cold" (*han*), a category distinct from "cooling," foods. The concept of cold was neither conspicuous nor clear in the minds of the Chinese to whom he spoke, but cold foods were believed to make one's stomach or body feel cold. This places them at the opposite end of the spectrum from certain heating foods, those, such as dog flesh, which are typically eaten in winter to warm one's body.

[10] For an excellent collection of recent articles on manifestations of the hot-cold food concept around the world, edited by Lenore Manderson, see the special issue of *Social Science and Medicine* (Vol. 25, No. 4, 1987) entitled "Hot-cold food and medical theories: Cross-cultural perspectives."

[11] Of interest in this regard are the striking similarities among the foods classified as hot or cold by the Vietnamese (Mathews and Manderson, 1981: 12-14), Malays (C. S. Wilson, 1971: 96-97; Laderman, 1983: 44-46), and Gujeratis (Pool, 1987: 390).

[12] "Congee" derives from the Tamil word *kanji*, meaning "boilings." The term is widely used in India to refer to the water in which rice has been boiled, which is the common starch of Indian launderers and also a food for invalids. Because of this, congee is used, on occasion, in India in the more general sense of liquid food given to invalids (Yule and Burnell, 1903: 245). In works on China, however, "congee" usually means "rice gruel," a watery porridge (*chu*; Cantonese: *jook* or *tsuk*) which may be eaten plain or contain other ingredients such as meat, fish, vegetables, or flavorings. Sometimes, however, the term congee is used in a more general sense for "gruel," which in North China is often prepared with millet, "millet congee." In this book, we use the term congee to refer not to the rice water of India, but to the Chinese dishes in question. In China, congee is commonly eaten for breakfast, though it may be consumed at other times as a snack or light meal, food for invalids, or even, with proper flavorings, as a sweet.

[13] For a discussion of Chinese love-philters, see Wallnöfer and von Rottauscher, 1965: 73-75.

[14] Among the London Chinese women studied by Tan and Wheeler (1983: 43, 44), *pu* foods were often made with alcohol, for alcoholic beverages such as rice wine, brandy, and beer, are regarded as repairing.

[15] The Chinese search for such items led to the establishment of trade that reached distant continents. This story remains little known, but, in some cases, Chinese demand for food and medicine was a significant element in threatening the extinction of certain species. In prehistoric and early historic times, when precipitation was heavier in China and forest extended far to the north of its present-day occurrence, the rhinoceros (*hsi*) was found in many parts of the country, even in North China. Remains of rhinoceros were found at An-yang in the second millennium B.C., and, judging from Shang oracle-bone inscriptions, it was both hunted and used as a sacrificial victim, though an uncommon one. By the T'ang period, the rhinoceros survived only in restricted areas south of the Yangtze River, especially remote areas of Hunan and parts of Nan-Yüeh. Horn was not the only rhinoceros product in demand: rhinoceros-hide armor, for one, was highly regarded in early China, and the people of Yüeh were said to use six different types of it (though old-fashioned, rhinocèros-hide armor was still used in T'ang times). At the same time, horn was the most significant of rhinoceros products. It was carved into elaborate cups, representations of gods, and objects of other sorts, with the craftsmen of Canton especially renowned for such carving. Rhinoceros horn was believed to be sensitive to the presence of toxins, and was used to counter arrow poisons, the bites of noxious snakes, and poisons of other kinds. It was said that cups of rhinoceros horn would sweat when poison was present, and that if one stirred a poisonous potion with a rhinoceros horn, it would froth and completely lose its toxicity. Rhinoceros horn had other medicinal virtues, too; prominent among them, it was *pu*, able to strengthen and invigorate (Soubeiran and Dabry de Thiersant, 1874: 47; Sowerby, 1940: 55, 172; C. A. S. Williams, 1941: 340; Burkhardt, 1953-58, 3: 142; Schafer, 1963: 83-84, 241-42, 260; 1967: 226-27; Wallnöfer and von Rottauscher, 1965: 72; I. H. Burkill, 1966, 2: 1926-28; Ware, 1966: 297-98; Pálos, 1971: 189; Read, 1976: Section 355; C. Li, 1977: 106, 107, 192; K. C. Chang, 1980: 139-41; 1986: 78-79, 211).

In T'ang times, rhinoceros horn was sent as tribute from parts of China and Indochina, and was also being imported to Canton by foreign merchants from Southeast Asia (Schafer and Wallacker, 1957-58: 225, Map 6; Schafer, 1963: 241-42; 1967: 77, 175, 194, 226-28; 1969: 83-84). In Sung times such trade continued, and Chinese merchants knew sources of rhinoceros horn in Indochina, Malaya, Java, India, and the horn of East Africa (Wheatley, 1959: 76, 77; I. H. Burkill, 1966, 2: 1926-28; Hirth and Rockhill, 1966: 15-16, 19, 46, 48, 67, 68, 77, 111, 116, 128, 149, 233).

Many other peoples across Southeast Asia, India, and Europe also valued rhinoceros horn (for many of the same uses as those of the Chinese: in carving, as an agent in identifying and countering poisons, and other medicinal uses). They also valued other rhinoceros products, with rhinoceros bezoar stone, a concretionary deposit of great rarity, even more valuable than horn, because it possessed magical qualities in addition to enjoying the medical virtues of rhinoceros horn (Watt, 1889-96, 6, Part 1: 490; I. H. Burkill, 1966, 2: 1926-28; B. Walker, 1968, 1: 48-49). Thus various peoples came to be involved in the trade in rhinoceros products. The ancient Romans, for example, obtained rhinoceros horn from Africa; then, after the beginning of the Christian era, they probably also obtained it from India and even Southeast Asia (I. H. Burkill, 1966, 2: 1926-27).

[16] Walnut-jujube paste is a Chinese health food, with the walnut kernels believed tonic to the brain, as noted on packages that are sold. When such packages are imported to the United States, however, that claim is taken off so as not to violate laws on truth in advertising (E. N. Anderson, 1988: 136).

[17] Research on Taiwan led one observer (Koo, 1973: 137) to suggest the possibility that red-colored foods, such as red crab, tomatoes, and carrots, may be classified as "nutritious" ("beneficial"?) because red is symbolic of good fortune, prosperity, and happiness. For more on symbolism of the color red, see footnote 1 in our chapter on "Eggs."

[18] For more information on the pangolin, see our chapter on "Flesh and Fowl," and on the raccoon dog, see the footnotes to that chapter.

[19] The Chinese believe that tiger flesh, bones, and skin are strongly protective, and capable of dispelling evil. Paintings and written accounts of the tiger have that capability, too, and some Chinese sages have recommended that persons wishing to be cured of persistent fever might do so by reading "treatises on tigers" (Cormack, 1935: 252). For details on medicinal uses of the tiger from the *Pen-ts'ao kang-mu* (sixteenth century A.D.), see Read, 1976: Section 351.

[20] For other Chinese concepts involved in food selection, see E. N. Anderson and M. L. Anderson, 1976: 150-53; Gould-Martin, 1978: 40-44; Mathews and Manderson, 1981: 12-13; Tan and Wheeler, 1983: 42-44; and E. N. Anderson, 1988: 188-89, 196.

[21] A Chinese woman who is not pregnant may also avoid soy sauce lest her skin darken, and this is particularly of concern when she has a wound that is healing, for the new skin, it is believed, will be darker than the old (Koo, 1973: 137-38).

[22] For general information on foods and festive occasions in China, see E. N. Anderson, 1988: 202-4; for Chinese festival dishes, see especially Feng, n.d.: 182-219, and also Tun, 1965: *passim* (Peking) and C. S. Wong, 1967: *passim* (Malaysia).

[23] For an excellent discussion of these and other non-subsistence uses of food in South China, see E. N. Anderson and M. L. Anderson, 1977: 366-80.

[24] Other portraits of the Kitchen God may be found in Cormack, 1935: opp. p. 144; Eberhard, 1952: 25; and Burkhardt, 1953-58, 2: opp. p. 1.

[25] In the Peking area, a festival sweet made of sticky rice is used for this purpose, pieces of it being cast into the fire along with the Kitchen God's portrait. In some parts of China, the Kitchen God's mouth may be daubed with opium so that he becomes sleepy, and unable to speak clearly; or his portrait may be dipped in wine to make him drunk so that he will be ejected and not permitted to speak of the family (Cormack, 1935: 144).

[26] Such practices were widespread in China, though there seem to have been differences from place to place and time to time. In former times, it was widespread practice to actually feed the deceased by placing food in its mouth (J. J. M. de Groot, 1964, 1: 29-30, 197, 228-31, 356-60; 2: 382).

[27] The actual burial could be put off for months, or even years, to enable selection of a lucky site and an auspicious day (Latourette, 1964: 538).

[28] One reads, for example, of the consort of one ruler being buried with a hundred pots of preserved flesh (J. J. M. de Groot, 1964, 2: 383).

[29] Each ancestor has his own tablet at home, "a rectangular piece of wood set in a base" on which the ancestor's name is inscribed and to which his spirit adheres during rites (Cormack, 1935: 103-4; H. C. Hu, 1948: 33-34). For interesting accounts of the role of such ancestral tablets, see M. Freedman, 1970 and H. D. R. Baker, 1979: 86-91 (general); Emily Ahern, 1973: 102-5, 116-48 (Taiwan); and H. D. R. Baker, 1968: 54-66, 106-8, 110-11 and H. G. H. Nelson, 1974: 267-73 (New Territories of Hong Kong). The tablets in the Taiwan village were made of wood, but those in Nelson's New Territories village were almost all of paper, not of wood, the change being relatively recent. Typically, according to Freedman, only three or four generations are included on tablets, but in the New Territories village many tablets went back six to eight generations, and a few went back much farther, one of them from the twelfth to the twenty-sixth generation.

[30] Cormack (1935: 90), writing of customs in the Peking area, says that for a young man or woman 15 or more years old, food offerings are made every morning and evening for an entire year after death. The offerings, made before the ancestral tablet, include a bowl of rice, chopsticks, and vegetables and meat, commonly things the dead person had particularly liked. Care was taken that the evening offering be done before it was dark, lest the spirit lose its way and cause the family difficulty.

[31] One reads of sacrificial food to household deities and ancestral spirits being taken and presented to a prospective mother to eat in the belief that this will bring her good fortune (Best, 1926: 570).

[32] This was the time of the spring festival called Ch'ing Ming ("Pure Brightness" or "Clear and Bright"), normally held each year on about the 5th of April, and sometimes called by Westerners the "Feast of the Dead," "Festival of the Tombs," or the "Grave-Sweeping Festival." Until the graves are cleaned and the ancestors honored at the time of Ch'ing Ming, no fires are lighted at home, which led to another of its names, "Festival of the Cold Food." Those acts completed, fires are re-lighted and festivities carried out. One special dish of this time is a pudding with nine layers, nine being a Chinese lucky number tied to "sacred power and ever-flowing life." A recipe for this pudding is given by Feng (n.d.: 191-95).

For more details on this festival, see S. W. Williams, 1895, 2: 252 (Kwangtung); Cormack, 1935: 188-90 and Tun, 1965: 26-27, 107 (Peking); C. S. Wong, 1967: 109-19 (Malaysia); and Eberhard, 1952: 113-27, H. D. R. Baker, 1979: 86, and Latsch, 1984: 46-54 (general).

[33] A lineage is a common descent group consisting of a number of men and their families. Its members have the same surname and ancestry; its center is a village or other settlement where its members, or many of them, live; and its worship is focussed on its ancestral temple. Many writers have called such common descent groups in China "clans," and some have called them "sibs" or "gens." H. D. R. Baker (1979: 49-70) uses the two terms "lineage" and "clan," but in different senses. By his usage, a "clan" is an "artificial kin group," a federation of lineages whose members are from a single geographical area and have a common surname, but which is formed primarily for increasing the social and

political influence of its member lineages. Baker observes that such a clan also may have an ancestral hall where it worshipped its assembled ancestors and performed other functions, but that the temple was usually located in a city and apparently only upper-class persons from its member lineages participated actively in its operations.

For accounts of the role of common descent groups in traditional China, including the veneration of ancestors and ancestral halls and their functions, see H. C. Hu, 1948; M. Freedman, 1970; and H. D. R. Baker, 1979.

[34] In Kwangtung, writes C. K. Yang (1961: 43), a person's membership in the lineage (clan) was demonstrated by his receiving a portion of the sacrificial flesh. If such meat was withheld, it signaled that the person had been excommunicated from the lineage because of a violation of its regulations.

[35] A Chinese ancestral spirit who does not receive such offerings is described not as "homeless" or "penniless," but as "hungry" (H. D. R. Baker, 1983).

[36] "Beiträge zur Geschichte von Vegetarismus und Rinderverehrung in Indien," *Akademie der Wissenschaften und der Literatur, Mainz, Abhandlungen, Geistes- und Sozialwissenschaften Klasse* 1961: 559-625.

[37] It has been observed (Welch, 1958: 88; D. H. Smith, 1968: 98) that it is improper to call Taoism a religion because it contains non-religious elements as well, among them the philosophical thinking of Lao Tzu; expeditions by sea to seek the Isles of the Blest; group sexual orgies; and secret societies dedicated to revolution.

[38] Chinese alchemy was not, like alchemy in Europe, a search for gold, but mainly for "a chemical elixir of immortality" (Welch, 1958: 96-97).

[39] Different lists have been given for the five, with Welch (1958: 91n) listing them as rice, millet, wheat, barley, and beans.

[40] There are other stories, as well, of Mencius' mother and her residential moves in the interests of her son. Legge (1895: 17-18), who presented a collection of such stories, observed that, though believed by the devout in China, some may be fabrications.

[41] After a sacrificial ritual had ended, the flesh, which was believed to possess special merit, might be divided among those who had attended. Eating pork from an autumn sacrifice to the ancestors was believed to make more likely the birth of male offspring and to help the sons of a family pass certain examinations and gain fame (Doolittle, 1865, 1: 229; Soothill, 1923: 136).

[42] This work, the "Essay on actions and retributions," may have been written in the eleventh century A.D. Like Gideon Bibles in the United States, copies of it are distributed without charge by monasteries and charitable organizations (Welch, 1958: 139-40; D. H. Smith, 1968: 110).

[43] For a photo of the sacred turtle pond in Penang, Malaya, at the "Temple of Paradise," described as "the largest and the most magnificent Chinese Temple in South East Asia," see C. S. Wong, 1967: Plate 8.

[44] For the tale of a devout vegetarian woman of Ming times who broke her vow of vegetarianism, took sick, and died, see Cormack, 1935: 161-66.

[45] For information on Buddhist vegetarian houses and vegetarianism in Malaysia, see C. S. Wong, 1967: 155-58.

[46] The seventh-century Chinese Buddhist pilgrim Hsüan-tsang noted that Hinayana Buddhists in India accepted the flesh of three animals as lawful food: wild geese, calves, and deer of a certain type. Hsüan-tsang, as a devout Mahayana Buddhist, would not eat such flesh, and other Mahayanists banned milk and dairy products as well (Watters, 1904-5, 1: 57; Banerjee, 1977: 59).

[47] According to one report from Ch'ing times, the cooks at Buddhist temples, in competing for customers, were not above improving the taste of their vegetarian dishes by adding a little chicken broth or chicken fat (Spence, 1977: 290). There are reports of even more serious violations involving the serving of meat and wine at Buddhist temples, but to balance these one finds accounts of foreigners being turned away for fear that they might smuggle meat into a temple and consume it there (Welch, 1967: 112-13).

[48] For a recipe for vegetarian ham, see H. J. Lin and T. Lin, 1972: 81; for vegetarian chicken, S. H. Huang, 1972: 111; and for "Mock Fish," with a potato taking the part of the fish, M. Gin, 1984: 97.

CHAPTER TWO

Major Regional Cuisines of China

"To be born in Soochow, to eat in Kwangchow,
to dress in Hangchow, and to die in Leouchow."

modern Chinese folk saying

Northern Cuisine
Eastern Cuisine
Western Cuisine
Southern Cuisine

China is widely acknowledged to have some of the finest food in the world, as reflected in the common expression in Southeast Asia that "the way to contentment is to live in an American house, to have a Japanese wife, and to employ a Chinese cook." Many contend, moreover, that of all Chinese food, Canton has the finest, a view reflected in the modern folk saying quoted above, which indicates where one may enjoy the best China has to offer: Soochow for its beauty, Hangchow for its fine silk, Leouchow for its excellent wood (used for coffins), and Kwangchow (Kuang-chou), or Canton, for its superb food. Whether or not Cantonese cuisine deserves primacy among the regional cuisines of China is, of course, a matter of dispute, as is the question of what constitutes a cuisine, and how many of China's regional cuisines qualify as "major" or "great." We turn first to a definition of "cuisine."

The Webster's definition of "cuisine" is "manner of preparing food" or "style of cooking." Elisabeth Rozin, in her book *Ethnic cuisine* (1983: xi-xvi), writes of cuisine in terms of differences among cultures in their culinary customs, their "rules or traditions about how to cook." She describes differences among cuisines in their essential foodstuffs, their methods of cooking, and their regular use of certain flavorings or combinations of flavorings. The latter, which Rozin calls "flavor principles," are a critical aspect of cuisine, ethnic markers that label a food as belonging to one cuisine or another.[1] They may vary not only by country but from one part of a country to another. Rozin thus writes of the cuisine of China, with its basic flavor principle of soy sauce, rice wine, and gingerroot, and of its regional cuisines, through additional flavorings, being variants of this.

Cuisine, in the sense used by Rozin, seems to have developed in China in ancient times. In *Food in Chinese culture,* K. C. Chang (1977b: 14) writes of a food tradition going back three thousand years or more in the literature, a tradition that enjoyed a special position in Chinese culture and entailed not just a distinctive collection of items used as food, but their preparation and consumption in terms of the fan-ts'ai principle and in agreement with beliefs about the significance of individual foods for human health and well-being. Chang sees that Chinese food tradition as persisting through time, changing but remaining basically the same. Other contributors to *Food in Chinese culture* credit specific historical periods, when outside culinary influences were especially strong, with having been particularly influential in the development of Chinese cuisine. Y. S. Yü (1977: 79-82) calls the Han period (206 B.C.-A.D. 220) one of culinary innovation, a time when major new edibles were introduced to China, one of experimentation with ordinary foodstuffs and receptiveness to the ideas of other peoples, one which was ultimately to form a unique episode in the history of Chinese cookery. Schafer (1977: 88, 118, 127-31), in the same volume, writes of the T'ang period (A.D. 618-907) as being characterized by innovations in agricultural production, of being a time of new foreign foodstuffs, particularly spices and aromatics; of dietary permissiveness and experimentation, at least among the upper classes; and of great interest in foreign foods. In his view, the T'ang period is to be credited with a revolution in food use which led ultimately to the abundance and variety of present-day Chinese cuisine. Freeman (1977: 144-45), in turn, emphasizes the importance of the Sung period (A.D. 960-1279), when striking improvements in agriculture and the expansion of trade created conditions of plenty that contributed to a significant development of Chinese cuisine.[2] He views Sung culinary development as an urban affair, centered in capital cities such as K'ai-feng and Hangchow whose inhabitants were exposed to foodstuffs drawn from throughout China and beyond, and were less bound by tradition, giving cooks broad scope for culinary experimentation. This calls attention to the broader proposition that cuisines are defined by superior dishes of a region, and that in China such dishes tend to be drawn from urban areas where the upper-classes and finer restaurants are concentrated. This should be borne in mind in reading the following discussion of China's regional cuisines.

Turning to the question of when regional cuisines may first be discerned in the Chinese record, Schafer (1977: 129-31) remarks that T'ang cuisine varied locally, but that little is known about it. It is clear, however, that differences between the north and south were widely recognized, for Schafer writes of the wonder and sometimes disgust northerners expressed toward an endless variety of southern dishes, which were characterized by strange flavor combinations such as fagara and oranges, salty and sour, flavor principles in the sense used by Rozin. In Sung accounts the matter becomes somewhat clearer, with writers referring to three major regional cuisines: those of the north, south [lower Yangtze], and Szechwan (Freeman, 1977: 168-70, 175).[3]

As for traditional China and its major regional cuisines, we note James Wei's suggestion that to qualify as a major regional school of cooking (regional cuisine), a respectable restaurant in the area should be able to provide to its dinner customers, as requested, over a hundred courses made from foods available locally (Hughes, 1972: 27). F. T. Cheng (1954: 55-57) suggests further that to qualify as major, a regional school must also have distinctive tastes and ways of preparing food, and he observes that while every province, and even every district, of China possesses its own food specialties, there are five major regional schools in China in terms of ways of cooking. The five he lists are identical to those provided by Wei: Shantung (including Peking and the north), Honan, Szechwan, Fukien, and Canton. Other Chinese, though they may not agree as to which regional schools are major, also commonly give them as five, a number which has had a special fascination for the Chinese since antiquity.[4] Writers of English-language books on Chinese food (many written by authors of Chinese descent), on the other hand, often speak of four major regional cuisines, and frequently list them as: (1) "northern" (sometimes "Peking" or "Shantung"); (2) "eastern" ("Shanghai," "lower Yangtze," etc.); (3) "western" ("Southwestern," "Szechwan," or "Szechwan-Hunan"); and (4) "southern"

("Canton").[5] In our account, we follow this four-fold system for the sake of simplicity, and consider China's regional cuisines in the order just given (Figure 19).

NORTHERN CUISINE

Northern Chinese cuisine is that of Peking and the provinces of Shantung, Hopeh, Shansi, Shensi, and Honan. This is a region of continental climate, with cold winters, hot summers, and low or relatively low average annual precipitation (20 to 30 inches a year in most of the North China Plain, whose North American climatic analogues are Kansas and Nebraska) (C. S. Chen, 1970: 35-36). Its growing season is far shorter than that of southern China, and double-cropping was found on less than half of all cultivated land. Precipitation, which occurs mainly in summer, varies considerably in amount from year to year to create problems of flood or drought, sometimes accompanied by famine. Its crops, most important among them wheat, millets, sorghum, and maize, are mid-latitude ones, and many of them, as well, are noted for their drought tolerance.

Northern Chinese cuisine was shaped mostly by the indigenous cooking of the region, especially that of Shantung and Hopeh. Significant in the development of regional cuisine was the area's geographic position with respect to the trade routes that since antiquity ran across Asia to the Near East and Europe, and which, as Berthold Laufer (1919) has demonstrated so well, brought to China a rich assemblage of food plants, spices and flavorings, and other edibles. Ultimately that route was to be the major one by which Islam penetrated China, the nation's northwest becoming the great center of Islamic population from which Moslems migrated to cities throughout the north to leave their mark on regional cuisine. The north's geographic position also exposed it to pastoral peoples, among them the Mongols, who lived in the arid lands to the north and northwest and regularly visited and sometimes conquered the region. Peking, as capital of China during most of the time since the great Mongol ruler Kublai Khan moved his capital there from Karakorum in the thirteenth century, was the nation's leading center of wealth and power, which gave it an unusual opportunity to attract some of China's best chefs, to obtain a broad range of dishes from throughout the land, and to support restaurants representing the major regional cuisines and some specialized ones (*i.e.*, Moslem) as well. Thus in talking of northern Chinese cuisine, one is dealing not only with indigenous dishes, but with those drawn from other parts of China, those created for the imperial court, and those deriving from Mongols, Moslems, and a Buddhist community which made its culinary contribution in vegetarian dishes. The last of China's northern conquerors were the Manchu, founders of the Ch'ing dynasty (A.D. 1644-1911), a people less advanced in culture and sophistication than the Chinese, and initially not particularly food-oriented. Over time, however, many Manchu rulers came to have a considerable interest in cookery and to exhibit an unusual interest in the creation of new dishes, with the result that the Ch'ing dynasty, especially the reign of Ch'ien Lung (1736-96), is regarded by some as the period when Chinese cooking reached its greatest heights, not achieved since in Republican or Communist China (Davis, 1966: 15; Em. Hahn *et al.*, 1973: 17).[6]

Its eclecticism has led some food writers to argue that Peking's cuisine is not "a true regional cuisine," that its sweet-and-sour dishes derive from Honan or that Shantung is the fount of the city's culinary greatness. Some food experts (F. T. Cheng, 1954: 55, 62; G. B. Miller, 1966: 16; Hughes, 1972: 27) thus speak not of Peking cuisine but of a "Shantung" or "Northeastern-Peking-Shantung" one. The latter view has a certain merit, for Shantung chefs

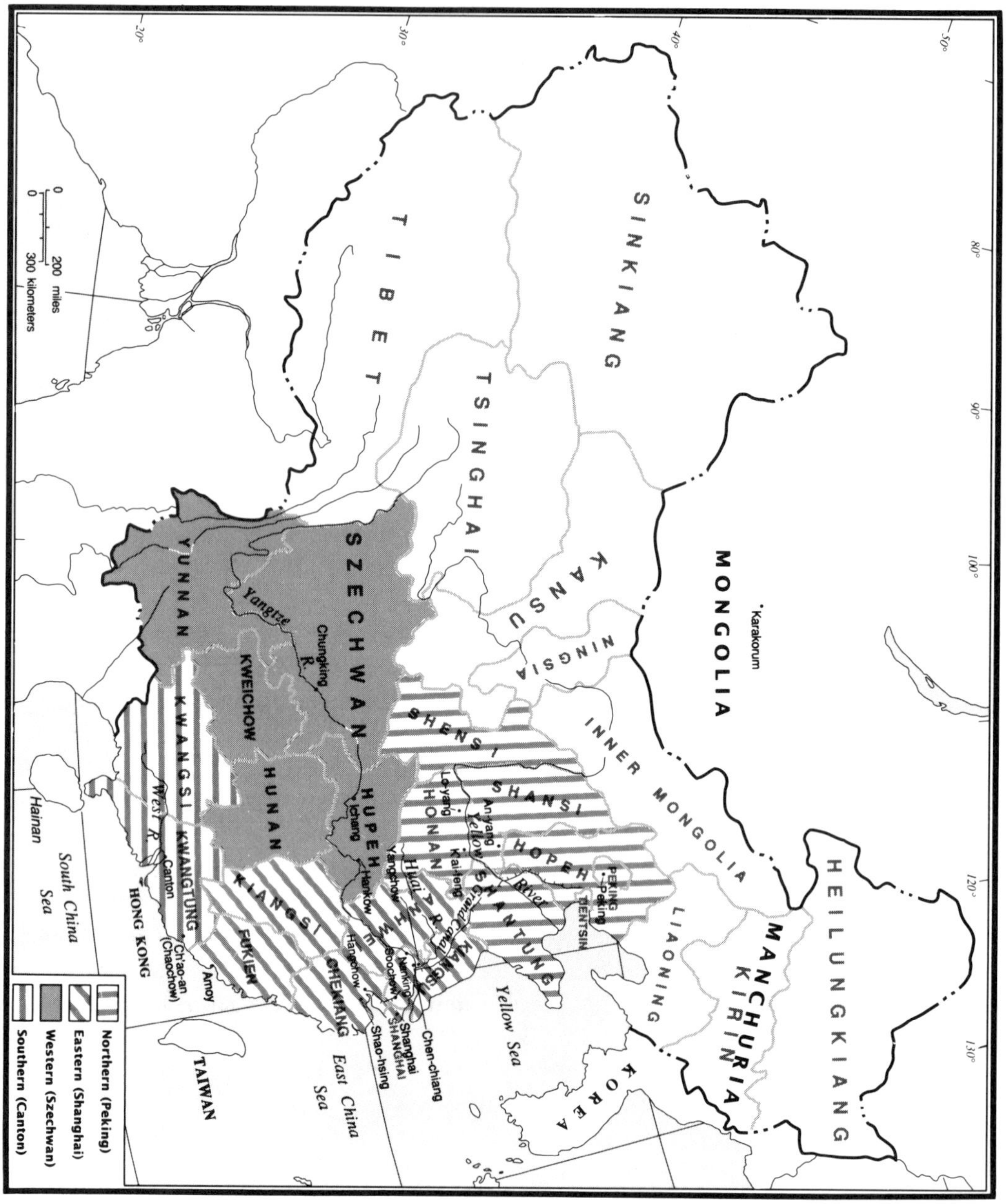

Figure 19. Main regional cuisines of China

dominated the royal kitchens in Ming times; they were prominent in Peking's best restaurants; and their influence was felt in Shanghai and other cities of eastern China as well (Sakamoto, 1977: 37; C. B. T. Lee and A. E. Lee, 1980: 89). At the same time, the cooking of Peking goes beyond that of Shantung to include a significant number of dishes drawn from other sources. Kenneth Lo (1971: 16), in his *Peking cooking*, writes of the city's many culinary debts, to surrounding Shantung and Hopeh, to more distant parts of China, and to the best chefs of dynasties (among them Chinese, Mongol, and Manchu) spanning nearly a thousand years. The result of these varied contributions, in his view, has been the creation of something new, and the best of all of China's culinary traditions today.

Basic to an understanding of northern Chinese cuisine is the importance of wheat, sorghum, millets, and maize, rather than rice. Various of these cereals are cooked and served as porridge or, in a fluffy form much like boiled rice, as a base for meat or vegetables. Maize is made into a bread, and wheat is used in preparation of many of the region's fried breads, steamed breads, and noodles; the stuffed dumplings for which the north is noted are wrapped in skins made of wheat flour; and its most famous dish, "Peking Duck," is eaten not with rice, but with pancakes made of wheat flour.[7] Elisabeth Rozin (1983: xv) characterizes Peking cuisine as often including soy bean paste, garlic, and sesame oil, in addition to the basic Chinese flavor principle of soy sauce, rice wine, and gingerroot. Also common among northern flavorings are vinegar; star anise; chives, leeks, scallions, and onions; sesame seeds; wine stock;[8] and sweet-and-sour sauce.[9] Overall, northern dishes are light, less spicy and less oily than those of Szechwan. Conspicuous in northern Chinese cuisine are celery cabbage (*Brassica pekinensis*),[10] pork,[11] lamb and mutton, and beef and fowl dishes as well. Fish are not abundant in most of the region,[12] and are often consumed in a dried or salted form (F. L. Chang and G. M. Fuchs, n.d.: 141-43; F. T. Cheng, 1954: 61-62; W. W. Chang *et al.*, 1970: 3-4; K. Lo, 1971: 16-17, 78; 1979a: 15-23; Claiborne and Lee, 1972: xviii; Em. Hahn *et al.*, 1973: 14-15; G. Z. Chu, 1975: 22, 55; F. Lin, 1975: 6; V. Y. N. Hsu and F. L. K. Hsu, 1977: 300-303, 309-11; Sakamoto, 1977: 37-39; C. B. T. Lee and A. E. Lee, 1980: 87-90; K. Lee and A. Branyon, 1984: 11-14; Y. K. Martin, 1984: 209-12).

Lamb and mutton are not used much or well liked by most Chinese because of their strong taste and the pronounced odor they impart to cooking utensils (G. Z. Chu, 1975: 55, 115, 123). Thus one finds few lamb and mutton dishes in the major cuisines of China. Such dishes, though still not many in number,[13] are more prominent in the north because of its proximity to arid regions where sheep are numerous, and to the influence of peoples (especially Mongols and Moslems) for whom lamb and mutton are dietary mainstays[14] and who have contributed the best-known Peking lamb dishes. The importance of such dishes in the several Peking Moslem restaurants of the 1930's was such that Chinese referred to them simply as "mutton houses" (Arlington and Lewisohn, 1935: 271). One type of lamb or mutton dish involves hot-pot (fire-pot) cooking, as in "Mongolian Hot-Pot" and "Mongolian Lamb Fire-Pot." For these dishes, the diner simmers strips of lamb or mutton, noodles, and other ingredients in broth in a charcoal-burning hot-pot[15] at the dining table, eating the flesh after dipping it in sauces of one sort or another. Sesame-seed rolls are commonly an accompaniment. Other well-known Peking dishes are barbecued lamb and beef which, like "Mongolian Hot-Pot," are prepared at the table and eaten with sliced onions or scallions, dips, and sometimes with pancakes or steamed buns.[16] Kenneth Lo (1971: 17, 26-27, 71-72) has written that despite Manchu rule in China for nearly three centuries, Manchu cooking had little impact on the cuisine of the Peking area except perhaps through the royal court, and the single Manchu dish we find in his *Peking cooking* is "Pheasant Hot-Pot."

Honan is a small, densely-settled province to the southwest of Shantung, much of it located between the valleys of the Yellow and Huai rivers, which places it within the Han Chinese heartland. Honan is covered in large part by loess soils, and is rich in early Neolithic sites. It was also of great importance in later times, the location of the Shang capital of An-yang, the Chou capital of Lo-yang, and the great trading city and Sung capital of K'ai-feng. This gave it an important early culinary advantage, one enhanced by its excellent location in terms of benefiting from overland plant introductions from the west.

In modern China, Honan enjoys a high culinary reputation in general, and some Chinese, as in early times, continue to list its cuisine as one of China's five major ones, and distinct from that of Shantung. Honan is widely recognized for its sweet-and-sour dishes,[17] especially "Yellow River Carp," China's best known fish dish and a favorite at banquets.[18] The general high regard in which this dish is held is reflected in James Wei's contention that to link Honan cuisine with that of Shantung and other parts of the north is to deny "luscious Yellow River Carp its legitimate segregated birthright" (Hughes, 1972: 27). The province of Honan is also noted for other dishes, among them "Bear's Paw"; "Walnut Kidney," a dish in which kidneys are cut to walnut-size, and quickly cooked in boiling oil so as to leave them tender but crisp; and dishes made with a highly-valued delicacy, a tree fungus called "monkey head" (*Hericium erinaceus*),[19] which has been likened it its delicate taste to the truffles of France (F. T. Cheng, 1954: 61-62; Howe, 1971: 146; Hughes, 1972: 27; Em. Hahn *et al.*, 1973: 11, 15-16; G. Z. Chu, 1975: 22; K. Lo, 1976: 45, 106; Sakamoto, 1977: 37-38; Liu and Bau, 1980: 177-79; Wittwer *et al.*, 1987: 294-95).

EASTERN CUISINE

The area of eastern Chinese cuisine as we define it centers on the lower Yangtze River Valley (though not all of the area is part of the Yangtze drainage system), and includes five provinces: from north to south, Kiangsu, Anhwei, Kiangsi, Chekiang, and Fukien. The lower Yangtze area has a long coastline along the East China Sea. It also has heavier precipitation than North China (usually 40 to 60 inches annually) and milder winters, which permit cultivation of a broader range of crops than in the north. This has led northerners, from antiquity through traditional times, to regard the region, correctly, as one of greater abundance than their own. The lower Yangtze also has extensive alluvial lowlands, with many waterways: the river and its tributaries, canals, swamps, and lakes. This provides the region with what may be a greater abundance of navigable waterways than found anywhere else in the world (Cressey, 1934: 283), and has contributed to the development of aquatic agriculture, fishing, and fish farming.

The region is also in a critical geographic position near the mouth of one of the world's great rivers, commonly called the "long river" by the Chinese, and "the main street of China" by Westerners because of its importance in the east-west transport of goods from Szechwan and the far interior to the East China Sea. The region is noted for its industry, commerce, and ocean shipping, with 10,000 ton ships reaching ports as far inland as Hangkow, six hundred miles up-river. The importance of the region in transport was enhanced in the seventh century when canals were opened to link Hangchow, south of the Yangtze in Chekiang Province, with the Yellow River of North China. It was further enhanced in the fourteenth century by opening of the Grand Canal (over a thousand miles long) from Peking to the Yangtze, and, late in the nineteenth century, by railway construction.

Besides the availability of a great variety of local foodstuffs, another significant factor in the region's culinary evolution is the fact that Shanghai is China's largest city and its most cosmopolitan. We hasten to add, however, that the emergence of superb cookery in the lower Yangtze long pre-dates Shanghai's rise to greatness following its being named a treaty port in the mid-nineteenth century. There were, and are, several cities of great culinary renown in the region, especially Hangchow and Nanking, each of which served as China's capital at various times, and Soochow, renowned as a center of wealth, beauty, and culture. Marco Polo visited Soochow and Hangchow (the latter he called "Kinsay" = *Ching-shih*=capital city, the former Sung capital) late in the thirteenth century. Soochow he viewed as a very large and illustrious city, one with many philosophers and highly skilled craftsmen, a place of considerable wealth and a center of trade and silk manufacture. Hangchow, which he described in greater detail, impressed him with its grandeur, its situation on the beautiful West Lake, and its population of over a million people. It was also an international port with a huge and varied trade with India and other foreign lands. Polo wrote of ten main markets and many lesser ones in the city. The main markets, held three days a week in huge squares, were each attended by 40,000 to 50,000 people and contained an unusual diversity of foodstuffs: game, domesticated fowl and mammals, fruit and vegetables, enormous quantities of plump and tasty fish, and spices, sugar, and rice wine. Polo also spoke of great buildings where at one time there might be a hundred different groups holding wedding feasts and banquets (Yule, 1921, 2: 181-218).

The reader will gather from the above that the lower Yangtze region has long enjoyed conditions widely regarded as favoring development of a major regional cuisine: wealth and sophistication, an abundance and variety of foodstuffs, and exposure to culinary ideas of other regions and countries. That it succeeded in achieving culinary excellence is clearly shown in the modern food literature, in which its cuisine is often described in superlatives.[20] Authors of one book (H. J. Lin and T. Lin, 1972: 112-13) speak of its "exquisite delicacy" gained by careful preparation. The taste in this region, they suggest, was shaped by gourmets who left nothing untasted, few combinations untried. Culinary techniques were perfected and considered ends in themselves, and concern with decoration was remarkable. Another food writer (Sakamoto, 1977: 105-6) focuses on the region's prosperity and cultural eminence, including the former capital status of Nanking and Hangchow, as contributing to creation of a region that "has been termed 'a heaven on earth.'" She credits Yangchow, a commercial center on the Grand Canal in Kiangsu, with China's best egg roll, Soochow with having excellent pastries prepared with centuries-old recipes, Chen-chiang (industrial and commercial center on the Yangtze near its junction with the Grand Canal in Kiangsu) with outstanding seafood dishes, and Hangchow with having the haute cuisine of the Orient. Some food writers (F. L. Chang and G. M. Fuchs, n.d.: 179; Hush and Wong, 1976: 51) say that this may be China's leading regional cuisine, the one closest to being a truly national one.

Unlike North China, the lower Yangtze was well suited, in its milder climate, greater abundance of water, and gentle terrain, for rice cultivation, and the region has become a major factor in China's rice economy. Rice is served at nearly all meals in the lower Yangtze area, though wheat is important, too, used in making certain noodles and dumplings. Rice is also widely used for stuffing poultry, as in "Eight Precious Chicken," whose stuffing, besides rice, may include lotus seeds, ginkgo nuts, melon seeds, chestnuts, mushrooms, barley, and bamboo shoots.[21] Fish and duck dishes are important, among them "Nanking Pressed Duck."[22] Pork and chicken dishes are common, including a good number of the slow-cooking kind. For example, whole chickens or ducks are sometimes steamed, along with other edibles, inside a squash or melon, and carried to the table that way. Vegetable and vegetarian dishes are also notable, such as Shanghai "Vegetable Rice," rice cooked with green vegetables.[23]

The most common word used to describe the taste of lower Yangtze dishes is "delicate." Sauces are common, but seasoning is typically mild, with salt and sugar or both used to accentuate flavors, which creates dishes usually richer and saltier than those of other regions, but less oily than western Chinese cooking. Soy sauce, too, is a common flavoring, and the region is famous for its red-cooked dishes (long, slow cooking with soy sauce), to which rice wine, sugar, gingerroot, and sometimes salt, scallions, and other flavorings are added.[24] Vinegar is also regularly used as a flavoring, as in "West Lake Vinegar Fish,"[25] and various types of vinegar are manufactured in the region, the most renowned being the thick, dark, sweetened "Chinkiang vinegar" from Chen-chiang in Kiangsu, used mainly as a dip and marketed throughout China and abroad. The region also produces the nation's premier rice wine, that of Shao-hsing (in Chekiang, not far from Hangchow), and rice wine is another common flavoring, used in "Drunken Chicken" and other "drunken" dishes (F. L. Chang and G. M. Fuchs, n.d.: 179-80; Joyce Chen, 1962: 51; H. J. Lin and T. Lin, 1972: 111-13; F. Lin, 1975: 7; Hush and Wong, 1976: 51; E. Chen, 1977: 83-86; Sakamoto, 1977: 105-7; K. Lo, 1979a: 139-46; C. B. T. Lee and A. E. Lee, 1980: 31-34; K. Lee and A. Branyon, 1984: 16-18).[26]

Fukien, which many consider as having a major cuisine in its own right, we include in the eastern region, but its geography is quite different from that of the lower Yangtze country. It is rainy (often 60 inches or more annually), subtropical, and, like southern Chekiang, has rough mountainous terrain and a coastline of submergence with rocky headlands, bays, and offshore islands. Its people cultivate rice and other food crops where they can, but topography severely limits agricultural possibilities, and the Fukienese have long looked to the sea and beyond, gaining fame for their fishing and sea trade, and becoming prominent among the Chinese settlers of Taiwan, Hainan, and Southeast Asia. Fukien cuisine tends toward broths, soup-like dishes, and stews. Such dishes typically contain noodles, whether of rice or wheat, and in Fukien noodles are said to be more varied in form and often more basic in diet than in any other part of China.[27] As for its soups, they are of unusual variety. At a Fukienese meal at home, it is said, at least two soups are provided; at banquets it is not unusual for a fourth of all dishes to be soups; and at one particular banquet, seven out of ten dishes were soups. Fish dishes are especially common in Fukienese cuisine, but chicken, duck, pork, and ham dishes are frequent as well. An unusual specialty is solidified blood. The solidified blood of pigs is ordinarily cut into pieces, which are stir-fried like soybean curd, and eaten with green Welsh onions; that of fowl is sliced thin and eaten with various dips. Fukienese cuisine is highly regarded in China for its simplicity and taste, its most distinctive flavoring being a sweet-flavored red paste made from rice-wine sediment and herbs, which is used in such dishes as "Braised Chicken in Wine Sediment Paste."[28] Another specialty is "swallow skin" (actually thin dough containing a good amount of meat), a wrapping for food which may then be steamed or added to soups such as "Shredded Swallow-Skin Meat-Ball Soup."[29] A specialty of Amoy, second largest city in the province, is *popia*, a thin pancake on which are placed: cooked filling; bean sprouts, seaweed, peanuts, and other side items; and condiments such as hot mustard and plum sauce. The pancake finally is wrapped around the edibles and eaten (F. L. Chang and G. M. Fuchs, n.d.: 117-40; F. T. Cheng, 1954: 61; G. B. Miller, 1966: 18; H. J. Lin and T. Lin, 1972: 124-136; Em. Hahn *et al.*, 1973: 16; F. Lin, 1975: 287-89; E. N. Anderson and M. L. Anderson, 1977: 356-57; E. Chen, 1977: 74-76; K. Lo, 1979a: 77, 131-35; C. B. T. Lee and A. E. Lee, 1980: 34; K. Lee and A. Branyon, 1984: 21-22; E. N. Anderson, 1988: 162, 164).

WESTERN CUISINE

The province of Szechwan is the great center of western Chinese cuisine, but as we define the region it also embraces the neighboring provinces of Yunnan, to the south, and Hunan, Hupeh, and Kweichow to the east and southeast. Szechwan is one of China's largest provinces, and the most populous, 99.7 million in 1982, or about ten percent of the nation's people. Much of Szechwan is in mountain or plateau, but relief in its eastern sections is lower, for the most part hilly, though rimmed by high mountains. This, and the characteristic color of its soils, has led the region to be called the "Red Basin," an area renowned for its fertility, but which suffers from a shortage of level land; thus on many of its hillsides, even on quite steep slopes, agricultural terraces have been constructed, the most extensive in China. To the west of the Red Basin is thinly-settled mountain and plateau country, one that adjoins the Tibetan Plateau, 12,000 feet above sea level, and whose inhabitants are mostly Tibetan. To the south and southeast are mountains and a high plateau, several thousand feet above sea level and cut by deep valleys, which covers much of Yunnan and Kweichow. Though elevations are lower and the upland more dissected in Kweichow, there is little level land for cultivation in either of these provinces (Cressey, 1934: 369-70). Both Szechwan and these neighboring provinces contain sizeable minority populations related linguistically to ones found in nearby areas of Southeast Asia. In 1982, Szechwan's minority population, mainly Tibetan and Yi (Lolo), amounted to about 3.7 million; Yunnan had quite varied minority groups numbering over 10 million, or nearly a third of its total population; and Kweichow, in turn, had nearly 6.9 million minority people, nearly a fourth of its total population (Banister, 1987: 298-99, 325).

The climate of Szechwan is subtropical, like that of northeastern Texas or adjacent Oklahoma: humid, moderate in precipitation (usually 35 to 45 inches annually, concentrated in the high-sun season), hot in summer but mild in winter because its mountains protect it from the cold air masses of the north. Its hilly but fertile "Red Basin," protected by encircling mountains from the biting, winter winds of the north, can produce two or even three crops a year. Irrigation and networks of canals existed in Szechwan already in pre-Christian times, and today it is one of China's most important agricultural areas. Rice is the most important crop, but the province's subtropical climate and reliable and moderate precipitation permit a broad range of mid-latitude crops (*i.e.*, wheat, oats, soybeans) as well as subtropical ones (oranges and sugar cane). The variety and abundance of Szechwan have long been extolled by other Chinese, who sometimes refer to the province as the "Heavenly State."

The Yangtze River has been Szechwan's major route of transport, linking the province with the heartland of China. At the same time, distance, mountain barriers, and the fact that the course of the river is broken by the famous Yangtze gorges between the cities of Chungking (Ch'ung-ch'ing) in Szechwan and Ichang (I-ch'ang) in Hupeh have created an isolation sufficient for Szechwan to develop a distinctive cuisine of its own, noted for its fiery spices.

How Szechwan cuisine came by its emphasis on strong spices has long been a matter of speculation. Some writers have observed that the regional climate tends to be hot and humid for much of the year, and that strong spices mask the smell of spoiling food, or induce sweating to help keep people cool and comfortable. Of possible significance, too, is the importance in Szechwan of preserved foods,[30] whether dried, salted, spiced, pickled in vinegar, or smoked; such preserved foods are strong and distinctive in taste, and commonly used in flavoring food. The question is whether their use predisposed people to like other strong flavorings and spices as well, part of an effort to make dull fare more palatable. Some writers have focussed on the region's distance from the main sources of salt along China's seacoast, arguing that, despite some production in the province, a scarcity of salt may have led its people to turn to strong spices as a flavoring alternative. Others have pointed to the region's geographic position adjacent to Southeast Asia and India, where piquant cuisine is the norm, as having had an impact on the cooking of China's southwest, especially in Yunnan where curries are common (F. L. Chang and G. M. Fuchs, n.d.: 247; Claiborne and Lee, 1972: xviii; Stallard, 1973a: 11; 1973b: 11; K. Lo, 1976: 65; 1979a: 196-97; Sakamoto, 1977: 225; C. B. T. Lee and A. E. Lee, 1980: 135; K. Lee and A. Branyon, 1984: 18, 20).

These explanations for the spiciness of Szechwan food recall those advanced to explain the preference for strong spices by other peoples of the world, whether today or in the past. One often reads, for example, that the widespread use of spices in medieval England was based mainly on their role as preservatives or usefulness in hiding the smell and taste of decaying flesh. Such statements have been challenged, however, by English-medieval-food historian Lorna Sass (1981), who insists that they are either wrong or exaggerated. She presents evidence that, at least in part, the heavy use of spices in medieval England derived from attempts to provide tasty fare on the numerous fast days of the time, which involved bans on meat and sometimes on milk, cheese, and eggs as well. She also notes the political competition among English aristocrats, and their efforts to impress others with their wealth, influence, and generosity by serving dishes flavored and colored with expensive spices imported from distant lands. Sass describes the close ties between cooking and medicine in medieval England, ties based on the conviction that food and spices can be effective in preventing and curing illness, a view reflected in the statement that a good cook is "half a physician." Of special interest to us, the medieval English, like the Chinese, believed foods and spices to be "hot" or "cold," "dry" or "moist," qualities passed on to the one who ate them. Though it remains uncertain why Szechwan and neighboring provinces developed their spicy cuisines, the matter is one that deserves as much attention from food historians of China as that given by Lorna Sass to the spice scene of medieval England.

Whatever its origin, it seems that the spiciness of Szechwan cuisine is not a recent development. Already in Sung times (A.D. 960-1279), there are suggestions that it was hot much as it is today (Freeman, 1977: 168-69). Its spices and flavorings at that time must have been quite varied, with fagara, or Szechwan pepper, most distinctive, but with ginger, cassia cinnamon, true pepper, star anise, fresh coriander, golden needles (dried lily buds), garlic and onions of various sorts, sesame oil and paste, soybean products, citrus fruits, rice wine, and vinegar among the items used in various combinations to create tasty dishes. Chili peppers must have been welcome additions to this assemblage when they arrived after the discovery of the New World. Chili peppers do burn the mouth in a way other spices of Szechwan may not have done, but as psychologist Paul Rozin has demonstrated (P. Rozin and D. Schiller, 1980; P.

Rozin *et al.*, 1982), people may dislike such burning on first tasting chili peppers but with repeated use most come to like, and even love, their taste and odor. That liking, according to Rozin, probably derives from several causes, among them the fact that eating chili peppers stimulates salivation, thereby enhancing flavor and facilitating the eating of bland, starchy foods. Whatever the reasons, chili peppers came to occupy an important dietary niche in Szechwan, Hunan, and other areas of western Chinese cuisine as well. In Szechwan, chili paste is the main form used, whereas in Hunan raw chili peppers are more commonly used than paste (K. Lee and A. Branyon, 1984: 19).

Today Szechwan cuisine is characterized not only by its pungent spices and varied seasonings and sauces, but by its many smoked dishes and liberal use of oil, especially sesame oil; its mixing of different flavors in a pleasing way in a single dish, for example in "Hot and Sour Soup," which is hot, sour, salty, sweet, and fragrant at the same time;[31] its use of several cooking processes in making a single dish; and its unusual emphasis on texture obtained through processes of cooking and use of such ingredients as mushrooms, bamboo shoots, and nuts (F. L. Chang and G. M. Fuchs, n.d.: 247-48; H. J. Lin and T. Lin, 1972: 103-10; Sakamoto, 1977: 225-26; K. Lo, 1979a: 196-201; C. B. T. Lee and A. E. Lee, 1980: 134-35; K. Lee and A. Branyon, 1984: 19-20). Szechwan cuisine includes many pork, chicken, and duck dishes, with "Camphor-wood and Tea-Smoked Duck," "Fragrant and Crispy Duck," and "Kung Pao Chicken" among those of greatest renown.[32] The name of the latter dish is said to derive from the title of an official (either a Szechwan governor or Peking emissary or exile) who had a special liking for it (C. B. T. Lee and A. E. Lee, 1980: 137; Y. K. Martin, 1984: 117). Along with "Twice-Cooked Pork" and "Fish-Flavored Pork," it is now among the favorite dishes in Szechwan restaurants in the United States. Szechwan is roughly a thousand miles inland, too far for it to have obtained fresh sea fish in the days before refrigeration. The Szechwanese did, however, import dried sea fish, to create tasty dishes from them as well as from the freshwater fish of their lakes and rivers, one of the latter being a spicy carp flavored with hot brown bean sauce, soy sauce, sesame oil, scallions, garlic, ginger, and other flavorings.[33] Also of note are Szechwan's bean curd dishes, most famous being the spicy and fragrant "Ma Po Bean Curd,"[34] long known elsewhere in China and in the 1930's a featured dish in a well-known Kweichow-style restaurant in Peking (Arlington and Lewisohn, 1935: 270).

While everyday dishes in Szechwan are characterized by spiciness, those served at formal dinners or feasts tend to be quite mild. This anomaly is usually explained in terms of northern influence, whether from northerners who migrated to Szechwan to evade political oppression (Sakamoto, 1977: 225) or from outside officials who governed the province and who, at grand affairs, served not the hot local foods but more subtle northern, and palace, dishes (K. Lo, 1979a: 196). Thus in Szechwan we are dealing not with one, but two styles of cooking: one, of outside origin, providing mild-flavored feast fare, and the other, of local origin, providing the spicy everyday food for which the province is noted. Kenneth Lo (1979a: 196) believes that the Szechwan peasantry is to be given credit for creating the spicy dishes that gained for the province its culinary reputation. If he is right, Szechwan contrasts notably with the north and the lower Yangtze Valley where, as we have seen, regional cuisine was influenced by the wealthy, by the presence of national capitals, and, especially in the north, by Mongol and Moslem cookery.

There seem to be few, if any, advocates of Kweichow or Hupeh as having major cuisines in their own right.[35] Hunan, in turn, has some distinctive dishes, and Hunan restaurants are popular in the United States, yet few, if any, food writers regard its cuisine as having sufficient culinary variety to qualify for major status. Lo (K. Lo, 1976: 65; 1979a: 200) argues that Hunan cuisine is best regarded as a marginal form of Szechwanese cuisine, and he and most other writers include it under "Szechwan" or western Chinese cuisine or link the two, as in "Szechwan-Hunan."

Anderson and Anderson (E. N. Anderson and M. L. Anderson, 1977: 354-55) have suggested that Yunnan may deserve recognition as having a major cuisine of its own. Yunnan is renowned throughout China for its excellent headcheese, and for hams that some Chinese consider the best in the world. It has an abundance of pork and fowl, with nuts of various kinds commonly used in meat dishes. As a region with some of China's richest forest reserves, Yunnan has a generous supply of game, fungi, and herbs for culinary use. It has a hot, spicy cuisine, like Szechwan; and curry dishes, like Southeast Asia to its south. Like Tibet and India, and in contrast to southern China in general, some of its people practice dairying and make yogurt, goat cheese, and fried milk curd. This use of dairy products has been stimulated by the province's Moslem minority, many of them descendants of soldiers of Kublai Khan who settled there in the thirteenth century (E. N. Anderson and M. L. Anderson, 1977: 354-55; K. Lo, 1979a: 201-2; K. Lee and A. Branyon, 1984: 20-21).

SOUTHERN CUISINE

Southern Chinese cuisine, as we define it, embraces the provinces of Kwangtung and Kwangsi. Its focus, however, is Canton, that great port city on the Pearl River eighty miles from the sea. As indicated in our Introduction, Kwangtung, most southerly province of mainland China, is subtropical and tropical with a year-round growing season. It has assured, ample rain; hot, humid summers; and mild winters with no frost—unusual climatic conditions for crop cultivation. At the same time, the province suffers from a dearth of level land, to which the Cantonese have responded by developing an intensive system of aquatic agriculture and fishing, and by seizing the excellent opportunities in overseas trade.

Some scholars hold that the Cantonese are the mixed descendants of the ancient Yüeh, yet we have found only scanty reference to what the Yüeh contributions to Cantonese cuisine might have been, or indeed to the more general question of the evolution of Cantonese cuisine. Schafer (1977: 129-31) did write that, in T'ang times, southern food and cooking, of the near south (Central China) and the far south (South China), was a source of endless wonder to the northern Chinese, who were intrigued by its variety. Their reactions to it were mixed, with some southern foods coming to be regarded as delicacies in the north, and others failing to be accepted because of indifference or, in some cases, revulsion. Schafer wrote little to suggest that the T'ang Chinese looked on Cantonese cookery as something special to be distinguished from that of the south in general. The same situation seems to have prevailed in the Sung period, when, according to Freeman (1977: 168-69), Kwangtung was a backwater whose cookery had not yet developed a distinctive style. Even in Yüan and Ming times, according to Mote (1977: 243), there seems to have been no awareness of Cantonese cuisine.

When, then, did Cantonese cuisine emerge, and what were the factors responsible for it? We do know that in Ch'ing times Canton was a place where Peking cooks might serve apprenticeships (Spence, 1977: 292), which demonstrates that the city had gained a national culinary reputation by then. Moreover, certain of the conditions commonly identified as leading to a major regional cuisine had been present in Canton long before. For one thing, it was a subtropical area which enjoyed an abundance and great diversity of local foodstuffs coming both from the land and sea. Another positive factor is the oft-noted interest that people of Kwangtung have in food and cooking, and their adventurousness in that regard. Still another factor, held by some to have led to the coalescence of the cuisine, was the wealth that developed in the great port of Canton and the culinary rivalry that emerged among its monied families. F. T. Cheng (1954: 57) places the latter among the most important factors in Canton's rise to high culinary rank. Its ascendancy was due not just to wealth, he insists, but to a competition for the best cooks and a rivalry in creating new dishes by the leisure class, members of wealthy merchant families. Other writers credit the fall of the Ming dynasty in 1644 with being a major factor in Canton's rise to culinary greatness. Some of them (G. B. Miller, 1966: 15; Em. Hahn *et al.*, 1973: 12) identify that period as one when many chefs from the imperial kitchens in Peking fled to Canton, bringing with them not only their own dishes and high standards of cookery, but recipes for excellent dishes they obtained along the route to the south. K. Joon (in J. D. Keys, 1963: xviii-xix) presents a somewhat different picture: of Manchu efforts to suppress adherents of the defeated Ming dynasty along the southern coast, and of Manchu officials, sent south to conquer and administer, bringing along their chefs from the north. He likens the situation to that of nineteenth century San Francisco when chefs of persons made wealthy in the gold rush came west and then remained in San Francisco. Returning to the situation in Canton, Joon observes that many chefs ultimately remained in the south, especially in Canton, where, drawing on previous culinary knowledge and a rich local variety of foodstuffs, they developed what is now called Cantonese cuisine.

Turning to modern Cantonese cookery, the characteristic that has most impressed both Chinese and foreign observers is its enormous variety, as perhaps most eloquently described by Lin and Lin (H. J. Lin and T. Lin, 1972: 137):

> "The exotic quality of Kwangtung cooking comes not from the taste but from the choice of material. It lies farther south [than] Fukien, along the sea coast. Mushrooms grew in the mountainous terrain. The sparrows and wild ducks were beaten out of the reeds, huge snails picked up. The squirming snake and slithering eel were highly regarded as food. Huge oysters were picked out of the sea, frogs, turtles and winkles from the watery rice fields, by the side of which grew quantities of green and leafy vegetables. Plump chickens and pigeons were raised specially for the table."

Today, with the Communist government's greater tolerance of culinary free enterprise, one finds an impressive array of live animals in Canton's large Ch'ing-p'ing market, and what seems to be an infinite variety of meat, fish, vegetables, spices, and other edibles—a range of foodstuffs perhaps not to be found anywhere else in China. Canton's restaurants, in their turn, can have a surprising range of dishes to offer. In the large Dai Tung restaurant in Canton, for example, our menu listed, as categories of "main dishes" (each with at least several to choose from): "Chicken"; "Duck"; "Pigeon"; "Sea and River Food"; "River Fish"; "Cuttlefish";

"Abalone"; "Frog"; "Pork"; "Beef"; "Steak"; "Hot-Pot"; "Cold Dishes"; "Soup"; "Shark Fins"; "Bird Nests"; and "Miscellaneous."

A second quality of Cantonese cuisine is the importance of fish, both fresh and salt water forms, as well as other seafoods, though there are many pork and fowl dishes as well.

In contrast to Szechwanese cuisine, that of Canton achieved major status without heavy use of chili peppers.[36] Elisabeth Rozin (1983: xv) writes of it as a cuisine in which garlic and fermented black beans are commonly added to the basic Chinese flavor principle of soy sauce, rice wine, and gingerroot.[37] Also common in Cantonese cooking is the use of herbs in sweetening and flavoring soups. Since some of these have medicinal value, other Chinese on occasion say that the Cantonese like to use medicine in preparing their dishes (F. T. Cheng, 1954: 60-61). The Cantonese often use ginger with fish and other seafoods to counter their fishiness, and vinegar, with crab and in other ways (E. N. Anderson and M. L. Anderson, 1977: 355; K. Lo, 1979a: 73; K. Lee and A. Branyon, 1984: 15). Other distinctive flavorings of Cantonese cuisine are oyster sauce, shrimp paste, and other seafood sauces or pastes;[38] small amounts of sugar;[39] and strong chicken bouillon used in soup and other dishes. Also common is the use of fruits, including subtropical ones, in dishes of flesh or fowl, as in "Orange Spiced Duck" (G. Z. Chu, 1975: 165-67), "Stir-fried Fillet of Beef with Mango" (Y. K. Martin, 1984: 96, 207), and "Pineapple Duck" (Sakamoto, 1977: 183).[40] Notable, too, is its broad range of finely sliced vegetables, its sauces (among them ketchup and Worcestershire sauce, commonly mixed together), and its mixing of seafood and meat in a single dish, for example "Casserole of Fish Head in Wine with Chicken Giblets and Pork Sweetbreads" (K. Lo, 1979a: 73, 116-17) and "Frog Legs with Winter Melon and Scallops" (F. L. Chang and G. M. Fuchs, n.d.: 102).

The overall intent in the Cantonese use of flavorings is to enhance the tastes of the principal ingredients of a dish; to create flavorful dishes interesting in texture, as through use of mushrooms and nuts; and to make them attractive in appearance and coloration. So as not to obscure the natural colors and flavors of other ingredients, the Cantonese use a soy sauce lighter in color than is usual in China. They also excel in quick methods of preparation, and are especially skilled in stir-frying,[41] for which they are renowned throughout China, although they use other such methods as well (*i.e.*, blanching, steaming, quick dipping, quick roasting). The resulting dishes may seem underdone to Americans and Europeans, but the Cantonese prefer them this way, as close as possible to their natural state. One reads of the outrage of a southerner if his food is overdone, and his happiness if chicken flesh near the bone remains a bit pink (Claiborne and Lee, 1972: xx; F. Lin, 1975: 8; Hush and Wong, 1976: 50-51; K. Lo, 1979a: 74; C. B. T. Lee and A. E. Lee, 1980; K. Lee and A. Branyon, 1984: 16). Of special note among Cantonese dishes are snack foods such as noodle dishes, wonton soup, and *tim-sam* ("dot hearts"; Mandarin: *tien-hsin*). Dim sum, as they are known in English, may be made of seafood, chopped meat or fowl, or vegetables wrapped in dough or thin wrappings and then usually steamed, deep-fried, or pan-fried to create spring rolls, pork-filled buns, and many other types (K. Lo, 1976: 66; 1979a: 76-77; E. N. Anderson and M. L. Anderson, 1977: 355; K. Lee and A. Branyon, 1984: 16-17, 91). Indeed, Canton is reputed to have over a thousand different kinds of dim sum, the best in China.[42]

In Canton today one gets dim sum at certain restaurants, usually at breakfast when they may be wheeled past a diner's table on heated carts. In pre-Communist times, however, they were served, ordinarily from 6 A.M. until noon, at tea houses, of which there were literally thousands, popular and well-attended, in Canton.[43] Some had several stories, and prices increased as one went up from the street level, with the top floor frequented by the well-to-do. In those days, boys carried the dim sum through a tea house on trays hung from their necks, and business was so heavy that waiters would not prepare bills. Instead, empty tea bowls and snack saucers were left on the table, and when a person was ready to depart, the waiter would quickly add them up and call out to the cashier the amount of the bill (Blofeld, 1985: 57-58). In Hong Kong and overseas, similar systems of reckoning are still found in places serving dim sum.

Altogether, people in China hold Cantonese cuisine in very high regard, as evidenced by the facts that hiring a Cantonese chef is viewed in China as comparable in prestige to hiring a French chef in Europe and the United States, and that Cantonese restaurants were popular and found throughout the country. There are, of course, many who have presented a case for Cantonese as China's finest regional cuisine, one of them gourmet F. T. Cheng (1954: 56), who writes that because of its broader base, with its chefs having created more dishes than those of any other regional cuisine, "it is only fair to put [it] at the top."

Cantonese cuisine in the United States derives not from the sophisticated cookery of Canton itself, but mainly from the country cooking of the Pearl River Delta districts from which the overwhelming number of our Chinese immigrants come. At the same time, that delta cooking is a variant of Cantonese cuisine; it has been the avenue through which most Americans approached Chinese cookery; and it has made us more familiar with dishes of this region than with those of any other part of China. There has long been controversy among Americans as to whether chop suey, still standard fare in many Chinese restaurants catering to the general public in the United States, is a true Cantonese dish or whether it was developed in America. Suffice it to say that one is unlikely to find chop suey on the menus of better restaurants in Canton; that it seems to have originated in the Toisan area south of Canton;[44] and that its name in Cantonese is *tsap-seui*, or "miscellaneous scraps." Chow mein, most familiar Chinese noodle dish in the United States, comes from the Cantonese *ch'ao-min*, "stir-fried noodles," but the noodles are prepared differently in America (S. F. Li, 1964: 211-12; G. Z. Chu, 1975: 46-47; E. N. and M. L. Anderson, 1977: 355-56; E. N. Anderson, 1988: 174, 177).[45]

Deserving attention from a culinary ethnologist, but unfortunately little studied, are the regional forms of Cantonese cuisine and the cuisines of other Kwangtung ethnic groups. One of the latter is the Hakka, whose cuisine has the stereotype of being a simple cuisine, but which, on the contrary, has sufficient variety for some to qualify it as "major." It is marked by a quite "sensitive way of combining the freshest of crisp vegetables, cut and mixed in many ways, and cooking them lightly to bring out the most delicate flavors. There is no heavy use of garlic, spices, or strong oils. Every part of the animals used is introduced into some dish; we had a most excellent dish of the spinal nerve cord of a calf, delicately stir-fried with the usual delightful vegetables" (E. N. Anderson, 1977: 8; E. N. Anderson and M. L. Anderson, 1977: 356). Another ethnic cuisine of Kwangtung is that of the Hokkien (Southern Min) peoples of Kwangtung, akin to that of Fukien but influenced by Cantonese. Then there is the cuisine of the Hainanese, linguistic relatives of the Hokkien, which also has certain similarities to that of Fukien (E. N. Anderson, 1977: 7; E. N. Anderson and M. L. Anderson, 1977: 356-57). None of these cuisines is well represented among American Chinese restaurants, though there are at least two restaurants in San Francisco that bill themselves as Hakka and more are opening nationally.

[1] The flavor principle of a dish may also affect its acceptability. For psychological perspectives on this question, see P. Rozin, 1978.

[2] "Cuisine," as defined by Freeman (1977: 144), is something beyond "manner or style of eating." He writes of it as "a self-conscious tradition of cooking and eating" that may emerge under stated historical conditions, as in China during Sung times.

[3] Mote (1977: 245), however, writes that accounts from Yüan and Ming times reveal a lack of awareness of major regional cuisines, though writers did recognize that dishes in the Yangtze Valley were more varied, less plain, that those of North China.

[4] As in the five elements, five cardinal points, five grains, five fruits or five tree-fruits, five domestic animals, five vegetables, five cereals, five tastes, five virtues, five colors, five sacred mountains, five regions, five rulers, five dynasties, five heroes, five demons, five classics, and many others (Mayers, 1874: 311-21; Ball, 1906: 485-87; Veith, 1949: 206; Burkhardt, 1953-58, 3: 119-21 and *passim*).

[5] Some argue that six or more of China's regional schools of cooking are "major schools."

[6] For a collection of Ch'ing court dishes, see Davis, 1966.

[7] A photograph of a Peking Duck dish with pancakes ("Mandarin pancakes" or "Peking doilies") may be found in Y. K. Martin, 1984: 122.

[8] Examples are Peking-style "Fish in Wine Sauce," for which a recipe and photograph can be found in Y. K. Martin, 1984: 123, 211; and the Hopeh and Shantung-style "Sliced Fish in Wine Sauce" and "Chopped Boiled Duck in Mustard and Thick Wine Sauce" (K. Lo, 1971: 80-81, 86).

[9] Sweet-and-sour sauce in the north, as is generally the case in China, is typically used with fish (E. N. Anderson, 1988: 173). Examples from northern cuisine are "Sweet-and-Sour Squirrel-Like Fish" and "Palace Sweet-and-Sour Fish Slices" (Sakamoto, 1977: 59-60, 78).

[10] For a northern-style recipe for celery cabbage ("Chinese Celery Cabbage in Cream Sauce"), and for other regional dishes that contain celery cabbage, see C. B. T. Lee and A. E. Lee, 1980: 15, and Y. K. Martin, 1984: 72-73, 123, 176-78, 212. Celery cabbage is also commonly preserved in garlic and salt.

[11] "Mu Shu Pork," now popular in the United States, is perhaps the best known of these.

[12] In a household budget survey in Peking in 1926-27, only 116 of 283 families were found to consume fish, and only a small number of these were lower income families (Gamble, 1933: 101-2).

[13] For a variety of northern-style Chinese lamb and mutton dishes, see F. L. Chang and G. M. Fuchs, n.d.: 148-49, 154, 157-59; W. W. Chang *et al.*, 1970: 365-69; H. J. Lin and T. Lin, 1972: 100-102; Sakamoto, 1977: 88-90, 97-101; K. Lo, 1979a: 43-50; 1979c: 186-94; and C. B. T. Lee and A. E. Lee, 1980: 106-10, 118-20, 125.

[14] It is fitting that among the various medicinal drinks of China and neighboring lands, there should be a "mutton wine" developed by the Mongols. Its preparation and use is described by Macgowan (1871-72).

[15] For excellent photographs of a Mongolian hot pot with ingredients ready or cooking, see Y. K. Martin, 1984: 71, and Tiger and Wolf, 1985: 224; for recipes for Mongolian hot-pot dishes, see F. L. Chang and G. M. Fuchs, n.d.: 146; F. Lin, 1975: 241-42, 275-76; Sakamoto, 1977: 97-99; K. Lo, 1979a: 44-45; 1979c: 191-92; C. B. T. Lee and A. E. Lee, 1980: 119-20; Gin, 1984: 157; and Y. K. Martin, 1984: 72-73.

[16] Recipes for northern-style barbecued beef or lamb may be found in W. W. Chang *et al.*, 1970: 366; K. Lo, 1971: 43-46; 1979a: 43-44; 1979c: 183-84, 192-93; and Sakamoto, 1977: 99-101.

[17] Some say that the sweet-and-sour flavor principle originated in Honan, though others see its origin elsewhere in China. Wherever the principle was first developed, the sweet-and-sour sauce of Honan and the north, made simply by mixing vinegar and sugar without tomato sauce or fruit, is, in the eyes of the Cantonese, lacking in refinement (K. Lo, 1979a: 17; C. B. T. Lee and A. E. Lee, 1980: 89; K. Lee and A. Branyon, 1984: 13). Traditionally the Cantonese did not like sweet-and-sour dishes very much, the main exception being fish. They, like other Chinese, are amused at the popularity among Westerners of pork and chicken prepared in this way; in fact, some Cantonese are now rejecting sweet-and-sour pork "because it is so thoroughly linked with the 'barbarians'" (E. N. Anderson and M. L. Anderson, 1977: 355; K. Lee and A. Branyon, 1984: 15; E. N. Anderson, 1988: 157, 173).

[18] A recipe for "Sweet-and-Sour Yellow River Carp" may be found in K. Lo, 1979a: 54; and 1979c: 265-66. For a superb photo of a related dish, "Peking Sweet-and-Sour Crispy Fish," together with a recipe, see Tiger and Wolf, 1985: 138-39. For other Honan sweet-and-sour dishes, see F. Lin, 1975: 140-41, 149-50 ("Sweet-and-Sour Squirrel-like Fish"); and Sakamoto, 1977: 60 ("Sweet-and-Sour Beef Liver Honan Style").

[19] Of interest here is the special dinner in Taiwan organized by James Wei. The host, a gourmet and enthusiast of Honan cuisine, lamented the fact that the dinner included no Szechwan smoked duck, Peking duck, Honan sweet-and-sour carp, or monkey head, after which the diners drank "a toast to monkey head" (Hughes, 1972: 27).

[20] Nor did its culinary attractions escape poets such Ku Tieh Ching, who wrote (H. J. Lin and T. Lin, 1972: 111) in the nineteenth century:

Soochow, the good place. Big beans
Pickled when the flowers bend low
And munched when tea is served.
Soochow, the good place—when the shad
Come by. River globefish
In spring we cook with ginger and chives.
Soochow, the good place. In summer
Plump fish dart about the river
Avoiding the fisherman's boat.
Purple crabs and red wine dregs
Make the autumn pass.
When waxy meats and sturgeon appear,
Carp and bream leap into the pot.

[21] K. Lo, 1979a: 141.

[22] A recipe for "Nanking Pressed Duck" may be found in W. W. Chang *et al.*, 1970: 271. For other eastern-style duck dishes ("Hangchow Steamed Spiced Duck," "Braised Duck Soochow Style," "Duck in Sauce Shanghai Style," "Nanking Salt Duck," "Red Cooked Duck," "West Lake Duck," etc.) see F. L. Chang and G. M. Fuchs, n.d.: 200-201, 203; W. W. Chang *et al.*, 1970: 268-82 *passim*; F. Lin, 1975: 78-81; Hush and Wong, 1976: 211-12; Sakamoto, 1977: 129-33; K. Lo, 1979a: 157-60; 1979c: 250; C. B. T. Lee and A. E. Lee, 1980: 44-46).

[23] K. Lo, 1979a: 141.

[24] Recipes for some of these ("Red-Cooked Pork Cubes with Braised Bamboo Shoots," "Red-Cooked Whole Shoulder of Pork," "Red-Cooked Fish," and "Red-Cooked Fish with Fresh Bean Curd") may be found in G. Z. Chu, 1975: 138-139, 178, 181. For "Red-Cooked Beef Tongue," see Gin, 1984: 155; and for "Red-Cooked Chicken," C. B. T. Lee and A. E. Lee, 1980: 35.

[25] A recipe for this dish is found in F. L. Chang and G. M. Fuchs, n.d.: 224, and, for carp, in F. Lin, 1975: 148 and Sakamoto, 1977: 144-46.

[26] Recipes for eastern-style "Drunken Chicken" may be found in F. L. Chang and G. M. Fuchs, n.d.: 198-99; W. W. Chang *et al.*, 1970: 233; F. Lin, 1975: 53-54; and K. Lo, 1979a: 157.

[27] One can only speculate on how the Fukienese came to such an emphasis on noodles. One notes, however, that in Sung times the growth of Fukienese population already required that rice be imported from other provinces, and that in Ming times continuing population increase, competition among crops for the best land, and a rise in rice prices led the poorest Fukienese peasants to stop consuming rice, with flour products and sweet potatoes becoming their staple foods (Rawski, 1972: 44, 51). Presumably the flour was consumed mainly in the form of noodles, which subsequently gained importance among the Fukienese population in general, wealthy as well as poor (E. N. Anderson, 1988: 80).

[28] For the recipe for this and other Fukien dishes made with wine sediment paste, see K. Lo, 1979a: 133-35; and K. Lee and A. Branyon, 1984: 219, 343.

[29] From K. Lo, 1979a: 132-33. For a similar Fukienese soup, see E. Chen, 1977: 72.

[30] Agronomists are said to have argued that Szechwan's dependence on spicy foods derives from its isolation and the fact that much of its land yields only one crop a year, which creates an unusual need for preservation with hot spices (Claiborne and Lee, 1972: xviii; K. Lee and A. Branyon, 1984: 16). This does not fit with Cressey's observation (1934: 314) that two crops a year are the norm throughout the Red Basin of Szechwan, with three crops not uncommon. Nor

does it fit with the results of the village survey (Buck, 1956: 80, 274), which found a little over two-thirds of Szechwan's cultivated land to be double-cropped, its index of double-cropping equal to the average for the Rice Region and far above the indices for all areas of the Wheat Region.

[31] H. J. Lin and T. Lin, 1972: 105. This soup, used widely in China today, some believe to have originated in Szechwan. It is made in many different ways. For generalizations about it, see E. N. Anderson, 1988: 158. For Szechwan-style recipes, see Sakamoto, 1977: 230; K. Lo, 1979a: 206-7; and Y. K. Martin, 1984: 202.

[32] Preparation of the first of these duck dishes included smoking the duck over a charcoal fire using a combination of tea leaves and camphor-wood leaves and/or camphor-wood twigs, chips, or sawdust. Recipes for preparing this dish may be found in Stallard, 1973b: 76-77; K. Lo, 1979a: 219-20; Gin, 1984: 118; and K. Lee and A. Branyon, 1984: 248-49. "Fragrant and Crispy Duck," also known as "Szechwan Duck" or "Szechwan Crispy Duck," is considered the Szechwan answer to Peking Duck, its crispiness deriving from deep-frying the duck, usually after long steaming. For recipes, see Em. Hahn *et al.*, 1973: 52; Field *et al.*, 1976: 92-93; Sakamoto, 1977: 252-54; K. Lo, 1979a: 218-19; C. B. T. Lee and A. E. Lee, 1980: 145; K. Lee and A. Branyon, 1984: 238-39; and Y. K. Martin, 1984: 200-201.

[33] Recipes for this dish will be found in F. L. Chang and G. M. Fuchs, n.d.: 281; W. W. Chang *et al.*, 1970: 160; and F. Lin, 1975: 147.

[34] The name of this dish is translated variously as "Pock-woman's bean curd," "Pockmarked wife's bean curd," "Grandmother pockmark's bean curd," and "Lady Ma's bean curd." There are various folk explanations for this name, with Lee and Lee (1980: 170) attributing the dish to a woman whose face was pockmarked, the wife of a famous chef who lived in Ch'eng-tu. Other explanations are listed by E. N. Anderson (1988: 168), and recipes for the dish may be found in F. L. Chang and G. M. Fuchs, n.d.: 286; W. W. Chang *et al.*, 1970: 433; F. Lin, 1975: 121-22; Sakamoto, 1977: 267; K. Lo, 1979a: 255; C. B. T. Lee and A. E. Lee, 1980: 170; and Y. K. Martin, 1984: 159.

[35] Kweichow's subordinate position may be explained by its being poor, thinly settled, and at the edge of Han Chinese influence through much of history.

[36] Chili is, however, used by the Cantonese in various dishes, as, for example, in "Hot Diced Pork with Bamboo Shoots" (F. L. Chang and G. M. Fuchs, n.d.: 101), and, along with soy sauce and other flavorings, in a favorite dip for prawns (K. Lee and A. Branyon, 1984: 15, 187, 335).

[37] As, for example, in the Cantonese dishes "Chicken with Black Bean Sauce" (F. L. Chang and G. M. Fuchs, n.d.: 89), "Clams in Black Bean Sauce," and "Soft-Shell Crabs with Black Bean Sauce" (K. Lee and A. Branyon, 1984: 166, 173).

[38] As in "Blanched Greens in Oyster Sauce" (H. J. Lin and T. Lin, 1972: 139), "Pork, Bean Curd and Peanuts with Oyster Sauce" (Sakamoto, 1977: 187-88), "Chicken Wings with Oyster Sauce," and that favorite in the United States, "Beef in Oyster Sauce" (K. Lee and A. Branyon, 1984: 202, 278).

[39] As in "Sweet-and-Sour Pork," another American favorite. For Cantonese recipes, see F. L. Chang and G. M. Fuchs, n.d.: 99, and Sakamoto, 1977: 193.

[40] Other fruits, or their juices, commonly added to Cantonese dishes are plums, lemons, and litchis, as in "Lemon Chicken," "Steamed Spare-Ribs with Plums," "Litchi Pork," and "Stir-Fried Roast Duck with Lichee and Pineapple" (H. J. Lin and T. Lin, 1972: 141; K. Lo, 1976: 66; 1979a: 96, 103).

[41] Stir-frying is rapid cooking by stirring and turning fine bits of food with a spatula in a little oil in a quite hot wok or pan. One advantage of stir-frying is that it uses little fuel, important in many parts of China where serious fuel shortages exist. It also has the advantage of sealing their juices within each of the foods being prepared.

[42] An excellent photo of several types of dim sum in serving trays is found in Emily Hahn *et al.*, 1973: 147, followed by recipes on pages 148-53.

[43] Blofeld (1985: 58-59), who knew the traditional circuit well, found not a single tea-house in a recent trip to Canton, a disappearance he ascribes to "official disapproval of unproductive leisure." There is, nevertheless, official pride in the quality of Canton's dim sum, with a brochure prepared by a government agency, the China International Travel Service, writing of it as the nation's best: distinctive in preparation, delicate in taste, and beautiful in color.

[44] For a different view on the origin of chop suey, see Em. Hahn *et al.*, 1973: 178-79.

[45] For a variety of Chinese chow-mein dishes, see K. Lo, 1979c: 98-109.

FOODS UTILIZED

CHAPTER THREE

Cereals and Pulses

"At frost time the weather is fine;
In brisk winds the tree leaves fall.
Now is the time to thresh the rice,
The flails making a wild noise."

Chinese poem (from Emily Hahn et al.)

Importance of Cereals and Pulses
Major Cereals and Pulses of Eurasian Origin
Rice
Soybeans
Common Millet and Foxtail Millet
Wheat and Barley
Broad Bean and Pea
Sorghum: An Early Arrival from Sub-Saharan Africa
Maize: America's Cereal Contribution
Minor Cereals and Pulses
Buckwheat, Rye, and Oats
Barnyard Millet
Job's Tears
Minor Pulses
Preparation and Consumption of Cereals and Pulses

IMPORTANCE OF CEREALS AND PULSES

In China, cereals and pulses dominate both diet and rural landscape, with roughly 90 percent of all calories deriving from them in traditional times, compared to nearly 40 percent in the United States (Table 1, and Maynard and Swen, 1956: 414). In addition, cereals and pulses together provided about 90 percent of total protein intake in China, and cereals alone, three-quarters of total intake (Maynard and Swen, 1956: 419). Cereals and pulses were also important sources of B vitamins and of various minerals. Their nutritional dominance in an area of dense population like China calls attention to the advantages they enjoy over other food crops: adaptability to a broad range of climatic and soil conditions; ease with which they can be transported and stored; large per acre yields of carbohydrate and protein; and the possibility that, by a proper balance of cereals and pulses, a diet may be achieved that is adequate in other nutrients (Pimental *et al.*, 1986: 42-43).

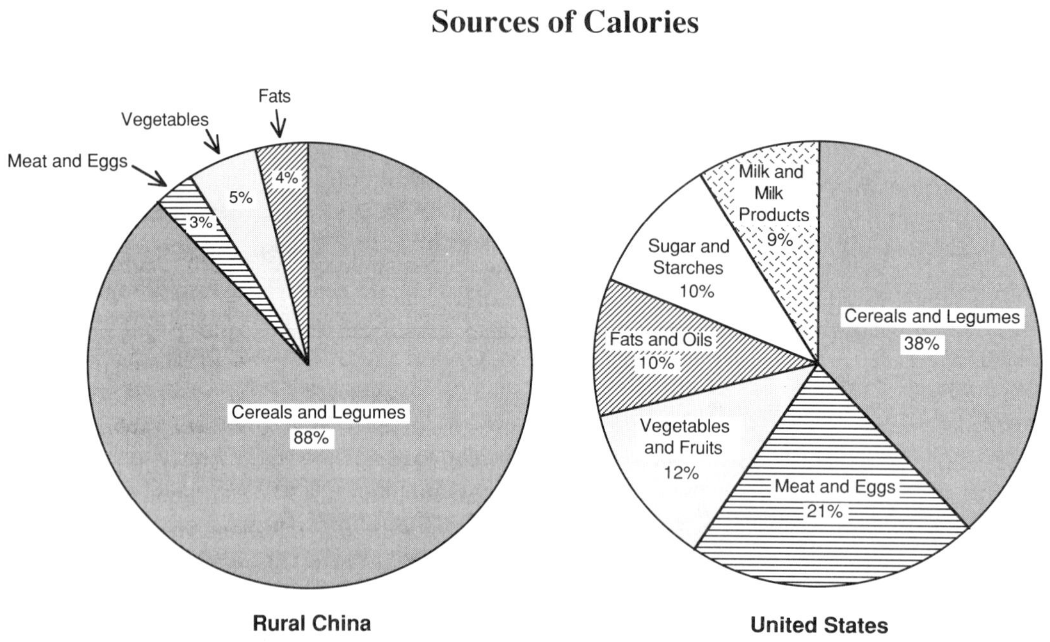

Table 1. Sources of calories in rural China and the United States (after Adolph, 1946b: 868)

Of all cereals, rice is the one popularly associated with China, and is the nation's most important cereal by far, with a production of 178 million metric tons in 1984. China has long rivalled India as the world's leading rice producer, and in recent years it seems to have passed India in production, China now accounting for more than a third of the rice production of the entire world (Wittwer *et al.*, 1987: 139-40).[1] In the Rice Region as defined by Buck in the village survey (Figure 20), roughly two-thirds of the total crop area was occupied by rice in 1929-33, and this supplied about two-thirds of total calorie intake. No other cereal or pulse approached rice in importance in the Rice Region, with wheat, the runner up, accounting for only 5 percent of the region's total calories (Figure 21 and Buck, 1956: 211).[2] Rice's dietary importance may be appreciated even more by noting that in Fukien it was the principal food at all three daily meals (Hurlbut, 1939: 78) and that in rural South China an adult male consumed 470 or 485 pounds of rice a year (C. K. Yang, 1959a: 54).[3] In the Double Cropping Rice Area, which is centered on Kwangtung (Figure 22), the dominance of rice was as great as, or greater than, in any other area of South China: of the double-cropped land, which in the 1929-33 survey amounted to two-thirds of all cultivated land, most involved one crop of rice followed by another, which supplied over three-fourths of total calorie intake (Buck, 1956: 85; Maynard and Swen, 1956: 413).

There were, nevertheless, striking regional differences in the relative importance of rice and other cereals and pulses in China (Figure 21 and Table 2), with contrasts between North and South China most often singled out by observers. In North China, the "Wheat Region," rice

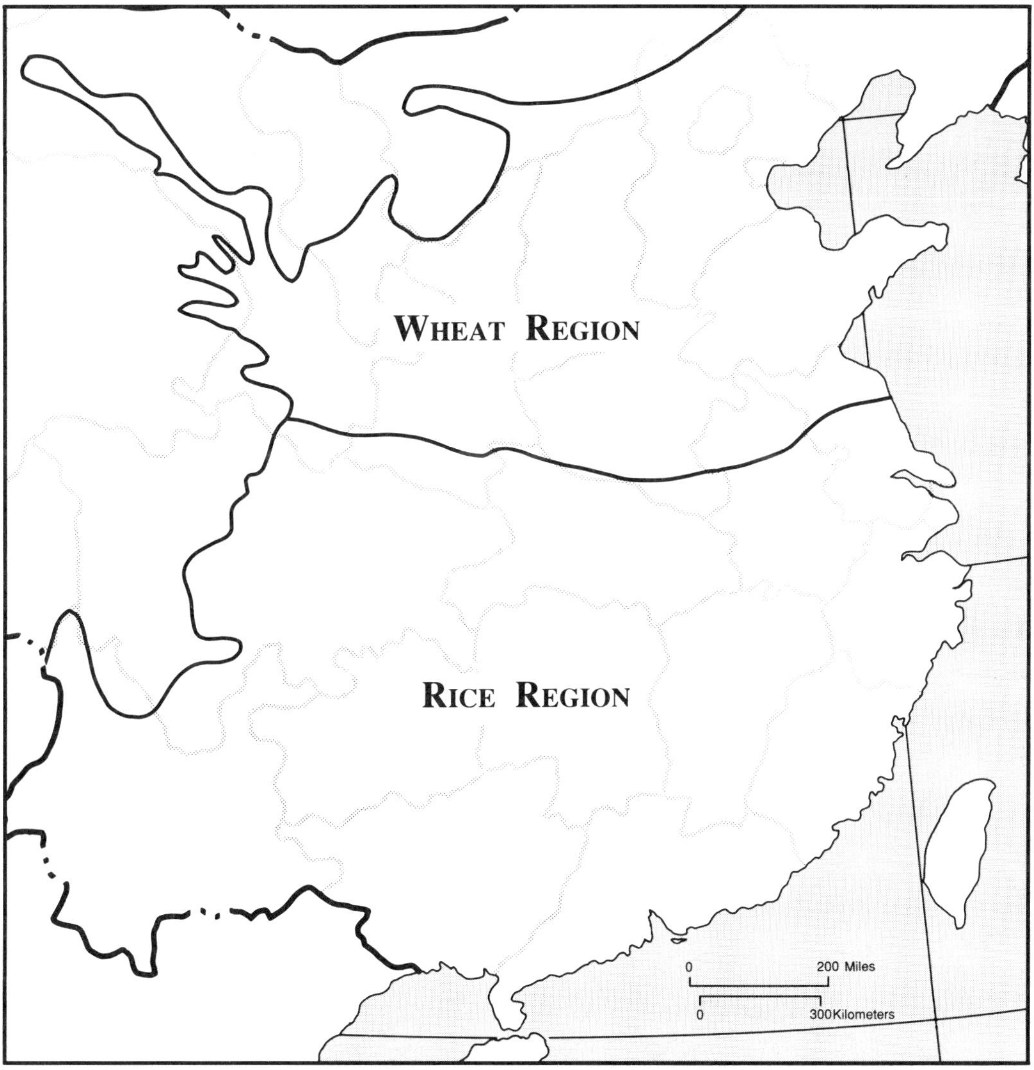

Figure 20. The two main agricultural regions of China (redrawn from Buck, 1956: 25)

was not important anywhere, and in the past many Chinese there ate it seldom, if at all (in the Wheat Region as a whole, it accounted for less than 1 percent of total calorie intake in 1929-33). Despite giving its name to the region, wheat contributed only about a fourth of total calorie intake in Wheat Region villages, with a number of other cereals and pulses playing significant dietary roles. In one or another area of North China, all of the following provided 3 percent or more of total calories: Italian millet, sorghum, maize, common millet, oats, soybeans, field peas, and mung beans.

Turning to individual cereals and pulses, we consider first those of major importance that were domesticated in Eurasia.

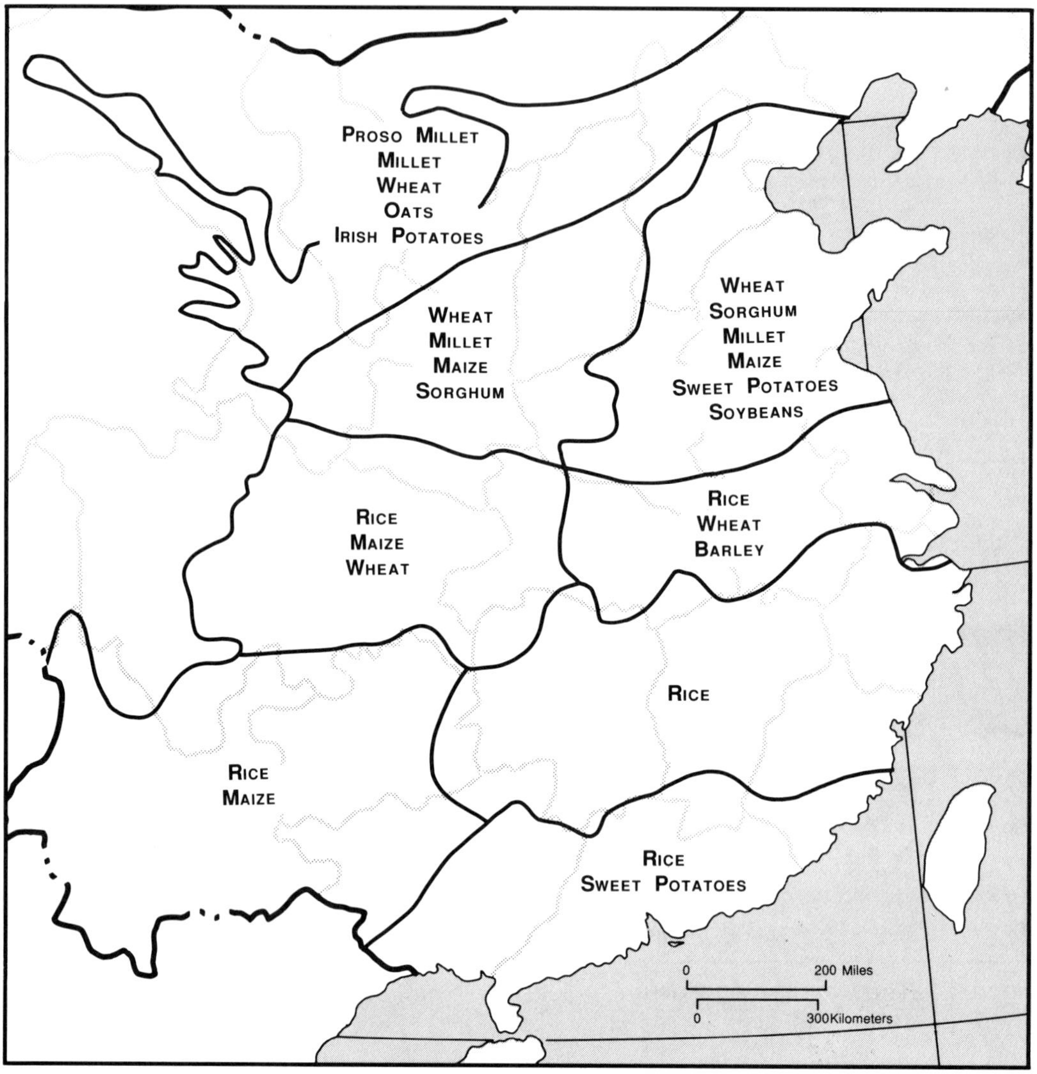

Figure 21. Crops providing five percent or more of total calories in each area of China, in order of importance (redrawn from Maynard and Swen, 1956: 410)

MAJOR CEREALS AND PULSES OF EURASIAN ORIGIN

Rice

There are two species of cultivated rice: *Oryza sativa* of Asian origin, and *O. glaberrima* of African. *O. glaberrima* was domesticated in West Africa, apparently independently of Asian rice. Its date of domestication is uncertain, though some, following Roland Portères, suggest that it may have been cultivated in the upper Niger Valley by 1500 B.C. There has long been

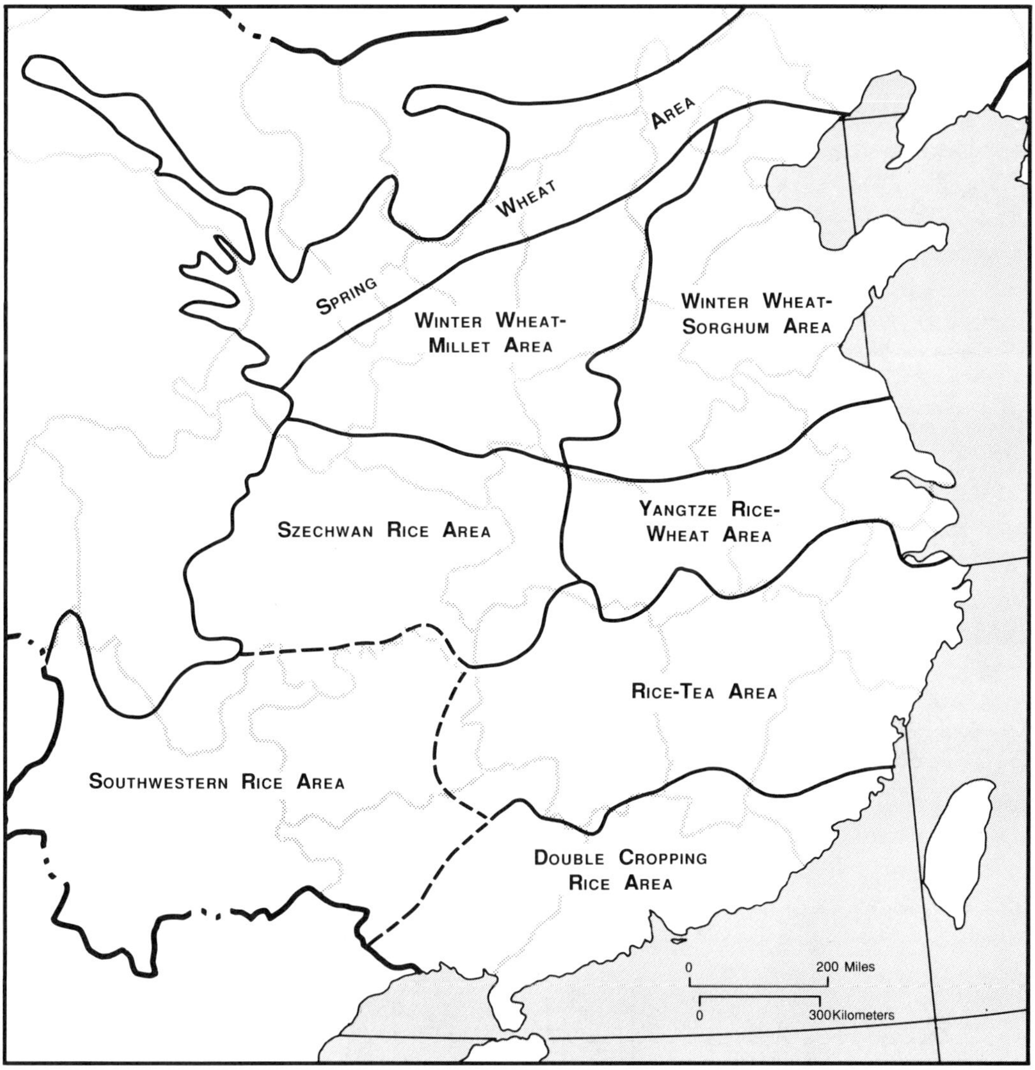

Figure 22. China's agricultural areas (redrawn from Buck, 1956: 27)

agreement that *O. sativa* was domesticated somewhere on mainland Asia in the zone from northern India to southern China, where the greatest genetic diversity of cultivated forms is found (Harlan, 1975: 220; T. T. Chang, 1976a: 101; 1976b: 430-32; 1976c: 143; 1983: 72; Glover, 1979: 26-28; Zeven and de Wet, 1982: 74; Higham, 1984; Zohary and Hopf, 1988: 81-82).[4] There are, however, differences of opinion as to the specific place where domestication occurred, with India, northern Thailand, and South China all having their advocates (Bray, 1984: 481-87). Contributing to the uncertainty is the humid, tropical climate of much of

Table 2. Percentage of calories supplied by various important crops: 17,351 persons, 2,727 families, 136 localities, 131 hsien, 21 provinces, China, 1929-1933 (from Maynard and Swen, 1956)

Regions and Areas	Number of Localities	Rice	Wheat	Millet	Kaoliang	Corn	Millet, Proso	Potatoes, Sweet	Barley	Soybeans
CHINA	136	35.0	14.4	10.4	7.5	7.5	2.8	2.6	2.0	2.0
Wheat Region	67	0.6	23.5	19.9	14.6	11.9	5.8	2.5	1.3	3.2
Rice Region	69	68.4	5.6	1.2	0.6	3.2	-	2.8	2.7	0.9
Wheat Region Areas:										
Spring Wheat	13	*	15.3	20.0	4.5	*	21.7	-	1.3	0.4
Winter Wheat-millet	21	1.0	26.7	23.2	11.5	17.4	4.1	0.3	0.8	2.2
Winter Wheat-kaoliang	33	0.6	24.7	17.9	20.5	13.0	0.5	4.9	1.7	4.9
Rice Region Areas:										
Yangtze Rice-wheat	22	57.8	12.8	1.2	1.6	2.1	-	1.8	7.4	0.7
Rice-tea	19	75.8	3.1	0.4	0.1	2.1	-	2.0	0.3	0.8
Szechwan Rice	6	56.9	5.3	-	0.1	14.0	-	3.1	0.3	1.9
Double Cropping Rice	11	76.9	0.5	0.1	-	-	-	8.2	1.2	0.6
Southwestern Rice	11	74.7	1.0	*	-	4.8	-	0.5	0.3	1.4

Regions and Areas	Mung Beans	Oats	Potatoes, Irish	Field Peas	Rice, Glutinous	Broad Beans	Soybean, Black	Other Grains	Other Legumes
CHINA	1.2	1.2	0.9	0.8	0.7	0.6	0.6	0.2	1.3
Wheat Region	2.0	2.2	1.7	1.3	-	0.2	1.2	2.3	1.4
Rice Region	0.4	0.1	0.1	0.3	1.5	1.1	*	1.3	1.3
Wheat Region Areas:									
Spring Wheat	0.4	10.9	7.3	3.2	-	1.0	0.7	8.6	0.8
Winter Wheat-millet	1.4	0.3	0.8	1.5	-	-	0.9	1.4	1.9
Winter Wheat-kaoliang	3.0	-	*	0.5	-	-	1.6	0.5	1.2
Rice Region Areas:									
Yangtze Rice-wheat	1.2	-	-	0.2	1.3	1.5	-	1.9	1.1
Rice-tea	*	*	0.1	0.1	1.8	0.5	0.1	1.7	1.5
Szechwan Rice	0.3	1.5	0.1	2.2	0.8	1.2	-	0.7	1.2
Double Cropping Rice	0.1	-	-	*	0.6	0.1	-	0.1	1.5
Southwestern Rice	*	-	0.3	0.1	2.4	2.1	-	1.3	1.1

* the amount is under 0.05

southern Asia, which leads rice and other plant materials to decompose quickly when exposed to the elements. The situation is made worse by the limited number of archeological excavations in much of the region, by problems of dating, and questions as to whether rice remains that do turn up are from wild or domesticated plants.

Close wild relatives of *O. sativa* have been found in this century in Yunnan; Kwangsi; Kwangtung; Hainan; Taiwan; and Kiangsu; and in antiquity they occurred as far north as the Po Hai (Gulf of Chihli) (T. T. Chang, 1983: 70-71). Until the mid-1970's, however, the earliest reliable evidence of what may have been domesticated *O. sativa* (from Central China, western India, and possibly northern Thailand) dated only from the fourth and third millennia B.C. (Nayar, 1973: 160-62; P. T. Ho, 1975: 61-64, 375; 1977: 440-41; T. T. Chang, 1976a: 99; 1976b: 430-31; Harlan, 1977: 371-72; Vishnu-Mittre, 1977: 572, 574-75; Glover, 1979: 19-35; Zohary and Hopf, 1988: 82). Then, in 1976, domesticated rice going back another millennium, to roughly 5000 B.C., was reported for the site of Ho-mu-tu in Chekiang Province, and that report since has gained credence with finds of rice in other Chekiang sites of comparable age (K. C. Chang, 1977a: 513; 1986: 197-99, 208-10; T. T. Chang, 1983: 72-73; Pearson, 1983: 124; Bray, 1984: 485; W. Chen, 1984: 49; Xin, 1984: 8).[5,6] Chekiang is well beyond cultivated Asian rice's center of genetic diversity and presumed hearth of domestication, where it would have been cultivated still earlier, but discoveries there do lend support to Ho's argument (P. T. Ho, 1975: 63-73) that China played a greater role in rice domestication than Vavilov and certain other botanists and historians had suspected. It seems prudent, however, to await further archeological evidence before accepting T. T. Chang's view (1983: 73) that China is probably to be credited with the first "true domestication" of rice (because of increased selection pressures brought on by cool weather and shortness of growing season, which made rice survival more dependent on humans).

O. sativa (*tao*) is divided by many into three major races which differ genetically and in geographical origins. One is *indica*, the earliest domesticated form known in both India and China, a tropical race which may have spread northward in China by two routes, one from Yunnan, the other from Indochina. The second race is *japonica*, a rice of temperate regions which probably separated from *indica* in southwestern China, and spread northward from there; thus, as Chinese botanists have noted, it would more properly be designated *sinica* since it goes back to roughly 5000 B.C. in Central China but reached Japan only about 300 B.C. The third race, *javanica*, seems to be a more recent form which evolved in Indonesia from tall, large-grained *indica* rice introduced from India (T. T. Chang, 1976a: 101-2; 1976b: 432-33; 1983: 72-73; Swaminathan, 1984: 82, 84, 90). *Javanica* has spread northward to the Philippines, Taiwan, and Japan, but the two common rices of mainland China are *indica* (in China, *hsien*) and *japonica* (*kêng*). In South China, including Kwangtung, *indica* is the major form, the dietary staple; in Central China, *indica* and *japonica* are of more equal importance; and in North China, *japonica* prevails (J. H. Shen, 1980: 9-13).[7]

The races *indica* and *japonica* differ in qualities of grain, including its length and starchiness, and in their cooking requirements and aroma. *Indica* is a long-grained form, and *japonica* (preferred form of Japan) has short, broad grains that, when cooked, are usually more sticky than *indica*. There are also many types of rice in China that are intermediate between *indica* and *japonica*. One also finds in China wetland and dryland forms of rice, early-season and late-season ones, and, perhaps of greatest importance to us, "glutinous"[8] and non-glutinous forms of both *indica* and *japonica* (Grist, 1953: 64-65; Bray, 1984: 478-80; Wittwer *et al.*, 1987: 143-46).

The development of rice varieties was favored by the Asian custom of harvesting the crop ear by ear, which better enables a harvester to recognize individual variation among plants and to select for future planting those that have useful qualities (Grist, 1953: 59-60). Among the rices having such qualities were the drought-resistant, early-ripening forms of *indica* that were selected from a rice introduced from Indochina at the start of the eleventh century. According to Ho Ping-ti (P. T. Ho, 1956-57: 200-201), the new forms permitted an expansion of rice cultivation in early modern China from the irrigated lowlands into relatively well-watered hill regions, which ultimately doubled the area under rice cultivation. Through this, and because their early-ripening quality assured the success of the double-cropping system for which China's rice area is famous, the new forms of rice contributed to a striking increase in the nation's food supply and population. Chinese farmers selected rice for other useful qualities as well, and today one finds forms of rice that can survive rapidly-rising water levels, as in flooding; forms that can be grown in saline coastal soils; and so forth. Altogether, according to Sun Han (in Wittwer *et al.*, 1987: 145), there are over 40,000 local rice cultivars in China.

The cultivation of rice in China is hard work, as reflected in such Chinese expressions as "Every grain of rice on your dish is won by the sweat of the brow" (Cormack, 1935: 133). Despite the effort involved, there are very good reasons why rice plays a dominant role in South China: it is high in calories produced per acre; superior to all starchy roots and tubers used in East Asia in the amount of plant protein it provides, and, in brown, undermilled, and home-pounded forms, to virtually all in amounts of thiamine, riboflavin, niacin, and iron.[9] Thus it excels in its ability to support dense human populations. Moreover, rice can also be processed simply; it stores well; it is digested readily; and it can be eaten for its own flavor or serve as a medium for taking on the tastes of spice and other foods with which it is prepared. As a result, it presents unusual possibilities for culinary creation, a fact that has been recognized by chefs and has contributed to many outstanding dishes in Cantonese and other regional cuisines of China (Leung *et al.*, 1973: 7, 11-15; E. N. Anderson and M. L. Anderson, 1977: 322-23).

Rice was the major cereal crop in South China wherever conditions suited it. Where they did not, rice's importance was reduced or it was replaced by other crops. In mountains, cold temperatures created conditions that favored barley or, on poor soils, buckwheat, over rice. On warm, steep slopes of low fertility, maize was favored. White potatoes replaced rice in moist, mountainous areas of interior South China. And in places both warm and sandy, sweet potatoes tended to be of most importance (E. N. Anderson and M. L. Anderson, 1977: 323).

Soybeans

The wild ancestor of the soybean is probably *Glycine soja* (= *G. ussuriensis*), a native of eastern Asia. Unlike the upright cultivated forms, it is typically a prostrate, vinelike plant. Domestication is believed by some authorities to have taken place in northeast China, the soybean's gene center and one of the world's greatest areas of production (T. H. Shen, 1951: 247; Leppik, 1971; Wittwer *et al.*, 1987: 184-86).[10]

Although certain experts formerly thought that the soybean was domesticated by about 3000 B.C., recent reviews (Porterfield, 1941b: 181; Hymowitz, 1970, 1976; P. T. Ho, 1975: 76-81; 1977: 451-54; K. C. Chang, 1977b: 28; Hymowitz and Newell, 1981: 276-77; Bray, 1984: 424, 511-12) report no convincing evidence of domesticated soybeans earlier than Chou times (twelfth century B.C.-221 B.C.).[11] The soybean did not spread across North China right away, but by the fourth and third centuries B.C. it was almost always mentioned as one of the two major food crops (the other being millet) (P. T. Ho, 1975: 79; 1977: 452). By the first century A.D. it was found not only in the north, but in Central China, South China, and Korea as well, and between that time and the period of European overseas expansion it was also introduced to Japan and Southeast Asia (Hymowitz and Newell, 1981: 277). Its spread cannot have been accomplished easily, for the plant is very sensitive to length of day and duration of frost-free period, and problems commonly arise when varieties are introduced to new regions (T. H. Shen, 1951: 248-49). Despite this, it has adapted to most of China, and in 1929-33 it was estimated (Buck, 1956: 211-12) to occupy roughly a tenth of the nation's total crop area. In 1936, its peak year, production amounted to 11.3 million metric tons annually, and its production in 1985 was 10.5 million metric tons (which placed it third in the world, behind the United States and Brazil). The nation's most important soybean area, in terms of acreage and production, is the northeast (especially Manchuria, Shantung, Kiangsu, Honan, Anhwei, and Hopeh) (T. H. Shen, 1951: 248-49; American Plant Studies Delegation, 1975: 73; Wittwer *et al.*, 1987: 183, 194-95). Indeed, in the mid-1940's, the soybean was said to occupy roughly a fourth of all cultivated land in Manchuria, to provide half of total farm income, and to contribute an even larger share of railway freight volume (Morse, 1947: 140). Much of the crop in Manchuria, then and now, is exported, whether to Japan, other parts of China, or elsewhere.

There are compelling reasons why the soybean (known today as *ta-tou*, "greater bean") should have gained importance in China. For one, it yields more usable protein per acre than other common cultivated plants, whether legumes, cereals, or others. It also contributes roughly four to eighteen times as much usable protein per acre as could be gained by producing milk, eggs, or meat. This means that the soybean is a very cheap source of protein. In addition, soy protein is of good biological value. The soybean also has a good store of certain B vitamins, calcium, phosphorus, and iron. Unlike most legumes, it is rich in oil. All this makes it an excellent crop for a country like China, with a shortage of pasture and arable land. The soybean's suitability is further enhanced by its nitrogen-fixing abilities which have led it to be included in various crop rotations, to the great long-term benefit of soil fertility in China. A final factor in the soybean's success was the Buddhist commitment to vegetarianism, which led monks to create tasty soybean analogues to flesh foods, thereby further encouraging use of the plant (A. K. Smith and S. J. Circle, 1972: 11-12).

In China, there are many varieties of domesticated soybeans, but they are divided into three main groups (yellow, green, and black) on the basis of seed color. The diversity is such that virtually every local area seems to have its unique varieties (C. V. Piper and W. J. Morse, 1923: 144). All, however, present similar problems for use. One is that without processing they are unpleasant, bitter in taste, and bring on considerable flatulence. Another is that they contain substances that inhibit digestion of their protein (Liener and Kakade, 1980). In ancient China these problems were solved through the development of simple processing techniques involving heat, soaking, grinding, sprouting, and fermentation.[12] As a result, people obtained not only readily digestible forms of soy protein, but products of differing textures and flavors that served quite different culinary uses (L. C. Hsu, 1974: 305). These will be considered later.

Common Millet and Foxtail Millet

In production of lesser millets, China today is second in the world only to India. The principal lesser millets of China, however, are of Eurasian origin: common millet (*Panicum miliaceum*) and foxtail millet (*Setaria italica*), whereas the principal ones of India are of African origin: bajra or pearl millet (*Pennisetum americanum* = *P. typhoideum*) and ragi or finger millet (*Eleusine coracana*).

Pearl millet and finger millet may first have been cultivated in sub-Saharan Africa between 3000 and 2000 B.C. (Zeven and de Wet, 1982: 129, 131). Pearl millet was known in western India by 1000 B.C. and possibly some centuries before (Brunken, de Wet, and Harlan, 1977: 173; Vishnu-Mittre, 1977: 573; Allchin and Allchin, 1982: 245, 264). Though archeological evidence is limited and unsatisfactory, it is suggested (Purseglove, 1976: 92; Vishnu-Mittre, 1977: 573, 575; Allchin and Allchin, 1982: 291-92)[13] that finger millet had also arrived in India by that time or earlier. Both of these millets are now cultivated in certain parts of Southeast Asia and China, but are of little importance and seldom mentioned in sources on China (Crevost and Lemarié, 1917: 82, 369, 392; Wester, 1924: 98, 152-53; I. H. Burkill, 1966, 1: 930-31; 2: 1717; American Plant Studies Delegation, 1975: 73; Bray, 1984: 437).

Common millet and foxtail millet, for their part, may have reached India somewhat earlier than the two African millets (Vishnu-Mittre, 1977: 576; Costantini, 1981: 275-77), and were far earlier elsewhere in Eurasia. Common millet was cultivated in China, Europe, and possibly Georgia and Iran, by the fifth millennium B.C., and cultivation of foxtail millet seems to be of comparable antiquity, at least in China (Renfrew, 1973: 99, 101-2, 204-5; Harlan, 1975: 240; 1977: 379-80; Zohary and Hopf, 1988: 78-81).[14] It is possible, according to Harlan, that common millet and foxtail millet were first domesticated in the West, and then spread to China; that domestication of these cereals occurred in China, with dispersal to Europe; or that they were domesticated independently in more than one region. Harlan (1975: 240-41; 1977: 380) is inclined to favor independent domestication. Many scholars (*i.e.*, P. T. Ho, 1975: 57-61; Mathon, 1981: 60, 62; Zeven and de Wet, 1982: 38), on the other hand, favor a Chinese domestication for both millets. In either case, Persia and India may have obtained common millet from China, for, as Laufer has observed (1919: 540, 565, 595), a Persian word for "millet" is similar to and may derive from its equivalent in Chinese, and one Sanskrit name for common millet (*cinaka*) means "Chinese."

Since 1963, it has been known that cultivation of *Setaria italica* in China goes back to roughly 5000 B.C. at the Yang-shao Neolithic site of Pan-p'o in Shensi, and remains of this millet have also been found in other sites of the Yang-shao culture (5000-3000 B.C.) in North China. Recently, moreover, discovery of carbonized millet seed at the still earlier Neolithic sites of Tz'u-shan (Cishan) in Hopeh and P'ei-li-kang (Peiligang) in Honan suggest that *Setaria italica* was cultivated in North China even earlier (*Recent archaeological discoveries in the People's Republic of China*, 1984: 9-11). Until recently, remains of *Panicum miliaceum* were reported for just one Yang-shao site, Ching-ts'un in Shansi (P. T. Ho, 1975: 57-58; 1977: 438; K. C. Chang, 1977a: 94-95, 95n; 1977b: 26; T. T. Chang, 1983: 66-67; H. L. Li, 1983: 30; Bray, 1984: 434). Now, however, K. C. Chang (1986: 90-92, 112) notes that *P. miliaceum* is reported for a second Yang-shao site, Chiang-chai in Shensi, and for pre-Yang-shao sites as well. Most scholars identify *P. miliaceum* as the cereal *shu*, mentioned on Shang oracle bones far more often than any other cereal. Many also believe that *chi*, another frequently mentioned cereal on oracle bones, was a starchy form of common millet, though others think that it was foxtail millet, *Setaria italica*. In either case, this evidence, together with that found in early Chinese writings, has led to a consensus that common millet and foxtail millet were very important crops in ancient North China (K. C. Chang, 1977a: 94-95; 1977b: 26-27; 1980: 146-48; Sinoda, 1977: 484; T. T. Chang, 1983: 65-70; Bray, 1984: 434-35).

Common millet is well adapted to dry, continental conditions. It requires less water than almost any other cereal not because it resists drought, but because it matures quickly, in 50 to 60 days after it appears (Arnon, 1972: 136). *Setaria italica* also has a short growing season, and is a hardy plant that can stand both wind and drought and grow in parts of China where other unirrigated cereals do poorly (Arnon, 1972: 137; American Plant Studies Delegation, 1975: 71; P. T. Ho, 1977: 438; C. Y. Hsu, 1980: 83-84).[15] Because of the ability of these millets to survive hot, dry conditions and because of their ancient importance in North China, most botanists place their domestication there and not elsewhere in the country. Fogg (1983), on the other hand, notes that aborigines on Taiwan have continued to cultivate *Panicum miliaceum* up to the present day and that *Setaria italica* may have been cultivated by ancient farmers in South China and Southeast Asia. He implies further that the latter may have been cultivated there as early or earlier than in the north. Over time, the two millets came into competition with other cereals and have gradually lost ground to them.[16] Nevertheless both *P. miliaceum* (*huang-mi*, for the hulled grain) and *S. italica* (*hsiao-mi*, for hulled grain),[17] but particularly the latter, continue to be grown as food crops in China, especially on unirrigated land. One writer in the 1930's, however, said that whereas *Setaria italica* was cultivated widely in China, particularly in hilly places, *P. miliaceum* was grown mainly in Shantung (Bray, 1984: 443).

Both of these millets are found in glutinous and non-glutinous forms. Though considered food for poor people, millet is well regarded for its taste and food value (Bray, 1984: 448). Judging from analyses provided by Leung *et al.* (1973: 5-10), *S. italica* and *P. miliaceum* are roughly comparable to other cereal grains in their content of protein, carbohydrate, vitamins, and minerals. *S. italica*, for example, is higher in protein (whole grain, 9.5 percent) than maize (9.1 percent) and rice (7.4-8.2 percent), but somewhat lower than sorghum, buckwheat, barley, and most kinds of wheat. It is higher in iron (5.5 mg./100 gm.) than almost all of the above cereals, and higher in thiamine (.43 mg./100 gm.) than all of them, though lower than some in other B vitamins. Adolph cautions, however, that *S. italica* protein has a low biological value, and that it also has a low protein efficiency ratio (grams of weight increase per gram of protein consumed by rats) (Adolph, 1951: 129; Leung *et al.*, 1973: 193-201). Protein efficiency, however, can be greatly increased if millet is consumed with meat or, as is sometimes done in North China, mixed with other cereals and with soybeans (Holmes, 1948: 219; Adolph, 1951: 129; 1956: 67). *S. italica* has come in for attention as a food crop in the People's Republic (Adolph, 1951: 129; American Plant Studies Delegation, 1975: 70-72), and in 1950 it even served as a standard of value in North China. One university in Peking, for example, set its "tuition fees at 100 half-kilos of millet per term," and salaries in that city were also established in terms of millet, with a senior professor paid "a salary of 1,200 half-kilos of millet per month" (Adolph, 1951: 129).

Wheat and Barley

Wheat (*Triticum* spp.) and barley (*Hordeum vulgare*) were among the most ancient of Near Eastern food crops, and were domesticated by about 8000 B.C. Wheat came to be more important as human food in classical times in that region, but in more ancient periods barley was the most plentiful and inexpensive cereal, the everyday food of the poor. Both cereals do well in cool weather and with moderate precipitation. In the ancient Near East, they were commonly cultivated under irrigation, as in the valleys of the Nile and Tigris-Euphrates. The advantages of barley are that it is fast-maturing, grows in drier situations and inferior soils, and has greater salt tolerance than wheat. It is better adapted to short growing seasons, as at the upper limits of agriculture in the mountains; and as it spread northward across Eurasia, it could grow at higher latitudes than wheat or most other cereals (Harlan, 1976: 93-95; 1977: 362-65; Zohary and Hopf, 1988: 52).

The story of wheat and barley in China has been ably told by Francesca Bray (1984: 459-77), and, as a result, we present it here in the barest detail. Wheat and barley are mentioned in oracle-bone inscriptions dating from late Shang times, and Ho (P. T. Ho, 1975: 74, 82, 352) believes that they were probably introduced to North China in the second millennium B.C. or a few centuries before. Some scholars (Schafer, 1977: 89; Sinoda, 1977: 484, 486; C. Y. Hsu, 1980: 84) have suggested that barley arrived before wheat. In either case, millets, well adapted to summer heat and rain, were already well-established crops in semi-arid North China, and wheat and barley, adapted to conditions of winter rain, did not readily fit into the system of unirrigated farming that prevailed. The picture ultimately changed with development of water-saving methods of cultivation, new races of wheat and barley adapted to local conditions, and a system of summer and winter crops. Prior to Han times, wheat is said to have been "a delicacy even among the aristocracy" (Bray, 1984: 459), but by that period, both wheat and barley were important winter crops in the north, well-suited to rotation with millet under a system of dry-farming (P. T. Ho, 1975: 73-76; Y. S. Yü, 1977: 71-72; C. Y. Hsu, 1980: 83-85; Z. Wang, 1982: 52; E. N. Anderson, 1988: 17).[18]

In traditional China, wheat and barley (along with such cereals of Near Eastern origin as rye and oats) were known collectively as *mai*: wheat as *hsiao-mai*, or "lesser *mai*," and barley as *ta-mai*, "greater *mai*" (F. P. Smith, 1469: 207, 443; Bray, 1984: 459). Both remained important cereals, typically grown in rotation with other crops. Wheat, however, is far more important than barley in the country as a whole. The 1929-33 village survey found nearly 40 percent of China's total crop area given over to these cereals (wheat, 29.2 percent; barley, 8.8 percent). It also found that wheat was consumed by nearly 80 percent of China's families in the survey, barley by nearly 20 percent. In terms of calories supplied by major staple crops, however, these cereals were far below rice (rice, 35 percent; wheat, 14.4 percent; barley, 2.0 percent) (Buck, 1956: 211; Maynard and Swen, 1956: 407, 413).

In China today, wheat ranks second in production among all cereals. Increases in production in recent decades, to 88 million metric tons in 1984, have tended to narrow the gap with rice. China is the world's largest wheat producer, exceeding both the Soviet Union and United States in that regard. Its yields per acre, as well, were much higher than those of the United States. Barley production at present is about 6.8 million metric tons (Wittwer *et al.*, 1987: 157-58, 209-10).

Broad Bean and Pea

The main pulses of Near Eastern or Mediterranean origin in China are the broad bean or horse bean (*Vicia faba*; *ts'an-tou* or "silkworm bean") and the pea (*Pisum sativum*; *wan-tou*).[19] Peas are among the earliest crops of the Near East, with domesticated peas found in Neolithic sites there dating from about 7000 to 6000 B.C. (Zohary and Hopf, 1973: 887; 1988: 96-97).

There is greater uncertainty as to where and when the broad bean was domesticated in the Near East/Mediterranean region; in archeological sites in Israel and Syria, remains of what may have been domesticated broad beans have been dated prior to 6000 B.C.; the case for domestication, however, is much stronger for the broad beans found in sites in various parts of the Mediterranean area and Central Europe at levels dating to the third millennium B.C. (Hanelt *et al.*, 1972; Schultze-Motel, 1972; Zohary and Hopf, 1973: 892-93; 1988: 104-7).

Both of these pulses were also included by the Chinese under the general terms *hu-tou*, "bean of the Hu" or "Persian bean," which suggests an introduction from Iran (Laufer, 1919: 305-7). Remains of what possibly were broad beans were found in a site of the Liang-chu culture in Chekiang (possibly the third millennium B.C. or even earlier) (K. C. Chang, 1977a: 180-81; 1977b: 28). Other evidence, however, suggests that the broad bean may be a much later introduction, possibly coming eastward over the early Silk Road, that ancient caravan route linking China with the West. Bretschneider (1882-95, 2: 35-36) says that the broad bean was the *hu-tou* introduced by Chang Ch'ien, famous Chinese envoy to Sinkiang in Han times, the second century B.C. Laufer (1919: 307-8), on the other hand, questions this identification and suggests that the broad bean was introduced no earlier than Yüan times. By Ming times, the broad bean was cultivated widely in China and apparently on a considerable scale in some places (Laufer, 1919: 307-8; Hanelt, 1972: 216, 221). The broad bean is little known in areas of China best known to Westerners, but is ahead of the soybean in "many mountainous, remote, or rainy parts of China," especially in western China (E. N. Anderson, 1988: 124). Overall, China has more acreage and production of broad beans today than any other country, over half the world's total in 1983 (Food and Agriculture Organization of the United Nations, 1984: 136). As for the pea, there is a tradition that it, too, was brought to China from the west by Chang Ch'ien (Y. S. Yü, 1977: 80), and C. Y. Hsu (1980: 83, 152, 217) includes it among the crops of Han China. Laufer (1919: 306), however, found convincing evidence of its cultivation only for Sui times (A.D. 581-618).

SORGHUM: AN EARLY ARRIVAL FROM SUB-SAHARAN AFRICA

Sorghum was domesticated somewhere in sub-Saharan Africa, though time and place of earliest domestication remain matters of speculation. One guess (Doggett, 1976: 114) is that sorghum was first domesticated there from about 4000 to 3000 B.C. It may have reached Oman about 2500 B.C. and the Indian subcontinent between 2000 and 1000 B.C., though it seems to have been unimportant there at that period (Vishnu-Mittre, 1974: 20, 21, 28; Doggett, 1976: 115; Harlan and Stemler, 1976: 474; Costantini, 1981: 273-77; Allchin and Allchin, 1982: 234, 264, 273, 278; Zohary and Hopf, 1988: 81).

Because a common early Chinese word for sorghum, *shu-shu*, means "millet of Szechwan," it has been suggested that sorghum was first introduced to China's southwest (P. T. Ho, 1975: 381; Bray, 1984: 449-50), probably overland from India and not by sea or over the Silk Road. There have, however, long been differences of opinion as to the antiquity of sorghum cultivation in China (Hagerty, 1940-41; W. Watson, 1969: 399). Sorghum is not mentioned in accounts of Han China; the name *shu-shu* does not appear in Chinese writings until roughly A.D. 300; and there are uncertainties about both the authenticity of the account and whether, at that time, it referred to sorghum (P. T. Ho, 1975: 381; Bray, 1984: 449). On the other hand, there are claims of sorghum remains being found in various Chinese archeological sites of the first millennium B.C. or much earlier (*i.e.*, the Yang-shao Neolithic site of Ching-ts'un in the

Yellow River area). These claims are suspect, and Ho Ping-ti, for one, writes that he knows of no certain botanical description of sorghum before the twelfth century A.D. Nevertheless, he and certain others believe that sorghum may have been introduced to parts of China, particularly Szechwan, much earlier than historical accounts suggest (W. Watson, 1969: 399; de Wet and Harlan, 1971: 132-33; P. T. Ho, 1975: 380-83; J. Harlan in P. T. Ho, 1975: 383; K. C. Chang, 1977a: 95; T. T. Chang, 1983: 79; Bray, 1984: 449-51; E. N. Anderson, 1988: 119). Whenever sorghum first arrived in China, it did not become important until quite late. Bray (1984: 451-52), for example, suggests that sorghum, now usually called *kao-liang* ("tall millet"),[20] probably remained unimportant through Yüan and Ming times and gained importance only as population pressure mounted during the eighteenth and nineteenth centuries.

Sorghum, like common millet and foxtail millet, is able to grow under conditions of summer heat and low precipitation, typical of its African homeland. In addition, it yields well on poor soils; it can survive floods; and some varieties can mature in a quite short growing season. It is true that people in North China prefer to eat millet rather than sorghum, that sorghum is a food of the needy, disdained and, if possible, not consumed (Bray, 1984: 451-52; E. N. Anderson, 1988: 119).[21] Yet sorghum provided not just seed for humans and animals to eat, but stalks that might be used as construction material, such as fencing, as well as fuel and fodder. In North China, where most of China's sorghum is produced today (especially the northeast), it was thus a major competitor for millet on marginal soils, and displaced it in many places.[22]

In time, however, sorghum itself was to be challenged by a new arrival, maize. The latter has always been a successful competitor in areas of China that are warm and have moderate precipitation. Today, moreover, hybrid forms of maize, capable of doing well under conditions of low rainfall and short growing season, are carrying the competition into arid areas. The result is that today maize is quickly replacing sorghum in large parts of China (E. N. Anderson, 1988: 119).

MAIZE: AMERICA'S CEREAL CONTRIBUTION

Maize (*Zea mays*) is an ancient domesticate of Meso-America, probably descended from the wild grass teosinte (*Zea mexicana*) (Galinat, 1971; Beadle, 1972; Harlan, 1975: 234-36), and was cultivated there by 5000 B.C. if not earlier (Harlan, 1975: 228-29; Bushnell, 1976: 118-19; Hammond, 1976: 123-24). Various suggestions have been made (Collins, 1909: 20-24; Ed. Anderson, 1943: 472-73; 1945: 90-91; Stonor and Anderson, 1949: 388-95; Jeffreys, 1957: 133-34; 1967; 1971; Carter, 1963; 1964; C. O. Sauer, 1963: 16-28; Vishnu-Mittre, 1966: 154-55; Vishnu-Mittre and Gupta, 1966; Sarkar *et al.*, 1974: 122; Dupuis, 1975; Needham and Lu, 1985: 61-62) that in pre-Columbian times maize may already have been cultivated in Africa or southern and eastern Asia, including China, but these suggestions are dismissed by most scholars. The prevailing view is that maize was first carried to Europe, Africa, and Asia following Columbus's discovery of America, and quickly became a staple cereal in regions as different as northern Italy and tropical Africa. Maize (*yü-mi*, "jade cereal"; *yü-mai*, "jade wheat" or "tribute wheat"; *yü-shu-shu*, "jade sorghum"; and other names) (E. Watson, 1923: 437; F. P. Smith, 1969: 464; Bray, 1984: 455) is believed to have reached China by one route or another by the mid-sixteenth century or even earlier (Laufer, 1907b: 235-41; P. T. Ho, 1955: 197; 1959: 187; Mote, 1977: 198; Bray, 1984: 455-56; Wittwer *et al.*, 1987: 171). By the late sixteenth century, it seems to have been a major crop at least in parts of China, for one observer reported that large quantities of it were paid as tribute to the emperor at that time (González de Mendoza, 1853-54, 1: 84; Boxer, 1953: 277-78). Indeed it is observed that, along with other introduced American crops (sweet potato, white potato, peanut), maize contributed to the great Chinese population explosion of Ch'ing times (P. T. Ho, 1959: 183-89; Spence, 1977: 263; E. N. Anderson, 1988: 89, 94). Early in the Ming period, China's population is estimated to have been about 60 million, virtually the same as at the beginning of the Christian era. Yet, aided by American plants, which arrived late in Ming times and became staples in Ch'ing times, population increased rapidly, to 431 million in 1851 (Spence, 1977: 263; Banister, 1987: 3-4).

The 1929-33 village survey found maize to be fourth among China's cereals and pulses in acreage cultivated (rice, 32.8 percent of total crop acreage; wheat, 29.2; millet, 10.9; maize, 9.6) and in calories supplied (Buck, 1956: 211; Maynard and Swen, 1956: 413). Its importance was far less in the Rice Region than in the Wheat Region, and in villages of the Double Cropping Rice Area it accounted for only 1.4 percent of total crop area, its calorie contribution being too small to report. Various observers (H. R. Davies, 1909: 311; E. H. Wilson, 1929: 335; E. N. Anderson and M. L. Anderson, 1977: 323) have noted that in Central and South China maize is cultivated mainly on land unsuited to rice, and becomes the staple crop only in hills and lower mountains where slopes are too steep for rice, "commonly so steep...that one wonders how the people manage to sow and reap the crop."

Gamble (1933: 79, 81-82), in a 1926-27 survey of household income and expenditures in Peking, found that maize was the principal cereal of families with the lowest-incomes, but that expenditures for maize decreased rapidly as family income rose, to make it a very minor item in the diet of families with high incomes. This fits with observations from other parts of the country: that in western China maize is a food of the poverty stricken (E. N. Anderson, 1988: 167, 184); that in a Chinese village in Yunnan, if one consumes maize it does not add to one's social status (Osgood, 1963: 162); and that in South China, the consumption of maize is regarded as suited only to children (Holmes, 1948: 132). The strongest statement of this sort is by Bray (1984: 458), who writes that the Han Chinese not only prefer other cereals (and even root crops) to maize, but that they have a clear distaste for it, with the average person rejecting it and even becoming ill from the odor of maize pancakes. She acknowledges that maize is a significant food in the mountains of Szechwan, Yunnan, and Kwangsi, but, like others, observes that this is especially among aboriginal groups, as is the case in Hainan and the northwestern margins of Kwangtung (Laufer, 1907b: 243-45; Legendre, 1911: 569, 570, 583; Graham, 1961: 18-19, 69; Bray, 1984: 458-59).

There has nevertheless been a recent striking increase in maize production in the People's Republic, from roughly 17 million to 72 million metric tons from 1956 to 1984. China is now the world's second largest maize producer, bested only by the United States, and it may account for nearly a third of all maize produced for human consumption outside of the developed world. Within China today, maize, a spring or summer crop, is the third-ranking cereal in production, surpassed only by rice and wheat. At the same time, it is estimated to provide only 12 percent of calories from cereals, for much of the increase in maize production is for animal feed. Cultivation is found principally in a belt extending from Manchuria to Yunnan, with Kwangtung and other southeastern provinces of little importance in this regard (Wittwer *et al.*, 1987: 169-81; E. N. Anderson, 1988: 17, 117, 118).

MINOR CEREALS AND PULSES

Buckwheat, Rye, and Oats

Two species of domesticated buckwheat are usually identified in China, one sweet (*Fagopyrum esculentum* = *F. vulgare*, common buckwheat; *ch'iao-mai* or *t'ien-ch'iao-mai*, "sweet buckwheat"), the other bitter and viewed as a bit toxic (*F. tataricum*, Tartar buckwheat; *k'u-ch'iao-mai*, "bitter buckwheat") (Hooper, 1929: 64; E. H. Wilson, 1929: 336). Tartar buckwheat, though bitter and lower yielding, is better adapted to high mountain conditions, and in Tibet its cultivation reaches the upper limits of agriculture, above 14,000 feet (E. N. Anderson, 1988: 120, 181). Domestication of Tartar buckwheat is commonly placed in western China, especially its mountainous regions, whereas domestication of common buckwheat is generally placed in Central Asia or western China (Vavilov, 1949-50: 21; Harlan, 1975: 72; Mathon, 1981: 58; Zeven and de Wet, 1982: 42, 100). Buckwheat is associated with barley today, and originally it may have been a weed in barley fields, or a wild plant growing nearby, domesticated because people came to recognize its virtues (E. N. Anderson, 1988: 120).

Even though China and its borderlands seem to be at or near the center of buckwheat domestication, the plant appears to have come into cultivation there quite late (Bertin *et al.*, 1971: 5). It is not listed among the cultivated plants of ancient China, whether found in archeological sites, reported in the literature, or cited by various authorities (P. T. Ho, 1969; 1975: 43-89; K. C. Chang, 1970: Table 1; 1977b: 26-27; 1986; Keng, 1974; Y. S. Yü, 1977; C. Y. Hsu, 1980: 81-91; H. L. Li, 1983; Bray, 1984: 429-552). Indeed, the first documentary accounts of buckwheat in China are said to date only from the fifth and sixth centuries A.D. (Campbell, 1976: 236),[23] though one suspects that it was cultivated long before in mountain areas and northern fringe locations not subject to detailed early description or extensive archaeologic excavation.

Turning to modern China, one notes that buckwheat is not singled out by Maynard and Swen (1956: 413) in the nutrition section of the 1929-33 survey, and presumably it is included in their category of "other grains." Quite minor as a source of calories in China as a whole, buckwheat was found in the survey to occupy 1.9 percent of China's total crop area, and 0.4 percent in the Double Cropping Rice Area (Buck, 1956: 211). The survey did find a few areas of China where it accounted for 3 percent or more of the total crop area, with the highest percentage, 6.8 percent, in the Spring Wheat Area, China's extreme north and northwest. Though it did not show up in the survey results, buckwheat is also important in China's cold mountain regions, where it is typically among the staple crops of both Han Chinese and minority peoples such as the Tibetans (E. H. Wilson, 1929: 337; E. N. Anderson, 1988: 120). It is also important in the Himalayas of India (Watt, 1889-96, 3: 310-12), for it can be cultivated at elevations as high or higher than barley (often as a spring crop in rotation with barley); it does well on quite marginal soils; it matures quite quickly; given good climatic conditions, it provides two crops in a season; and it demands little care other than planting and harvesting (E. H. Wilson, 1929: 336-37; E. N. Anderson, 1988: 120, 181).

Rye (*Secale cereale*), a weed in ancient fields of wheat and barley, finally came to be cultivated in its own right. Remains of rye in Neolithic sites are scarce, and the place of its earliest deliberate cultivation is uncertain, with suggestions ranging widely, among them the region from Turkey to Turkestan and Afghanistan (Evans, 1976: 110); Russia (Helbaek, 1971: 275-77); and northern Europe (Bertin *et al.*, 1971: 2). A recent review of the evidence by

Zohary and Hopf (1988: 69-71) points to Turkey and nearby Armenia as the most likely place of rye origin, with earliest deliberate cultivation in Anatolia in Bronze Age times and possibly much earlier. Though rye may have reached Japan and Korea in the sixth or seventh centuries A.D., its main modern area of cultivation is northern, central, and eastern Europe (Bertin *et al.*, 1971: 2). There it was made into the rye breads for which that region is noted, and which came to be popularized in the United States in large part by Jewish immigrant bakers. As for China, most authors simply don't mention rye, not Laufer, Buck, Hosie (describing cereals of Manchuria), or the authors of *Food in Chinese culture*. Bretschneider, on his part, thought that rye was not cultivated in China (de Candolle, 1886: 370-71). There is, however, evidence to the contrary: E. H. Wilson (1929: 335), a naturalist who travelled mainly in Yunnan, Szechwan, and Hupeh, writes of rye, which he identified as *Secale fragile*,[24] as cultivated on a very small scale in the mountains; the American Plant Studies Delegation (1975: 54-55), though they did not see rye cultivated or visit the places in question, wrote of small areas of rye in upper elevations in certain provinces of the west; and Bray (1984: 463, 463n), though she says nothing about its cultivation in her book on Chinese agriculture, gives a name for it, *hei-mai*, "black *mai*" (*mai* is a collective term for fall-sown grains of Near Eastern origin, especially wheat and barley).

Oats (*Avena* spp.) were also weeds in ancient fields of wheat and barley in the Near East and certain of them came to be cultivated, though at what date remains uncertain. In the modern world, only the common white oat, *Avena sativa*, a hexaploid, is an important cereal, though other species are of minor importance, especially the hexaploids *A. byzantina*, red oat, and *A. nuda* (= *A. sativa*, convar. *nuda*), large-seeded naked oat. By one view (Holden, 1976: 88), when wheat and barley spread across Europe, weed oats went along, and there, about 1000 B.C., they were first cultivated in their own right.[25] As for China, oats were cultivated already in Han times (C. Y. Hsu, 1980: 82, 83, 221, 222, 223, 225; Bray, 1984: 240). Some (de Candolle, 1886: 373; Holden, 1976: 87-89) say that the early oats of China were *A. nuda*, presumably introduced from an east European center of domestication. In any case, oats (*ch'iao-mai*, "sparrow grain"; or *yên-mai*, "swallow grain") (F. P. Smith, 1969: 59, 164; Bray, 1984: 463) were to remain quite a minor crop in China. The 1929-33 village survey found only 1.8 percent of total crop area in China given over to oats (Buck, 1956: 211), which accounted for only 1.2 percent of total calorie intake (Maynard and Swen, 1956: 413). Oats were of somewhat greater importance in the Wheat Region (3.0 percent of total crop area), especially in the Spring Wheat Area of the northwest (11.7 percent). In the Rice Region, however, they were of little importance (0.4 percent of crop area), though more so in Szechwan (1.8 percent). Judging from accounts of Szechwan, however, they are little grown by Han Chinese, but cultivated on a significant scale only by Tibetans and tribal groups in upland areas, and indeed Legendre says that among the Lolo living in the mountains of western Szechwan they are the main food (Legendre, 1911: 570-71; E. H. Wilson, 1929: 335).[26] Pollard (1921: 198-99), writing of the Nosu (Black Lolo) or upland tribal groups in southwestern China in general, observed that oatmeal mixed with cold water is among their principal dishes; that men may journey for days or weeks at a time consuming no other food, merely stopping along the way to mix the oatmeal; and that certain of them may consume two, three, or more pounds of such oatmeal at a sitting.

In the Double Cropping Rice Area, the village survey reported no oat cultivation or consumption, but a map by Bertin *et al.* (1971: 3) does show scattered cultivation there.

Barnyard Millet

Though, as we have seen, the two lesser millets from Africa, pearl millet (*Pennisetum americanum*) and finger millet (*Eleusine coracana*), reached India by 1000 B.C. or before, the history of their spread to China is little known. Bretschneider (1882-95, 2: 285) reports that *E. coracana* is identified in classical works, and it is also shown on a recent map of cereals (Bray, 1984: 437) as occurring as far east as coastal China and on Taiwan. Pearl millet, in turn, is mentioned as yielding less than *Setaria* millet in contemporary China, and is little known and of quite minor significance there (American Plant Studies Delegation, 1975: 73). Our other sources are silent on the presence of these millets in China, which at best are quite minor crops far overshadowed by the *Setaria* and *Panicum* millets.

More often mentioned is a millet cultivated widely in East Asia, commonly known as barnyard millet, Japanese millet, or Japanese barnyard millet. This domesticate is often grouped with an Indian domesticated millet as *Echinochloa frumentacea* (formerly *Panicum frumentaceum*). Yabuno (1962: 304; 1966), however, found significant genetic differences between the East Asian cultivated species and that of India. Thus he suggested a new term for the East Asian domesticate, *E. utilis*, while retaining *E. frumentacea* for the Indian one. In his view, we are dealing with two cytologically different species, that of India deriving from one wild ancestor, *E. colona*, and that of East Asia from another, *E. crus-galli*.

The genus *Echinochloa* consists of roughly twenty species of grasses, some annual and some perennial, that occur widely in warm regions of the Old and New Worlds. Some are aggressive invaders of habitats disturbed by humans. Some serve as fodder, and may be cultivated as such. Young shoots may be eaten as human food. Stems of some are used as thatch or, for at least one species, are chewed as a sweet, crushed, with the sweet extract drunk, made into candy, or fermented into liquor. Seed of some species is also collected as food, especially at times of food scarcity. And two (*E. colona* and *E. crus-galli*) have been cultivated as cereals in ancient Egypt, early Europe, modern sub-Saharan Africa, or in ancient and modern southern and eastern Asia, including China where *E. crus-galli* is sometimes called *pai* or *pai-tzŭ* (Bretschneider, 1882-95, 2: 160-61; Watt, 1889-96, 6, Part 1: 7-10; E. H. Wilson, 1929: 336; Dalziel, 1937: 526-27; Ames, 1939: 104; Bailey, 1949: 148; Schnell, 1957: 101, 153-54; I. H. Burkill, 1966, 1: 902-3; F. P. Smith, 1969: 304-5; P. M. Smith, 1976; de Wet, 1977: 18-19; Read, 1977g: 32; de Wet *et al.*, 1983). What historical connection, if any, there may have been among African, European, and Asian areas of *Echinochloa* domestication, is not known.

Various wild species of *Echinochloa* are weeds in paddy fields, and in Southeast and East Asia there are forms, including types of *E. crus-galli*, that mimic the rice plant throughout its period of growth, creating a weeding problem for rice cultivators in both lowland and upland fields (I. H. Burkill, 1966, 1: 902-3; Yabuno, 1966: 286-92, 314-15).[27] One of the most successful, *E. oryzoides* (= *E. oryzicola*), drops most of its seed before the rice harvest, though some may be harvested with the rice; this provides a chance of its seed being sown with rice seed the next season or, in any case, a new growth is assured when next the old rice field is cultivated (Yabuno, 1966: 288; de Wet *et al.*, 1983: 283). *E. crus-galli* is found also in fields of millet and wheat, on occasion comprising 30 percent of the threshed millet in China; it is bitter but edible and, when food is in short supply, it may be consumed as a cereal (F. P. Smith, 1969: 304-5). Domesticated *E. frumentaceum* in India is said to be the fastest maturing of all millets, ready for harvest within six weeks and providing an early and cheap source of food for the poor (Watt, 1889-96, 6, Part 1: 9, 10). One wonders whether barnyard millet had a pattern of domestication similar to that of oats, with a troublesome weed in stands of preferred cereals coming to be grown in its own right because of advantages it enjoyed at some times and places.

We seem to be dealing not, as some have implied, with a single species and hearth of domestication, but with two distinct species that may have quite different histories, *E. colona* domesticated in India or Africa and *E. crus-galli* in East Asia. This is in keeping with statements by Yabuno (1966: 318-19) and de Wet *et al.* (1983: 283-84). The latter conclude that *E. colona* (presumed ancestor of *E. frumentacea*) was likely domesticated in India in the not too distant past, and *E. crus-galli* in East Asia by 2000 B.C. or somewhat earlier. They place the domestication in Japan, whereas Harlan (1975: 72) had suggested eastern China. In either case, *E. colona* is more a grass of tropical regions and *E. crus-galli* of temperate ones (I. H. Burkill, 1966, 1: 903; de Wet *et al.*, 1983: 284).

B. E. Read (1977g: 32) notes that steamed *E. crus-galli* is remarkably tasty. It is also good nutritionally, as shown by Leung *et al.* (1973: 6-7) who report the whole grain to be roughly comparable to rice in calories, thiamine, and niacin, but higher in protein, riboflavin, calcium, and iron.

Job's Tears

Job's tears (*Coix lacryma-jobi*)[28] is a tall grass (it can grow four to six feet in height) whose genus is in the same tribe (*Maydeae*) as maize. Today Job's tears occurs wild and cultivated in Asia, Africa, and other parts of the world, but it probably originated in tropical Asia (Watt, 1904: 192, 203-5; Vallaeys, 1948: 254-57; I. H. Burkill, 1966, 1: 638).[29] It is well-known around the world for the use of its attractive seeds, whether from wild or domesticated plants, in necklaces, bracelets, and rosaries. In its Asian homeland, in addition, certain of its varieties have long been grown for food, though only one is important in this regard: the variety called *ma-yuen*, which has large, sweet-tasting seeds with soft shells, easy to hull.

The exact place of earliest cultivation of Job's tears is uncertain, but it was somewhere in the zone from Indochina and the Malay Archipelago to eastern India (Vavilov, 1949-50: 29; H. L. Li, 1970: 12, 16; 1983: 51; Koul, 1974: 64; Arora, 1977: 360; Mathon, 1981: 70; Zeven and de Wet, 1982: 53). Some have said that the plant was already cultivated in the period from 2000 to 1000 B.C. in India (Schaaffhausen, 1952: 217; Koul, 1974: 64; Mathon, 1981: 70), and Job's tears is also commonly regarded, along with rice and *Setaria* millet, as one of the earliest cultivated cereals of Southeast Asia. There has even been speculation that its domestication in Southeast Asia may have preceded that of rice (H. L. Li, 1970: 12; 1983: 51; Harlan, 1977: 381; Bray, 1984: 322).[30]

Though today Job's tears is important in relatively few places, it is grown for food to some extent for in all countries from India to Indochina, Malaysia, Indonesia, the Philippines, China, and Japan (Watt, 1889-96, 2: 497-98; 1904: 192-93, 223-25 and *passim*; 1908: 393-95; Crevost and Lemarié, 1917: 86-88; Wester, 1924: 17-18; Vallaeys, 1948: 260-63; Lebar *et al.*, 1964: 118; I. H. Burkill, 1966, 1: 638; Jain and Banerjee, 1974: 40-41; Koul, 1974: 64-65; Harlan, 1977: 381). In the Southwest Pacific it is found, self-sown, farther eastward than *Setaria* and other millets of Eurasian origin (Bray, 1984: 437); its seed is ordinarily used for making necklaces, though in New Guinea there are reports of its use as food, but to a minor degree and not on a regular basis (Massal and Barrau, 1956: 43). In India, by contrast, Job's tears is a regular crop, even a staple, among certain tribal peoples in the hills and mountains of the northeast and in the Central Provinces (Madhya Pradesh), though it is virtually unknown as a food plant to the general Indian population (though people may collect and eat the seed of wild *Coix* at times of food shortage) (Watt, 1889-96, 2: 497-98; 1904: 223-24; 1908: 394-95; S. Roy, 1960: 134; Lebar *et al.*, 1964: 46; Hutton, 1968: 60, 98; 1969: 77, 93; Jain and Banerjee, 1974: 40-41; Whyte, 1974: 212; Arora, 1977: 359-65 *passim*).[31]

Turning to China, we note an interesting tradition about Job's tears relating to the famed Han general Ma Yüen (14 B.C.-A.D. 49), who was sent to suppress a revolt against Chinese rule in what today is Tonkin. Perhaps learning from local practice, Ma ate Job's tears to survive the malarial miasmas of that hot, tropical region,[32] and when he returned to China he carried with him a large store of the seed, intending to cultivate it. In fact, he is credited by the Chinese with introducing Job's tears to their land, presumably the afore-mentioned soft-shelled variety which is named for him (Mayers, 1874: 149-50; Bretschneider, 1882-95, 3: 383; Watt, 1904: 224; Couling, 1917: 262; Schaaffhausen, 1952: 217; Schafer, 1967: 132; 1977: 89; F. P. Smith, 1969: 122). If this tradition is true, we are dealing with a Nan-Yüeh contribution to China's food scene. In any case, Job's tears are mentioned in the *Huang-ti su-wen, The Yellow Emperor's classic of internal medicine*, whose text in large part existed during Han dynasty times, with much of it earlier (Veith, 1949: 9, 183). Along with soybeans, millets, rice, and hemp, Job's tears were also found in many earthenware vessels uncovered in Han tombs near the ancient capital of Lo-yang in North China (Y. S. Yü, 1977: 71), the associations suggesting that they were used as food. In the fifth century, T'ao Hung-king reported the plant to be commonly cultivated in middle China, but noted that its seed was used in medicine (Bretschneider, 1882-95, 3: 383). In T'ang times, seed from both wild and cultivated forms served now and then as a dietary supplement to the major cultivated cereals (Schafer, 1977: 89). And in the sixteenth century Li Shi-chen reported Job's tears to be grown on a considerable scale, with the seed of one form made into gruel, or pounded into flour and eaten as food or used to make a fermented beverage (Bretschneider, 1882-95, 3: 383-84).

Though most of our sources on traditional China fail to mention Job's tears (*i-i, i-mi, i-i-jên,* etc.) there are reports of it being grown on a small scale everywhere in central and western parts of the country (E. H. Wilson, 1929: 336; Graham, 1961: 19) and in Manchuria (Hosie, 1904: 180). It is also reported to be grown by the Yami of Taiwan (Fogg, 1983: 100) and by other shifting cultivators in China (Bray, 1984: 99). Though sometimes eaten simply as food (as gruel or added to soup or sweets) by the traditional Chinese, the seed of Job's tears was consumed by them mainly with medicinal purposes in mind (Hosie, 1904: 180; Watt, 1904: 225; E. H. Wilson, 1929: 336; Vallaeys, 1948: 259; I. H. Burkill, 1966, 1: 639). Today, if E. N. Anderson (1988: 120) is correct, no one eats Job's tears in South China except, once in awhile, persons well versed in medicinal matters. Job's tears are regarded as nutritious and restorative; soothing; capable of relieving pain, reducing fever, and cleansing the bowels; and effective for urinary and pulmonary problems.[33,34] One reads that Chinese doctors may prescribe a soup or congee made of the hulled seed. Such seed was also made into a refreshing drink for invalids, or was fermented to obtain an alcoholic beverage considered effective against rheumatism. It was roasted and popped, to provide a product (*chao-i-mi*, or "roast Job's-tears grain") similar to popped rice. It was also milled into a flour which was considered excellent for sick persons. In view of the above, it is no surprise to learn that seed of Job's tears was found in Chinese drug stores and was exported for medicinal as well as food purposes, its physical resemblance to pearl barley being such that Chinese customs reports sometimes incorrectly designated it as such (Soubeiran and Dabry de Thiersant, 1874: 98; Bretschneider, 1882-95, 3: 382-85; Hosie,

1904: 180-81; Watt, 1908: 396; Couling, 1917: 262; Hooper, 1929: 40; E. H. Wilson, 1929: 336; R. S. Harris *et al.*, 1949: 30; Schaaffhausen, 1952: 218; F. P. Smith, 1969: 122; J. D. Keys, 1976: 45-46; Perry, 1980: 164).

The success of rice in competition with Job's tears derives from rice's important advantages over the latter cereal, among them faster growth, higher yields, better keeping qualities, and greater ease of milling with mortar and pestle (Schaaffhausen, 1952: 219-220; Arora, 1977: 365).[35] Though some have written of the nutritional superiority of Job's tears over other cereals, we cannot discern a clear superiority over rice, at least before rice has been milled (brown rice). The whole hulled seeds of rice and Job's tears are roughly equivalent in the food energy they provide. Those of Job's tears are reported (Platt, 1962: 6-7; Leung *et al.*, 1973: 5, 7-8) to have far more protein (12 to 14 gm./100 gm.) than rice (7.4 to 8.2 gm./100 gm.), but rice protein is much higher in nutritive value (Food Policy and Food Science Service, Nutrition Division, FAO, 1970: 38-41). Job's tears seem higher than most forms of rice in calcium and perhaps in thiamine, but rice is richer in certain other minerals and vitamins. One clear advantage of Job's tears may be its ability to survive attacks by insects, including locusts (Koul, 1974: 65), a trait that perhaps led to its inclusion in the fifteenth-century famine herbal written by Chou Ting-wang (Read, 1977g: 32).

Minor Pulses

Turning to the minor pulses, one must acknowledge that not one of them comes close to soybeans in its dietary contribution. Nevertheless, some minor pulses were widely consumed in the Double Cropping Rice Area (Maynard and Swen, 1956: 407) and as a group their dietary contribution was quite significant. For these reasons, it seems essential that we identify as many of them as possible, and ascertain their dietary role. That task, however, has not proved easy, for the lesser pulses are many and often are overlooked by observers. The task is made more difficult because pulses in great variety are also imported. Herklots (1972: 226) writes that before World War II large amounts of dried beans arrived in Hong Kong from North China, Korea, Southeast Asia, and India. "Sacks of beans," he writes, "green, and brown, and red, and white, short and fat, long and thin, I could only name a fraction of them." Though incomplete, we present our results below.

By far the greatest number of minor pulses cultivated in the Double Cropping Rice Area today derive from a zone stretching from China and Japan through Southeast Asia to India. Traditionally most of that region was non-dairying, which may have encouraged the domestication and use of legumes as alternatives to milk products in providing dietary protein and calcium. Two of the minor pulses—the adzuki bean (*Phaseolus angularis*) and velvet bean (*Stizolobium hassjoo* = *Mucuna hassjoo*)—are believed by authorities to have been domesticated in South China, elsewhere in China, or in Japan.[36] The domestication of others is usually placed in Southeast Asia (sometimes including South China or parts of it) or India: the mung bean (*Vigna radiata*), sword bean (*Canavalia gladiata*), mat or moth bean (*Phaseolus aconitifolius* = *Vigna aconitifolia*), and red or rice bean (*Phaseolus calcaratus* = *Vigna calcarata*).[37] Two others, the hyacinth or lablab bean (*Dolichos lablab* = *Lablab niger*) and the black-eyed pea or cowpea (*Vigna unguiculata* = *V. sinensis*), are believed by most to have come from Africa.

Here we would direct attention briefly to the mung bean (*lü-tou* or "green bean") the only of our minor pulses to be identified in tables of the 1929-33 village survey, as "green beans" from its Chinese name. Even this bean, which is usually green but sometimes yellow or brown in color, occupied an insignificant percentage of crop area in the Rice Region as a whole (0.4 percent) and even less in the Double Cropping Rice Area (0.05 percent). In the Wheat Region, however, the mung bean occupied more acreage (5.2 percent) and was cultivated on more farms (16 percent) that any pulse other than soybeans (Buck, 1956: 211-12, 215).

The adzuki bean (*ch'ih-tou*), whose major varieties have dark red, brown, yellow, or black seed, is not identified as such in the village survey. Yet it is said to be an important crop in China, especially in the Yangtze Valley. It is also said to be important in Japan, sixth among all crops there. An interesting aspect of the adzuki bean is that, along with certain other red beans (E. N. Anderson, 1988: 125), it plays a major role among Chinese sweets, its most common use being in a sweetened paste[38] used as filling for sweet dishes such as "Red Bean Paste Steamed Buns" (F. Lin, 1975: 301) and "Red Bean Fool," a popular Cantonese pudding made of red beans and rice (Y. K. Martin, 1984: 96, 208). It is also interesting that the adzuki bean is regarded as the most "heating" of all beans and thus is employed in treating problems deriving from excess "cold," such as difficult labor, danger of miscarriage, failure of a mother to secrete milk, constipation, and kidney disease (Sacks, 1977: 9, 10, 13-14).

Also of interest is *Vigna unguiculata*, subspecies *sesquipedalis*, the "yard-long bean" or "asparagus pea or bean," whose pods are one to three feet in length (Figure 23) (Herklots, 1972: 264; Zee and Hui, 1981: 55). Though *V. unguiculata* comes from Africa, this subspecies is not traditional there. It seems to have been developed in India and is widely found there and in East Asia (Steele, 1976: 183). Its unripe pods, though somewhat tougher than string beans, are cut into sections, stir-fried, and eaten as a green vegetable, usually mixed with other vegetables (Zee and Hui, 1981: 55; E. N. Anderson, 1988: 125). A good recipe for yard-long beans with sliced pork or beef, ginger, garlic, soy sauce, salt, and pepper, may be found in H. L. Chung and J. C. Ripperton (1929: 45).

It is clear that when finally the New World pulses found their way to China, they were faced with a formidable assemblage of well-established competitors. Partly as a result of this, the common (haricot, french, string, or kidney) bean (*Phaseolus vulgaris*) and the lima bean (*Phaseolus lunatus*) did not assume a role as important as they otherwise might have.

PREPARATION AND CONSUMPTION OF CEREALS AND PULSES

The Chinese prefer polished rice despite its diminished nutritional value, and the use of polished rice has greatly increased in modern times (see "Thiamine Deficiency and Beriberi" in our nutrition chapter).[39] The non-glutinous forms of rice, after milling, are the common, everyday rice (*fan*, "cooked rice") consumed in China, usually after boiling in water without salt being added. Rice water left over after boiling the rice may be consumed as a cooling drink, both to allay thirst and in terms of the hot-cold system of food classification. Leftover boiled rice among the Cantonese is commonly made into fried rice prepared with chopped vegetables and meat; a similar, better-liked fried rice is that of the Eastern Cuisine of China, which is made by slowly sautéing rice along with other ingredients, to provide a dish both fluffy and with a somewhat firm coating. The Chinese may use rice for the deep-fried rice crust that makes the serving of "Sizzling Rice Soup" so dramatic. They also grind non-glutinous rice, in water or without water, into a flour used for making noodles, wrappings, cakes, sweets, and such (G. Z.

Figure 23. Bundles of yard-long beans (*Vigna unguiculata* subspecies *sesquipedalis*) in marketplace (Ch'eng-tu, 1987)

Chu, 1975: 71; F. Lin, 1975: 329-31; K. Lee and A. Branyon, 1984: 82, 84; Y. K. Martin, 1984: 14, 228; E. N. Anderson, 1988: 114, 173-74). Glutinous forms of rice (commonly, when hulled, called *no-mi*), though they may be used in ways similar to non-glutinous rice, tend to serve more specialized uses. They are essential in rice-wine manufacture, with an estimated 6 percent of all glutinous rice going to this purpose (C. P. Li, 1948: 199; T. H. Shen, 1951: 201-2). Like other rice, glutinous rice may be made into vinegar. Glutinous rice, alone or with other rice, is also commonly used in congee, the soft-rice porridge (*chou*; Cantonese: *chuk* or *jook*) served for breakfast or late-night suppers or eaten by old people or persons in poor health. Congees may vary in preparation, thickness, and ingredients, with a favorite congee of Canton, "Sampan Congee," including various kinds of seafood, sometimes chicken, eggs, and peanuts, and with onion, garlic, and ginger as common flavorings (W. W. Chang *et al.*, 1970: 416; K. Lo, 1972: 57-59; 1976: 119, 122; 1979c: 88, 90; G. Z. Chu, 1975: 70; M. Gin, 1984: 67). Glutinous rice, whether as whole grains or flour,[40] is also used in the puddings and pastries common on festive occasions, and in other dishes where the qualities of stickiness and ease of shaping are important (R. S. Harris *et al.*, 1949: 32; T. H. Shen, 1951: 201; W. W. Chang *et al.*, 1970: 416, 418-19; Wittwer *et al.*, 1987: 143). In fact, the use of glutinous rice in sweets is of such prominence that one of its Chinese names is "sweet" rice (rice used in sweets, not itself sweet in taste) (G. Z. Chu, 1975: 70, 283). One of the best known desserts made with glutinous rice is "Eight-Precious Rice Pudding," "Eight-Treasure Rice Pudding," or "Eight-Jewel Rice Pudding," a steamed dish decorated with nuts, seeds, and dried, honeyed, or candied fruit that is served with a sweet syrup, particularly at the time of Chinese New Year (Feng, n.d.: 173; R. S. Harris *et al.*, 1949: 24; Em. Hahn *et al.*, 1973: 172-74; Stallard, 1973b: 178; K. Lo, 1976: 137; 1979c: 329; Y. K. Martin, 1984: 186, 224). Glutinous rice is also commonly used as stuffing, as in poultry dishes such as "Chicken Stuffed with Glutinous Rice" and "Duck Stuffed with Glutinous Rice" (C. B. T. Lee and A. E. Lee, 1980: 190-91; Hom and Steiman, 1981: 284; Y. K. Martin, 1984: 112-13). It may be mixed with ingredients such as pork and dried plums, wrapped in bamboo leaves, and steamed (Tiger and Wolf, 1985: 70-71). Glutinous rice is made into balls such as the Hupeh "Pork Balls with Glutinous Rice" ("Pearl Balls") and the Cantonese "Fried Glutinous Rice Chicken Rolls" (W. W. Chang *et al.*, 1970: 324, 418-19). Glutinous rice may also be fermented, and the fermented product used as a flavoring or food. One such product, which originated in or is popular in eastern and western Chinese cooking, is "wine rice" in which yeast and other fungi, sold commercially in a wine yeast ball, are the agents of fermentation. Wine rice, which may be stored in a refrigerator for weeks or months, has a pleasant odor and is quite soft, juicy, sweet, and has a low alcohol content. It enjoys a special place, as a strengthening food, in the diet of Chinese mothers after they have given birth.[41] It may be eaten as a snack or dessert, for example by boiling in water with sugar and eggs. Or it may be used as a tasty ingredient (commonly in or with sauces) in cooking, for example in the Szechwan dish "Spicy Shrimp with Wine Rice," which also contains ground pork, soy sauce, gingerroot, garlic, hot red peppers, Shao-hsing rice-wine, and other spices and flavorings (R. S. Harris *et al.*, 1949: 25; H. L. Wang and C. W. Hesseltine, 1970: 574-75; G. Z. Chu, 1975: 194-95, 300; F. Lin, 1975: 248-49, 298, 328; Hom and Steiman, 1981: 42, 289-93).

Of special interest here is the practice of popping rice in China. One recent Western observer (pers. comm., Gwendolyn T. Schwabe) reported rice being popped and sold by a lone Han Chinese street vendor in the city of Lan-chou in Kansu. The popped grain was a treat for children, and was eaten directly, without oil, salt, or other flavorings. She was later told by a Chinese from Harbin that the practice of popping grains (rice, maize, and sorghum) is most common in the northeast.[42] In China's northeast, popped grain may be prepared at any season, and in big cities is sold as a snack, sometimes with an added sweetener, at stands along the street or near bus terminals. Early in this century, Sammons (1918: 182) wrote of popped rice being produced on a commercial basis in a Shansi district called Tsingyuan (= Chin-yüan?). This was a local trade, but sufficiently well-enough organized that there was a guild of rice poppers which at one time had 130 members. The process followed was quite elaborate, and ended up with popped rice that was either plain or colored red or yellow. Popped rice, according to Sammons, was used in sweets or as food for invalids or the aged. He also said that plain popped rice was probably produced in two other places in Shansi and in Honan as well.

In East and Southeast Asia, soybean products are of great dietary importance, and essential elements in vegetarian diet. A great variety of them are found (for which, see C. V. Piper and W. J. Morse, 1923: 218-59; Morse, 1947: 142-43; A. K. Smith, 1949; Uphof, 1959: 171; I. H. Burkill, 1966, 1: 1099-1103; A. K. Smith and S. J. Circle, 1972: 12-16), with Hymowitz and

Newell (1981: 280-84) providing brief histories of the principal ones, their origin and spread.[43] In China, young soybean pods may be cooked, or the seeds removed and prepared in soups and other dishes. Children is some parts of China also carry paper bags of soybeans, in their pods, to school as snacks (Wittwer *et al.*, 1987: 189).

Mature soybeans, because they are toxic and taste unpleasant, are rarely eaten raw, or by simple cooking or roasting, but instead are processed in various ways to improve their nutritiousness and edibility. One product well known to the nutritionally-aware in the West, is soybean milk (*tou-fu-chiang*); in traditional times this product was not normally used in feeding infants, but was consumed by others, for example old people who might take it in place of tea, or by mothers to stimulate their milk secretion. Use of soybean milk for infants did increase among the Chinese, encouraged by the example of the Western use of cow's milk. At the same time, the dispensing of soybean milk for infants at health stations in Peking early in this century was described as just as atypical as that of dispensing cow's milk (Guy, 1936: 440; Platt, 1936: 418; Platt and Gin, 1938: 345, 351; Jelliffe, 1968: 58). E. N. Anderson and M. L. Anderson (1977: 342) speak of a further, quite recent, development: the use of soybean milk becoming quite popular and spreading, together with soft drinks. Soybean milk may be prepared at home, but soybean-milk factories are also widespread, and in some cities in traditional China such milk was delivered early in the morning to the homes of regular customers, as well as being sold by street vendors (Kale, 1937: 136-40, 158; Holmes, 1948: 232; A. K. Smith and S. J. Circle, 1972: 16).

Another common soybean product in China is soybean-milk skin, a surface film skimmed from boiling soy milk which is then dried. This product, after soaking, serves as a wrapper in cooking or it may be sliced into strips or small pieces and added to soups or cooked in other dishes such as "Stir-Fried Pork with Bean Milk Skin" (C. V. Piper and W. J. Morse, 1923: 246; R. S. Harris *et al.*, 1949: 32; A. K. Smith and S. J. Circle, 1972: 379, 445; G. Z. Chu, 1975: 239-40, 298-99; F. Lin, 1975: 98-99, 120-21).

Bean curd or tofu (*tou-fu*) is a product said to have been known in Han times, though literary evidence suggests that it originated late in T'ang times or early in Sung.[44] It is made from soy milk with precipitating agents such as vinegar or gypsum, and is an essential Chinese food, especially critical in vegetarian diets.[45] Bean curd is low in cost but high in protein of good biological value. It contains substantial amounts of lysine and certain other essential amino acids; it is easy to digest; and it is rich in minerals, especially calcium,[46] and in certain vitamins. The dietary role of tofu in East Asia has been compared to that of bread in Western diet. In Japan, there are said to be 38,000 tofu shops compared to 19,000 bakeries in the United States; those shops annually provide each Japanese with about seventy tofu cakes, each twelve ounces or so in weight. In China, the average consumption of tofu is much higher than in Japan (Shurtleff and Aoyagi, 1975: 16).

Fresh bean curd is bland in taste, and may be used without further processing, whether eaten alone, stuffed, or served in soups or dishes with other ingredients such as rice, vegetables, meat, and fish. It may be eaten in a sweetened form, for example, after mashing, as a filling for steamed bread. Fresh bean curd may, as an alternative, be deep-fried before use or sale. Moisture may be removed from bean curd by compression and drying, as in pressed bean curd or soy sauce-pressed bean curd. Bean curd may also be processed in other ways, for example fresh bean curd may be pressed, fermented, brined, and aged to form white or red bean-curd cheese (*su-fu*), which has the consistency of Western cheese and is commonly employed in small amounts as a flavoring, as in congee (Hesseltine and Wang, 1967: 284-86; H. L. Wang and C. W. Hesseltine, 1970: 572-74; G. Z. Chu, 1975: 272, 279, 281, 290-91; F. Lin, 1975: 310-12; K. Lo, 1979a: 260-61; Hom and Steiman, 1981: 89-93).

Indeed it is in fermented products, such as this, that soybeans find their greatest use in China (E. N. Anderson, 1988: 123). Most important among these is soy sauce (*shih-yu* or *chiang-yu*), whose manufacture probably began during Chou dynasty times (C. P. Li, 1948: 170; K. C. Chang, 1977: 31). Traditionally, it was fermented naturally from soybeans, flour (usually wheat), salt, and water, but it varies greatly in color, thickness, taste (from salty to nearly sweet), and overall quality.[47] In addition, there are flavored soy sauces, as with mushrooms or shrimp (Hom and Steiman, 1981: 40), and there are many other fermented soybean products, among them black beans (*tou-shih*),[48] yellow beans, and a range of pastes and sauces.[49,50] Soy sauce and most other fermented products are flavorings or relishes used alone or in combinations to create a broad range of dishes. The nutritional contribution of soy sauce and other ferments, however, should not be overlooked, for fermentation not only improves the digestibility of soy protein, but contributes vitamins and other nutrients as well (E. N. Anderson and M. L. Anderson, 1977: 340).

Among other significant soybean products are soybean oil (*tou-yu*), which is used not only for cooking but for soap-making, lubrication, and other purposes; and soybean flour, which is added to other types of flour in making various foods, such as noodles and steamed bread. Also to be noted is processed soybean protein, whose use is expanding as a substitute for meat in processed foods, as well as in preparing meat analogues (Wittwer *et al.*, 1987: 189).

Wheat kernels in modern China are not eaten whole, whether parched or cooked.[51] Nor are they parched and broken into fragments like the bulgur of Turkey and nearby areas. Instead, among the Han Chinese they are milled into flour which is usually made into steamed dumplings (small or large, unfilled or filled) or into noodles. The Chinese also use wheat flour for pancakes; fried or boiled dough-strips; wrappings, as for spring rolls; cakes; and other edibles (Gamble, 1933: 84-85; Cormack, 1935: 134; E. N. Anderson and M. L. Anderson, 1977: 337; V. Y. N. Hsu and F. L. K. Hsu, 1977: 300; Hom and Steiman, 1981: 294-322; E. N. Anderson, 1988: 54, 112, 117-18, 178).

We feel obliged to mention one attractive and frequently-noted type of cake, which, if one story is true, played a significant role in Chinese history. We are referring to "moon cakes," about which there are different stories of origin (Eberhard, 1968: 89). In traditional China, they were prepared and eaten at the time of the "Moon Festival" or "Mid-Autumn Festival," a festival in which household women may offer such cakes to the moon, the cakes later being eaten by members of the family. Moon cakes were also commonly exchanged as gifts with friends, thereby serving to strengthen ties with them. Moon cakes are made of gray or "moon-colored" flour; they are circular in shape, like a full moon; sweet; filled with stuffing; and may be stamped with the emblem of the deity and decorated with inscriptions or floral or other motifs, such as that of the rabbit, which is associated with the moon (for excellent color photographs of moon cakes, see Em. Hahn *et al.*, 1973: 165, and Tiger and Wolf, 1985: 73). Because they require elaborate preparation using special equipment, moon cakes are commonly bought in shops or restaurants rather than made at home. In North China, according to Hsu and Hsu, moon cake fillings are of just two types, a paste of brown dates or one of white sugar, but in South China fillings of considerable variety are used (among them lotus jam, sweet sesame seeds, nuts, fruits, sweet beans, preserved melon, ham, and others) (Feng, n.d.: 211; Cormack,

1935: 180-81; Eberhard, 1952: 97-108; Burkhardt, 1953-58, 1: 53-54; 2: 71-72; J. D. Keys, 1963: 94; Latourette, 1964: 595; Tun, 1965: 65-67; C. S. Wong, 1967: 147-48; Em. Hahn *et al.*, 1973: 162, 164; V. Y. N. Hsu and F. L. K. Hsu, 1977: 310; and Tiger and Wolf, 1985: 73).[52]

As to the story referred to above (Tun, 1965: 66; C. S. Wong, 1967: 147n; Em. Hahn *et al.*, 1973: 162), which is more of a legend, it is said that during a rebellion of the Chinese against their Mongol rulers in the fourteenth century A.D. messages calling for rebellion against the Mongols were baked into moon cakes at the start of the Moon Festival, and this led to an uprising and to the eventual overthrow of the Yüan dynasty. This moon-cake story bears a striking similarity to one about India's Sepoy Rebellion, and the role of chapattis in signaling an uprising against another foreign ruler, the British.

Chinese noodles (*mien*) are commonly made of wheat and are of various forms. Ordinarily they are prepared without eggs, though in Kwangtung people like egg noodles as well.[53] In South China, the making of noodles and buns is a commercial activity, and families purchase these items in shops or elsewhere (E. N. Anderson and M. L. Anderson, 1977: 337). The practice of separating wheat bran, starch, and gluten for food and medicinal uses is also found in China, with wheat gluten commonly used in preparing meat analogues for consumption by vegetarians (E. Watson, 1923: 174; E. N. Anderson, 1988: 118, 124).

As told by E. N. Anderson (1988: 117), only among Turkic and Iranian groups in China's far northwest (Sinkiang) is true bread, presumably made with local yeast or sourdough leavening, the most important way in which wheat is consumed. Such bread, after baking on the inside walls of an oven, is an inch or so in thickness, and often circular in shape and sprinkled with sesame seeds (Figure 24). This is, of course, an extension into China of Near Eastern practice, and indeed in Sinkiang the bread in question is usually known as *nan*, its Persian name. The Chinese, having had contact with Inner Asia and Iran since antiquity, prepare similar baked goods, too, but they are of quite minor dietary importance in China. Already in T'ang times a small, sesame-covered bread was sold by Central Asian bakers in Chinese cities, and it seems to have been a new introduction. Its descendant, most common of such baked goods in present-day China, is *shao-ping* ("roast cakes"), which are square, thin rolls, about six inches across, which are sprinkled with sesame seeds. Frequently meat is inserted in them, and they serve as sandwiches, much in the way Armenian pita bread is used in California (E. N. Anderson, 1988: 54, 117, 178). Western influence, especially in cities with large Western communities such as Shanghai, has also encouraged the use of baked yeast bread, but Gamble (1933: 84) wrote that in Peking in the 1920's such bread was eaten only by families that were well off.

As for barley, it was mainly milled into flour and used in traditional China for making large, flat cakes. In South China, the making of barley cakes is common, especially in western China and the margins of Tibet. In Tibet itself, barley is the principal food crop, used for making *tsamba*, roasted barley flour, that Tibetan staple which may be eaten in buttered tea or in other ways. *Tsamba* is common today not only in Tibet but in certain neighboring areas, including northwestern China, where the name *tsamba* has been modified to *tsan-pa* by the Chinese (Bray, 1984: 462-63).[54] In China, barley was also sometimes cooked in one way or another, alone or in congee or other dishes;[55] it was ground and used as a ferment in making beer and distilled sorghum liquor; and it was used for making malt sugar (S. W. Williams, 1895, 1: 772; Hosie, 1904: 178; F. P. Smith, 1969: 388; Em. Hahn *et al.*, 1973: 120; E. N. Anderson and M. L. Anderson, 1977: 338; E. Tropp, 1982: 553; E. N. Anderson, 1988: 181). One also reads of

Figure 24. Baked wheat bread being removed from oven among the Uighur of Turfan (photo by Stephen F. Cunha)

malt sugar and broths made from barley in T'ang China (Schafer, 1977: 89, 108), and of dried barley of various sorts serving in Chou and Han China as food for travellers and soldiers, and as an everyday food as well (Y. S. Yü, 1977: 76).

In China, sorghum seed is ordinarily consumed as porridge. In some places, especially Manchuria, it is also pearled, to create a product which, when cooked, is better-tasting (though less nutritious because the outer seed coat has been removed) (E. N. Anderson, 1988: 119). As

Hosie (1904: 175-76) described the preparation of pearled sorghum in Manchuria, the winnowed grain is first placed on a circular stone base, the husks broken by a stone roller, and then separated from the grain. Then the grain is placed in a pot of water and boiled (without salt or other flavorings) for sixty minutes or so. By that time, it is soft and has increased in size three to four times. Then, Hosie wrote, the mushy substance is served in bowls and, with boiled vegetables, eaten with chopsticks, in the same manner as rice. Since sorghum was Manchuria's leading cereal, an ordinary servant ate two pounds of sorghum daily, and a very active man was reputed to consume twice that amount. People in China may also mill sorghum into flour, which, after mixing with water and salt, is baked into large, thin cakes which may be used as a wrapping for a chopped, spiced mixture of meat and vegetables, to create a food reminiscent of a Mexican tamale (Powell, 1945: 99). Sorghum is also used for making distilled liquor (for which see our chapter on "Beverages"), and sweet sorghum serves as a sweetener (see our chapter on "Spices and Flavorings").

Common millet is mainly used in modern China for making alcoholic beverages. Foxtail millet, whether glutinous or non-glutinous, may be made into porridge, but only the glutinous forms serve in making alcoholic beverages or steamed millet cakes, commonly flavored with jujubes (S. W. Williams, 1895, 1: 772-73; American Plant Studies Delegation, 1975: 71; Fogg, 1983: 97-98; Bray, 1984: 443; Wittwer *et al.*, 1987: 206; E. N. Anderson, 1988: 120).

In China's southwest, especially among Tibetans and upland tribal groups, oats are roasted and milled into oatmeal, which is often eaten after mixing it with cold water; oats may also be cooked and eaten whole; or they may be made into oat cakes (S. W. Williams, 1895, 1: 772; Legendre, 1911: 570-71; Pollard, 1921: 198-99; E. H. Wilson, 1929: 335).

All other minor cereals (except wheat) may be made into porridge as well, or they may be added to other dishes. Various writers (Hosie, 1904: 179; E. H. Wilson, 1929: 335-36; Osgood, 1963: 162) have noted that maize is also dried, ground, and made into cakes or alcoholic beverages. Thick maize cakes are described as a regional specialty in China's west and northwest (E. N. Anderson, 1988: 166, 184).[56,57] Immature corn on the cob is also prepared as a vegetable, commonly in other dishes, and boiled green corn was sold at street stands in a Chinese city early in this century (Headland, 1914: 197; E. H. Wilson, 1929: 335). Mature corn on the cob may be boiled or roasted. Maize grains, whether whole or cracked, may be mixed with rice (Osgood, 1963: 162; E. N. Anderson, 1988: 119), or prepared in soup[58] or other ways; and cornstarch may be used as a thickening agent. In addition, maize is made into noodles (E. N. Anderson, 1988: 119, 166).

Despite having a variety of ways of preparing maize, the Chinese have failed to adopt ways of cooking maize with alkali, found among many American Indian groups, which increases the availability of niacin and improves the balance among essential amino acids and the nutritional quality of its protein (Katz, Hediger, and Valleroy, 1974).[59] One of these ways, which involves lime treatment of maize kernels, as in making tortillas, provides the additional benefit of significantly increasing the amount of calcium and phosphorus (Bressani *et al.*, 1958: 772). It also increases their availability by combining with the phytic acid found in the outer coat of maize grains, which otherwise would tie up considerable amounts of the calcium and phosphorus in insoluble compounds and limit their availability (E. N. Anderson, 1988: 119).[60] Peoples of southwest China do treat buckwheat flour with lime in making noodles (E. H. Wilson, 1929: 337), but maize products in China, lacking such treatments, are not as nutritious as they might be, and also are inclined to be thick, dull, and mediocre as compared to American Indian maize products (E. N. Anderson, 1988: 119).

As for buckwheat, it may be cooked alone or added to other dishes. It may be milled into flour which is be boiled as gruel, or baked into large, flat cakes, or processed into buckwheat noodles which are popular in certain places, such as North China and mountainous areas of the southwest (S. W. Williams, 1895, 1: 772; Hosie, 1904: 179-80; Legendre, 1911: 570; E. H. Wilson, 1929: 336-37; E. N. Anderson and M. L. Anderson, 1977: 338; Y. K. Martin, 1984: 15, 228; E. N. Anderson, 1988: 118, 120).

The pulses may be consumed in various ways. Fresh pods of pulses are eaten (usually boiled or stir-fried) either with the young seeds or after those seeds have been removed.[61] Especially notable in this regard are the sugar pea (*Pisum sativum*, var. *saccharatum*), yard-long bean, and common (or string) bean, but other minor pulses are eaten in this way as well. String-bean pods and those of certain other pulses may be sun-dried, before or after cooking, and stored for use when few vegetables are available. The main object in other cases is the seed, whether young or mature, fresh or dried. Seed may also be roasted for use as a snack. It may be stir-fried or cooked whole, or it may be candied, salted, fermented or ground into flour or made into bean or pea curd or noodles, or processed and consumed in other ways. This is well illustrated in the uses of the mung bean. Young mung bean pods, with seeds, may be cooked as a vegetable. Mung bean seeds may be cooked and prepared as curd. They may be ground into a flour and made into what we call "cellophane noodles" (also "bean thread," "long rice," or "Chinese vermicelli").[62] After soaking, cellophane noodles are translucent and require only brief heating or cooking. They have little flavor and are not usually eaten alone, but with other foods whose flavors they take on. Their contribution is the unusual texture and appearance they give to soups and other dishes. Mung bean seed is also commonly germinated, and the sprouts are regarded as the best of all such sprouts,[63] added to stews or soups; cooked as a vegetable, whether boiled or stir-fried; or consumed in other ways. Mung bean seeds are also boiled in sugary water and served as a snack or sweet, and seedlings, too, may be candied. Indeed, sweet porridge made from mung beans or red beans is the most common dessert or sweet for many Chinese families. Leaves and young shoots of some minor legumes, for example the pea, may also be cooked and eaten, but leaves and pods of certain others, such as the lima bean, may be toxic (E. H. Wilson, 1929: 337; Ames, 1939: 57; Read and Wagner, 1940: 265; M. Yang, 1945: 35; J. D. Keys, 1963: 97; I. H. Burkill, 1966, 2: 2273; Herklots, 1972: 225, 247, 248, 251, 260; G. Z. Chu, 1975: 104-5, 132, 283; F. Lin, 1975: 312-13; Zee and Hui, 1981: 49, 51, 53, 54; Y. K. Martin, 1984: 15, 23, 144-45, 228, 233; E. N. Anderson, 1988: 124-25).

This calls attention to the widespread problem of legume bitterness and toxicity, which can affect not only leaves but other plant parts, including seeds (Herklots, 1972: 227-28). Such toxicity is common among wild legumes and may have originated as a protective device against insects and other creatures. Some domesticated legumes may have undergone selection by humans to reduce or eliminate their toxicity. Others, however, including some minor pulses consumed in Kwangtung (for example, sword bean, hyacinth bean, common bean), are toxic in a raw or undercooked state when eaten by humans or laboratory animals. The toxicity is eliminated by proper cooking before consumption.

Of special concern in terms of toxicity are broad beans which, usually in genetically-sensitive persons (deficient in Glucose-6-Phosphate Dehydrogenase or G6PD), can bring on the disease favism. Males are much more likely than females to develop favism (among Kwangtung in-patients, a sex ratio of 7 males for each female was reported by S. F. Chung [1965: 881]). Adults usually recover spontaneously and quickly from an acute stage of hemolytic anemia lasting one to two days and involving loss of color, fatigue, labored breathing, back or abdominal pain, and high body temperatures and chills. Favism in young children (2 to 4 years of age) is more serious than in adults, and the chances of mortality are much greater. One study, for example, reported a mortality of 6 to 8 percent among children under 6 years of age.

Best known among Mediterranean and Near Eastern peoples, favism is common as well in China (Du, 1952; S. F. Chung, 1965; Mager *et al.*, 1980: 266-67).[64] Writing of Kwangtung, S. F. Chung (1965) describes favism as basically a problem of children and most common in March and April when the harvest creates an abundance of fava beans. The most serious attacks of favism result from consuming raw beans, whereas dry or cooked beans ordinarily bring on moderate symptoms. Some studies have found that inhalation of broad-bean pollen can bring on favism, though others have failed to confirm this (Mager *et al.*, 1980: 266). Of special interest are reports of favism in nursing infants who have eaten no fava beans. Presumably the toxicity derives from the milk of mothers who have consumed large amounts of the beans (S. F. Chung, 1965: 880-81; Mager *et al.*, 1980: 267).

Species of the genus *Lathyrus* are found in China and have long been used as medicine and scarcity food (F. P. Smith, 1969: 232; Read, 1977g: 34, 49; Perry, 1980: 229). Fortunately, however, the grass pea (*Lathyrus sativus*) seems not to be cultivated in China (Mitchell, 1971: 31), though this pulse is widely grown for food and fodder in Eurasia from Western Europe through the Indian subcontinent. Probably domesticated in southwest Asia, its seed, whether from wild or cultivated plants, was found in association with wheat and barley in western Iran (Tepe Sabz) at levels dating from about 5200 to 4000 B.C. (Hole *et al.*, 1965: 106; Mitchell, 1971: 31-32). Its seed was also found in several other Neolithic sites in the Near East and Mediterranean area, and seems to have been domesticated there by the late Neolithic and Bronze Age (Zohary and Hopf, 1988: 110). The grass pea was among the earliest pulses cultivated in India, beginning at least in 2500 to 1800 B.C., and is widely cultivated there today (Watt, 1889-96, 4: 590-94; Vishnu-Mittre, 1974: 25, 28). It is fortunate that it did not become a crop in China, despite its hardiness, drought-resistance, and ability to tolerate waterlogging more than other pulses, and despite its high calcium content and a protein content as high or higher than that of all of the minor pulses considered above (Platt, 1962: 11; Arnon, 1972: 240-41; Leung *et al.*, 1973: 16-22 *passim*), since it can cause a paralytic disease, lathyrism, a major contributor to the population of crippled beggars in Indian cities (Watt, 1889-96, 4: 592-94; Mitchell, 1971; Padmanaban, 1980).

[1] According to a fairly recent estimate by Swaminathan (1984: 83), China produced nearly a fourth (23.9 percent) of the world's rice, and India, somewhat more (27.6 percent). Sun Han (in Wittwer *et al.*, 1987: 139-40), noting striking recent increases in Chinese rice production, claims that China now produces more rice than India, or more than 36 percent of the world's total.

[2] Some observers have questioned the data presented by Maynard and Swen in the village survey. One concern has been the failure to compensate for the Chinese tendency to present the bright side of things. This may have led to the study of villages within a region that were better off than normal, that were not representative. It may also have led people to give data to those conducting the survey that looked more favorable than was actually the case. As for South China, figures for rice may have been too high for some areas, and figures for secondary staples too low.

[3] This is very close to the "grain" allotment (about 43 pounds a month) for an adult male factory worker in Peking in 1984 (Bennett, 1984: 24).

[4] Noting that in highland regions of Southeast Asia rice is often cultivated as a dry-land crop by shifting cultivators, some have argued that such cultivation preceded wet-land cultivation in the lowlands, a view, however, not in accord with the fact that upland rice plants and their seed are both more advanced and less varied than lowland types (T. T. Chang, 1976b: 431).

[5] For rice in early Central Chinese sites, see K. C. Chang, 1986: 199, 214-15, 254 and *passim*. Sun Han (in Wittwer *et al.*, 1987) notes that the rice in one of these sites, Lo-chia-chiao, is slightly earlier than that of Ho-mu-tu.

[6] There has also been a subsequent report of rice, confirmed as domesticated by T. T. Chang, in levels of a site in central Indonesia that may go back to 4000 B.C. (Glover, 1979: 19-24).

[7] Since antiquity, most Chinese rices have been cultivated in shallow paddies in the lowland, which means that China, unlike Southeast Asia, lacks rice varieties strongly adapted to growth on dry land or in deep water (T. T. Chang, 1983: 74).

[8] Though called glutinous, such rice does not contain gluten. Its stickiness is presumed to derive from the dextrin, maltose, and starch found in the endosperm (Bray, 1984: 478).

[9] The iron of rice and most other cereals, however, is poorly absorbed (Lynch *et al.*, 1984: 42).

[10] Some Chinese writers place its domestication in other parts of China (Wittwer *et al.*, 1987: 184-86). And Ho (P. T. Ho, 1975: 77-79) argues convincingly for domestication not by the Chinese, but by proto-Tungusic tribes in Manchuria or its borderlands. If he is correct, they, and not the Chinese, must be given credit for one of East Asia's greatest contributions to the crop wealth of mankind.

[11] In *Food in Chinese culture*, K. C. Chang (1977b: 28), citing Chinese authorities, wrote that paleographic evidence of soybeans dates to the Western Chou period (*c.* twelfth century B.C.-771 B.C.), and archeological remains to the Spring and Autumn period (722-481 B.C.). In his *Shang civilization* (K. C. Chang, 1980: 148), he does not include the soybean among cultivated plants of Shang, though he notes that one Chinese authority believes that a Shang character, commonly believed to represent rice, actually represented the soybean. In his latest edition of *The archaeology of ancient China* (1986: 362), however, he states, but without citing sources, that "the Shang and Chou...planted soybeans..., but this we know only from inscriptions and texts." If this is so, the earliest soybean cultivation in China would date from Shang times, 18th to 12th century B.C.

[12] The perfection of techniques of soybean fermentation is placed in late Chou or early Han times (E. N. Anderson, 1988: 42).

[13] These suggestions rest in part on reports of finger millet at Hallur in South India at levels that may date from early in the second millennium B.C. Harlan, however, believes the identification of finger millet to be wrong (Vishnu-Mittre, 1977: 575, 575n).

[14] A form of *Setaria* is also believed to have been cultivated as a cereal in Tamaulipas, Mexico, in the fourth millennium B.C., but cultivation declined and finally ceased after maize was introduced late in the third millennium B.C. (Callen, 1967: 535-37).

[15] For an interesting account of the range of *Panicum* and *Setaria* forms in China, which are "suited to almost any conditions of soil and climate," see Bray, 1984: 442-43.

[16] The story of this competition is told by Bray, 1984: 434-48.

[17] For information on the confusing, often inconsistent, use of terms for common millet and foxtail millet in China, see Bray, 1984: 440.

[18] Millet and rice, however, seem to have been better-liked and of higher prestige in Han China than were wheat and barley (Y. S. Yü, 1977: 72-73).

[19] Older sources identify several species of *Pisum*, among them *P. sativum*, the garden pea, and *P. arvense*, the field pea. Today the genus is regarded as having just two species, *P. sativum* and *P. fulvum* (D. R. Davies, 1976: 172; Zohary and Hopf, 1988: 93). The species *sativum* is divided by some into six subspecies, among them the field pea, *P. sativum*, var. *arvense*, and the garden pea, *P. sativum*, var. *hortense* (Zeven and de Wet, 1982: 97).

[20] *Kao-liang*, sorghum's most common name today, first appears in an account dating from early in the fourteenth century (Bray, 1984: 449).

[21] That preference is reflected in Gamble's 1926-27 survey of family incomes and expenditures in Peking (1933: 79, 83-84), which found that city families, in striking contrast to rural families of North China, consumed very small amounts of sorghum, less in amount not only than millet, but than wheat, rice, and maize as well.

[22] For a map of sorghum production and further information on sorghum in present-day China, see Wittwer *et al.*, 1987: 202, 204-6.

[23] According to Victor Hehn, buckwheat reached Germany early in the fifteenth century. It came to be called *buch-weizen* ("beech wheat") in Low German because its seed resembled beech-nuts. In South Germany, another name is used for buckwheat: *heiden-korn* ("heathen cereal"), now often *heide-korn* ("heath cereal"). Names for buckwheat in Polish, Czech, and Magyar have a similar meaning, "heathen" or "pagan." A third German name is *tater-corn*, "Tartar cereal," which is similar to names for buckwheat in Czech, Little Russian, Magyar, Finnish, and Estonian, and may be taken to indicate that it came from the Tartars, meaning not just the Volga and Crimean Tartars, but from Mongolian peoples in general. On the other hand, Hehn notes that the gypsies were also known in Europe as "heathen" and "Tartars," and he thinks that these terms may have been referring to them. Russian names for buckwheat (*grecha*, *grechukha*, and similar ones) mean "Greek cereal," but used in the sense of "cereal from the south" or "foreign cereal." In the lower Volga, however, the Russians have a name for buckwheat, *dikusha* ("wild cereal"), which may mean "uncultivated cereal" or "cereal deriving from wild people" (nomads), and in addition some there call buckwheat by its Tartar name, *kurluk* (Hehn, 1885: 386-88). The evidence presented by Hehn is somewhat confusing, but, to this writer, it suggests that buckwheat is a late arrival in Europe from the east or southeast.

[24] Cultivated rye (*S. cereale*) is believed by many botanists to derive from the wild perennial *S. montanum* ("mountain rye"), which is found from Morocco and Spain to Turkey, Iraq, and Iran. *S. montanum*, in turn, some believe to derive from *S. fragile* (= *S. silvestre*), an annual wild species found from Hungary to western Siberia and the Pamir and Alai mountains. This would mean that *S. fragile* is "phylogenetically the oldest species" of *Secale* (Mathon, 1981: 125; Zeven and de Wet, 1982: 91-93). And, if Wilson was correct in his identification, we would therefore be dealing not with cultivation of *S. cereale* but of an ancestral form of rye.

[25] For an excellent discussion of oats, wild and cultivated, their use in the early Near East and Europe, and their domestication, see Renfrew, 1973: 87-98.

[26] E. H. Wilson observed (1929: 335) that the Han Chinese prefer *A. nuda*, whereas Tibetans and tribal people of the highland southwest prefer another oat, which he identifies as "*A. fatua.*" *A. fatua* is a common wild oat in the Old World, a hexaploid which, along with *A. sterilis*, the wild red oat, some botanists (but by no means all) hold to be one of the wild forms from which cultivated hexaploid oats evolved (Renfrew, 1973: 87; Zeven and de Wet, 1982: 109-10, 152; Zohary and Hopf, 1988: 71-74). Among Tibetans and tribal people, grain of both sorts of oats is processed into oatmeal or cooked whole.

[27] *Echinochloa* can be an extremely serious weed in irrigated crops in the Central Valley of California, where it is known as water grass. In some other areas in the United States this weed is known as barnyard grass (pers. comm., James Street).

[28] The common English name for the plant is simply a translation of the species name assigned to it by Linnaeus. Its genus name comes from the Greek, but how Linnaeus arrived at the species name is uncertain. Somewhere along the way the shape of the seeds probably led to them being called "tears," but the plant is not included in *Plants of the Bible* (Moldenke and Moldenke, 1952) and we know of no evidence of a historical connection with the Old Testament figure. Likely the species name derives from the Arabs, who were located along the plant's route of diffusion from the east, who are credited with introducing it to Spain and Portugal, and who first called *Coix* "David's Tears" and later, perhaps by substitution, "Job's Tears." In recent times, the Visayan name for the plant, *adlay* or *adlai*, has also come into general use for Job's tears (Watt, 1889-96, 2: 492; 1904: 191; 1908: 392-93; Schaaffhausen, 1952: 217; I. H. Burkill, 1966, 1: 639; Jain and Banerjee, 1974: 38; Koul, 1974: 63).

[29] Though Job's tears had spread into the South Pacific in pre-Columbian times, it was not found in the American tropics before the Europeans arrived (Merrill, 1954: 246, 295).

[30] On the other hand, Glover (1979: 18, 21, and *passim*), in a review of prehistoric plant remains in Southeast Asia, mentions specimens of domesticated rice for various early sites (pre-1000 B.C.), but just a single specimen of Job's tears (for pre-3000 B.C. Timor). The Job's tears seed in question, moreover, was pierced, suggesting that it was used for decoration rather than food.

[31] Because of this and other reasons, Watt (1908: 395) argued that domesticated Job's tears may have been spread in India by invading "Mongolians," by which he presumably meant tribal peoples (Sino-Tibetan, Mon-Khmer, and Tai in speech) who came from Southeast Asia. We would note, however, that there are reports in the early Sanskrit literature of *gavidhuka or gavedhuka*, commonly identified as Job's tears, *Coix lacryma-jobi*, being used as food or medicine. Patañjali (?200 B.C.-?A.D. 500), for example, mentions it as an inferior grain made into porridge (Prakash, 1961: 88, 260); and the *Charaka Samhita* (Charaka: ?A.D. 80-180?) cites its use, whether alone, in an infusion, or boiled as a cereal, to treat weakness and emaciation (P. Ray and H. N. Gupta, 1965: 58). Though Apastamba (?500 B.C.-A.D. 200) said that *gavidhuka* was uncultivated (Prakash, 1961: 260) and Macdonell and Keith (1912, 1: 223) identify the plant as *Coix barbata*, a grass species, the accounts do show that some sort of *Coix*, whatever its identity, was used as food and medicine in early India.

[32] In modern Tonkin, *Coix* seeds (perhaps from *C. lacryma-jobi*, var. *ma-yuen*) are believed to eliminate miasmas and cleanse "unwholesome water" (Perry, 1980: 164), both of which were prominent, in Chinese thinking, as possible causes of malaria (Wiens, 1954: 338-39).

[33] The plant root and leaves also serve medicinal purposes.

[34] Arora's description (1977: 362-65) of the uses of cultivated Job's tears by tribal peoples in northeast India reveals a variety lacking in accounts of its use among the Han Chinese in traditional times (though we suspect that China's non-Han minority groups who cultivate the plant are more similar to the Indian groups). In any case, the Indians in question often eat the seed directly, after its hull has been removed, in the way we eat peanuts. The seed may be boiled like rice, and is sometimes eaten mixed with rice. It may be pounded and made into beer or, with sugar added, fried to make a sweet dish. On rare occasions, pounded seed is even made into a thick paste used to prepare leavened bread. Whole seed also serves as poultry feed, foliage as fodder for pigs, and leaves for thatching. The seed made into beads, however, is derived from wild forms of *Coix* found locally.

[35] There is, however, a superior variety of Job's tears, developed in the Philippines, which is reported to have, among other good qualities, a shorter growing season. It is also said to provide higher yields than rice (Schaaffhausen, 1952: 221-25).

[36] For drawings of many of these, see Herklots, 1972: 225-72; for information on their cultivation and use, especially in Southeast Asia, see Herklots, 1972: 225-72, and I. H. Burkill, 1966: *passim.*

[37] The mung bean has been identified in India at archeological levels dating from 1660 to 1440 B.C. (Vishnu-Mittre, 1974: 24).

[38] A recipe for red bean paste is provided by F. Lin, 1975: 300.

[39] For Chinese medicinal use of rice and other cereals and pulses, see Soubeiran and Dabry de Thiersant, 1874: *passim*; I. H. Burkill, 1966: *passim*; and Hooper, 1929: 29, 40, 54, 64, 74, 97, 136, 137-38. For a list of sources on such use, see Read, 1936: *passim.*

[40] E. N. Anderson and M. L. Anderson say that in South China a thin clear noodle is occasionally made from rice, but that rice flour is used in that region "for cosmetics as much as for food" (1977: 337).

[41] Wang and Hesseltine (H. L. Wang and C. W. Hesseltine, 1970: 575) have observed that one of the fungi used in the fermentation of wine rice, *Rhizopus chinensis*, produces an antibacterial compound found to minimize infection and stimulate growth in animals with an inadequate diet. They suggest that this must be taken into account in explaining the nutritional value of wine rice.

[42] Job's tears and amaranth seeds are also popped, at least in parts of China. For more details on the popping of amaranth seed, see our section on amaranth in the chapter on "Other Vegetables."

[43] For the food composition of individual soybean products from China and neighboring lands, see R. S. Harris *et al.*, 1949: 35; Platt, 1962: 10-11, 28-29; and Leung *et al.*, 1973: 19-21.

[44] E. N. Anderson, 1988: 69. Tradition holds that bean curd was first made by Liu An (d. 122 B.C.), though it is observed that the literary evidence is insufficient to support that view (C. P. Li, 1948: 170-71; Y. S. Yü, 1977: 81).

[45] For descriptions of traditional methods and technology of tofu-making in China, see C. V. Piper and W. J. Morse, 1923: 234-44; Hommel, 1937: 105-9; and C. P. Li, 1948: 179-80.

[46] One way of precipitating soybean milk in making tofu in China is by the use of gypsum or calcium sulfate, which makes the end product, the preferred form of tofu, an important source of calcium (Adolph, 1939: 332; H. L. Wang and C. W. Hesseltine, 1970: 572).

[47] For the traditional methods of making soy sauce in China, see C. P. Li, 1948: 174-79, and for Kwangtung, E. H. Groff (1919).

[48] For the ancient way of preparing such fermented soybeans, see C. P. Li, 1948: 171-74.

[49] Shurtleff and Aoyagi (1976: 227-30) provide a good description of Chinese soybean pastes and sauces and similar fermented products. In addition, those authors and Shibasaki and Hesseltine (1962) provide a wealth of information on miso, the Japanese fermented soybean paste, its preparation and use.

[50] An indication of the versatility of the Chinese is their using soybean paste and other Chinese ingredients in developing a version of the saté sauce of the Malays. For details of this, see E. N. Anderson, 1988: 123-24.

[51] Y. S. Yü (1977: 79) writes that in Han China the poor always prepared their wheat by boiling and that, presumably as a result, this was considered inferior food. S. W. Williams (1895, 1: 772), writing of China in the last century, however, does not mention wheat as being cooked in that manner, but speaks only of wheat flour made into cakes, dumplings, and such, but not into baked bread. The cereal grains he mentions as being cooked whole are rice, maize, buckwheat, oats, barley, and *Setaria* millet. V. Y. N. Hsu and F. L. K. Hsu (1977: 300-301), in describing the cooked grains (*fan*) of modern North China, identify rice, sorghum, maize, and millet as being prepared in that way, but not wheat.

[52] Recipes for moon cakes are given by I. C. Chang, 1970: 187, 205-6.

[53] Such egg-less noodles have encountered problems in the United States, where use of the term "noodles" is restricted to products that contain eggs (Piper, 1970: 8-9). Egg-less noodles, as a result, are sometimes labelled "imitation noodles" or "alimentary paste" in this country.

[54] Bray (1984: 462-63) observes that there are also traditional Chinese names for *tsamba*, and that such names must have been used in antiquity, when the Chinese may also have prepared *tsamba* of wheat, as is done sometimes in modern times in China's northwest and in its Inner Asian provinces.

[55] Pearled barley, according to E. N. Anderson (1988: 120), seems to have come into use recently, and is consumed only as a medicinal broth. In South China, it has taken the place of Job's tears in that role, and indeed pearled barley is known by a name formerly used for Job's tears.

[56] Such cakes seem to be different from the maize flour buns (*wo-thou*) that Bray (1984: 458), writing in general terms of the Han Chinese, says nobody eats "if they can avoid it."

[57] Since China's aboriginal groups of the south and southwest are often singled out for their use of maize, we would note that the Li of Hainan boil and eat maize, often mixed with millet but seldom whole (Odaka, 1950: 25); that boiled maize meal is the staple of tribal people, such as the Lolo, Miao, and Ch'iang, who live at lower elevations in the mountains of southwest China (Graham, 1961: 18); and that the Lolo not only boil maize into a porridge but cook it "in cakes under the cinders" (Legendre, 1911: 570).

[58] A modernized recipe for such a soup, "Velvet Corn Soup," is given by M. Field *et al.*, 1976: 21.

[59] Interestingly, such alkali processing of maize is found primarily among American Indian groups who have a high dietary dependence on maize (with maize comprising one-third to over two-thirds of total diet); among American Indian groups for whom maize is of minor dietary importance (below one-third of diet), on the other hand, processing with alkali is rare (Katz, Hediger, and Valleroy, 1974: 770). Use of such methods would, of course, be justified in China, especially in areas where maize is of major dietary importance. In general in China, however, maize is of minor dietary importance, and its people are no different from those American Indian groups who consume little maize and do not prepare it with alkali.

[60] Briggs and Calloway (1979: 252) write that this action of phytic acid is considered to be of minor significance in practical terms as long as enough absorbable calcium and phosphorus remains to satisfy physiological requirements. The key words in the above sentence, as it relates to China, are "as long as enough absorbable calcium and phosphorus remains," for, as we shall see in our chapter on nutrition, amounts of dietary calcium in China are low and sometimes inadequate.

[61] One such dish is "Dry-fried Four-Season Beans," a Szechwan dish of green beans that are deep-fried with shrimp, Szechwan preserved vegetable, green onions, garlic, gingerroot, soy sauce, and other flavorings (Y. K. Martin, 1984: 75, 203).

[62] For a detailed account of the processing of mung beans into cellophane noodles in Manchuria early in this century, see Hosie (1904: 184-86).

[63] Soybeans, broad beans, and black-eyed peas (cowpeas) are also germinated for sprouts, but are less tender. Bean sprouts in general contain good amounts of vitamin C and riboflavin and fair or good amounts of iron (Read and Wagner, 1940: 210, 212; B. S. Platt, 1962: 15; Leung *et al.*, 1973: 34, 55, 67).

[64] The incidence of G6PD deficiency is highly variable among the Mediterranean and Middle Eastern peoples tested, from 0.2 percent for Ashkenazi Jews and 0.7-3 percent for Greeks from Crete and the mainland, to 53 percent for Kurdish Jews (males). Among Chinese, the incidence is 5.5 percent (Mager *et al.*, 1980: 270).

CHAPTER FOUR

Roots and Tubers

"Yam and taro brought back by every man,
Bin and kitchen filled richly every day..."

Su Kuo (translated by Edward H. Schafer)

Historical Sketch
Root Crops of Eurasian Origin
- Yam
- Taro
- Radish
- Carrot and Turnip
- Aquatic Roots and Tubers
  - Old World Arrowhead
  - Chinese Water Chestnut
  - Lotus Root
- Minor Roots and Tubers: Great Burdock, Kudzu, and Others

Root Crops of American Origin
- Sweet Potato
- White Potato
- Manioc
- American Yam Bean
- Notes on Other American Root Crops

Sago: Another Starchy Food

Before turning to individual roots and tubers and their dietary role in traditional China, we shall briefly sketch their history as food crops there. Our sketch is of necessity general and in considerable measure speculative because remains of the important root crops, yam and taro, have not been found in ancient Chinese archeological sites (K. C. Chang, 1977b: 28). It is clear, however, that we are dealing with a long history of domestication, selection, adaptation, competition, and displacement of food crops. Sometimes the main competition was among root crops themselves, and sometimes between root crops and cereals or other food plants.

HISTORICAL SKETCH

Cultivation of root crops in Southeast Asia, including southernmost China, may date back to 8000 B.C. or before, with the yam (*Dioscorea* spp.) and taro most important among those used for food. The wild yam was a plant of well-drained soils, whereas wild taro was found along the margins of swamps or in similar aquatic situations. At first, these plants may have been cared for in a casual way. If so, this changed over time, most notably for taro which came to be cultivated intensively, and probably, some say, in the irrigated agricultural terraces whose remains are widespread in Southeast Asia.

As the yam spread northward in Asia, new wild forms were domesticated or old ones selected for the less tropical conditions that prevailed. In early times the yam was cultivated north of the Yangtze, as well as in southern Korea, and in all of Japan's large islands except Hokkaido (Bertin *et al.*, 1971: 13). Nevertheless, in China, as in the Pacific region (I. H. Burkill, 1951: 445-46; 1966, 1: 648), the yam failed to spread as widely as did taro. Taro does require much water for growth but, given this, it has great adaptability to other conditions. In addition, taro of certain races can be stored as long as several months (I. H. Burkill, 1966, 1: 648). Adaptability and storability favored the diffusion of taro, and by pre-Columbian times it had spread into a great diversity of environments: from rainforest to desert, from tropics to mid-latitudes. It found its way westward to Egypt, tropical Africa, and Mediterranean Europe, northward across China and on to Japan, and eastward across island Southeast Asia, Melanesia, and Polynesia as far south as New Zealand and as far north as Hawaii (Massal and Barrau, 1956: 6-11; I. H. Burkill, 1966, 1: 647; Plucknett, 1976: 11; Cable, 1984: 28-30). By selection, "dryland" or "upland" forms of taro were developed which could grow without always being in water (Murai *et al.*, 1958: 99), and could thus compete in a different niche from that of the "wetland" forms of earliest cultivation.

The exact time of taro's northward spread in China remains uncertain, but by Han times it was a staple crop in the middle Yellow River area, cultivated in pits, fertilized, and watered as needed (Y. S. Yü, 1977: 76; S. H. Shih, 1982: 25-27, 41, 42). In an exchange perhaps of equal value, *Setaria* millet—domesticated in the North Chinese hearth—spread at some unknown early date to South China and Southeast Asia, to compete with dryland roots and tubers. One can still find among aboriginal groups in South China, elements of what may have been a widespread ancient pattern. One such group is the Yami, aborigines who live on an island south of Taiwan. The Yami do not cultivate rice, but instead their favorite crops are taro, in its wetland form (their staple), and *Setaria* millet, a dryland crop. Because they are unable to cultivate enough wetland taro, the Yami also grow dryland taro, yams, and other roots and tubers (Kano and Segawa, 1956: 11, 140-55).

Elsewhere in South China, however, taro and yams have a long history of losing ground to cereals, especially rice. Though the story of this competition remains obscure, some scholars (Haudricourt and Hédin, 1943: 153; Haudricourt, 1962: 41n; Barrau, 1965a: 341) have argued that initially rice may have been a weed in the gardens of taro cultivators.[1] Barrau (1970: 500) has suggested further that the domestication of rice was carried out not by indigenous Southeast Asian cultivators, whose focus was on vegetatively-reproduced crops such as taro, but by immigrants from the north who knew and used cereals and may have brought *Setaria* millet with them. An alternative possibility is that Southeast Asian taro cultivators did indeed domesticate rice and consumed it on a minor scale, though it did not initially yield enough to provide serious competition with their root crop staples. This may have changed when rice in small amounts reached the north, where cereal cultivation was well established. Perhaps in the north people better appreciated the potential of rice when it was introduced, then selected it for non-shattering seeds, yield, and other useful qualities. One might speculate that such new forms then spread to South China to make rice for the first time a strong competitor with root crops there.

One notes that in the South Pacific islands where taro has recently been losing ground in the diet of native peoples, this has been mainly because other foods are more convenient to use (rice, wheat flour, manioc), store well and can be cooked with less fuel (cereals), or are easier to cultivate (sweet potatoes and manioc) (Parkinson, 1984: 219, 222). Yields are involved to a degree, but under conditions of intensive cultivation and irrigation, taro can equal rice (and manioc) in energy production (Parkinson, 1984: 216). Yams, for their part, are costly to produce. As Coursey has observed, they need good soil, staking, and considerable attention throughout their growth; in addition, as much as a quarter of the yams harvested must be saved for future planting. In West Africa, where yams sell for much more than other carbohydrate foods, one can obtain an equal amount of calories from rice at a half to a fifth the price. This partly reflects costs of production and partly demand, for yams in West Africa are a food of the rich, a prestige food that, when consumed in large amounts, confers status on a family. In various other yam-cultivating areas, yams are also highly regarded, usually far more than substitute foodstuffs (Coursey, 1967: 141-42; 1983: 564).

That was not the case among the Han Chinese, who did not confer prestige on the yam. On the contrary, they far preferred rice. On Hainan in T'ang times, for example, the Han Chinese are said to have despised the aboriginal diet of yams and taro; though of necessity they ate them, this was only because rice was in very short supply (when ships did not arrive, it was said, "a grain of rice was like a pearl") (Schafer, 1970: 81, 93). Their position unprotected by taste preference, yams were subject to other competitive factors. Coursey (1967: 141-44), in considering the modern gains of rice against yams in various parts of the tropics, ascribes this not to the unquestioned nutritional superiority of rice, but to a growing appreciation of the taste of rice, and to its lower cost and ease of preparation. The latter factors have been critical in the minds of urban populations who cannot readily obtain traditional foods at low price from relatives who remain farmers. For South China, one can only speculate about the combination of factors that were most critical to the success of rice over yams and taro.

Rice competed for space not only against irrigated taro but against unirrigated crops, for there were upland forms of rice that were alternatives to upland taro, yams, and *Setaria* millet. The cultivation of rice by irrigation involved the transplanting and care of individual plants, which offered unusual possibilities for selection as compared to other cereals. Concern with individual plants may also have helped farmers in recognizing the food potential of certain weeds. In this regard, one notes the present-day occurrence of wild water chestnut, *Eleocharis dulcis*, as a weed in rice fields of the Tonkin Delta, and wild Old World arrowhead, *Sagittaria sagittifolia*, as a weed in nearly all Chinese paddies (Blasdale, 1899: 9; Hodge, 1956: 50). One also notes that China has a range of domesticated aquatic plants unmatched anywhere else in the world.

By the fifth and sixth century A.D., the main roots and tubers cultivated for food in China, judging from the lists provided by H. L. Li (1969) and K. C. Chang (1970), were as follows. In the north, taro, though less important than in the south, was the major carbohydrate source among roots and tubers. Also grown were the turnip, whether developed in North China itself or introduced from Central Asia, the Mediterranean, or Europe; and the radish, from the Near East or the Mediterranean. In Central and South China these were joined by yams and aquatic roots and tubers. The carrot reached China from the west in the thirteenth or fourteenth century. And following discovery of the Americas, New World roots and tubers also found their way to China.

Many scholars believe that the sweet potato, an American domesticate, spread to Polynesia in pre-Columbian times, perhaps, according to D. E. Yen (1971; 1974: 329; 1976: 43), before the eighth century A.D. Others (*i.e.*, Brand, 1971), however, are convinced that it was introduced to Polynesia only after discovery of the Americas, and that only then did it reach Southeast Asia and China, and by two routes: across the Pacific via the Philippines, carried by Spanish ships, and from the Caribbean and eastern South America across the Atlantic to Europe, and then, by Portuguese vessels, around Africa and across the Indian Ocean (Conklin, 1963: 133; I. H. Burkill, 1966, 2: 1266-67; Bertin *et al.*, 1971: 11; D. E. Yen, 1976: 43). Supporting the case for a post-Columbian introduction of the sweet potato to China are local records suggesting that it was first introduced in the mid-sixteenth century or shortly afterwards to Fukien, by sea, and Yunnan, perhaps by land from India (Laufer, 1929: 245-46; P. T. Ho, 1955: 193-94).

It was not long before the sweet potato gained an important position in the south. This was aided by certain official actions, as in Fukien Province in A.D. 1594 when a serious crop failure, brought on by drought or hurricane, led the governor, Chin Hsueh-tseng, to urge farmers to cultivate sweet potatoes extensively so as to avert famine.[2] His role in fostering sweet potato cultivation was so well recognized that even today in China the sweet potato is sometimes designated as the "*Dioscorea* of Governor Chin." The sweet potatoes in question were obtained from Luzon, where there was a considerable Fukienese community (Laufer, 1929: 244; P. T. Ho, 1955: 193-94; Spence, 1977: 262).

The sweet potato reached Okinawa in 1605 and Japan in the latter part of the seventeenth century (Laufer, 1929: 245). In Japan today, the sweet potato provides higher yields per acre in upland areas than any other crop, and nationally it is second only to rice in importance (Schery, 1972: 492). In Kwangtung and other parts of South China, its importance in hills and other areas not suited to paddy cultivation is based on a similar yield advantage over other food crops. It is favored, too, because it is simple to cultivate and produces a crop quickly, much more so than yams (I. H. Burkill, 1966, 2: 1268; Herklots, 1972: 439). Because of these advantages, today it occupies 70 percent of all the area in China given over to root crops (Wittwer *et al.*, 1987: 229).

We turn now to individual roots and tubers and their dietary role in traditional China, dealing with them in the rough historical sequence of their appearance.

ROOT CROPS OF EURASIAN ORIGIN

Yam

The rhizomes or tubers of most wild yam species are toxic, and indeed some have been used as sources of fish poison or poisons of other sorts (Coursey, 1967: 148-50). The early human use of wild yams as food—and domestication of yams itself—may have been facilitated, according to the hypothesis of Timothy Johns (pers. comm.), by the detoxifying effect of earth eating, geophagy, a practice widespread both in China and Southeast Asia (Laufer: 1930: 111-43).[3] By Johns' hypothesis, such earth absorbs yam alkaloids and other toxic substances which then pass through a person's intestinal tract with limited ill-effect. In any case, early cultivators selected for non-toxic forms of yams, though some domesticated species still have such toxicity, which must be destroyed by cooking or removed by soaking in or leaching with water.

Figure 25. Tuber of the yam *Dioscorea alata* (from Bois, 1927-37, 1: 482)

The first detailed information on yam cultivation in southeastern China is provided by Chi Han in his fourth century flora, where he describes a purple-skinned yam cultivated on Hainan Island and by groups in "the South Seas" (Southeast Asia). Chi Han attributed the extreme longevity of certain Southeast Asian people to their eating this yam rather than grain. The yam in question, *kan-shu*, has been identified by H. L. Li (1979: 41-43) as *Dioscorea esculenta*, the lesser Asiatic yam,[4] a high-yielding Southeast Asian domesticate with small tubers but free of toxicity and very palatable.

Yams of other species have also been cultivated or collected from the wild in ancient and modern China for use as food or medicine (Bretschneider, 1882-95, 2: 193-94; 3: 415; F. P. Smith, 1969: 150-51). Most common of the cultivated food yams are *D. alata*, the greater yam (Figure 25); *D. opposita*, the Chinese yam; and its relative *D. japonica*. Like *D. esculenta*, *D. alata* is a tropical species originally from Southeast Asia. It is said (Bertin *et al.*, 1971: 13) to have the largest tubers and best keeping qualities of any food yam. As a result, it spread in early times from Southeast Asia far across the Old World tropics, from Polynesia to Madagascar and

East Africa. Today it is even more widespread, the best known and most widely disseminated of all food yams.

D. opposita and *D. japonica*, on the other hand, are warm temperate species not cultivated in the lowland tropics of Southeast Asia, but widely grown in China and Japan in places much colder than most cultivated yams can tolerate. Both were introduced by the Japanese to the Hawaiian Islands and are still cultivated there. And both were introduced to Western Europe during the potato blight of the mid-nineteenth century as alternatives to the white potato; though not cultivated there to any extent today, one still occasionally finds *D. opposita* grown in gardens as a curiosity (I. H. Burkill, 1966, 1: 834; Coursey, 1967: 56; 1983: 560). *D. opposita* was also used in East Asia to provision whaling ships operating in the north Pacific (I. H. Burkill, 1966, 1: 834). At sea, its keeping qualities and antiscorbutic activity were useful to whalers just as those of more tropical yams were to early Polynesian seafarers, to Portuguese on their return trips from India, and to Europeans in the African slave trade (I. H. Burkill, 1966, 1: 827; Coursey, 1967: 16-19, 166-167, 169). *D. japonica*, which some botanists view as a kind of *D. opposita*, is included by Coursey with the food yam species (Coursey, 1967: 71, 138; 1976: 71). In China, however, it also serves pharmaceutical uses, giving rise to one of its names, *shan-yao*, "mountain drug" (Bretschneider, 1882-95, 2: 193; F. P. Smith, 1969: 150).

Coursey, in his book on yams (1967: 130-34), found no recent data on yam production in mainland China, with the principal international data source, the FAO, usually combining statistics on yams with those for sweet potatoes. Recent FAO data (1983) does separate yams from sweet potatoes, but unfortunately it lacks information on China. Coursey did estimate (1964) a production of five million metric tons annually for the region embracing South Asia, Southeast Asia, East Asia, and the Pacific. Though China, including Taiwan, is responsible for only a minor share of this, Coursey nevertheless judged it considerable. Yam cultivation is found today principally on Hainan and in the tropical parts of Kwangtung (Yu Youtai in Wittwer *et al.*, 1987: 230-31).[5] The nutritional value of yams is principally as a source of carbohydrates, though when eaten in large amounts (1 kg. daily), as common in West Africa, they can provide significant amounts of protein and most daily requirements of thiamine and vitamin C (Coursey, 1967: 168-70; 1983: 571-73).

Taro

Taro (*Colocasia esculentum* = *C. antiquorum*; *yü*) (Figures 26-27), also known by other names such as cocoyam and dasheen, is an edible aroid, a starchy root crop and food staple in various parts of the tropics today. It was domesticated in ancient times either in western India or Southeast Asia. There are many races of taro in Southeast Asia and the Pacific and they vary greatly in their utility; though some are important human foods, others are fed mostly to pigs (I. H. Burkill, 1966, 1: 649). And many now are ornamentals grown indoors in regions otherwise too cold for them.

Though people of high-Chinese culture have long preferred rice to taro, it remains to be explained how their views of taro have changed over time. Taro was among the foodstuffs listed on bamboo slips found in the Han tomb of a noble lady near Ch'ang-sha, Hunan (Y. S. Yü, 1977: 57), which suggests that taro was a food acceptable to her. Taro was also sent as tribute to the royal court in T'ang times (Schafer and Wallacker, 1957-58: 222; Schafer, 1977: 92), and was among the imperial foodstuffs shipped on the Grand Canal to Peking in Ming times (Mote, 1977: 215), which further suggests that it was an acceptable food to persons of high standing.

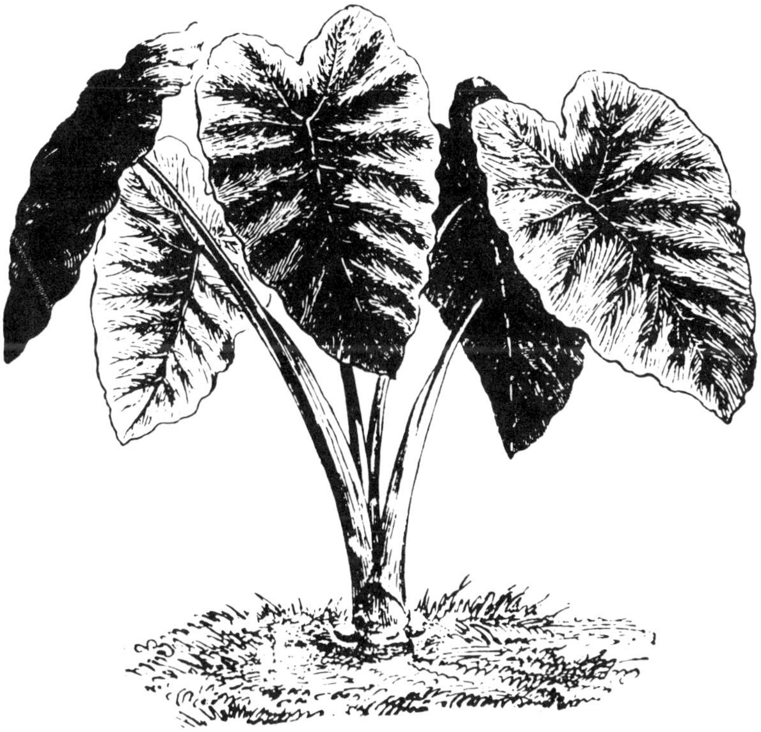

Figure 26. Taro (*Colocasia antiquorum*) (from Bois, 1927-37, 1: 522)

Yet Eberhard (1968: 97-98) writes that high-civilization Chinese regarded taro solely as food for the poor, a view that he supports with an account of a man who planted rice "as a sacrifice for his parents" (to offer in sacrifice to?), but for himself, only taro. There is also the case of the T'ang poet Po Chü-i who, when headed for China's southeast, was warned of the diet of that region, including taro soup, and many other items not customary among respectable North Chinese (Schafer, 1977: 130).

In modern times, taro is a minor crop in the Double Cropping Rice Area, one which in the 1929-33 village survey occupied just 1 percent of the total crop area there (Buck, 1956: 212). Early races of taro mature rapidly, with as little as four months from transplanting to harvest. Other races can take eight months or even up to a year. Taro grows at shallow depths and can be obtained without much digging. By contrast, tubers of many yam species require deep digging, and forms that grow near the surface may be toxic or, for a few species, bear thorny roots (I. H. Burkill, 1966, 1: 648-49, 824). Taro has an additional advantage over yams, a triple value: not only are its corms good sources of readily-digested starch, but, for certain races, the stems can be used in salads and the young leaves as green vegetables (Porterfield, 1944: 11; Massal and Barrau, 1956: 34; Murai *et al.*, 1958: 89-90; I. H. Burkill, 1966, 1: 649; Parkinson, 1984: 215). In China, however, the leaves are used much more as fodder or feed (Buck, 1956:

Figure 27. Taro tuber (from Bois, 1927-37, 1: 523)

238). One disadvantage of taro is that corms and leaves are acrid. They may irritate the hands and, unless properly prepared and cooked, they irritate the mouth and throat (I. H. Burkill, 1966, 1: 648, 649; Steinke *et al.*, 1984: 226). It may be this quality to which Schafer (1967: 182; 1977: 92) is referring in saying that one or more kinds of taro cultivated in T'ang China were poisonous and had to be leached before use.

Problems of acridity or toxicity have not prevented taro leaves, stem, and seed from serving medicinal purposes. All three are officinal, with seed, though viewed as somewhat toxic, taken for indigestion, flatulence, and problems of parturient women (Bretschneider, 1882-95, 3: 414; F. P. Smith, 1969: 123). We have found little on the symbolism of taro except that in Fukien, because of its dietary importance, it represents fertility. In this regard, one reads of a Fukienese grandmother placing a bunch of taro, with a large taro in the middle and a group of small taros clinging to it, beneath her grandson's bridal bed (Y. H. Lin, 1947: 40).

Taro corms and cormels may be prepared as food in various ways, by baking, steaming, or boiling, and care must be taken that cooking be sufficiently long to eliminate the toxic elements (Herklots, 1972: 410-11). The leafstalks may be prepared by chopping into small pieces, then boiling in water until tender (H. L. Chung and J. C. Ripperton, 1929: 27-28). I. H. Burkill (1966, 1: 649) says that because of the irritants in the leaf-tissues, double boiling is desirable.

One finds a considerable range of taro dishes in China. Among the main Cantonese taro dishes are "Taro Eel," a pot that includes, besides eel and taro, garlic, salted black beans, ginger, "sherry" (rice wine),[6] ham, pork or chicken soup, scallions, salt, and vegetable oil; and "Taro Root Duck," a duck stuffed with taro root and simmered in water with mushrooms, ham, ginger, scallions, and more taro, until tender, with "sherry" added shortly before serving (W.

W. Chang *et al.*, 1970: 211, 279). From the Hui-chou region of Anhwei comes another main dish which includes taro; this dish, "Hui-chou Pot," includes celery cabbage, dried bamboo shoots, slices of taro, fried bean curd, red-cooked meat and chicken, along with egg wraplings and noodles, all arranged in layers in a large pot and cooked over a slow fire (B. Y. Chao, 1972: 181-82). What may merit the top prize among taro dishes, however, is an unusual and striking banquet dish from Szechwan called "Taro Seafood Nest," which consists of scallops, shrimp, and lobster served in a nest of taro, potatoes, and rice noodles (K. Lee and A. Branyon, 1984: 189-90).[7] Taro may also be used in appetizers and desserts.

Taro as prepared in the above dishes is quite tasty, a long way from the dish, found in Canton and elsewhere in South China, that was described by Anderson and Anderson (E. N. Anderson and M. L. Anderson, 1977: 330; E. N. Anderson, 1988: 177). This was a steamed dish consisting of slices of taro between slices of bacon, one that was starchy, oily, and insipid. This, for the Andersons, was a rare case, a Chinese dish they disliked. It seems, moreover, that they are not alone in their reaction: though taro is fairly common in markets in California, few people of European extraction seem to have acquired a taste for it.

Taro corms are high in carbohydrates and calories (94 calories/100 gm. raw edible portion). In this regard they are similar to the principal starchy root crops of present-day Southeast Asia: yam, sweet potato, and manioc (97-135 calories) (Leung *et al.*, 1973: 12-15). Though low in protein (2.2 gm./100 gm. raw edible portion), taro tubers have more protein than those starchy root crops (yams, *D. alata* and *D. esculenta*, 1.5-1.9 gm.; sweet potato, 1.0-1.2; manioc, 1.0). They are not especially high in calcium, phosphorus, or iron, but when consumed in large quantities, taro is a good supplier of these minerals, and of thiamine and niacin. Taro tubers are poor sources of vitamins A, C, and riboflavin (Murai *et al.*, 1958: 95-96), but the leaves are excellent sources (Leung *et al.*, 1973: 70). Altogether, the above justifies Parkinson's conclusion (1984: 219) that taro, like other aroids, is excellent nutritionally.

Research and development efforts with taro and other edible aroids have been minimal when compared with wheat, maize, and rice, and far below those even for such tropical root crops as manioc, sweet potatoes, and yams (Chandra, 1984: ix). With a serious and continuing effort, taro and other aroids could gain a more important dietary role in the Asian tropics, one that better reflects their nutritional worth. One can only speculate what the future holds for them in South China.

Radish

It is commonly supposed that the garden radish (*Raphanus sativus*) was introduced to China from the West in ancient times, perhaps about 500 B.C. (H. L. Li, 1969: 257; Banga, 1976: 60), and that the Chinese then developed their own large, elongated form (*Raphanus sativus*, var. *longipinnatus*; *lo-po*) commonly known in the United States by its Japanese name, daikon. The usual American radish (called "small *lo-po*") is also found today in parts of China influenced by the West (E. N. Anderson, 1988: 129).

In the fifth and sixth centuries A.D., the radish was a root crop of much less importance than the turnip, a position that today is reversed (H. L. Li, 1969: 257). Indeed, Anderson and Anderson (E. N. Anderson and M. L. Anderson, 1977: 327-28) group radishes with cabbages and mustards as "by far the most important" of China's minor crops.

In China, the radish is mostly a winter crop, and in the Double Cropping Rice Area, according to the 1929-33 survey, 0.6 percent of all cropland was in winter radishes (Buck, 1956: 212). Radishes are eaten in a much greater variety of ways than in Europe. Roots may be used fresh or preserved, as by drying or with salt, soy sauce, or vinegar, sometimes with added items such as garlic and chili. They are commonly cut up and added to soup or meat dishes. They may be used in seafood dishes, such as "Stir-Fried Icicle Radish with Clams" (Field *et al.*, 1976: 37). They may be carved for decoration. Seedlings may be cooked as green vegetables. Leaves and flowering tops may be prepared as vegetables. Seeds, too, may be used, as for medicinal purposes (H. L. Li, 1969: 257; I. H. Burkill, 1966, 2: 1899-1900; Herklots, 1972: 139, 143; Hom and Steiman, 1981: 32, 98; Zee and Hui, 1981: 65; Harrington, 1984: 117-19; E. N. Anderson, 1988: 128-29). The roots of the Oriental radish are regarded as "cold" (E. N. Anderson, 1970a: 209), and nutritionally they are an excellent source of vitamin C. Raw greens and seedlings are even better in this regard, as well as being good to excellent sources of calcium, vitamin A, and riboflavin. The greens, in addition, contain good amounts of iron and niacin (Leung *et al.*, 1973: 62).

Carrot and Turnip

Most wild forms of the carrot genus (*Daucus*) occur in the area from the Mediterranean to southwestern Asia, but there has long been uncertainty about where and when carrots (*Daucus carota*) were first cultivated. It is agreed that cultivated carrots fall into two principal groups, the Asian or anthocyanin carrots, typically with purple or yellow roots, and the European or carotene carrots, mostly with orange roots. The greatest diversity of anthocyanin carrots occurs in Afghanistan, where Banga and others believe these forms were first cultivated, the yellow forms as genetic variants of the purple ones. Banga suggests that anthocyanin carrots from Afghanistan spread westward in the tenth century A.D. or perhaps earlier, that they reached Spain in the twelfth century and northwestern Europe from the thirteenth or fourteenth centuries onward. The first acceptable reports Banga found of orange-colored carrots in Europe date only from the seventeenth century and later, and he concludes that they were developed through selection in Holland. As for China, anthocyanin carrots (including varieties that were reddish-purple, red, orange, or yellow) seem to have arrived in the north about the same time they arrived in northwestern Europe (in Yüan times, according to Laufer), and then spread to Central and South China. Since the carrot is a cool-weather vegetable, in temperate areas it may be cultivated in spring, summer, and fall, but in sub-tropical lowland areas of Kwangtung usually in fall and winter (Laufer, 1919: 451; Banga, 1957: 54-55; 1963: 13-21, 29; 1976: 291-92; Herklots, 1972: 431-33; Zeven and de Wet, 1982: 86, 101, 119, 162; E. N. Anderson, 1988: 68, 129).

In China, carrots (*hu-lo-po, hung-lo-po*) may be eaten in soup, in stir-fried dishes, such as the stir-fried "Hot Cabbage, Carrot, and Radish Salad" (K. Lo, 1974: 160). They may be cooked as a vegetable, or prepared in other dishes such as the Ch'ing-dynasty court dish "Fried and Cooked Carrots, Pork, Soy Beans, and Bean Curd" (Davis, 1966: 119), the Hunanese "Tripe Simmered with Vegetables" (Stallard, 1973a: 114, 116), and the Cantonese "Crab Meat with Carrots" and "Carrots with Shrimp Sauce" (W. W. Chang *et al.*, 1970: 198, 385). Carrots are also commonly shaped into decorative forms, which are sometimes added to dishes and sometimes used as ornaments (Zee and Hui, 1981: 58).[8]

Known especially in the West as an excellent source of vitamin A, the Chinese regard the carrot as strengthening and very "cooling," to be avoided when the body is already in a cold state but very effective against conditions thought to be brought on by an excess of heat, such as digestive upsets and diarrhea. It is thus no surprise that books on Chinese medicine often list carrots as excellent for the digestive tract, and note that the seed is recommended for chronic dysentery. Since cooking in boiling water makes carrots even more "cold," carrot soup is still more effective against "hot" illnesses (Soubeiran and Dabry de Thiersant, 1874: 199-200; Roi, 1955: 248; E. N. Anderson and M. L. Anderson, 1968: 24; 1976: 148; F. P. Smith, 1969: 147; E. N. Anderson, 1970a: 209; 1988: 129, 203; Koo, 1973: 137, 141; Perry, 1980: 415). According to Pillsbury (1978: 13, 20), however, some Chinese believe that the "hot" color of carrots lessens their coldness, and sets them apart from other "cold" root vegetables. Their color also makes carrots especially suitable for the Chinese New Year, when red dominates, and when, by tradition, they are consumed more than at other times (Herklots, 1972: 432).

Many, even most, of the "turnips" referred to in Western accounts of China have really been radishes (E. H. Wilson, 1929: 341; E. N. Anderson, 1988: 128). Nevertheless, the turnip, too, has been cultivated in China since early times. *Feng*, a vegetable repeatedly mentioned in the *Shih-ching* (early Chou times), and apparently important and under cultivation, has been identified by various authorities as the turnip, *Brassica rapa*. Whether or not that identification is correct,[9] the turnip (whether called *B. rapa*, *B. campestris*, var. *rapa*, or otherwise), has been identified in the *Ch'i-min yao-shu* (sixth century A.D.) and in various later accounts of China.

In China today, the turnip, like the carrot and Oriental radish, is regarded as "cold" (Koo, 1973: 141; Yeung *et al.*, 1973: 198; E. N. Anderson and M. L. Anderson, 1976: 148). This is so, says Pillsbury (1978: 13), "because they grow in the dark, damp earth." The raw roots are not outstanding nutritionally in any way, but are a good source of vitamin C. Raw greens, in turn, are good to excellent sources of calcium and vitamins A and C, as well containing good amounts of iron. Turnips may be used fresh; dried; or preserved, as in brine, commonly with chili pepper. In Shantung it is when turnips are plentiful or begin to rot in storage, that they are likely to be sliced, boiled, and dried for future use (M. C. Yang, 1945: 32, 35), such dried turnips keeping for several months. When needed, they are first soaked, then they may be added to stir-fried or slow-cooking dishes, as with pork and snow peas (G. Z. Chu, 1975: 280). Fresh turnips, too, may be prepared in various ways. One reads of "Fried Turnips," a Ch'ing court dish consisting of sliced turnips that are browned, then simmered with cornstarch in a mixture of chopped pork, onion, ginger, and soy sauce that has been prepared in water and sesame oil (Davis, 1966: 121). Fresh turnips may also be cooked as a vegetable; or they may be included in dishes or soups, commonly with some strong-flavored meat, such as beef or lamb, or its broth (K. Lo, 1974: 97-98; 1976: 44). Turnips preserved with salt or brine are normally employed only in small amounts, as a flavoring: in soup, congee, and other dishes, often with pork or beef (G. Z. Chu, 1975: 290; Sakamoto, 1977: 32; C. B. T. Lee and A. E. Lee, 1980: 299), and one reads of preserved turnips served in North China with bowls of millet porridge (Winfield, 1948: 67).

Aquatic Roots and Tubers

Of special interest among tuberous plants used in South China are the aquatics Old World arrowhead, Chinese water chestnut, and lotus root, all primarily sources of carbohydrates. These species regularly enter into systems of rotation with irrigated rice in southeastern China.

Figure 28. Old World arrowhead (*Sagittaria sinensis*) (from Herklots, 1972: 494)

Old World Arrowhead

This elephant ear (*Sagittaria sagittifolia*; *tz'ü-ku*) is an aquatic, perennial herb found growing wild in marshy and swampy areas of Europe and Asia and in nearly every pond and paddy field in China (Henry, 1893: 27).[10] Named for its arrowhead-shaped leaves (in Latin, *sagitta* is "arrow"), in many places the plant is an ornamental. In China, and especially in Japan, however, it is also cultivated in ponds and irrigated fields for its edible underground corms. The Japanese and Chinese varieties are sometimes designated respectively *S. japonica* and *S. sinensis* or *S. chinensis* (Figure 28). The plant is very prolific and from each corm planted, eight or more runners are sent out, with a new corm forming at the end of each runner. Unless corms are dug up from time to time, arrowhead plants may spread over large areas. The corms, which are yellowish in color, weigh less than an ounce each, and are firm, like sweet potatoes, and delicate in flavor (H. L. Chung and J. C. Ripperton, 1929: 45; Porterfield, 1940b: 45, 47).

Arrowhead is believed to be an ancient domesticate of South China, but part of a system of aquatic farming known in the Yangtze area since Neolithic times (K. C. Chang, 1976: 17; H. L. Li, 1983: 43, 44). In modern China, it is cultivated widely but particularly in the south, where it is planted in spring. In the Double Cropping Rice Area, it was third in importance among the starchy root crops (after sweet potato and taro) in the 1929-33 survey (0.2 percent of total crop area). It has also been carried overseas by the Chinese, to Malaya, Tahiti and Moorea in the

Society Islands, and Hawaii, for cultivation as a food crop (H. L. Chung and J. C. Ripperton, 1929: 45; Massal and Barrau, 1956: 37; I. H. Burkill, 1966, 2: 1976).

Arrowhead corms are not eaten raw, in which state they are believed to cause weakness and other health problems (Porterfield, 1940b: 47; F. P. Smith, 1969: 389). Ordinarily, the corms are baked, steamed, or boiled, like white potatoes. One reads of them being peeled, chopped, and cooked in red-cooked dishes with pork, soy sauce, ginger, and other ingredients (H. L. Chung and J. C. Ripperton, 1929: 45-46; K. Lee and A. Branyon, 1984: 60, 264), or made into a cooked salad (Harrington, 1984: 198). They are also processed into starch (Porterfield, 1940b: 46). The stalk of arrowhead, too, may be steamed and eaten, but the leaves are acrid (though applied externally for sores, bites, and itching) (F. P. Smith, 1969: 389).

Arrowhead corms are valuable nutritionally mainly for the carbohydrates they provide. In this, they compare well with yams and taro, and are superior to them in protein (5.0 gm./100 gm. raw edible portion), iron (2.6 mg./100 gm.), and niacin (1.4 mg./100 gm.).

Chinese Water Chestnut

This water chestnut (*Eleocharis dulcis*; *pi-ch'i*; *p'o-chi*; *ma-t'i* or, in Cantonese, *ma-t'ai*, "horse's hoof"; etc.) is sometimes called the "true water chestnut" to distinguish it from the fruit of the aquatic *Trapa bicornis*, which is also called "water chestnut." Both of these, but especially *T. bicornis*, present a health hazard, for encysted larvae of the intestinal fluke *Fasciolopsis buski* can be transmitted to humans who eat them (Barlow, 1925: 57-60, 89-90; Faust, 1929: 32).

Eleocharis is a genus of the sedge family, *Cyperaceae*, in which are found the papyrus of Egypt and a host of other species around the world that are useful to humans, for example for grazing and thatch, with some providing edible plant parts as well. *E. dulcis* itself grows wild in fresh and brackish waters in many parts of the Old World tropics. Its corms are collected for food in some places in that broad region, but in others they are overlooked. It was in southeastern China and Vietnam that *E. dulcis* first came to be grown for food, with domesticated forms now having corms far larger and sweeter than typical in the wild. *E. dulcis*, as a plant of moist places and marshy areas, is cultivated in paddies in China, often in rotation with rice or other aquatic crops. As a tropical plant, it is cultivated more in warm areas of the south, but is common as far north as the Yangtze River and some grows even in the Peking area (40°N.) (Hodge, 1956: 49-50, 53, 56, 63-64).

Under the best conditions, including liberal use of night soil and other fertilizers, the Chinese water chestnut is an excellent producer. Corm yields under normal field cultivation are between a half and three-quarters of a pound per square foot of cultivated area. For a single plant, there is a record yield of twenty pounds of corms in one season. Corms vary greatly in size, with larger ones (one and a quarter inches or more in diameter) marketed, and smaller ones kept for seed, for use at home, or for feeding farm animals or wild birds (Hodge, 1956: 56, 59).

Corms are planted, initially in a nursery bed, in spring. In most parts of China, they are typically harvested in late fall, though availability can be extended to March or April by leaving them in the ground. One reads (Gray, 1878, 2: 194-95) of women wading in ponds to gather water chestnuts, and, when their baskets are full, going to a nearby town or village to hawk them. Corms used locally may be stored for several months in crocks kept in cool places. From time to time, they are removed for inspection, and decaying ones disposed of (Hodge, 1956: 59-60).

Water chestnut corms may be consumed as a vegetable or made into starch or flour, a white substance that looks and feels like chalk. There are two seemingly distinct varieties, one (Cantonese: *hon-ma-t'ai* or *Kweilin-ma-t'ai*) eaten fresh or cooked, and the other (*sui-ma-t'ai*) used mainly for starch. The very best water chestnuts for eating fresh come from the Kweilin area of Kwangsi Province, thus the second term, *Kweilin-ma-t'ai*. Freshly peeled corms are eaten raw, especially during the winter and spring when fruit is scarce or unavailable. It is quite common for guests to be offered a plate of them. Fresh corms, peeled and on bamboo skewers, are a regular item of sale by vendors on streets and trains. Fresh peeled corms, usually sliced, diced, or shredded, may be added to various dishes (such as soups, salads, meats, and fish) to which they impart their notable crispness. Examples of Cantonese dishes of this sort are "Beef and Water Chestnut Fun Goh," "Shrimp with Water Chestnuts," and "Stir-Fried Minced Pork with Water Chestnuts and Lettuce Leaves" (W. W. Chang *et al.*, 1970: 78, 182, 296). Minced water chestnuts may be used for puddings or made into fritters or cakes of which one type, *ma-t'ai-ko* in Cantonese, can regularly be obtained from sidewalk vendors or in tea houses. Water chestnut corms may also be steamed or boiled as a vegetable. Made with a sauce of sugar, butter, and flour, they are said to be quite like sweet corn in taste and appearance. Those water chestnuts intended for starch are taken for milling and processing to simple thatch shelters erected near paddies or streams. The starch can be used as a thickening agent or for coating food, as in deep-frying. This starch is believed by the Chinese to be nutritious and good for the digestive system. The corms are given in large amounts to treat children who have accidentally swallowed coins (Porterfield, 1938c: 32-33; 1951: 18-19; Hodge, 1956: 50, 53, 60, 62; I. H. Burkill, 1966, 1: 921-22; F. P. Smith, 1969: 399; Herklots, 1972: 491-92; Hom and Steiman, 1981: 41; Tropp, 1982: 578; Gin, 1984: 176). Also of interest, water chestnut corms contain the antibiotic puchin, which, however, is destroyed even by moderate heat (Hodge, 1956: 62).

Chinese water chestnuts are a delicacy, of value nutritionally mainly as a source of carbohydrates (16 gm./100 gm. edible portion) (Leung *et al.*, 1973: 54).

Lotus Root

The "Chinese" or "Indian" lotus (*Nelumbium speciosum = Nelumbium nelumbo = Nelumbo nucifera*; *ho*), often confused with the Egyptian lotus (*Nymphaea* spp.), is a perennial water plant native to southern and eastern Asia. Its domestication is placed by Vavilov (1949-50: 24) and certain other botanists in China, or, possibly, India. In China, it is found growing in marshes and ponds, such ponds being quite numerous around Canton. Since antiquity it has also been cultivated in China and other parts of Asia for food, medicine, and for its beautiful flowers (Figure 29). It is one of the few aquatic plants mentioned in the *Shih-ching*, that North Chinese classic which dates from early Chou times (*c.* 1000-500 B.C.) (Keng, 1974: 402-3), and it may have been cultivated much earlier in the Yangtze Valley (H. L. Li, 1983: 43) and areas to the south.

Lotus stalks and roots likely were also in common use as foods among the Vedic Indians (Prakash, 1961: 22), and later the lotus was highly regarded by Hindu poets, and enjoyed strong religious associations with Hinduism and Buddhism, symbolizing, among other things, the vulva, fertility, divine birth, spiritual power and life, purity, beauty, and perfection (Dymock, 1889b: 81-83; Hastings, 1908-26, 8: 142-44; B. Walker, 1968, 2: 216; Lessing, 1976: 141-44; Stutley and Stutley, 1977: 160-61; Kinsley, 1986: 21, 22, 32, 62, 173). These associations are found in Chinese Buddhism as well (Arlington and Lewisohn, 1935: 329; Sowerby, 1940:

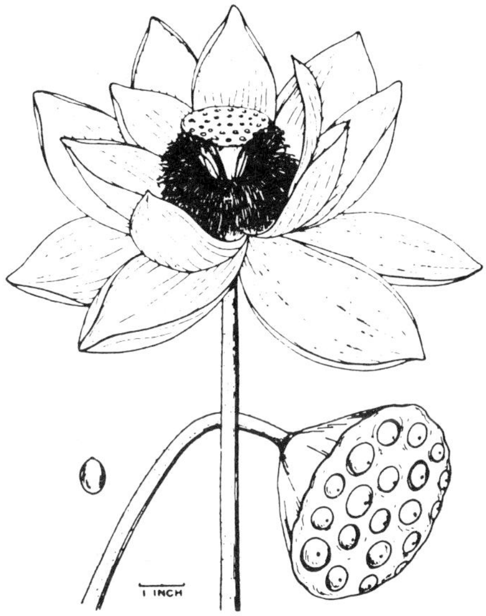

Figure 29. Flower, seed capsule, and a seed of the lotus (*Nelumbo nucifera*) (from Herklots, 1972: 493)

130-31, 136-37; Schafer, 1963: 128; C. S. Wong, 1967: 79; Em. Hahn *et al.*, 1973: 65-66; Lessing, 1976, 141-44), with the lotus a sacred plant grown near Buddhist temples and widely represented in Buddhist and non-Buddhist art and crafts (Figure 30). The Buddha is commonly depicted as sitting on a lotus blossom, as are various other Buddhist deities and the Goddess of Mercy of the Chinese (Cormack, 1935: 197-98). In a ceremony observed by F. L. K. Hsu (1983: 28-29) in Yunnan in 1942, the chief priest,[11] who was simulating a Buddhist monk or Buddha, wore a lotus-petaled hat, sat on a seat rimmed by lotus petals, and as he walked in procession, a subordinate placed in his footsteps a series of tiny lotus lamps. The lotus flower is also symbolic of feminine beauty in China, and the small feet of Chinese women were referred to as "Golden Lotus Flowers" (Arlington and Lewisohn, 1935: 329). The lotus was sacred to Taoists, and Confucian scholars, too, were liberal in praising it (C. S. Wong, 1967: 79). The lotus was grown in the waters of the Forbidden City in Peking, and Chinese poets wrote of the beauty of its flower, of visiting lotus gardens, and of picking lotuses from boats (Schafer, 1963: 127-30).

The lotus also was prominent at certain festivals including "All Soul's Day" (Chung Yüan Day), a beautiful autumn celebration, Buddhist and ancestral, when graves of the ancestors were cleaned and sacrifices of food made to them. Among other activities in Peking, boys, starting at twilight, went along the streets with lanterns made of lotus leaves or lotus flowers, and in the evening people placed lighted candles on lotus leaves[12] and set hundreds of them adrift on rivers or on a lake to guide the spirits of individuals who had drowned (Arlington and Lewisohn, 1935: 91, 329-30; Cormack, 1935: 190-95; Eberhard, 1952: 129-33; Tun, 1965: 60-62, 111).

Figure 30. Chinese statuette of an aspiring Buddha holding a lotus flower and standing on a lotus-flower pedestal (from Porterfield 1941: 281)

Most important of the edible products of the lotus are its "root" (actually a rhizome, called *ou*) and seed (*lien-tzŭ*) (Figures 29, 31), though, according to Wilson (E. H. Wilson, 1929: 342), they were costly luxuries used only by well-to-do persons. Young lotus leaves may be cooked as greens or added to other dishes;[13] or full-grown leaves, which are quite large and commonly sold in a dried form, may serve to wrap meat, fowl, fish, rice, or stuffing that is to be steamed or baked, which adds a special fragrance to the food they enclose.[14] The petals of the flower (*lien-hua*), too, may be used to garnish salad or soup, or they may be used in other ways, as in "Fried Lotus Flower" (K. Lo, 1974: 169).[15] The root may be used fresh, dried, or canned, and is eaten in a variety of ways, for example sliced and stir-fried as a crunchy vegetable, added to soups and meat dishes, made into salad, stuffed with rice and steamed, preserved in syrup, candied, pickled, or processed into starch, a thickening agent.[16] The seed, a delicacy, may be

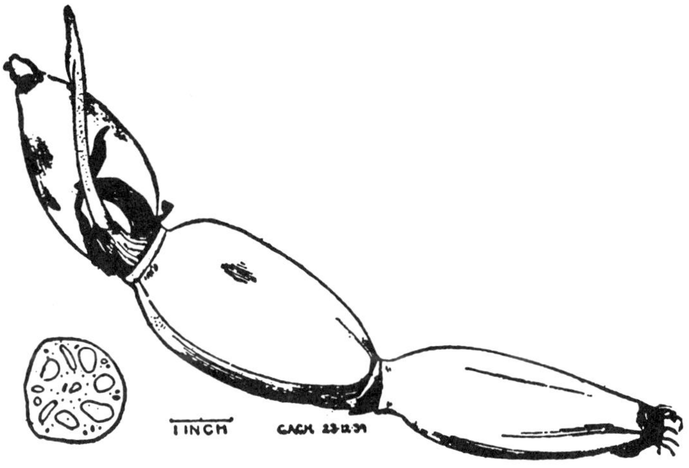

Figure 31. Lotus rhizome (from Herklots, 1972: 493)

eaten raw or dried as a dessert or snack, for which it is sold on the streets. It is also available in cans. It may be roasted, boiled, candied, pickled, or ground into flour, and is commonly made into a tea or soup[17] or added to dishes of one sort or another for its delicate flavor (Bretschneider, 1882-95, 2: 216-17; Blasdale, 1899: 17-18, 39-40; H. L. Chung and J. C. Ripperton, 1929: 53; Hooper, 1929: 92-94; Arlington and Lewisohn, 1935: 329; Read *et al.*, 1940: 26; Porterfield, 1941a: 283-85; 1951: 10-11; I. H. Burkill, 1966, 2: 1565; F. P. Smith, 1969: 279-80; Herklots, 1972: 492-94; K. Lo, 1972: 247; Em. Hahn *et al.*, 1973: 65-67; Hom and Steiman, 1981: 36; Zee and Hui, 1981: 63; Harrington, 1984: 189-90). Lotus root is valuable nutritionally mainly for its vitamin C content. Lotus seeds, when dried, are good to excellent sources of carbohydrate, protein, calcium, iron, and certain B vitamins (Platt, 1962: 14-15; Leung *et al.*, 1973: 26-27, 54).

The lotus is said to be "cold" because water is cold and the lotus grows in watery habitats (Pillsbury, 1978: 13). Many virtues are ascribed to the lotus, and it serves an unusually broad range of medicinal uses, with its products finding their way into pharmacies not only in China but in overseas Chinese communities as well. Of special note, however, is the belief that lotus products impart purity and strength to the body, improve virility, increase fertility, and slow the aging process (Soubeiran and Dabry de Thiersant, 1874: 219-20; Blasdale, 1899, 17-18; Hooper, 1929: 92-94; Porterfield, 1941a: 283-85; F. P. Smith, 1969: 278-81; J. Y. P. Chen, 1973: 229; Em. Hahn *et al.*, 1973: 66). Thus it seems especially appropriate that at the time of New Year, lotus seed is included among the fruits and seeds eaten in a clear effort at enhancing fertility, and that the name for lotus seed bears a similarity to words meaning "many sons" (E. N. Anderson and M. L. Anderson, 1977: 380). A symbolic role is also played at weddings by the lotus root. Here, however, it is the root's pronounced stickiness, which continues to hold its parts together when eaten, that symbolizes the hope that nothing will separate the joyful newly-weds (E. N. Anderson and M. L. Anderson, 1977: 330; E. N. Anderson, 1988: 202).

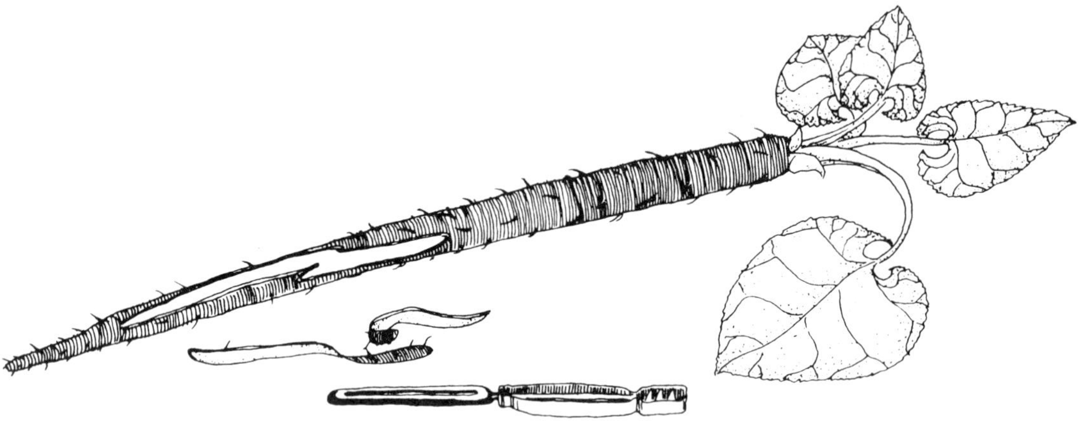

Figure 32. Great burdock (*Arctium lappa*) (from Harrington, 1984: 114)

Minor Roots and Tubers: Great Burdock, Kudzu, and Others

Of lesser importance than the aquatics just considered are five minor roots and tubers: (1) great burdock, found especially in temperate parts of China; (2) the root usually known in the United States by its Japanese name "kudzu"; (3) "Chinese artichoke"; (4) an aroid, *Amorphophallus rivieri*; (5) the parsnip; and (6) the beet (*Beta vulgaris*). We consider them briefly here.

The general term "burdock" is applied to a group of troublesome weeds (*Arctium* spp.) native to temperate regions of Europe and Asia. One species, great burdock (*Arctium lappa* = *Lappa major*; *niu-p'ang, niu-p'ang-tzŭ, ta-li-tzŭ*) (Figure 32) came to be cultivated in China and Japan for its leaves and roots. Its place of domestication, according to both Vavilov (1949-50: 21, 24) and Harlan (1975: 74), is temperate China, with Vavilov including it in a list of traditional Chinese food plants. Apparently, however, use of the plant was once much greater there. In his famous materia medica of the sixteenth century A.D., Li Shih-chen writes that in ancient China great burdock was cultivated in fertile soils and its leaves and roots used as food. Though very nourishing, he added, it was little used in his day (Bretschneider, 1882-95, 2: 173). Perhaps for this reason, many of our sources fail even to mention the plant. Herklots (1972: 462) does refer to its cultivation on Taiwan, but especially by Japanese, who call it *gobo*. And others report it grown in Hawaii and consumed there by Orientals (H. L. Chung and J. C. Ripperton, 1929: 49-50), or served in Oriental restaurants in the United States (Harrington, 1984: 113).

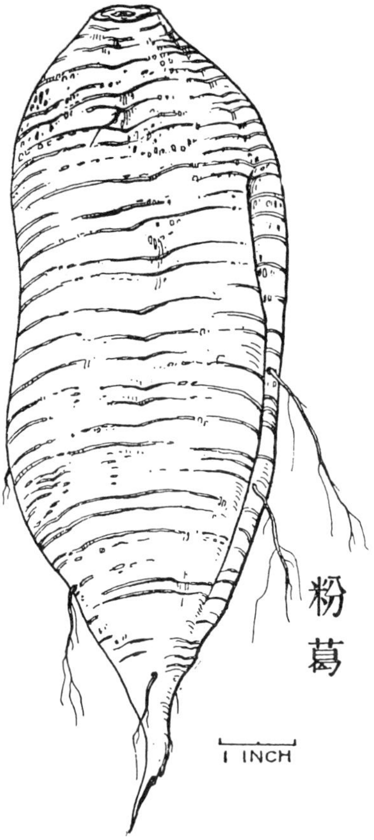

Figure 33. Root tuber of kudzu (*Pueraria lobata*) (from Herklots, 1972: 469)

Both the tender leaves and stems of great burdock can be consumed, but the root is of greatest dietary importance. Young roots can be peeled and eaten raw, but mature ones, after peeling, may be soaked or scalded, then cooked in various ways and served alone or with other foods (H. L. Chung and J. C. Ripperton, 1929: 50; Harrington, 1984: 114-15).

Besides the food uses mentioned above, great burdock root, stem, and seed are all officinal. The seed has hooked spines which gave the plant one of its popular names, *wu-shih* or "evil fruit"; yet the seed is valued for its diuretic and other medicinal qualities and is exported to Malaya for drug use by Chinese (Bretschneider, 1882-95, 3: 172-73; Hooper, 1929: 14; F. P. Smith, 1969: 46).

Nutritionally, great burdock is a source of carbohydrates. The root in a raw state is a good source of calcium and thiamine, though much thiamine is lost in boiling (Leung *et al.*, 1973: 11).

The yam bean kudzu (*Pueraria lobata, Pueraria thunbergiana*, or, formerly, *Pachyrrhizus thunbergianus*; *ko*) is an herbaceous perennial vine, sometimes cultivated for its edible root (Figure 33) and other useful products. Originating in South China or Southeast Asia, it spread eastward at an early date to New Guinea and parts of Oceania, and northward to Korea and Japan (K. C. Chang, 1970: Table 1; Harlan, 1975: 74, 221; Shurtleff and Aoyagi, 1977: 12;

Mathon, 1981: 78; Zeven and de Wet, 1982: 40, 58). In recent centuries, it has spread to other tropical and subtropical parts of the world as a purple-flowered ornamental and cover crop excellent in retarding soil erosion. It may also be cultivated as forage or fodder, for kudzu is nitrogen-fixing, can do well in poor soils, and its hay is said to be of equal value to that of alfalfa (Porterfield, 1951: 13; Herklots, 1972: 470).[18] Its cultivation was encouraged in the American South by the Soil Conservation Service in the days of the New Deal. By 1945, 500,000 acres of kudzu were cultivated there. It readily escaped cultivation and spread so rapidly as to be considered a common weed, a pest creating landscapes reminiscent of the tropics (Agricultural Research Service of the U.S.D.A., 1971: 236-37). One reads of kudzu in the American South that cannot be killed; of a plant that may grow a foot a day or a hundred feet in a summer, one hard to kill with ordinary herbicides or burning mainly because of the store of starch in its taproot. One reads of back roads flanked with kudzu; of a vine that will engulf a deserted vehicle in just a few weeks; of one that runs up billboards, poles, and unoccupied structures until they are clothed in green; of warnings to naughty children that they will be cast into the kudzu and lost there; of a folk saying in Georgia that in areas where kudzu grows one must shut the windows at night to make certain that it doesn't come into the house; and of kudzu as an example of the dangers of government intervention (Tanner *et al.*, 1979: 400-401; Maharidge, 1985: 1; Pasour, 1985).

One is unlikely to have heard such tales of kudzu in its place of domestication in Asia, for kudzu served uses, as fiber and food, that kept it under control. Along with the yam, taro, and other aroids, it was among the earliest roots and tubers domesticated there for human food. With those plants, or even before them, it apparently spread across Oceania for use as food and fiber, and indeed may have been a staple food in some places there. Though cultivation of kudzu for food continues in East Asia, Southeast Asia, and India, in Oceania it has ceased almost everywhere, apparently continuing only in highland New Guinea and New Caledonia (Barrau, 1956: 89-91, 151; 1965b: 282-85; J. B. Watson, 1964; 1965: 296, 297-98, 300-301; 1968: 268-69).[19] The plant is found wild elsewhere in Oceania as far east as Samoa and Niue, in many places an apparent escapee from earlier cultivation. Wild roots may be eaten at times of food scarcity or as a ceremonial food or chiefly delicacy, and wild plant fiber used for fishing lines, netting, and rope (Barrau, 1965b: 282-85). In the New Guinea highlands, the decline of kudzu as a food crop is attributed to competition with the sweet potato after the latter arrived from the New World (J. B. Watson, 1964; 1965: 300-301), though we do not know whether its decline elsewhere in Oceania can be attributed to this cause.

In modern China, kudzu is cultivated in both north and south, including Kwangtung. It is commonly grown in the same plots as taro and ginger, with stakes erected on which the vines can climb (Herklots, 1972: 470). The above-ground portions of kudzu are believed to be somewhat toxic, having emetic qualities (Bretschneider, 1882-95, 3: 309; F. P. Smith, 1969: 299), and thus the root is the part of the plant normally used for food. Young kudzu roots, which are shaped somewhat like sweet potatoes, may be cooked as vegetables, but their covering is tough, and the flesh both tough and fibrous. Thus it is mature roots (which may be up to three feet long and 80 pounds or more in weight) that are normally used for the excellent, edible starch they provide. Or kudzu stems may be processed into a fiber that is woven into a strong, yellow-colored grass cloth (*ko-pu*, or "*ko* cloth") for summer dress. Such cloth has been used since ancient times in China, and was the most notable textile produced in the tenth century A.D. in the empire of Min, an independent Fukienese state which sent large amounts of it as tribute to the Chinese court (Bretschneider, 1882-95, 2: 208-10; 3: 308-10; Schafer, 1954: 63, 63n, 64, 69; Keng, 1974: 399; E. N. Anderson, 1988: 25, 26). In the Orient, kudzu fiber from vines and roots is also processed into a fine quality paper, and is made into shoes, fishing nets, basketry, and other products.

Kudzu starch, in the form of flour (*ko-fên*), is marketed widely in eastern Asia from Japan to Indochina. Starch, roots, and other plant parts are exported from Canton and other Chinese ports to overseas Chinese communities, including those of Malaya and the United States. Fine, sweet in taste, and odorless, the starch may be used as a thickening agent, for example in soup or pastry (Bretschneider, 1882-95, 2: 208-10; 3: 308-10; Blasdale, 1899: 25; H. L. Chung and J. C. Ripperton, 1929: 52; Hooper, 1929: 119; Porterfield, 1938b: 203-5; 1951: 13; I. H. Burkill, 1966, 2: 1870; Herklots, 1972: 468-70; Zee and Hui, 1981: 64).[20] Kudzu starch can now be found in many health food stores in the United States but, though it contains modest amounts of calcium, phosphorus, and iron, it is of value, nutritionally, mainly as a source of carbohydrates (Leung *et al.*, 1973: 13).

Medicinally every part of the kudzu plant may be used for treating skin rashes. Its leaves may also be applied to cuts to stop bleeding, and its flowers and seeds taken for "alcoholic excess." Shoots may be applied to boils, or to mouth sores in children. Seeds and root are used to treat dysentery. And the root is believed to satisfy thirst, act against fevers and diarrhea, and is also prescribed for colds, influenza, insect and snake bites, and to counter poisons (Hooper, 1929: 119; F. P. Smith, 1969: 299).

The multi-faceted use of the kudzu plant in the Orient has not escaped American observers, and some have urged broader use of kudzu as a solution to the environmental problems the plant has created in the American South. Several years ago, at least one firm in Atlanta, following the Oriental process, was already producing a kudzu paper intended for artists, one that resists mold and doesn't fade or become stiff. Among other suggested uses of kudzu in the South, its thick vines could be burned along with coal in producing electric power, which would reduce atmospheric pollution brought on at present by the burning of high-sulphur coal; its starch could be employed as a substrate in making ethanol and a lysine-rich bakers' yeast; and a fermentation process might be used for separating kudzu fibers for use in textiles (Tanner *et al.*, 1979).

A third minor but widespread root crop, known as "Chinese artichoke," is *Stachys affinis* (= *S. tubifera* = *S. sieboldii*; *ts'ao-shih-ts'an, kan-lu-tzŭ*, etc.) (Figure 34). An ancient North Chinese domesticate (Vavilov, 1949-50: 21; K. C. Chang, 1970: Table 1; Harlan, 1975: 73), this nettle-like member of the mint family (*Labiatae*) bears elongated white tubers unusual in form, divided into internodal segments which can number a dozen or more. Tubers quickly discolor in the atmosphere, and if they are not to be used promptly they may be stored in moist sand. After washing, they are boiled for a few minutes, then fried. Their taste, says Herklots (1972: 470-71), is excellent. Pickled, they are a very good source of calcium and possibly of iron (Leung *et al.*, 1973: 35).

Seldom mentioned in the literature but also cultivated in China is the aroid *Amorphophallus rivieri* (= *Amorphophallus konjac*) (Figure 35). This plant, called *mo-yü* in China, is sometimes grown as an ornamental in the United States, where it is known as "devil's tongue" or "snake palm." We mention it here because, like taro, *A. rivieri* is an ancient aroid which has continued in use as a food crop to the present day, with E. H. Wilson (1929: 342) observing that it was cultivated on a small scale everywhere in the Yangtze Valley. We have found no report on the nutritional value of *A. rivieri*, but the corm of a related cultigen (*A. campanulatus*) is com-

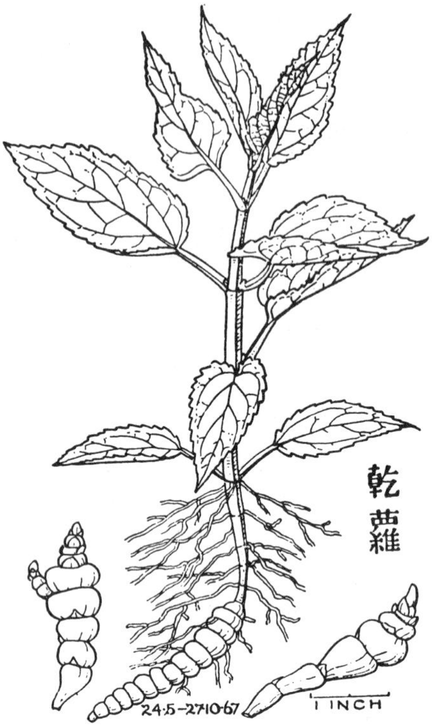

Figure 34. Chinese artichoke (*Stachys affinis*) (from Herklots, 1972: 471)

parable to taro nutritionally (Leung *et al.*, 1973: 12, 14). That cultigen seems to yield as much or more than taro when correctly fertilized and irrigated. It also has the capability of growing in areas too dry for most types of taro; and it enjoys unusual resistance to insects and disease (Sakai, 1983: 30). Perhaps *A. rivieri* possesses some of the same qualities, and has persisted in cultivation because of them.

A. rivieri is of a genus native to the Old World tropics which consists of nearly a hundred species. Of these, there are two main domesticated ones. Both originated in southern or eastern Asia, and both are cultivated there primarily for their corms, which are used as food and, in some places, as medicine. One is *A. campanulatus*, the elephant yam or elephant-foot yam, probably domesticated in India (Sanskrit: *kanda*), and now cultivated for food in gardens from India to Indochina and Malaysia; it may have been introduced as a food plant to Melanesia and Polynesia in ancient times, but is seldom eaten there today. The second species is *A. rivieri*, believed to have been domesticated in China or Indochina. Less tropical than *A. campanulatus*, it is cultivated for food from Indochina to Japan, where it was introduced in the tenth century A.D., and where, in 1979, 100,000 tons of corms were produced (Watt, 1889-96, 1: 225-27; Bailey, 1935, 1: 276-77; Vavilov, 1949-50: 21, 27; Massal and Barrau, 1956: 6-7; I. H. Burkill, 1966, 1: 139-41; Herklots, 1972: 403-4; Zeven and de Wet, 1982: 49, 71; Sakai, 1983: 53, 56-57).

Figure 35. *Amorphophallus rivieri* (from Crevost and Lemarié, 1917: 147)

The corms of wild *Amorphophallus* and of some cultivated forms contain poison which must be eliminated by leaching or cooking before the corms are consumed. In China and Japan, fresh corms of *A. rivieri* may be prepared directly as food or they may be dried and made into a flour. In Japan, most of the crop is dried and made into such flour, which is largely used for making a gelatinous food called *konnyaku* (Sakai, 1983: 57, 59, 65, 67). The Chinese way of flour preparation seems to be similar. According to I. H. Burkill (1966, 1: 139), the tubers are sliced; then, to eliminate poison, the Chinese carefully wash the slices, which are then dried, pulverized, and used for making a food that looks like bean curd. Wilson (E. H. Wilson, 1929: 342) writes only that the tubers are ground with water and made into a curd-like substance.

Though we have found only a partial analysis of them, it seems that the fresh corms of *A. rivieri* are only a fair carbohydrate source. The prepared food *konnyaku* as made in Japan and Hawaii is of questionable nutritional value though it is a good source of calcium and has been found to reduce serum cholesterol levels significantly in a high cholesterol diet (Sakai, 1983: 30, 68-73).

The parsnip (*Pastinaca sativa*) is believed to have been domesticated in Europe, the Mediterranean area, or the Near East (Vavilov, 1949-50: 36; Mathon, 1981: 35; Zeven and de Wet, 1982: 163) but it is of quite minor importance in China and few of our sources even mention it. Wilson (E. H. Wilson, 1929: 342-43), however, observed it in cultivation in Yunnan, Hupeh, or Szechwan, noting that its roots are quite thin and that the entire plant is prepared and consumed. Use of the beet (*Beta vulgaris*) for sugar will be considered in our chapter on "Spices and Flavorings," but the beet is used in China as a vegetable as well, with beet greens or the spinach beet (var. *cicla*) usually mentioned (Laufer, 1919: 399-400; Wilson, 1929: 345; Read *et al.*, 1940: 16; Herklots, 1972: 154-55, 428-30; Zee and Hui, 1981: 9).

ROOT CROPS OF AMERICAN ORIGIN

Sweet Potato

For persons familiar only with Chinese restaurant fare in this country, it may come as a surprise to know that China produced over half of the world's sweet potatoes (*Ipomoea batatas*; *kan-shu*, "sweet tuber"; *fan-shu*, "foreign tuber"; and by other names) in the late 1930's and early 1940's, and that it remains the leading producer today, with an output of 90 to 100 million tons each year (Cooley, 1948: 84-85; Wittwer *et al.*, 1987: 228). Indeed, per capita consumption in China in 1935 amounted to 61 pounds versus 19 pounds for the United States (Cooley, 1948: 84-85). As for the Double Cropping Rice Area, in the 1929-33 village survey sweet potatoes were cultivated on 54 percent of all farms (Buck, 1956: 212-15; Maynard and Swen, 1956: 407), though such cultivation was concentrated on salty or sandy soils and steep slopes, places unsuited to rice (Cressey, 1934: 342; Spence, 1977: 262). Yet overall, sweet potatoes occupied 12 percent of all crop area; 84 percent of all farm families consumed them; and the percentage of total calorie intake provided by sweet potatoes in the Double Cropping Rice Area (8.2 percent) dwarfed that for all other areas of China (0-4.9 percent) (Maynard and Swen, 1956: 413).

At the same time, one should not overlook other places of high sweet potato consumption outside the Double Cropping Rice Area, for they are grown in many parts of the country. Szechwan is one major area of sweet potato production, and they are cheap and plentiful there (Jenner, 1968: 40; Wittwer *et al.*, 1987: 228).[21] They are also produced in considerable amounts in the valleys of the Huai and Yellow rivers; and at Ting Hsien, Hopeh Province, in the 1920's, they were the principal food of all income groups, more important than millet and all other cereals together (V. Y. N. Hsu and F. L. K. Hsu, 1977: 301; Wittwer *et al.*, 1987: 228).

Despite their widespread use as food, sweet potatoes have always been unpopular in China, and nearly everywhere they are considered the most marginal of foods. One observer (E. N. Anderson, 1988: 122) writes of sweet potatoes as pig feed among well-off families; of people consuming them only when their food situation is hopeless; and, in places where sweet potatoes are cultivated, of determining a family's income in terms of the sweet potato's dietary importance. Other writers (Buxton, 1929: 113; Cressey, 1934: 342; T. H. Shen, 1951: 210; Jenner, 1968: 40; C. B. T. Lee and A. E. Lee, 1980: 99, 303) confirm the observation that in China sweet potatoes are a food of the poor. For example, at Taitou, a village in Shantung studied by Martin Yang, sweet potatoes were eaten by poor people all year long and at all meals. As families increased in affluence, sweet potatoes and millet became the staples, then millet and wheat, and, finally, wheat alone. Sweet potatoes were so important overall in Taitou, however, that outsiders called its people "sweet potato eaters," an offensive term because it was a mark of poverty (M. C. Yang, 1945: 32-35).

In China, sweet potatoes may be consumed fresh, following harvest, and are usually prepared by steaming or boiling (Holmes, 1948: 143; E. N. Anderson, 1988: 122). They may be packed in straw or sand to delay their going bad (I. H. Burkill, 1966, 2: 1268). Or, because tubers of most races spoil within a few months, they may be sliced and sun-dried, as they are in Japan (E. H. Wilson, 1929: 340; Schery, 1972: 493; E. N. Anderson, 1988: 122). In Fukien at the fall harvest, entire hillsides may be covered with slices of sweet potatoes drying in the sun.[22] This "sweet potato money," as it is designated, can be kept for use through the winter when rice is expensive (Hurlbut, 1939: 77-78). In Fukien, and among Chinese of Fukienese origin on Taiwan, sweet potatoes are shredded into strips and dried in the sun for use through the year (Holmes, 1948: 38; Diamond, 1969: 18-19). The dried slices are prepared for eating with cereal grain if that is available (E. N. Anderson, 1988: 122), but for Fukien one reads of dried, shredded sweet potatoes cooked into a substitute for rice, called "sweet potato rice" and used by very poor people (Holmes, 1948: 38). In Chinese houses, according to a person who served as a missionary in Fukien (Holmes, 1948: 142), one finds huge containers of sweet potato rice, gray and impalatable, on the stove. Martin Yang (1945: 32) notes that in Shantung dried sweet-potato slices are either boiled, or ground into a meal that is added to other flour in preparing bread or noodles. E. H. Wilson (1929: 340) writes of such sweet potato noodles in the Yangtze Valley. And others (Holmes, 1948: 32, 131-32) mention such noodles, made of sweet potato flour or sweet potato flour and eggs, as everyday food in the Foochow area of Fukien. I. H. Burkill (1966, 1: 1269), in turn, notes that in Malaya the Chinese make dried sweet-potato slices into a thirst-quenching tea.

Sweet potatoes are also a common snack in China, a steamed sweet potato being quite satisfying on a cold day (Holmes, 1948: 132; B. Y. Chao, 1972: 18; Stallard, 1973a: 122). Roasted or boiled sweet potatoes are prepared by street vendors in Peking and Tientsin, and sold as hot, inexpensive snacks or meals in the cold weather from October onward (Headland, 1914: 197; Gamble, 1933: 106; Cormack, 1935: 132-33; Tun, 1965: 83-84; V. Y. N. Hsu and F. L. K. Hsu, 1977: 302, 306).[23] Sweet potatoes may also be prepared as an appetizer in other ways, as in the Cantonese "Sweet Potato Balls," for which tubers are peeled, quartered, simmered until soft, mashed, mixed with sugar and flour, and shaped into small balls that are deep fried for a few minutes until golden brown (W. W. Chang *et al.*, 1970: 84). This is similar to a popular Peking snack called "Sweet Mashed Mountain Potatoes," a dish that came originally from the rural south. In making this dish, sweet potatoes are peeled, cut into pieces, boiled, drained, and mashed; sometimes a bit of boiled, mashed chestnut may be added to the mashed sweet potato; then sugar is added, and the mixture beaten until light, then stir fried for a few minutes and served hot (K. Lo, 1971: 105). Sweet potatoes may be eaten as a dessert, as a vegetable, or cooked with rice (Zee and Hui, 1981: 61). There are even main courses that feature sweet potatoes, as in the dish from Szechwan (where sweet potatoes are a regional specialty) called "Steamed Beef with Sweet Potato" (W. W. Chang *et al.*, 1970: 344; H. J. Lin and T. Lin, 1972: 24-25). Despite the perception of sweet potatoes as food of the poor, main sweet-potato dishes were found even in the imperial court, for example "Fragrant Duck Braised with Sweet Potatoes" (J. D. Keys, 1963: 26, 47). Sweet potatoes may also, as among the Li people of Hainan, be made into a distilled liquor (Odaka, 1950: 28).

Sweet potatoes are high in calories, more so even than white potatoes. Though they contain only a small amount of protein (about 1 gm./100 gm. raw edible portion), it is of good biological value (Adolph and Liu, 1939: 337; Leung *et al.*, 1973: 14). They contain varying amounts of minerals and vitamins, with the yellow varieties, which are most important nutritionally, very rich sources of beta-carotene, a precursor of vitamin A.[24] Because intake of vitamin A is low in China, sweet potatoes can play a critical role in this regard. In the words of E. N. Anderson (1988: 122), the sweet potato "may have saved many million pairs of eyes in the four hundred or so years since its introduction." At the same time, he notes, the Chinese favor the whiter forms of sweet potatoes, which are not good sources of vitamin A. The difference is striking: tubers of the pale variety were found to have a beta-carotene equivalent of only 35 mcg./100 gm. raw, edible portion, whereas those of the yellow variety had 1680 mcg./100 gm. (Leung *et al.*, 1973: 14).

The leaves of sweet potatoes are good sources of riboflavin (0.16-0.35 mg./100 gm.), among other vitamins, and one reads of the tops, young tops, or leaves as human food in Malaya, Indonesia, the Philippines, and India (Watt, 1889-96, 4: 479; Wester, 1924: 100; Ochse and Bakhuizen van den Brink, 1931b, 1: 159; Caldwell and Enoch, 1972: 311). They do, however, contain hydrocyanic acid and, if eaten in large amounts, can cause death (I. H. Burkill, 1966, 2: 1269). In China they, and sometimes the entire plant, are used as pig feed (Herklots, 1972: 444; Zee and Hui, 1981: 61), but we have not determined the extent to which they serve as human food.

White Potato

The white or Irish potato (*Solanum tuberosum*; *yang-yü, ma-ling-shu*, and other names), a domesticate of highland Peru, was cultivated in Taiwan by A.D. 1650 and on the mainland of China by A.D. 1700 (Laufer, 1938: 70, 71). Its introduction is credited to the Dutch, and in South China is still called *ho-lan-shu*, "Holland tuber" (Couling, 1917: 456). Its spread in China in the eighteenth and nineteenth centuries, however, is attributed in good measure to efforts by French Catholic missionaries (E. N. Anderson, 1988: 94, 121-22). Early in this century, for example, E. H. Wilson observed (1913, 2: 58) that during a great famine sixty years before, Catholic missionaries had introduced the white potato to the upper Yangtze area.

Today, China accounts for nearly a fifth of the world's potato acreage and a tenth of its production, about 30 million metric tons annually, a figure well below that of the sweet potato (Wittwer *et al.*, 1987: 234, 237). That the potato did not come to occupy the dominant dietary role in China that it gained in such European countries as Germany and Ireland (Salaman, 1949), is explained by Laufer (1938: 69-70) in terms of the Chinese preference for rice, in turn based on sound economic reasons and an intuitive sense of its superiority both in food value and as a producer of calories. Cultivation of the white potato, Laufer observed, was usually on a small scale and carried out especially by indigent folk in the mountains and among aboriginal groups. In Laufer's view, there were simply too many alternatives. However foreigners might encourage the cultivation of potatoes,[25] most Chinese continued to look down on them as food for the poor, and seldom ate them. Even the poor might pass them by, as with the 73 lower-income families in a survey in Peking in 1926-27, only ten of whom bought white potatoes (Gamble, 1933: 107). This fits with Laufer's observation that the history of the white potato in China has been local, sporadic, and isolated rather than national.

One cannot, however, dismiss the white potato as a food in China,[26] for it is grown in all provinces, and is an excellent backup crop at times when there is a failure of the grain crop. In the north, it has come into favor as a vegetable capable of ready storage in winter (Wittwer *et al.*, 1987: 235, 237), and in some places, for example in parts of mountainous Yunnan and Szechwan, it is even a staple (Hosie, 1897: 48, 51, 59, 149, 175; E. H. Wilson, 1929: 340-41; Laufer, 1938: 75-78; see also the map of Bertin *et al.*, 1971: 12). Even in lowland areas of South China, where they are simply another vegetable, white potatoes do well as a winter crop, and in Kwangtung, Fukien, and Kwangsi farmers are able to get two potato crops and two rice crops a year (Herklots, 1972: 458-60; Wittwer *et al.*, 1987: 236).[27]

Potatoes are commonly eaten by the Chinese as a vegetable: fried, baked, roasted, boiled (with or without their skins), stewed, or otherwise. They may be diced and added to soups. They may be used in stews and curries such as the western Chinese "Curried Beef with Potatoes" or in these and other dishes of foreign origin (G. Z. Chu, 1975: 72; C. B. T. Lee and A. E. Lee, 1980: 153; Zee and Hui, 1981: 67). And they may be made into potato flour, a thickener (Y. K. Martin, 1984: 17, 229). Despite the low regard in which potatoes are held, moreover, there are some interesting dishes that show them to be compatible with traditional Chinese cuisine. There is, for example, the Cantonese "Red Bean Curd Cheese Pork with Potatoes," made with red bean curd cheese, bite-size pieces of boneless pork butt and potatoes which are simmered in water with sherry, ginger, soy sauce, and sugar (W. W. Chang *et al.*, 1970: 312). There is also "Potato Meat Balls, Shanghai Style" (E. Chen, 1977: 126-27) and North Chinese "Duck with Potatoes" in which the duck, its skin rubbed with soy sauce, is browned, basted with a sauce made of water, soy sauce, brown bean sauce, red bean curd, star anise, sherry, salt, sugar, and pepper, then simmered with cubed potatoes in what sauce is left (C. B. T. Lee and A. E. Lee, 1980: 99). In addition, potatoes may be made into dry slices or into noodles (Wittwer *et al.*, 1987: 237), and there are unusual Chinese potato sweets such as "Sugared Potato Chips" and "Potato Cream" (K. Lo, 1974: 170; 1979c: 326-27).

Manioc

Manioc (*Manihot esculenta*) is the third major root crop domesticated in the Americas (along with white potato and sweet potato). I. H. Burkill (1966, 2: 1437) writes that manioc may have come from Manila to the extreme south of China, and then spread to Java. Bertin *et al.* (1971: 10), in their map of manioc spread and cultivation, provide dates of its spread: to the Philippines in the 1600's and to Taiwan in the 1800's. Though manioc proved a great success in tropical Africa and in parts of the Pacific and South and Southeast Asia, in China it is of quite minor importance compared to the sweet potato, its highest annual production there, early in the 1980's, amounting to six million tons (Wittwer *et al.*, 1987: 229). Most of our sources fail even to mention manioc, though it is cultivated in the far south, particularly Hainan (E. N. Anderson, 1988: 122), in Pearl River Delta villages, and on Taiwan. E. N. Anderson (1988: 171) mentions it to illustrate the range of food items used by the Cantonese, but Herklots (1972: 449) writes of it as seldom seen in Hong Kong, though he says that it may be marketed in villages there. I. H. Burkill (1966, 2: 1437-38) has noted the role of Chinese in Malaya in cultivating manioc and processing tapioca, and others have observed that tapioca is a common thickening agent in China (Y. K. Martin, 1984: 229). Blasdale (1899: 25-27) even reports manioc roots imported from China for sale in San Francisco. Roots were sometimes boiled and

Figure 36. Yam bean (*Pachyrrhizus erosus*) (from Herklots, 1972: 451)

eaten as vegetables there, but they were used mainly for obtaining a starch which, because of a supposed resemblance to ginseng (*Panax ginseng*), was believed to have medicinal value.

American Yam Bean

The plant that concerns us here is the yam bean *Pachyrrhizus erosus* (= *P. angulatus*) (Figures 36 and 37), known in China as *ko* or *fan-ko*, and sometimes confused by Westerners with the Oriental yam bean *Pueraria*, which is also known as *ko*. *Pachyrrhizus erosus* is a twining, climbing legume of Mesoamerican origin, known in the United States by its Mexican name, jicama.[28] It is cultivated for its large, turnip-like tuberous root (each plant bears a single one). Carried to the Philippines by the Spanish, *P. erosus* presumably spread from there to

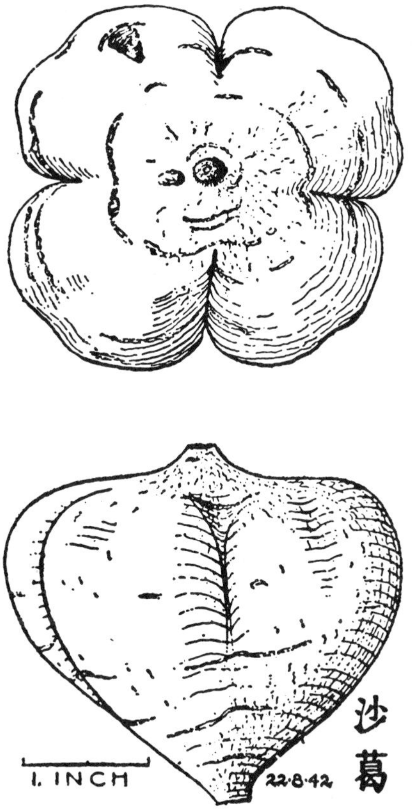

Figure 37. Yam bean tuber (from Herklots, 1972: 450)

mainland East Asia, including Indochina and China. Today it is widely cultivated in the latter country, where it may be trained on bamboo trellises, and it has also run wild there.

Usually it is the young roots which are consumed directly, whether raw or cooked. They may be served whole, or cut into pieces or grated. They may be added to other dishes. Or they may be eaten as a separate item, for example the crisp raw slices served with sharp chili sauce which are common snacks in South China. Tough old roots, on the other hand, are made into starch, and were imported from Canton by San Francisco Chinese for this purpose. Though we have not heard of the practice in China, certain other peoples are reported as eating the young pods of *P. erosus*, though they are said to cause diarrhea. Mature pods, seeds, and leaves, which are toxic and narcotic, are not consumed. Because of these qualities, seeds of *P. erosus* may be employed as fish intoxicants; and the stems, at least in Fiji, are used for making fish nets (J. Smith, 1882: 444; Blasdale, 1899: 23-25; Crevost and Lemarié, 1917: 133; E. H. Wilson, 1929: 341; Bailey, 1935, 3: 2425-26; 1949: 572-73; Porterfield, 1939b: 108; 1951: 12-13; I. H. Burkill, 1966, 2: 1647-48; Herklots, 1972: 449-52; Schery, 1972: 460, 506; Harlan, 1975: 76; P. M. Smith, 1976: 314; Zeven and de Wet, 1982: 174, 193; E. N. Anderson, 1988: 125-26). This calls to mind Carl Sauer's hypothesis that early man's interest in intoxicants for stupefying fish, and fiber for fish nets, led to unusual familiarity with wild plants, and a predisposition to domestication. Jicama may first have been used by primitive fishermen in the American tropics as net fiber and fish intoxicant, then domesticated for its edible root.

Nutritionally the yam bean root is not outstanding in any way. It is far lower in carbohydrate (10-13 gm./100 gm. raw, edible portion) and food energy (46-55 calories/100 gm.) than the yam, taro, kudzu, great burdock, manioc, sweet potato, and white potato (which, in food energy, provide from 82 to 135 calories/100 gm.). Nor is the yam bean root outstanding nutritionally in any other way (Platt, 1962: 8-9; Leung *et al.*, 1973: 11-15, 75).[29]

Notes on Other American Root Crops

Of minor importance in South China are the American plants arrowroot (*Maranta arundinacea*) and Jerusalem artichoke (*Helianthus tuberosus*) (Herklots, 1972: 433-35, 465). Arrowroot rhizomes may be eaten boiled or roasted, or starch may be extracted from them. In Hawaii, young rhizomes, after boiling, are eaten by Orientals. Mostly, however, Oriental people there cultivate arrowroot for its starch, which they use at home (H. L. Chung and J. C. Ripperton, 1929: 46). Arrowroot starch has little nutritional value other than as a source of carbohydrates.

SAGO: ANOTHER STARCHY FOOD

Though not a root or tuber, sago is included here because, like many roots and tubers considered above, it is starchy, and because there seems to be no other appropriate place for it. Sago, or palm starch (*hsi-ku*; Cantonese: *sai-kou*),[30] comes from the pith of the stems of various palms and cycads. Though produced in China's south today, it is far less important than in T'ang times, when, according to Schafer, sago, consumed with milk, was a basic food, and when, in Canton, sago flour contended with cereal flour in preparing cakes (though it did not surpass rice). The arenga palm (*Arenga pinnata* = *A. saccharifera*), though far more significant for the sugar it provided, was China's leading source of sago in T'ang times, though other species supplied sago as well (Schafer, 1967: 176; 1977: 92-93).[31]

The arenga palm's primary center of genetic diversity may be the Malay Archipelago, and its secondary center India, with its domestication usually placed in the Southeast Asia/Pacific region (Harlan, 1975: 75; Mathon, 1981: 64; Zeven and de Wet, 1982: 61, 78). It was described by Chi Han in the fourth century A.D. as growing in what now is northern Vietnam (H. L. Li, 1979: 90-92), and its cultivation in China today (Kwangsi, Kwangtung, and Hainan) (S. C. Lee, 1935: 155) is a northward extension of its Southeast Asian distribution. In addition to sugar and sago, our sources on China mention fermented toddy as an edible product the tree provides (Schafer, 1977: 92-93, 121). One suspects that other edibles obtained from the arenga palm in Southeast Asia and India may also be consumed in China: fresh (as well as fermented) toddy, arrack, and vinegar from its juice; a vegetable from its "cabbage"; and a preserve or snack from its seed (Watt, 1889-96, 1: 302-4; Crevost and Lemarié, 1917: 140-41, 347, 350-51; Wester, 1924: 103; Ochse and Bakhuizen van den Brink, 1931b, 3: 554-56; Massal and Barrau, 1956: 34; Uphof, 1959: 32; I. H. Burkill, 1966, 1: 231-38; Schafer, 1967: 175-76; 1977: 92-93, 109; H. L. Li, 1979: 90-92).

According to E. Watson (1923: 518), sago is ordinarily marketed in two processed forms: as grains or as translucent globules called "pearl sago." The grain, or inferior, form of sago, was imported to China in gunny sacks in the 1920's, whereas the latter, or superior, form was imported in boxes (E. Watson, 1923: 518). Chinese merchants have long been active in the sago trade of Southeast Asia, and interestingly, Burkill (1966, 2: 1486) credits nineteenth-century Chinese in Sumatra with developing the process for making pearl sago, a product which holds its shape when cooked.

We have found no nutritional analysis of "sago flour" or "sago meal" from the arenga palm, but those of the sago palm *Metroxylon sagus* are quite similar to manioc starch and corn starch in calories provided (about 350 per 100 gm./edible portion), and, like them, it is quite low in protein (0.5-1.4 gm./100 gm.), far lower even than highly-polished rice (6-7 gm./100 gm.). Nor are such sago products outstanding in any mineral or vitamin (B. S. Platt, 1962: 6-9; Leung *et al.*, 1973: 7-8, 12, 13, 14-15).

[1] This view is discounted by E. N. Anderson (1988: 15).

[2] Given sufficient water in early stages of growth, sweet potatoes are quite drought resistant (T. H. Shen, 1951: 210).

[3] Geophagy is a deliberate practice usually involving earths obtained locally, but sometimes from considerable distances and through well-established trade channels. The practice is found in many parts of the world, in the United States especially among southern Blacks, who seem to have brought the custom with them from Africa (Vermeer, 1986). The earths consumed by geophagists may consist of coarse, medium, or fine material, but fine clay is most commonly used. The practice of geophagy, though it occurs among children and adult males, is especially common among pregnant women, a fact that has aroused special interest in its possible benefits (*i.e.*, nutritional and pharmaceutical) and detrimental effects. Because geophagists around the world use a wide range of earths quite different in composition, however, it is difficult to reach valid general conclusions about the merits of the practice. That it may be useful as a pharmaceutical is suggested by the clay of Uzalla, Nigeria—the main West African clay traded for eating purposes; this clay is said by villagers to act against diarrhea, and indeed it has a mineral composition strikingly similar to that of the clay in Kaopectate, that widely-used antidiarrheal agent (Vermeer and Ferrell, 1985). There is also the observation (Vermeer, 1986) that, at least among some peoples, it may provide nutrients useful to the pregnant woman. At the same time, various studies reveal that geophagy may contribute to such conditions as iron and zinc deficiency, intestinal parasitism, and intestinal blockage, and create serious public health problems (Danford, 1986).

With respect to China, Laufer (1930: 111-26) found considerable evidence of geophagy in the literature, and identified three different ways in which people have consumed earths: as medicine; in Taoist magic; and as famine food. The latter use, he observed, was very rare in China, the final resort of starving people whose other possibilities of obtaining food had been exhausted.

[4] Bretschneider held that the *kan-shu* mentioned by Chi Han was the sweet potato. Most botanists, however, now believe that the plant in question was in fact a yam (I. H. Burkill, 1966, 2: 1266-67; H. L. Li, 1979: 43).

[5] Herklots (1972: 415-28) provides much useful information on the cultivation and use of yams in the Hong Kong area.

[6] Though sherry is known in China, rice wine rather than sherry is customarily used in cooking. English-language cookbooks, from which this recipe derives, often substitute sherry or dry sherry for rice wine. In our account, we follow recipes as given, but put sherry in quotes where rice wine would ordinarily be used in China.

[7] For other taro dishes, see F. L. Chang and G. M. Fuchs, n.d.: 206, 335 ("Pork with Taro" and "Fried Taro Dumplings"); and I. C. Chang, 1970: 206 ("Taro Pudding").

[8] See, for example, Tiger and Wolf, 1985: 31 for a dragon with a head and tail carved of carrot.

[9] Various authorities (Keng, 1974: 400-401; K. C. Chang, 1977b: 28; H. L. Li, 1983: 35) have identified the word *feng* in the *Shih-ching* as the turnip, but E. N. Anderson (1988: 26) questions that identification, suggesting instead that the term referred to Chinese cabbages.

[10] Some writers consider *S. sagittifolia* to be the same species as *S. latifolia*, the American arrowhead. Corms of the latter were collected for food by North American Indians and repeatedly mentioned by observers, among them Lewis and Clark in their expedition to the Pacific coast. Corms of *S. latifolia* are quite like the corms of Old World arrowhead; they can be substituted for them in recipes; and in Chinese markets in the United States, they, as well as corms of Old World arrowhead, are sold (Blasdale, 1899: 8-9; Porterfield, 1940b: 45; 1951: 16; Harrington, 1984: 199).

[11] The chief priest and other priests performing the ceremony ordinarily wore Taoist ceremonial garb and were known as Taoist priests. For the purposes of this ceremony, however, the chief priest took on the "role of a Buddhist monk or Buddha" (F. L. K. Hsu, 1983: 28).

[12] In some places in China, according to Cormack (1935: 191), candles are not used, but small reed lights, which are placed inside watermelon shells and set adrift on canals and rivers.

[13] They may even be added to cooked cereal, as in the Ch'ing court dish "Lotus Leaf and Rice Porridge" (Davis, 1966: 207).

[14] There are many recipes for such steamed dishes, among them "Pork and Lotus Leaves," a Ch'ing court dish (Davis, 1966: 107); "Lotus Leaf-Wrapped Duck," from Peking (K. Lo, 1979a: 32); "Steamed Rice in Lotus Leaves," from Shanghai or eastern cuisine (E. Chen, 1977: 148); and "Carp Wrapped in Lotus Leaf" and "Chicken and Glutinous Rice in Lotus Leaves" from Canton or the south (K. Lo, 1972: 132-33; E. Chen, 1977: 78-79).

[15] The so-called "lotus-leaf buns," a usual accompaniment to "Szechwan Fragrant and Crispy Duck" and most duck dishes other than "Peking Duck," do not contain lotus leaves, but are so named because they look like them (K. Lo, 1979c: 252; Y. K. Martin, 1984: 200-201).

[16] For recipes to "Lotus Root Soup" and "Sliced Lotus Root Braised with Pork," see J. D. Keys, 1963: 92; and K. Lo, 1979a: 27-28. For a Cantonese recipe for "Stir-Fried Beef with Lotus Roots," see W. W. Chang *et al.*, 1970: 304-5. In addition to dishes such as these, there are lotus root chips, made much like potato chips and used widely in China; lotus root cakes from Canton (W. W. Chang *et al.*, 1970: 73); lotus root custard from Shantung (K. Lo, 1972: 168-69); and lotus root sweet-and-sour salad, from Hangchow, Soochow, and Shanghai (G. Z. Chu, 1975: 109-10; E. Chen, 1977: 141-42).

[17] These may be soups as we know them, such as "Lotus Seeds in Chicken Broth" (J. D. Keys, 1963: 93), or sweet soup, a dessert usually served as a welcome change between courses at a dinner. For a Shanghai recipe for such lotus seed soup, see K. Lo, 1979a: 181. Both dried and candied lotus seed are used in this soup, along with glazed fruit, sugar, cream, milk, and corn flour. For simpler recipes, see G. B. Miller, 1966: 753-54; G. Z. Chu, 1975: 230; and K. Lo, 1979c: 325.

[18] According to H. L. Li, *Pueraria phaseoloides*, a relative of kudzu, may be the forage crop for horses, "the horse-fattening herb" mentioned in the fourth century flora of Kwangtung and Southeast Asia. In China this species is much more limited in distribution than kudzu, being restricted to coastal areas of the subtropical southeast (H. L. Li, 1979: 62). It continues to be cultivated or collected from the wild in modern Southeast Asia, where it is used mainly for fiber and medicine. It has also been planted experimentally as a green manure on plantations. And to a small degree in some places, it is used for its edible roots (I. H. Burkill, 1966, 2: 1869-70).

[19] In New Guinea kudzu is found quite widely in a wild state, but at the time J. B. Watson wrote in the late 1960's (1968: 268-69, 277-79) there were no reports that it was cultivated there except in the Central Highlands. Judging from statements by native informants in the highlands, Watson concluded that kudzu was an ancient food plant of some importance. He raised the possibility that kudzu had been introduced to New Guinea even before taro or yams, and that kudzu cultivation had ceased in many parts of New Guinea because of competition with taro and yams "or some other combination of food sources." In the Central Highlands, however, kudzu survived such competition, and has continued to play a subsistence role up to the present time. Kudzu's role is a minor one today, but Watson observed that at least in some places the plant is more than just a "famine or forage" food.

[20] Shurtleff and Aoyagi (1977) have written the most complete account we have found of the dietary uses of kudzu, with a focus on those of Japan. They give excellent information on how to make kudzu powder, and on its use as a thickener for sauces, soups, congees, noodles, jelled salads, desserts, confections, beverages, and other dishes. They provide information, as well, on the preparation and use of kudzu noodles, and food uses of its leaves, shoots, flowers, and roots. In addition, they describe in detail the use of kudzu as medicine and of its fiber in making cloth, as well as presenting a useful background account on kudzu in a historical and ecological perspective.

[21] For a recent map showing where sweet potatoes are cultivated in China, see Wittwer *et al.*, 1987: 228.

[22] For Ch'ing China, there is a description (Spence, 1977: 265) of yet another way of processing sweet potatoes, whether for storage or direct consumption. Sweet potatoes were skinned and steamed, then passed through a strainer to remove their fiber. And finally they were rolled into strips or dried into cakes.

[23] V. Y. N. Hsu and F. L. K. Hsu (1977: 302) write of such hawkers selling sweet potatoes by the ounce or catty. At one end of a carrying pole, a hawker would have a basket of uncooked sweet potatoes, and at the other end, a portable oven in which layers of sweet potatoes were suspended over a charcoal fire.

[24] For additional information on sweet potatoes, see Adolph and Liu, 1939 (value as a famine food); T. H. Shen, 1951: 210-12 (cultivation and breeding); and M. C. Yang, 1945: 35 (processing).

[25] At Macao, potatoes had been cultivated in paddy fields as a winter crop since about 1800, but the harvest went mainly to its Portuguese residents (Laufer, 1938: 69).

[26] Other species of *Solanum*, native to the Old World, also provide food and/or medicine in China and Southeast Asia (I. H. Burkill, 1966, 2: 2078-86). For food, the most important by far is *S. melongena*, eggplant, to be considered later. As medicine, *S. dulcamara* and *S. nigrum* are of particular note for China. The former is a small, prickly shrub which grows wild in the south. Its berries and root are added to mixtures used to treat swelling and sores, and its roots serve also in treating malaria and fever (F. P. Smith, 1969: 411-12). Dried stems of this plant are imported to Malaya by Chinese for medicinal use (I. H. Burkill, 1966, 2: 2078). *S. nigrum* serves some of the same medicinal uses as *S. dulcamara*. In addition, the cooked shoots or seeds are boiled and consumed as a tonic and for other purposes (F. P. Smith, 1969: 413).

[27] For more information on white potatoes and their cultivation in contemporary China, see Wittwer *et al.*, 1987: 233-40.

[28] There is a certain amount of confusion in the literature between *Pachyrrhizus erosus* and one of its domesticated relatives of South American origin, *P. tuberosus*, which is also called jicama.

[29] For medicinal uses of *P. erosus* in China and elsewhere in Asia, see I. H. Burkill, 1966, 2: 1648; and Perry, 1980: 221.

[30] In the fourth century A.D., Chi Han called the arenga palm *kuang-lang*, a name still used for sago in present-day China (E. Watson, 1923: 517-18; H. L. Li, 1979: 90-91). Writing of Sumatra in the thirteenth century, however, Chau Ju-kua used the word *sha-hu* for the sago, which, as Hirth and Rockhill observe (1966: 61, 65), is *sagu* in the Malay dialects of the western Malay Archipelago for both the sago palm (*Metroxylon sagus*) and its meal. E. N. Anderson (1988: 122) writes that the name for sago in China's south is borrowed from that Malay word. Also of possible relevance, however, are some present-day Indochinese names for the sago palm: *sà-cù* in Cambodia, and *sa-khu* in Laos (Crevost and Lemarié, 1917: 140).

[31] The meal of the arenga palm is considered inferior to that of the sago palm (Schafer, 1967: 176; 1977: 93; H. L. Li, 1979: 91-92).

CHAPTER FIVE

Other Cultivated Vegetables

> "My new Province is a land of bamboo-groves:
> Their shoots in spring fill the valleys and hills...
> For two farthings, I buy a whole bundle.
> I put the shoots in a great earthen pot
> And heat them up along with boiling rice...
> All the time I was living at Lo-yang
> They could not give me enough to suit my taste.
> Now I can have as many shoots as I please;
> For each breath of the south-wind makes a new bamboo!"
>
> *Po Chü-i (translated by Arthur Waley)*

Cabbages and Related Plants
Mallow: An Early Competitor
Other Green Leafy Vegetables
- Spinach
- Water Spinach
- Amaranth
- Malabar Nightshade
- Spinach Beet and Swiss Chard
- Watercress
- Lettuce
- Water Shield
- Alfalfa
- Sow Thistle
- Some Additional Greens

Squashes, Gourds, Cucumbers, and Other Cucurbits
Onions and Their Relatives
Fungi
Lily Bulbs
Bamboo Shoots
Wild Rice
Celery
Water Drop-Wort
Tomato
Eggplant
Okra
Asparagus

What we designate here "other cultivated vegetables" are considerable in number and variety, and here we focus on the more important or interesting of them.[1] Most fall into five categories: the cabbages and related plants; other leafy vegetables; squashes, gourds, cucumbers, and other cucurbits; onions, leeks, and chives; and fungi. The largest number of vegetables in these categories originated in China or elsewhere in Eurasia, the great exceptions being squashes and gourds, which are of New World origin. In addition to vegetables fitting into the five categories, there are many others, including several of dietary significance. Some of these are Oriental in origin and, at least until recent times, were unfamiliar in Western cuisine: bamboo shoots, lily bulbs, wild rice, and water drop-wort. Others, long familiar in the West, are celery, asparagus, eggplant, okra, and tomato, and of these only the latter two come from outside Eurasia: okra from Africa, the tomato from the New World.

In terms of nutrition, many of the vegetables in question are of critical dietary importance, and in terms of the of the hot-cold system, most are cold, and some, especially watercress, are very cold. Such very-cold vegetables should not be given to young babies or eaten by mothers after delivery or when nursing. They should be eaten only in small amounts by old people. They are, however, regarded as very effective in treating hot illnesses (Koo, 1973: 141; Yeung *et al.*, 1973: 198; B. Chang, 1974: 437, 438; E. N. Anderson and M. L. Anderson, 1976: 148; E. N. Anderson and M. L. Anderson, 1977: 368; M. Mathews and L. Manderson, 1981: 12-13; Tan and Wheeler, 1983: 41, 44; Wheeler and Tan, 1983: 52, 54-55; E. N. Anderson, 1988: 128, 191).

CABBAGES AND RELATED PLANTS

To appreciate the range of cabbages found in China, one need only skim the account of them by Herklots (1972: 182-224). That author first distinguishes the Asiatic forms from the common European forms of *Brassica oleracea* (which include European cabbage, kale, brussels sprouts, broccoli, savoy cabbage, cauliflower, and kohlrabi). Wending his way through the considerable differences of opinion among botanists, he then divides the Asiatic forms into five species, four of which he identifies as Chinese:

(1) *Brassica juncea*, mustard cabbage (*chieh-ts'ai* or *chiai-ts'ai*, "mustard greens") (Figure 38),

(2) *Brassica alboglabra*, "Chinese kale" (*chieh-lan-ts'ai* or *chiai-lan-ts'ai*) (Figure 39),

(3) *Brassica pekinensis*, celery cabbage (*pai-ts'ai*, "white leafy vegetable," or *Tientsin pai-ts'ai*), common in North China (Figure 40), and

(4) *Brassica chinensis* (also called *pai-ts'ai*), common in South China (known, after its Cantonese name, as bok choy in the United States) (Figure 41).

When one adds the several varieties into which each species is divided and the multitude of local names for them, the complexity is bewildering.

As for the antiquity of *Brassica* cultivation in China, the evidence most often cited in recent decades (K. C. Chang, 1970: 182; 1977b: 29; P. T. Ho, 1975: 87n, 108; H. L. Li, 1983: 35; E. N. Anderson, 1988: 12) is that from the Yang-shao site of Pan-p'o, not far from Hsi-an (Sian) in Shensi Province, where *Brassica* seeds were found in pottery jars, apparently stored for planting. This would seem to date cultivation of some form of *Brassica* to the fifth millennium B.C.[2] Recently, moreover, K. C. Chang (1986: 92) noted a report that remains of "oil cabbage" were uncovered in Hopeh at Tz'u-shan (Cishan), a site dating to the sixth millennium B.C. or

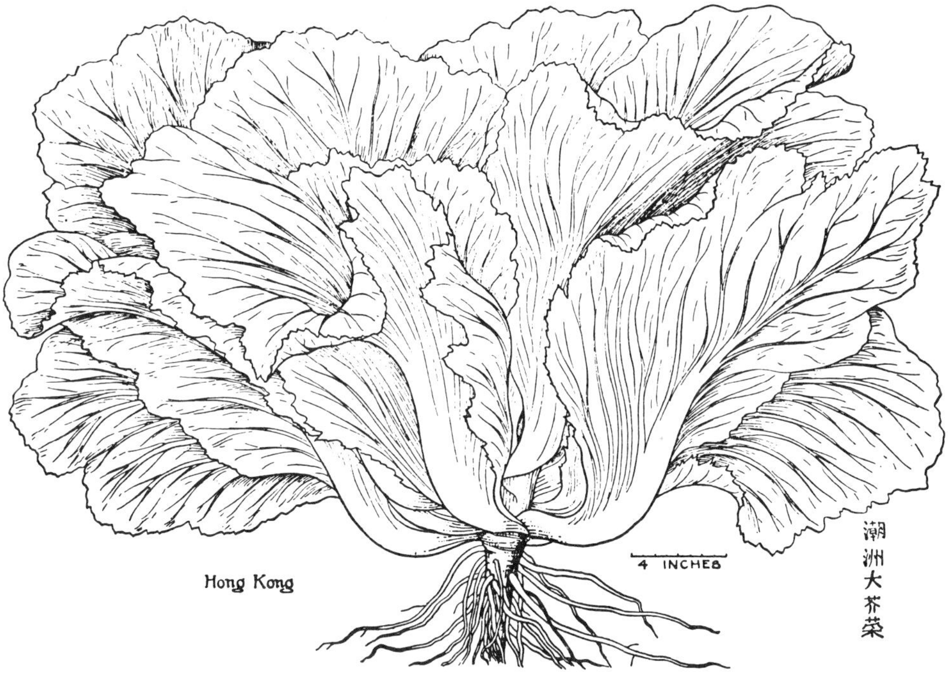

Figure 38. Mustard cabbage (Swatow mustard) (*Brassica juncea* var. *rugosa*) (from Herklots, 1972: 215)

even a bit earlier. Whatever the actual date of their initial cultivation, Brassicas seem to have been among the earliest cultivated food plants of China.

As for *B. oleracea*, its domestication and development of its common forms occurred in Europe or the Near East (I. H. Burkill, 1966, 1: 366; Schery, 1972: *passim*; Harlan, 1975: 70; K. F. Thompson, 1976: 49-51; Snogerup, 1980: 121-22, 131-32), and, though it has been claimed that one or another of those forms reached China in ancient times, we have uncovered little firm evidence bearing on the matter.[3,4] Turning to modern China, we note that Herklots (1972: 182-89) and Zee and Hui (1981: 15-18) give Cantonese names for European cabbage, and observe that broccoli is gaining popularity in Hong Kong because it fits well in Chinese dishes. The other European forms of *B. oleracea*, Zee and Hui observe, are cultivated in that city on a quite small scale. Anderson and Anderson (E. N. Anderson and M. L. Anderson, 1977: 327; E. N. Anderson, 1988: 172, 209), in a similar vein, write of European cabbage, broccoli, cauliflower, and kohlrabi as minor, recently-introduced Western crops to China's south or southeast, and note that at least some of them are more commonplace in Kwangtung than elsewhere in the country.

The importance of cabbage in traditional North China is well illustrated in a household survey of 73 lower income families in Peking in 1926-27. All families were found to use cabbage, commonly celery cabbage, and cabbage was by far their leading vegetable, accounting for a fourth of total expenditures for vegetables (Gamble, 1933: 104). In South China, the cabbages and their relatives are harvested in large amounts and can be obtained throughout the

Figure 39. Chinese kale (*Brassica alboglabra*) (from Herklots, 1972: 194)

year. Their dietary importance is well illustrated by the observation (E. N. Anderson and M. L. Anderson, 1977: 328) that a very large bowl of rice, a good quantity of bean curd, and a plate of cabbage are the standard everyday fare of South China. Celery cabbage grows in the cold season; Chinese kale, in the hot season; and one type or another of Chinese white cabbage, throughout the year.

The leaves and stems of Brassicas may be consumed fresh, as in stir-frying or in soups, or they may be pickled.[5] In addition the seed of certain varieties may be used as a spice in sauces, dips, or as a source of oil.

The greatest nutritional importance of cabbages in China is as providers of vitamins A and C, and calcium and iron (H. C. Hou, 1937; T. J. Chu and B. E. Read, 1938; Kung *et al.*, 1938; E. F. Yang and M. Y. Dju, 1939; Fan and Hou, 1953; Caldwell, 1972: 315). *Brassica chinensis*, principal cabbage of South China, and *B. juncea*, mustard cabbage, are clearly superior both to *B. pekinensis*, mainly of the north, and European cabbage (*B. oleracea*, var. *capitata*) in all four nutrients (Leung *et al.*, 1973: 37, 38). Cabbages, like other greens, can also be significant sources of riboflavin in countries such as China, where riboflavin intake is far below international standards (Caldwell and Enoch, 1972: 309), and again *B. chinensis* and *B. juncea* are the clear winners over the other two Brassicas (.13 and .14 mg./100 gm., raw leaves, edible portion, vs. .06 and .07 mg./100 gm. for *B. pekinensis* and *B. oleracea*, var. *capitata* respectively).

Figure 40. Celery cabbage (*B. pekinensis* var. *cylindrica*) (from Herklots, 1972: 197)

MALLOW: AN EARLY COMPETITOR

Though today the cabbages are China's most important leafy vegetables, in ancient North China, according to H. L. Li, mallow (*Malva verticillata*; *k'uei* or *tung-k'uei*) was the principal one. Li says that it retained its position at the time of the *Ch'i-min yao-shu* (sixth century A.D.), but that after T'ang times it lost favor and was gradually replaced by *Brassica chinensis* (H. L. Li, 1959: 141-42; 1969: 256; 1983: 35-36; Schafer, 1977: 91). At the close of the sixteenth century, Li Shi-chen wrote of mallow as seldom cultivated and little eaten (Bretschneider, 1882-95, 3: 193). And at the present time, its cultivation as a vegetable is uncommon, though Chinese in many parts of the country still know mallow in the wild, and on occasion grow it as an ornamental. One should not, however, overlook the fact that in certain places, such as Szechwan and Yunnan, local cultivation of mallow for food may be on a considerable scale. According to S. Y. Hu (1956: 23-24), hundreds of acres of mallow were grown as a winter crop near his home in Ch'eng-tu, Szechwan, and it was sold in markets, to be eaten as a vegetable or in soups.[6]

Though not the equal of *B. chinensis* in niacin, ascorbic acid, or calcium, a related Chinese mallow species, *M. sylvestris*, has twice the protein, thiamine, and riboflavin, and is far superior in vitamin A and iron (Leung *et al.*, 1973: 37, 54). If *M. verticillata* has a food value similar to

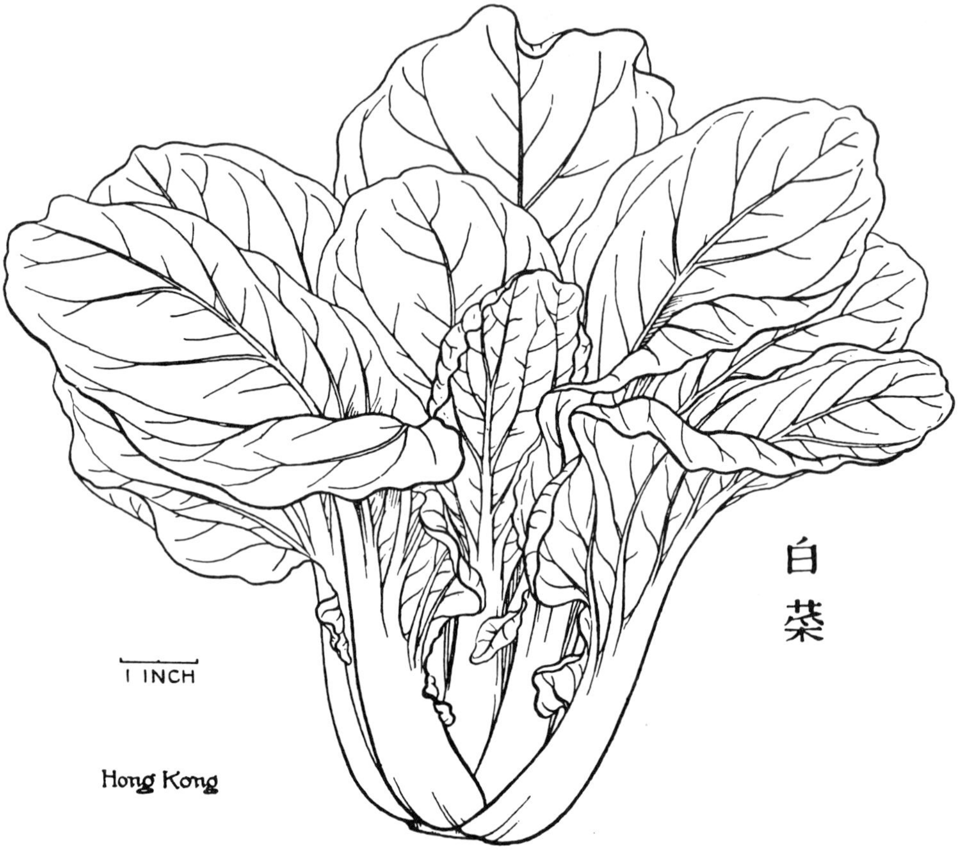

Figure 41. Chinese white cabbage (*Brassica chinensis* var. *chinensis*) (from Herklots, 1972: 202)

that of *M. sylvestris*, the decline of its cultivation for food[7] in China must have come about for reasons other than nutritional worth. H. L. Li (1969: 256) explains that decline in terms of the development of technology for extracting oil from vegetable seeds. Before that development, he contends, mallow was favored in cooking because it was sticky, moist, and readily prepared. With vegetable oil available, those qualities were no longer essential, and mallow gradually lost favor because it was a perennial, less readily cultivated than leafy vegetables that were annual or biennial. E. N. Anderson (1988: 27), notes, however, that cabbage today is ordinarily steamed or boiled, and that oil is not needed to cook it, implying that the explanation for mallow's decline must be sought elsewhere.

OTHER GREEN LEAFY VEGETABLES

This category includes some plants (water spinach, amaranth, and Malabar nightshade) that were domesticated in China, Southeast Asia, or India, and others (spinach, Swiss chard, spinach beet) of temperate Asian or European domestication. The tropical Asian plants do well under subtropical conditions and are good providers of summer greens in South China, whereas the temperate Asian and European plants provide a welcome supply of greens from late fall to spring.

Spinach

Spinach (*Spinacia oleracea*) is a native of southwestern Asia (P. M. Smith, 1976: 304). De Candolle (1886: 98-100) believed it to have been domesticated in Persia, whereas Vavilov (1949-50: 32) placed its domestication to the north, in Central Asia. Laufer (1919: 398) believed spinach to have been domesticated no earlier than about the sixth century A.D. In any case, it spread from the Persian realm rather late. Spinach was not known to the ancient Greeks or Romans. The Arabs seem to have introduced it to Spain by the eleventh century A.D., and it spread from there to other parts of Europe (de Candolle, 1886: 98-100; Laufer, 1919: 395-96). Nor was it an ancient plant in India, Southeast Asia, or China. Laufer (1919: 392-98), after reviewing the historical documents, concluded that spinach was first introduced from Iran to Nepal, and that then, in the seventh century A.D., it spread from Nepal to China. By his interpretation, spinach was then a novelty in Nepal, and was sent to the emperor of China in response to a request that tributary nations send their finest vegetable commodities. Another view (F. P. Smith, 1969: 417) is that spinach seed was introduced to China from India by a Buddhist monk in T'ang times, thus explaining its use as fasting food by Buddhist clergy in China. Whatever the route by which they obtained it, the Chinese have a variety of spinach with narrower, more pointed, tender, and tasty leaves than our type of spinach, and they prefer it to ours (Herklots, 1972: 160; E. N. Anderson, 1988: 127).

Spinach is a plant of temperate rather than tropical areas. In tropical Africa, for example, spinach can only be cultivated at high altitudes (Vickery and Vickery, 1979: 48). And in Malaya, where spinach was introduced by Europeans, there were trial plantings in both lowland and hill areas, but the plants quickly developed disease (I. H. Burkill, 1966, 2: 2103). In the subtropics of coastal South China spinach can be cultivated only in winter, and is ordinarily available in Hong Kong from mid-October to mid-May (Herklots, 1972: 161; Zee and Hui, 1981: 31).

In China, spinach is most commonly prepared in soups (E. N. Anderson, 1988: 127),[8] though it may be stir-fried with such flavorings as salt, soy sauce, ginger, garlic, and sugar.[9] It may be combined with noodles, bean curd products, or other vegetables (most commonly mushrooms) in vegetarian dishes, for example in "Stir-fried Spinach in Bean Curd 'Cheese' Sauce" (Y. K. Martin, 1984: 220). It may be used as a wrapping, as in vegetarian "Spinach Balls" (K. Lo, 1979c: 118). It is also common in shrimp or meat dishes such as Shanghai "Stir-fried Spinach with Dried Shrimp" and "Fresh Ham with Spinach" (W. W. Chang *et al.*, 1970: 338, 399). And it may be served cold in salads (after the spinach has been boiled, however briefly) such as Peking "Spinach Salad with Mustard Dressing" (Joyce Chen, 1962: 168),[10] or be used as a colorful base on which to serve a dish.

Spinach is rich in vitamins A and C, and a fair source of riboflavin. It also contains abundant iron and calcium, though most is in forms that are not readily absorbed (Kung *et al.*, 1938: 311, 312, 314; E. F. Yang and M. Y. Dju, 1939: 484, 486; Snapper, 1965: 10-11; Briggs and Calloway, 1979: 277).

Spinach, both leafy parts and roots, is believed to be antivinous, carminative (effective in expelling gas from the alimentary canal and relieving flatulence and intestinal pain), thirst-quenching, and "cooling" (F. P. Smith, 1969: 417). Because of the latter quality, the Sung Chinese are said to have viewed spinach as especially good for northerners but not for southerners, for the latter were already "cold" from eating "water food" such as fish and turtle flesh (Laufer, 1919: 393).

Water Spinach

Another popular green vegetable widely used in Central and South China is "water spinach," sometimes called "swamp cabbage" or "Chinese cabbage" (*Ipomoea aquatica* = *Ipomoea reptans*; *yung-ts'ai*; *wêng-ts'ai*; or *k'ung-hsin-ts'ai*, "hollow-hearted leafy-vegetable"). In appearance, the plant resembles its relative the sweet potato, but does not bear tubers. In the Old World it occurs in a wild state in tropical areas from western Africa through Asia to Australia. In a domesticated one, it is found in various parts of South and Southeast Asia (Watt, 1889-96, 4: 477-78; Ochse and Bakhuizen van den Brink, 1931b, 1: 159-61; Dalziel, 1937: 439; I. H. Burkill, 1966, 2: 1270; Edie and Ho, 1969: 32), but its greatest area of cultivation is China's southeast coast.

It is uncertain where *Ipomoea aquatica* was first cultivated, though India and China have both been suggested as possibilities. The first record of its cultivation, however, occurs in Chi Han's fourth century flora of Kwangtung and nearby areas of South China and Indochina (Edie and Ho, 1969: 32, H. L Li, 1979: 71). Of special interest in Chi Han's account are the floating reed rafts on which water spinach was cultivated. Such floating gardens, which the Chinese called "frame fields" or "zizania fields" (after Manchurian wild rice, *Zizania latifolia*, which was also cultivated on them), were described in a Yüan account as occurring "more or less everywhere" in southeastern China, and in a Ch'ing account as found in the Yangtze and Huai area and on Lake Tien in Yunnan.[11] The Chinese floating gardens seem to have been overlooked by many Western scholars, though certain "floating gardens" in other parts of the world (*i.e.*, the *chinampas* of Aztec Mexico[12] and the *rádh* of modern Kashmir) have attracted considerable attention (H. L. Li, 1979: 15-17; Bray, 1984: 119-21).

In modern South China, water spinach is of two types. One is a dry-land type that can better withstand cold and is cultivated in fields in spring and early summer in a manner similar to other leafy vegetables. The preferred, more common form, however, is a wet-land type usually grown in fields, often former paddies, that are bordered by raised banks to retain the water let in as the plants mature (Edie and Ho, 1969: 34-36). Water spinach occupies an important place in South Chinese diet because many other leafy crops, such as Chinese cabbage and lettuce, do poorly in the heat of summer.[13] In addition, wet-land water spinach can be grown with little effort. It can be harvested at various times according to market demand. And in coastal areas its aquatic habit and manner of growth help protect it from damage by storms, including hurricanes (Edie and Ho, 1969: 34).

It is the tender parts of the water spinach, including the stems, leaves, and petioles, that are eaten by humans, the coarser parts being fed to pigs or other domesticated animals. Water spinach, which is mild in taste, is prepared in various ways, like spinach. It may be washed, chopped into small pieces, and fried in peanut oil with garlic, then served directly with salt or soy sauce (H. L. Chung and J. C. Ripperton, 1929: 25). That preparation may be modified, for example by substituting shrimp sauce for soy sauce or salt and, as a final step, boiling the spinach and flavoring in water (Herklots, 1972: 159). In Shanghai, according to Read and Wagner (1940: 264), the Chinese prepare it with pork.

Water spinach is a good source of vitamins A and C., and a fair one of riboflavin. It contains significant amounts of iron, but available iron is much lower than that of spinach and many other leafy vegetables. It also contains abundant calcium, but we have found no studies of how much can be absorbed (E. F. Yang and M. Y. Dju, 1939: 484; Caldwell, 1972: 315; Caldwell and Enoch, 1972: 311; Leung *et al.*, 1973: 72). Water spinach has a laxative effect, and has been used in cases of poisoning, difficult labor, and for other medicinal purposes in China or elsewhere (Watt, 1889-96, 4: 476; I. H. Burkill, 1966, 2: 1271; F. P. Smith, 1969: 219-20; Perry, 1980: 106). On the other hand water spinach is reputed to transmit intestinal parasites.

Amaranth

Wild amaranths (plants of the genus *Amaranthus* or *Amarantus*) are found widely in tropical and subtropical regions of the world, and since antiquity their green leaves, and in some cases their seeds, have been collected and eaten by humans. Many amaranth species, moreover, have displayed an unusual capacity to fit into the disturbed habitats created by humans (J. D. Sauer, 1957: 11; 1967: 103-4). As a result they have become successful pioneers, fellow-travelers which have sprung up around settlements and cultivated plots. Various of them came to be domesticated by humans in one part of the world or another, for use as potherbs (greens), as ornamentals or dyestuffs, or for their grain.[14]

In the Old World, the earliest food use of domesticated amaranths was as greens, a use mentioned in sources on China from ancient times onward (Bretschneider, 1882-95, 2: 66-67; 3: 404-7). Known in English as "Chinese spinach," the most-frequently mentioned Chinese amaranth species (often called *hsien* or *hsien-ts'ai*) are identified as *A. tricolor*, *A. mangostanus*, or *A. gangeticus* (Figure 42). In her provisional key to edible amaranths, Feine-Dudley (1981: 35-37) lists the latter two under *Amaranthus tricolor* L., which would indicate that they are synonymous with or are forms of *A. tricolor*. Other botanists may separate the three, in one way or another, into two or more species or varieties, but however one classifies them, domestication of the amaranths in question is usually placed in India or Southeast Asia (de Candolle, 1886: 101; Vavilov, 1949-50: 27; H. L. Li, 1970: 13, 16; 1983: 48; Zeven and de Wet, 1982: 49). *A. mangostanus* is identified as a minor cultivated plant in North China by the fifth and sixth centuries A.D. (H. L. Li, 1969: 253), and today it is listed as a cultivated plant, a hot-weather leafy vegetable, in China (and in countries stretching from India to Japan) (Uphof, 1959: 20; Choudhury, 1976: 195-96).

There are various types of Chinese spinach which differ in the color and form of their leaves (Read and Wagner, 1940: 262-63; S. Y. Hu, 1948: 30-31; Herklots, 1972: 150; Zee and Hui, 1981: 3). They may be stir-fried as a vegetable or used in soup, as in the Cantonese "Spinach Soup" (Sakamoto, 1977: 31-32, 168-69; Zee and Hui, 1981: 3).

Amaranth greens have a nutritional value similar to that of other green leafy vegetables. Since, when fresh, they commonly contain much dry-matter, amaranth may provide two to three times the nutrients of an equal amount of other greens (National Research Council, 1984: 43). Most are good sources of riboflavin, and good or excellent sources of vitamins A and C, folic acid, calcium, and iron in an available form (Read *et al.*, 1940: 16; Caldwell, 1972: 315; Caldwell and Enoch, 1972: 310; Leung *et al.*, 1973: 30-31; Devadas and Saroja, 1980). Though amaranth greens have oxalate and nitrate levels comparable to those of other green leafy

Figure 42. Chinese spinach (*Amaranthus gangeticus*) (from Herklots, 1972: 151)

vegetables, these do not reduce their excellent nutritional value to any significant degree (Der Marderosian *et al.*, 1980: 31, 39-41).

In recent years, there has been growing interest in the amaranths as neglected plants of great potential in feeding the world's people. Of special note in this regard are the principal grain amaranths, *A. hypochondriacus, A. cruentus, and A. caudatus*, which bear large inflorescences and many tiny seed. In pre-conquest times these were essential food plants of the Aztecs and various other American Indian peoples (National Research Council, 1984: 1-5). Amaranth seed can be cooked as porridge. It can be parched or popped and eaten as a snack. The popped seed can be consumed as cold cereal, or can be mixed with honey to make a confection. The seed can be ground into flour for making flat breads such as tortillas. Though deficient in gluten, amaranth flour can also be mixed with wheat flour in making raised breads and similar baked goods. And there are other uses as well.[15]

Amaranth seed has more protein (about 16 percent) than common varieties of rice (7 to 10 percent), maize (9 to 10 percent), and other common cereals (National Research Council, 1984: 5, 7). Whereas the proteins of cereals in general are viewed as unbalanced because they are low in the amino acid lysine, amaranth protein averages more than twice the lysine of rice and maize protein, and as much as that of milk (National Research Council, 1984: 5, 7). Though average U.S. yields of grain are much lower for amaranth (900-1900 kg./ha.) than for rice (5040 kg./ha.), maize (6350 kg./ha.), and wheat (2130 kg/ha.), it is believed that they can be improved substantially through breeding and improved cultivation (Cole, 1979: 289). Indeed, in California and India yields of 3000 kg./ha. or more have already been reported (National Research Council, 1984: 12).[16]

The grain amaranths were all domesticated in the New World and seem to have come into cultivation in Eurasia and Africa only in post-Columbian times, especially in the eighteenth and early nineteenth centuries (Pal and Khishoo, 1974: 132).[17,18] Most important of the introduced species is *A. hypochondriacus*, which by the early nineteenth century had become a staple food crop in certain hill and mountain regions of the Indian subcontinent. Since then, it has spread widely not only in that region, but across Tibet to western China, and on to Manchuria and eastern Siberia (I. H. Burkill, 1966, 1: 128; J. D. Sauer, 1967: 115-17; 1976: 6). In South and Central China, grain amaranths are reported as cultivated only in the interior: Yunnan, Szechwan, Kweichow, Hopeh, and Manchuria (J. D. Sauer, 1967: 117, 137; 1977: 20, 23-24). The grain amaranth *A. caudatus* is reported always to be grown as a subsidiary crop in maize and cotton fields, whether along their margins or in places within a field where insects have destroyed plants. Its young shoots, which are not as tender as those of vegetable amaranth species, may be eaten by people as a green. Usually, however, plants are permitted to mature, after which they are dried and threshed, with leaves used as pig feed and seed as human food. In western China, the seed, known as "millet from heaven," is popped and, with maltose or dipped in syrup, is made into a tasty candy prepared in large amounts at the time of Chinese New Year (S. Y. Hu, 1948: 32-34; Needham and Lu, 1985: 62n).

Malabar Nightshade

This perennial creeping and twining herb (*Basella rubra*; *lo-k'uei*) (Figure 43), also called Ceylon or Malabar spinach, is another source of leafy greens. In a wild form, *B. rubra* is found in Africa and Asia and may be native to both continents (I. H. Burkill, 1966, 1: 310). In a domesticated form, it is widely cultivated in tropical regions of the New and Old Worlds, among them Africa, India, Southeast Asia, the South Pacific islands, and China (Watt, 1889-96, 1: 403-4; Ochse and Bakhuizen van den Brink, 1931b, 1: 75-76; Dalziel, 1937: 38; Massal and Barrau, 1956: 35; Winters, 1963: 195-96; I. H. Burkill, 1966, 1: 310; Herklots, 1972: 152; Schery, 1972: 512-13). The domesticated form has two varieties, one with green leaves, var. *alba* (sometimes, *B. alba*), and the other with reddish ones. The plant spreads and climbs vigorously, its stems on occasion growing to ten or fifteen feet in length. It may be cultivated in beds and allowed to run over the ground, or it may be trained or allowed to grow on bamboo fences or trellises; in Bengal, it is said (Watt, 1889-96, 1: 404), nearly every village has a hedgerow covered with *B. rubra*.

According to most authorities, Malabar nightshade was probably first domesticated in India. It was known in China, however, by the early fourth century A.D., and was cultivated there at least by the fifth or sixth centuries A.D. (Laufer, 1919: 324-25; H. L. Li, 1979: 71). Its early cultivation in China was both for its leaves, used as food, and its berries, whose red juice served as a rouge or purple dye (Laufer, 1919: 324-25). These are still major uses of the plant in India, Southeast Asia, and China.[19] The leaves and shoots of Malabar nightshade are eaten like spinach, often in soup or stews.[20] They seem to be a very good source of vitamins A and C, and a fair source of riboflavin and calcium (Winters, 1963: 198; Caldwell, 1972: 315; Caldwell and Enoch, 1972: 310; Leung *et al.*, 1973: 72), though we have not determined how much calcium is readily absorbed. The leaves also have a mild laxative property and serve

Figure 43. Ceylon or Malabar spinach (*Basella rubra*) (from Ochse and Bakhuizen van den Brink, 1931b, 1: 76)

various medicinal uses (Watt, 1889-96, 1: 404; I. H. Burkill, 1966, 1: 310; F. P. Smith, 1969: 66; Herklots, 1972: 154).

Spinach Beet and Swiss Chard

Beta vulgaris is a species that includes several plants (red beet, sugar beet, Swiss chard, and others) cultivated for their leaves and/or roots. Varieties of *Beta vulgaris* were domesticated in a broad region centered on the Mediterranean, but extending northward in Western Europe and eastward in western Asia (de Candolle, 1886: 58-59; Vavilov, 1949-50: 34, 36; Harlan, 1975: 69). De Candolle observes that there is no common name for the plant among Indo-Europeans and that they probably did not cultivate it when they began their migrations across Europe. As for India, authorities differ as to whether the Sanskrit term *palankya* referred to a type of *Beta* or not, but I. H. Burkill (1966, 1: 323) thinks *Beta vulgaris* was found in northern India from "remote" times.

Whenever it may have reached India, *Beta vulgaris*, a temperate land crop, may not have been cultivated on a regular basis in Southeast Asia, where it does best at high altitudes, until the time of European contact and settlement. In modern Indonesia and Malaya, for example, the plant is cultivated entirely or almost entirely for Europeans (Ochse and Bakhuizen van den

Brink, 1931b, 1: 102-3; I. H. Burkill, 1966, 1: 323). Its varieties are known among natives of those countries, moreover, as "*bit*," a term apparently derived from English or Dutch. In Indonesia, the plant does not normally produce seed, and is cultivated there with seed imported from Europe.

One variety or another of *Beta vulgaris* reached China by T'ang times. Whether the initial introduction was made by sea, as Laufer suggests (1919: 399-400), or overland, is not certain. In the earliest T'ang description Laufer found, however, the leaves were mentioned as food and, for southerners, the sprouts, which were steamed, "the dish being very fragrant and fine" (Laufer, 1919: 400). Beetroot is cultivated in modern Hong Kong, but does not seed there because it requires long days; thus seed must be imported each year (Herklots, 1972: 428-30).

The variety that mainly concerns us here, however, is *Beta vulgaris*, var. *cicla*, which includes spinach beet, whose leaves are completely green; and Swiss chard, seakale beet, or silver beet, whose leaves have white mid-ribs (Herklots, 1972: 155). In present-day Hong Kong, says Herklots, the local green form seeds readily, whereas European and American cultivars neither flower nor seed. This suggests that through breeding the plant has become acclimatized to South China. Herklots also notes that these are ideal vegetables for Hong Kong, that seed is sown from mid-September to mid-February, that the plant grows well and with little care, and provides greens from mid-October to May.

The leaves of Swiss chard have moderate amounts of carotene and ascorbic acid but much less than water spinach, Malabar nightshade, and true spinach. They have fair amounts of riboflavin. And they contain abundant iron and calcium but we have not established how much of these can readily be absorbed. We have found no information on the nutritional value of spinach beet.

Watercress

Plants related to watercress (*Nasturtium* spp.), most of them weeds, occur widely around the world in streams, swamps, and other low-lying areas (Bailey, 1935, 3: 2895). Some scholars identify the *ting-li*, a minor cultivated vegetable mentioned in the *Ch'i-min yao-shu*, as one of these, *Nasturtium indicum* (H. L. Li, 1969: 257; Schafer, 1977: 91).[21] E. N. Anderson (1988: 68, 138), in turn, writes of cresses used in Sung times and in some places on a small scale today.

An early medicinal and salad plant of the Mediterranean region, watercress (*Nasturtium officinale = Rorippa nasturtium-aquaticum*) was introduced to Hong Kong about 1890, whence it spread to the mainland. A temperate land aquatic crop, it is now an important winter crop in South China, reproduced not by seed but by cuttings. Rather than being used in salads as in the West, it is cooked as a green, usually in soup or with meat dishes (Herklots, 1972: 143-48; E. N. Anderson and M. L. Anderson, 1977: 330-31; Zee and Hui, 1981: 30; E. N. Anderson, 1988: 128). Its leaves contain far less carotene than most of the green leafy vegetables considered above, but they do have abundant calcium and ascorbic acid, and moderate amounts of iron and riboflavin (Caldwell, 1972: 316; Leung *et al.*, 1973: 73). We would mention again that watercress is "very cold" in the hot-cold system, outstanding for treating conditions that are hot, but a great danger to persons who are already in a cold state (E. N. Anderson, 1970a: 209; Koo, 1973: 171; Yeung *et al.*, 1973: 198; E. N. Anderson and M. L. Anderson, 1976: 148; 1977: 368; Tan and Wheeler, 1983: 41; E. N. Anderson, 1988: 128, 191).

Lettuce

The genus *Lactuca*, to which lettuce belongs, consists of roughly 100 species of annual and perennial herbs, most of them indigenous to the Old World. Though there are serious problems of identification, certain native species of *Lactuca* seem to have long been used in China as medicine and food (Soubeiran and Dabry de Thiersant, 1874: 169; E. H. Wilson, 1929: 345; F. P. Smith, 1969: 230-31; Read, 1977g: 18, 56, 59; Perry, 1980: 96), and even to have been cultivated in early North China (Bretschneider, 1882-95, 3: 409-10). The species most often singled out as cultivated for food in early times is *Lactuca denticulata* Maxim., though it is no longer cultivated in China today (S. H. Shih, 1962: 3; K. C. Chang, 1970: Table 1; 1976: Table 1; H. L. Li, 1970: 9; Zeven and de Wet, 1982: 34). Still cultivated, however, is *L. indica* L., a species native to southern China and raised for its leaves, which may be cooked as a vegetable, used in soup, or taken as medicine; this perennial, which grows to a height of four feet or more, is also cultivated in southern and eastern Asia (Malaysia, Indonesia, Indochina, southern Japan) (Crevost and Lemarié, 1917: 173; Ochse and Bakhuizen van den Brink, 1931b, 1: 135-37; I. H. Burkill, 1966, 2: 1316; Herklots, 1972: 137-38; Perry, 1980: 96; Zeven and de Wet, 1982: 34). Known in English as "Indian lettuce," *L. indica* was believed by Vavilov (1949-50: 27) to have been domesticated in India or Burma. Mathon (1981: 74), on his part, is less specific, placing it in southern Asia (Asie méridionale). And Zeven and de Wet (1982: 34) note that many varieties of *L. indica* are found in China and place it in their Chinese-Japanese center of domestication. This fits with I. H. Burkill's observation (1966, 2: 1316) that *L. indica* was brought into Malaysia by the Chinese, and leads one to suspect that we may be dealing with a South Chinese domesticate that has largely been ignored in the history of Chinese agriculture.

By far the most important *Lactuca* species in China, however, is lettuce itself, *Lactuca sativa*, a species embracing all of our better-known forms of salad lettuce: crisphead, butterhead, cos (romaine), and leaf. In China, leaf lettuce is the most popular form and is more a cooking lettuce, whereas head lettuce, though it may be cooked as a vegetable, is also, in a recent imitation of Western practice, used as a garnish (Herklots, 1972: 134; E. N. Anderson, 1988: 127).

According to most botanists, *L. sativa* derives directly from prickly lettuce (*L. serriola* Torner = *L. scariola* L.), a wild plant of Europe, North Africa and the Near East, northern India, and parts of Inner Asia.[22] *L. sativa* is believed to have been domesticated in the Mediterranean or Near East; some believe that it was already cultivated in Egypt by the mid-fifth millennium B.C., initially, it has been suggested, to obtain oil from its seed (Vavilov, 1949-50: 34, 36; Lindqvist, 1960: 339-40; I. H. Burkill, 1966, 2: 1316; Harlan, 1975: 70; Ryder and Whitaker, 1976: 39; Mathon, 1981: 31; Zeven and de Wet, 1982: 150; Zohary and de Wet, 1988: 170). The ancient Egyptians considered lettuce to be an aphrodisiac,[23] a food stimulating sexual desire, whether because of the pointed, erect growth of the early Egyptian form, because of its green color or freshness, or because, when squeezed, it gave forth a milky fluid.[24] Even in present-day Egypt, there is a folk belief that if you eat a great deal of lettuce, you will have many offspring. In any case, lettuce was linked to the deity Min, god of fertility and harvest, and was used in offerings and depicted as a backdrop to the image of the deity (Darby *et al.*, 1977, 2: 675-80; Harlan, 1986: 4-10). Turning to China, H. D. R. Baker (1983) notes that lettuce is offered in fertility rites, but ascribes this to its name which, at least in Cantonese, "sounds like 'the birth vegetable.'" He likens this to the serving of fish and oyster dishes at New Year feasts because their names sound like those of desirable states: "fish" like "surplus," and

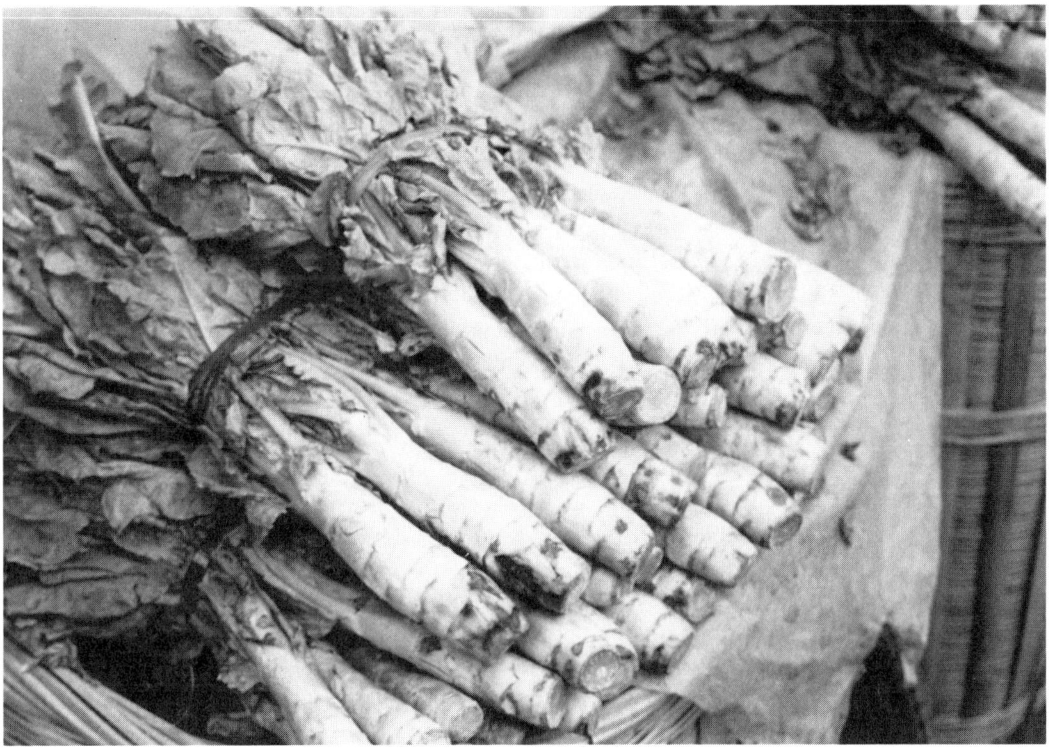

Figure 44. Pile of Stem or Asparagus lettuce (*Lactuca sativa* var. *angustana) (Canton, 1987)*

"oysters" like "happy events." We have found no other evidence of a link between *Lactuca sativa* (*pai-chü*; or *sheng-ts'ai*, "raw vegetable")[25] and sex and reproduction in China, and no certainty that it was known there in very ancient times.[26]

If one accepts de Candolle's widely-cited opinion (1886: 96), lettuce was introduced to China from the West at a fairly late date, probably no earlier than T'ang times. He relied on Bretschneider for support, but Laufer (1919: 400-402), in a careful review of Bretschneider's statements and the Chinese literature, concludes only that superior forms of *Lactuca* may have arrived in China from the West during the T'ang period.

Whenever lettuce arrived, China is credited by Vavilov (1949-50: 24) with the development of one variety, stem or asparagus lettuce (Figure 44), *Lactuca sativa*, var. *angustana* Irish (= var. *asparagina* Bailey; *wo-chü*), which is grown widely in China, the principal region of its cultivation (Bois, 1927-37, 1: 301; Helm, 1954: 105; Lindqvist, 1960: 328, 340).[27] This variety is raised primarily for its thick, fleshy stems which can reach a height of two to three feet or more,[28] and which may be eaten raw or cooked by stir-frying or in soup (F. P. Smith, 1969: 229; Zee and Hui, 1981: 25; E. N. Anderson, 1988: 127).

Salad made with uncooked vegetables was not consumed in traditional China, for raw salads were dangerous and had little appeal to most Chinese; instead, Chinese salads were customarily made of parboiled or stir-fried vegetables and served either hot or cold (G. B. Miller, 1966: 703; K. Lo, 1974: 149-64; G. Z. Chu, 1975: 99; Tropp, 1982: 284; E. N. Anderson, 1988: 127). At the same time, the Chinese use lettuce in a range of ways that go well beyond the salad uses so prominent in the United States.[29] Lettuce may be stir-fried much like cabbage or spinach, though, because of its tenderness, it is cooked for a shorter time. Stir-fried lettuce may be prepared "plain": simply and quickly, sometimes with mushrooms or other vegetables along with spices and flavorings (K. Lo, 1974: 20-23). As an alternative, it may be prepared in bean curd or meat dishes, as in "Sliced Pork Stir-Fried with Lettuce" (K. Lo, 1979c: 148-49), "Veal Cutlet with Lettuce and Scallions," and "Lettuce and Fermented Bean Cake" (W. W. Chang *et al.*, 1970: 370, 392). Lettuce may be quickly stir-fried, then steamed as a separate dish, or it may be steamed without stir-frying in other dishes, as in "Clear-Simmered Bean Curd with Lettuce and Cellophane Noodles" (K. Lo, 1974: 146). Reminiscent of Belgian cooking are China's lettuce soups, such as the Cantonese "Fish with Lettuce Soup" and "Minced Chicken with Lettuce Soup" (C. B. T. Lee and A. E. Lee, 1980: 240-42). Such soup should be made with fresh, crisp lettuce, and consumed shortly after preparation to avoid its becoming limp.

Lettuce, stem lettuce and other forms, is quite low in calories and protein, and of nutritional value mainly for the modest amounts of certain minerals and vitamins it provides. Stem lettuce (leaves and stems) seems to be a fair to good source of vitamin C and can be a fair source of calcium, whereas the green leaves of other types of lettuce are fair sources of calcium and vitamins A and C (Platt, 1962: 14-15; Leung *et al.*, 1973: 39, 53).

Water Shield

Water shield (*Brasenia schreberi* = *B. peltata*; *shun* or *ch'un*) is an aquatic herb, native to Asia, Australia, North America, and Africa, which is sometimes cultivated for its edible leaves in China, as well as in Japan (Uphof, 1959: 58; Zeven and de Wet, 1982: 34). H. L. Li (1970: 10, Table 1; 1983: 43-44) places its domestication in southern China, and notes that it may have been cultivated in the Yangtze valley since Neolithic times. Various scholars observe that it has also been a food in North China since antiquity, already mentioned, as *mao*, in the *Shih-ching* (early Chou times) (Keng, 1974: 403; H. L. Li, 1983: 45). E. Biot, translator of the *Chou-li* (likely late-Chou times), includes water shield among the "seven vegetable pickles" offered to the emperor (the others were salted turnips, malva, bamboo sprouts of two types, a form of *Allium*, and Chinese celery) (Bretschneider, 1882-95, 2: 188, 189n, 220). Water-shield stew is also said to have been "a famous delicacy" in Chou writings (E. N. Anderson, 1988: 27), and in T'ang times, water shield was made into a tasty soup (Schafer, 1977: 91).

Turning to more recent times, one notes the Chinese belief that water shield is cooling, anthelmintic, and antidysenteric, among other things (F. P. Smith, 1969: 73; Perry, 1980: 288). At the same time, F. P. Smith (1969: 73) reports the view that continued consumption of water shield is bad, damaging to the stomach, teeth, hair, and bones, and that "in the seventh month" eating water shield may bring on cholera. He justifies certain of these beliefs by observing that the plant grows in dirty ponds and streams, and is eaten raw or slightly cooked. Soaking the leaves in vinegar may have reduced whatever danger exists, but nutritionally vinegar-soaked young leaves are not outstanding in any way, and quite modest in the amounts of minerals and vitamins they contain (Leung *et al.*, 1973: 73).

Alfalfa

The alfalfa species *Medicago sativa* (*mu-hsü*) evolved in a region centering on the Caspian Sea, from which it spread as a wild plant to adjoining areas. That included some of the steppe lands of Inner Asia from which pastoral nomads ventured forth repeatedly to pillage and to conquer. Indeed, the domestication of alfalfa in western Asia was as a forage crop (first plant species to be grown as forage), and likely came about because of the growing importance of horses and invention of horse-drawn chariots early in the second millennium B.C. Alfalfa was introduced to Greece in the fifth century B.C., from which it spread to Italy and other parts of Europe (Zeven and de Wet, 1982: 97). According to Chinese tradition, alfalfa arrived in China in Han times by the overland route across Central Asia (Laufer, 1919: 208-19), and since that time it has been grown as a forage crop, especially for horses in North China. In addition, however, its young shoots and leaves have served in China as a vegetable. In the *Ch'i-min yao-shu*, a full chapter was given up to the matter of alfalfa as a vegetable; it was good for that purpose when young and tender, and, when mature, was excellent for horses (H. L. Li, 1969: 259). Alfalfa was also consumed as a vegetable in T'ang times; its cultivation was encouraged in Yüan times largely for use at times of famine; and it continues to be eaten as a vegetable today, though, because it is nearly without taste, it is said not to be a favorite (Bretschneider, 1882-95, 3: 403-4; Laufer, 1919: 216-17; Read *et al.*, 1940: 16; F. P. Smith, 1969: 260-61; Schafer, 1977: 91; E. N. Anderson, 1988: 128).

According to Herklots (1972: 176-77), alfalfa was never found as a vegetable in Hong Kong prior to 1945, after which time it was introduced by immigrants from Shanghai. Tips of alfalfa shoots in bunches then came to be found in markets. "Lightly cooked," Herklots observed, they are "a good addition to meat dishes," and their nutritional value better than that of spinach. They are, nevertheless, considered "cooling," too much so, according to F. P. Smith (1969: 261), for one to eat very often.

Sow Thistle

Sow thistle (*Sonchus* spp.) belongs to a genus of Old World plants, most of them weedy, but some with such excellent foliage that they have become ornamentals. In addition, *Sonchus* species are used as food, fodder, or medicine in various parts of the Old World. *S. oleraceus* and *S. arvensis* are the two species regularly mentioned as food plants in the literature on China. Usually they seem to have been collected from the wild, but Bretschneider (1882-95, 3: 410) reported that in Peking, besides growing wild, *S. arvensis* was cultivated and its leaves eaten.

The eating of *Sonchus* goes back to antiquity in China, and, though it doesn't seem to have been a regular in haute cuisine, one does read of a dog flesh and *Sonchus* dish in Han times (Keng, 1974: 402; K. C. Chang, 1977b: 28; Y. S. Yü, 1977: 58; E. N. Anderson, 1988: 27, 42). There is, however, considerable confusion caused by terms used in the early Chinese literature. *T'u* was an early name which in general referred to *Sonchus*, but it was applied to various species of *Lactuca* as well. In a similar vein, *k'u-ts'ai* ("bitter vegetable"), used for certain species of *Lactuca* and *Cichorium*, could also refer to *Sonchus* (Bretschneider, 1882-95, 2: 33-4, 177-79; 3: 407-10; Laufer, 1919: 400-401; F. P. Smith, 1969: 230, 414).

The Chinese are not the only people for whom species of *Sonchus* serve as food. *S. jacquinii* (= *S. congestus*?) is the *lachuza de pastor*, "pastor's lettuce," whose leaves may serve as salad in the Canary Islands (Bailey, 1935, 3: 3189). In Java, the young shoots and leaves of *S. arvensis* and *S. aster* may be made into a salad, perhaps after slight steaming, or they may be cooked or steamed and served with rice (Ochse and Bakhuizen van den Brink, 1931b, 1: 141; I. H. Burkill, 1966, 2: 2086). *S. oleraceus*, which is said to have leaves much less bitter than those of *S. arvensis*, was grown as a pot-herb in early Europe (H. M. Burkill, 1985: 493). In areas closer to China, *S. oleraceus* is reported to be used a vegetable on Java and in the Philippines (Ochse and Bakhuizen van den Brink, 1931b, 1: 145; I. H. Burkill, 1966, 2: 2086); the South Sea islands, especially New Caledonia (Massal and Barrau, 1956: 36); Japan (Bretschneider, 1882-96, 2: 33; 3: 410; F. P. Smith, 1969: 230); and Kashmir (Watt, 1889-96, 6, Part 3: 275). *Sonchus oleraceus* is said to be high in vitamin C (Read, 1977g: 55).

Some Additional Greens

The group of Chinese green vegetables considered above is broad by American standards, but, in addition, the Chinese use many other leaves or young shoots, whether from cultivated plants or collected in the wild, which are commonly eaten in soup. E. H. Wilson (1929: 345), on the basis of lengthy travels in Yunnan, Szechwan, and Hupeh, identified a dozen species besides those we have already considered, and other observers mention still more. Some of these are New Zealand spinach (*Tetragonia expansa*); shepherd's purse (*Capsella bursa-pastoris*); garland chrysanthemum (*Chrysanthemum coronarium*); aster (*Aster trinervius*); Chinese box-thorn (*Lycium chinense*); garden sorrel (*Rumex acetosa*); parsley (*Petroselinum sativum*); purslane (*Portulaca oleracea*); cocklebur (*Xanthium strumarium*); perilla (*Perilla ocymoides* = *P. frutescens*), whose seed is a source of oil; white goosefoot, common pigweed, or fat hen (*Chenopodium album*);[30,31,32] and plantain (*Plantago major*)[33] (E. H. Wilson, 1929: 345; Read and Wagner, 1940: 262-67 *passim*; F. P. Smith, 1969: 92, 250, 313, 335, 348, 461; Herklots, 1972: 127-224 *passim*, 488-89; Zee and Hui, 1981: 1-33 *passim*; E. N. Anderson, 1988: 26-27, 41, 50, 105, 127, 128). Of special note as greens are the leaves of certain trees. One is the Chinese cedar *Cedrela sinensis* (= *Toona sinensis*; *ch'un* or *hsiang-ch'un*), an ornamental and timber tree whose wood resembles that of the true cedars (*Cedrus*). *Cedrela sinensis* is found widely in China in a wild or cultivated state. Its tender shoots and young leaves are picked in springtime, and then salted or cooked as greens (E. H. Wilson, 1929: 300, 345; S. C. Lee, 1935: 697-700; Read and Wagner, 1940: 266; F. P. Smith, 1969: 100; Zee and Hui, 1981: 32). A second is the elm (*Ulmus* spp.), certain species of which provide leaves which have been used as greens. Ancient Chinese authors also mention the inner bark of the elm being ground into a sticky meal and eaten, and elm fruits being boiled with rice. Bretschneider, writing in the nineteenth century, spoke of such ground bark as a food common among mountain people in North China, and other people, too, consumed elm leaves, bark, and fruit at times of severe food shortage (Bretschneider, 1882-95, 2: 128-29, 366; F. P. Smith, 1969: 448; Read, 1977g: 48; E. N. Anderson, 1988: 27, 41).

SQUASHES, GOURDS, CUCUMBERS, AND OTHER CUCURBITS

The cucurbits, members of the family *Cucurbitaceae*, are found in great variety in China, where they are known by the general term *kua*. Anderson and Anderson (E. N. Anderson and M. L. Anderson, 1977: 329) place them in a category of plant foods of lesser importance, in

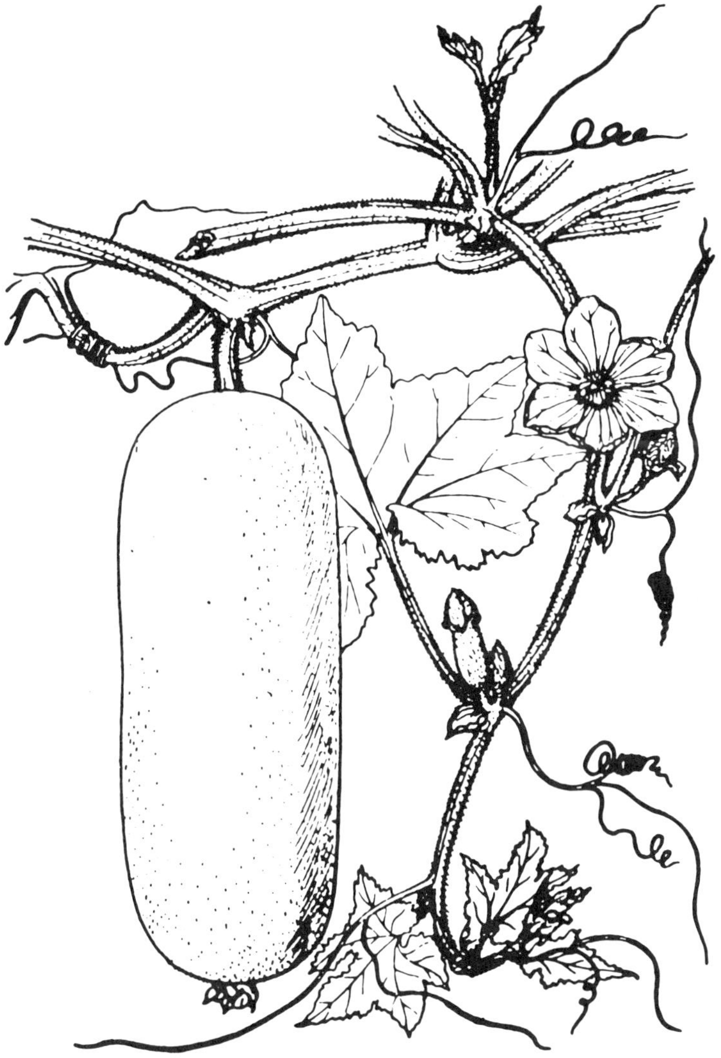

Figure 45. Wax gourd (*Benincasa hispida*) (from Herklots, 1972: 308)

which none is of major importance but all serve culinary needs. It is their flesh that is most commonly eaten, in some cases raw but normally stir-fried or boiled or pickled; though moist and flavorful, the flesh usually has little food value except for vitamins and minerals. The bitter cucumber (*Momordica charantia*), for example, is very high in vitamin C; and winter squash (*Cucurbita maxima*) has good amounts of vitamin A (T. J. Chu and B. E. Read, 1938: 250, 254; Leung *et al.*, 1973: 32, 68). In addition, the leaves of luffa and bottle gourd are very high in iron and vitamin A (E. F. Yang and M. Y. Dju, 1939: 485-86; Leung *et al.*, 1973: 38, 48, 49).

For the full range of cucurbits cultivated in China, one should consult Herklots (1972: 273-349) and I. H. Burkill (1966: *passim*). Here it should suffice to single out several of the better known ones widely used as vegetables: (1) the wax gourd, white gourd, or Chinese preserving melon (*Benincasa hispida* = *B. cerifera*; Figure 45) (*tung-kua*, "winter *kua*" or

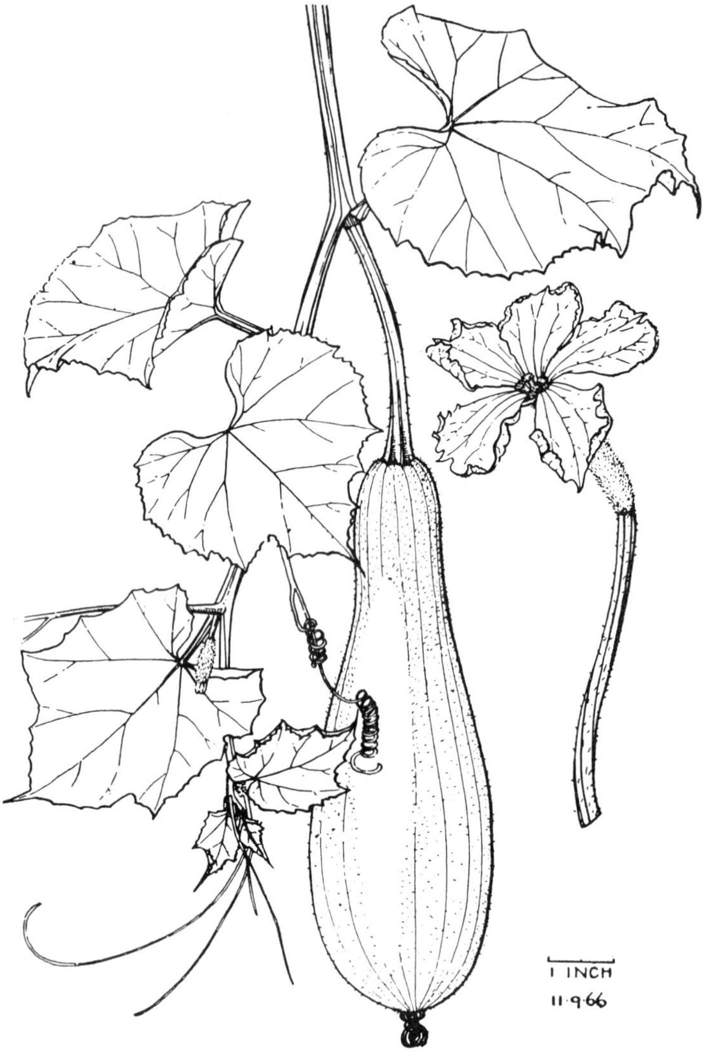

Figure 46. Bottle gourd (*Lagenaria siceraria*) (from Herklots, 1972: 325)

"winter melon"), which may bear very large fruit, up to 100 pounds in weight when eaten at maturity; (2) bottle gourd (*Lagenaria siceraria*; Figure 46); (3) bitter cucumber, bitter melon, bitter gourd, or balsam pear (*Momordica charantia*; Figure 47); (4) snake gourd or serpent gourd (*Trichosanthes cucumerina* = *T. anguina*; Figure 48); (5) luffa or loofah (*Luffa acutangula* and *L. cylindrica*; Figures 49 and 50); and (6) cucumber (*Cucumis sativus*). Wherever in, or beyond, the Old World they were domesticated, all of the above have been cultivated in China or southern Asia since ancient times (I. H. Burkill, 1966: *passim;* H. L. Li, 1969; K. C. Chang, 1970: Table 1).[34] The squashes and pumpkins (*Cucurbita maxima*, *C. pepo*, and *C. moschata*), on the other hand, did not reach China from the New World until the sixteenth century A.D. or later.[35]

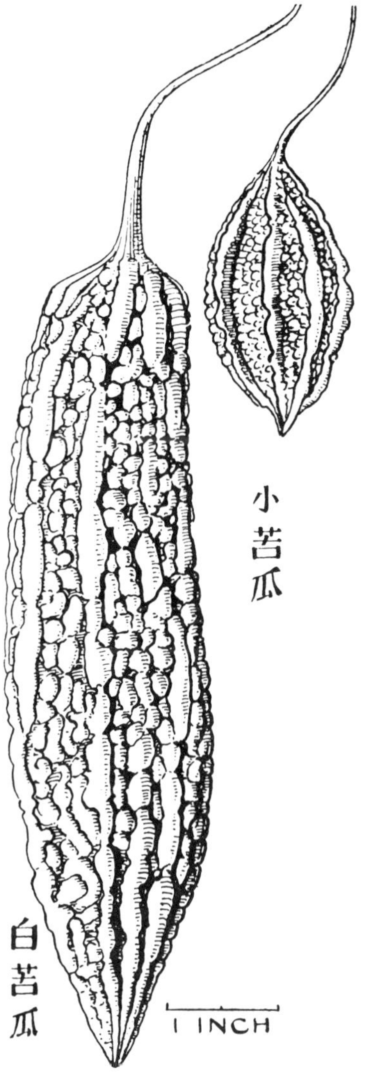

Figure 47. Bitter cucumber (*Momordica charantia*) (from Herklots, 1972: 336)

Some of the above are worthy of special note because of additional, sometimes unusual, uses. *Luffa cylindrica* (*ssŭ-kua*), for example, may be cooked as a vegetable when young, though like *L. acutangula* it has a purgative quality. As it matures, that quality becomes more pronounced and *L. cylindrica* may be used medicinally. Or the fibrous sponge-like skeleton that develops with age, commonly a foot or two in length, may be dried to serve various household uses, giving rise to the names "dishcloth gourd" or "towel gourd" (Blasdale, 1899: 30-31; Porterfield, 1943; I. H. Burkill, 1966, 2: 1393-94; F. P. Smith, 1969: 248-49). The bitter cucumber (*k'u-kua* or "bitter *kua*"), in turn, is of value less as a vegetable in its own right than for the flavor it gives to other dishes, such as salads and stir-fried meat or shrimp (Blasdale, 1899: 30; J. Y. P. Chen, 1973: 229; E. N. Anderson and M. L. Anderson, 1977: 329). It may be made into soup. It may also be stuffed, as in "Steamed Stuffed Bitter Melon," made of bitter

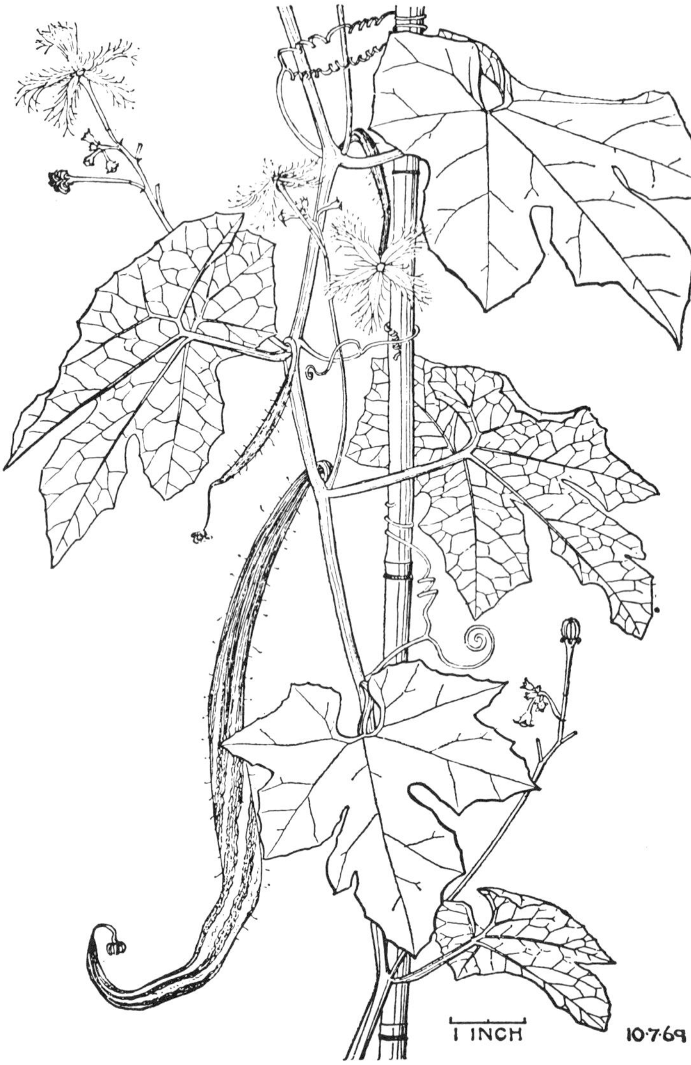

Figure 48. Snake gourd (*Trichosanthes cucumerina*) (from Herklots, 1972: 349)

cucumber slices whose insides are removed and stuffed with pork, water chestnuts, gingerroot, green onion, rice wine, soy sauce, and other ingredients, then steamed, and served with a sauce (Hom and Steiman, 1981: 83-86). Bitter cucumber, especially when mature and quite bitter in taste, is also a tonic medicine and, in a dried state, is found in Chinese drug stores (Hooper, 1929: 90; F. P. Smith, 1969: 265; Perry, 1980: 117; Hom and Steiman, 1981: 83). Since it is also "cold," even "dangerously cold," in the hot-cold system of classification, it is given for migraine headaches but may be banned for pregnant women (Koo, 1973: 141; Ling *et al.*, 1975: 129; Gould-Martin, 1978: 43).

The flesh of winter melon may be preserved. As with certain other cucurbits, its flesh may be prepared in various ways, but is often made into soup, for example the Cantonese "Winter Melon-Fish Maw Soup" (G. Z. Chu, 1975: 97-98) and "Crab and Winter Melon Soup" (E. Chen,

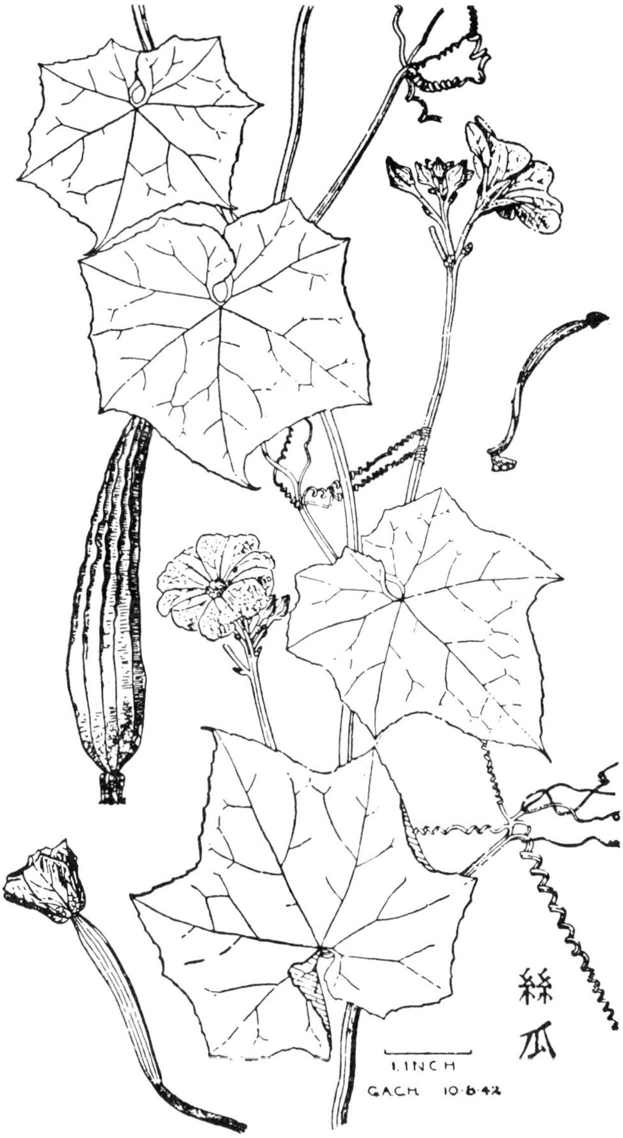

Figure 49. Angled luffa (*Luffa acutangula*) (from Herklots, 1972: 327)

1977: 69). The general dish of "Winter Melon Soup," despite its name, is made not only with melon flesh, which is cooked and cut into pieces, but with ham, chicken stock, and other ingredients (Em. Hahn *et al.*, 1973: 107). Since, when mature, the winter melon is very large and has a hard rind, it may, when cut and hollowed out, also serve as a soup kettle, and sometimes may even be decorated with designs cut into its outer surface, as illustrated in the book by Emily Hahn (Em. Hahn *et al.*, 1973: 66, 67). Soup is prepared in the melon by slow steaming, for hours, and is then served with the melon as soup tureen. Because of the effort needed in preparation, such soup is normally made only for large dinners or on special occasions. Certain restaurants in Hong Kong and elsewhere in China, however, do specialize

Figure 50. Smooth luffa (*Luffa cylindrica*) (from Herklots, 1972: 332)

in such a soup, sometimes called "Winter Melon Pond," which may be made with duck, ham or pork, mushrooms, bamboo shoots or water chestnuts, and other delicacies (F. L. Chang and G. M. Fuchs, n.d.: 74-75; Howe, 1971: 160-61; E. N. Anderson and M. L. Anderson, 1977: 329; E. Chen, 1977: 65; Sakamoto, 1977: 169-70; C. B. T. Lee and A. E. Lee, 1980: 244-45; Hom and Steiman, 1981: 330-32; Zee and Hui, 1981: 34; Y. K. Martin, 1984: 65). Also worthy of note is the use of melon and watermelon as fruit, to be considered in our chapter on "Fruits," and the eating of cucurbit seeds, which will be considered in the chapter on "Edible Nuts, Nut-like Fruits, and Seeds."

ONIONS AND THEIR RELATIVES

Allium is a large genus, with over 500 species indigenous to the northern hemisphere, usually found in open areas such as grassland and desert (H. A. Jones and L. K. Mann, 1963: 25, 26). There seem to have been three major areas of *Allium* domestication, all in the Old World: (1) Central Asia, (2) China and its neighbors, and (3) the Near East-Mediterranean. Cultivated in China and neighboring countries and considered by many botanists to have originated in the Chinese-Japanese region are the spring, Japanese bunching, or "Welsh" onion (*A. fistulosum*; *ts'ung*),[36] and *A. chinense* (= *A. bakeri*; *ch'iao*). The spring onion has been the principal garden onion in China since ancient times, and *A. chinense* is a vegetable onion which is often pickled and eaten as a snack. In the United States, *A. chinense* is sometimes known by its Japanese name, *rakkyo*, but it has no agreed-upon English name, and the different names applied to it (*i.e.*, "shallot," "leek," and "scallion") lead to considerable confusion. Some botanists also credit East Asia with the Chinese leek (*A. ramosum*); Chinese garlic (*A. macrostemon*); another small onion (*A. ledebourianum*); and possibly the Chinese chive (*A. tuberosum*; *chiu-ts'ai*) (Vavilov, 1949-50: 24; H. L. Li, 1969: 257; 1970: 16; Herklots, 1972: 387-401; McCollum, 1976: 187; Mathon, 1981: 53; Zeven and de Wet, 1982: 32-33).

The other domesticated forms of *Allium* found in China are believed by most scholars to have been introduced from Central Asia or the Near East and Mediterranean: the common onion (*A. cepa*; *yang-ts'ung* or *hu-ts'ung*, "foreign onion"; *hui-hui ts'ung*, "Moslem onion"), shallot (usually *A. ascalonicum*), leek (usually *A. porrum*), common chive (*A. schoenoprasum*),[37] garlic (*A. sativum*; *suan* or variations of *suan*), and giant garlic or rocambole (*A. scorodoprasum*; sometimes *ta-suan* or "large garlic"). Some, however, suggest that an indigenous wild garlic, var. *pekinense*, was domesticated in North China and that the cultivated garlic of China and Japan derives from it (H. L. Li, 1969: 257; Zeven and de Wet, 1982: 33). Whether or not this is so, garlic was known in China from ancient times and was identified by the simple root word *suan* (Laufer, 1919: 302), which may indicate a considerable antiquity.

Anti-*Allium* forces have been at work in the Old World since antiquity. There are claims that onions and garlic were considered unclean and avoided by priests in ancient Egypt, though the evidence is inconsistent and unclear (Darby *et al.*, 1977, 2: 660, 661-62). For ancient India, on the other hand, the evidence for such views is more plentiful. Brahmins, other upper-caste Hindus and respectable persons were enjoined to reject various *Allium* species, such as onion, garlic, and leeks, along with certain other foods, prominent among them objectionable flesh such as pork and that of carnivorous birds. The concern here was with maintaining ritual purity and avoiding pollution, with avoiding Alliums because they were unclean.[38] Some suggest that their impurity derived from the perception of them as being similar to flesh.[39] One example is found in the *Bower Manuscript*, an Indian Buddhist medical treatise written in the fifth century A.D. (*Bower Manuscript*, I, 10-12; or Hoernle, 1893-1912: 11), which indicated that Brahmins should not eat garlic because it originated from a living being, which gave it an evil odor as well. Specifically, the treatise recounts a story of the deity Vishnu severing the head of the lord of the demons. Drops from the demon's throat fell on the ground and these were the first garlic. By this tradition, garlic derived from an enemy of the gods and thus was evil. In Samkhya, an early school of orthodox Hindu philosophy, onions and garlic, like mushrooms, beef, pork, and certain other foods, were believed to have the quality of *tamas* ("darkness"). *Tamas* is a sensual, animal quality that brings on ignorance, sloth, fear, and lewdness in humans. Foods that have *tamas*, moreover, are cold, dull, rotten, stale, and "made up of the leavings of others" (B. Walker, 1968, 2: 264, 344), qualities unlikely to encourage people to eat them. A similar explanation is that onions and garlic were forbidden to Brahmins and celibates because they had an aphrodisiac effect (B. Walker, 1968, 2: 215).

In India, anti-*Allium* sentiment occurred also among Jains[40] and Buddhists. Though one cannot determine whether Indian Buddhists simply adopted views already prevalent among other Indian religious and philosophical groups, that may have been the case, for the ban seems to have been less strong and widespread among them than among Hindus.[41,42]

Turning to China, where anti-*Allium* sentiment is also found, there is a strong suspicion that Buddhists were responsible for introducing it. One reads (Laufer, 1919: 303) of Kumarajiva (A.D. 344-413), a famous Indian Buddhist scholar brought to China by the emperor, who fostered Indian ideas there and translated Buddhist works. In one of these, the *Brahmajala-*

sutra, various forms of *Allium* were included among the *wu-hun*, "five vegetables of strong odor" forbidden to Buddhist clergy (de Groot, 1896: 42; Laufer, 1919: 303), and in T'ang times, such "lists of five" usually included them (Schafer, 1977: 132). Alliums are among foods banned to Chinese Buddhists even today (Welch, 1967: 112), and there is at least one Buddhist vegetarian dish especially fitting for monks because it excludes garlic, scallions, and ginger, items believed to stir the passions and reduce one's likelihood of attaining nirvana (Y. K. Martin, 1984: 218). One also finds *Allium* bans among minority peoples in China influenced by Buddhism, for example the Mongolian-speaking Monguors of the Kansu-Tibetan border, among whom there is a sect, "those who fast," who do not eat garlic or meat or drink "wine" (Schram, 1954: 117).

"Lists of five" banned items in China are not restricted to Buddhists, but are also found among Taoists and geomancers, some of whom avoid Alliums too (Welch, 1958: 132; F. P. Smith, 1969: 28). Perhaps this is what led Doolittle (1865, 2: 183) to write, apparently in a general sense, that garlic, onions, and scallions are forbidden to Chinese vegetarians. Interestingly, items included on the lists of five were also believed to be insect repellents, for one reads of a Han midsummer custom, which survived in traditional China, of putting the five vegetables of strong odor, including onions and garlic, on red cords hung on house doors and gates to repel harmful insects (Bodde, 1975: 291, 302-3).

As in ancient Egypt and India, religious influences seem to have had limited impact on the use of the Alliums in China, for they were important in all periods of Chinese history. One reads (Y. S. Yü, 1977: 75-76) that in Han times garlic, scallions, and leeks were common foods; and that these were eaten more often by the poor though the wealthy used them as well (there was an imperial greenhouse in which out-of-season vegetables were raised, and among them were leeks and scallions). In T'ang times, the leaves and bulbs of Alliums were liberally used in cooking and were strongly endorsed as heating and tonic; garlic was often included in cooked dishes; the bulbs of *A. chinense* were pickled; and the spring onion served as a seasoning and was highly regarded as a means of countering the winter cold (Schafer, 1977: 91-92). In Ming times, one traveller said that customary meals in Shantung inns consisted only of rice, steamed bread, and garlic (Mote, 1977: 243). And in modern Kwangtung, onions and their relatives continue to be important and well-regarded, often adding character to what otherwise would be indifferent dishes. Some consider garlic and certain onions as "warm" or "hot," and one study found them commonly prescribed as postpartum food (Koo, 1973: 141; Yeung *et al.*, 1973: 198, 200; E. N. Anderson and M. L. Anderson, 1976: 147; M. Mathews and L. Manderson, 1981: 12, 14).

Onions and garlic are described as having little nutritional value (White and Selvey, 1974: 186). One should not, however, dismiss out of hand the nutritional merits of all *Allium* species, for some (for example chives, Chinese chives, leeks, spring onions) contain significant amounts of calcium and vitamins A and C (H. C. Hou, 1937; Kung *et al.*, 1938; Kwoh *et al.*, 1941; Fan and Hou, 1953; Leung *et al.*, 1973: *passim*).

There is also the question of what role onions and garlic may have had in fostering, or hindering, the growth of intestinal parasites in humans. I. H. Burkill (1966, 1: 99), for one, states that several wild species of *Allium* and, to a lesser degree, some domesticated species, have anthelmintic properties. In this belief, they are used in various forms as anthelmintics in Southeast Asia. This is in keeping with widespread and ancient Old World folk beliefs about the medicinal value of garlic and other *Allium* species (for China, see F. P. Smith, 1969: 25-28;

for passing references on such uses, Malay Peninsula: I. H. Burkill, 1966, 1: 99-103; India: Ainslie, 1826, 1: 150-51, 603; 2: 475-76 *passim;* Watt, 1889-96, 1: 169-174; general, especially the classical Mediterranean: H. Phillips, 1827, 2: 15-27; ancient Egypt: Darby *et al.*, 1977: 656-63). Many physicians as well have testified to their effectiveness medicinally, and some, to their vermifugal qualities. On the other hand, one leading physician and nutritionist says (pers. comm., W. C. Darby) that he knows of no scientific experiments to support the view that the Alliums are in fact effective as anthelmintics. R. C. Robertson (1936), writing of the Shanghai area, suggests a quite different role in including onions (*A. cepa* and *A. fistulosum*) among plants that commonly transmit helminths. They are grown in soils fertilized by night soil, human excreta. Along with other vegetables, they may be transported to market in sampans that previously had been used for carrying night soil. They may be washed in contaminated water. And they are commonly eaten raw.

FUNGI

In their fascinating book *Mushrooms, Russia, and history*, Valentina and R. Gordon Wasson (1957) identified an area of mycophobes, mushroom haters, around the North Sea, especially in England, and one of mycophiles, mushroom lovers, in eastern Europe, particularly Russia. In the former area, the Wassons found that mushrooms were burdened by a negative literary tradition of fear and disgust. As a consequence, wild mushrooms were little used and instead, it was the domesticated form (*Psalliota campestris* = *Agaricus campestris*), the indifferent-tasting "slattern of the mushroom family," that was dominant in cuisine. By contrast, people in the area of mycophilia exhibited love for, interest in, and knowledge of, wild mushrooms. Seasonal expeditions were organized in Russia to collect mushrooms in woods and forests, and a naughty child might be punished by being denied the privilege of going mushrooming.

How, one might ask, would the Chinese be regarded in terms of mycophobia and mycophilia? Unfortunately, the Wassons only touch on China in their book, though they do write (1957, 2: 333-34) of a "cult of the mushroom in the Chinese cuisine" and appreciate what they call "the fungal hierarchy" that exists in it. That the Chinese belong among the mycophiles is clearly indicated in their literature. In the Chinese classics, for example, a certain fungus growing on soft earth is highly pleasing, the "longevity mushroom," "ten-thousand-year mushroom," or "plant of immortality" (*Ganoderma lucidum*) which dispels "earthy vapors" and leaves "a heavenly atmosphere"; today this fungus is highly regarded as a medicine (I. H. Burkill, 1966, 1: 1061; F. P. Smith, 1969: 271; B. Liu and Y. S. Bau, 1980: 170-71; Wang Lianzheng in Wittwer *et al.*, 1987: 306; Yu Youtai in Wittwer *et al.*, 1987: 296-97; E. N. Anderson, 1988: 133). In Sung times, a monograph on wild edible mushrooms was written (Sinoda, 1977: 490), which is a further testimony to Chinese mycophilic commitment. There is also clear evidence of mycophilia in other historical and contemporary documents, so much so as to challenge the view that the Russians and eastern Europeans may be the leading Old World mycophiles.

The general name for mushrooms in China is *ku*, but fungi are also classified as to habitat: those that grow on soft earth (*chih*); those on hard ground (*chün*); and still others (*êrh*), on wood. Overall the Chinese recognize a broad range of fungi, with more than a hundred and twenty species listed in one recent work (B. Liu and Y. S. Bau, 1980), as having useful medicinal qualities. Even poisonous ones may be utilized, in some cases after they have been burned to ashes, for treating a broad array of ailments (F. P. Smith, 1969: 273-74; B. Liu and Y. S. Bau, 1980: *passim*).

As mentioned briefly in our discussion of garlic, ancient Hindus looked on mushrooms as growing from unclean substances, and they were forbidden to Brahmins and other respectable persons (Prakash, 1961: 55, 115, 128, 164). We have found nothing to indicate that such ideas were transmitted from India to China, or that Chinese are reluctant to consume mushrooms because of religious sensibilities.

Turning to the cultivated fungi of China, we note first that various of them are grown as medicine, food, or both (B. Liu and Y. S. Bau, 1980: *passim*). Among those of dietary importance, perhaps best known to Westerners is the Chinese padi-straw mushroom (*Volvariella volvacea*; *ts'ao-ku* or "straw mushroom") (Figure 51), so named because it is commonly grown on rice straw. This mushroom is ordinarily used fresh, an expensive delicacy in China and in certain areas of Southeast Asia where Chinese have settled (S. T. Chang, 1977: 374). Yet despite its popularity today, the Chinese padi-straw mushroom is not mentioned in early sources. Indeed it is only in the early nineteenth century that one finds in local annals the first account of what seem to be padi-straw mushrooms, produced at the Nan-hua Buddhist Monastery of northern Kwangtung, a small institution but well-known and considered sacred since T'ang times. After reviewing relevant historical accounts, S. T. Chang (1977: 374-76) concludes that cultivation of this mushroom may have been started at that institution. Buddhist monks, he notes, are vegetarian, and mushrooms one of their best-liked foods. Those of the Nan-hua Monastery, moreover, lived in an isolated area, and thus ordinarily cultivated their own vegetables. One of them, he suggests, likely discovered the method of cultivating *Volvariella* mushrooms on straw, and in time their excellence came to be widely recognized. An account published in 1875 mentions four boxes of Nan-hua mushrooms being sent each year as tribute to the royal family. Another report of about the same date indicated that villagers of northern Kwangtung were also cultivating them, using the monks' method of cultivation. Today the padi-straw mushroom is cultivated widely in Hong Kong and elsewhere in South China (Herklots, 1972: 505-6; B. Liu and Y. S. Bau, 1980: 154),[43] for in growing it, higher temperatures and more moisture are needed than for the European mushroom.

Also of dietary importance are fungi that grow on trees. First among those mentioned by Anderson and Anderson (E. N. Anderson and M. L. Anderson, 1977: 332; E. N. Anderson, 1988: 133) is the *hsiang-ku* or *hsiang-chün* ("fragrant mushroom") or *tung-ku* ("winter mushroom"), the Japanese *shiitake* (*Lentinus edodes*), which is important in many dishes, particularly if they are vegetarian.[44] There are also the "earlike wood fungi," species of *Auricularia* (*mu-êrh*, "wooden ear" or "tree ear") or *Tremella*. The Chinese may consume more Auricularias than any other people, and, in any case, demand for them as food is much greater than in Europe (I. H. Burkill, 1966, 1: 270),[45] and in the nineteenth century they were even imported from Hawaii (Thrum, 1914).[46] The earlike wood fungi occur naturally, often after summer or fall rains, on various broad-leaf forest or garden trees such as oak, willow, and poplar; or they may be cultivated on logs out-of-doors or in a saw-dust mixture in-doors (B. Liu and Y. S. Bau, 1980: 43-53). Such cultivation was known at least by T'ang times, when one account said "Put the steamed bran on logs; then, covered with straw, wood ears will grow" (S. T. Chang, 1977: 374).[47] Such cultivation is found today especially in western China (Szechwan, Hunan, and Yunnan) (Y. K. Martin, 1984: 229, 230). Even with cultivation, however, the domestic supply

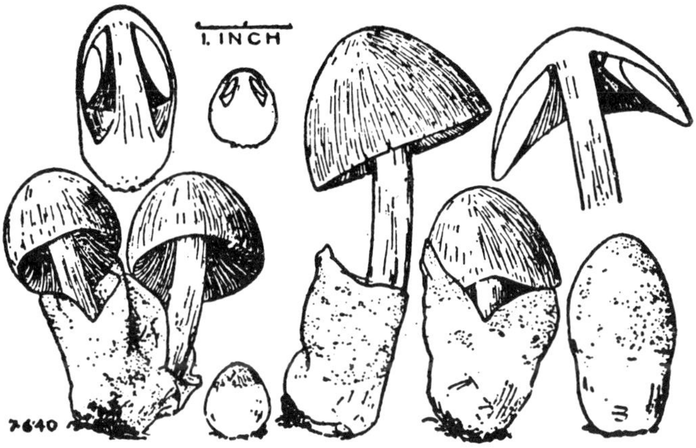

Figure 51. Chinese padi-straw mushrooms (*Volvariella volvacea)* (from Herklots, 1972: 506)

of *mu-êrh* has been insufficient and, though China does export *mu-êrh*, it must also import them.

Ear-like wood fungi have been likened to the truffles of France because they enhance flavor and contribute a pleasing texture to dishes. In China they are generally used in a dried form and require soaking, and may be shredded so that their texture will affect the entire dish. Ear-like wood fungi are common ingredients, often along with "golden needles" (dried flowers of the day lily), in soups and vegetable dishes, such as "Eight-Treasure Vegetarian Assemblage" (Y. K. Martin, 1984: 153), and in other dishes as well (Joyce Chen, 1962: 12; J. Y. P. Chen, 1973: 226, 229, 232; G. Z. Chu, 1975: 92, 127-28, 280; F. Lin, 1975: 317; E. N. Anderson and M. L. Anderson, 1977: 332; Sakamoto, 1977: 23, 33; Hom and Steiman, 1981: 31, 67-68; K. Lee and A. Branyon, 1984: 87; Y. K. Martin, 1984: 16, 229, 230; E. N. Anderson, 1988: 133). Some of these bear names that may sound strange to a Western ear, for example "Ginger Soup with Pork and Wood Ears" (Y. K. Martin, 1984: 56-57), "Quick-Fried Ox Kidney with Wood Ears" (K. Lo, 1979a: 241), and "Clouds and Mushrooms," a Szechwan dish that includes cloud ears along with other fungi, noodles, ham, crab meat, and flavorings (Sakamoto, 1977: 277). In addition to their food uses, the earlike wood fungi serve medicinal ones, with some regarded as outstanding tonics (F. P. Smith, 1969: 170, 272; J. Y. P. Chen, 1973: 232, 235; B. Liu and Y. S. Bau, 1980: 44-52).[48]

There are also other famous mushrooms in China, and, though we cannot consider them all, we cannot pass over "monkey head" (*Hericium erinaceus*). This highly-valued banquet delicacy, also known as "hedgehog fungus," was mentioned in our chapter on "Major Cuisines of China." Its standing among China's great delicacies (bear's paw, bird's nest, shark's fin, and bêche-de-mer or sea cucumber) is revealed in the expression "mountain delicious the monkey head, and seafood delicious the sea cucumber" (Wittwer *et al.*, 1987: 294). In traditional times, this rare mushroom, most common in China's northeast, was collected in mountain forests, but recently Chinese scientists have succeeded in cultivating it (B. Liu and Y. S. Bau, 1980: 177-79; Wittwer *et al.*, 1987: 294-95). A second species singled out by some food writers is *yin-êrh*, "silver ear" (also "white tree ear"; *Tremella fuciformis*), which is valued as a tonic (is *pu*), and is sometimes included in soups and vegetarian dishes or made into a sweet dessert soup, such as "Silver Fungus in Syrup" (G. Z. Chu, 1975: 231, 295; F. Lin, 1975: 317; B. Liu and Y. S. Bau, 1980: 48; Wittwer *et al.*, 1987: 296; E. N. Anderson, 1988: 194). A third fungus of special dietary interest is one which is added to rice in Fukien, the result being a rice of attractive red hue and somewhat pungent taste, a "distinctive marker of Fukienese cuisine" (E. N. Anderson, 1988: 117, 164).

The European mushroom (*Psalliota campestris* = *Agaricus campestris*; among its names, *pai-chün*, "white mushroom," and *yang-chün*, "foreign mushroom"), both fresh and dried, is much used by the Chinese, who ship them south from the northern provinces (I. H. Burkill, 1966, 2: 1843). On Taiwan they are cultivated on a large scale, and in Hong Kong on a small one (Herklots, 1972: 506-7; E. N. Anderson, 1988: 133).

LILY BULBS

Many Westerners think of lilies, species of the genus *Lilium*, only as attractive ornamentals, yet bulbs of various lilies have long served as human food in eastern Asia and other parts of the world. In Japan, for example, the bulbs of various wild *Lilium* species are collected in the forest by both Ainu and Japanese, and processed and consumed in various ways ("Lily flowers and bulbs used as food," 1889: 117-18). In addition, several species of lilies (a general term for lilies in China is *pai-ho*, "hundred together," referring to the many scales forming the bulb), including what may be the oldest lily domesticate, the tiger lily (*L. tigrinum*; *chia-pai-ho*), are believed to have been brought into cultivation in China from ancient times onward for their edible bulbs (Vavilov, 1949-50: 24; H. L. Li, 1959: 115-16; 1970: 11, 16; K. C. Chang, 1970: Table 1; 1976: 18; Mathon, 1981: 59; Zeven and de Wet, 1982: 41). Today the tiger lily and other species of *Lilium* are grown for food not only in China, but in Japan, Korea, and Vietnam as well. Such bulbs, which one Western observer likens to parsnips in flavor, were valued products in traditional China, significant both in domestic trade and as exports to overseas Chinese communities. For example, lily bulbs (identified as from *Lilium brownii*, *L. tigrinum*, and other species) were imported from China and sold fresh or dried in the Chinatowns of San Francisco and other American cities, more a delicacy than an ordinary vegetable (Blasdale, 1899: 20-21; Porterfield, 1951: 25-26). Lily bulbs may be dried (*pai-ho-kan*) or eaten fresh (*hsien-pai-ho*). They may be baked or grated (lily flour = *pai-ho-fên*), and they are often added to soup (Read and Wagner, 1940: 217). They may also be given as a tonic with chicken broth, and they have other medicinal uses as well (Soubeiran and Dabry de Thiersant, 1874: 107; F. P. Smith, 1969: 240; Perry, 1980: 237-38). Nutritionally, fresh lily bulbs (*L. lansifolium* = *L. tigrinum*?) (Synge, 1980: 135-36) are notable only as a fair source of ascorbic acid, but dried ones (*L. tigrinus*) are high in calories, iron, and fairly high in ascorbic acid (Harris *et al.*, 1949: 34; Leung *et al.*, 1973: 54).

BAMBOO SHOOTS

The term "bamboo" is applied to various genera of the grass family whose species commonly provide long, woody canes suitable for construction purposes. In California, the bamboo plant is generally perceived as an attractive ornamental, but sometimes as aggressive and spreading out of control. The strength and utility of its canes may also be recognized, but whatever their uses, they are few compared to those found in the Orient. The role of bamboo there is so important that, even in a study focussing on food, we would be remiss if we failed to mention it. Schafer (1967: 178) writes of an early legend in Kwangtung, Kwangsi, and northern Vietnam that portrayed bamboo as "a gift of the gods," a story reminiscent of that of the infant Moses, of a woman finding a boy floating in a bamboo stem, and of his becoming the hero of aboriginal peoples, a "bamboo king" worshipped throughout the region.

In ancient China, bamboo (*chu* or *chu-tzŭ*) was likely grown as far north as the Yellow River, probably on a small scale. In T'ang times, it cannot have been abundant in the north, as suggested by the delight of Po Chü-i (A.D. 772-846), quoted at the beginning of this chapter (Waley, 1962: 116), with the plentiful bamboo shoots of his new province (Szechwan?) compared to the northern capital city of Lo-yang. Today bamboo is widely cultivated from the Yangtze Valley southward, particularly in the coastal provinces from Kiangsu to Kwangtung (T. H. Tsien, 1985: 59). Bamboos raised commercially in southern China are not propagated from seed, but solely by vegetative means, by dividing and/or cutting culms, rhizomes, or branches of existing plants (McClure, 1938: 4-8). Bamboo is a rapidly growing plant, and all parts of it, including leaves, canes, shoots, and roots, are used in China, and in ways so striking in their variety as to lead to it being called a "universal provider." How many its uses may number has long been a matter of conjecture among Westerners, with the Jesuit Nicolas Trigault, writing in the seventeenth century, suggesting six hundred altogether, a conservative estimate. Besides serving as an ornamental, bamboo may be employed as construction material for bridges; for rafts; masts, ropes, and fishing and boating equipment (with its shavings used as caulking); for water-wheels, agricultural implements, fencing, basketry, traps, snares, and cages; and for yokes, beds of ox-carts, and other equipment for land transport, including the carrying pole, the basic way for humans to transport goods. It may be employed in building a house, thatching its roof, and providing its furniture, blinds, matting, and household goods of other sorts. It may be made into personal items such as hats, raincoats and umbrellas, sandals and shoe soles, combs, and pipes for opium or tobacco.[49] It may be fashioned into toys, musical instruments, boxes, works of art, and into pens and paper.[50] It may provide items used in ritual, such as divining sticks and blocks, mats on which devotees kneel, and the foundation material of incense sticks. If sufficiently wide, its leaves may be used as wrappings for dumplings to be steamed. And finally, of primary concern to us, its shoots may provide food for humans (S. W. Williams, 1895, 1: 358-60; Ball, 1906: 72-75; E. H. Wilson, 1929: 293-94; Porterfield, 1933: 177-83; McClure, 1956: 335-55; E. N. Anderson, 1988: 176).

Bamboo shoots were significant items in Chinese diet in ancient times (K. C. Chang, 1977b: 28). In Han times they were mentioned frequently among the foods served at feasts (Y. S. Yü, 1977: 67), and for T'ang times Schafer (1977: 88) includes them among "the old reliable foodstuffs." They were used in varied ways indeed. One reads, for example, of honeyed bamboo shoots, a delicacy; of "pickled or salted" ones; and of bamboo shoots, along with vegetables and live frogs, in a soup made in South China. There was even a separate department set up on the grounds of the T'ang palace to care for the bamboo gardens that supplied succulent shoots for the royal table. In addition, the court received bamboo shoots of special quality from distant places at one time of year or another (Schafer, 1977: 90-91, 108, 131, 134).

There are several genera and many species of bamboo in China, with thirty-three species listed in one index of Chinese plants. Bamboo shoots used as human food in China come from many of these species, and differ in time of appearance, edibility, need of special preparation, and importance from one locality to another.[51] In eastern China, the best quality shoots are said to come from species of the genus *Phyllostachys* (especially *P. edulis*, *P. bambusoides*, and *P. mitis*), which is important both in China and Japan (*P. edulis* is likely the most important single species in supplying bamboo shoots in China as a whole). In southern China, species of *Bambusa* (especially *B. arundinacea* and *B. vulgaris*), a genus of warmer and rainier parts of Asia, are also important sources of shoots (E. H. Wilson, 1929: 294-96; McClure, 1938: 14; Porterfield, 1938a: 282).

Bamboo shoots are buds that may, depending on the type of bamboo, grow from the base of older canes or from an underground stem or rhizome. Shoots are cut from the mother plant when young, just as they start to appear at the surface of the ground. The ground around emerging bamboo may be mounded up or emerging shoots may be covered with baskets to prevent exposure to the sun, which turns them bitter and fibrous.[52] When cut, a shoot is conical in form and covered with tough, overlapping leaves that are removed to expose a tender core, its edible part (H. L. Chung and J. C. Ripperton, 1929: 2-3; Porterfield, 1938a: 282; 1951: 20-21; Zee and Hui, 1981: 7).

Bamboo shoots (*chu-sun* or *tung-sun*) are preferred fresh in China, and precautions, such as packing them in mud, are taken to preserve their crispness. Their protective leaves removed, fresh shoots are commonly parboiled in water to remove any hydrocyanic acid they may contain.[53] They are then cut, chopped, or shredded into slices, strips, or other pieces. These may be prepared in various ways such as steaming, boiling, or stir-frying; they may be added to soups, stews, or other dishes. For storage or transport to distant markets, moreover, they may be salted, pickled, sugared, canned, steeped in oil, or dried.

The processing of dried bamboo shoots as reported by Porterfield (1951: 21) extended over two days, and involved several steps: washing the newly-dug shoots; removing sheath leaves; boiling in salt water, then drying the shoots in an oven; rolling, pounding, slicing them; and then drying the slices further in an oven. Finally the slices were graded according to tenderness, after which they were ready for packing and storage or shipment. For every hundred pounds of fresh shoots, there were only ten pounds of dried ones.

Some provinces of China, among them Fukien, Chekiang, and Szechwan, were noted for their bamboo cultivation and output of dried bamboo shoots. Within those provinces, moreover, there were districts where people were dependent largely on such production for their livelihoods. As reported in 1930 for one such district in Chekiang, processing of bamboo shoots was carried out in small workshops which operated for about two months a year. Each workshop turned out from about 400 to 1500 pounds of dried shoots, which were shipped not just to nearby regions, but to Tientsin, Peking, Mukden, Hong Kong and other ports of South China, and even to Singapore (Porterfield, 1938a: 284-85).

Analyses of the shoots of many species of bamboo shoots from different localities show them not to be outstanding in any way. They averaged 91 percent water, 4.5 percent digestible carbohydrate, 2.6 percent protein, 0.3 percent fat, and provided 27 calories of food energy per hundred grams (Young, 1954: 384; Leung *et al.*, 1973: 32-33).[54]

WILD RICE

The Indians of eastern North America collected the seed of wild rice (*Zizania aquatica*, also "water rice," "Indian rice," "Canada rice") in swampy areas for food. This wild rice has come into modern American diet as a delicious, though expensive, substitute for true rice (*Oryza sativa*), or as an additive to give true rice a special flavor and texture. Wild rice (*Zizania*) is also found in various parts of eastern Asia from Siberia on the north to Malaya and Assam on the south (Herklots, 1972: 498; Thrower and Chan, 1980: 24; Mathon, 1981: 10-11). Herklots (1972: 498) says that the wild rice of eastern Asia is of the same species, *Z. aquatica*, as that of North America. Thrower and Chan (1980: 20-21), on the other hand, insist that it is a distinct species, *Zizania caduciflora* Turcz. (= *Z. latifolia* [Griseb.] Stapf = *Hydropyrum latifolium* Griseb.), but like *Z. aquatica* a hydrophytic, perennial grass sometimes called "Manchurian water-rice."

Manchurian wild rice, more commonly referred to as *Zizania latifolia*, is not only wild, but is cultivated in Japan, Korea, China, Annam, and Malaya (Crevost and Lemarié, 1917: 172; Uphof, 1959: 390; Thrower and Chan, 1980: 24).[55] Its place of domestication is believed to be China, with most authorities favoring its northern or central areas (Vavilov, 1949-50: 24; H. L. Li, 1970: 12; Harlan, 1975: 73; Mathon, 1981: 10-11, 64; Zeven and de Wet, 1982: 39), and with H. L. Li (1983: 43) suggesting that it may have been grown in the Yangtze Basin since the Neolithic period. In some places in the Yangtze and Huai areas, as we have seen above, it was once grown in floating gardens. Today its cultivation in China is most common in the Yangtze Valley and areas to the south.

In ancient China, *Z. latifolia* was used for its seed. In the *Chou-li* ("Ritual of the Chou dynasty," probably late Chou), what seems to have been *Zizania* (*ku*) is included, along with rice, wheat, and millets, among the six grains of China. It was also included in such a list in the *Li-chi*, "Book of Rites" (Han period). H. L. Li believes that it was probably never widely "or intensively cultivated as a grain." Moreover, by at least the eleventh century A.D., its seed had ceased to be used except in periods of scarcity, and that remains the case today (Bretschneider, 1882-95, 2: 138-39, 158-59; 3: 350-52; F. P. Smith, 1969: 211; H. L. Li, 1983: 44-45). Thus the seed of wild *Zizania* rice, an expensive delicacy ("wild rice") in the United States today, is not normally eaten in modern China.

When *Z. latifolia* is cultivated today, very few, if any, plants flower and seed (Thrower and Chan, 1980: 23), for they are grown for their stems (culm) and young inflorescences, eaten as vegetables. Such a vegetable, according to H. L. Li (1983: 44-45), is first mentioned in herbals dating from Sung times, though one Sung writer sought to identify it in the *Erh-ya*, a work probably dating from the second century B.C. By itself, the stem is tough, and requires the growth of a parasitic smut fungus to turn it into a tender delicacy, *chiao-pai* or *chiao-sun* in Mandarin, or *gau-sun* or *kaau-sun* ("sticky shoot") in Cantonese. Thrower and Chan (1980: 20) note considerable confusion in the literature about the role of the host, *Z. latifolia*, and the fungus, *Ustilago esculenta*, in creating gau-sun. Some observers have suggested that the end product is largely fungus, but, in contrast, Thrower and Chan say that gau-sun consists mostly of the tissue of the host, *Z. latifolia*, swollen by growth of the fungus.

It has been said (Thrower and Chan, 1980: 20) that gau-sun is probably "the only product of a plant host and a fungal parasite that is systematically cultivated and eaten as a vegetable."[56] The fungus does not grow on other plant species and cannot even grow on dead tissue of *Z. latifolia*, and it remains a mystery how the association of host and parasite originated (Thrower and Chan, 1980: 20, 24). One suggestion (H. L. Li, 1970: 12) is that the pathological condition may have come about when a northern plant was introduced into more humid areas to the south; another (Thrower and Chan, 1980: 25), that human selection was involved in fostering the association between plant and parasite.[57] Even in the former eventuality, humans would have had to recognize the value of infected stems, still another example of Chinese sensitivity to a broad range of food possibilities.

As described by Thrower and Chan (1980: 23) for the New Territories, Hong Kong,[58] *Z. latifolia* was cultivated in fields usually covered with about four inches of water. After a field was hoed and fertilized, cuttings were planted in the mud, March and April being the best months. Fertilizer was added periodically, and in about four or five months the plant was ready for harvesting, done by hand at intervals of a week or two. In harvesting, both the swollen part of the culm, including the leaf sheath, and the normal-sized part of the culm nearby, were taken from the plant, and marketed in bunches of a dozen or so. The mycelium of the fungus remains from year to year in the rhizome and culm, so that each newly-formed shoot is infected. Indeed it is such infected plants that farmers want for cultivation, and they throw uninfected ones away, with the result that the latter are rarely found in gau-sun fields.

In Cantonese dishes, gau-sun is used, like bamboo shoots, to provide texture, and indeed another of its Cantonese names means "white bamboo shoot." When prepared, the outer green layer of a stem is first removed, leaving a product nearly white and a bit spongy. The stems, which have a pleasant sweet taste, may be consumed raw or cooked, with cooked ones favored (Bretschneider, 1882-95, 3: 351; Read and Wagner, 1940: 218; F. P. Smith, 1969: 210; Herklots, 1972: 499). Stems are first softened by parboiling, then prepared with dishes to which they are added. They may be sliced and stir-fried with meat and vegetables, or sliced, fried with other ingredients, and then simmered in a stock or broth for ten to twelve minutes (Thrower and Chan, 1980: 24). As E. H. Wilson (1913, 2: 63) observed, "from a European standpoint it is really very good eating."

One analysis of the food value of gau-sun (Zee and Hui, 1981: 33) shows every 100 gm. to contain 93 gm. of water; little protein (1.5 gm.), fat (0.1 gm.), or carbohydrate (4 gm.); and quite small amounts of minerals and vitamins.

The stems of gau-sun, as well as the rhizomes and seeds, serve various medicinal uses, for they are believed to be "very cooling" (Read and Wagner, 1940: 218; F. P. Smith, 1969: 210-11; Perry, 1980: 172).

CELERY

Celery (*Apium graveolens*; *ch'in* or *ch'in-ts'ai*) is found in a wild state in moist places in North Africa and temperate areas of Europe and Asia. Both Vavilov (1949-50: 36) and Zeven and de Wet (1982: 118) place its domestication in the Mediterranean region. It may have been used in Europe initially as a medicine, then as flavoring, and finally as food (Herklots, 1972: 127-28; Schery, 1972: 512). As for the earliest cultivation of celery in China, considerable confusion exists because terms used for celery have also been applied to other plants such as parsley, cress, and water drop-wort (Bretschneider, 1882-95, 2: 183; F. P. Smith, 1969: 42-43; E. N. Anderson, 1988: 128). H. L. Li (1969, 1970) does not include celery among the food plants of early China. It was likely first cultivated there several centuries after the beginning of the Christian era, with Laufer (1919: 402) speculating that a vegetable *hu-ch'in* mentioned in a seventh-century account may have been celery (though he acknowledged the possibility that it was parsley).

Whatever the date of its first cultivation there, celery is used by the present-day Chinese, its leaves and stalks sometimes added to soups and other dishes. F. P. Smith (1969: 43) writes of it as a common vegetable in China, but E. N. Anderson (1988: 128) observes that celery is far more prominent in American Chinese cooking in which it is a low-cost filler, and that many Chinese gastronomes are dissatisfied with such dishes. Celery is a good or fair source of calcium, and vitamins A and C (Leung *et al.*, 1973: 39). The Chinese regard it as "cooling," soothing, tonic, and an aid to digestion; and celery juice, pressed from the stalk, is a common medicine (F. P. Smith, 1969: 43).

WATER DROP-WORT

A plant used in ways similar to celery is water drop-wort (*Oenanthe javanica* DC = *O. stolonifera* Wall.; *shui-ch'in* or "water-*ch'in*") (Figure 52), which is sometimes referred to incorrectly as Chinese or Oriental "celery."[59] It is a herbaceous perennial that grows in ditches, marshes, and other wet areas. Both wild and cultivated forms are utilized. Domestication of the plant seems to have occurred in South China and it was a common vegetable, even a delicacy, in ancient China (H. L. Li, 1970: 10, 16; 1983: 45). It was cultivated in paddies then, as it still is, and it was of moderate dietary importance (H. L. Li, 1969: 254, 258; Schafer, 1977: 91).

Today, the leaves and stems of the plant may be used in various ways. They may, for example, be added as flavoring to soup. Either raw or cooked, they may be consumed with rice and other foods or they may be pickled. In nutritional value, water drop-wort contains good amounts of vitamin A, calcium, and iron and a fair amount of riboflavin (Leung *et al.*, 1973: 73).

Several wild species of *Oenanthe* are highly toxic, especially their roots and seeds (I. H. Burkill, 1966, 2: 1604), and certain kinds of *Oenanthe stolonifera* are also reported to be toxic (F. P. Smith, 1969: 290).

TOMATO

The tomato (*Lycopersicon esculentum*; *fan-ch'ieh*, or "foreign eggplant") was domesticated in the New World, with Mesoamerica and Peru both possible places of origin (Rick, 1976: 269). The tomato was first carried to the Old World following discovery of the Americas, and the present-day tomatoes of Europe and the United States derive almost completely from those initial introductions (Rick, 1976: 269).

The tomato requires a distinct warm season and good drainage, but it adapts well to a variety of soils and climates and under some conditions can provide enormous harvests of 50 tons per hectare (Rick, 1976: 268). It is a very good source of vitamins A and C. And it is tasty, colorful, and can be added with good results to a broad range of dishes. For all these reasons, the tomato ultimately gained importance in many parts of the world.

There was, however, a curious delay in using tomatoes as food among people of the Mediterranean lands, northern Europe, and areas of overseas European settlement. The plant was initially cultivated in Europe as a curiosity, or for the beauty or supposed aphrodisiac qualities of its fruit which led to its being called "love apple." Its use as food, however, was delayed because the plant was recognized as a relative of the deadly nightshade and was feared to be toxic. It was probably with this knowledge that it was described as "rank and stinking" in

Figure 52. Water drop-wort (*Oenanthe stolonifera*) (from Crevost and Lemarié, 1917: opp. p. 184)

England. In any case, after some centuries the tomato gradually gained popularity as food, initially in eighteenth century Spain, then Italy and France, and finally, in the nineteenth century, in Britain (Ainsworth-Davis, 1931: 122; Norman, 1972: 281-82; C. A. Wilson, 1973: 346).

As for China, we have found nothing to suggest a similar fear of using the tomato as food. Yet, though its use was encouraged by the presence of foreign enclaves along China's coast, the tomato was accepted slowly after its introduction in the sixteenth century, and even today, according to one account by E. N. Anderson (1988: 94, 131), the tomato remains largely restricted to the cooking of urban Cantonese. In another statement, however, he writes (E. N. Anderson and M. L. Anderson, 1977:328-29) that tomatoes and chili peppers were responsible for changing the very character of cooking in South China, greatly altering the taste of dishes and improving nutrition by providing abundant supplies of vitamin A and C, as well as minerals. Tomatoes were especially favored in the South, he wrote, because they could bear nearly all year under the mild climatic conditions that prevailed, and in addition they provide high yields and are simple to cultivate.

The chili peppers will be considered in our section on spices, for the Chinese, like us, place them with pepper. Despite their importance in American Chinese dishes, sweet peppers are quite uncommon in China except in Hong Kong and other places subject to strong Western influence. They have not, however, become part of ordinary cookery (E. N. Anderson, 1988: 131-32), and for this reason we will not consider them further.

EGGPLANT

Eggplant (*Solanum melongena*; *ch'ieh*) was domesticated in India or Southeast Asia. The Persians seem to have introduced it from India to Africa, and the Arabs to Spain (I. H. Burkill, 1966, 2: 2081). In early China it was sometimes called "Malayan melon" or "Malayan purple melon" (F. P. Smith, 1969: 412; Schafer, 1977: 93), suggesting that it may have been introduced from Southeast Asia. It was cultivated in what is now Kwangtung in the fourth century A.D. (H. L. Li, 1979: 68-70). Though the eggplant is often regarded as an annual, it is in fact a perennial which, where not killed by frost, can grow into a strong plant up to eight feet high. Apparently in early Kwangtung it was grown as a perennial known as the "eggplant tree," and it continued to be grown in this way at least until the seventeenth century (H. L. Li, 1979: 69-70).

Several varieties of eggplant were found in traditional China, and they contrasted in shape, color, and date of maturity. Some were very large, a foot in length and over two pounds in weight (E. H. Wilson, 1929: 344). Even in early China several varieties seem to have been known, including one with fruit "the size of a marble." They were eaten raw or cooked and, as in later times, may have been pickled as well. Eggplant seems, however, to have been of minor importance in early North China, whereas today it is a very common crop and major summer vegetable in the north as well as the south (H. L. Li, 1969: 256; 1979: 68-70; Schafer, 1977: 93).

The most popular forms of eggplant in South China today are long and narrow, often purple-red, white, or green in color (Herklots, 1972: 381-82; Zee and Hui, 1981: 47; K. Lee and A. Branyon, 1984: 60-61) (Figure 53). Because they are picked younger than our eggplant, usually they are not as bitter in taste. China has considerable variety in its eggplant dishes, both in ways of preparation (steamed, boiled, fried, or otherwise)[60] and flavorings. As one would expect, some have flavorings that mark them as Chinese or as from one of China's regional cuisines, for example the hot dishes "Ginger Juice Eggplant," "Sweet-and-Sour Eggplant" (W. W. Chang *et al.*, 1970: 389-90), "Red-Cooked Eggplant" (B. Y. Chao, 1972: 152; K. Lo, 1974: 64-65), and "Eggplant, Szechwan Style," a stir-fried dish prepared of cooked eggplant, with garlic, gingerroot, sesame oil, and vinegar as flavorings and served with a spicy sauce (F. Lin, 1975: 206).[61] Eggplant may also be served in cold dishes such as a steamed or boiled "Eggplant Salad," a dish of both the present-day and of the Ch'ing dynasty court (F. L. Chang and G. M. Fuchs, n.d.: 308, 352; Davis, 1966: 118; G. Z. Chu, 1975: 110-11). It may also be pickled.

Despite the tastiness and variety of eggplant dishes, the vegetable's nutritional contribution was insignificant, since cooked eggplant is 93 percent moisture and 5.6 percent carbohydrate, and its vitamin and mineral content is low (Leung *et al.*, 1973: 44). Though the eggplant has

Figure 53. Chinese eggplant in market (Ch'eng-tu, 1987)

been called "mostly flavor and water" (White and Selvey, 1974: 157), it was, nevertheless, well regarded in China. One reads that wealthy families in Sung China "had a passion for obtaining the very first eggplants of the season" (Freeman, 1977: 155), and that Ming royalty made temple offerings of eggplant to family ancestors (Mote, 1977: 217). Besides its food uses, eggplant had certain medicinal ones, though prolonged use was believed to bring on digestive troubles and to damage the uterus (F. P. Smith, 1969: 412).

OKRA

Okra (*Abelmoschus esculentus = Hibiscus esculentus*) is generally believed to be of sub-Saharan African domestication. Joshi and Hardas (1976) argue that this has not conclusively been demonstrated and that the plant, in part or entirely, may have originated in Asia, perhaps India. On the other hand, there is no evidence that it was an ancient food plant there (de Candolle, 1886: 190). It is not mentioned by Prakash (1961) in his account of the foods of ancient India, and it does not have a Sanskrit name (de Candolle, 1886: 190), or at least not one on which authorities agree. All this, together with the existence of a single name for okra in modern Indian languages, has been taken to indicate a late introduction, perhaps following the start of the Christian era (de Candolle, 1886: 191). According to I. H. Burkill (1966, 1: 1184), okra's Malay name is of Hindi origin. This suggests that it arrived from India, but late. As for China, okra is not included in any of the sources on ancient foods and food plants we have consulted (H. L. Li, 1969; 1979; K. C. Chang, 1977b). When and how it was introduced to China remains a mystery, though a route by way of Southeast Asia seems likely.

Okra recipes are absent in the cookbooks we consulted, and we have found nothing to indicate how okra is prepared in China. As elsewhere, the immature pods are presumably eaten uncooked, boiled, or fried as vegetable, or they are sliced and added to dishes such as soup or stew (Herklots, 1972: 360). We have found nothing to indicate whether, as in Africa, Arabia, India, or Southeast Asia (Watt, 1889-96, 4: 239-40; I. H. Burkill, 1966, 1: 1185), the seeds of okra are added to coffee or made into a drink substituting for coffee. As in India and Southeast Asia, however, okra flowers, seeds, and roots are viewed as demulcent and diuretic. They are prescribed in cases of difficult labor and as soothing applications to burns, scalds, sores, and wounds (F. P. Smith, 1969: 205).

Okra has a nutritional value much like string beans, not outstanding in any one nutrient, but a significant source of a few, especially calcium and ascorbic acid.

ASPARAGUS

The place of domestication of asparagus (*Asparagus officinalis*; *lung-hsü-ts'ai*, or "dragon's-whiskers vegetable," *lu-sun*, or *chiao-sun*),[62] is given as Eastern Europe or the Mediterranean area, where it was grown by the Romans in pre-Christian times (Vavilov, 1949-50: 36; P. M. Smith, 1976: 315; Mathon, 1981: 23; Zeven and de Wet, 1982: 157-58). Originally asparagus was regarded as a medicine, though later it became a favorite vegetable, and was introduced to, and often become naturalized in, many parts of the world. For China, we have not found *A. officinalis* on any list of early food plants. It was only after the occupation of Cochin China by France that *A. officinalis* was introduced to Indochina (Crevost and Lemarié, 1917: 154). With reference to China, Anderson (1988: 129) writes that it is "a very recent introduction," and Herklots (1972: 165), that the asparagus of North China was introduced, not native.

Asparagus is marketed in China in a dried, salted form, as well as in a fresh one (Read and Wagner, 1940: 218). Fresh asparagus may be prepared in various ways. It is commonly stir-fried, whether directly or after parboiling, alone or as part of a main dish.[63] As an alternative, it may simply be simmered along with other ingredients of a dish. It may, after parboiling, be made into a cooked salad.[64] Or it may be added to soup.[65]

There are also wild species of the genus *Asparagus* in eastern Asia, and some of them, too, have long been sources of medicine or food. For China, the most frequently mentioned of these is "shiny asparagus" (*Asparagus lucidus*), a perennial herb found in Indochina, China, Taiwan, Korea, and Japan. In the case of *A. lucidus* (*mên-tung* or *t'ien-mên-tung*), the plant part of greatest interest to humans is not the stem, but the tuber, which in China is of value principally for medicinal purposes (Bretschneider, 1882-95, 2: 67-68; 3: 311-14; Hooper, 1929: 19-20; F. P. Smith, 1969: 55-56; J. D. Keys, 1976: 56-57; Perry, 1980: 234-35). The tubers of *A. lucidus* were in Chou Ting-wang's fifteenth-century famine herbal, but, according to Read, they must be washed to remove their bitterness before use. The cores of the tubers may then be boiled; or they may be dried in the sun and cooked with honey (Read, 1977g: 29).[66] They may also be preserved in sugar as a sweet (F. P. Smith, 1969: 56).[67] There are reports of *A. lucidus* being cultivated in China, at least in some places (Shantung, Peking area, Chekiang), but mainly or entirely for medicinal use of its tubers, some of which were exported (Hooper, 1929: 20). However in Japan, given by certain botanists (Vavilov, 1949-50: 24; Mathon, 1981: 54) as its place of domestication, *A. lucidus* is cultivated for food purposes, its tubers being boiled and consumed as a vegetable (Uphof, 1959: 38).

[1] For a more complete accounting of Chinese vegetables, see Herklots, 1972; Zee and Hui, 1981; and Wittwer *et al.*, 1987: 253-76.

[2] Both Ho Ping-ti (P. T. Ho: 1975: 16-18) and K. C. Chang (1977a: 84, 120, 485; 1986: 87, 110-11) place the Pan-p'o phase of the Yang-shao culture in the fifth millennium B.C.

[3] Schafer (1977: 90) does touch on the antiquity of European forms of *B. oleracea* in China in noting that kohlrabi was cultivated in T'ang times. Indeed E. H. Wilson (1929: 341), in his excellent account of the upper Yangtze Valley early in this century, mentions no European forms of *B. oleracea* other than kohlrabi, saying that the Szechwanese grow very fine ones.

[4] White mustard (*Sinapis alba = Brassica alba*), whose origin is placed in the Mediterranean area by Zeven and de Wet (1982: 108), seems to have been in China by the middle of the first millennium A.D. (Laufer, 1919: 380; S. H. Shih, 1962: 49; H. L. Li, 1969: 257 Schafer, 1977: 112).

[5] For a recipe to "Pickled Chinese Mustard Greens," see Hom and Steiman, 1981: 88.

[6] For a Chinese way of preparing malva in soup or as a vegetable, see S. Y. Hu, 1956: 30. For a Jurchen recipe for "Mallow Cold Soup," see Franke, 1975: 174. The latter recipe includes some unusual ingredients, such as wild red currants (*Schizandra chinensis*), "shredded half-bred ducks' eggs, mutton tongue, kidneys, intestines, meat with skin from head and feet, all shredded." The Jurchen (Juchên) were a Tungusic tribal people who conquered northern China early in the twelfth century A.D. and established a dynasty called Chin.

[7] Malva leaves, seed, and other plant parts continue to be used medicinally (S. Y. Hu, 1956: 24; F. P. Smith, 1969: 256-57; Perry, 1980: 255).

[8] For a recipe to "Spinach Soup" (made with pork, onion, and celery), see I. C. Chang, 1970: 229; for "Meatball with Spinach Soup" (with ground pork and cellophane noodles), see F. Lin, 1975: 223-24; for "Beef Ball Soup with Spinach" (with minced beef, chestnuts, and onion) and "West Lake Spinach Soup" (with ham and chicken), see K. Lo, 1979a: 78-79, 148.

[9] Recipes for stir-fried spinach will be found in Joyce Chen, 1962: 173; K. Lo, 1971: 161; 1974: 21-22; 1979c: 111; B. Y. Chao, 1972: 144; F. Lin, 1975: 188; Hush and Wong, 1976: 146; and C. B. T. Lee and A. E. Lee, 1980: 231-32. A more elaborate stir-fried spinach dish is one from Canton: "Spinach with Crab Meat Sauce" (K. Lee and A. Branyon, 1984: 167).

[10] For "Spinach Salad with Dried Shrimps," see F. L. Chang and G. M. Fuchs, n.d.: 309, and B. Y. Chao, 1972: 144.

[11] According to Bray (1984: 119), the first mention of such floating gardens in China, in a poem about the Yangtze, dates from about A.D. 300.

[12] Some, but by no means all, early observers described the *chinampas* of Mexico as floating gardens that would be moved from place to place. If their accounts are true, floating *chinampas* survived from the time of the Spanish conquest until early in the nineteenth century. Today, however, Mexican *chinampas* are all non-floating, and most scholars dismiss the early accounts of *chinampas* which floated (Gibson, 1964: 321, 557).

[13] The importance of water spinach in summer is well illustrated in a survey conducted in Kwangtung and Fukien in 1935. In July, 95 percent of all families consumed water spinach, a percentage higher than for any other leafy vegetable, including kale (85 percent) and cabbage (65 percent). In the previous January, by contrast, none consumed water spinach, but 99 percent consumed cabbage, 87 percent kale, and 84 percent spinach (Ta Chen, 1940: 90).

[14] See Senft *et al.*, 1981 for a comprehensive bibliography on the amaranths.

[15] See National Research Council, 1984: 36-38 for other food uses of amaranth grains, and Cole, 1979: Appendix B for recipes involving amaranth seed and greens.

[16] For further discussion of the nutritional value of grain amaranths, see Cole, 1979: 275-79; Senft, 1980; and National Research Council, 1984: 34-35.

[17] Needham and Lu (1985: 62), however, write of literary allusions to grain amaranths in Szechwan and elsewhere in China in the tenth century A.D., but add that the antiquity of such amaranths in China is unknown.

[18] The use of amaranths as dyestuffs also seems to have been entirely a New World phenomenon in pre-Columbian times (J. D. Sauer, 1950: 412).

[19] The dye may be used by the Chinese as liquid or powder, for coloring foods and sweets.

[20] For recipes, see H. L. Chung and J. C. Ripperton, 1929: 17.

[21] Others have identified *ting-li* not as *N. indicum*, but as some plant whose seeds were medicinal (Bretschneider, 1882-95, 2: 55; F. P. Smith 1969: 155-56). In any case, *N. indicum* is a weed in China today; and I. H. Burkill (1966, 2: 1560) reports medicinal use of its seed by the Chinese, who obtain it from Indochina.

[22] Lindqvist (1960: 345-47), however, argued against the widely-held theory that *L. sativa* is descended directly from *L. serriola*, and favored instead a hypothesis involving origin of cultivated lettuce by hybridization involving other species of *Lactuca*.

[23] The Egyptian view was not held by the Greeks, who, on the contrary, considered lettuce to be an antiaphrodisiac, to bring on impotence (Darby *et al.*, 1977, 2: 678; Harlan, 1986: 8-9).

[24] It was this milky fluid that led to the plant's Latin name, *lactuca*, and to its French one, *laitue*. Our word "lettuce," which some early English authors call "lectuce," is probably a corruption of the Latin form (H. Phillips, 1827, 1: 315).

[25] Lettuce was called "raw vegetable" because of a statement in the *Pen-ts'ao* that it should be eaten raw, with vinegar and salt, rather than being cooked (F. P. Smith, 1969: 229).

[26] F. P. Smith (1969: 229) writes that the *Pen-ts'ao* states that *Lactuca sativa* was brought to China in Han times. Laufer (1919: 401-2), however, notes that the *Pen-ts'ao* statement in question, which relies on an eleventh century work, does not specify a date, Han or otherwise.

[27] Seeds of asparagus lettuce were obtained from western China in 1938 by the W. Atlee Burpee Company, which subsequently marketed them in the United States under the name "celtuce" (Lindqvist, 1960: 326; Herklots, 1972: 137).

[28] The stems we saw in South Chinese marketplaces in May, however, averaged about ten inches in length.

[29] Besides food uses, lettuce and its seed serves medicinal uses, for which see Soubeiran and Dabry de Thiersant, 1874: 169; F. P. Smith, 1969: 229-30; and Perry, 1980: 95.

[30] *Chenopodium album* is in the same family as the beet, but a closer relative of quinoa (*Chenopodium quinoa*), long a major food crop, raised primarily for its seed, among Indian groups of the Andes (Bois, 1927-37, 1: 402-4; I. H. Burkill, 1966, 1: 529).

[31] The Chinese Buddhist monk Fa-hsien, on returning to Shantung from a pilgrimage in India early in the fifth century A.D., is said to have been certain that he was actually in China when he once more saw the *li-ho* vegetable, which Bretschneider (1882-95, 2: 261-62) and Read (1969: 104) identify as *Chenopodium album*, a common weed in North China. One wonders about the matter, however, and why Fa-hsien did not observe *li-ho* in India. For *C. album* is also a common weed in many parts of temperate and tropical India, where its seed and greens are collected and eaten; in addition, it is cultivated by some peoples there, and in northern Burma and Annam (Watt, 1889-96, 2: 265-66; Bois, 1927-37, 1: 399; I. H. Burkill, 1966, 1: 529; Read, 1977g: 52, 61). The greens of *C. album* (like those of certain other *Chenopodium* species) have also been used for food in more distant parts of the world (*i.e.*, in the Hebrides of Scotland, among Indians of the American Southwest, and gold seekers in California) (Bois, 1927-37, 1: 397, 398-401; I. H. Burkill, 1966, 1: 529-30; Renfrew, 1973: 170). *C. album* seeds have also been widely eaten, and finds of them in various prehistoric sites in Europe suggest that they were collected for food there (Bois, 1927-37, 1: 398-401; Renfrew, 1973: 170). Zeven and de Wet (1982: 149) think it likely that *C. album* was also cultivated in Neolithic Europe.

In China, *C. album* is included among famine foods listed in the fifteenth-century famine herbal compiled by Chou Ting-wang (Read, 1977g: 52, 61). It also has medicinal uses in China (Read, 1969: 104; Perry, 1980: 77), but we do not know whether any of China's peoples have cultivated *C. album* on a regular basis for food or medicine. *C. album* leaves are reported to be high in calcium, beta-carotene, riboflavin, and vitamin C, and the seeds, high in protein (16.1 gm./100 gm.) (Leung *et al.*, 1973: 48; Read, 1977g: 52, 61).

[32] Also mentioned in the *Shih-ching* (early Chou times) and possibly used as vegetables in ancient China, was *fan*, which has been identified as mugwort (*Artemesia vulgaris*; in modern times: *ai*) or a related species, *Artemesia sieversiana*; and *p'u*, identified as the common cattail (*Typha latifolia*) or grass-leaved sweet flag (*Acorus gramineus*) (Keng, 1974: 402, 403; K. C. Chang, 1977b: 28; Read, 1977g: 29; H. L. Li, 1979: 53; S. Y Hu, 1980: 5, 100). In modern times, the cattail is found in many places in China, sometimes grown for its tender shoots, which serve as a vegetable (F. P. Smith, 1969: 447; H. L. Li, 1983: 44, 45). In modern Malaya, where they are used by Chinese, sweet flag (*Acorus gramineus* and *A. calamus*; "Acorus reed" = *ch'ang-p'u*) serves primarily for medicinal purposes, and mugwort (*Artemisia vulgaris*) for flavoring food as well as medicinal purposes (I. H. Burkill, 1966, 1: 34-38, 247-48). Medicinal uses for sweet flag and mugwort are also reported for China (F. P. Smith, 1969: 12-13, 52-54; Perry, 1980: 34), where both seem to be famine foods as well (Read, 1977g: 21, 29-30). Of special interest is the fact that since very early times, sweet flag has been an auspicious plant in China, one used by Taoists in an effort to achieve immortality (H. L. Li, 1979: 53-54); and that in modern times, sweet flag and mugwort are hung up at home to exorcise evil spirits at the time of "Dragon-Boat Festival." It is believed that the pointed, dagger-shaped leaves of sweet flag are weapons of which spirits are particularly afraid (Cormack, 1935: 165, 166; F. P. Smith, 1969: 13, 53).

[33] This plant should not be confused with the cooking banana (*Musa paradisiaca*), which is also called "plantain." *Plantago major* is a common household weed of Eurasia, now dispersed widely around the world. Since antiquity, its seeds, leaves, and roots have served medicinal uses among many peoples from Europe to East Asia (Watt, 1889-96, 6, Part 1: 285; I. H. Burkill, 1966, 2: 1797-98; Renfrew, 1973: 180). In China, the plantain grows not only around houses, but along roadways, which led to one of its names, *chê-ch'ien-ts'ao*, or "cart-track plant." It seems always to have been collected mainly from the wild, but there are also reports of it being cultivated, and with Zeven and de Wet (1982: 42) placing *P. major*, var. *asiaticum* in their Chinese-Japanese hearth of domestication. As elsewhere, *P. major* in China has primarily served medicinal purposes, and today both the plant, fresh or dried, and its products may be found in Chinese drug stores at home and abroad (Hooper, 1929, 109-10; McClure and Hwang, 1934: 24; I. H. Burkill, 1966, 2: 1797-98; F. P. Smith, 1969: 335; Perry, 1980: 316-17).

At the same time the food use of *Plantago major*, mentioned in the early Chinese literature (but likely unimportant), today is called minor or rare (Bretschneider, 1882-95, 2: 99, 256-57; 3: 210-12; F. P. Smith, 1969: 335; Keng, 1974: 402; Read, 1977g: 10; Zeven and de Wet, 1982: 42; E. N. Anderson, 1988: 27). One wonders, however, whether some selection for tenderness of leaf may have taken place, for Read (1977g: 10), in reference to plantain as a vegetable, writes that Chinese forms of plantain have thin leaves that are more edible than those of the common plantain found in other lands. The seeds of the plantain are very rich in vitamin B_1 (thiamine), and the Chinese make them into a jelly highly regarded for its cooling properties in summer (Hooper, 1929: 110; E. H. Wilson, 1929: 346; I. H. Burkill, 1966, 2: 1798; Read, 1977g: 10). Its raw leaves, in turn, are rich in calcium, beta-carotene, and riboflavin (Leung *et al.*, 1973: 60).

[34] The bottle gourd may be one of the earliest cultivated plants in the world, and was known in great antiquity in both the Old and New Worlds. In China, seeds of *Lagenaria siceraria* were uncovered in the early Neolithic site of Ho-mu-tu in Chekiang, which dates from roughly 5000 to 3500 B.C. (H. L. Li, 1983: 48; K. C. Chang, 1986: 210).

[35] In Cantonese, *C. moschata* is *faan-kwa* ("foreign gourd") or *naam-kwa* ("southern gourd") (Herklots, 1972: 284).

[36] Though this onion is called "Welsh," it does not derive from or have anything to do with Wales. The term, instead, seems to be a corruption "of the Anglo-Saxon *welise* and the German *welsche*," meaning "foreign" (H. A. Jones and L. K. Mann, 1963: 41; Herklots, 1972: 396; Schery, 1972: 503). According to de Candolle (1886: 68), this onion was not known in ancient Greece or Rome, and seems to have come to Europe by way of Russia only in the Middle Ages. Laufer (1919: 303) thinks that this may have been the cultivated *Allium* mentioned in the Chinese literature in the seventh century A.D. as deriving from the West, which would be at odds with the view that it is a Chinese domesticate.

[37] When Western chives were first mentioned in the Chinese literature, in the fifth century A.D., they were designated *hu-swan*, "garlic of the Hu" or "Iranian garlic," thus revealing a Western origin (Laufer, 1919: 302).

[38] In the *Code of Manu* (?600 B.C.-A.D. 300?) there is a long list of foods forbidden to Brahmins and other upper castes. Among the array of objectionable foods were many of animal origin. Also included were garlic, leeks, and onions, which were listed with mushrooms and other plants deriving from unclean matter. Should a twice-born man knowingly eat such foods he would lose his social standing and become an outcast (*Code of Manu* V. 5, 19-20; or Bühler, 1886: 170-72). A similar list of forbidden *Allium* occurs in the *Apastamba dharma-sutra* (?500 B.C.-A.D. 200), but here "red garlic" is specified (*Apastamba dharma-sutra* I. 5. 17. 26; or Bühler, 1882: 63). Judging from the *Mahabharata*,

the great epic poem of the Hindus, onions and garlic were apparently not consumed by honorable people (*Mahabharata, Karna-parva* [200 B.C.-A.D. 500] 44.17 and *Mahabharata, Anusasana-parva* [A.D. 200-400] 91. 38, 39; or P. C. Roy, 1919-30, 5: 111, 10: 156). Similar prohibitions were found in a Hindu code of law of comparable age (*Yajñavalkya Smriti* [A.D. 100-300] I. 171, 176; or Vidyârnava, 1918). There is also a relevant statement in the chronicles of the kings of Kashmir written by Kalhana, a twelfth century poet-historian who was probably a Brahmin of Buddhist leanings. Kalhana wrote of a Hindu king who removed, from consecrated land in the vicinity of a new shrine, those Brahmins who ate garlic (*Rajataramgini*, I. 341-43; or Stein, 1900, 1: 51). That these prohibitions were widespread in India, apparently among Hindus and Buddhists alike, is indicated by two famous Chinese Buddhist pilgrims of the seventh century A.D. The first, Hsüan-tsang, commented that in India onions and garlic were "little used and people who eat them are ostracized" (Watters, 1904-5, 1: 178). The second, I-ching, wrote that people in India eat no onions of any sort because they cause pain when fasting, harm the stomach, progressively weaken the body, ruin the eyesight, and enhance disease (Takakusu, 1896: 44-45, 137-38).

39 One is Prakash (1961: 221), who attributes to Kalhana, the twelfth century writer, the statement that some people rejected onions because, when cut, they resembled flesh, and garlic because it came from abroad. We have been unable to locate those statements in the writings of Kalhana cited by Prakash. In keeping with those statements, however, are Crooke's observations (1885: 41; 1926: 291) that certain Hindu castes refuse to eat onions because of a belief that they turn into meat, that the name for garlic means "root of the barbarian," and that the plant seems to be a recent introduction to India.

40 Garlic was not to be accepted by Jain monks or nuns when begging (*Acaranga Sutra* II. 1. 8. 13; or Jacobi, 1884: 110). It was, however, simply one of a long list of foods not to be accepted, apparently solely because of *ahimsa* considerations (concern that insects or other forms of life might be in the forbidden foods).

41 There is a story in an early Buddhist collection of tales, legends, riddles, and puzzles, the *Jataka* (*c.* 450 B.C.), indicating widespread use of garlic by Buddhists. The story is of Buddhist nuns who regularly went to a farm owned by a lay brother where they obtained a supply of garlic, two or three handfuls each. On one holiday, their garlic supply exhausted, the nuns went to get some, but the manager of the farm did not have any at home. So he directed the nuns, led by Sister Fat Nanda, to a field where they could pick some. The sisters took so much garlic, however, that the angry farm manager complained of their greed. Word of this reached the Buddha, and he condemned them, too. Significantly, however, he objected only to their greed, not to their garlic eating, which suggests no strong objection to the practice (*Jataka* I. 136; or Cowell, 1969, 1: 292-93).

In the *Vinaya Pitaka*, early regulations set down for the governing of Buddhist monastic life, there is a story of a monk who had eaten onions and sat apart at a meeting where the Buddha was preaching (*Vinaya Pitaka, Kullavagga* V. 34. 1; or Davids and Oldenberg, 1885: 153-54). The monk did this so as not to annoy his fellow monks, presumably with bad odors. When the Buddha learned of this, he banned the eating of onions by monks so as to enable them to participate fully in such meetings. The rule seems to have been a pragmatic one, involving an offense (*dukkata*) that could bring on ill or suffering. Nothing in the account suggests that onions or other *Allium* were viewed as unclean. Indeed, the Buddha is quoted as permitting monks to eat onions for medicinal purposes (*Kullavagga* V. 34. 2). This is in keeping with the use of garlic in various formulae in an Indian Buddhist medical treatise, the Bower manuscript (*Bower Manuscript* I. 1-43; Hoernle, 1893-1912: 10-15 *passim*).

42 Though onions and garlic are important foods in India today, the ancient attitudes have not been eliminated altogether. One reads in R. M. Martin's account (1838, 1: 121) of Bihar and Patna, for example, that though Moslems consume onions, among Hindus they are used only by the lowest [castes]; that most Hindus use garlic, which is purer than onions; but that Brahmins and Rajputs reject both vegetables. Watt (1889-96, 1: 171) writes that strict Bengali Hindus consider onions "as objectionable, and rarely if ever eat them." And similar attitudes are found today among the Brahmins of Benares (pers. comm., D. O. Lodrick).

43 For information on cultivation of *Volvariella* mushrooms in southeastern China, see S. T. Chang, 1965, and in the Canton area, see Benemerito, 1935-36, and Espino, 1935-36.

44 For Chinese vegetarian dishes with mushrooms, see especially W. W. Chang *et al.*, 1970: 376-404; K. Lo, 1974; and Y. K. Martin, 1984: 217-18.

45 Cormack (1935: 123-24), whose observations focus on the Peking area, reports the use of wood fungi at the mourning ceremony, held on the third day following a funeral, when food is offered and paper money burned for the spirit of the dead person. At that time, two white-flour cakes are also taken to the grave. Each is split apart, and a bit of *mu-êrh* inserted. The cakes are then set at the head of the coffin, one on each side, in the hope that the wood of the coffin may open like a door and permit the soul of the deceased to come and go.

[46] The story of Hawaii as a supplier of *Auricularia* mushrooms to China has not been fully told, but starting in 1852 the Hawaiian mushroom called *pepeiao akua* (*Hirneola polytricha* = *Auricularia polytricha*, the so-called "hairy jew's ear"?), which grows on decaying wood of the tree *Aleurites moluccana*, began showing up on lists of goods shipped to San Francisco. Collecting, sun-drying, baling, and shipping was entirely in the hands of Chinese, who initially obtained their supply entirely from Oahu but subsequently from other islands as well. In later years, shipments were directed to China, a steadier and more profitable market than San Francisco, with exports in 1864, the peak year, amounting to 368,000 pounds. Production declined in subsequent years and ceased altogether in 1888 (Thrum, 1914: 201-3). This is nevertheless an interesting parallel to other Chinese efforts at seeking delicacies overseas, some of which will be considered in our chapter "Four Chinese Delicacies."

[47] One modern method of cultivating *mu-êrh* involves chopping down small oak trees, cutting them into poles six to ten feet in length, and leaving them on the ground to rot for a year. Then, their wood quite rotten, the poles are made into a "shed-like structure" from which the *mu-êrh* grows spontaneously, to cover the structure in two years or so. Then, when the wood is fully rotten, *mu-êrh* is no longer produced (I. H. Burkill, 1966, 1: 270).

[48] For which, see B. Liu and Y. S. Bau, 1980, and Wittwer *et al.*, 1987: 293-99.

[49] In ancient Kwangtung and South China, fibers of one kind of bamboo, perhaps *Lingnania chungii* or some other species of its genus (formerly included in *Bambusa*), were used for making cloth. Such bamboo cloth is unknown in modern China (Schafer, 1967: 178; H. L. Li, 1979: 135-36).

[50] Bamboo, because of its long fibers and modest cost, has been important in paper manufacture in China since mid-T'ang times, perhaps at first as a substitute for hemp and rattan. Kwangtung is the first place mentioned in connection with bamboo paper manufacture, which probably began there (T. H. Tsien, 1985: 59-61). For the process of making bamboo paper, see C. P. Li, 1948: 162-64, and Y. S. Sung, 1966: 224-30.

[51] There is also a report for one district of the seed of bamboo being collected and eaten, and its quality was said to be similar to that of rice (A. Henry, 1893: 51).

[52] In tests made of one Chinese species of bamboo, *Sinocalamus beecheyanus*, in which such precautions were not taken, the strongly unpleasant smell and taste of the shoots could be eliminated only by such lengthy cooking as to make the shoots unsuited for consumption (Young, 1954: 384).

[53] Tests in the United States found raw shoots of various bamboos, among them three of the important species above (*P. edulis*, *P. bambusoides*, and *B. arundinacea*) to have an unpleasant, pronounced acrid taste, though cooking and one or two changes of water removed nearly all of the unpleasantness (Young, 1954: 380-84). It is not clear whether, in these tests, precautions were taken to keep the emerging shoots from exposure to the sun, which can make bamboo shoots acrid. Even with such care, however, some types of bamboo require special treatment before use. I. H. Burkill (1966, 1: 298) has noted that in Malaya treatment may consist of storage in mud for a few days. He also mentions that the Chinese parboil certain types of bamboo shoots and then soak them awhile before cooking them for the last time.

[54] For medicinal uses of bamboo, see especially F. P. Smith, 1969: 63-65, and Perry, 1980: 162-63, 169.

[55] Chinese immigrants probably introduced cultivation of the plant to Malaya, for there is just a single report of wild *Z. latifolia* there, and Perak, where it is cultivated, has had large numbers Chinese settlers for a century or more (Thrower and Chan, 1980: 24).

[56] Fungi are, of course, widely eaten in their own right around the world, and are important in making certain human foods or drinks, among them bread, beer, grape wine, and some kinds of cheese. In China, they are also essential in the preparation of important indigenous edibles, such as fermented soybeans and soy sauce, as well as in unusual ones, such as caterpillars infected with *Cordyceps* fungus (see our section on "Insects" in the "Flesh and Fowl" chapter). The claim here, however, is that the case of gau-sun is unique in that it is both systematically cultivated and consumed as a vegetable.

[57] Thrower and Chan (1980: 25) also suggest that humans selected the plant for stems that were less woody. In support of this, they noted finding in abandoned fields of *Z. latifolia* certain plants that had returned to a primeval state, smaller and far more woody than cultivated forms. They also observed that importers of gau-sun from China regard the plants of the Canton and Shanghai areas to be different cultivars.

[58] Gau-sun is imported to Hong Kong, but not cultivated there. It was, however, grown on a small scale in the New Territories until 1969, when its cultivation ceased (Herklots, 1972: 498; Thrower and Chan, 1980: 24).

[59] There has also been inconsistency of usage in China. If we read our sources (Bretschneider, 1882-95, 2: 71-72, 183-84; 3: 396-98; F. P. Smith, 1969: 42-43, 289-90; Schafer, 1977: 91; H. L. Li, 1983: 44, 45; E. N. Anderson, 1988: 26, 128) aright, in early times the term *ch'in* was used (though not applied to it alone) for water drop-wort, that native plant also sometimes called *shui-ch'in* ("water-*ch'in*") in contrast to celery, which today may be designated *han-ch'in* (dry-soil *ch'in*) or *hsi-ch'in* ("western-*ch'in*").

[60] The variety in preparation is sometimes illustrated in the names of eggplant dishes, such as "Fried Eggplant Cakes" (made of eggplant slices soaked in batter, then fried), "Steamed Eggplant with Minced Meat," "Barbecued Eggplant" (W. W. Chang *et al.*, 1970: 387-90), and "Deep-Fried Stuffed Eggplant" (K. Lee and A. Branyon, 1984: 302). For five vegetarian eggplant dishes, see K. Lo, 1974: 31, 36, 42, 58, 64.

[61] For other Szechwan and Hunan-style eggplant dishes, see W. W. Chang *et al.*, 1970: 387, 389; Stallard, 1973a: 123-25, 129; G. Z. Chu, 1975: 203-4; Hush and Wong, 1976: 193; Sakamoto, 1977: 275-76; C. B. T. Lee and A. E. Lee, 1980: 169-70; Hom and Steiman, 1981: 82; Gin, 1984: 98-99; K. Lee and A. Branyon, 1984: 299-303; and Y. K. Martin, 1984: 150-51.

[62] *Sun* means "shoots," and *chiao-sun* is the same name used for the shoots of wild rice (*Zizania latifolia*), which bear a certain resemblance to asparagus.

[63] Examples of dishes in which asparagus is prepared in this way are Cantonese "Beef in Oyster Sauce" (Y. K. Martin, 1984: 129-130), Shanghai "Abalone with Asparagus and Turkey," and "Szechwan Pork with Asparagus" (W. W. Chang *et al.*, 1970: 144, 299).

[64] A recipe for such a salad, flavored with soy sauce, sugar, and sesame oil, is found in Field *et al.*, 1976: 42.

[65] For example, "Chicken Soup with Asparagus Shoots and Bean Curd" (W. W. Chang *et al.*, 1970: 102).

[66] Schafer (1977: 91) writes of *A. lucidus*, not identified as to state of domestication, providing "leaves" (cladophylls?) consumed as greens in T'ang China.

[67] The tubers of *A. lucidus* preserved in sugar have long been used as a cough medicine in Annam (Crevost and Lemarié, 1917: 154).

CHAPTER SIX

Seaweeds and Other Algae

Their Importance in East Asia
Types of Seaweed Eaten
A Note on Edible Microscopic Algae in Traditional China
Beginnings of Seaweed Cultivation in China
Marine Algae in Chinese Diet
Marine Algae and Human Nutrition
  Minerals
    Iodine
    Calcium
    Iron
  Vitamins
  Protein
  Carbohydrates
Medicinal Uses

"More vile and worthless than seaweed," an expression used by Virgil and Horace, is taken by certain authorities (W. A. P. Black, 1953: 32; Zaneveld, 1959: 89) as symptomatic of an ancient and widespread European indifference to seaweeds, macroscopic marine algae, as human food. In that continent, the authorities note, people ate seaweed primarily in areas of poverty and at times of food shortage, and today, though seaweed is consumed in scattered parts of the continent, such as the British Isles, its dietary use is of quite minor and declining importance. Though seaweed is utilized on a considerable scale in Europe, this is mainly for fertilizer and for feeding animals rather than people (Chapman and Chapman, 1980: 30-55).

THEIR IMPORTANCE IN EAST ASIA

The situation is quite different in East Asia, which is part of the world's greatest area of seaweed eating, or phycophagy, which stretches from Southeast and East Asia across the Pacific (Leber, 1977). Within that area, Japan is among the lands best-endowed with seaweed resources. Seaweed use is ancient in Japan, and seaweed is a well-liked food, even a delicacy. The Japanese are also the largest and most sophisticated consumers of seaweed and seaweed foods in the East Asia-Pacific region. This is even more so in modern times than in antiquity, for since the eighteenth century the Japanese have cultivated seaweed, not just collected it. They have also carried out more research on seaweed than any other people in the East Asia-Pacific area, and probably in the world as a whole. As a result, we refer regularly in this account to Japan and Japanese research, for in East Asia Japan is the greatest source of seaweed knowledge, much of it relevant to China.

China's coast extends across 25 degrees of latitude, and its seaweeds derive from two quite different floristic regions, the North Pacific and the Indo-West Pacific. Most of its seaweed species are temperate, subtropical, or tropical ones, but in the Yellow Sea there are fair numbers of cold temperate species and even a small number of lower boreal ones (Tseng, 1983: i). Overall, however, China is considered poorly endowed with seaweed resources as compared to Japan because of the high temperatures of China's seas in general, a sea bottom in large part unfavorable for seaweed growth, with such growth further hindered by sediment and fresh water carried by China's great rivers (Kirby, 1950: 203-4). It is claimed that the Chinese taught the Japanese how to use seaweeds and that they use a greater number of seaweed species as food and medicine than the Japanese or any other people (Xia and Abbott, 1987: 341, 351). At the same time, China's development of its seaweed potential lagged far behind that of Japan, with seaweed cultivation not practiced in China until the 1940's, two centuries after it began in Japan.

There is no question, however, that, since antiquity, the Chinese have held seaweeds in high regard, which is suggested by the colloquial name *hai-ts'ai*, "sea vegetable," commonly applied to them. By one ancient account, some algae were viewed as "a delicacy for the most honorable guest, even for the king himself" (Porterfield, 1922: 299). The word for algae, in turn, was used metaphorically as a high compliment; and brightly-colored representations of algae on the roof beams of houses protected against fire because, as water plants, they could repel the fire spirit (Porterfield, 1922: 300).

Because China's domestic supply of seaweed did not meet demand, a sizeable import trade developed. One reads, for example, that in the early and mid-nineteenth century, it was customary for Chinese junks to carry dried seaweed from Malaysia to China, and that in 1935-37 China imported an average of roughly 326,000 tons of seaweed annually, mainly from Japan (Kirby, 1950: 203). In China's southeast, Hong Kong was a major factor in the seaweed trade: in 1956 it imported several hundred tons or more of seaweed from coastal areas of China, from Korea, and from Japan. Some was for local consumption and some for re-export, along with ten tons of local seaweeds going largely to Indochina and Malaysia (B. T. Chiu, 1958: 133).

We have found little to indicate what per-capita consumption of seaweeds may have been in traditional China. There is, however, a recent estimate by Xia and Abbott (1987: 341) that in China as a whole 110 million pounds of seaweed, fresh and dried, are consumed annually. Assuming a total Chinese population of a billion, the annual per capita seaweed consumption would be 50 grams. Seaweed consumption in coastal China, of course, is much higher than the national average, and Xia, judging from personal experience, says that an average family of four living in coastal Shantung may eat a *jin* (=1.1023 lb.) of seaweed on each of two or three occasions a week (Xia and Abbott, 1987: 341), a very high level of consumption indeed.[1]

TYPES OF SEAWEED EATEN

Seaweeds, in the sense of macroscopic marine algae, belong to many genera but only three phyla: *Rhodophyta* ("red algae"), *Phaeophyta* ("brown algae"), and *Chlorophyta* ("green algae"). In China as a whole, over a hundred seaweed species are used economically in some way, and at least 74 as human food or medicine (B. T. Chiu, 1956: 171; Tseng, 1983: i; Xia and Abbott, 1987: 341, 351). At the same, there, as elsewhere in East Asia, two forms of seaweed are consumed in larger amounts than all others together: the large brown algae known as kelp (especially *Undaria pinnatifida*, the Japanese *wakame*; and *Laminaria* spp., the Japanese *kombu*), and the red algae called purple laver (*Porphyra* spp., the Japanese *asakusa-nori*) (Tseng, 1944: 37-38; Chapman and Chapman, 1980: *passim*; Arasaki and Arasaki, 1983: 21-22).

For China's southeast, the most complete study of economically-important seaweeds we have uncovered is that of C. K. Tseng (1935), which is based on a three year survey in Kwangtung.[2] Altogether Tseng identified sixteen genera or species that were economically important in the main coastal areas of the province. Some were used as fertilizer. Some served as pig feed. A few were sources of mucilage, paste, or agar-agar (a dried gelatinous seaweed extract used for food and other uses).[3] And still others were used solely as food or medicine, or as food or medicine along with other uses such as those just mentioned.

Tseng identified five species of seaweed as being of major economic importance, and listed food uses for three of the five. One of the three, *Eucheuma gelatinae*, a red alga, served industrial purposes, as well as being made into agar-agar at home. In East Asia, agar-agar is mainly used in food preparation, and is ordinarily sold in bundles of long, white, thin dry strips.

The other two types of seaweed of major economic importance for which Tseng listed food uses were *Porphyra* and *Sargassum*. There seem to have been two main species of *Porphyra* used in Kwangtung: *P. suborbiculata* and *P. tenera*. When fresh, these *Porphyra* are dark purple in color, and are known in Chinese as *tzu-ts'ai*, "purple vegetable." Collected in traditional China with bamboo rakes from coastal rocks, they are especially plentiful from January to March. Usually they are sun-dried and sold in flat, circular sheets, and are highly regarded as vegetables and soup ingredients (Tseng, 1933: 60; 1935: 99-100). *Sargassum* species (most important is *S. fusiformis*) are mostly used as fertilizer or pig feed.[4] *Sargassum* may, however, also be prepared as human food or made into a sort of "tea." For food, plants are collected in spring before fruiting and are cooked fresh, especially for use in soup. For "tea," they are sun-dried and prepared much like real tea, as a cooling drink (Tseng, 1933: 58-59; 1935: 100; B. T. Chiu, 1956: 171).

Food uses were also indicated by Tseng for certain seaweed of minor economic importance. Among them were *Gracilaria confervoides* and another species of that genus, *Monostroma* sp., *Hypnea musciformis*, *Eucheuma edule*, and *Grateloupia ligulata*. Green laver or sea lettuce (*Ulva lactuca*) was listed both as pig feed and a "cooling" medicine for humans, and elsewhere (Tseng, 1983: 258) as human food; and *Hypnea cervicornis* was food both for pigs and humans.[5] These, like all seaweed considered above except *Porphyra*, were collected in spring or summer.

Not to be overlooked are the large brown algae, *Laminaria japonica* and related species, which traditionally were imported in a dried, sometimes powdered, form. *Laminaria japonica*, commonly known in China as *hai-tai*, "sea belt" or "sea ribbon,"[6] occurs naturally in cooler seas and even in Japan comes almost entirely from the northern island of Hokkaido (Chapman and Chapman, 1980: 15, 73). Judging from literary sources, *Laminaria* were first imported to China from Korea in the fifth century A.D. Over the past several centuries, however, it was Japan which provided most *Laminaria* products used in China, and Japanese *kombu* was sold widely in China, even in places thousands of miles form the coast (Tseng, 1944: 38; T. H. Cheng, 1969: 219-21).[7] It was eaten particularly when green vegetables were scarce, an inexpensive food and well-liked when properly cooked, as with pork and soy sauce (T. H. Cheng, 1969: 219).

A NOTE ON EDIBLE MICROSCOPIC ALGAE IN TRADITIONAL CHINA

Though by our definition they are not seaweeds, certain microscopic algal species from the phylum *Cyanophyta* ("blue-green algae") were also traditional foods in East Asia. We mention them here because as algae they seem to fit best with their relatives, the seaweeds. One edible species, a terrestrial form, is the highly-valued delicacy *fa-ts'ai* ("hair vegetable"), commonly identified as *Nostoc commune* or *Nostoc commune*, var. *flagelliforme*,[8] which grows in black tangled clumps that look like hair. It is found in parts of the interior, such as Mongolia, Kansu, Shansi, and Szechwan, especially the arid northwest where it is said to grow in desert springs, mountain streams, and in places where lowland areas are flooded (Blasdale, 1899: 46; Uphof, 1959: 249; J. Y. P. Chen, 1973: 232; Wittwer *et al.*, 1987: 257; E. N. Anderson, 1988: 133; Cost, 1988). Kansu is said to be the main source of this alga today (Cost, 1988), and it is traded widely across China and exported to overseas Chinese for use as food and medicine.[9] The alga is first soaked and cleaned to remove extraneous material. Then it may be prepared in various ways: for example in soup; by stir-frying or sautéing and then being added to a dish, such as the Cantonese "Duck with Hair Seaweed" (W. W. Chang *et al.*, 1970: 278); or by boiling to make a gelatinous substance that serves as a thickening agent and also provides a slippery texture and delicate flavoring, as in the traditional New Year's dish "Dried Oysters and Hair Seaweed."[10] Cantonese use of hair vegetable at the time of New Year is explained in terms of its name resembling their greeting at that season, which has an approximate meaning of "Have a prosperous New Year" (Cost, 1988: 3). Hair vegetable, as is also true of seaweeds, is common in Buddhist vegetarian dishes, with one such dish, given by Cost, also including bean curd, golden needles (dried lily buds), mushrooms, and other vegetables (Blasdale, 1899: 46; J. Y. P. Chen, 1973: 232; G. Z. Chu, 1975: 284; Madlener, 1977: 123, 207; Y. K. Martin, 1984: 24, 25, 218, 221; E. N. Anderson, 1988: 135; Cost, 1988: 3). Another edible blue-green alga, which resembles *Nostoc*, is *Brachytrichia quoyi*. This, however, is a marine plant, blue-green or blackish-green in color, which is common on intertidal rocks along coasts of the South China Sea and Yellow Sea, and is readily observed and gathered (Arasaki and Arasaki, 1983: 87; Tseng, 1983: 40). Xia and Abbott (1987: 351) say that it is the only of the marine blue-green algae commonly consumed as food in China,[11] and is especially abundant and of good size on Taiwan. *B. quoyi* may be stored by shaping it into balls, pressing out the water, and letting it dry in the sun for a day; when used, it is first soaked, and then consumed in various ways, as with fish and pork, or, with sugar added, as a candy (Xia and Abbott, 1987: 351).

BEGINNINGS OF SEAWEED CULTIVATION IN CHINA

In the mid-1920's, *Laminaria japonica* was first discovered growing near the port of Dairen in Manchuria. It is believed that spores had inadvertently been carried from Hokkaido to Dairen on the bottoms of lumber rafts and the cargo vessels that towed them. In any case, over the next few decades *Laminaria* became naturalized in the Dairen area, and what *Laminaria japonica* was produced in mainland China derived largely from such wild stands. In the early 1940's, the cutting off of *Laminaria* imports from Japan led the Japanese to try cultivating *Laminaria japonica* in Shantung Province, and later, following the Japanese evacuation, the Chinese continued that effort (T. H. Cheng, 1969: 219-21; Arasaki and Arasaki, 1983: 76).

After the People's Republic was established, moreover, there was a great expansion of *Laminaria* cultivation, with large plantings at various places as far south as Kwangtung Province. Cultivation was by two methods. The first, and by far the most important, involved growing seaweed on floating rafts. The second entailed planting on the sea floor. In either case, the success of plantings was helped by a new, higher-yielding strain of *Laminaria japonica* developed for the warm coastal waters of China. Indeed the expansion of *Laminaria japonica* cultivation to China's southeast coast came to be regarded as one of the People's Republic's greatest successes in aquaculture (T. H. Cheng, 1969: 221-28), with production increasing from 12 tons, dry weight, in 1952 to about 275,000 tons in 1975 (Tseng, 1981: 127).

The actions of the People's Republic aimed at encouraging cultivation of marine algae were part of a broader commitment to increasing consumption of aquatic foods. One project involved *Chlorella*, a unicellular, fresh-water green alga that can provide phenomenal yields of protein per acre (Tamiya, 1960: 384-85). *Chlorella* was widely cultivated in the early 1960's for use in making biscuits and sweets, and as a fertilizer and pig feed. Another project was to develop artificial propagation of *Porphyra*, that other important seaweed (T. H. Cheng, 1969: 215). By 1980, there was commercial cultivation of *Porphyra* spp. (7200 tons, dry weight, produced in 1979-80) and, on a much smaller scale, of certain other marine algae (*Undaria pinnatifida*, *Eucheuma gelatinae*, and *Gracilaria verrucosa*) (Tseng, 1981: 143, 148-50). Overall, it has been estimated that cultivation and harvest of *Laminaria* and *Porphyra* now employ a labor force of a quarter of a million people (Xia and Abbott, 1987: 342).

MARINE ALGAE IN CHINESE DIET

Some types of seaweed were too expensive to be eaten regularly by the poor. Others, however, were relatively modest in price, especially near the coast and at times of abundant supply. In Hong Kong in the late 1950's, for example, *Porphyra suborbiculata*, the principal seaweed eaten, sold retail, presumably in a dried form, for about 70 cents U.S. a pound, but *Ulva lactuca*, consumed by poorer people, could be had for only 21 cents U.S. a pound (B. T. Chiu, 1958: 132).

As for preparation, seaweed (whether fresh, dried, or powdered) may be used as a flavoring or major ingredient in dishes prepared in various ways, among them red cooking, stir-frying, steaming, and boiling (Xia and Abbott, 1987: 343-51). It may be added to soup, such as the Cantonese "Seaweed Soup," made of dried laver ("purple seaweed") along with "sherry" (rice wine), ginger, scallions, and an egg (W. W. Chang *et al.*, 1970: 114).[12] Seaweed may be used in fish dishes, such as Ningpo "Green Seaweed and Fillet of Sole" (G. Z. Chu, 1975: 179-80), and in vegetable dishes and ones of other sorts.[13] One of these involves mixing seaweed with meat or chopped vegetables for use as a filling or stuffing for dumplings (Xia and Abbott, 1987: 344). Dried laver (*Porphyra*) sheets, and perhaps other seaweeds, may also serve as a wrapping, as in "Minced Shrimp in Laver Rolls" (K. Lee and A. Branyon, 1984: 124).

Strips of agar-agar may be soaked, boiled, and used as a gelatin (with sugar or other flavors added),[14] commonly eaten in China as a cooling dish in summer, and sold along the street by mobile canteen vendors as a dessert to follow the simple meals they provide poor workers. Agar-agar strips may also be used as a thickening agent in soups and sauces. Or they may be softened by washing or soaking in cold water, then prepared as "Agar-Agar Salad," commonly with ham or chicken and other ingredients (W. W. Chang *et al.*, 1970: 448-49; G. Z. Chu, 1975: 105-6, 269; F. Lin, 1975: 198-99; Hush and Wong, 1976: 24, 204; Xia and Abbott, 1987: 342).[15] Agar-agar also serves as a cheap substitute for birds nests in bird's nest soup (Chapman and Chapman, 1980: 193).[16]

MARINE ALGAE AND HUMAN NUTRITION

European indifference to marine algae as human food may have contributed to a tendency to overlook or belittle their food value. To counter this tendency, some Western seaweed enthusiasts have started using the term "sea vegetable" instead of "seaweed." Whatever general acceptability the new term may gain, nutritional research over recent decades seems to justify its use, for there is now a consensus among experts that, at least in some respects, marine algae can make a significant dietary contribution.

Minerals

Fresh and dried seaweed, for one, are outstanding sources of minerals, including both macrominerals and trace minerals or trace elements (Ishibashi and Yamamoto, 1960; Yamamoto and Ishibashi, 1972; Lontoc *et al.*, 1982).

Iodine

Of the minerals in seaweed, iodine is the one most often mentioned, for seaweeds are very rich in iodine, and seaweed consumption is believed responsible for the virtual absence of endemic goiter in those parts of East Asia near the sea.[17] In Japan, iodine intake is said to be more than double that of any other people (McClendon, 1933: 99), for an individual commonly eats 10 gm. or more (dry weight) of seaweed at a single meal. Seaweeds provide such an abundance of iodine in the Japanese diet that it is unnecessary to iodize table salt (Arasaki and Arasaki, 1983: 46). Salt spray in coastal areas of Japan is rich in iodine, and this is said to account for the much higher than average iodine content of Japanese cereals and fruits (Yamagata and Yamagata, 1972: 84-85). Seaweeds are used as fertilizer in Japan and this, too, may contribute to the higher iodine content of food products there. One wonders, as well, what impact the burning of kelp to obtain iodine, practiced in Japan on a large scale from 1900 at least until 1929, may have had, for one recalls the observation (W. A. P. Black, 1953: 36) that with the end of kelp burning in Western Europe[18] there was a great decrease in the iodine content of the air over Central Europe, presumably to the detriment of humans and other animals. In any case, goiter incidence is quite low on Japan's four main islands (Honshu, Kyushu, Shikoku, Hokkaido), a problem mainly in certain mountainous interior regions (McClendon, 1933; 1939: 85, 91, 100; Kelly and Snedden, 1960: 182-84).

As for traditional China, by far the most important seaweeds imported for food are *Laminaria*, which are unusually rich in iodine.[19] Seaweed consumption acted not only to prevent goiter there, but where goiter did occur, especially in the interior provinces, seaweeds or their extracts were used to treat the condition. In the fourth century A.D., for example, the great alchemist and physician Ko Hung wrote of an alcoholic extract made from seaweed, to be taken as a goiter treatment for a month in spring and a month of summer. The T'ang scholar Wang T'ao (eighth century A.D.), in turn, gave thirty-six goiter prescriptions, of which twenty-seven contained seaweed (H. C. Hou, 1938: 351-52).

Calcium

The calcium content of many types of seaweed, especially when dried, is high,[20] and one might expect people in coastal China to benefit greatly from this in nutritional terms. There is, however, a question of availability; is the calcium in forms that can be absorbed by humans? If availability were high, seaweed would be richer in calcium than virtually all other foods; yet research on this question is incomplete (Arasaki and Arasaki, 1983: 45). Johnston (1972) did make an attempt to determine calcium availability for two species of Japanese marine algae, with interesting results: for *Porphyra yezoensis* virtually all calcium seemed to be available, whereas for *Heterochordaria abietina*, 55 percent seemed to be. The finding by Japanese researchers that calcium deficiency was absent among people in coastal Japan, where seaweed consumption was high, but common in inland regions (J. Y. P. Chen, 1973: 233) may therefore be significant, and relevant for China as well as Japan.

Iron

It has long been recognized that the total iron content of most seaweeds is very high, higher than most other foods (Read and How, 1927: 101, 104, 105-6). One seaweed eaten in China and Japan, *Monostroma latissimum*, has an iron content (278 mg%, dry weight) which Johnston (1972: 430-31) called extraordinary. There remain, however, questions about the availability of the iron in seaweeds. Johnston (1972: 434), in his study of two other marine algae, judged only 37 percent of the iron in *Porphyra yezoensis* and 54 percent in *Heterochordaria abietina* to be available.

Vitamins

Various writers (Norris *et al.*, 1937: 432; Rao, 1965: 21-22; Leber, 1977: 8; Chapman and Chapman, 1980: 95-97; Arasaki and Arasaki, 1983: 44) regard marine algae in general as good sources of vitamins. All dried seaweeds used in China on which we have information contain considerable beta-carotene, a precursor of vitamin A, and most have fair to good amounts of vitamins B_1, C, riboflavin, and niacin. *Porphyra tenera*, one of China's most important edible seaweeds, stands out as an unusually good source of vitamin A, with roughly ten times that of fresh spinach. In addition to vitamins B_1, C, and riboflavin, it has good amounts of B_6 and B_{12}. There are, however, considerable variations, by time of year and place of growth, in both the vitamin and mineral content of seaweeds.

Protein

Seaweeds also differ in their protein content by season and habitat, as well as by age and from one plant part to another. Overall, the crude protein, dry weight, of algal species consumed as food in southeastern China ranges from about 8 to 48 percent (Johnston, 1972: 429-30; Chapman and Chapman, 1980: 89-93; Arasaki and Arasaki, 1983: 39; Portugal *et al.*, 1983: 168). Two algal species stand out in this regard: the seaweed *Porphyra tenera*, with from 19.5 to 47.7 percent protein, and the terrestrial alga *Nostoc commune*, with 21 percent. The range of crude protein reported for *Porphyra tenera* is roughly equivalent to that, dry weight, of common beans (25 percent), peanuts (27 percent), whole cow's milk (27 percent), soybeans (41 percent), beef (45 percent), and whole chicken eggs (48 percent) (Briggs and Calloway, 1979: 99). The question remains as to what the digestibility and biological value of this protein may be. This is a matter on which there has been little research but persistent differences of opinion. Research has demonstrated, however, considerable species differences among seaweeds in digestibility of their protein; also that some species have protein of low digestibility, whereas others have protein of high digestibility (75 percent or more) (Bender *et al.*, 1953; W. A. P. Black, 1953: 36; Larsen and Hawkins, 1961: 523; Johnston, 1972: 434).

For algae eaten in South China we have found few studies of protein digestibility. Kimura (1952-53: 179), in a study of three types of Japanese seaweed also used in South China (*Laminaria japonica, Porphyra tenera*, and *Undaria pinnatifida*), found protein digestibility of 57 to 69 percent, within the range for green leaves of land plants mentioned by Briggs and Calloway (1979: 99) (cabbage, 35 percent; mustard greens, 60 percent; kale, 85 percent). Johnston (1972: 434), in turn, reports digestibility of 95 percent for *Porphyra yezoensis*, which is found in coastal areas as far south as Chekiang (Tseng, 1983: 46); this compares well with peanuts (89 percent) and soybeans (90 percent).

The protein of *P. tenera* and other seaweeds eaten in Japan has also been found to be balanced in amino-acid composition, better than land plants and similar to egg white protein (Arasaki and Arasaki, 1983: 39-40).[21] Tamiya (1960: 381-82) reported animal studies with dried sheets of *Porphyra tenera* (Japanese: *asakusa-nori*) which found their protein both high in digestibility and their nutritional value comparable to that of horsemeat.

Carbohydrates

When in a dried state, seaweeds of types eaten in South China contain about 40 to 75 percent carbohydrate. That carbohydrate, however, is complex and for long it was widely believed that most of it was unavailable or of minimal use in nutritional terms (Chapman and Chapman, 1980: 90-93). Even after many years of study with humans and other animals, results are inconclusive. For three common types of seaweed eaten in China and Japan (*Laminaria* sp., *Porphyra tenera*, and *Undaria pinnatifida*), the percentages of carbohydrate digestibility in humans reported by one researcher ranged from 75 to 92 percent, but by another, only 25 to 51 percent (Arasaki and Arasaki, 1983: 46-47). Percentages for other Oriental seaweeds (of the genera *Ulva, Enteromorpha, Hypnea*, and *Gracilaria*) averaged from 9 to 34 percent (Chapman and Chapman, 1980: 94).

The above research was done with persons of contrasting ethnic backgrounds and over varying lengths of time, and one cannot ignore the question, which has recurred over the years, of whether some people around the world have inherited a greater ability to digest seaweed carbohydrates. Present thinking inclines to the view that the ability to digest seaweed is an acquired trait, that regular consumption of seaweed in time brings on intestinal changes—in flora or enzymes—that facilitate seaweed digestion (Leber, 1977: 10; Madlener, 1977: 25-26). If such adaptation does occur, people who regularly eat seaweed, such as those of Japan and coastal South China, would be better able than other groups to digest seaweed carbohydrate. The final answer to the question, however, awaits further research.

MEDICINAL USES

Seaweed is among the foods the Chinese regard as "cold," one proscribed to pregnant women in Hong Kong, and sufficiently cold as to pose a great risk for weak persons and a remarkably strong remedy for persons suffering from excess heat (E. N. Anderson, 1970a: 209; Koo, 1973: 141; Wheeler and Tan, 1983: 53). It is thus no surprise that in Hong Kong one finds seaweeds of various kinds not only in Hong Kong food stores, but in all herbalist shops as well (B. T. Chiu, 1954: 39).

Scientific studies have indeed found certain substances derived from seaweed of species or genera used in South China to have useful or potentially useful medicinal qualities. Some seaweeds exhibit pronounced antimicrobial or antiviral actions, as, for example, in inhibiting the growth of influenza B virus and mumps virus in embryonated eggs (Gerber *et al.*, 1958; Aubert *et al.*, 1979). Others have anticoagulant, antitumor, or other qualities (Schwimmer and Schwimmer, 1955: 13-21; Baslow, 1969: 56-85; Burkholder and Sharma, 1969; Nisizawa, 1979: 244-47; Chapman and Chapman, 1980: 233-38; Arasaki and Arasaki, 1983: 53-60), some of which were known to practitioners of Chinese traditional medicine.

Most important among seaweeds gathered for medicinal purposes in Kwangtung is the red alga *Digenea simplex*, source of a traditional drug called *tse-koo-ts'oi* in Cantonese. In Hong Kong there are warehouses devoted exclusively to storing *Digenia simplex* brought from Pratas Island, 140 miles to the southeast (B. T. Chiu, 1954: 39). There, in a large lagoon, divers collect the seaweed, with production of 300 tons a year in the mid-1930's (Michanek, 1979: 208). After it was collected, the seaweed was sun-dried, packed in jute bags, and shipped either to Hong Kong or, by Japanese collectors, to Taiwan (Kirby, 1950: 287).

Though the time and place of earliest use of *tse-koo-ts'oi* are not known, this drug was used in both China and Japan as an anthelmintic that had no bad side effects (Kirby, 1950: 287; Takemoto, 1978: 2). The Chinese have great confidence in this drug, and it is found in pharmacies throughout the country. Researchers have confirmed its effectiveness, especially against ascarids (Baslow, 1969: 69; Michanek, 1979: 207; Chapman and Chapman, 1980: 234; Arasaki and Arasaki, 1983: 56-57).

The active anthelmintic agent in *Digenea simplex*, isolated by Japanese scientists in 1953, was first called digenic acid, but now is known as kainic acid. An adult taking only 5 to 10 mg. of kainic acid experiences a drop of from 40 to 70 percent in the number of parasitic intestinal worms (Baslow, 1969: 70). Indeed, kainic acid is so effective as an anthelmintic that it is credited with eradicating ascariasis in Japan following the Second World War (Takemoto, 1978: 2). Its effectiveness against ascarids and roundworms is based on its ability to paralyze them, a quality that has led kainic acid, like other excitatory amino acids, to be used as a tool in neurobiological research (Shinozaki, 1978: 17).

Another use of the drug *tse-koo-ts'oi* is as a laxative for children believed to have an infection called "womb poison" (Tseng, 1935: 96). Though we have found no research on the laxative qualities of *tse-koo-ts'oi*, laxatives deriving from agar-agar and kelp (alginates) have been found both effective and non-irritating in treating constipation (Mulinos and Glass, 1953; *Handbook of nonprescription drugs*, 1982: 72-73). Their gentleness and suitability for infants is based on their action as bulk-forming laxatives which dissolve or swell in the intestinal fluid to facilitate passage of the intestinal contents without interfering with absorption of nutrients.[22]

Among the other medicinally important seaweeds in traditional southeast China were the brown algae *Sargassum fusiformis* and *Laminaria japonica. S. fusiformis* is a common drug plant used not only in treating goiter but as a "cooling" agent that cleans the blood, brings blood pressure to normal, and is effective in cases of apoplexy and glandular infection (Tseng, 1933: 58; 1935: 100). One notes that Japanese researchers have found extracts of *S. thunbergii*, a related *Sargassum* species also used in China, to have antibiotic qualities against certain bacteria (Levring *et al.*, 1969: 369-70; Chapman and Chapman, 1980: 236). *L. japonica*, besides its use in treating goiter, was employed in China to reduce high blood pressure and to treat edema and menstrual problems. In this regard, one notes that research with food products or other substances derived from various species of *Laminaria* and/or *Sargassum* has found them to possess antihypertensive, antilipemic, and anticoagulant qualities (Baslow, 1969:

72-73, 79-81; Abe and Kaneda, 1972; J. Y. P. Chen, 1973: 232; Chapman and Chapman, 1980: 236-37; Arasaki and Arasaki, 1983: 53-54).[23]

In a review of antitumor substances of marine origin published a decade ago, C. P. Li *et al.* (1974: 97-98, 119, 121) observed that the oceans have remained largely untapped as a source of new antitumor agents. Noting that some promising antitumor substances had already been identified, they urged an expanded investigation. Most antitumor substances of marine origin at that time were more or less impure, and only one clinically useful drug had been developed. That drug, moreover, derived from study of the sponge, not from seaweed. In contrast to the practitioners of scientific medicine, traditional healers of China had long had extracts of various seaweeds, including *Laminaria* and *Sargassum*, that they used in treating cancer (Yamamoto *et al.*, 1974: 543).

Relevant here is the recent hypothesis advanced by Teas (1981) concerning possible links between seaweed consumption and the incidence of breast cancer. Teas notes that the incidence of breast cancer in Japan is only a sixth that in the United States, but increases in the children of Japanese immigrants. While acknowledging the possible role of dietary fat as an etiological factor, she finds the dietary fat hypothesis lacking. She suggests that attention be given to foods, notably seaweed, that are widely consumed in Japan but not in the U.S., and that might act as anti-carcinogens. She cites recent studies (Yamamoto *et al.*, 1974; 1977) that found extracts of three species of *Sargassum* or *Laminaria* to have an anti-tumor effect in laboratory mice.[24] The extracts were highly effective in inhibiting the growth of implanted sarcoma-180 solid tumor, with over half of the mice in one study experiencing a complete regression (Yamamoto *et al.*, 1974: 543-44). Teas observes that seaweed consumption is high in Japan, sufficient to be of significance in this regard. Moreover, the two areas of lowest breast cancer rates in Japan are also important in seaweed production (and presumably in consumption). Though correlations between high seaweed production and low breast cancer rate are not found everywhere in Japan, Teas notes that other factors must be involved in the etiology of breast cancer and that a perfect inverse correlation would not be likely. Several possible mechanisms by which seaweed may act against breast cancer are indicated. It is further noted that if seaweed is indeed effective in preventing breast cancer, it may be effective as well against colon and prostate cancer, which seem to have a similar geographic occurrence and may have a similar etiology.

[1] If such a family ate one *jin* of seaweed on each of two occasions weekly through an entire year, per capita daily consumption, by our calculation, would be roughly 35 grams. It is difficult to believe that such high consumption is characteristic of any sizeable part of coastal China, for one study of three districts in Japan found seaweed consumption to average 7.3 grams daily per capita (Teas, 1981: 604).

[2] For a recent account of edible seaweeds for China as a whole, see Xia and Abbott, 1987: 344-52.

[3] Two species with industrial uses might be singled out here: the red algae *Gloiopeltis furcata* and *Eucheuma gelatinae*. *Gloiopeltis furcata*, besides being edible, provides a paste used as sizing in the Canton silk industry. In Swatow it is also a hair paste used by ladies. Thus one of the vernacular names for it in China is "glue vegetable." From *Eucheuma gelatinae*, a mucilage is obtained in Canton for mounting paintings, maps, and lanterns. *E. gelatinae* and *Gelidium amansii*, another red alga, are also sources for the commercial manufacture of agar-agar (Tseng, 1933; 1935: 97-99; 1983: 66, 92, 112; B. T. Chiu, 1954: 39; 1956: 171).

Gelidium amansii, it might be noted, is also the most important source of agar-agar manufactured in Japan, and *Eucheuma gelatinae* and related species are additional sources in Japan and Southeast Asia (Zaneveld, 1959: 110-12, 115; I. H. Burkill, 1966, 1: 970-71, 1080-81; Levring *et al.*, 1969: 227-29, 233-37). Japan is the largest producer of agar-agar in the world. For a history of its use there and in other parts of the world, as food and in medicine, dentistry, cosmetics, and other ways, see Humm, 1947; and Chapman and Chapman, 1980: 148-93.

[4] In some places in Kwangtung, various types of *Sargassum* are collected, usually when they are cast ashore in summer, to be used as a low-cost fertilizer. They are applied either in a fresh or sun-dried form to coconut trees and coffee shrubs (Tseng, 1935: 100). In the Amoy area, Sargassums are used as fertilizer mainly for sweet potatoes and peanuts (Tseng, 1933: 59). One judges from the account of Chapman and Chapman (1980: 44-45) of the manurial value of other types of seaweed, that Sargassums, like the others, are relatively high in potash and nitrogen, but low in phosphorus; that the potash is readily utilized by plants, whereas the nitrogen is made available only slowly; and that their high potash content suits seaweed especially well for fertilizing cabbage and root crops.

[5] For other seaweeds eaten in China, see Tseng, 1933; 1983; Kirby, 1950: 204-5; B. T. Chiu, 1958; Zaneveld, 1959; and Johnston, 1966.

[6] For details of *Laminaria* production, see Levring *et al.*, 1969: 351-52; Bardach *et al.*, 1972: 801-13; Chapman and Chapman, 1980: 72-81; and Arasaki and Arasaki, 1983: 71-77.

[7] *Laminaria* products are also traded far across southern Asia. I. H. Burkill reports (1966, 2: 1322-23) that the Chinese of Malaya import a medicine made from *L. saccharina*, that one of its uses is for treating goiter, and that this medicine reaches even remote markets of the Punjab in India.

[8] Other species of *Nostoc* are also consumed in China (Xia and Abbott, 1987: 351), and, it seems, at least one non-*Nostoc* species of terrestrial blue-green algae as well (Chapman and Chapman, 1980: 69; Arasaki and Arasaki, 1983: 87, 94).

[9] In San Francisco Bay area markets, it is sometimes marketed as "Black Moss."

[10] For one recipe to a dish prepared in this fashion, as well as a photograph of "hair seaweed" and of the dish itself, see Y. K. Martin, 1984: 25, 187, 221. For a recipe with fewer ingredients and quite different preparation, see Madlener, 1977: 207. For still other recipes, see Cost, 1988.

[11] Tseng (1983: 32), however, also identifies *Calothrix crustacea*, a marine blue-green alga that grows on intertidal rocks, as a common food.

[12] For other recipes to Chinese seaweed soup, see J. D. Keys, 1963: 1-3; I. C. Chang, 1970: 231; G. Z. Chu, 1975: 91-92; K. Lo, 1979c: 72-73; and M. Gin, 1984: 41.

[13] For other uses of seaweed in Chinese cookery, see Davis, 1966: 113 ("Steamed Pork, Kelp, and Green Onions," a Ch'ing dynasty court dish that came from Honan Province); K. Lo, 1979b: 39 ("Buddhist's Vegetable Ensemble for the Gods" made of seaweed, golden needles, noodles, and other ingredients); Y. K. Martin, 1984: 211 ("Seaweed," a northern deep-fried vegetable dish made from a special form of seaweed not found elsewhere); F. L. Chang and G. M. Fuchs, n.d.: 353 (a seaweed, ginger mix used in a meal to balance hot, spicy dishes); and H. J. Lin and T. Lin, 1972: 132 (fried seaweed used as a side dish, described as providing "an indescribably delicious toasted flavour," and served with popia, the thin Amoy crepes mentioned earlier).

[14] Examples are the sweets "Almond Curd" (Davis, 1966: 144), "Jelly Dessert" (made of agar-agar, sugar, and rice-flour water) (I. C. Chang, 1970: 156), and "Agar-Agar in Syrup" (prepared by simmering the agar-agar in sugary water until it nearly dissolves, then cooling it, cutting it into pieces and serving it in a brown sugar syrup) (H. J. Lin and T. Lin, 1972: 194).

[15] For a recipe for "Abalone and Agar-Agar Salad," a banquet dish in China, see G. Z. Chu, 1975: 101-2.

[16] Agar-agar, it should be noted, is not a seaweed product of particular nutritional importance. The commercial form of agar-agar in Japan contains from 67 to 76 percent galactose, but very little of this carbohydrate can be digested; agar-agar does contain small amounts of protein and fat; its food value, however, is mainly as roughage and as a jelly that protects food or makes liquids more solid for readier consumption (Chapman and Chapman, 1980: 182, 192-93).

[17] Preparation of kelp by the traditional Chinese method of soaking in brine and sun-drying, however, results in significant loss of iodine. One study found that kelp soaked in brine for eight days lost about 45 percent of its iodine (T. H. Cheng, 1969: 219).

[18] In Western Europe, "kelp" factories, in which brown seaweeds (especially *Laminaria* spp. and *Saccorhiza polyschides*) were burned on a large scale, were first established in France and Britain in the seventeenth and eighteenth centuries A.D. Such kelp factories subsequently developed in Norway, Sweden, Denmark, Spain, and Germany. The product was an ash rich in sodium and potassium salts, used in glazing pottery and in the manufacture of glass and soap. Later, iodine was the object of such kelp burning. With development of other sources for these products, however, kelp

burning in Western Europe declined (Taft, 1965: 212-13; Levring *et al.*, 1969: 346-49; Chapman, 1970: 25-27, 40-41; Chapman and Chapman, 1980: 18-19).

[19] For comparisons of the iodine content of *Laminaria* and other marine foods, see Read and How, 1927: 101; T. H. Tang *et al.*, 1936; Grimm, 1952: 319-21; and Arasaki and Arasaki, 1983: 46.

[20] Worthy of special note here is *Nostoc commune*, the fresh water alga, which has more calcium and iron than many seaweeds (J. Y. P. Chen, 1973: 227, 232).

[21] For other material on the amino acid content of seaweed protein, see D. G. Smith and E. G. Young, 1955; Larsen and Hawkins, 1961: 524; Rao, 1965: 18-19; Schmid, 1967: 43-44; and Portugal *et al.*, 1983: 169-70.

[22] The ancient Okinawans considered the dugong to be unusually long-lived, and attributed this to the sea mammal's feed of sea grasses and red algae, including *Digenea simplex*. In an effort to increase their own longevity, they made and ate a porridge made of *Digenea simplex* (Arasaki and Arasaki, 1983: 57).

[23] For other medicinal uses of seaweed in China and Japan, see particularly Kirby, 1950: 287-91 *passim*; Schwimmer and Schwimmer, 1955: 13-21 *passim*; and Arasaki and Arasaki, 1983: 53-60.

[24] In these studies, those species whose extracts proved very effective inhibitors were *Sargassum fulvellum* and *Laminaria angustata*. *L. japonica* extracts had only a minor inhibitive effect. Anti-tumor effects have also been reported by other, mainly Japanese, researchers for derivatives of other seaweed: *Sargassum confusum*, *S. horneri*, *S. hemiphyllum*, *S. tortile*, *Laminaria cloustoni*, *Gloiopeltis complanata*, and *Codium pugniformis* (Yamamoto *et al.*, 1974: 543; Arasaki and Arasaki, 1983: 54).

CHAPTER SEVEN

Fruits

"The loquats are now ripe,
splendent as gold and pearl."

Su Shih (translated by Edward H. Schafer)

Importance of Fruit in Chinese Agriculture and Diet
Kwangtung as a Producer of Fruit
Fruits of Chinese or Southern Asian Origin
- Citrus Fruits
  - History and Variety
  - Sour Oranges, Sweet Oranges, and Mandarin Oranges
  - Pummelo
  - Citron
  - Lemon and Lime
  - Kumquat, Calamondin, Trifoliate Orange, and Ichang Orange
  - Wampee
- Litchi
- Longan
- Loquat
- Persimmon and Its Relatives
- Peach, Apricot, and Chinese Plums and Cherries
- Chinese Jujube
- Chinese Olive
- Some Lesser Fruits of Chinese Origin
  - Chinese Quince
  - Chinese Haw
  - Chinese Gooseberry or Kiwifruit
  - "Human Face"
  - Chinese Strawberry
  - Japanese Raisin Tree
  - Chinese Mulberry

Additional Fruits from Tropical Asia
- Banana and Plantain
- Carambola
- Myrobalans
- Other Fruits of Tropical Asian Origin

Some Introductions from the Near East, Europe, or Africa
- Common Pear and Apple
- Melons
- Pomegranate
- Date
- Fig
- Olive
- Grape

Fruits from the New World

IMPORTANCE OF FRUIT IN CHINESE AGRICULTURE AND DIET

In traditional China, the percentages of total crop area devoted to fruit cultivation were low as compared with Japan and Western countries. By one estimate (1929), China's percentage (0.9) was half that of the United States (1.8) and Britain (1.9), a third that of Germany (2.7), and less than a tenth that of Japan (9.7) and Italy (14.4) (Buck, 1956: 205). As a rule, fruit orchards in China were small and scattered, often just a few trees on the edge of a vegetable garden; fruit cultivation was thus not central to the thinking of most farmers; and T. H. Shen (1951: 217) writes of it as "an avocation" for them. As for trade in fruit, dried fruits could readily be shipped to distant markets in traditional China, but without refrigerated facilities, shipment of many fresh fruits was difficult or impossible. This meant that types of fruit whose cultivation was restricted to a single climatic zone in China were little used in a fresh form, or even known, in other zones. If such fruits were brought in, it was usually for gourmets, well-to-do persons, or for special occasions rather than everyday consumption by ordinary folk. Most people in North China, as a result, seldom if ever ate subtropical fruits, as is well illustrated by the story (V. Y. N. Hsu and F. L. K. Hsu, 1977: 311), perhaps apocryphal, of Wu Chün-sheng, a Manchurian warlord who visited Peking in the 1930's and was entertained at a great feast. After the main courses had been consumed, Wu faced a large bowl of fruit in the middle of the banquet table. Taking a banana, which was unfamiliar to him, he ate it peel and all. His host, to demonstrate proper procedure, selected another banana and, in a conspicuous manner, peeled it, then ate it. To save face, the ingenious Wu took a second banana and ate it as before, but only after commenting "I always eat these things with the peel on."

One type of cultivated fruit, berries, is notable for its absence or limited notice in Chinese cookbooks and accounts of Chinese fruit and food. It is uncertain why berries should not have come under greater cultivation in China, for wild berries, among them raspberries, blackberries, blueberries, and strawberries,[1] some of excellent quality, are found widely in China and serve medicinally and sometimes as food.[2] One factor, noted by Meyer (1911: 49), is that the Chinese appear not to like them, in the same way they don't like other soft fruits. Another factor, suggested by Anderson and Anderson (E. N. Anderson and M. L. Anderson, 1977: 332-33) for fruits and nuts in general, may be that berries provided too small a return for the labor and space they required. In any case, this is a notable instance in which the Chinese failed to develop a potential food resource.

Buck observed (1930: 357) that in China fruit is not regarded as part of a meal, or even as food. If someone does eat fruit, he said, it is done to while away the time, "to play" as farmers might say. Thus per capita consumption of fruit in China was very low, with T. H. Shen estimating (1951: 217) that average per capita consumption was about thirty pounds per year compared to 156 pounds in the United States. Such consumption was even lower for poor people, with a survey of household incomes and expenditures of 283 families in Peking in 1926-27 finding that poor families consumed virtually no fruit. The survey also found that expenditures for fruit were made not in winter, but mainly in summer and early fall, when fresh fruits were available, and that they increased at the time of the Autumn Festival (Gamble, 1933: 61, 69, 74). Lang (1946: 84), in turn, noted that poor workers and coolies in Peking in the 1930's ate fruit, like meat and rice, solely on holidays and at religious affairs.

Also contributing to low fruit consumption was the view, reported for western China where fresh fruit was abundant, that eating such fruit brought on illness (Jenner, 1968: 40). This view may have been based on the fact that in the hot-cold classification of foods, most fruits are "cold" and that pregnant or nursing women and others inclined toward bodily coldness must take care about the fruit they consume. Indeed, many pregnant or nursing Chinese women avoid eating fruit, with watery fruits (melon), sour fruits (lemon or lime, pineapple), and bananas conspicuous among those commonly mentioned. Babies are somewhat on the hot side of things and foods that are neutral to somewhat cooling are excellent for them. Most fruits, however, are too "cold," and are not tolerated well by young babies. Thus London Chinese mothers were found not to include fruit in the diet of infants less than three months old, with some mothers not giving their infants fruit until they were five to six months old (B. Chang, 1974: 438; Ling *et al.*, 1975: 129; M. Mathews and L. Manderson, 1981: 12, 13; S. P. Tan and E. Wheeler, 1983: 45; Wheeler and Tan, 1983: 52, 53, 54, 55). London Chinese women also noted that pineapple is sour and bananas slippery, and that both, if consumed, can bring on miscarriage; bananas, they held, can also lead to a baby with jaundice and muscle cramps (Wheeler and Tan, 1983: 53).

There were, however, certain influences that favored greater fruit consumption. South Chinese returnees from Southeast Asia, apparently affected by their experiences abroad, consumed fruit regularly at meals (Ta Chen, 1940: 98). Western influence, in turn, influenced the use of more fruit among wealthy Chinese (Lang, 1946: 99), and foreigners contributed to an abundance of good fruit (including strawberries, pears, grapes, and bananas) in Peking markets (Read, 1921: 5).

KWANGTUNG AS A PRODUCER OF FRUIT

In 1947, Kwangtung was estimated to have produced 370,000 metric tons of fruit, nearly twice that of any other province in South China (T. H. Shen, 1951: 217). In the country as a whole, however, that production was exceeded or approached in three other provinces: Hunan (940,000 metric tons), Shantung (856,000), and Hopeh (342,000), neighbors within, or largely within, the Winter Wheat-kaoliang Area, and producers of mid-latitude fruits. The situation was very different in China's southeast, where it is possible to grow fruits from the mid-latitudes, subtropics, and tropics. As a result, in Kwangtung one finds fruits from all four of the climatic belts of eastern Asia identified by H. L. Li (1970), as well as from India, the Near East and Europe, Africa, and the Americas. One must admit that some of them were not well suited to southeastern conditions. Yet whatever the difficulties, there were places, favored by microclimate or other local conditions, where a fruit might find a niche in which it could survive and even do well. Examples of this are the mangosteen, a delicious fruit from tropical Asia reported to be grown on Hainan, and the date palm, from the arid Near East, which found a niche in Kwangtung. Other examples are the pineapple, from the New World tropics, and the watermelon, from sub-Saharan Africa. Not to be overlooked are the mid-latitude fruits, such as the peach, pear, persimmon, and plum, though they are better suited to conditions in North and Central China.

Before considering the rich variety of fruits found in the southeast and elsewhere in China, we would caution the reader about English names used for Chinese fruits. In some cases, English-speaking people have named a Chinese fruit after one familiar from their homeland because of a perceived resemblance in flavor or appearance. One should not, however, assume that such a fruit has a close botanical relationship to the similarly-named Western fruit, for it may come from a quite different family and have a different history of domestication and use. One example of this is the "raisin tree," which is no relative of the grape vine and produces an edible stalk rather than a dried fruit. Another is the "Chinese olive," which comes from a tree of a different genus, *Canarium*, from that of the European olive, *Olea*. To compound the confusion, the same English name may be used for quite different Chinese fruits, with "Chinese gooseberry" applied to two fruits quite unlike in appearance that come from plants of different genera. The reader should therefore exercise more than normal care in considering Chinese fruits.

Since it has occupied more acreage in Kwangtung than any other fruit, we turn first to citrus.[3]

FRUITS OF CHINESE OR SOUTHERN ASIAN ORIGIN

Citrus Fruits

History and Variety

In an article on the origins of fruit cultivation in the Old World but with a focus on the Near East and Mediterranean, Zohary and Spiegel-Roy (1975) do not mention citrus fruits at all. This is because, despite their importance today, most of that region was beyond the range of wild *Citrus*, a genus native mainly to tropical and subtropical areas of South, Southeast, and East Asia (Tolkowksy, 1938: 3-5; Swingle and Reece, 1967: 358). It is in the latter three areas that authorities believe the various forms of *Citrus* to have first been cultivated. Harlan (1975: 75), for one, sees all major domesticated forms as originating in Southeast Asia or South China: sweet orange (*Citrus sinensis*), sour orange (*C. aurantium*), mandarin orange or tangerine (*C. nobilis*, *C. reticulata*, and others), lemon (*C. limon*), lime (*C. aurantifolia*), grapefruit (*C. paradisi*), citron (*C. medica*), and pummelo or shaddock (*C. decumana* or *C. grandis*). This has contributed to South China becoming a place of great genetic diversity in citrus, with many unusual wild and cultivated forms. An American plant explorer (Meyer, 1911: 43, 45), for example, writes of eighty different types of edible oranges (including mandarin oranges) in southeastern coastal China and on its offshore islands.

Today the production of citrus fruit in the People's Republic is 1.8 million tons a year (1985), far below that of apples (3.8 million tons) and somewhat less than that of pears (2 million) (Wittwer *et al.*, 1987: 283, 284, 286). In traditional times, citrus was grown as a commercial crop as far north as southern Shensi Province, but China's leading citrus region by far was southeast China, with about 5.6 million citrus trees in 1949, including Taiwan (3.1 million); Kwangtung, with Hainan (1.49 million); Fukien (600,000); and Kwangsi (400,000). The principal types of citrus grown in the region were mandarin orange, sweet orange, and pummelo. Second in importance was the upper Yangtze region, with about 1.6 million citrus trees, including Szechwan (1.485 million), Yunnan (100,000), and Kweichow (40,000). And third of the major regions was the lower Yangtze, 628,000 trees, including Hunan (310,000), Chekiang (160,000), Kiangsi (138,000), Hupeh (10,000), Kiangsu (5,000), and Anhwei (5,000). Fruits were mostly consumed locally, but some were sent to other parts of China or exported, with five ports in Kwangtung in 1933 shipping the equivalent of 433,000 boxes of fruit, each weighing 70 pounds (Webber, 1943b: 102; Burke, 1967: 140-43).

Citrus is usually cultivated on level sites, as in deltas, valleys, or on hill terraces; and manures, among them night soil, are applied as needed. Since commercial citrus cultivation in China is found principally in areas with annual precipitation from 40 to 60 inches which typically occurs throughout the year, irrigation is not ordinarily needed. In the southeast, much commercial citrus is produced in coastal regions that have a high ground water table which limits the depth to which roots can grow, requires the digging of drainage ditches, and limits the trees' profitable period to ten to fifteen years. Typically such orchards are located next to paddies, and rice, vegetables, or melons may be cultivated among the trees. Orchards in south coastal regions are also subject to hurricanes and are commonly protected by bamboo windbreaks (C. C. Hu, 1930-31; Condit *et al.*, 1937-38; Webber, 1943b: 101-2; Burke, 1967: 143-45). Citrus growers of the southeast may attempt to protect their trees further by use of tree-nesting ants to destroy harmful insects. This practice, known already in the fourth century A.D., is said to be the earliest known case of controlling plant pests through use of a biological agent (H. T. Huang, 1986: 531).[4]

Sour Oranges, Sweet Oranges, and Mandarin Oranges

There is evidence (Tolkowsky, 1938: 100-101; Andrews, 1961: 45-46) of orange cultivation in Italy in the first centuries of the Christian era, but the earliest description of the sour orange (*C. aurantium*) in Europe dates only from the thirteenth century, and the first reference to the sweet orange (*C. sinensis*), from the fifteenth century (Webber *et al.*, 1967: 6-12). The mandarin orange (*C. reticulata*), in turn, does not appear to have reached Europe until the nineteenth century.

In China, by contrast, there is clear evidence that oranges have been cultivated since antiquity. The earliest work on the botany of China's southeast, the fourth century A.D. monograph by Chi Han, reports that in Han times there was a salaried official, the Minister of Oranges, whose job it was to present oranges to the royal court. In the succeeding period, a governor from what today is Vietnam presented the emperor with a single "stalk" with seventeen oranges on it, an auspicious symbol that brought congratulations from the royal entourage (H. L. Li, 1979: 118-20). In T'ang times, the fruiting of mandarin trees in the imperial gardens, apparently indoors, led to "formal congratulations to the monarch on his divine charisma" (Schafer, 1977: 90). It is understandable that the Chinese have given the world the earliest surviving monograph on the orange, Han Yen-chih's *Chü-lu*, translated as "Monograph on the oranges of Wên-chou, Chekiang" (Hagerty, 1923). In this twelfth-century work, one reads not only of sour, sweet, and mandarin oranges, but of the related kumquat and trifoliate orange. Drawing both on his own observations and earlier literature, Han Yen-chih presents a picture of diversity, identifying many varieties of oranges. He also displays sophisticated knowledge of planting, grafting, cultivation techniques, irrigation, disease control, and of methods of harvesting, storage, and use. One reads of oranges as highly-regarded fruit, with elders tasting the first mandarin oranges of the harvest, and of the sweet orange extolled in poetry and sent as tribute to the royal court (Hagerty, 1923: 18-19). An awareness of the Chinese fondness for oranges may have brought on the interesting tradition, reported for seventeenth century Java, that when a Chinese ship arrived in the port of Bantam, a representative of the ruler brought a gift of oranges along with two umbrellas (I. H. Burkill, 1966, 1: 582).

In South Chinese folk religion, orange and red-colored fruits, including mandarins and other oranges, are common ritual offerings, and one also reads of small mandarin trees in homes at the time of Chinese New Year (E. N. Anderson and M. L. Anderson, 1977: 377, 378; E. N. Anderson, 1988: 137). This is because their colors are religious and magical, life-sustaining, capable of warding off evil spirits and assuring good luck. The ritual associations of the orange in China need further investigation,[5] and here we would only note that they may have roots in the cultures of the early non-Han peoples of South China (E. N. Anderson, 1988: 136).

The Chinese people are noted for their propensity to gamble and for the rich variety of their gambling methods, and even oranges have not escaped their notice. When one sees several people surrounding a fruit peddler's supply of oranges in Hong Kong, according to Ball (1906: 582) writing early in this century, likely it is because they are gambling with the fruit. This is done by placing bets on the number of seeds an orange contains. Each player first gets to look carefully at the orange, to estimate the number of its seeds, and then to place his bet. When all players have done so, the fruit seller peels the orange, opens each segment, and carefully counts the seeds. The person who "guesses right" wins three times the amount of his bet, and the next two closest win double theirs.

Of all the areas of South China today, the Double Cropping Rice Area may be best suited for orange cultivation, for it has fertile alluvial soils in lowland areas and a climate much like that of Florida. In the 1929-33 village survey (Buck, 1956: 209, 212), oranges accounted for more than a tenth of the total acreage devoted to fruits in the Double Cropping Rice Area, an amount greater than for any other fruit. S. W. Williams, writing in 1895 (1: 774), describes the streets of Canton and Amoy in winter as lined with produce, including golden oranges in baskets, piles, and stalls, in colorful contrast to the other fruits and vegetables. Turning to forms of oranges, we note that the earliest report of marmalade made from sour oranges may be Chinese, and today several kinds of sour oranges (*suan-ch'êng*) are cultivated in China, especially in Chekiang and Fukien but as far south as Ch'ao-an (Chaochow), near Shan-t'ou (Swatow) in Kwangtung. They are, however, grown mainly as a rootstock for other citrus, though the fruits, produced in quite small amounts, may be candied, made into an excellent marmalade, or consumed in other ways. There is also a dwarfish variety, *daidai*, which is grown in Kiangsu and Chekiang specifically for its fruit, valued as a medicine, and its fragrant flower buds which are dried and added to tea leaves (Tanaka, 1929a: 341-42; C. C. Hu, 1930-31: 544; 1934: 79; Swingle and Reece, 1967: 377).

In world perspective, the sweet orange is the major citrus species. Nevertheless in traditional China, the sweet orange (*t'ien-ch'êng*), though more important than the sour orange and found in many forms, was of minor significance compared to the mandarin. In a 1928-29 survey of the four most productive citrus regions of Kwangtung, C. C. Hu (1930-31: 502) found 58 to

100 percent of the annual production to be of mandarins and only 0 to 42 percent of sweet oranges.

Mandarin trees are adaptable to a wider range of climatic conditions than other types of cultivated citrus. They are better able to stand heat than most other types. They also tend to be more cold resistant than other citrus trees of commercial importance, though the fruit, mainly because it is small and thin-skinned, is more readily damaged by cold than are other oranges (Hodgson, 1967: 497). As a result, mandarins are grown in all of China's citrus-producing regions: South China, West China, Southwest China (Yunnan), Central China, and Northwest China (Burke, 1967: 142).

Though the mandarins are the largest and most varied group of cultivated citrus, their fruit possess the common characteristics of small to medium size; shape like a flattened globe; a loose, readily removed skin, yellow or reddish orange in color; and fruit segments that are separated without difficulty. The mandarin is sufficiently different from other oranges as to lead Swingle (Swingle and Reece, 1967: 370, 380-81) to place it in a separate species, *Citrus reticulata*, distinct from that of the sweet orange, *C. sinensis*, and the sour orange, *C. aurantium*. Plant scientists, geographers, and archaeologists have suggested South China, Southeast Asia, or eastern India as the place of earliest domestication of the mandarin oranges. Hodgson (1967: 497) argues for northeast India as their probable home. In that area, there occurs a primitive, related form sometimes called the "Indian wild mandarin" (*C. indica*); also found there are highly developed forms of mandarins that are absent elsewhere, as well as many mandarin hybrids. The mandarin was cultivated in very early times in Southeast Asia and East Asia, with both of those regions important in developing mandarin forms.

In the early fourth century A.D., Chi Han had already distinguished red and yellow forms of mandarins (*chü* and *kan*, names still in use) in the south, one of them the famous variety known in present-day Canton and the West as Ponkan or "Chinese honey orange" (H. L. Li, 1979: 118-20). Accounts from T'ang times mention highly-prized varieties of common mandarins and King mandarins, the latter (formerly *C. nobilis* Loureiro) believed to be a natural hybrid, perhaps between mandarin and sweet orange (Hodgson, 1967: 505-6; Swingle and Reece, 1967: 380-81). Mandarins seem to have been grown even in the gardens of the imperial palace far to the north in Shensi, apparently in some sort of greenhouse (Schafer, 1977: 90, 95-96), a situation reminiscent of the orangeries or orange-houses of ancient Rome and later Europe (Tolkowsky, 1938: *passim*; Webber, 1943a: 17; Webber *et al.*, 1967: 13, 14; Swingle and Reece, 1967: 376; R. Ray and L. Walheim, 1980: 5-7).

Many mandarin varieties are grown in South China today, of which two deserve special note. One is the afore-mentioned Ponkan, which produces large fruit with loose skin and tender, pleasant-tasting, juicy flesh, and which can be picked in October. It is said to be the finest mandarin produced in China, in East Asia, and perhaps in the entire world. The Cheokan, known better as Tankan, is an ancient form which originated in southeastern China. Another of the Orient's best citrus fruit, it has qualities of both the Ponkan and sweet orange, and may be of hybrid origin. Its fruit is smaller than that of the Ponkan, but is sweet, juicy, and has a deep orange flesh. Its fruit matures later than the Ponkan and may be left on the tree for harvesting from February to April (Tanaka, 1928-29: 122; 1929a: 343-45; 1929b: 37-43; C. C. Hu, 1930-31: 543-44; Burke, 1967: 143-44; Hodgson, 1967: 520-21).

In parts of South China in traditional times, fresh oranges might be stored in earthenware jars for weeks or months, with decaying ones removed periodically (Condit *et al.*, 1937-38: 69). In early China, there were more ingenious and likely more effective means of storage. One involved placing oranges on a bed of rice straw on the earthen floor of a tightly-sealed store-room, inspecting the fruit periodically and removing those that had gone bad (Hagerty, 1923: 31-32). Another involved burying oranges in a pit in the floor of a farmhouse and sealing the pit with loam to keep the temperature steady and cool enough to permit them to keep through the winter (Laufer, 1919: 230-31). There is even one account of people burying in the earth the orange-laden branches of a living tree for several months, then digging them out and eating the fruit. If the orange was a bit damaged, the skin might be removed and the fruit smoked and baked in an oven, and then consumed, sometimes with sugar or honey to improve its taste (Hagerty, 1923: 32-33).

In traditional China, besides being consumed fresh or after storage in jars, oranges were preserved in honey or sugar, or their peels were sun dried, later to be soaked and used as a flavoring, particularly in spicy dishes from Szechwan and Hunan (Hom and Steiman, 1981: 33; Tropp, 1982: 557). We note, in this regard, a "Tangerine-Peel Chicken" dish from that region (Sakamoto, 1977: 259), as well as Cantonese "Roasted Duck with Orange Peel" (W. W. Chang *et al.*, 1970: 269), and the imperial court dish "Tenderloin of Pork Braised with Orange Peel" (J. D. Keys, 1963: 63).[6] The fruit itself might be added to dishes, as in a Szechwan dish made of tangerines or oranges and "silver ears" (jelly fungus) (W. W. Chang *et al.*, 1970: 457). Oranges might also be used in sweets, among them a sweet drink, "Orange Tea" or "Sweet Orange Tea," often served, whether hot or cold, between courses at a dinner. The intent was to help the guests clean their mouths so as to ready themselves for the forthcoming dishes (G. B. Miller, 1966: 748-49; K. Lo, 1979c: 328).

Oranges, like other citrus fruits, are very good sources of vitamin C, with the rind having more ascorbic acid than the pulp. Fresh oranges are also good sources of folic acid. Some, but not all, authorities report mandarin oranges (or tangerines), including King mandarins, to have significantly less ascorbic acid than either sour or sweet oranges (Su and Liu, 1938: 43; Leung *et al.*, 1973: 91-92; R. Ray and L. Walheim, 1980: 168-69; Ting, 1983: 222-26). Besides being used as food, the orange, including its fruit, peel, seed, leaves, and flowers, serves various medicinal uses (Hagerty, 1923: 33-34; Hooper, 1929: 35-38; F. P. Smith, 1969: 111-17; J. D. Keys, 1976: 176-77).

Pummelo

The pummelo (pumelo or pomelo) or shaddock (*Citrus grandis*; *yu*) is an evergreen tree widely believed to be an ancient Southeast Asian domesticate. From Southeast Asia, it seems to have spread north to China in very ancient times (Tanaka, 1929a: 340), but west to India, the Near East, and Europe much later, reaching Spain only in the twelfth century A.D. (Bonavia, 1888: 220-23; Tolkowsky, 1938: 26, 34-35, 146-48; Hume, 1957: 73). The pummelo tree bears round, sometimes flattened or pear-shaped fruit resembling grapefruit in color but with a thicker rind, but pummelos are much more varied than grapefruit in their characteristics. Some are relatively small trees, others large. Some have fruit the size of a large grapefruit; others have fruit ten to twenty pounds in weight, larger than any other citrus. Some are seedy, some almost seedless. The pulp of some forms is reddish in color, that of others whitish or yellow, and it may range in taste from quite acid to insipid, orange-like sweetness. Most Asian pummelo varieties are of little value as fresh fruit, but some are highly regarded. In the latter category are certain delicious Thai pummelos that have been exported to Malaysia and coastal China, and some Chinese varieties, among them ones of presumed Thai ancestry (G. W. Groff, 1929-30: 442; Bailey, 1935, 1: 782; Hume, 1957: 73-76; Hodgson, 1967: 533-39; Swingle and Reece, 1967: 382-83; R. Ray and L. Walheim, 1980: 97-101).

The pummelo we bought in Canton markets were the size and color of a large grapefruit, but pear-shaped, and sold for 35 to 50 cents U.S. per pound. It took an effort even for a knowledgeable vendor to remove the rind (about 3/8 of an inch thick) by hand and to separate the segments of the fruit within. The flesh of these pummelo was whitish and coarse; it resembled grapefruit in flavor, but was sweeter and contained numerous large seeds. We preferred by far the small, delicious local oranges.

By tradition, a certain Captain Shaddock, master of an East India trading vessel, was responsible for introducing the pummelo to the West Indies, probably in the seventeenth century. The earliest references to the grapefruit, from the mid-eighteenth century, are from that same region, and many believe it to have originated there. Botanists hold that the grapefruit, which is closely related to the pummelo, derived from it by mutation or as a natural hybrid, perhaps with the sweet orange. Some regard the grapefruit as a variety of pummelo, others place it in a separate species, *C. paradisi* (Tolkowsky, 1938: 265-67; Tanaka, 1954: 10-11, 18-19; Hodgson, 1967: 533-39; Swingle and Reece, 1967: 382-84; Webber *et al.*, 1967: 14-15; Sinclair, 1972: 1-2; International Board for Plant Genetic Resources, 1982: 3).

Though the ancestry of the grapefruit is obscure, clearly it is a late form that was known in China only in the eighteenth century or later. The pummelo, on the other hand, is an ancient Chinese fruit, and of the two it remains the common one there today. The grapefruit's Chinese name, *p'u-t'ao-yu*, is a translation meaning "grape pummelo" (E. N. Anderson, 1988: 187). The fruit is said to be "not well liked in China" (Wittwer *et al.*, 1987: 286), and we observed no grapefruit in Canton or Ch'eng-tu markets, though we did see them in certain Hong Kong markets that handled imported fruits and vegetables. Pummelo fruit is less juicy than grapefruit, but its segments can easily be separated and their membrane removed. With little if any loss of juice, the segments can then be placed in dishes for eating, with sugar added if necessary. Unlike grapefruit in the United States, the sweet edible pummelo in traditional China was not a breakfast fruit but a dessert, eaten in segments and not as half a fruit. Pummelo pulp might also be dried and then preserved or candied; and pummelo rind might be parboiled and cooked as a vegetable, for example in a meat or shrimp-egg stew (Zee and Hui, 1981: 81), or it might serve to give tea leaves a good scent (Holmes, 1948: 39).

Pummelos are sensitive to cold, and even in South China most major producing districts are in lowland coastal areas. One is the Chang-chou region of Fukien, home of the famous "Amoy pummelos," which are readily separated into segments which are sweet and juicy (Meyer, 1911: 44; "Production of fruit in Fukien Province," 1928: 1033). The Pearl River Delta is another, and indeed it has been argued that the finest pummelos come from such fertile deltas subject to the tidal ebb and flow of salty water (G. W. Groff, 1929-30: 442). A pummelo hybrid, it is true, is grown in the Yangtze Valley, but as an ornamental, its fruit being inedible (C. C. Hu, 1930-31: 543, 544). Pummelo fruit, because of their thick rind, will keep for about six months, from October to April (Wittwer *et al.*, 1987: 286). Even in traditional China, they were shipped far to the north, with pummelos abundant in Shanghai, served even in hotels catering to foreigners (Meyer, 1911: 44; "Production of fruit in Fukien Province," 1928: 1033).

The pummelo, like other citrus fruits, is high in vitamin C, some Oriental varieties having more of this vitamin, by weight or volume, than most oranges, and than grapefruit (Su and Liu, 1938: 43, 50; Read *et al.*, 1940: 23; Mustard, 1952: 33; Swingle and Reece, 1967: 383; Leung *et al.*, 1973: 83, 91-92, 96).

Besides use as food, the pummelo is a common table plant (McClure, 1933: 138), and its fruit serve as colorful decorations, especially at the Chinese New Year festival for which thousands are imported annually to California (Hume, 1957: 74-75; R. Ray and L. Walheim, 1980: 8). Pummelos are among the ritual offerings in the folk religion of modern South China, and they are viewed as having magical qualities. Pummelo leaves or skins are soaked in water to make a ceremonial bath to ritually cleanse a person and repel evil (E. N. Anderson and M. L. Anderson, 1977: 377; E. N. Anderson, 1988: 136-37). One reads that in a Chinese wedding a ceremonial bath in such pummelo-leaf water is an essential element in the proceedings (Burkhardt, 1953-58, 1: 86); that in certain places near Canton newborn infants are bathed in such pummelo-leaf water to purify them; and that pummelo leaves added to the water in which a Chinese child bathes on New Year's Eve are believed to cleanse its soul for a year (R. Ray and L. Walheim, 1980: 7-8). The pummelo fruit, peel, leaves, flowers, and seed also serve other purposes, including medicinal ones, in China and Southeast Asia (Hooper, 1929: 36; McClure and Hwang, 1934: 16; I. H. Burkill, 1966, 1: 578; F. P. Smith, 1969: 113-14).

Citron

Its Early Cultivation in India, the Near East, and Mediterranean Area

The citron is a thorny shrub or small tree that may flower and fruit at any season and is more sensitive to frost than any major cultivated form of citrus. Though this greatly limits the region within which it may have been domesticated, the time and place of earliest citron cultivation remains a mystery. Citron seems to have been in a domesticated state in both India and Persia by at least the middle of the first millennium B.C. Some authorities have suggested India as its place of origin, others southern Arabia, which would make its place of domestication farther westward in Asia than that of any other major form of citrus. This may account for citron being the earliest citrus fruit in the Mediterranean region, and apparently the only form known in Europe for hundreds of years after it was introduced (Andrews, 1961). Some scholars credit the Greeks with its introduction to the Mediterranean region. By that hypothesis, the Greeks found the plant growing in Iran during the conquest of western Asia by Alexander the Great about 325 B.C., which led them to call it the "Median apple" or "Persian apple" (Tolkowsky, 1938: 39-70). Others (Isaac, 1959: 181; Andrews, 1961: 35-37; Zohary and Hopf, 1988: 160) suggest a much earlier introduction of the citron to Mesopotamia, Palestine, other parts of the Fertile Crescent, or Cyprus.

In either case, its spread across the Mediterranean lands is credited mainly to the Jews, who came to use the citron, or *etrog*, as a vital element in religious ceremony, in the Feast of Tabernacles. According to Midrashic commentators, the citron fruit symbolized the human heart, and its use in the ceremony atoned for sins of the heart (Abrahams, 1908-26: 144; Isaac, 1959: 183; Goor and Nurock, 1968: 155). It was also medicinal, among other things increasing a man's potency and leading a pregnant woman to give birth to a son (Goor and Nurock, 1968: 166-67). Perhaps it was such beliefs that led the Jews of Roman days to decorate their bridal chambers with citrons (Tolkowsky, 1938: 19). One is dealing, as Isaac has noted (1959), with a clear example of the role of religion in the diffusion of a domesticated plant. Wherever in the

Mediterranean region Jewish settlements were established—in Sicily, North Africa, and elsewhere—the citron, the "goodly tree" or "tree of loveliness," came to be cultivated, if at all possible.

There is considerable confusion about the identity of citrus fruits mentioned in ancient Indian texts or depicted in Indian art. Textual problems stem from use of the same Sanskrit word or words for more than one type of citrus fruit, from changes in usages over time, as well as from inconsistencies in modern Indian and European usage (Bonavia, 1888, 1: 214-44; Tolkowsky, 1938: 23-24; Hayes, 1966: 199-200, 204). Problems of identifying fruits represented in art arise from the occurrence in India of a considerable diversity of citrus fruits, including forms not known in the Western world and which do not fit clearly into categories, such as citron and lemon, with which Westerners are familiar (Bonavia, 1888, 1: 63-72; Hayes, 1966: 203-9). An additional problem is created by the many natural hybrids in India with citron and lemon qualities, and by the fact that to some extent most lemon-like fruits in India have characteristics of citron (Hodgson, 1967: 558). A particularly troublesome linguistic problem is presented by the Sanskrit word *jambira*, which some translate as citron, others as lemon. In fact, insists Bonavia (1888, 1: 61), it is nearly impossible to determine from Sanskrit accounts which of the many forms of citrus is intended by *jambira*. In Hindi today, a similar term is used for the "rough lemon," a fruit different from the type of lemon common in the Mediterranean region, which is still relatively rare in India. Indeed Hodgson (1967: 571-72) puts the rough lemon into a category of "fruits resembling the lemon," but a distinct species, *Citrus jambhiri*.

Whatever the uncertainty, it is clear that the citron or rough lemon, or both, played a role in Hindu and Buddhist religion. In Hinduism, the citron, as a phallic symbol, is said to be sometimes depicted in the hand of the elephant-headed god Ganesa (Getty, 1971: 17-18). As for Buddhism, Tolkowsky (1938: 24) has found citron "unmistakably recognizable" in at least three scenes from the lives of the Buddha depicted in one of the famous Ajanta Caves at Hyderabad in central India. Those caves, among them ones used as monasteries, chapels, and galleries, were carved into the face of a cliff over a period from about 150 B.C. to A.D. 650, and are renowned for their wall paintings, some by Hinayana, and others by Mahayana, Buddhists (B. Walker, 1968, 1: 17-19). The religious role of citron or rough lemon is more clearly shown in the case of Kubera (Kuvera), a Vedic earth spirit and later lord of the earth's treasures who was adopted by the Buddhists. Kubera is often depicted as golden, and in one form always with a particular fruit in his hand (Tolkowsky, 1938: 23, Plates IV and V; Getty, 1962: 159), though some authorities call it a citron and others, a lemon. On Java, where both Hinduism and Buddhism were introduced and widely practiced in the first millennium A.D., Kubera is depicted apparently with the same fruit, which again is called a lemon (Daweewarn, 1982: 110-11) or citron (Goor and Nurock, 1968: 155).

Introduction of Citron to China; Its Uses There

In China, one of the earliest references to domesticated citron (known by a name, *kou-yüan*, associated with it until Sung times; now, *hsiang-yüan*) seems to be that in Chi Han's fourth century flora. The citron in question had fruit with a shape normal for citron in the West, and highly fragrant. Chinese women, the account said, cut them into the form of flowers or birds, then simmered them in honey for use as decorations on the banquet table. Chi Han writes of citron being highly valued by foreigners, and of ten jars of citron being sent to the emperor as tribute from western countries in the previous century (H. L. Li, 1979: 127-28). Though the

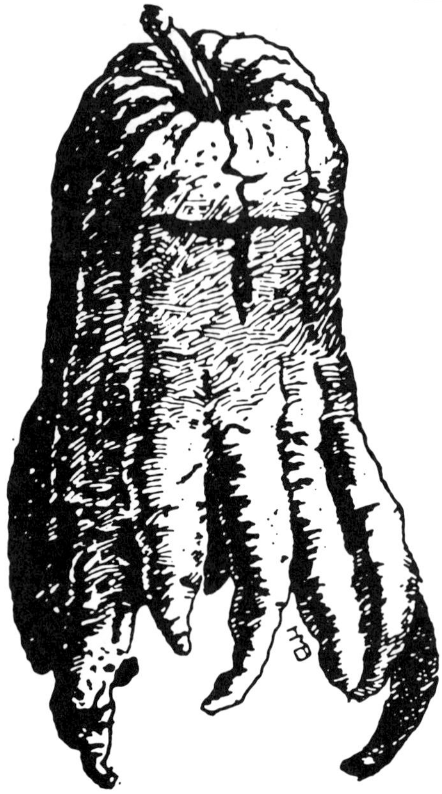

Figure 54. Buddha's-hand citron (*Citrus medica* var. *sarcodactylis*) (from Bailey, 1935,1: 781)

above suggests that citron may have been introduced to China not long before, there is nothing in the account to indicate that it was associated with religions of Indian origin.

Of special interest in this regard is the fingered citron (*C. medica*, var. *sarcodactylis*), a highly-valued variety known in China as *fo-shou* ("Buddha's hand"), in which the carpels of the fruit naturally separate and grow so as to look like fingers (Figure 54).[7] Not depicted in early Buddhist art in India, the Buddha's-hand citron is nevertheless believed to have originated there and been brought to China by Buddhist monks. Though not mentioned by Bretschneider in his account of the plants of ancient China, it was already important in the tenth century A.D. in the empire of Min, an independent Fukienese state (Schafer, 1954: 67).

In traditional China, citron was cultivated from the Yangtze Valley southward (H. L. Li, 1979: 128), with the Buddha's hand the most common variety (F. P. Smith, 1969: 114). Buddha's-hand citron was a popular pot plant and its fruit was symbolic of happiness, wealth, and longevity. The tree and its fruit were highly-prized gifts, especially at New Year, for they are believed to bestow good fortune on a household. The fruit was also a very common art-motif in China. It was often represented in jade or ivory carvings, or, with stems, leaves, and flowers, depicted on lacquered wood panels and in other ways (Meyer, 1911: 45; McClure, 1933: 138; Tolkowsky, 1938: 5-6, 8-9; Sowerby, 1940: 140; C. A. S. Williams, 1941: 51; Burkhardt, 1953-58, 1: 112). Of special note is the plate described and depicted by Yetts (1912: 27, Fig. 20), which was rich in fruit symbolism: the peach for longevity; the pomegranate for numerous offspring; and the Buddha's-hand citron for happiness and wealth. The Buddha's-hand citron was a sacrificial offering commonly presented in temples or at the shrines of household gods at New Year and other times (Giles, 1900: 26; Tolkowsky, 1938: 8-9; C. A. S. Williams, 1941: 51; E. N. Anderson, 1988: 136). Buddha's-hand citron also serves as an altar offering in Indochina (Crevost and Lemarié, 1917: 202-3), which fits with the suggestion (E. N. Anderson, 1988: 136) that the ritual role of citrus fruit in China may derive from non-Han groups in the south. Though Kubera is a common figure in Chinese Buddhist temples, we have found nothing to indicate whether he is associated with the citron in China.

The principal uses of citron fruit are as an ornament, a source of a pleasant fragrance, and medicine and food. One reads of people carrying Buddha's-hand citron in their hands or placing them on tables for their strong, delicious odor; of their being used to perfume clothes when pressed; and of washing fine linen in citron juice (F. P. Smith, 1969: 114). The rind is of greatest food value, and may be eaten fresh or preserved, sometimes by drying, often as "candied peel" (E. Watson, 1923: 366). Various plant parts serve as drugs, with peel and flowers, the principal commercial products, sold in pharmacies as stimulants, tonics, or for other purposes (E. Watson, 1923: 365-66; Hooper, 1929: 37; I. H. Burkill, 1966, 1: 579-80; F. P. Smith, 1969: 114; J. D. Keys, 1976: 176).

Lemon and Lime

There has long been confusion surrounding the lemon in China, for the English word "lemon" has been applied to two different forms of *Citrus*, both of which probably originated in India or Southeast Asia. One is the true or common lemon, *C. limon*. The other is the Canton lemon, *C. limonia*. Many foreigners have stated that "the lemon" was a recent introduction to China, a view Laufer (1934) proved to be erroneous. Tanaka (1929a: 342-43), distinguishing between the two forms, says that the Canton lemon reached China in Sung times, and the true lemon in Yüan times. As various writers have observed, however, few true lemons are grown in China, though occasionally they are cultivated in Canton and elsewhere (Tanaka, 1929a: 343; C. C. Hu, 1930-31: 543, 544; Swingle and Reece, 1967: 373). Canton lemons, on the other hand, are very common in the Canton region, but rare in other parts of China.[8] The fruit of the Canton lemon is small to medium in size, yellowish to reddish orange in skin color, quite sour in taste, and resembles a mandarin in shape, looseness of skin, and orange juicy pulp. Quite likely it originated in India (where one of its names is *rangpur*), possibly as a hybrid between the mandarin orange and rough lemon (*C. jambhiri*). A supposed hybrid origin has led Swingle and Reece (Swingle and Reece, 1967: 364, 368-69) and some others to reject its being designated a species. In any case, it spread from India to Southeast Asia and then on to China and Japan (where it is known as the "*hime* lemon"). In both India and the United States, it is sometimes also referred to as "lime" (*i.e.*, "rangpur lime," "mandarin lime," and "marmalade lime"), a usage leading to further confusion, this time with the true limes, *Citrus aurantifolia* (Hodgson, 1967: 532-33).

In the late 1920's Tanaka (1929a: 340) observed that the true lime, *C. aurantifolia*, was not found in China, and that the scattered reports of it there were in error. A more recent account (Burke, 1967: 144) indicates that the Mexican (West Indian) lime (a form of *C. aurantifolia*) has been introduced both to the mainland and Taiwan and is now cultivated to some degree.

Kumquat, Calamondin, Trifoliate Orange, and Ichang Orange

So far as is known, the kumquat first reached Europe from China only in the mid-nineteenth century, long after many other citrus species, and it seems to have arrived in the United States about the same time (Hodgson, 1967: 580; Ting, 1983: 203).

The kumquats are known in China by various names, among them *chin-chü*, "golden orange"; *chin-kan*, "golden mandarin"; or *chin-tou*, "golden bean." Their English name, in turn, comes from the Cantonese. The kumquats are placed by Swingle and Reece (1967: 326-32) in the genus *Fortunella*, like *Citrus* one of the six genera of true citrus. They identify four species of *Fortunella*, all of them shrubs or small trees native to southern China, and they note, as well, various *Fortunella* hybrids. Southern China is believed to be the place of earliest kumquat domestication.

The principal cultivated species of kumquats are the oval kumquat (*F. margarita*) and the round kumquat (*F. japonica*), but the Hong Kong wild kumquat (*F. hindsii*) is also sometimes cultivated. In subtropical and warm temperate areas, the two common kumquats, though evergreen, are semidormant during fall, winter, and spring, and grow only in the heat of summer. This apparently contributes to the kumquat's unusual ability to survive the winter cold, an ability greater than that of any species of the genus *Citrus* (Hodgson, 1967: 581). In China this has contributed to the kumquat becoming one of the most important citrus fruits cultivated in the lower Yangtze Valley, a region that in some years is subject to winter frost and snow (Burke, 1967: 142, 143, 144). Meyer (1911: 44) reports that kumquats are marketed in Peking, and one presumes that the fruit is shipped north from the Yangtze region.

The kumquat fruit is small, orange in color, with a somewhat sour pulp but a thick, tender, sweet rind. It is the only citrus fruit normally eaten whole. In China kumquats are sometimes grown as ornamentals, whether outdoors or as indoor pot plants. The fruit may be used as a garnish, or it may be eaten fresh, made into jelly or conserve, or dried. The fruit may also be put in sacrificial dishes. It may be candied or preserved in sugar syrup or honey. A dwarf kumquat plant covered with ripening fruits may be a gift to friends or guests, hence one of its early names in China was *chi-k'o-ch'êng*, or "give-guest orange." Sometimes a dwarf pot kumquat may be placed on the dinner table so diners may eat its fruit between courses. Kumquats are also believed to be of medicinal value, serving as a stimulant or deodorant, or in relieving flatulence or countering inflammation (Hagerty, 1923: 16; Tanaka, 1929a: 340, McClure, 1933: 138; Hume, 1957: 90-95; I. H. Burkill, 1966, 1: 575; Hodgson, 1967: 580-83; F. P. Smith, 1969: 115; Schafer, 1977: 113; R. Ray and L. Walheim, 1980: 108).

The calamondin or calamondin orange (*Citrus mitis*, *C. microcarpa*, or *C. madurensis*) is a fruit tree of Chinese origin. Swingle and Reece (1967: 337) regard it as probably a natural hybrid between a kumquat and a loose-skinned sour mandarin orange and not a valid species, whereas Tanaka retains its specific rank. Though widely cultivated in tropical Southeast Asia, including the Philippines (where it is known as *calamonding*), it is very hardy and tolerant of cold, and bears fruit at almost all seasons. The fruit, reddish-orange in skin color, is small but abundant, and a tree in fruit is striking in appearance. The orange-colored flesh may be used for marmalade, or the juice sweetened and made into a drink. The tree also serves as an ornamental, sometimes as a table plant, a potted plant brought to the table (Ochse and Bakhuizen van den Brink, 1931a: 131-32; McClure, 1933: 138; I. H. Burkill, 1966, 1: 580;

Hodgson, 1967: 531; R. Ray and L. Walheim, 1980: 111, 113, 156). Both the calamondin and the kumquat are excellent sources of vitamin C (Mustard, 1952: 33).

We mention the trifoliate orange (*Poncirus trifoliata*; *chih*), another ancient Chinese domesticate, even though it is a medicine rather than a food, as an example of the variety of citrus in the Chinese hearth. The trifoliate orange is a spiny shrub or small tree, a true citrus fruit related to the genera *Citrus* and *Fortunella* yet different from them in significant ways. For one, it has a compound leaf consisting of three leaflets. For another, it, alone of all true citrus, is deciduous. As a result, it is able to grow in areas subject to frost and snow, far north of the tropical and subtropical areas in which other true citrus are found. In eastern Asia it is widely grown in central and northern China, in Korea, and in Japan where it has been cultivated at least since the eighth century A.D. The plant also grows in other parts of China, including South China, and in Vietnam.

In China, the trifoliate orange is a hedge plant or ornamental with small fruit, greenish or dull lemon in color, and fragrant, but with little juice, bitter and inedible. It is used primarily in a sliced, sun-dried form as a bitter, aromatic drug. The plant is also used in China to provide valuable rootstocks, as for certain mandarin oranges, kumquats, and dwarf pot citrus. And in Japan, the United States, and elsewhere it serves similar uses as an ornamental and hedge plant, and, together with some of its hybrids with *Citrus* species, as hardy rootstocks, resistant to both heat and cold, for ordinary citrus trees. In the United States it has long been hoped that new varieties of citrus plants might be developed with the ability to survive severe frost conditions. Starting in 1892-93, the U.S. Department of Agriculture began a serious decades-long effort in Florida aimed at developing hybrid citrus that were hardy, and it was expected that *Poncirus trifoliata* would be the parent to provide that quality. Hybrids of *P. trifoliata* with *Citrus* were obtained, those with *Fortunella* with much more difficulty. The fruit of the hybrids has generally proved to be sour and bitter but edible, and the hope remains that further experimentation may lead to a citrus of commercial value that may grow much farther north than present commercial varieties (Meyer, 1911: 45; Hooper, 1929: 115; Tanaka, 1929a: 339; C. C. Hu, 1930-31: 543; 1934: 75; Hume, 1957: 28-34, 168-71, 172; I. H. Burkill, 1966, 1: 583; Burke, 1967: 145; Hodgson 1967: 584-88; Swingle and Reece, 1967: 344-49; F. P. Smith, 1969: 116-17; J. D. Keys, 1976: 181; R. Ray and L. Walheim, 1980: 157).

Also of interest as breeding stock for hardy new forms of citrus is the Ichang (I-ch'ang) orange or Ichang papeda (*Citrus ichangensis*; *I-ch'ang-kan*), a wild shrub or small tree found in west-central and southwestern China. This is the northernmost, and most cold-resistant, evergreen species of the citrus group, able to take considerable winter cold and grow at elevations 3,000 to 5,000 feet above sea level. It has hybridized with cultivated species of the genus *Citrus*, and a few hybrid forms are cultivated in China for their fruits, used as fragrance or food. One, the yuzu (sometimes *C. junos*), seems to have been cultivated in China since antiquity, and today is even more commonly grown in Japan for its sour fruit and as a rootstock (Tanaka, 1929a: 339; C. C. Hu, 1930-31: 544; 1934: 79; Bailey, 1935, 1: 784; Swingle and Reece, 1967: 388-92).

Wampee

A distant relative of the true citrus but member of the orange subfamily, is the fruit known to foreigners in China as "wampee." The wampee (*Clausena lansium*; *huang-p'i*) is a domesticate of South China cultivated to a small extent as far north as the Yangtze Valley (Hooper,

1929: 38; K. C. Chang, 1970: Table 1; H. L. Li, 1970: 11, Table 1). The fruit is small, the size of a pigeon egg, usually roundish and covered with fine short hairs. Yellowish in skin color, the fruit has a thin peel and tender, whitish pulp, aromatic and delicious, like a grape or litchi. It is produced on a shrub or tree which may reach a height of twenty feet or more. Though only a remote relative, it can serve as a rootstock for *Citrus*.

Clausena is more widespread than any other genus in the orange subfamily. Species of *Clausena* are native to New Guinea, Australia, southern Asia, and Africa, and several of these bear edible fruit. One, *C. dentata*, var. *dulcis*, the Indian wampee, is the object of large-scale fruit collecting, but *C. lansium* is the only species to be grown commercially. In South China, there are many varieties of the latter species, their fruits differing in size, shape, time of ripening, and taste—whether sour, sub-acid, or sweet (Swingle and Reece, 1967: 209-12, 216-18; R. Ray and L. Walheim, 1980: 158-59). Such variety suggests that we may be dealing with an ancient domesticate, the object of careful selection by humans over a long time span.

Wampee fruit, besides being eaten fresh, is made into a jam, or is dried. The fruit and root serve medicinal uses in China, and the root and seed are sold in Chinese pharmacies in Malaya. In Indochina, the Chinese buy from the Annamese unripe wampee fruit, cut in half and sundried, for use in treating bronchitis. In China the fruit is considered strengthening to the stomach, cooling, and anthelmintic (Crevost and Lemarié, 1917: 214; Hooper, 1929: 28; I. H. Burkill, 1966, 1: 585-86; F. P. Smith, 1969: 117-18).

The fruit seems to be a good to excellent source of vitamin C and a fair one of niacin (Mustard, 1952: 33; Leung *et al.*, 1973: 99).

Litchi

The litchi (*Litchi chinensis = Nephelium litchi*; *li-chih*) (Figures 55-56) is a subtropical, evergreen fruit tree, valued also as an ornamental and as a source of cabinet wood. It tends to bear every other year, flowering from February to April and providing clusters of fruit from April or May through August. The litchi is found wild on Hainan, and is believed by most authorities to have been domesticated in South China; it has, however, been cultivated in Tonkin since antiquity and continues to be widespread there and varied in form (Crevost and Lemarié, 1917: 221-22; Chun, 1921: 237; H. L. Li, 1979: 114). In the fourth century A.D., Chi Han described it as a large tree, fifty to sixty feet high, or more, with dense green foliage and red fruit "as big as hen's eggs, with a yellowish black seed resembling a ripened lotus seed, and a white flesh resembling lard—sweet and juicy like the pomegranate" (H. L. Li, 1979: 113-14). Its outer covering is rough and hard, inedible, in many varieties crimson in color when it reaches maturity, but turning gray-brown as it dries. The edible part of the litchi is its white flesh, which is firm, and of unusual, delicate flavor. It is highly prized, most famous of all the fine fruits of China's south. In the words of T'ang poet Po Chü-i (Hart, 1954: 114) "the red fruit of the litchi is as precious as the pearl," and in those of Edward H. Schafer (1967: 188) it is "a jewel among fruits in China."

Though the litchi tree thrives in the humid, rainy climate of Kwangtung, it does not do well everywhere in China or elsewhere in southern Asia. For one thing, it requires much soil moisture. For another, it is quite sensitive to frost, especially when young, which has limited its northward extent in China (cultivation extends along the southeast coast only as far north as Foochow in Fukien). Furthermore, under conditions of dry heat and wind when fruit are maturing in summer they may be damaged and spoil, which has restricted the litchi's spread in

Figure 55. Litchi (*Nephelium litchi*) (from Bailey, 1935, 2: 1892)

India. And in Southeast Asia the litchi's success has been limited by its need for a cool, dry season to fruit well (Meyer, 1911: 47; Popenoe, 1920: 317-19; I. H. Burkill, 1966, 2: 1572; Hayes, 1966: 325-26; P. M. Smith, 1976: 321). Nevertheless, where conditions are satisfactory, the litchi may bear for a century or longer, reaching its best production in two or three decades. Indeed in Kwangtung and Fukien, there are venerable litchi trees, reputed to be a thousand years old or more, which still yield fruit ("Fruit cultivation in Kuangtung," 1936: 718; Rosengarten, 1984: 301).

One can only guess at the antiquity of litchi cultivation in South China, but when the Han Chinese conquered Nan-Yüeh they found litchis grown there, and developed an immediate taste for the fruit. There is a record of litchi trees being taken in the late second century B.C. from what today is Vietnam to the Chinese capital of Ch'ang-an in the north. The emperor built a new palace there named *Fu-li* after the litchi, and over a hundred litchi trees and other rare plants from the South were grown there. All litchi trees but one died within a few years, and that one, though it neither flowered or fruited, became a great favorite of the emperor. When,

Figure 56. Litchi fruit on tray

without warning, it, too, wilted and died, dozens of caretakers are said to have been executed. Litchi fruits continued to be sent to the royal court as tribute from the south, and on a considerable scale. Many men carrying litchis, wrote Chi Han, "died of exhaustion along the highway...indeed a great peril of the people" (H. L. Li, 1979: 114).

In T'ang times, fine litchis were cultivated in Fukien and Szechwan, but the best came from Nan-Yüeh (Nam-Viet), especially the country west of Canton. They continued to be sent north to the royal court, where tradition singles out the Lady Yang, consort of the emperor Hsüan Tsung but a native of Szechwan, for her fondness of litchis. That fondness was so well recognized that the court musicians even composed a new song, "Litchi fragrance," presented in her honor at the winter palace. Tradition holds further that each year fresh Canton litchis, the finest of them all, were sent to her by post horse. That the latter may be an exaggeration of history is suggested by the medieval writer who observed that fresh litchis could not stand such a trip, for in the summer heat "they would deteriorate after a single night" (Schafer, 1967: 188-89; 1970: 46).

Whatever the form and condition of the litchis received by the T'ang royal court, the general populace of the north probably had little appreciation of the delicate flavor of the fresh fruit, for it would not have been available to them. The abundance of fresh litchis in Nan-Yüeh, on the other hand, seems to have eased the pain of exile for some persons who were banished there. Certainly this was the case with the exiled Sung poet Su Shih (Sun Tung-p'o) who even said that litchis might make one adjust to permanent banishment. His fondness for litchis did not, however, lead him to excess: he consumed no more than three hundred a day, a small number compared to the thousand a day that, according to him, certain others consumed (Popenoe, 1920: 312). His moderation may have been based on fear of illness, for when raw litchis are eaten in sufficient quantity they are said to bring on fever and nosebleed (F. P. Smith, 1969: 281).

Further indications of the general status of the litchi in China may be found in its prominence in poetry and art, and in the fact that the first monograph on any fruit written in China (A.D. 1059) was Ts'ai Hsiang's "Treatise on Litchi" (Hayes, 1966: 323; H. L. Li, 1979: 114). Its position is also revealed in the elaborate festivals, already held in Canton in the ninth century A.D., honoring the ripening of the fruit (Schafer, 1967: 189; 1970: 46).

A fruit tree so esteemed might be expected to spread early and far, yet the litchi does not seem to have reached much beyond its homeland in South China and Indochina until quite late. It seems first to have been cultivated in Burma and eastern India only late in the seventeenth century or soon after, and to have reached the subtropical regions of Australia, New Zealand, Africa and the Americas, including Hawaii, California, and Florida, only in the eighteenth or nineteenth centuries. China, however, remains the world's principal producing region, especially the provinces of Kwangtung, Fukien, and Szechwan, and India is probably next in production (Watt, 1889-96, 5: 346-47; Popenoe, 1920: 314-16; Hayes, 1966: 323-24; Cavaletto, 1980: 469).

There are many varieties of litchis in China, of which G. W. Groff (1921) has described a good number. They differ from one another in many ways, including fruit color, size, shape, and flavor. The fruit of many varieties can be eaten fresh, that of others not. The author was in Hong Kong in May at the beginning of the litchi season. Vendors were selling fresh litchis, from Singapore as well as from China, the latter more expensive and far tastier. We were assured that there are still better varieties of litchis that mature later, in the summer.[9] Litchis may also be canned, dried in the sun or oven, pickled, or made into other preserves or alcoholic beverages. Today one finds for sale in Hong Kong a sweet litchi liqueur called "Ming," a name deriving from a title (Ming Huang, or "Luminous Illustrious") given to Hsüan Tsung, the T'ang emperor who made such an effort to supply his consort with litchis from Canton. Some insist that canned litchis are even tastier than fresh ones. Dried litchis, as well as canned ones, are shipped in large amounts from Kwangtung, a testimony to the high regard in which Chinese, at home and overseas, hold them (Popenoe, 1920: 316-17; "Fruit cultivation in Kuangtung," 1936: 717-18; Porterfield, 1951: 28-30; Hayes, 1966: 331; Cavaletto, 1980: 470-71, 476). To many non-Chinese, however, dried litchis, seem little, if at all, better than raisins.

Two studies (Koo, 1973: 141; Yeung *et al.*, 1973: 198) report that litchis, unlike most fruits, are considered "hot" in the hot-cold classification, which fits with the belief that when the raw fruits are eaten in quantity, they bring on fever and nosebleed (F. P. Smith, 1969: 281). One wonders whether the "hotness" of litchis is based on the fact that the fruit is red on the outside. In any case, litchis were forbidden to pregnant Chinese women in London, but they could not explain why (Wheeler and Tan, 1983: 53). Whether consumed fresh or dried, litchis are viewed as thirst-quenching and nutritious (F. P. Smith, 1969: 282). They may also be prepared in desserts and other dishes, among them sweet-and-sour dishes such as Cantonese "Litchi Duck" and "Litchi Snowball Chicken," and the Fukienese "Sweet and Pungent Litchi Pork" (W. W. Chang *et al.*, 1970: 267, 280; K. Lo, 1979c: 163: C. B. T. Lee and A. E. Lee, 1980: 55). Litchi leaves are also made into a tea. Tests have found litchis to be an excellent source of vitamin C (Leung *et al.*, 1973: 87; Cavaletto, 1980: 474-75). Besides serving as food, litchi fruit, seed, and other plant parts serve medicinal uses (Hooper, 1929: 94; I. H. Burkill, 1966, 2: 1573; F. P. Smith, 1969: 282; J. D. Keys, 1976: 183).

Longan

The longan (*lung-yen* or "dragon's eye"; *Euphoria longana = Nephelium longana*) (Figure 57) is a fruit tree related to the litchi, and some have placed it in the same genus. It occurs wild in Ceylon, parts of southern and eastern India, Burma, and South China (de Candolle, 1886: 315; Watt, 1889-96, 5: 348; S. C. Lee, 1935: 787; Hayes, 1966: 332; E. H. Walker, 1976: 683).

In Bengal today, the longan is cultivated to a small degree for its fruit and as a shade and ornamental tree (Hayes, 1966: 332). Watt (1889-96, 5: 348), however, gives no Sanskrit name for the longan, nor does Prakash (1961) include it among the fruits of ancient times. The longan is also cultivated in Thailand and Malaysia, but is said to have been introduced to Malaysia by the Chinese only "some centuries ago" (de Candolle, 1886: 315; I. H. Burkill, 1966, 2: 1573). In China and Indochina, on the other hand, longan cultivation is ancient and important today. If an account of much later times is given credence, King Chao T'o of Nan-Yüeh, whose capital was P'an-yü (modern Canton), presented longans to the first Han emperor, Kao Tsu (who ruled 206-195 B.C.) (H. L. Li, 1979: 122). Other accounts report that longans, along with litchis, were sent by fast horse from Kwangtung to the Han court (Y. S. Yü, 1977: 80); that at the royal court longans were presented to a visiting dignitary, "the Khan of the Huns"; and that Wen Ti, a Wei emperor (A.D. 220-226) ordered that longans be presented to him as annual tribute (H. L. Li, 1979: 122). The longan is described by Chi Han in the fourth century A.D. as growing in what today is Vietnam (H. L. Li, 1979: 122), and indeed in modern Annam and Tonkin the name for the longan is related to its name in China (Crevost and Lemarié, 1917: 222). The suggestion has been made (Popenoe, 1920: 326) that the longan had been introduced to South China from India. The above evidence, however, leads this writer to side with those authorities (Vavilov, 1949-50: 25; K. C. Chang, 1970: Table 1; H. L. Li, 1970: 11, Table 1; Harlan, 1975: 75) who favor South China or Southeast Asia (Vietnam) as the likely place where the longan first came into cultivation.

In modern times, the longan is cultivated overseas, far beyond its original homeland, in, for example, southern Florida and southern California where it does well and bears abundant fruit if located in places free of severe frost (Popenoe, 1920: 326-27).

The longan is a tree that may live to a great age, as much as 400 years, and it continues to bear many fruit even when 80 to 100 years of age (Zee and Hui, 1981: 89). A smaller tree than the litchi, the longan also has a smaller fruit, yellow or brown in color. Though the longan resembles the litchi in certain ways, the Chinese value its fruit, even of the best varieties, far less.

Figure 57. Longan (*Euphoria longana*) (from Crevost and Lemarié, 1917: 223)

The longan, like the litchi, is a tropical and subtropical species. Though the longan is said to be less sensitive to cold than the litchi, apparently it grows no farther north, roughly to the latitude of Foochow in Fukien. In Kwangtung, it matures in August, later than the litchi, for which it is sometimes called "slave of litchi," following the litchi as a slave follows its master (Meyer, 1911: 47; Bois, 1927-37, 2: 139-40; "Fruit cultivation in Kuangtung," 1936: 718; Ochse *et al.*, 1961, 1: 724-26; H. L. Li, 1979: 122).

The longan is a commercial fruit in both Kwangtung and Fukien. In Kwangtung the fruit is usually sold fresh, but there, and in Fukien, it may be dried by sun or fire, or canned for shipment elsewhere in China or abroad. Traditionally dried longans, which taste like raisins, were available at Chinese New Year in the northernmost cities of China. They might be eaten directly or added to other dishes for their flavor. Canned longans are held by some people to be the equal of ordinary canned litchis in flavor, and they are often eaten as a dessert, whether alone or with loquats or other fruits, as in "Dragon Eye and Litchi Dessert," a Shanghai recipe (W. W. Chang *et al.*, 1970: 457).

The longan tree serves a variety of non-food uses in China as well. Its timber is hard, good for furniture and construction. Its leaves and flowers are medicinal, and exported for such purposes from Canton (Bretschneider, 1882-95, 3: 429; F. P. Smith, 1969: 282). Its seed contains a saponin, and in China is used to wash one's hair, or is ground for use as a styptic (I. H. Burkill, 1966, 2: 1573; J. D. Keys, 1976: 184). Its fruit is believed to be effective medicinally, particularly as a tonic, and is used as such, in one form or another (I. H. Burkill, 1966, 2: 1573; F. P. Smith, 1969: 282; J. Y. P. Chen, 1973: 229-30; J. D. Keys, 1976: 184). Among other uses, it is given to debilitated or anemic persons, and to women who have recently given birth. Indeed, like many dried fruits, its flesh contains significant amounts of iron (Leung *et al.*, 1973: 87 and *passim*). And its usefulness against fatigue seems confirmed in one laboratory study with rats (J. Y. P. Chen, 1973: 229-30). The fresh and canned fruit, like the litchi, is valuable nutritionally, mainly as an excellent source of vitamin C (Leung *et al.*, 1973: 87).

Loquat

The loquat, *Eriobotrya japonica* (= *Photinia japonica* = *Mespilus japonica*; *p'i-p'a*),[10] is an evergreen fruit tree of warm temperate rainy areas with mild winters. Today it is cultivated widely in South and Central China (Meyer, 1911: 35; Chun, 1921: 170-71; H. L. Li, 1983: 46), and is believed to have been domesticated there at an early date, probably in western Hupeh or Szechwan (H. L. Li, 1983: 46). The loquat has been cultivated in Japan since antiquity, and its species name, *japonica*, reflects an earlier belief that it was indigenous there. The tree has spread to Southeast Asia and India, though it does not fruit well in tropical lowland areas that are too hot. It has also spread to more distant overseas areas, among them the Mediterranean region, California, and the American South (Watt, 1889-96, 3: 257; Popenoe, 1920: 252-55; Bois, 1927-37, 2: 310-13; Ochse *et al.*, 1961, 1: 721-23; I. H. Burkill, 1966, 1: 952-53; Hayes, 1966: 404).

Several varieties of cultivated loquats are found in China, and they differ considerably in flavor, color, and other qualities. The fruit is oval or pear-shaped, with green skin when immature, turning pale yellow to orange at maturity. Its flesh ranges from white to orange. It has large seeds, usually from one to five. And the best of the Chinese varieties, the "white loquat," is of excellent taste (Meyer, 1911: 35).

The fruit is ready for picking in summer, sour when immature but sweeter as it reaches maturity. Loquats present a marketing problem, however, for they are readily bruised and require care in handling. Most fruit is consumed fresh, but loquats may also be canned or made into preserves of other sorts. Canned loquats may be used as dessert, alone or with other ingredients, as in "Almond Bean Curd with Loquats" (W. W. Chang *et al.*, 1970: 456); they may also be added to chicken dishes or main dishes of other sorts.

The vitamins most studied in the loquat are A and C. The fresh fruits are poor sources of vitamin C and have a considerable range of vitamin A, depending on variety and color. Fresh fruit with deep orange skin color can be a significant source of vitamin A. In mineral content loquats are noted especially for their relatively high potassium (roughly 210 mg./100 gm. pulp) and low sodium (under 1 mg./100 gm. pulp) which has led them to be suggested as a tasty source of potassium for persons receiving diuretics (Y. T. Chang and H. B. Collier, 1936: 438; C. H. Lee *et al.*, 1941: 68; Fan and Hou, 1953: 129; Leung *et al.*, 1973: 87; Shaw, 1980: 485-88).

Loquat fruit, flower, and inner bark all serve medicinal purposes, but their leaves are most important in this regard. Bitter in taste, they are used to control or prevent vomiting and cough and for other internal or external conditions. Loquat leaves were enough in demand to be exported from Canton, Amoy, and other Chinese ports in the last century (Bretschneider, 1882-95, 3: 428; F. P. Smith, 1969: 164-65; J. D. Keys, 1976: 169).

Persimmon and Its Relatives

Background

Persimmons are species of *Diospyros*, a genus of the *Ebenaceae*, a family renowned for its tropical Asian and African members that since antiquity have provided ebony-wood to the high civilizations of the Old World.[11] Several species of *Diospyros* also provide edible fruits, with three of them well-known for their contribution to human diet. The three are indigenous mainly to temperate and sub-tropical regions of Asia or North America. The "common persimmon," *D. virginiana*, is native to the eastern woodlands of the United States, a tree with excellent fruit but not much cultivated. The "date-plum" of the Mediterranean, *D. lotus*, is native to an area stretching from southeastern Europe to northwest India, China, and Japan, and is cultivated in certain places, among them southern Europe, northwest India, and China (Watt, 1889-96, 3: 146-47; Chun, 1921: 272; Bois, 1927-37, 2: 432; S. C. Lee, 1935: 890; I. H. Burkill, 1966, 1: 845; H. L. Li, 1983: 35). Both of the above species not only provide fruit but may be used as rootstock for the third, and most important commercial species, *D. kaki* L. (= *D. chinensis* Blume = *D. roxburghii* Carr). This is the "Japanese persimmon," long important in Japan, China, and Korea, and introduced to the United States in the nineteenth century.[12]

The Persimmon in China

D. kaki (*shih*) is often said to be indigenous to a region stretching from northeast India and China to Japan (Watt, 1889-96, 3: 145; Bois, 1927-37, 2: 420-21, 425; I. H. Burkill, 1966, 1: 844). For China, S. C. Lee (1935: 889-90) reported wild *D. kaki* (var. *sylvestris*) from Yunnan northward to Szechwan and eastward to Kiangsu and Chekiang, and H. L. Li (1983: 34-35) writes of it in North China and Korea. Some recent research, however, suggests that this persimmon arrived in Japan from China only in the eighth century A.D. (Itoo, 1980: 442).

Pollen from some sort of *Diospyros*, whatever its species and state of domestication, has been found in Neolithic sites in North China (H. L. Li, 1983: 34). Persimmon fruit, in turn, was identified among the food remains in a Han tomb at Ma-wang-tui, near Ch'ang-sha in Hunan (Y. S. Yü, 1977: 58); and it was also among fruits mentioned in the older Chinese classics (Bretschneider, 1882-95, 2: 315-16; 3: 426; S. H. Shih, 1962: 54). Because of this and because of its genetic diversity there, some botanists (Vavilov, 1949-50: 25; H. L. Li, 1970: 9, 16; Zeven and de Wet, 1982: 36) have come to view China as the place where *D. kaki* was domesticated, with its northern or central regions usually favored. If so, "Chinese persimmon" would be a more appropriate name than "Japanese persimmon." By way of caution, however, we note that Meyer (1911: 14), in extensive travels to investigate tree fruits and nuts in northern China and Manchuria, only once found a few trees that might have been the ancestral *D. kaki*,[13] and that other scholars (P. M. Smith, 1976: 306; Mathon, 1981: 71) argue that the domesticated *D. kaki* may derive from a wild form found from Assam to Indochina. If so, its center of domestication would be well to the south, perhaps not even in China.

In any case, *D. kaki* is cultivated widely in China today from the far south (Yunnan, Kwangtung) northward to Manchuria (S. C. Lee, 1935: 888; "Fruit cultivation in Kuangtung," 1936: 716; T. H. Shen, 1951: 216),[14] and it is also cultivated in certain upland regions of northern India and Burma (Watt, 1908: 498; Randhawa and Nath, 1959: 47; I. H. Burkill, 1966, 1: 844; Regmi, 1982: 95) as well as in Indochina, especially Tonkin (Crevost and Lemarié, 1917: 257).

There are many cultivated varieties of *D. kaki* in China, always grown by budding or grafting on seedlings of the date-plum, *D. lotus* (Meyer, 1911: 11; Dorsett and Dorsett, 1928: 2).[15] The fruit may vary in color from yellow to red. It may be seeded or seedless, ovoid in shape or flattened. Trees of one variety in North China were estimated to become thirty to fifty feet tall, and of another, the tallest variety, sixty to eighty feet. The average orchard tree there was said to be seventy-five to a hundred years old, but one could find trees of some varieties that were hundreds of years old and still highly productive (Meyer, 1911: 11-16; Dorsett and Dorsett, 1928: 2-3). Certain valleys of North China were entirely devoted to persimmons, though, since they are slow growing, the trees were commonly interplanted with annual crops or other fruit trees or bushes for the first twelve years or so. Persimmon fruit ripens in fall but was usually picked before fully ripe. Though the fruit of at least one variety may be eaten while still immature and hard, like an apple, that of other varieties at that stage is usually quite astringent. The astringency, caused by the presence of tannin, affords protection against birds that might eat the fruit (Nakayama and Chichester, 1963: 73). To remove such astringency before sale, the Chinese of the northeast boil water in a container, remove it from the fire, and soak the fruit overnight. Another way is to set the fruits in piles, cover them with dry grass, rice husks, or other litter, and let them ripen for eight to ten days (Dorsett and Dorsett, 1928: 6).[16]

Near Peking, many persimmons picked in the fall were also stored in large open-air beds usually twelve feet wide and any length desired. A bed had trenches beneath or was raised on timbers or logs which permitted cold air to circulate under the fruits and provided drainage for rain or melting snow. A mat of sorghum stems was laid across the trenches, or over the logs or timbers, and fruits piled on the mat in layers, but no more than six deep. A bed was then protected by placing sorghum stems and other litter around it, and the fruits were covered with a reed-grass mat that might be tied or weighted down against the force of the wind. After the winter cold froze the fruit, the bed was covered further with up to two feet of grass or other litter. These persimmon open-air beds were impressive landscape features in some valleys near Peking, with stored persimmon fruit in just one small valley estimated to number over ten million (Dorsett and Dorsett, 1928: 7-10). Such frozen persimmons could be stored for months, or in some years until late March.[17] Fruits were transported from orchard or cold-storage yard in a fresh or frozen state to the railroad or to markets. Sometimes they were carried in baskets borne by a man, otherwise by pack donkey or camel. Persimmons were very popular in northeastern China, and people in Peking could buy them at moderate cost in one of the many fruit stands in marketplaces or along the streets of the city. In Peking at the height of winter, Chinese would place frozen persimmons in cold water for slow thawing. They could be eaten when still somewhat frozen, like sherbet, or allowed to thaw completely when they were little different from newly-harvested fruit and were enjoyed as such. Europeans and Americans there usually liked persimmons, too, and would eat cooled fruit with a spoon as refreshment, or consume them at breakfast much like grapefruit in the United States (Meyer, 1911: 12, 14; Dorsett and Dorsett, 1928: 7-11; H. J. Lin and T. Lin, 1972: 88). Persimmons of the less-juicy varieties, instead of being frozen or eaten fresh, may be preserved in a three-step process of steaming, flattening, and, finally, sun-drying. There is a certain resemblance between dried persimmons and dried figs, and indeed the Portuguese name for them means "red figs." The dried Chinese persimmons we purchased were covered with a fine, white, sweet powder of the sort one sometimes finds on dried Greek figs, and the flesh was somewhat similar to that of figs in its brownish color, firmness, and sweetness. The identity of the fruit, however, was unmistakable, revealed by their persimmon shape and size; their color (an orange tinge, pronounced on the inner side of the skin); an occasional flat, elliptical seed; and their distinctive persimmon flavor. Dried persimmons are delicious raw and were commonly eaten that way in traditional northeast China. Alternatively, they were stewed like dried prunes or apricots, or made into a compote.

Judging from an account of persimmon processing in Fukien ("Production of fruit in Fukien Province," 1928: 1026-29), procedures in that rainy, humid region of the southeast differ somewhat from those described above for northeast China. In Fukien, persimmons of one small-fruited variety, "monkey persimmon," are commonly mixed with water, filtered, and the resulting "persimmon oil" marketed in barrels. The plum-sized fruit of another variety are often processed into a dry form called "persimmon balls." This is done by first peeling them, then exposing them to the sun by day and keeping them in baskets at night. After five days of such treatment, they become red, and, over the next two weeks, black. At that stage, they are put in gunny sacks for about two weeks more, during which time a white coating forms on the fruits. After this, they are shipped to market. Fruit of a third persimmon variety, "pearl persimmon," is made mostly into "persimmon cakes." This is done by peeling the fruits and then, over the days of drying, kneading them bit by bit into flat cakes. If the weather is sunny and wind is from the north, drying is done entirely in the sun. If rainy or if the wind is from the south, cakes may have to be baked on a fire, the dryness of the final product determining how storable it is.

Also somewhat different in the accounts of Fukien we have considered are the processes for eliminating astringency. One large variety of persimmon, "egg-long persimmon," is picked before maturity, in which state it is astringent and inedible. The astringency is eliminated by one of three methods: steaming, smoking, or storing in sealed vessels. In the first method, steaming, two hundred persimmons or so are placed in a large jar. Its mouth is then stuffed with straw and it is turned upside down in a pottery container of water. The container with the inverted jar is put on a fire and the water brought to a boil. The steaming continues for twenty to thirty days, with the water changed every other day. After such steaming, persimmons are not only sweet, but they can stand shipping and will keep for several months. In the second method, smoking, bamboo crates of persimmons are placed on a specially-made brick structure in which a fire of chestnut shells is built, and the entire affair then covered with a wooden container. After the fire goes out, the smoke is said to continue to affect the persimmons. This procedure takes a day and a night, after which the fruits are sweet and edible. Though better tasting than steamed persimmons, smoked persimmons keep for just four or five days, and most are consumed locally. The third process involves packing the newly-harvested persimmons in large wooden containers that have been coated inside and out with mud to make them air-tight; each will hold from 250 to 400 pounds of fruit. By the time it reaches Foochow by boat, the fruit is sweet and edible. This process clearly involves autoasphyxiation, with the fruit itself exhausting the oxygen supply in the sealed vessels. The effectiveness of oxygen deprivation as a means of eliminating persimmon astringency may be an old one in China, but it was not until the early years of this century that American growers tried processing persimmons in this way (D. Fairchild in Gore, 1911: 5). In China, moreover, there were methods of eliminating astringency in addition to those mentioned previously: by soaking persimmons in lime-water for twenty-four hours, putting drops of an alkali solution into the fruit, or hastening its ripening by inserting bamboo splinters into the fruit alongside the stem ("Production of fruit in Fukien Province," 1928: 1028; I. H. Burkill, 1966, 1: 844; F. P. Smith, 1969: 152).[18]

Fresh persimmons can be a good source of ascorbic acid and a fair one of vitamin A. Dried they are a fair source of vitamin A, and contain good amounts of iron (Embrey, 1921: 430; Platt, 1962: 20-21; Leung *et al.*, 1973: 94; Itoo, 1980: 458). The Chinese believe the fruit to possess various medicinal properties (Soubeiran and Dabry de Thiersant, 1874: 188; F. P. Smith, 1969: 152-53; Perry, 1980: 130). They also believe that certain processed persimmons are generally more effective medicinally than fresh ones, and may be needed for specific ailments. One form of processed fruit, called *hung-shih*, is prepared by putting an unripe persimmon in a container along with leaves to aid its ripening. When ripe, it is said to be "as sweet as honey" and to be soothing, effective against fevers, and antivinous.[19] Also used medicinally are skinned, dried persimmons, which are believed to be much more effective, useful in ways, for example as an anthelmintic and restorative, beyond those of the *hung-shih*. Also used as medicinals are dried and dried and smoked persimmons, and persimmons preserved in water or salt. There is a persimmon sweet, *shih-kao*, made by mixing rice and dried persimmons and steaming the mixture until ready for eating;[20] it is prescribed for children with dysentery and other forms of fluid discharge (F. P. Smith, 1969: 152-53). In addition, unripe persimmons are used to make a varnish for waterproofing paper hats and umbrellas (Chun, 1921: 272-75).

The Date-Plum in China

In China, the date-plum (also called wild persimmon or Caucasian persimmon), *D. lotus*, is found wild and cultivated from Manchuria to Yunnan, and it, too, may have contributed to the development of the domesticated *D. kaki* (S. C. Lee, 1935: 890; H. L. Li, 1983: 35). In the north, a few *D. lotus* trees are commonly grown in a persimmon orchard to obtain stock for budding or grafting or for their small, edible fruit, and some are found in scattered places in the mountains. Wild *D. lotus* are usually small trees or even shrubs, but in favorable spots they may grow into sizeable trees. *D. lotus* produces abundant fruits which, when ripe, are usually blackish in color and the size of big cherries, with large seeds and little pulp. Though the Chinese recognize the similarity of the tree to the persimmon, they liken the fruit to the jujube and commonly call it by the name *hei-tsao*, "black jujube," or *suan-tsao*, "sour jujube" (the latter term is also used for the wild jujube *Zizyphus sativa*, var. *spinosa*) (Henry, 1893: 47; Laufer, 1919: 435; Dorsett and Dorsett, 1928: 2; F. P. Smith, 1969: 153). When ripe, the fruits may be eaten by those who collected them, frequently children. They are also found in large amounts, either fresh or dried, in markets, where they are bought especially by poor people (Meyer, 1911: 15-16; Dorsett and Dorsett, 1928: 2, 3).[21]

The fruit of *D. lotus* is believed to dispel evil forces and, if eaten over a period of time, to "give a pleasing appearance to the countenance, and strength to the body." It is also regarded as effective in reducing fevers and in other ways (Hooper, 1929: 53-54; F. P. Smith, 1969: 153; Perry, 1980: 130). We have found no nutritional analysis of this fruit.

Peach, Apricot, and Chinese Plums and Cherries

These plants are species of *Prunus*, a large genus of trees and shrubs indigenous to cool temperate areas of the world, but believed to be of Central Asian origin. Many of its species are found in China (S. C. Lee, 1935: 603; Watkins, 1976b: 243-44), and China, along with Central and western Asia, Europe, and North America, is one of several centers of origin of cultivated *Prunus* species. Western and/or Central Asia are credited by Watkins (1976b: 243-45) with the sweet cherry (*P. avium*), European dwarf or ground cherry (*P. fruticosa*), sour or pie cherry (*P. cerasus*), blackthorn (*P. spinosa*), cherry plum or myrobalan plum (*P. cerasifera*), and damson plum (*P. insititia*); Europe, with the European plum (*P. domestica*); and North America, with the sand cherry (*P. pumila*), western sand cherry (*P. besseyi*), and American plum (*P. americana*). China, on its part, is credited with the peach (*P. persica*); the Manchu, Korean, or bush cherry (*P. tomentosa*); and the Chinese or Japanese plum (*P. salicina*). In addition, Watkins calls western China the primary center of origin of the common apricot (*P. armeniaca*), with western Asia a secondary center of origin. Not mentioned by Watkins, but also cultivated in China, are other *Prunus* species, most notably the Japanese apricot (*P. mume*) and the Chinese cherry (*P. pseudocerasus*). Most important as contributions to the outside world, however, are the peach and apricot, and we turn first to them.

Peach

Domestication of the peach (*t'ao*) is usually placed in temperate China, with most authorities favoring northern and/or western China, especially its hill and mountain regions, including Tibet (H. L. Li, 1970: 16; Harlan, 1975: 73; D. R. Harris, 1981: 29; Mathon, 1981: 61; Zeven and de Wet, 1982: 42). The ancient Chinese looked to the west as the home of the peach, and specifically the K'un-lun Shan, those ranges of Central Asia (H. L. Li, 1959: 62). Whatever the exact time and place of its domestication, the peach has been known in China since great antiquity, with remains found in archeological sites in Chekiang that may date as early as the fourth or fifth millennium B.C. (K. C. Chang, 1977a: 180-81; 1986: 254; Xin, 1984: 10). Today the peach is a typical crop of northern China, but in the south, except for upland regions of the southwest, it is flowering rather than fruit varieties that are more commonplace (E. N. Anderson, 1988: 134), and in Kwangtung, at least in the mid-1930's, peaches were not among major fruit exports ("Fruit cultivation in Kuangtung," 1936: 718).

The peach tree has been surrounded with greater mysticism in China than any other plant, cherished for its delicious fruit and beautiful flowers, important in symbolism and ritual, and frequently mentioned in literature and depicted in art (H. L. Li, 1959: 57). Because of this, Schafer wrote (1977: 93-94), the peach deserves first place in any list of classical Chinese fruits. The traditional Chinese viewed the peach as symbolic of long life and immortality, and various deities and illustrious persons of the past are depicted as carrying the *shou-t'ao*, "peach of immortality." This fruit, believed to bestow immortality on all who ate it, came from a mythical tree—one whose fruit ripened only after three thousand years—at the palace of Hsi Wang Mu, Queen Mother of the West, in the legendary mountain home of the Chinese (C. A.

S. Williams, 1941: 316; Burkhardt, 1953-58, 1: 112; 2: 169-70; J. J. M. de Groot, 1964, 1: 56). The Monkey God was believed to have stolen the peach of immortality from Hsi Wang Mu, and Chinese artists came invariably to depict monkeys as holding peaches (Sowerby, 1940: 46, 137; Burkhardt, 1953-58, 2: 42, 169-70). Shou Hsing, God of Longevity, is also depicted in Chinese art as carrying a peach, and often as emerging from one (Couling, 1917: 426; Sowerby, 1940: 140; C. A. S. Williams, 1941: 316; Burkhardt, 1953-58, 1: 54; 3: 149; H. L. Li, 1959: 58). The peach of immortality is represented in cookies and pastries which serve as presents at birthday parties (H. L. Li, 1959: 57-58). A gift of porcelain or some other form of art decorated with peaches, in turn, carries the wish that the recipient enjoy a long life (Skinner, 1911: 214). Taoists, who were dedicated to the search for longevity and immortality, came to consume peach gum in an effort to gain immortality, and to include the peach in their elixirs of life. Folk belief credited certain ancient Taoist priests with gaining immortality from consuming such elixirs (Couling, 1917: 425; H. L. Li, 1959: 58).

The Chinese also considered peach wood (*t'ao-fu*) protective against evil spirits, who held the peach in awe. In ancient China, peach-wood bows were used to shoot arrows in every direction in an effort to dispel evil. Peach-wood slips or carved pits served as amulets to protect a person's life, safety, and health (Couling, 1917: 426; Burkhardt, 1953-58, 1: 181; Bodde, 1975: 304-6). In Han China, peach-wood seals, with a purpose similar to that of the amulets, were set up on the gates and doors of a house (Bodde, 1975: 302, 304). Carved peach-wood figurines of two guardian deities, Shen Shu and Yü Lü,[22] were also set up alongside doors at the time of New Year to protect against evil spirits; as a result, one Han account reads, "the buildings in the capital are made tranquil and pure; everywhere a good state of affairs prevails" (Bodde, 1975: 85). Associated with the deities, and for additional protection against evil spirits, one found representations of tigers, and, in each deity's hand or nearby, rush ropes. Apparently following Han times, such peach-wood figurines came to be replaced by representations of the two deities painted on flat boards of peach wood. In later times, representations of the guardian deities came to be painted on paper, and, still later, printed on paper. Finally, in T'ang times, the old deities were replaced by new guardians dating from that period (J. J. M. de Groot, 1964, 6: 956-62; Bodde, 1975: 85, 127-38).

Another aid in fighting evil spirits were peach-wood wands. The *Li-chi* (Han period) reported that the emperor went to the funeral of a minister escorted by a sorcerer carrying a peach-wood wand to keep bad influences away. Since that time, peach-wood wands have remained an important means of exorcism in China. At times of epidemic, a peach twig might be soaked in warm water and used for washing one's body. And peaches themselves were made into decoctions used to expel evil spirits or prepared as a soup consumed at New Year (Bretschneider, 1882-95, 2: 274-75; 3: 423; J. J. M. de Groot, 1964, 1: 41-42; 6: 958-62; Bodde, 1975: 133).[23]

In modern China, little mysticism is associated with the peach, but on New Year's Eve people sometimes suspended a peach spray over their gates, and used peach branches to sprinkle water on the ground around a house as protection against evil spirits (C. A. S. Williams, 1941: 316; H. L. Li, 1959, 58-59). Until recently, strips of peach-colored paper continued to be placed as charms on the gates or doors of Chinese houses at the time of New Year, and left to age until replaced a year later to assure continued protection and good luck (J. J. M. de Groot, 1964, 6: 959-60; Bodde, 1975: 136). In addition, the Chinese, always sensitive to beauty, not only depicted its blossoms in art, but cut flowering peach branches and brought them home for decoration on festive occasions. The time of Chinese New Year, in February, is both the time when the peach blooms and the favorite time for weddings. Thus the peach was regularly alluded to in association with marriage (C. A. S. Williams, 1941: 316-17). It symbolized springtime and marriage as well as immortality, and its blossom a bride or the face of a beautiful woman (Thiselton-Dyer, 1889: 154; Ball, 1906: 285; Sowerby, 1940: 137; C. A. S. Williams, 1941: 316-17; H. L. Li, 1959: 59).

Many varieties of peaches in China are solely ornamental, grown in pots or in gardens for their attractive double-flowered blossoms (H. L. Li, 1959: 59-61). Peach blossoms, like other peach parts, enjoyed the ability to dispel evil spirits. They served to cure various ills, being considered diuretic, effective against intestinal worms, and quieting (Soubeiran and Dabry de Thiersant, 1874: 271; J. J. M. de Groot, 1964, 6: 958; F. P. Smith, 1969: 357). Other peach products, whether fruits, kernels, leaves, gum, or bark, were also used medicinally, as befitting a plant with such strong, ancient supernatural associations (Soubeiran and Dabry de Thiersant, 1874: 271; Hooper, 1929: 117-18; F. P. Smith, 1969: 356-57; Perry, 1980: 345).

Meyer divided China's peach varieties into three groups: cling, honey, and peento (*pien-t'ao*, or "flat peaches;" *P. persica*, var. *platycarpa*).[24] The cling group does best in places with hot, somewhat dry summers and fairly cold, dry winters. The honey group is found principally in Central and South China where summers are hot and humid, winters mild and rainy. The peento peaches seem to do well everywhere, but their fruits are small, and trees of this group are cultivated less than those of other groups except in areas of the south where other peaches do not do well. Deserving special note is the fei peach of North China, considered the finest of all cling varieties, pale-yellow in skin, sweet, fragrant, and outstanding in flavor. The fei peach is outstanding, as well, in its ability to be stored and shipped, in one case being carried to market eight days in baskets suspended from a carrying pole. Fei peaches each commonly weigh more than a pound; they cost about two-thirds the normal daily wage of a field worker in China; and they were among the annual tribute sent to the imperial court in Peking early in this century. There are also several Chinese races of peaches with red flesh, some so red as to resemble beets. And in Manchuria there was a cultivated peach which, though its fruits were small, was able to survive winter temperatures that sometimes reached -40°F. (Meyer, 1911: 16-19).

The peach is one of China's important fruit contributions to the Western world, with some major commercial varieties in the United States introduced directly from China. Long before discovery of the Americas, however, the peach had spread widely across Eurasia. In Southeast Asia, it spread as an ornamental, for the tree may not fruit in the lowland tropics or, if it does, it usually yields poor-quality fruit.[25] The westward spread of the peach, on the other hand, seems to have been primarily for its fruit. Literary sources suggest that it was introduced to Persia after the time of Alexander (Laufer, 1919: 539). In the first century A.D., Pliny (*Natural history* XV. 11, 13), called the peach "Persica" and said that it came from Persia and that it was exotic even in Greece and Asia Minor. He also noted that the peach had been introduced to Rome only shortly before from Rhodes, whence it came from Egypt.[26] For India, there is no evidence that the peach was cultivated in ancient times. There is, moreover, the report by Hsüan-tsang (seventh century A.D.) that the peach, known as *cinani* or "Chinese fruit," had been introduced by hostages from China's western borderlands in the time of Kanishka (?A.D. 100-162?) (Watters, 1904-5, 1: 292-93; Laufer, 1919: 540). The peach tree has not, moreover, been a great success in India. Late in the last century it was widespread in the northwestern Himalayas, growing near almost all villages at up to 10,000 feet in elevation, but its fruit was small and green, the trees commonly neglected and the fruit not eaten (Watt, 1889-96, 3: 449; 6, Part 1: 349-50). The tree is also cultivated to a limited extent in other parts of the subcontinent, but in some places its fruit is of poor quality (Hayes, 1966: 444).

Returning to China, peaches, usually quite green, were eaten as such or were prepared in desserts, preserved, or made into alcoholic beverages or vinegar (Soubeiran and Dabry de Thiersant, 1874: 271; E. Watson, 1923: 572; R. H. Cobbold, cited by Holmes, 1948: 145; F. P. Smith, 1969: 358; Schafer, 1977: 113). They were also made into sweet soups, among them the Ch'ing imperial "Peach Soup" made of peeled, seeded, mashed peaches, sugar, cornstarch, water, lard, and rose essence, and which might be served hot or cold (Davis, 1966: 147). And peach kernels, sometimes mistaken for almonds, were also used in confectionery (Hooper, 1929: 118).[27]

Apricot

Apricots (*Prunus armeniaca = Armeniaca vulgaris*; *hsing*) are found wild or spontaneous in certain mountainous areas of Central Asia centered on the T'ien Shan, and in many parts of northern China, as well as in Korea and eastern Siberia (Bretschneider, 1882-95, 2: 293; 3: 421-22; Meyer, 1911: 20; S. C. Lee, 1935: 603; Evreinoff, 1954: 435; H. L. Li, 1959: 156). De Candolle (1886: 215-18), after reviewing the evidence, concluded that China was the apricot's place of domestication, and some present-day botanists agree. The symbol for apricot is found among inscriptions on the Shang oracle bones (18th-12th century B.C.) (H. E. Gibson, 1937: 308, 309; K. C. Chang, 1980: 146). The apricot is also mentioned in various of the Chinese classics as an ornamental and fruit tree; and, like the peach, it was believed to possess magical powers (Bretschneider, 1882-95, 2: 293; 3: 421-22; Sowerby, 1940: 137; H. L. Li, 1959: 156-57; 1983: 34).

In traditional China the apricot was cultivated on a large scale in the north, but a farmer usually had just a few apricot trees rather than an entire orchard. China has many apricot varieties, and their fruits differ in size and color, whether orange, yellow, red, red and white, or white-spotted. The Chinese almost always eat apricot fruit unripe, relishing even those that are still small and hard. Apricots, whether ripe or somewhat ripe, may also be stewed in various ways with sugar or honey. And the fruit may be cut in half and dried for use during the winter season (Meyer, 1911: 20-21; H. L. Li, 1959: 156-57).[28] We take up the use of its kernels in a later section. Though not listed among Kwangtung's major commercial fruits, apricots are grown there and elsewhere in the south, and even in native gardens in Tonkin where one of their names is *hanh* (Crevost and Lemarié, 1917: 241), apparently from the Cantonese *hang*.

The apricot, like the peach, seems to have been late, about the beginning of the Christian era, in reaching Greece and Rome (Hehn, 1885: 320-22; de Candolle, 1886: 215; André, 1981: 80). Laufer (1919: 539) has suggested that possibly it arrived in Iran in the second or first century B.C., then spread westward by way of Armenia, which led to its Greek name and ultimately its species name, *armeniaca*.[29] At least one observer believed the wild apricot is indigenous to the Himalayas of India's Punjab, but this is a minority view among botanists of India (de Candolle, 1886: 216-17; Watt, 1889-96, 6, Part 1: 345). In any case, it seems to have been late in coming into cultivation there, first clear mention of it—as planted "here and there...from Kashmir on" (toward China?)—apparently being that of Hsüan-tsang (seventh century A.D.) (Watters, 1904-5, 1: 177-78). Even in modern times, Kashmir is the only place in India where apricots are of much significance (Hayes, 1966: 444).

Chinese Plums and Cherries

According to E. H. Wilson (1929: 305), the species from which the European plum (*P. domestica* = *P. communis*) derives, has not been reported in China and likely is not found there. There are, however, related species native to China that have come into cultivation for their fruit and/or as ornamentals. Indeed the two trees sometimes called "plums" that are most often mentioned as cultivated in China were both domesticated there. Probably most widespread is *Prunus mume* Sieb. and Zucc., a shrub or small tree usually known in China as *mei* (Cantonese: *mui*)[30] though, as has often been noted, it lacks a satisfactory English name. Though commonly called "plum" or "flowering plum" in English, botanically it belongs with the apricots; and though called "Japanese apricot," it was not domesticated in Japan but introduced from China. H. L. Li suggests that it may have been domesticated in Shensi Province in North China, where it was wild, at least in the past. Today it is reported as wild or spontaneous in western Hupeh, Honan, Szechwan, Kansu, and on Taiwan (Bretschneider, 1882-95, 3: 422; E. H. Wilson, 1929: 305; Evreinoff, 1954: 434; H. L. Li, 1959: 48-49; 1963b: 284), and it is cultivated widely in China. There are many varieties of *P. mume*, both wild and cultivated. Some of the latter are grown for their flowers, others for their fruit, which is globose in form (resembling but usually smaller than those of the apricot, *P. armeniaca*), yellow or green in color but often red on one side. The fruit has firm, sour flesh, usually inedible or barely edible. Thus it is ordinarily used after processing, as by pickling, drying, or smoke-drying, as a snack, flavoring, or medicine. Half-ripe, smoke-dried fruit are called *wu-mei* ("black *mei*"), and green fruit pickled in brine, then dried, are known as *pai-mei* ("white *mei*") or by other names. The latter product is a regular pickle or relish at Chinese dinners, especially with meat dishes. *Mei* are also made into sauce, and fresh *mei* into a sweet, or, after pressing, into a refreshing drink (sour *mei* may also be used for making such a drink). One also reads of *mei* being made into soup, and of other food uses, some to be considered below (Bretschneider, 1882-95, 2: 294-96; Dudgeon, 1895: 17; Hooper, 1929: 117; F. P. Smith, 1969: 355-56; E. N. Anderson and M. L. Anderson, 1977: 334; E. N. Anderson, 1988: 134).

Mei fruit, along with other plant parts, also serves a range of medicinal uses, most often in the form of "black plums" which are used against intestinal worms, diarrhea and dysentery, colic and flatulence, and in other ways (Bretschneider, 1882-95, 3: 422; Hooper, 1929: 117; F. P. Smith, 1969: 355-56; Perry, 1980: 344).

One cannot leave the *mei* without noting that since antiquity it has been highly-valued in China as an ornamental, whether as a pot plant or in gardens or temple grounds where trees hundreds of years old may be found. The *mei* is esteemed both for its attractive white or pink flowers and the form in which it grows, especially the gnarled branches typical of aged trees. Since antiquity it has been a great favorite of Chinese poets, writers, and artists, and a common art motif, as well, on all sorts of items of everyday use. Since it blossoms in spring before the snow and ice have disappeared, the *mei* is linked in Chinese paintings with the pine and bamboo as symbolic of hardiness, purity, and longevity, and was chosen as the national flower by the Republican Government of China (Yetts, 1912: 20; Sowerby, 1940: 137; H. L. Li, 1959: 48-56).

Often confused by Westerners with the *mei* is the Chinese or Japanese plum (*Prunus salicina* Lindl. = *P. triflora* Roxb.), known usually as *li* in China, where it is both wild and cultivated. This is a small tree with white blossoms, whose primary gene center is in North China where it has been known since antiquity and where, with peach, apricot, jujube, and chestnut, it was among the five renowned fruits of antiquity (Veith, 1949: 206; H. L. Li, 1959: 157; Zeven and de Wet, 1982: 44). Japan is its secondary gene center and *P. salicina* is widely cultivated there, though not indigenous to that land. Though the tree itself is hardy, it flowers early and, as a result, is especially subject to frost damage (Roach, 1985: 143). The fruit may be consumed fresh, pickled, or in other ways.

One cannot be certain which of the above fruits—*mei, li,* or both—are involved, but plums may be ingredients in main dishes, such as the Cantonese "Steamed Spare-Ribs with Plums" (K. Lo, 1979a: 103). They may also be made into desserts, such as Hunan "Stuffed Prunes," in which the kernels are removed from the prune seeds and stuffed in the pitted fruit (Stallard, 1973a: 170). Plums may also be made into alcoholic beverages or into plum sauce, that thick spicy sauce so popular as a dip in American Chinese restaurants. In Canton, plum sauce serves a more specific purpose, both to flavor, and to counter the greasiness of, flesh. It may simply be an ingredient in cooking, as in the Cantonese dishes "Chicken in Plum Sauce" and "Pork in Plum Sauce" (W. W. Chang *et al.*, 1970: 261, 308). It may be used, mixed with oil and perhaps other ingredients, as a dip,[31] especially for roast or deep-fried duck. It may be spread on the pancakes used to wrap pieces of duck or other food for eating, as hoisin sauce is commonly used for Peking Duck.

Early in this century, Meyer (1911: 23) wrote that the sweet cherry (*P. avium*) was apparently not found in northern and eastern China, but noted two other cherries with edible fruit found in the north or in the Yangtze Valley. These two cherries, both of Chinese origin, he identified as the Chinese sour cherry (*P. pseudocerasus* Koids.?),[32] which the Chinese call *ying-t'ao*, and the Manchu or Chinese bush cherry (*P. tomentosa* Thunb.), *shan-ying-t'ao*. The Chinese sour cherry, which may be cultivated for fruit and as an ornamental, has delicious fruit though usually sour and somewhat small. It has the advantage over other cherries of flowering and fruiting very early in the season (Meyer, 1911: 23). The Manchu cherry, the most common cherry in North China, is found there both wild and cultivated. In a wild form, it is a thick shrub found in dry, rocky places in the mountains of Manchuria, Korea, and North China, and has quite small, red fruit. When cultivated, as in mountains near Peking, its growth is more open, and its fruit larger (Bretschneider, 1882-95, 2: 299-300; Meyer, 1911: 23-24). Other species of cherries or plums may also be cultivated on a small scale. One, producing a small, sour, red fruit like a plum or cherry, is the *yü-li, Prunus japonica* Thunb., which is cultivated in some parts of China, and its kernel used as medicine and its fruit sometimes eaten (Bretschneider, 1882-95, 2: 296-97; 3: 509-10; Hooper, 1929: 117; F. P. Smith, 1969: 355). Another species, mentioned by botanists (Bailey, 1935, 3: 2827; Rehder, 1940: 458; Vavilov, 1949-50: 24; Uphof, 1959: 297; Mathon, 1981: 61; Zeven and de Wet, 1982: 44) as cultivated, fruit-producing, and of North Chinese origin, is the Simon plum or apricot plum, *P. simonii* Carr., whose fruits commonly are bitter but sometimes quite pleasant in taste. Cherries may be consumed in various ways in China, whether eaten fresh, made into compote, jelly, jam, or vinegar, or preserved in honey or in other ways (Waln, 1935: 69; C. A. S. Williams, 1941: 65). In Chinese symbolism, the cherry represents beautiful women or women in general (Sowerby, 1940: 137; C. A. S. Williams, 1941: 65).

Figure 58. Chinese jujube (*Zizyphus sativa* Gaertn.) (from Bois, 1927-37, 2: 105)

Chinese Jujube

The Chinese jujube, Common jujube, or Chinese date, *Zizyphus* [or *Ziziphus*] *sativa* Gaertn. (= *Z. jujuba* Mill. and *Z. vulgaris* Lam.; *tsao*) (Figure 58), is a small tree, ten to thirty feet tall, suited to areas with somewhat low precipitation and cold, dry winters.[33] Though found in Kwangtung (S. C. Lee, 1935: 798), that region is too humid and rainy for it to do well. On the other hand, it is well suited to and important in North China, from which its fruits are shipped south in considerable amounts. In the north, the jujube is cultivated everywhere except in places with very cold winters. It prefers a porous, somewhat alkaline clay soil like North China's loess, but also yields well on highly alkaline or compacted soils, and seems quite tolerant of neglect. In the north, cultivated jujube may be found as individual trees, in small groves, or in orchards roughly ten to twenty acres in size (Meyer, 1911: 35-36).

Most jujube fruits are plump and brown when fresh, but wrinkled and reddish when dried. They are of many varieties, however, and they differ in shape, size, color, and other characteristics. There is a "seedless" jujube from Shantung (its seeds are so soft as to be hardly noticeable when eaten) whose fine, sweet fruit was sent as an annual gift to the royal court in Peking. Such seedless fruit is said to be very tasty when cooked with rice or millet as a compote. When dried, it may be eaten with peanuts in the way we eat peanuts mixed with raisins. The Chinese also eat other varieties of jujubes: fresh or dried, preserved in honey or sugar, or smoked (Bretschneider, 1882-95, 2: 119; Meyer, 1911: 36-39; Read *et al.*, 1940: 25; Porterfield, 1951: 26). Dried jujubes may keep indefinitely if stored in a dry place. When used, they may be boiled in water to make a compote, as in "Cooked Dried Fruits," a Ch'ing court dish which also contained dried persimmons, fresh lotus root, sugar, and rose essence (Davis, 1966: 226). They may be prepared in other desserts, such as a date cake of Soochow said to be "reminiscent of English plum pudding" (Sakamoto, 1977: 159). Or they may be an ingredient in soups and slow-cooking dishes of other sorts, such as the Cantonese "Braised Duck with

Chestnuts and Chinese Dates" (Sakamoto, 1977: 186). One reads of jujube wine in T'ang times (Schafer, 1977: 121), and there are non-food uses of the fruit as well (Porterfield, 1951: 27). There is also a rare jujube called "dragon's claw" (*Zizyphus sativa*, var. *tortuosa*), which has gnarled branches and is a highly-valued ornamental (Meyer, 1911: 40).

Wild jujube (*Zizyphus sativa*, var. *spinosa*; *suan-tsao* or "sour jujube") are usually shrubs, but sometimes trees twenty to thirty feet tall. They are found widely in North China where they grow on city walls or in isolated, stony places of little or no use for cultivation. Though cultivated jujube have no spines, the wild plant has curved ones that will readily tear clothes or impale themselves in flesh, which has led to widespread use of its branches as fencing. Fruit of the wild plant, though small and sour, is valued as food, gathered by old women and children and made into a preserve or paste (Meyer, 1911: 40).

The jujube is one of the earliest fruits of North and Central China, its importance in early agriculture there comparable to that of the fig and date in early western Asia (H. L. Li, 1983: 33-34). Jujube remains have been identified (state of domestication and other details lacking), along with those of other food plants, in early agricultural sites dating from the fourth and fifth millennia B.C. or before: Eh-kou in Honan (K. C. Chang, 1986: 92) and Ho-mu-tu in Chekiang (T. K. Cheng, 1982: 11; Xin, 1984: 8, 10). The character for jujube is inscribed on Shang oracle bones (E. N. Anderson, 1988: 20). There are also many ancient Chinese literary references to the jujube, wild or cultivated, including some from the Chou period (Bretschneider, 1882-95, 2: 305-9; 3: 424; Keng, 1974: 403-4; H. L. Li, 1983: 33). In Han times, the best jujube orchards were found in southern Shansi Province (H. L. Li, 1983: 33).

Vavilov (1949-50: 24, 32, 34) looked not only to China, but to Central Asia (including northwest India) and the Near East as a center or secondary center of jujube domestication. Most authorities, on the other hand, place its domestication solely in China (de Candolle, 1886: 196; H. L. Li, 1970: 16; Harlan, 1975: 73; Mathon, 1981: 64; Zeven and de Wet, 1982: 42), for in China its genetic diversity is greatest, with forty-five varieties or races identified already in the *Ch'i-min yao-shu*, written in the sixth century A.D., and with several hundred today, most of them from North China (S. H. Shih, 1962: 55; H. L. Li, 1983: 33).

The story of the spread of the Chinese jujube plant is not well known, but it probably reached Rome at the beginning of the Christian era. Jujubes have since become somewhat common in the Mediterranean area, where they are harvested in fall and either sold fresh in markets or dried on trays for export (de Candolle, 1886: 194-96; Bois, 1927-37, 2: 104).[34] The English name "Chinese date" derives from the fact that the dried jujube fruit, though ordinarily reddish in color, is shaped like the date of the Near East, has a long pit like the date, and a taste reminiscent of it. The Chinese recognized these similarities, too, for when they came to know the date, they called it, among other things, "foreign jujube" and "Persian jujube" (Laufer, 1919: 385; Schafer, 1967: 176; H. L. Li, 1979: 124).

The jujube, wild and domesticated, has a long history as a medicine in China, involving use of fruit, pit, leaves, root, wood, and bark (Soubeiran and Dabry de Thiersant, 1874: 250; Bretschneider, 1882-95, 3: 424, 503-4; F. P. Smith, 1969: 466-67; Perry, 1980: 339). Jujube fruit, among other things, is considered highly tonic and nourishing (F. P. Smith, 1969: 466; E. N. Anderson, 1988: 136). It is also considered very "heating" and may be included in soup prescribed for a mother during confinement (Yeung *et al.*, 1973: 198; Yee, 1975: 69). Nutritional analysis of Chinese jujube shows that, when dried, it is high in carbohydrate, contains good amounts of iron and riboflavin, and fair amounts of calcium and thiamine. The smoked fruit appears to have similar food value, and the raw fruit, in turn, is a good source of ascorbic acid (Read, 1940: 58-59; Leung *et al.*, 1973: 85). Overall it compares quite well in nutritional value with both the fig and date.

Chinese Olive

This is one of the Chinese fruits and vegetables named by Europeans and Americans for the products at home they most resemble. In fact, the "Chinese olive" derives from evergreen trees that are species of *Canarium*, a genus indigenous mainly to the Asian and African tropics. This genus is quite distinct from *Olea*, that of the cultivated olive (*O. europea*) of the Mediterranean area and Near East, which was described in T'ang times as a tree of Persia and Syria whose fruit was pressed to obtain a cooking oil used like sesame oil in China.

Turning back to *Canarium*, various of its wild species provide useful products, among them wood; an aromatic, oil resin known as elemi, used mainly in varnish, lacquer, and printing ink; and edible fruit and seed. Certain species may sometimes be planted partly or largely for their edible fruit or seed. One is *C. schweinfurthii*, the bush-candle tree, incense tree, or African elemi of West Africa, whose name derives from its aromatic resin which serves as an illuminant, fragrance, and for other purposes, but whose fruit and seed are also eaten (Dalziel, 1937: 315-16). Others, whose seeds are their most important products, are *C. commune*, the kenari-nut or Java almond tree of Malaysia, and *C. ovatum*, the pili-nut of the Philippines (Watt, 1889-96, 2: 94-95; Crevost and Lemarié, 1917: 214-15; Bois, 1927-37, 2: 99-100; Uphof, 1959: 71; I. H. Burkill, 1966, 1: 429, 433-34, 436).[35]

As for China, two *Canarium* species, *C. album* and *C. pimela*, are cultivated for food in the south (Kwangsi, Kwangtung, Fukien, Szechwan, and elsewhere), and both are also cultivated in Indochina (Porterfield, 1951: 27-28; I. H. Burkill, 1966, 1: 430; Zeven and de Wet, 1982: 34, 51). *C. album* seems first to have been described in China in the third and fourth centuries A.D. as *kan-lan*, still one of its names today (another common one being *ch'ing-kuo*) (Laufer, 1919: 417n; H. L. Li, 1979: 120-22). *Kan-lan* is similar to its name in modern Annam, and for this and other reasons, Laufer (1919: 417n) has argued that *C. album* was probably domesticated in Indochina, with the cultivated plant then spreading into southern China.[36] On the other hand, *C. album* is found wild today on Hainan (Fenzel, 1933: 137, 168; H. L. Li, 1983: 52), and, according to Porterfield (1951: 27), it is known only in the wild on that island. Whether on these grounds or others, certain authorities (Harlan, 1975: 73; Mathon, 1981: 55; Zeven and de Wet, 1982: 34) have placed its domestication in China, rather than Indochina. Authorities differ, as well, about the precise place of domestication of *C. pimela* (*wu-lan*), whether in Indochina or South China (Haudricourt and Hédin, 1943: 146; K. C. Chang, 1970: Table 1; H. L. Li, 1970: 11, Table 1; Zeven and de Wet, 1982: 51). In either case, one is dealing with the same general region, in antiquity the homeland of the Nan-Yüeh. Thus the Chinese olive likely represents yet another of the Nan-Yüeh contributions to China's wealth of domesticated plants, one which provided those seafarers not only with food but with a superior varnish and sealing-oil for boat hulls (Schafer, 1954: 67; 1967: 197-98; 1977: 97). The Chinese olive was among the southern plants brought north for planting at the palace of Han emperor Wu Ti when Nan-Yüeh was conquered in 111 B.C., and in subsequent periods its fruit was sent north as tribute (Laufer, 1919: 417n; Schafer and Wallacker, 1957-58: 218; H. L. Li, 1979: 120). In modern China its fruits, in one form or another are popular throughout the country.

The fruit of *C. album*, yellow in skin color when ripe, is known to Westerners in China as "white olive," and that of *C. pimela*, which is purple black, as "black olive." Both resemble European olives and, like them, they may serve as a side dish or flavoring. Like European olives, they have an acrid, unpleasant taste when fresh, though this differs from one variety to another. They may be eaten fresh after processing to make them palatable. They may be salted by soaking in salt water, then sun-dried. Or they may be preserved in sugar, syrup, or honey, or processed with licorice. The seed kernels are also eaten, and those of the black olive, which are considered best, are said to resemble walnuts in taste. In South China, *Canarium* seed kernels are well-liked in cooking though used on a small scale (Blasdale, 1899: 43-44; Meyer, 1911: 48; E. Watson, 1923: 473; "Production of fruit in Fukien Province," 1928: 1023; Porterfield, 1951: 28; F. P. Smith, 1969: 89-90; Schafer, 1977: 97, 131; Zee and Hui, 1981: 79; E. N. Anderson, 1988: 136, 137, 172).

The seeds of *C. album* are high in fat and, like many other seeds, are also high in protein. The semi-dried, salted fruit is also fatty and contains good amounts of iron and calcium (Read *et al.*, 1940: 56-57; Porterfield, 1951: 28; Leung *et al.*, 1973: 91). In addition both species of Chinese olives, especially their fruit and seed, serve medicinal purposes in China (Porterfield, 1951: 28; F. P. Smith, 1969: 89-90; J. D. Keys, 1976: 91; Perry, 1980: 63).

Some Lesser Fruits of Chinese Origin

Chinese Quince

The Chinese quince (*mu-kua* = "tree *kua*" or "tree melon or gourd"), the form usually found on fruit stands in China, has an aroma resembling that of the common quince of Europe, *Cydonia oblonga* (= *Cydonia vulgaris*). This has contributed to the two being placed by some botanists in the same genus, the Chinese quince usually as *Cydonia sinensis*. Other botanists, however, insist that the two belong in different genera, *Cydonia* and *Chaenomeles*, the Chinese fruit commonly classified as *Chaenomeles sinensis*.[37] By either classification, we are dealing with yet another Chinese domesticated fruit that, though independently domesticated, had its parallel in the West.[38]

The Chinese quince, mentioned already in the Chinese classics (Bretschneider, 1882-95, 2: 300-301; 3: 425), is a shrub or small tree of Central China, where it has been grown mainly in the Yangtze Valley, particularly Anhwei (F. P. Smith, 1969: 363). It is less common in western China (E. H. Wilson, 1929: 307) and in the south, though the southeast is said to have produced excellent quinces in T'ang times (Schafer, 1967: 183). In northern China, Meyer (1911: 32) found the Chinese quince to be rather uncommon, noting that it was raised on a commercial scale only south of central Shantung.[39]

Chinese quince fruit, which can weigh ten pounds, serves in China as ornament and fragrance, a medium for decorative carving,[40] and as a medicinal[41] and food. Early in this century Meyer (1911: 32-33) wrote that in North China only foreigners preserved the fruit, as in jam and jelly; that the Chinese there never ate quinces; and that instead they were only used by people to add their pleasant fragrance to a room. E. H. Wilson (1929: 307), writing in more general terms, mentions only their use as ornament, fragrance, and medicine. Other accounts, however, give clear evidence of food uses in China, saying that quinces were eaten in T'ang times (Schafer, 1977: 94); that in Sung times the people of K'ai-feng ate preserved quinces (Freeman, 1977: 168); and that in Ch'ing times they were made into a wine, although an inferior one (Spence, 1977: 277). In contemporary times, they continue to be used in alcoholic beverages, such as the white distillation known as "Tiger Tendon and Quince Wine" (W. W. Chang *et al.*, 1970: 503; Em. Hahn *et al.*, 1973: 146). In addition, the quince is much used as a candied fruit (H. L. Li, 1959: 125). We have found no analysis of its nutritional value.

Chinese Haw

In China, the small edible fruits of various species of hawthorn (*Crataegus*)[42] have been consumed from antiquity onwards (Bretschneider, 1882-95, 2: 111, 135-36, 306; Freeman, 1977: 155; Mote, 1977: 217; and Schafer, 1977: 94).[43] Some fruits are gathered from the wild, and some, usually red in color, picked from the several species that are cultivated (E. H. Wilson, 1913, 1: 34; 2: 29). In traditional China, the hawthorn was a typical fruit of North China's Winter Wheat-kaoliang Area (Buck, 1956: 30). Chinese hawthorn plants and fruits, commonly called "haws" or "haw apples" in English,[44] are known in Chinese by various names, commonly modifications of *cha*, a generic term for hawthorn.

The main hawthorn species cultivated for fruit is the "Chinese hawthorn," *Crataegus pinnatifida* (commonly *shan-cha*, "mountain haw"),[45] a shrub or small tree placed by botanists in the Chinese or Chinese-Japanese center of origin (Vavilov, 1949-50: 24; Mathon, 1981: 57; Zeven and de Wet, 1982: 43) or more specifically in North China (K. C. Chang, 1970: Table 1; 1976: Table 1; H. L. Li, 1970: 9, 16; 1983: 35). A second cultivated species is *C. hupehensis* of Hupeh Province (E. H. Wilson, 1913, 1: 34; 2: 29; 1929: 34, 307; Zeven and de Wet, 1982: 43). *C. pinnatifida*, usually a thorny shrub and indigenous to North China (Honan, Shansi, Hopeh, Shantung), Manchuria, Korea, and eastern Siberia, is exceptionally hardy and able to take much drought and cold, leading to its use as a rootstock for North China's large-fruited cultivated varieties. Cultivated *C. pinnatifida* can also take much heat, and is thus well suited to conditions in North China, where one finds entire orchards of it, all trees given careful attention. Usually its fruit is bright red, somewhat hard, and normally pleasantly sour in taste, though it differs from one variety to another in acidity as well as size. Fruits of the best varieties, which come from Shantung Province, are only a little sour, comparable to crab apples in size, and they are marketed throughout China (Meyer, 1911: 33-34; F. P. Smith, 1969: 130). The fruit of *C. hupehensis*, also grown in orchards, is described as red, not quite an inch in diameter, and "of insipid flavor" (E. H. Wilson, 1913, 2: 29; 1929: 307).

Haw apples from *C. pinnatifida* are widely used in China in making sweetmeats, jellies, or jams regularly sold in shops. They may also be glazed with sugar, with fruits commonly strung on straws, bamboo slips, or long skewers made of wood, and sold by street vendors. In contemporary China, the fruit is also preserved in juice, syrup, or wine (Headland, 1914: 173, 180; Chun, 1921: 176; F. P. Smith, 1969: 130; Sakamoto, 1977: 71; Wittwer *et al.*, 1987: 287). There is also a haw apple candy made in the form of thin, quarter-sized disks packed in a cylinder (Kan, 1978: 220) apparently much like our Necco Wafers. Haw apples are also made into jelly cake, as in the Ch'ing dynasty imperial dish "Steamed Lotus Seeds with Haws Jelly Cake" (Davis, 1966: 148). The fruit of one somewhat sour, dry-fleshed variety may also be sliced, dried, and stored for consumption in winter, and then stewed, made into a compote with sugar, or brewed into a tea regarded as a purifier of the blood and body (Meyer, 1911: 33). The juice or fruit of this and various other species of haws, among their other medicinal uses, are also considered blood stimulants, aids to digestion, antidiarrheal and antiscorbutic agents, and effective against hypertension and arteriosclerosis (F. P. Smith, 1969: 131; Perry, 1980: 342; Wittwer *et al.*, 1987: 287). The fruit of *C. pinnatifida* are high in vitamin C, and contain fair to good amounts of iron and calcium. They are also an excellent source of potassium (327 mg./100 gm.) (Leung *et al.*, 1973: 84; Wittwer *et al.*, 1987: 287), nearly as good as bananas, and should thus be valuable in replacing excessive potassium lost through severe diarrhea.

Chinese Gooseberry or Kiwifruit

The Chinese gooseberry (*Actinidia chinensis*)[46] is a significant, though recent, contribution to the store of fruits in the Western world. Exports of fruit from New Zealand to the United States in 1960 led an exporting company to coin a new name, "kiwifruit" (now often shortened to "kiwi," the name of flightless New Zealand birds of the genus *Apteryx*), which has become widely accepted. In fact, however, the fruit had been introduced to New Zealand only in the first decade of this century (Schroeder and Fletcher, 1967: 81; Ferguson, 1984: 3).[47]

The genus *Actinidia*, according to one recent classification (1983), consists of more than fifty species of climbing or straggling woody vines, many of which have edible fruit (Ferguson, 1984: 4-5). The genus is indigenous to a region stretching southward from Sakhalin Island and eastern Siberia through Japan and China to Malaysia and northeastern India, but the uplands of southwestern China are where the genus evolved (Ferguson, 1984: 9). Of the many species found in China, two, *A. arguta* and *A. chinensis*, have been singled out by H. L. Li (1952: 4) as of economic importance there, long used for their edible fruit collected from the wild but not cultivated.[48] Both of these, often called *yang-t'ao* ("goat's peach" or "sheep's peach")[49] in China, are widespread species, found from northeastern Asia and Japan to western China. *A. arguta*, however, is particularly common in North China, including Manchuria, and the more important *A. chinensis*, which is more sensitive to frost, is most common in the Yangtze Valley, a region of mild winters (Meyer, 1911: 49; H. L. Li, 1952: 1-5, 54). *A. chinensis* was first cultivated there in 1900, when it was planted by foreign residents of Ichang; subsequently it was called "Ichang gooseberry" throughout the Yangtze Valley (E. H. Wilson, 1929: 311). Today large quantities of Chinese gooseberries (over 100,000 metric tons annually) are collected from the wild in China (Ferguson, 1984: 3). In addition, the plant is now grown in many parts of the country, and is undergoing selection for larger fruit and higher vitamin C content (Wittwer *et al.*, 1987: 289-90).

The fruit of *A. chinensis* is the size of a small chicken egg, roundish or oval in shape, and has a thin skin, brown and often hairy. Its flesh is green in color, contains many tiny black seeds, and is tender and delicious in flavor. Its similarity in taste to the gooseberry has often been noted, but among fruits important in the United States the strawberry is perhaps most like it in taste. We have found nothing to indicate whether the Chinese preserved this fruit in the past, but it was used by missionaries in China to make an excellent jam.[50,51]

It is beyond the scope of our work to trace the spread of *Actinidia chinensis* and its relatives to the Western world. We would note, however, that *A. chinensis* was introduced to cultivation in England and the United States, as to New Zealand, in the first decade of this century; that today the leading producers of cultivated kiwifruit are New Zealand, the United States, France, Japan, and Italy; and that the fruit is marketed in various forms, and made into jam and even an alcoholic beverage (E. H. Wilson, 1913, 2: 32; Schroeder and Fletcher, 1967: 82; Ferguson, 1984: 3; Vietmeyer, 1987).[52] Consumption of kiwifruit in the United States has been increasing rapidly in recent years, but about half of the California crop is exported, with Japan the largest purchaser. The kiwifruit we observed in Hong Kong were from California, and on returning home we found New Zealand kiwifruit in certain California markets. The kiwifruit has been found to be an exceptionally rich, but variable source of vitamin C, containing at least two times the ascorbic acid of citrus fruit and ten times that of apples (Randoin and Boisselot, 1941; Matheson, 1951; Platt, 1962: 34; Ferguson, 1984: 43-44).

"Human Face"

Another edible fruit cultivated in Kwangtung is the *yan-min* (Cantonese for "human face"),[53] which comes from an evergreen tree (*Dracontomelum dao = Dracontomelon dao*) indigenous from South China and Vietnam to the Philippines and Indonesia (Tanaka, 1976: 266; H. L. Li, 1979: 132; Perry, 1980: 17). This is a member of a small genus of trees found in southeastern Asia and the Pacific, several species of which provide, among other things, edible fruit (Crevost and Lemarié, 1917: 231; Bois, 1927-37, 2: 164; I. H. Burkill, 1966, 1: 873).[54]

Dracontomelum dao was known to Chi Han in the fourth century A.D., though he lacked enthusiasm for its fruit which he described as tasteless, but somewhat palatable when cooked in honey (H. L. Li, 1979: 132). Though Chi Han did not indicate whether the tree was cultivated in southeastern China or Vietnam in his day, today it is commonly found in cultivation around Canton. It is not, however, cultivated in the Philippines (Wester, 1924: 69), and we have no information that it is cultivated elsewhere. Today, according to Zee and Hui (1981: 88), the small, dark green, quite sour fruit of *D. dao* may be prepared as a vegetable or as a preserve. One delicious dish prepared from fresh fruit, they say, is made by steam cooking it with oil and salt along with chili pepper and preserved soybeans. It also has certain medical uses (Perry, 1980: 17).[55]

Chinese Strawberry

This is yet another instance of a Chinese fruit being given an English name because of its resemblance to a fruit of the West. In this case, the fruits come from quite different cultivated plants, the Western fruit from perennial herbs, mainly *Fragaria ananassa*, the Chinese one from an evergreen shrub or small tree, *Myrica rubra* (= *M. nagi* DC), a wax myrtle. Commonly called *yang-mei*,[56] this plant is a Chinese domesticate, probably originating in east-central China, likely, according to H. L. Li (1983: 46), in hill areas of northern Chekiang. It is common as a cultigen today in Kwangtung and eastern China, and is also grown in Taiwan, Japan, and Korea (Meyer, 1911: 45; Chun, 1921: 57; H. L. Li, 1979: 117; Mathon, 1981: 59). Its lack of tolerance of cold conditions is clearly revealed in Japan: the Myricas do well in the southern islands, but the quality of fruit declines with increasing latitude, and cultivation ceases at about 34° or 35°N. (Ikeda, 1909: 38-39).

Bois (1927-37, 2: 513) writes that *M. rubra* appears to have been cultivated in China since the greatest antiquity, but we know of no documentation for that statement. It is true that remains of *M. rubra* fruit were among the foods discovered in a Han tomb at Ch'ang-sha that dates from the second century B.C. (Y. S. Yü, 1977: 56), but nothing is said about whether they were from wild or domesticated trees. It is also true that Chi Han described the fruit in his fourth century flora (H. L. Li, 1979: 117) of the southeast, but unfortunately he wrote only of gathering it in the mountains, with nothing to indicate cultivation. Our conclusion is that the antiquity of strawberry tree cultivation may be uncertain, but there is no question that its fruit has long been consumed as food in China.

The fruit of the Chinese strawberry tree is ovoid or globose in shape, and may range in color from white to dark red or purple. Its size varies from that of a cherry to that of a good-sized plum, and its taste, from quite sour to sweet (Meyer, 1911: 46). The sour taste may be eliminated by soaking the fruit in salted water or by boiling it with sugar ("Production of fruit in Fukien Province," 1928: 1033). The Chinese distinguish three main varieties by the color of their fruits: purple, red, and white, with the purple most highly regarded (F. P. Smith, 1969: 275). The fruits, which may be found in Shanghai markets seasonally, may be eaten fresh or cooked; or they may be salted or preserved in other ways, or made into a drink (Chun, 1921: 57). Early accounts also mention a "wine" made from wild strawberries, one, however, served only to "noblemen and distinguished guests" (H. L. Li, 1979: 117; Needham and Lu, 1986: 446). Chinese strawberry fruit, like that of some other wax myrtle species, has an edible seed, which, along with the fruit, may serve medicinal purposes (F. P. Smith, 1969: 275; Perry, 1980: 278). We have uncovered no analysis of the food value of either seed or fruit.

Japanese Raisin Tree

For want of a more appropriate place for it, we would mention here *Hovenia dulcis*, a deciduous tree known in English as the "Raisin tree" or "Japanese raisin tree," but which more accurately should be called "Chinese raisin tree." This tree *(chih-chü)*[57] is found from Hopei, in the north, to Kwangtung, Kweichow, and Yunnan, in the south (S. C. Lee, 1935: 806-7), and today is widely planted in gardens and along roadsides, mainly as an ornamental. It was mentioned in the *Shih-ching* ("Book of Odes"), a work dating from early Chou times (*c.* 1000-500 B.C.) (Keng, 1974: 404). Though initially cultivated in North China, it spread in early times to Japan and the lower slopes of the Himalayas of India (H. L. Li, 1963a: 249; 1983: 35).

The part of the tree eaten in China, Japan, and India is not the fruit, but the peduncle, the stalk that supports the fruit. As the fruit matures, the peduncle becomes fleshy and thick, reddish brown, sweet or subacid, with a taste reminiscent of a dry raisin (Watt, 1889-96, 4: 301-2; Chun, 1921: 242-43; Bois, 1927-37, 2: 106-7), thus the name "raisin tree" (Figure 59). The peduncle was not only eaten like a fruit in China, but it (and other plant parts) served medicinal uses, as in warding off the bad effects of drinking alcohol (Hooper, 1929: 74; E. H. Wilson, 1929: 311; F. P. Smith, 1969: 209; Perry, 1980: 340).[58]

Chinese Mulberry

One feels uneasy about placing the white mulberry (*Morus alba; sang*)[59] among Chinese "also-rans" of any sort, for it is a tree of great importance and antiquity in China, essential to the development of the silk industry there. Domesticated in North China, the tree came to be a conspicuous landscape element, and to occupy a major place in economic life there. Mulberry cultivation and silk production spread southward over time, to gain a position of great importance in other places, among them Kwangtung; and silk became a critical item in trade, both domestic and foreign. The white mulberry's essential role as provider of the leaves on which silkworms feed, meant that for *M. alba* the leaf and its tenderness were the essential concerns in selection. Though the Chinese could have selected *M. alba* for tastiness of fruit, we have no information that in fact they did so.[60,61] The case was different with *M. nigra*, the black mulberry of the ancient Near East and Mediterranean, which in early times was of use primarily for its fruit. In that region, mulberries were consumed fresh, whether raw or stewed, or made into preserves. It was also common to dry mulberries for future consumption,[62] and in

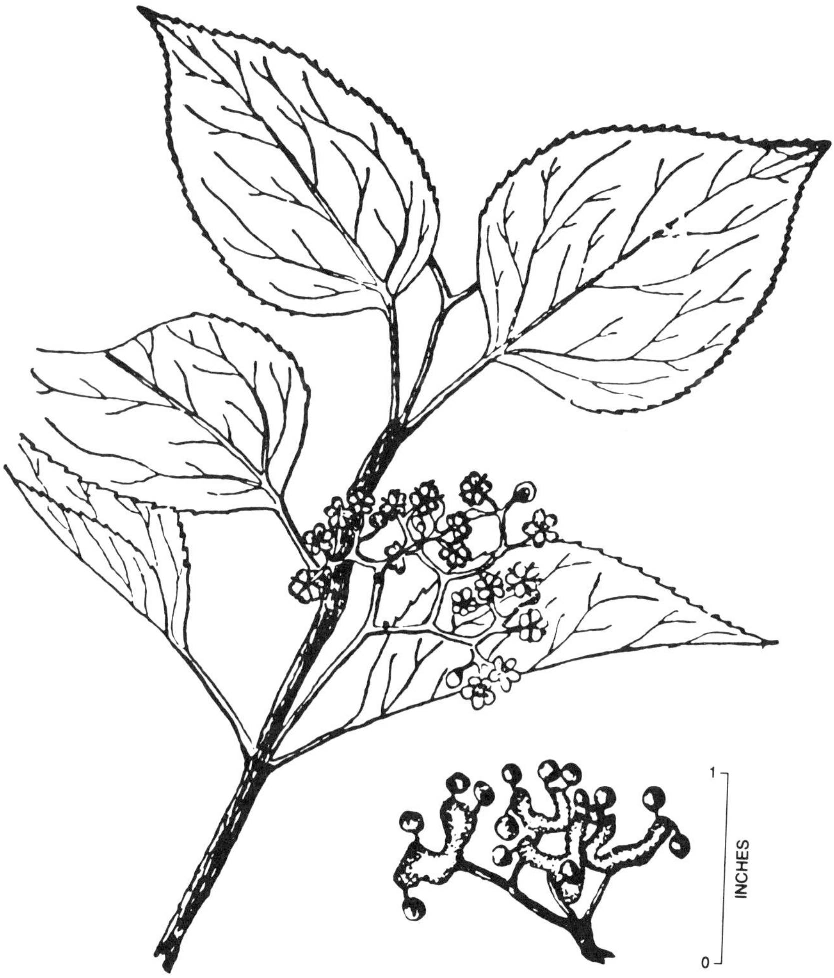

Figure 59. Japanese raisin tree (*Hovenia dulcis*) (redrawn from S. C. Lee, 1935)

Afghanistan, dried mulberries were even ground into flour and made into bread (Watt, 1889-96, 5: 281). The leaves of *M. nigra*, on the other hand, proved inferior to those of *M. alba* for getting quality silk from the silkworm when, in medieval times, that insect was introduced to Europe from the east (Hehn, 1885: 290-93; de Candolle, 1886: 149-53).

Returning to the question of the mulberry as human food in China, we note that the fruit of *M. alba*, though sweet, has been described by some as insipid, devoid of taste, and even as hardly edible to humans;[63] and that references to Chinese eating mulberries are scanty in our sources. E. N. Anderson (1988: 20, 28, 41, 66, 68) does include the mulberry among fruit sources in China from ancient times onward, but provides little detail. Turning to traditional China, we note various references to the fruit of *M. alba* having a dietary role. S. W. Williams (1895, 2: 10) wrote of the mulberry being grown as a shade and fruit tree in North China.[64] S.

C. Lee (1935: 429) wrote of *M. alba* fruit being "eaten by the peasants" in China. Adam Grainger listed mulberries, with other fruits, among the foodstuffs of Szechwan (Graham, 1961: 19). In the Chinese village of Kao Yao in Yunnan, Osgood (1963: 148, 164) observed that boys and girls would pick mulberries from the tree and eat them raw, but adults would not, saying that they brought on stomach pain. Meyer (1911: 48) wrote of a white mulberry variety near Peking that bears large, white, sweet tasting fruit that was sold, though "sparingly." Bretschneider (1882-95, 2: 329; 3: 499) in one place says that the Chinese eat mulberry fruit even though it is insipid, and elsewhere he reports that mulberries, presumably dried, were exported from Ning-po and Canton in 1889 for medicinal purposes. Medicinal use of the fruit and other plant parts of *M. alba* has also been reported by others (I. H. Burkill, 1966, 2: 1523; F. P. Smith, 1969: 267-68; Perry, 1980: 275), as, for example, "mulberry fruit prepared as a jam for medicinal use."

The overall conclusion is that though it was of quite minor importance, the fruit of *M. alba* had both dietary and medicinal uses in China. There would seem to have been no great nutritional gain or loss if the Chinese had not consumed such fruit, which is notable mainly for the good amounts iron it contains (Leung *et al.*, 1973: 89).

ADDITIONAL FRUITS FROM TROPICAL ASIA

Banana and Plantain

Though Alexander von Humboldt, relying on early Spanish accounts and native traditions, argued that bananas of some forms were cultivated in the Americas in pre-Columbian times, that idea has long since been rejected (de Candolle, 1886: 304-11; N. W. Simmonds, 1966: 313). Most telling among evidence to the contrary, wild members of the genus *Musa* are not found in the Americas, but in Asia from southern and eastern India through Southeast Asia to New Guinea (N. W. Simmonds, 1966: 308-13; 1976a: 211-13; de Langhe, 1969: 57-58). Ancient hunting and gathering groups of that region must, like its modern peoples, have used wild banana plants for several purposes: fiber for nets and fabric; leaves as wrapping material and thatch; and male buds, flowers, and inner sheaths of the pseudostem as vegetables.[65] The fruit, however, contained hard seed and thin flesh, and was likely of little use to them.

Though the time of earliest cultivation of the banana is unknown, Carl Sauer (1952: 24-27) regarded it as a cultigen of great antiquity, among the earliest of Southeast Asian plant domesticates. The domesticators of the banana probably first cultivated it to have a ready supply of several useful products, with its seedy fruit likely the least among them (N. W. Simmonds, 1962: 132). Perhaps its fiber was of greatest interest, for, as Crawfurd (1820, 1: 207-8) has observed, there may have been a period in Malaysia when most clothing was made of banana fiber, an intermediate step between an earlier predominance of bark cloth and a later one of cotton clothing.

According to N. W. Simmonds (1962: 136-41; 1966: 28-29, 308-9; 1976a: 212-13), the initial step in the evolution of bananas with edible fruit came about in Malaysia with selection of diploid strains of *Musa acuminata* for fruit seedlessness (sterility) and parthenocarpy (ability to develop a mass of edible pulp without pollination). When the cultivated *M. acuminata* spread from Malaysia to surrounding areas, including the Asian mainland from India to Indochina and South China, it hybridized with the wild *Musa balbisiana*, to give rise to polyploid forms with edible fruit. Though edible diploid bananas are still widespread in Southeast Asia, most present-day cultivated bananas are triploids, more vigorous than the diploid forms and with faster-growing and larger fruit. Some have fruits that can be eaten fresh, usually called "bananas" in English and commonly designated *Musa sapientum*, whereas others have larger, starchy fruits that must be cooked before use, called "plantains," *Musa paradisiaca*.

Bertin *et al.* (1971: Map 14) give the first half of the second millennium B.C. as the hypothetical date for cultivation of the banana in Indochina and parts of China's extreme south, including the margins of what today is Kwangtung. The northward spread of the plant to its present limits they place in the category "date of introduction unknown." Whatever the time of that spread, it was not known in earliest times in North China, for that region was far to the north of producing areas.[66] Moreover, the first banana product known in the north was not its fruit but its fiber. Banana fiber came to be called there by a general term, *chiao*, already used for hemp and other plant fibers. The banana fiber was made into cloth, described as thin and of excellent quality, which continued to be important in Chinese commerce as recently as the eighteenth century. After that time, its position, like that of other hand-made textiles, was undermined by machine-made imports (Schafer and Wallacker, 1957-58: 234; H. L. Li, 1979: 34-35).

Only when the Chinese pushed southward do they seem to have become familiar with the banana plant itself. One reads—though in accounts written much later—that in 111 B.C., after the defeat of Annam, bananas were transplanted in the north at Hsi-an (Sian), the national capital, though most soon died (Reynolds and Fang, 1940-41: 166; H. L. Li, 1979: 33). Like banana cloth, the plant and its fruit were also called *chiao*, with additional words to distinguish among the various kinds of bananas and banana products (Reynolds and Fang, 1940-41: 166). Chi Han, in his fourth-century flora, for example, used the term *kan-chiao*, "sweet banana," for both banana and plantain; *niu-ju-chiao* and *yang-chiao-chiao* for certain of their varieties; and *chiao-ko*, "*chiao*-linen,"[67] for cloth made from banana fiber (H. L. Li, 1979: 32-33).

As for so many cultivated plants, the Nan-Yüeh seem to deserve credit for the introduction of fruit bananas to China, for their homeland was strategically placed on the edge of the Southeast Asian hearth of domestication. One judges that in the fourth century and succeeding centuries, bananas remained curiosities of the far south. Chi Han said only that they grew in the administrative units of Chiao (centered on what today is Vietnam) and Kuang (centered on present-day Kwangtung) (H. L. Li, 1979: 32-33). And Chia Ssu-hsieh, writing of agriculture and economic life in the Yellow River area in the sixth century, mentions bananas only as rarities (S. H. Shih, 1962: 54). It was not long, however, before they became a favorite of poets, who wrote of their fruits, flowers, and, even more, of their leaves. In later centuries, when banana cultivation reached its northern limits in the Yangtze Valley, the plant grew well but did not flower or, if it did, failed to yield mature fruit (Reynolds and Fang, 1940-41: 169; Schafer, 1967: 187-88; F. P. Smith, 1969: 270). Some fruit was brought in from warmer areas to the south, an item of trade or tribute, as in T'ang times when bananas were sent to the royal court from coastal Fukien, Kwangtung, and Annam (Schafer and Wallacker, 1957-58: 217). Ordinary northern folk of that time, however, seldom saw bananas.

We have little information on the role of the plantain in traditional China, except that many varieties were found in the south; that they were available most of the year; and that they formed a considerable part of the diet of poor people (S. W. Williams, 1895, 1: 774; Ball, 1906: 298). As for the banana, in 1936 it was the most abundant fruit in the central districts of Kwangtung, where it was cultivated along rivers nearly everywhere. Its three main varieties came largely from the Canton area and Pearl River Delta: *ta-chiao*, which had large fruit;

lung-ya-chiao, with straight fruit; and *hsiang-ya-chiao*, with small, curved ones. Fruit production of the latter variety was by far the most important (accounting for 24,000 metric tons, or nearly 90 percent of all fruit coming from these three varieties), and was shipped in large quantities not just to Canton and Hong Kong, but to Shanghai and Tientsin, far to the north ("Fruit cultivation in Kuangtung," 1936: 716-17).

As for use of bananas in China, one notes that in the fourth century they were not only eaten fresh, but sometimes were preserved in honey (H. L. Li, 1979: 32). One also finds "Honeyed Bananas" and "Candied Banana Fritters" among contemporary dishes, but made from fresh bananas that are fried, then dipped into a hot, sugary syrup (G. B. Miller, 1966: 742; S. H. Huang, 1972: 159). Other contemporary banana dishes are the Cantonese "Banana Pancakes" (W. W. Chang *et al.*, 1970: 459), and two in quite a different spirit, one from Taiwan, the other from Hunan. The Taiwan dish is called "Stuffed Bananas," and consists of sandwiches made of a shrimp paste spread between banana slices, the sandwiches then coated with a floury mixture and deep-fried (S. H. Huang, 1972: 178). The Hunan dish is "Sautéed Bananas with Sweet Bean Paste" (Stallard, 1973a: 168-69). Most of our sources, however, are silent on the ways in which bananas and plantains are prepared.

Bananas and plantains are largely water and carbohydrate, with the latter readily digested if the fruit is ripe or properly cooked. Both types of fruit are low in protein, usually even more so than taro and yam, but are fair to good sources of vitamins B_6 and C. Bananas are also a relatively good source of potassium, and plantains a fair one of vitamin A (Murai *et al.*, 1958: 86-87; Platt, 1962: 18-19; Leung *et al.*, 1973: 13, 77; Forsyth, 1980: 270-71; Marriott and Lancaster, 1983: 101-4). Medicinally, the fruit of bananas, which the Chinese regard as very cooling, is used in many ways, and so are its other plant parts (Soubeiran and Dabry de Thiersant, 1874: 127; Reynolds and Fang, 1940-41: 176-78; F. P. Smith, 1969: 269-70; Perry, 1980: 277-78).

Two relatives of the banana, both of considerable economic importance, deserve mention here because they illustrate so well the different ways in which the *Musaceae* serve man. One is *Musa textilis*, abacá or Manila hemp, a domesticate of the Philippines used not for food but for fiber.[68] The second is *Ensete ventricosum*, a domesticate not of Southeast Asia but highland Ethiopia, one used in a range of ways much like the earliest domesticated bananas of Malaysia. The species of *Ensete*, once included in the genus *Musa* but now commonly placed in a separate genus of the *Musaceae*, are found wild not just in Ethiopia and other parts of the African tropics and subtropics, but in a belt stretching from southern India to Burma, Indochina/South China, the Philippines, eastern Indonesia, and New Guinea (N. W. Simmonds, 1962: 25-26, 28-29). Though the leaves and fiber of wild *Ensete* are used in many parts of that region, only one species is regularly cultivated for food. This is *E. ventricosum* (formerly *Musa ensete* or *Ensete edule*), known as *inset* among the Sidamo, prominent among Ethiopian groups who cultivate it, and as "ensete" or "false banana" in English. It is not its fruit, however, but other plant parts (especially the pseudostem and young shoots) that are normally used as food, cooked fresh like vegetables or after fermentation underground for weeks or months. Nutritionally, "ensete bread," a flat bread made of the fermented pulp, is a good source of carbohydrate, comparable to other starchy tropical products such as manioc flour. This has led areas of ensete monoculture in Ethiopia to have among the highest rural densities of population anywhere in that country (Smeds, 1955; Simoons, 1960: 89-99; Bezuneh and Feleke, 1966).

Knowing the importance of ensete as a cultivated plant in southern highland Ethiopia, the presence of wild *Ensete* in Indochina and southernmost China, and the ingenuity of native peoples there in their food quest, one may wonder whether they made any use of *Ensete* for food. Unfortunately we have found nothing to indicate such food use in Kwangtung or neighboring areas of China. The only one of our authorities to mention *Ensete* in an Indochina or China context are Crevost and Lemarié (1917: 271, 279-80), who write of two other species of *Musa*, reported in Yunnan and forested regions of Indochina, that are ornamentals "in the manner of, or analogous to, *Musa ensete*." One has an upper trunk reputed to be edible, and the other has pseudostems whose heart is consumed in Cambodia as a vegetable. Such uses of wild bananas are common in the region (N. W. Simmonds, 1962: 132; 1966: 266-67), but our authorities do not say that *Ensete* itself was eaten in this way in Indochina or China. Even if it were, its importance as a vegetable would be quite minor, and in any case there is nothing to suggest processing of the sort employed in highland Ethiopia. Presumably dietary carbohydrates were more readily obtained from other plants—among them yam, taro, plantain, and banana—that grew well in the lowland, without the need for such elaborate processing.

Carambola

The carambola (*Averrhoa carambola; yang-t'ao*, a name also applied in China to species of *Actinidia*), is a small tree, usually fifteen to thirty feet tall when full grown. Its fruit is three to five inches long, its skin green when unripe and burnished yellow or yellow-brown when ripe. Oval or elliptical in shape, it has ribs, usually four or five, running lengthwise, which give it a star-shape in cross section (Figure 60), and led to one of the carambola's early names, *wu-han-tzŭ*, or "fruit of five ridges." The fresh carambolas we consumed in Canton were sweet in taste, and with a crispness reminiscent of an apple but with flesh less firm and more juicy.

Some botanists say that the carambola is native to Malaysia, but H. L. Li (1979: 127) insists that it is indigenous to a region from southern China through Malaysia to eastern India. When the Portuguese reached India, they found it cultivated there, and adopted its local Malayalam name, *carambola*. When they reached Malaya, they found it cultivated there, too, but under a different name, *belimbing*, which is also used for a related fruit tree, the cucumber tree or bilimbi (*Averrhoa bilimbi*) (I. H. Burkill, 1966, 1: 271-73). The carambola is grown in Indochina as well, known there by still different, and quite varied, names (Crevost and Lemarié, 1917: 199). Such a diversity of names would be expected among peoples familiar with a plant since great antiquity, and this may indeed be the case, for in "Nan-hai" (the South Sea area in general?) it has been known since at least the fourth century A.D. (Bretschneider, 1882-95, 2: 318n; H. L. Li, 1979: 127). The cultivated carambolas are of two types, sour and sweet. The former, more like the wild form in taste, seems to be found principally in India, whereas some of both types, sour and sweet, are found in Malaysia, Indochina, and China (Crevost and Lemarié, 1917: 199; Bois, 1927-37, 2: 63-64; Ochse and Bakhuizen van den Brink, 1931a: 92-93; Kennard and Winters, 1960: 33; Hayes, 1966: 397; H. L. Li, 1979: 127; Zee and Hui, 1981: 78). This suggests that one must look to the latter region, and not to India, for ennoblement of the carambola, if not for its domestication.

In China, the carambola is cultivated in Fukien, Kwangtung, and Kwangsi, but is little known in the north (F. P. Smith, 1969: 59; Wittwer *et al.*, 1987: 290). In the last century, Gray (1878, 2: 194) wrote of many orchards of carambola trees in Kwangtung, especially along the Pearl River between Canton and its outport, Huang-p'u (Whampoa), a dozen miles away. In the

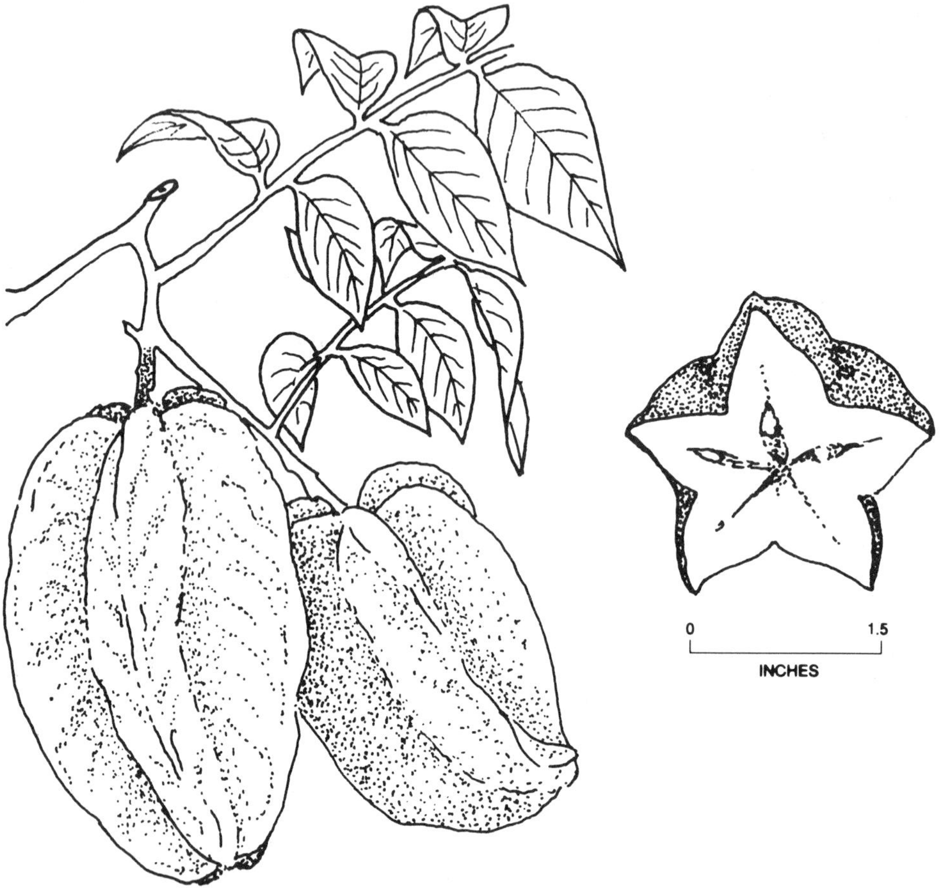

Figure 60. Fruit of the carambola (*Averrhoa carambola*) (drawn by Shelby Hall)

fall, as the fruit ripened, birds descended on the carambola orchards and, to frighten them away, boys were hired to beat gongs. Without the din of the gongs, Gray concluded, half the fruit supply would be lost.

The English, encountering the carambola in India, China, and elsewhere, have sometimes referred to it as "gooseberry" (Chinese gooseberry, country gooseberry, Coromandel gooseberry) after a favorite preserve fruit of their homeland. In China, too, it may be canned for export or used in other preserves. Unripe carambola may also be eaten as a vegetable, commonly pickled. Ripe fruit, depending on the variety, may be eaten out of hand or fresh in salads or drinks. It may be made into an alcoholic beverage, or used for juice or in other ways. It may also be stewed, and served with sugar added if needed. It may be cut cross-wise and dried, the dried fruit having a star shape and a taste reminiscent of a dried pear.

Nutritionally, the carambola, like the bilimbi, seems notable mainly as a rich source of vitamin C (Mustard, 1952: 33; Leung *et al.*, 1973: 78, 79). In early Chinese medicine, the fruit was believed to increase one's saliva, to quench thirst, and to lower fever (F. P. Smith, 1969: 60). One reads (Holmes, 1948: 292) that in Foochow a bride was given a carambola to eat before departing in her wedding chair for the house of her husband. This is because the carambola is believed to be anti-diuretic,[69] and eating one enables the bride better to stand the many hours she must be in public, exposed to the joking and ribaldry of wedding guests, and without an opportunity to urinate. In Indochina, carambola flowers are used as an anthelmintic or in other ways (Crevost and Lemarié, 1917: 199; Perry, 1980: 300-301). And in Indonesia and Malaya, too, the fruit and other plant parts are used as medicine, though not as much as *A. bilimbi* (I. H. Burkill, 1966, 1: 274; Perry, 1980: 301).

Myrobalans

One cannot leave a discussion of fruits that came to China from Southeast Asia and India without considering the myrobalans, for these small astringent fruits, celebrated for their miraculous powers, have been significant trade items from those regions since antiquity. The position of the myrobalans in China has been likened to that of ginseng, for both were rejuvenators of outstanding reputation (Schafer, 1977: 97). Just as one finds ginseng wine in China, one finds, at least in the north in T'ang times, a myrobalan wine (Schafer, 1963: 146).

Various myrobalan species were involved in trade from Southeast Asia and India to China, but most renowned were the *triphala*, "three fruits," of early India: the emblic (*Phyllanthus emblica = Emblica officinalis*), chebulic (*Terminalia chebula*), and belleric (*Terminalia bellerica*). By Indian belief, the goddess Sri (Lakshmi), who was associated with prosperity and good luck, would come to a person after he had eaten myrobalans and performed certain rites. Myrobalan fruits also occupied a major place in the Ayurvedic medicine of India (Dutt, 1922: 161-64, 226-29), especially valued for strengthening the body, improving health, reducing the effects of aging, and prolonging life. Consuming a myrobalan drink, it was believed, would permit a sage to regain his youth, mental ability, and physical vigor, and assure his survival in good health for many hundreds of years (B. Walker, 1968, 2: 215-16). One wonders whether the drink in question may have been the famous Hindu tonic preparation, also called *triphala* (for which, see Yule and Burnell, 1903: 609, Dutt, 1922: 162, 163-64; and Chopra *et al.*, 1965, 1: 380).

This preparation was also known in China, perhaps introduced by Indian monks or traders, at least by T'ang times. The myrobalan trade from India and Southeast Asia to China, however, goes back much farther in time, probably into pre-Christian times. The planting of myrobalan trees in China, at least of certain species, is also much earlier, with chebulic myrobalans grown in Canton in the third century A.D., their presence at early Buddhist temples or monasteries further suggesting an Indian connection (Schafer, 1977: 191; H. L. Li, 1979: 92-94).[70]

We focus here on the emblic, which seems to be the only one of the three myrobalan species which was significant from a food perspective in traditional China. The emblic, a deciduous shrub or tree, is indigenous to a region stretching from tropical India and Ceylon through Southeast Asia, possibly including parts of South China. Emblic remains, whether wild or cultivated, were found at Navdatoli in peninsular India at levels dating to the first half of the second millennium B.C. (Allchin and Allchin, 1982: 267). It was Indians, too, who had the greatest influence in spreading its use,[71] and its Sanskrit names (*amalaka* or *amalaki*) (Laufer,

1919: 551; Prakash, 1961: 266; Schafer, 1963: 145, 1977: 96) accompanied the fruit in the early trade that developed with Southeast Asia and China. In Malaysia, for example, one of its names today is *malaka* (Ochse and Bakhuizen van den Brink, 1931b, 2: 286; I. H. Burkill, 1966, 1: 935),[72] and among its names in early China, *an-mo-lo* or *an-mo-lê* (Laufer, 1919: 378; Schafer, 1977: 96; H. L. Li, 1979: 130). In all three regions, moreover, the fruit has been used as a food, whether raw as a flavoring, or as a pickle, or with sugar as sweetmeat or preserve (Watt, 1889-96, 6, Part 1: 218-21; Bois, 1927-37, 2: 463; Ochse and Bakhuizen van den Brink, 1931b, 2: 287; Prakash, 1961: *passim*; Chopra *et al.*, 1965, 2: 766; I. H. Burkill, 1966, 1: 935; Atkinson, 1980: 718; and Zee and Hui, 1981: 96).

In China, the earliest mention of the emblic seems to be that of Chi Han in the fourth century A.D., who wrote of it growing at Chiu-chên in Annam. He described its fruit as similar to a plum, and "yellowish-green in color." The emblic tree seems to have been found in the Canton area in T'ang times,[73] and today the tree is quite common around Canton and in southern China, including Hong Kong where it is a wild shrub. Its fruit (now called *yu-kan-tzŭ* in South China) is quite sour even when mature, but it is eaten nevertheless, whether directly or sweetened, or as jelly, pickle, or other preserve (H. L. Li, 1979: 130; Zee and Hui, 1981: 96). Raw emblics are one of the best-known natural sources of vitamin C, having at least five to nine times the vitamin C content of raw oranges.[74,75,76] Most of the vitamin C, however, seems to be lost in preservation. Emblic fruit and other plant parts also continue to serve various medicinal uses in present-day China (Perry, 1980: 149).[77]

Like the emblic, the fruits of chebulic (*ho-tzŭ* or *k'o-tzŭ*) and belleric myrobalans serve as dyestuffs, tanning substances, and medicinals in India, Southeast Asia, and China. These uses, moreover, far overshadowed food uses of the fruit. For India, the detailed account of Watt (1889-96, 6, Part 4: 18-22, 24-36) contains abundant references to non-food uses of fruit and other plant parts of the belleric and chebulic myrobalans, but scant mention of food uses: eating of kernels or occasional eating of *T. chebulica* fruit by humans. In a similar vein, Hayes (1966) in his *Fruit growing in India* includes a discussion of *Phyllanthus emblica* but fails to mention *T. bellerica* or *T. chebulica*. For Southeast Asia, I. H. Burkill (1966, 2: 2174-79) mentions no food uses of either fruit, except to say that *T. bellerica* kernels are edible, though dangerous because of their intoxicating or narcotic effect. And in traditional China, too, it is the medicinal and non-food uses of *T. bellerica* and *T. chebulica* fruit, seeds, and leaves that are emphasized by our sources (E. Watson, 1923: 204-5; Smith, 1969: 430-31).[78]

Other Fruits of Tropical Asian Origin

Among southeastern China's other tropical fruit trees of Asian origin, two deserve special mention: the mango (*Mangifera indica*) and the jackfruit (*Artocarpus heterophyllus = Artocarpus integrifolia*).

The genus *Mangifera* is indigenous to an area stretching from India to New Guinea, including Hainan and the Lei-chou Peninsula, with *M. indica* itself believed to have been domesticated in the India-Burma area (Vavilov, 1949-50: 27; Mukherjee, 1953: 131; 1972; Singh, 1976: 8; Zeven and de Wet, 1982: 71). The date of mango domestication is unknown, but 2000 B.C. has been suggested. Whatever it may actually have been, the mango was carried eastwards across Southeast Asia by Buddhist monks starting perhaps 400 to 500 B.C. (Singh, 1976: 8). It was not mentioned, however, by Chi Han in his fourth century flora. And even as late as T'ang times, it received scant notice in Chinese literature, and even then seems to have been a foreign fruit, known in India and Cambodia. It next appears in accounts from Sung times (A.D. 960-1279), but only as a medicinal (Schafer, 1967: 190-91). In the centuries since, however, the mango has become an accepted food in China, and is cultivated widely in the southeast, including the Hong Kong and Canton areas (F. P. Smith, 1969: 259). Its antecedents are clearly shown by its Chinese name *mêng-kuo* (Cantonese: *mon-go*), which derives from the Malay *manga* (or *mangga*), which in turn comes from the Tamil *mangas*. Another Chinese name for the mango (*an-lo-kuo*), perhaps coming by way of Persia, derives from the Sanskrit *amra* (Laufer, 1919: 552; I. H. Burkill, 1966, 2: 1426-27; F. P. Smith, 1969: 259; Zee and Hui, 1981: 93).

China's 1979 mango production was estimated to be 240,000 metric tons, a substantial one but dwarfed by that of India, the world leader, which produced 9,300,000 metric tons (Stafford, 1983: 403). Mango fruit, a good source of vitamins A and C when ripe and raw, is usually eaten fresh in China, but is also prepared in other ways, for example as jelly or jam (Leung *et al.*, 1973: 88; Zee and Hui, 1981: 93). A blending of Old World and New World ingredients, and a dessert both tasty and colorful, is the orange-yellow and white Taiwan dessert "Mango-Tapioca Pudding" (S. H. Huang, 1972: 180).

The jackfruit is a large tree related to the breadfruit (*Artocarpus altilis*), that staple food of some islands of traditional Oceania. Jackfruit cultivation, however, was centered in South and Southeast Asia before the time of European overseas expansion. Unknown in the wild, it may have originated in India, where it has been cultivated since antiquity, and Indians were likely responsible for its introduction to Malaysia (I. H. Burkill, 1966, 1: 255-56; Zeven and de Wet, 1982: 77). We don't know when it reached South China. It was not mentioned by Chi Han in his fourth century flora, yet, if we read Schafer (1967: 196-97; 1969: 46) aright, by T'ang times the jackfruit was cultivated in the Canton area. Whatever the time of its arrival, the jackfruit grows in that area today, and elsewhere in Lingnan and in Yunnan as well (F. P. Smith, 1969: 54). One can only speculate on its route of travel from India to China, but one notes *panasa* among its names in Sanskrit; *p'o-na-so* in Persian; and *panas* in northern Sumatra (Yule and Burnell, 1903: 440; Prakash, 1961: 270; I. H. Burkill, 1966, 1: 255; F. P. Smith, 1969: 54). It may have come that far on Indian and Persian ships, then on to China, with F. P. Smith (1969: 54) writing that its Chinese name, *po-lo-mi* (Cantonese: *bo-lo-muk*), is its Sanskrit name given in Chinese characters.[79]

Jackfruit is among the largest of all domesticated fruits, reaching a weight of 40 pounds or more in Hong Kong. It is barrel-or pear-shaped (Figure 61), its covering warty and dark green when immature but becoming brown at maturity. Its yellow smelly pulp is the edible portion, and may be cooked unripe as a vegetable, or eaten ripe as a fruit. It may also be dried or preserved in other ways, as in a candied form. Its seeds can be eaten as well, and, after roasting or boiling, are said to be similar to chestnuts (Kennard and Winters, 1960: 32; I. H. Burkill, 1966, 1: 256; C. A. Thomas, 1980: 158; Zee and Hui, 1981: 77).[80] In many parts of the tropics, people have not liked the taste and texture of jackfruit, which I. H. Burkill (1966, 1: 256) calls unpleasantly sweet and lacking in distinctiveness. Of perhaps more relevance to us, mature raw jackfruit and its seeds do not seem to be outstanding nutritionally in any way (Leung *et al.*, 1973: 26, 85). Also to be weighed, however, is the fact that in India production of jackfruit is greater than of any other fruit except the mango, and that in India's south it is said to provide more food per acre than any other fruit (Hayes, 1966: 385). Perhaps such productivity is less significant in China, with its broad range of alternative food crops, but the jackfruit is a

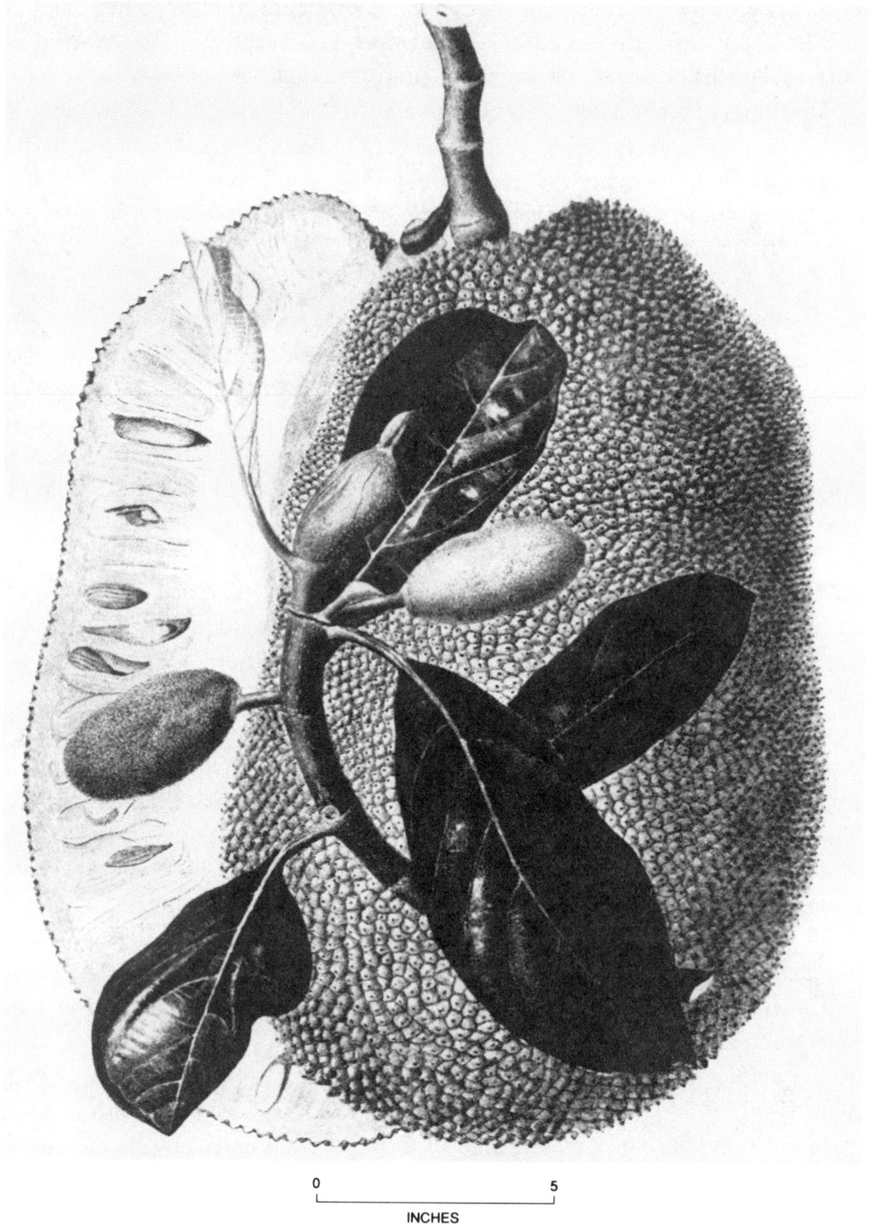

Figure 61. Jackfruit (*Artocarpus heterophyllus*) (from Ochse and Bakhuizen van den Brink, 1931a: opp. p. 69)

multi-purpose plant, and one should take seriously Thomas's suggestion (1980) that the jackfruit is deserving of more study in the hope of improving it nutritionally and in other ways.

SOME INTRODUCTIONS FROM THE NEAR EAST, EUROPE, OR AFRICA

So far we have dealt almost entirely with fruits of Oriental origin, those of China, Southeast Asia, and India, for they dominate the fruit scene of Kwangtung. Here, however, we focus on fruits originating in the far west (the Near East and Europe), most of which appear to have been what Berthold Laufer calls "Sino-Iranica," transmitted by the Persians, "those great mediators between the West and the East" (Laufer, 1919: 188-89). Most important of these fruits are the pear, apple, melons, pomegranate, fig, and grape, but we would note briefly, as well, two others, the date and olive, which, at best, had quite limited success in traditional China.

Common Pear and Apple

The common pear and common apple are fruits of temperate lands, and some botanists place them in a single genus, *Pyrus*, the apple usually classified as *Pyrus malus*, the pear as *Pyrus communis*. Other botanists place them in the separate genera *Pyrus* and *Malus*, with the pear in the former genus and the apple in the latter, usually as *Malus pumila* (formerly *M. communis*). We follow the latter usage, in placing the two in separate genera.

There is wide agreement among botanists that the primary center of origin of the two genera, *Pyrus* and *Malus*, is a region stretching westward from Central Asia and the Himalayas of India to Iran, the Caucasus, and Asia Minor. Some botanists, however, argue for an origin farther to the east, even, in the case of *Pyrus*, as far as western China (Watkins, 1976a: 248). Certain individual species originated in that primary Central Asia-Asia Minor center, some to the west of it, and still others to the east, as in China. Turning to cultivated forms, origin of the common apple, *Malus pumila*, is usually placed in the region from Central Asia to the Balkans or Europe in general,[81] and that of the common pear, *Pyrus communis*, in Europe or western Asia, especially Iran and Turkey (Wilcox, 1958-62: 639-40; Harlan, 1975: 69-70; Mathon, 1981: 36; Zeven and de Wet, 1982: 159, 160; Zohary and Hopf, 1988: 153, 156). Both species, however, are self-incompatible and each crosses without difficulty with related species within its genus (Watkins, 1976a: 248). *M. pumila* and *P. communis* have been cultivated in China since antiquity, but in addition China has indigenous species of *Pyrus* and *Malus* that have come into cultivation or contributed to the development of cultivated forms. Altogether China accounts for over 20 percent of the world's production of pears at the present time, and roughly 10 percent of its apples (Wittwer *et al.*, 1987: 283, 284).

Pear

Eight species of *Pyrus* are identified by Rehder (1940: 401-6)[82] as indigenous to China, where they are known generically as *li* or *t'ang*. Three are commonly mentioned as cultivated for their fruit: *P. ussuriensis* Maxim., *P. pyrifolia* (Burm.) Nakai (= *P. serotina* Rehd.), and *P. bretschneideri* Rehd. The primary gene center of *P. ussuriensis*, the Ussuri pear, is northeastern China and along the Ussuri River of southeastern Siberia; it is better adapted to winter cold than other pears; and it is widely cultivated in North and Central China (Zeven and de Wet, 1982: 44). The fruit of the wild form (*t'ang-tu-li*) is described as small, globular, greenish-brown in color, covered with tiny warts, and pleasant in taste (Bretschneider, 1882-95, 2: 304). The primary gene center of *P. pyrifolia*, known in China as *sha-li* ("sand pear") and in English as the Chinese sand pear, Chinese pear, or Japanese pear, is in the uplands of North and Central China. The cultivated *P. pyrifolia* is drought-resistant but less able to stand cold than *P. ussuriensis*. The fruit of the sand pear is of good size but sandy, tough, and lacking in flavor;

despite this, it is excellent for preserves (de Candolle, 1886: 233; Bailey, 1935, 3: 2869; Zeven and de Wet, 1982: 44).[83] The primary gene center of *P. bretschneideri* is in North China, where it was domesticated and is the most common cultivated pear, known as *pai-li* ("white pear").[84] Its fruit is described typically as having "hard, crisp, white sweet flesh" (Zeven and de Wet, 1982: 44).

There has, however, been considerable hybridization among species of pears in China, and there are great differences in fruit among the many varieties found there.[85] These include contrasts in color of skin, shape and size of fruit, and keeping quality, with one pear, the *yu-li* ("oil pear") of Shansi, capable of storage for over a year. There are also differences in juiciness and other qualities of flesh. At maturity, some pears have sweet flesh; others, tart. Some at maturity are soft; others, hard. Hard varieties of pears, such as the "crunch pears" that have attracted attention in the United States in recent years,[86] are actually favored in China over soft varieties. Meyer (1911: 25-30), who described the pears of North China, considered the *ya'rh-li* the best of the hard pear varieties; large, juicy, and sweet, fruit also ship readily and keep well, and thus are available in North China all winter long. In Meyer's opinion, the best of North China's soft pears is the "quince pear," whose fruit, which looks and smells like a quince, is juicy, mellow and quite large, commonly over a half pound in weight. With gentle handling, it can be stored for most of a year and stand shipping as well. Meyer also had a high regard for the "Peking pear," called *pai-li*, whose fruit is usually small, globular, yellowish-white, and sweet, but does not keep well. Though its identity remains uncertain, one notes as well the pear Marco Polo reported seeing in the thirteenth century in markets of the Sung capital, which he called "Kinsay" (the modern Hangchow): white in flesh, fragrant, and of tremendous size, weighing up to ten pounds each (Yule, 1921, 2: 202). A century before Marco Polo's observation, the Chinese scholar Lu Yu wrote of seeing excellent pears as large as a *sheng* (a measure of capacity two-thirds of a liter in size) in his travels up the Yangtze (C. S. Chang and J. Smythe, 1981: 171). Such oversize pears have been reported by other Chinese authors as early as Wei dynasty times (third century A.D.) (E. N. Anderson, 1988: 68).

The pear is often described as a fruit of North and Central China. At the same time, the Chinese have made persistent efforts to introduce pears to the south. Already in T'ang times, Schafer reports (1967: 183), there were pears of fine quality in the far southeast—present-day Kwangtung, Kwangsi, and northern Vietnam. In modern times, *P. ussuriensis* Maxim. or a related species is reported as cultivated in Annamite gardens in Tonkin (Crevost and Lemarié, 1917: 243), and it has even been introduced to Malaya and Singapore, to which it is poorly suited (I. H. Burkill, 1966, 2: 1879). The success of the pear was greater in the south of China, and in the southeast. Though shipments arrived from the north, some local pears were exported as well. Indeed in Kwangtung in the early 1930's pears accounted for about 3 percent of fruit exports by volume ("Fruit cultivation in Kuangtung," 1936: 721).

The story of the introduction of foreign pears to China and the spread of Chinese pears abroad is yet to be told. Here we would note B. Y. Chao's statement (1972: 20) that "the American pear" has recently been introduced to China where it is called the "*Yen-t'ai* (Chefoo) pear" for the Shantung port where it is grown. We would also note Bretschneider's brief comment (1898: 242) that the Chinese sand pear was called *salli* (from the Chinese *sha-li*?) in the Calcutta area. The suspicion is that this pear may have been shipped by sea to Calcutta from China in the nineteenth century. Yet these might not be the first Chinese pears known in India. For according to a tradition reported by Hsüan-tsang (Yüan Chwang; seventh century A.D.), the initial introduction of the pear to India (second century A.D.?) was from China. Thus in India it was called *cinarajaputra* ("crown-prince of China"), a name still in use (Laufer, 1919: 540-41). There are also certain names for the common pear in Kashmir and/or the Punjab (*li*, *tang*) (Watt, 1889-96, 6, Part 1: 374) that resemble the general Chinese names for pear, *li* and *t'ang* (F. P. Smith, 1969: 365).

Besides eating pears fresh, one reads, for T'ang times, of pears being steamed and of an alcoholic beverage being made from them (Schafer, 1977: 116, 121). Steaming of pears is still practiced in modern times, as in steamed pears in or with honey and liqueur. There is also a popular Chinese dish that originated in Yunnan, "Honey Pear Ham," a sweet and savory ham made with pears (G. B. Miller, 1966: 743; K. Lo, 1974: 171-72; 1979c: 165, 327). Pears, like other fruits, also have medicinal uses, as in pear syrup, used for relieving sore throat and cold symptoms (Soubeiran and Dabry de Thiersant, 1874: 266-67; Hooper, 1929: 120-21; F. P. Smith, 1969: 362-65; Perry, 1980: 345; E. N. Anderson, 1988: 198).

Apple

The genus *Malus* has been divided by Wilcox (1958-62: 638-39) into thirty-three species, exclusive of hybrid ones. Of these, *M. pumila*, a tree indigenous to southeastern Europe and to Asia as far east as the Altai Mountains, is regarded by most authorities as the main parent of the European varieties of cultivated apples, which, as indicated above, are commonly designated *M. pumila*, too. Many other species of *Malus* in Europe and Asia, however, are likely to have made genetic contributions to modern cultivated apples. *M. sylvestris*, the wild crab apple tree indigenous to a region from western Europe to western Asia, overlaps the natural distribution of *M. pumila* and is widely regarded as most important among the contributing species.[87]

The apple (*p'in-kuo* or *p'ing-kuo*) is China's leading fruit crop in terms of production (3.8 million tons in 1985) (Wittwer *et al.*, 1987: 283), and China is rich in apple species, with twenty, wild and domesticated, identified by T. T. Yü and C. L. Yen (1956) and Carlone (1957). Some are cultivated mainly as rootstock for other apple species, some are raised for their edible fruits or leaves (used for making tea),[88] and some for their attractive flowers. Of special note among the "flowering crab apples" (collectively known as *hai-t'ang*) is *M. spectabilis*, the "Chinese flowering apple" or "Chinese crab apple," first flowering crab apple to be cultivated in China, already in T'ang times (H. L. Li, 1959: 121-25; 1983: 35). This apple is usually a small tree with beautiful red flowers, for which it became popular with poets and was grown in the grounds of the imperial palace in Peking. Its fruit, like that of other crab apples, is also eaten, particularly in a candied form (E. N. Anderson, 1988: 135). The fruit production of apples of good eating quality, however, derived almost entirely from cultivated forms of *M. pumila* and *M. asiatica* (= *M. prunifolia*, var. *Rinki*),[89] and, to a small extent, from *M. micromalus* (Carlone, 1957: 302). These three species are all found in the Kansu corridor and North China, with *M. asiatica* and *M. pumila* extending southwest to Yunnan as well (Carlone, 1957: Fig. 8). *M. asiatica*, "the Chinese apple," has yellow or red fruit, smaller (up to two inches in diameter) than that of *M. pumila*, but similar in taste. *M. micromalus*, believed by some to be of hybrid origin (*M. baccata* x *M. spectabilis*), has yellow fruit about an inch in diameter (Rehder, 1949: 271; Carlone, 1957: 294, 296; H. L. Li, 1959: 124-25).

In traditional China, the apple, because one of its names sounds like the Chinese name for "peace," was symbolic of peace. Gifts of apples or their ceremonial use could express a wish for peace, as at the wedding of the last Manchu emperor in 1922, after the Chinese republic had been established (Cormack, 1935: 75-76; C. A. S. Williams, 1941: 18). In a similar vein, bowls of apples might be placed on Chinese banquet tables to symbolize peace, with the fruit sliced and eaten when the meal is finished (G. Z. Chu, 1975: 45-46). Apples might also be prepared in other ways, as in the apple dessert from North China called "Spun Apples," "Peking Glazed Apples," or "Peking Drawn-Thread Glazed Apple." This is made by dipping apple slices in a batter, then deep-frying them and coating them with syrup. Sesame seeds may be added to the syrup or sprinkled on afterwards. The warm slices may then be dropped into a bowl of ice water, causing the syrup to form a solid glaze, after which they are quickly served; or they may be taken immediately to the table along with a bowl of ice water, with each diner dipping his still-warm apple slice in the water for a moment before he bites into it (K. Lo, 1971: 171-72; Em. Hahn *et al.*, 1973: 174-75; Field *et al.*, 1976: 105; C. B. T. Lee and A. E. Lee, 1980: 128).[90]

Though we have found little information on them, we would also note two species of apples indigenous to China's southeast that are sometimes cultivated for their fruit. One is *M. formosana*, native to Taiwan, Hainan, and South China, and the other *M. melliana*, native to many of China's eastern provinces from Anhwei to Kwangsi, and to Yunnan (T. T. Yü and C. L. Yen, 1956: 110; Carlone, 1957: 284, 301; H. L. Li, 1963b: 274-76). Kwangtung itself was not good apple country, and its apples were not very popular in China's markets ("Fruit cultivation in Kuangtung," 1936: 722).[91]

Melons

Both the watermelon, *Citrullus lanatus* (= *C. vulgaris*; *hsi-kua*, "western gourd") and melon, *Cucumis melo* (*t'ien-kua*, "sweet gourd"), are common crops in China. Domestication of the watermelon is usually placed in Africa. Earliest cultivation of the melon, in turn, is generally placed in Africa or the Near East, but there are secondary centers of diversity as well in other regions, including China (Burkill, 1966, 1: 567, 706; Harlan, 1975: 69; Mathon, 1981: 28, 42; Zeven and de Wet, 1982: 36, 82, 89, 124; Zohary and de Wet, 1988: 167-68). Seed identified as that of *Cucumis melo* has been identified at Ch'ien-shan-yang, a site near the mouth of the Yangtze River dating from the third and fourth millennia B.C. (K. C. Chang, 1986: 254). That, however, is a disturbed, unreliable site, and it seems prudent to withhold judgment on the melon seed found there. Nevertheless it is known that *Cucumis melo* did reach China in antiquity, for it was a major crop in North China in the fifth or sixth century A.D. (H. L. Li, 1969: 254-55). The watermelon, in turn, reached North China in the twelfth century A.D., though the Chinese knew it earlier in Central Asia (Laufer, 1919: 438-45), which led them to call it "western gourd" (F. P. Smith, 1969: 110).

In China today, there are many varieties of *Cucumis* melons, which are sometimes identified by the locale in which they originated, one of these being the renowned Hami melon (*Ha-mi-kua*) of Sinkiang (E. N. Anderson, 1988: 132). Use of melon seed will be considered in our next chapter, but here we would note that *C. melo* leaves were used as a vegetable in ancient China, and that the fruits of both of these melons may be eaten fresh, with most people preferring to eat fresh watermelon with salt (I. C. Chang, 1970: 177). They may also be consumed medicinally or in other ways. Of special note in this regard is the Oriental pickling melon, *Cucumis melo*, var. *conomon*, whose fruit are consumed after cooking or pickling, but not raw (H. L. Li, 1969: 254-55; F. P. Smith, 1969: 110-11; 134-35; Herklots, 1972: 301-5, 314-18; E. N. Anderson, 1988: 132). A watermelon may also be cut and hollowed out to form a container for steaming food, as in the Shanghai dish "Watermelon Chicken" (W. W. Chang *et al.*, 1970: 263-64), and in one of the imperial court, "Boneless Duck Steamed in Whole Watermelon with Button Mushrooms" (J. D. Keys, 1963: 45).

As pointed out at the beginning of this chapter, the melons, like most fruits, are "cold," to be avoided by persons who are in a cold state; watermelon is singled out by certain of our sources as being "very cold," perhaps because of its watery nature, and to be avoided by Chinese women during pregnancy (Yeung *et al.*, 1973: 198, 200; Ling *et al.*, 1975: 129; Wheeler and Tan, 1983: 53). Interestingly, however, Linda Koo (1973: 141), who lived on Taiwan for seven years, places red watermelon among Chinese "hot" foods, though she includes watermelon rind and yellow watermelon among "cold" ones. One wonders whether, at least on Taiwan, the red color (hot) of watermelon flesh is viewed as offsetting the coldness watermelon would otherwise be expected to have.

Pomegranate

Most clearly associated with Persia itself is the pomegranate (*Punica granatum*; *shih-liu*, or "stone tumor"). Not only does this arid-land shrub or small tree grow wild in Persia and neighboring areas (Kurdistan, Afghanistan, and Baluchistan), but historical and philological evidence supports the idea of its origin in that region (de Candolle, 1886: 237-40; Zohary and Hopf, 1988: 151). The pomegranate was grown in the famous hanging gardens of Babylon, and in the Old Testament it was mentioned as cultivated in Israel, apparently abundant in some places. It has also been cultivated in other parts of the Mediterranean area since antiquity, and is found in a spontaneous state[92] there. The cultivated pomegranate was mentioned in the *Odyssey*, and seems to have come to Greece from Persia by way of Syria (de Candolle, 1886: 239; Seymour, 1963: 213, 340). It was cultivated in Egypt by the reign of Thutmose I (1530-1520 B.C.) or earlier (Darby *et al.*, 1977, 2: 742). Apparently it was carried by the Phoenicians to their colony at Carthage. Some say that it was introduced to Rome from Carthage, giving it one of its Latin names, *Malum punicum*, "Punic apple," whereas others believe the name meant only that in Rome the finest quality pomegranates came from Carthage.

To the Jews, the pomegranate was symbolic of "sanctity, fertility, and abundance," and as such it was depicted on temple designs and coins, royal robes and priestly regalia, with its tufted crest a model for the royal crown (W. Smith, [192-]: 543; Goor and Nurock, 1968: 70-75). The Jewish perception of the pomegranate was shared by various other ancient peoples of the region (Muthmann, 1982).[93] In the religious rites of Syria and Phoenicia, the pomegranate was so important that *rimmon*, the name of its fruit, was part of the name for the Sun God, Hadad-Rimmon. In Greece the pomegranate tree and fruit were cult objects symbolizing fruitfulness and production, "death and destruction." It was associated with the cult of Hera (identified by the Romans with Juno), wife and sister of Zeus, cultivated near her temples and used in cult ceremonies. In one of her temples she is depicted as holding a pomegranate in one hand (Hehn, 1885: 180-84), and at Pelusium in Lower Egypt, Zeus was shown holding a pomegranate "as a mystical symbol" (Darby *et al.*, 1977: 744). On Rhodes, pomegranate blossoms were included in the royal coat of arms (Moldenke and Moldenke, 1952: 191). In Persia, the fruit formed the head of the king's scepter (Moldenke and Moldenke, 1952: 191), and in his invasion of Greece, ten thousand of Xerxes' élite infantry, apparently part of the imperial bodyguard, carried spears, each decorated at its lower end with a pomegranate of gold or silver (Herodotus *History* VII. 41; Grant, 1897, 2: 100). Ultimately the pomegranate came to symbolize the hope of an everlasting life in early Christianity (Goor and Nurock, 1968: 73).

Some say that the pomegranate is wild or spontaneous in India (Watt, 1889-96, 6, Part 1: 369; 1908: 909; Dutt, 1922: 167; Mathon, 1981: 37). Laufer (1919: 285-86), however, has noted that its Sanskrit name (*dadima, dalima*) derives from the Persian (**dulim[a]*); he says that it was not mentioned in Vedic, Pali, or early Sanskrit accounts; and he presumes it to have been introduced to India in the early centuries of the Christian era. It is true that modern Indian pomegranate fruit does not compare in quality with that Persia or of Afghanistan (from which pomegranates have been imported). Yet one reads that the tree (*dadima*) was referred to in the *Prajñapana Sutra* and *Acaranga Sutra* (*c.* 500 B.C.?) (Prakash, 1961: 73); that it was mentioned by Panini (?350 B.C.-250 B.C.?) and its fruit named by Patañjali (?200 B.C.-?A.D. 500) (Agrawala, 1963: 217); and that Charaka (?A.D. 80-180?) wrote of its juice as a medicinal (P. Ray and H. N. Gupta, 1965: 56). Though it is possible that some of the above were later insertions into the texts in question, one suspects that the pomegranate may have been known in India somewhat earlier than Laufer suggested. Prakash (1961: 2, 267-68), in fact, suspects it to be of far greater antiquity there.

On the basis of linguistic and historical evidence, Laufer (1919: 280, 282-86) concludes that from India the pomegranate spread to Malaysia and Cambodia, reaching the latter country at least by the beginning of the sixth century A.D. One variety may also have reached Yunnan from India, but Chinese names for pomegranate strongly point to an introduction overland, not directly from Iran but by way of Iranian peoples, late in the third century A.D.[94] This was before it was reported in Cambodia, and would seem to rule out an initial introduction to China by the sea route.

The pomegranate is a drought-resistant desert plant which yields well when irrigated. It does best in places with dry summers and cool winters, but will grow and provide fruit in locales with quite different climatic conditions. These include areas of heavy precipitation, as in India, the Philippines, and Florida, but fruit produced under such conditions is soft and juicy, not of the best quality in terms of eating, storing, or shipping. The tree is, moreover, severely damaged by temperatures any lower than about 8° to 15°F., and freezes to ground level with zero temperatures (Hodgson, 1917: 174; Hayes, 1966: 408). In view of the above, much of China would be ruled out as providing good conditions for pomegranate growth, the far north for its low winter temperatures, the southeast for its heavy summer rains.

Yet after it arrived in China, the pomegranate gradually spread, and early in this century E. H. Wilson (1929: 308) wrote of it as widespread, common in warmer sections of China, and even naturalized in some places. Of its several cultivated varieties in early China, some had sweet fruit, some sour fruit, and still others bitter fruit (Laufer, 1919: 279-82; F. P. Smith, 1969: 361). Meyer (1911: 46), in writing early in this century of Shantung, noted that the pomegranate was grown in gardens in "quite a number of different varieties." Some were dwarf forms only a few feet tall which yielded only small fruit, whereas others were fifteen to twenty feet high with fruit weighing a pound or more. They also differed in color of fruit, that of some varieties having red flesh, and others, white. Wilson (1929: 308) wrote of Chinese pomegranates as barely edible. The fruit, however, was not of primary concern, for most Chinese pomegranates were grown only as ornamentals, typically with double flowers varying in color from dark red to white. This is similar to the situation in India, where the pomegranate is popular in part as an ornamental with colorful flowers, but is "nowhere of much commercial importance" (Hayes, 1966: 408).

The pomegranate did, however, serve medicinal uses in China, with various plant parts used, among them fruit, peel, and root (Hooper, 1929: 120; F. P. Smith, 1969: 361; Perry, 1980: 331). As in Southeast Asia and India, in China pomegranate parts were anti-diarrheal, anti-dysentery agents or anthelmintics, the latter use being prominent in the Near East as well (Watt, 1889-96, 6, Part 1: 369-70; Dutt, 1922: 167; I. H. Burkill, 1966, 2: 1872-75; Goor and Nurock, 1968: 73-74; Darby *et al.*, 1977: 2: 744). As for other uses of pomegranate, Laufer writes (1919: 286) that unlike the case in the ancient and modern Near East and Mediterranean, he knew of no use of pomegranate wine among the Chinese. He and others (Hirth and Rockhill, 1966: 177; Schafer, 1967: 251) did, however, cite reports from T'ang times of a wine on Hainan that was fermented with pomegranate flowers.

The pomegranate in China is considered a Buddhist symbol, and pomegranate fruits, flowers, and stems are a favorite art motif (Sowerby, 1940: 140). As in the Near East and Mediterranean, in China the pomegranate is also a symbol of fertility, and in addition it represents the continuation of the family, the fruit's many seeds standing for "numerous offspring" and "many noble sons" (Yetts, 1912: 27; Burkhardt, 1953-58, 1: 30, 120; 2: 16; Eberhard, 1968: 130). Chinese women who desired children might offer pomegranates to the Goddess of Mercy, whose temple porcelain was decorated with representations of the fruit (Skinner, 1911: 223). One also reads of pomegranates as esteemed wedding presents or as playing a role at wedding dinners (Laufer, 1919: 287). Burkhardt (1953-58, 1: 118), for example, tells the story of a procession carrying furniture to the future home of a couple soon to be married in Hong Kong. The most highly-valued item, a modern toilet, was at the head of the procession but, since no article might be empty, it was sent brimming with pomegranates and eggs. Both of these, as symbols of fertility, were especially well-suited to the occasion. One also reads of a Hong Kong funeral procession, in which carts carrying the wives of a deceased's grandchildren carry a special emblem, a red pomegranate flower (Burkhardt, 1953-58, 1: 99), presumably because it represents the future survival of the family.

In passing, it should be noted that the symbolism of the pomegranate in China is strikingly similar to that found in the modern Near East and Mediterranean. When an Arab bride dismounts at the tent of her bridegroom, for example, she is given a pomegranate which she "smashes on the threshold," then throws the seeds inside (Laufer, 1919: 286-87). A Turkish bride, in turn, throws a ripe pomegranate to the ground, with the number of seeds falling out taken as an indication of the number of children she will bear (Skinner, 1911: 221-22).

Date

The date palm (*Phoenix dactylifera*; *wu-lou-tzŭ*) is not native to China, and it remains a matter of controversy as to when dates and the date palm first arrived. If one follows Laufer (1919: 203, 385-91) and Schafer (1963: 121-22; 1967: 176), the earliest Chinese references to the date are as a fruit of Persia, with one of its names meaning "Persian jujube." The fruit would seem first to have been imported to China in T'ang times, with the tree arriving then as well. The latter may have been carried to North China overland by envoys from the Caspian Sea area in A.D. 746, but, if so, it was unsuccessful in acclimatizing there. It did better in the Canton area, where it was reported growing already in the ninth century A.D.

Bretschneider and H. L. Li have suggested earlier dates for introduction of the tree, basing their views on Chinese textual references to what may be the date palm. One is in Chi Han's fourth century flora which describes a tree called *hai-tsao*, "ocean jujube," which, according to others, may have been the palm-like *Cycas revoluta*, commonly called "sago palm" (Laufer, 1919: 388; F. P. Smith, 1969: 140-41; H. L. Li, 1979: 123-24). Whatever the date palm's actual time of arrival, today it is cultivated only in southernmost China.

Fig

Ficus is a very large genus found in tropical and subtropical parts of both the Old World and New World. About a hundred species, most of them shrubs, are found in South China (S. C. Lee, 1935: 447). None of the Chinese species, however, was to become as important a provider of food as the fig, *Ficus carica*, was in the Near East and Mediterranean area. The fig is a deciduous tree, tolerant of summer heat and productive in arid and semiarid regions, if, where rainfall is insufficient, irrigation is provided. It is, however, sensitive to frost and low winter temperatures. In Europe, as a result, fig cultivation usually stops at the foot of the Alps and other mountains that border the Mediterranean basin to the north (de Candolle, 1886: 295-96).

Though there are uncertainties in taxonomy, wild or feral figs closely related to *Ficus carica* are found in an area stretching from Afghanistan and Iran to Ethiopia and the Mediterranean area, and as far as the Canary Islands (de Candolle, 1886: 295; Zohary and Spiegel-Roy, 1975: 324). Most authorities look to the eastern Mediterranean or southwest Asia (Turkey, Syria, Iraq, south Arabia, or Iran) as the place of earliest fig cultivation (Condit, 1947: 9; Moldenke and Moldenke, 1952: 105; Goor and Nurock, 1968: 46; Harlan, 1975: 69; Zohary and Spiegel-Roy, 1975: 324; D. R. Harris, 1981: 31; Mathon, 1981: 30; Zohary and Hopf, 1988: 144-45). The fig seems to have been used as food in great antiquity in that region, for pips resembling those of *Ficus carica* (but only once the flesh of a fig) were found at certain early Neolithic sites in the eastern Mediterranean, the earliest dating to 7000 B.C. or before (Hopf, 1969: 356, 357; Zohary and Hopf, 1988: 144). Fig remains, whether pips or whole fruit, have also been found in late Neolithic levels at sites in that region (Zohary and Spiegel-Roy, 1975: 324; Zohary and Hopf, 1988: 144-45). One cannot tell whether the figs mentioned above were from wild or domesticated trees, but there is clear evidence of fig cultivation in Mesopotamia, Syria, Palestine, and Egypt in the third millennium B.C. (Goor and Nurock, 1968: 48; Darby *et al.*, 1977, 2: 708-9). In ancient Egypt, the fig was among the most frequently represented fruit in tombs. It was also cultivated in ancient Greece,[95] and was one of the most important plants of the Bible, mentioned altogether fifty-seven times (E. W. G. Masterman in Hastings, 1910: 262-63; W. Smith, 192_: 13; Moldenke and Moldenke, 1952: 103-4; Goor and Nurock, 1968: 46-69). Mount Olivet was renowned for its fig trees, and the fig was linked with the vine as a symbol of peace and prosperity, as shown in the biblical expression "And Judah and Israel dwelt safely, every man under his vine and under his fig tree" (*I Kings* 4: 25).[96]

The spread of *Ficus carica* in eastern India and Southeast Asia would not have been facilitated by climatic conditions, for the fig does not do well in the humid tropics. Such conditions, however, do not prevail everywhere in India, yet cultivation of *F. carica* there seems to have begun quite late (Gode, 1941-42: 127-29, 134-35; 1943).[97] The names for it in Sanskrit and Hindi (Watt, 1889-96, 3: 347; Condit, 1947: 9) apparently derive from the Persian *anjir*, and suggest an introduction from Persia. That its Malay name also derives from Persian suggests that the fig, likely dried, came to be known in that region through Persian sailors on Arab ships (I. H. Burkill, 1966, 1: 1022).

For China, there is no certain evidence of *Ficus carica* until the T'ang period (Bretschneider, cited by de Candolle, 1886: 296; Laufer, 1919: 410, 412-13). By the sixteenth century, figs (*wu-hua-kuo* or "flowerless fruit") were obtained from trees cultivated in the lower Yangtze area, Yunnan, Chekiang, Kiangsu, Hupeh, Hunan, Fukien, and Kwangtung (Laufer, 1919: 410).[98] Early in this century, however, Meyer (1911: 47) could still write that figs, found out-of-doors in warmer parts of China, were "sparingly grown" and seemingly "not great favorites of the Chinese." Today there are few cultivated fig trees in Hong Kong, but small plantings of figs are found as far north as Shanghai and Nanking, in the Yangtze Valley.[99] The fig grown in the Yangtze Valley is described as a small bush with fruit far smaller and of lower quality than that of Persia (F. P. Smith, 1969: 174). To the north, winter cold, which may drop well below zero, limits fig cultivation out-of-doors. Where figs are found in North China, they are exotics, grown mainly in pots and tubs and protected in winter (Meyer, 1911: 105; Condit, 1947: 95).[100]

Dried figs, a gentle laxative, are said to be found only in Chinese medicinal stores (Zee and Hui, 1981: 90), and it has been suggested (Read, 1929: 434) that the Chinese adopted the fig tree because of the laxative value of its fruit. One is somewhat surprised that the fig was not embraced more heartily by the Chinese, for in its homeland it symbolized fertility and propagation, and served as a phallic symbol. Such imagery was apparently found even in Japan, where "figs of sugar candy in the form of a phallus" were sold at a Shinto shrine in Kyoto during an autumn festival (Condit, 1947: 5). Yet we have found nothing of this sort in the literature on China, though other members of the genus *Ficus*, the largest trees in India, have played roles in religion there.

One is the peepul tree (*Ficus religiosa*), the sacred fig-tree of India,[101] which can grow to an impressive size, with wide-spreading branches that provide shelter from the sun. Tradition holds that it can also live to an advanced age.[102] To Hindus, it is by far the most revered tree, meritorious to plant, sinful to cut down, but potentially harmful magically and to be kept away from a house. It is said to be inhabited by deities, among them Brahma, Vishnu, and Shiva. It is worshipped as the home or symbol of a deity, or in its own right. Vows may be made at the peepul tree. The deity Vishnu was born under a peepul, and today Hindus perform ceremonies under peepul trees to dedicate them to Vishnu, who appears when appealed to by a worshipper. Ritual may involve circumambulation of the tree while magical formulas are recited, sprinkling the tree with water, and making offerings to it (Watt, 1889-96, 3: 359-60; Crooke, 1926: 407-8; Bosch, 1960: 67-69; B. Walker, 1968, 1: 358; 2: 498; Stutley and Stutley, 1977: 27). In former times, it was said, a woman, in an effort to become fertile, would circumambulate a peepul tree while naked, whereas in modern times she winds a cotton thread around its trunk many times (Crooke, 1926: 408), apparently with similar intent. More relevant to China, the peepul was the tree under which the Buddha gained illumination, which led to its being called the *bodhi* or *bo* tree, "the tree of enlightenment." As such, the tree was introduced from India to areas where Buddhism spread, its Chinese name, *p'u-t'i*, deriving from the Indian *bodhi*. As to its date of introduction to China, Sampson found no mention of this other than for the Canton area, to which, early in the sixth century, an Indian Buddhist priest brought a bodhi plant by sea and planted it at the "Bright Filial Piety Temple" (Sampson, 1869c: 103; Samagalski and Buckley, 1984: 200-201). One also reads of a bodhi plant sent by an Indian king to a T'ang emperor in A.D. 641 (Schafer, 1963: 122-23). It is thus understandable that certain of its uses in China, too, suggest an introduction by Buddhists (Read, 1929: 435). Commonly grown in temple grounds,[103] the tree was symbolic of the Buddha and his gift to mankind. It was also a Buddhist art motif, as in stone carvings where it was represented in a conventionalized style (Sowerby, 1940: 143).

A second fig tree of religious importance in India is the banyan tree *F. bengalensis*, a wide-spreading tree that can reach great age and a height of seventy to a hundred feet. It is often mistaken for the peepul in writing and art. Unlike the peepul, however, it has branches that send down roots to give birth to daughter axes and provide the tree with a considerable circumference. One reads of such a banyan tree in India with a circumference of two thousand feet, large enough to provide shade for seven thousand people (Watt, 1889-96, 3: 344). Like the peepul, the banyan tree has been involved in Hindu belief and ritual since antiquity, though its sanctity may be later than that of the peepul tree. It, too, is associated with Hindu deities (Watt, 1889-96, 3: 346; Crooke, 1926: 407; Bosch, 1960: 69-70; B. Walker, 1968, 1: 357-58), with the Chinese Buddhist pilgrim Hsüan-tsang in the seventh century A.D. describing one in front of a Hindu temple, and around it many bones of persons who had sacrificed themselves there (Watters, 1904-5, 1: 362-63; B. Walker, 1968, 1: 358). Buddhists, as well, hold *F. bengalensis* to be sacred (Grünwedel, 1901: 181n; I. H. Burkill, 1966, 1: 1019).

As for China, it has its own species of banyan trees[104] found, according to Sampson (1869b: 72), as far north as the Yangtze area. They have been admired and, at least in some cases, were believed to have healing and protective qualities. One reads, for example, of banyan branches suspended over a door to keep contagion away, and of votive offerings made to the tree itself (Burkhardt, 1953-58, 2: 115, 117, 152, 167; J. J. M. de Groot, 1964, 1: 45, 67). Chi Han in the fourth century A.D. described a form of banyan tree, sometimes of enormous size, commonly planted in what today is Kwangtung and Kwangsi, though he did not mention a ritual role for it. H. L. Li (1979: 80) identifies this tree as *Ficus retusa* (*yung-shu*), a related species found from India to China, Malaysia, and New Caledonia. In traditional South China, sacred trees were usually of this species. In the last century, Sampson (1869b: 72) wrote of *Ficus retusa* as a shade tree near virtually all public buildings in South China, along streets in Hong Kong, and at nearly all rural ferry landings in Kwangtung. We do not known whether *F. bengalensis* itself was carried by Buddhists or others from India to China, but banyan trees of some type have long been grown in Buddhist temple grounds in China. Indeed one of Canton's most important shrines, "Temple of the Six Banyan Trees," derives its name from the trees that grew there in the eleventh or twelfth century A.D. (Samagalski and Buckley, 1984: 200).

Olive

Though species of *Olea* are indigenous to a region stretching from South Africa to Australia and New Zealand, the olive (*Olea europea*) itself was domesticated in the eastern Mediterranean area. Present archeological evidence points to its earliest cultivation, in the fifth or fourth millennium B.C., in Lebanon and Palestine. Many botanists believe that the olive derives from the oleaster, *O. oleaster* (= *O. europea*, var. *oleaster*), whereas others hold the oleaster to be an escapee from cultivation and look to other wild species, especially *Olea chrysophylla*, as possible or likely ancestral forms (Renfrew, 1973: 131-34; Zohary and Spiegel-Roy, 1975: 319-21; Mathon, 1981: 105-9; Zeven and de Wet, 1982: 117; Zohary and Hopf, 1988: 131-36). Though the olive tree seems to have been introduced to India in antiquity, we have found no indication that *Olea europea* was successfully cultivated for its fruit in early India.[105] In the fourteenth century A.D., Ibn Batuta wrote that there were no olives anywhere in India (Yule,

1866, 2: 486), and early in the last century Ainslie (1826, 1: 268) could still write that the olive was not cultivated there.[106]

Though it is possible that olive fruit (called *ts'i-t'un*, from the Persian *zeitun*) was imported to China from Persia in T'ang times, there is no convincing evidence of that or of the tree itself being introduced (Laufer, 1919: 415-19; Schafer, 1963: 148). Ibn Batuta noted that there were no olive trees (*Olea europea*) in China (Yule, 1866, 2: 486), and even at the end of the nineteenth century, S. W. Williams (1895, 1: 775) was able to say the same. The few olive trees that had been introduced to southern China before 1949 by Europeans were largely unnoticed and unproductive, and the carefully-done maps of *Olea europea* distribution we consulted (Bertin *et al.*, 1971: Map 17; Mathon, 1981: 106) do not show such olives anywhere in China, though they do show them in Japan, introduced early in this century.[107]

Starting in the 1960's, the Chinese government, interested in expanding production of edible oil, started large-scale introductions of olive seeds and seedlings, initially from Albania, in the hope of identifying and selecting trees suitable to China's climatic conditions. The extent of the government's commitment was made clear in 1964 when China's Prime Minister planted a symbolic olive tree in Yunnan. By 1980, about twelve million olive trees had been planted, largely in central and southern China including Kwangtung, but most were quite young (under five years of age) and not yet producing (*China: Development of olive production*, 1980: 4-8, 104). It will be interesting to see the results, for in various places in India olives have been grown successfully since early in the nineteenth century, yet remain unproductive. Though they flower, their fruits rarely set, or when they do, they drop off before maturity. This problem seems to derive from a lack of sufficient cold in winter, which is needed for encouraging differentiation of flower buds (Watt, 1889-96, 5: 485; Hayes, 1966: 421). In Malaya, Java, and the Philippines, too, the olive will grow but does not fruit (Wester, 1924: 136-37; I. H. Burkill, 1966, 2: 1607). South China, most of which has a subtropical climate much like northern India in temperature regime, might well pose similar problems to olive cultivation. And the north presents a different problem, that of an extended period of freezing in winter.

Grape

The genus *Vitis* is large and widespread, with many wild species found in North America and Eurasia, especially in the eastern sections of those continents (Mathon, 1981: 111). In China, wild grapes of various *Vitis* species have been eaten since antiquity,[108] but it was not until the domesticated grape, *Vitis vinifera*, arrived from the west that grape cultivation began (Laufer, 1919: 226-27).

The cultivated grapes of the world are all believed to derive from *Vitis vinifera*, a species which, according to Olmo (1976: 295), was probably domesticated about 4000 B.C. somewhere in a belt stretching from Afghanistan to the south shores of the Black Sea. Archeological evidence of grape cultivation has been found for the Early Bronze Age (*c.* 3000-2000 B.C.) or earlier in Iran, Anatolia, Syria, Palestine, Jordan, Egypt, and Greece, but such cultivation seems to have spread to the western Mediterranean only after 1000 B.C. (Bertin *et al.*, 1971: Map 18; Renfrew, 1973: 127; Zohary and Spiegel-Roy, 1975: 321-22; Costantini, 1979b: 91, 93; Jarrige and Tosi, 1981: 125; Mathon, 1981: 111; Zohary and Hopf, 1988: 140-42).

Grapes and wine are mentioned in early Indian literature, but it is not always clear whether they came from cultivated *Vitis vinifera* or wild grapes, of which four species in India and Burma yield edible fruit (Laufer, 1919: 239; Hayes, 1966: 333). As a result, authorities differ as to the antiquity of grape cultivation in India, with late first millennium B.C. a reasonable choice on the basis of literary evidence.[109] Such a date would have been long after domestication of the grape, and two millennia or more after it was cultivated in the eastern Mediterranean area. Its spread to China, however, was even later, with tradition having it introduced from Central Asia by the famous Han traveller Chang Ch'ien late in the second century B.C. (Sampson, 1869a: 50; Bretschneider, 1882-95, 3: 438; Laufer, 1919: 221; Y. S. Yü, 1977: 80; Needham *et al.*, 1980: 137; Needham and Lu, 1986: 163-64).

The grape, its Chinese name (*p'u-t'ao*) presumably deriving from an Iranian word (Laufer, 1919: 225; H. Huang, 1980: 461), was cultivated at royal residences in or near Ch'ang-an, the Han capital, and grape wine made from Han times onward (Hulsewé, 1979: 52, 135-36). Even in the fifth century, however, the grape remained unimportant, and grape wine, "a rare and exotic drink" (Schafer, 1963: 142). This changed somewhat at the start the T'ang period, a time of Chinese expansion into the grape-growing areas occupied by Turkic and Iranian peoples of Central Asia. Among the T'ang tribute from those regions were raisins of three different kinds ("dried," "parched," and "crinkled"); "grape juice," likely a grape syrup; and grape wine. New varieties of vines found their way to the heart of China, and certain regions in northern China, especially Kansu and the T'ai-yüan district of Shansi, became important in grape production. Grapes were eaten fresh or made into raisins. Their juice was made into vinegar. And, perhaps, with improved knowledge of the vintner's art, better wine was made of domesticated grapes.[110]

Grape cultivation and wine production, however, were never to gain the importance they enjoyed in Mediterranean lands. In the late-thirteenth century, Marco Polo reported that T'ai-yüan, capital of Shansi Province, had "many excellent vines, supplying great plenty of wine" that was shipped throughout China.[111] He noted, however, that this was the only place in all China where wine was produced (Yule, 1921, 2: 13).[112]

With the arrival of Westerners in China, new forms of grapes were introduced to China by sea, from both Europe and the Americas. It was common for missionaries to have grapevines in their church grounds, and a good number of missions had their own wine cellars (H. Huang, 1980: 461). Nevertheless the map of Bertin *et al.* (1971: Map 18) shows that even at the end of the nineteenth century there were only scattered areas of grape cultivation, mostly in North and Central China, none in Kwangtung. In the last decade of that century, a Chinese winery, initially called the Chang Yü Pioneer Wine Company, was established in Shantung, and was subsequently responsible for introducing, from Europe and the United States, over a hundred varieties of table and wine grapes. After the Japanese invasion, additional grapes were introduced to North China and southern Manchuria, and these were still in existence after the Second World War (H. Huang, 1980: 461).

Though in the 1929-33 village survey the grape was listed among typical fruits of the Wheat Region, details about production were not given (Buck, 1956: 30, 40). It is said, however, that in 1949, before establishment of the People's Republic, China had about 16,000 acres in grapes, with an annual production of less than 50,000 metric tons (H. Huang, 1980: 461). Today grape acreage is believed to be double or triple that amount, and grape production, of many different varieties and places of origin, is about 360,000 tons. China's main producing area is Sinkiang (from which fresh grapes and raisins are shipped to many of China's larger cities), and certain other parts of China north of the Yangtze are also important (H. Huang, 1980: 461, Figure 1; Wittwer *et al.*, 1987: 286-87).[113]

In traditional China, the grape was as highly regarded as it was in the West, and there were several varieties with white and purple fruit, with central Shansi famed for its purple ones. H. Huang (1980: 462-64) has identified many of the cultivated varieties of *Vitis vinifera* found in China today. Perhaps of greatest interest to us are those known in the region since antiquity, which bear colorful names. One of these is the *lung-yen* ("dragon's eye"), a late-ripening, drought-resistant, highly-productive variety widely grown in Manchuria and North China. Its fruit are pink to purple in color and neutral in taste; they keep and ship well, and serve as table grapes and for making wine and champagne. Another variety is "cow's nipple," a mid-season table grape, with handsome, long, light yellow fruit that are sweet and subtle in taste but do not keep or ship well. Others are *wuhebai* or *ak kishmish* ("sultania" or "Thompson seedless"), a grape known in Sinkiang since antiquity; and, of minor importance, "red cock's heart," "white cock's heart," and "small bottle."

In older areas of grape cultivation in China, which are principally in the hill country of the north, grapes are nearly all grown on pergolas (vine arbors). For some decades, however, trellised vineyards have been constructed there, particularly on the edges of settlements. Whether grown on pergolas or trellises, the severe winter cold in North China has led to the practice of removing vines and covering them with earth in the winter. That practice is not found in areas south of about latitude 35°N., or in all of Shantung Province, parts of which enjoy a coastal location and milder winter (H. Huang, 1980: 462-65).

Such burying of pergola-grown grape vines in winter, as described for traditional North China (Sampson, 1869a: 54; Meyer, 1911: 40-42), involved permitting the vines to develop quite lengthy main leaders. After cold weather arrived in October, they were carefully taken down, pruned, and tied in long bundles. Then pits were dug from one or four to six feet deep, sufficiently wide for a man to work in easily, and long enough to take a bundle of vines. Each bundle was placed in a pit, covered with stalks of sorghum and mats, and then covered with a few feet of earth. A small number of holes were left to provide air during periods of warm weather. When a blizzard developed, however, the air holes were plugged so as to protect the vines from the cold. The root of the vine was also protected by placing mats around it and covering the entire affair with earth. The result was that in winter such a Chinese vineyard appeared empty, giving few clues that in summer it would be providing an abundant harvest of grapes.[114]

In addition to the cultivated *Vitis vinifera*, there are wild *Vitis* species whose edible berries are collected in China, and, in some cases, even made into wine.[115] A few are occasionally cultivated. Most notable of the latter is *Vitis amurensis*, found in North China, Manchuria, and eastern Siberia, a vigorous plant which grows over shrubs and trees and forms dense thickets. Its fruit is abundant, though small and somewhat sour. More important, perhaps, the plant is able to stand winter temperatures of -40°F. or more (Meyer, 1911: 42; H. Huang, 1980: 464).[116]

One reads of sugared grapes put on bamboo slivers and eaten (Headland, 1914: 180). However, according to Meyer (1911: 42) raisins were not common in China, though there were some from Mongolia that were small, greenish white, and seedless. The Chinese did, however, keep grapes over long periods of time in other ways, and have done so since antiquity. The *Ch'i-min yao-shu* (sixth century A.D.), for example, describes the storing of fresh clusters of grapes in pits inside a farmhouse, with each pit then filled and/or covered with earth to maintain an even temperature. Under these conditions, grapes remained fresh through an entire winter (Laufer, 1919: 230-31; S. H. Shih, 1962: 60). S. W. Williams (1895, 1: 775) and F. H. King

(1927: 303) both mention such storage of grapes in China around the turn of this century. King did not observe the practice first-hand or find any foreigner with knowledge of the methods involved. He could learn from his interpreter only that the fruit was wrapped in paper and stored in earth cellars kept dry and at a constant temperature. And he was told that it was common for such grapes to be kept into July. A somewhat similar form of grape storage, reported by Meyer (1911: 41), involved carefully placing grapes in containers and keeping them in basements kept cool with ice, some for nearly a full year.

In Chinese medicine, wild grapes—fruit and other plant parts—serve a variety of uses. The root, vine, leaves, and fruit of *Vitis vinifera* serve such uses as well, with the fruit, among other things, viewed as strengthening, cooling, and diuretic (F. P. Smith, 1969: 457-58; Perry, 1980: 435). Fresh grapes are not outstanding nutritionally in any way, but dried they contain good amounts of iron, and are a fair source of energy and calcium (Leung *et al.*, 1973: 83).

FRUITS FROM THE NEW WORLD

Following discovery of the New World, various tropical fruits of Mexican/Central American, Caribbean, or South American origin also found their way to southeastern China. Most often mentioned are the pineapple (*Ananas comosus*), papaya (*Carica papaya*), and guava (*Psidium guajava*, the guava, and *P. cattleianum* = *P. littorale*, the strawberry guava),[117] but also present today are others, among them the sapodilla (*Manilkara achras* = *Achras zapota*) and Annonaceous fruits such as the sugar apple or custard apple (*Annona squamosa*).[118]

The pineapple, which was probably introduced from Malacca to Macao by the Portuguese, is first mentioned in Chinese accounts dating from the beginning of the seventeenth century. From Macao, it reached Kwangtung and Hainan, then Fukien and Taiwan, where it was already known by the middle of that century. The pineapple also seems to have reached China by another route: from Burma to Yunnan (Laufer, 1929: 251).

In traditional China, pineapple growing and canning was carried out on a considerable scale, the fruit being eaten in large amounts by rich and poor alike (Laufer, 1929: 251). The pineapples (Cantonese: *bo-lo*) one sees in Kwangtung markets today are quite small in size compared to those of Hawaii, and they may be eaten raw or cooked. Pineapple has gained a significant role in China as an ingredient in certain sweet-and-sour dishes, especially with chicken and duck. Sweet-and-sour dishes are more widely used in Canton and the south, where fruit and their juices are more regularly used in cooking (K. Lo, 1979c: 222). How well pineapple has been incorporated into Cantonese cuisine is indicated by its inclusion in a colorful and tasty dish, "Dragon-Boat Shrimp with Pineapple and Sweet-and-Sour Sauce," which is prepared in celebration of the dragon-boat races in the Pearl River at Canton (Sakamoto, 1977: 204). These are held at an annual festival found mainly in Central and South China, which, according to popular tradition, originated in Hunan in the latter part of the first millennium B.C. (S. W. Williams, 1895, 1: 147-48, 816; Ball, 1906: 222-23; Cormack, 1935: 157-69; Eberhard, 1952: 73-75; Latsch, 1984: 59-60).[119] Unfortunately for the people of Kwangtung, pineapple is not notable for its nutritional value.

Turning to the papaya, we would note first the term *kua*, a general term embracing cucurbitas, but also certain non-cucurbit fruits, among them the papaya, which are seen as resembling melons. Papaya is sometimes called *mu-kua*, "tree melon," in China, but since this term is also used for the quince (*Chaenomeles sinensis* and *Cydonia oblonga*), it can cause confusion. The papaya is also known as *fan-mu-kua*, "foreign tree melon"; *fan-kua*, "foreign melon"; or by other names (F. P. Smith, 1969: 93; S. Y. Hu, 1980: 27, 55-56, 73, and *passim*; Zee and Hui, 1981: 80, 108; E. N. Anderson, 1988: 80, 132-33). The papaya, which is common in China's southeast, may be consumed ripe as a fresh fruit or cooked immature as a vegetable (Zee and Hui, 1981: 80). Despite the recency of its introduction, the Chinese well understand the papaya's good qualities, including its usefulness as a meat tenderizer (F. P. Smith, 1969: 93). In nutritional terms, papaya is notable for its unusually high vitamin C content (73 mg./100 gm., ripe raw fruit, edible portion), which is well above that of lemon, lime, pummelo, and orange (32-57 mg./100 gm., raw fruit) (Leung *et al.*, 1973: 86-87, 91-92, 93, 96).

Meyer (1911: 47) noted that guavas (identified as *P. cattleianum*) are never observed in North China, and seldom in Central China, "and then only as pot plants." Farther south, he said, others have reported them as growing "in regular plantations." In the late 1920's, guavas (species not given) were grown near the mouth of the Min River in Fukien, and sold in the fruit stalls of Foochow ("Production of fruit in Fukien Province," 1928: 1035; Holmes, 1948: 39). Today guavas are marketed more widely than in Meyer's time, for E. N. Anderson writes in general terms that, like the papaya, the guava (identified as *P. guajava, fan-shih-liu*)[120] is very well-liked in coastal Chinese cities. Sometimes, he notes, it is called "women's dog meat" because, like dog meat, it is "heating," and is used, instead of dog meat, for winter warmth by women (E. N. Anderson, 1988: 137, 191). The strawberry guava is a good source of vitamin C (20 to 32 mg./100 gm., edible portion, raw fruit), and the common guava, *P. guajava*, is an exceptional one (132 mg./100 gm.) (Leung *et al.*, 1973: 84), far richer in this regard than the papaya. It may be consumed raw or preserved, and its juice made into a widely-used drink (Zee and Hui, 1981: 99).

[1] In a recent visit we found strawberries in Hong Kong and Ch'eng-tu markets, though the Hong Kong strawberries were imported. The cultivated strawberry *Fragaria ananassa*, however, is seldom mentioned in accounts of traditional Chinese food or fruit, not once, for example, in Meyer's excellent account of fruit and nut cultivation in China.

[2] Wittwer (1987: 290), for example, writes about "blueberries" (*Vaccinium uliginosum*; bog bilberry), found in large amounts in the northeast, whose fruits are gathered and made into wine.

[3] For developments over the past thirty years in citrus production and breeding in China, see W. C. Zhang, 1981a, b.

[4] Called by Groff and Howard (G. W. Groff and C. W. Howard, 1924-25) "the cultured citrus ant of South China," in fact it is the well-known "red tree ant" of the Old World tropics and subtropics. Chi Han in the fourth century had reported that the people of Chiao-chih, which today is the Hanoi area of Vietnam, sold, in their marketplaces, bags containing nests of ants fastened to twigs or leaves. Without such ants, he said, mandarin fruit would be injured by other insects (H. L. Li, 1979: 118-19). Other Chinese writers from the ninth century onwards have also reported the use of citrus ants, whether in Lingnan in general, in Kwangsi or Kwangtung, or, in one case, possibly in Szechwan (H. T. Huang, 1986: 531-35).

In modern Kwangtung, the citrus ant is still viewed as essential to successful citrus cultivation. If a district has a climate too cold for the ants in winter, nests of dormant ants are brought in from outside areas shortly before spring commences. Some, as seems to have been the case in early Vietnam, make a regular business of raising ants or collecting wild nests for sale to fruit growers (G. W. Groff and C. W. Howard, 1924-25).

With the arrival of warm weather, an ant colony becomes active, drives out ants of other species, and weaves a new, larger nest in the citrus tree. Because food is scarce early in the season, an orchardist places dog intestines, a delicacy, or silkworm larvae, in the tree as food, which discourages the ants from leaving. To assist them in reaching other trees, he builds ant-bridges of bamboo, high enough above the ground so as not to interfere with orchard work. Orchardists of Kwangtung believe the ants to be effective against the stink-bug insect (*Tesseratoma papillosa*), a serious pest of citrus, and against boring and flower-eating insects. Without such ants, they insist, their fruits would be few and blemished. Acknowledging their effectiveness in destroying caterpillars and larger insects, Groff and Howard also noted their failure to kill scale insects (which are tended by the ants for their edible excretion) and, sometimes, plant lice—both serious threats to citrus trees. W. E. Hoffman, who published the results of his field studies of the citrus ant in 1936, made similar observations, and concluded that in balance citrus ants were likely more harmful than beneficial. While acknowledging that the citrus ant is not perfect as a pest control agent, Chinese researcher Chhen Shou-chien suggested that it might be useful in areas lacking chemical insecticides and proposed further research to find ways to help the ant's performance. Since Chinese orchardists, who have used citrus ants for nearly two millennia, remain convinced of their effectiveness, it seems prudent to withhold judgment in the matter (H. T. Huang, 1986: 537-39).

Also worthy of mention is the use of another red ant, *Tetramorium guineense*, in southern Fukien for control of sugarcane pests, and of a black ant, *Polyrhachis* sp., in paddy fields and elsewhere. The use of *Polyrhachis* is modern, that of *Tetramorium* of unknown antiquity. Both, however, may have come to be used in imitation of the citrus ant (H. T. Huang, 1986: 543-44).

[5] In the Yangtze region, orange trees are commonly used in "tree marriage," a fertility ceremony which Eberhard (1935; 1968: 174) believes to be related to an earth-god cult.

[6] For the preparation of home-dried orange and tangerine peel, see Tropp, 1982: 557-58.

[7] When it arrived in China from Mexico, after discovery of the New World, the chayote (*Sechium edule*), whose fruit bears a vague resemblance to the Buddha's hand citron, came to be called "Buddha's-hand gourd" (Herklots, 1972: 339; E. N. Anderson, 1988: 132, 187).

[8] For a history of lemons in Kwangtung, see I.-N. Shiu, 1933.

[9] Among the best varieties of litchis reaching Hong Kong in April and May are the colorfully-named "March red" and "jade purse," and in June and July, "black leaf" and "princess's smile."

[10] The loquat is called *p'i-p'a* because its leaves are similar in shape to those of the Chinese guitar, *p'i-p'a*. Its English name, loquat, derives from its name in Cantonese, *lu-chü* (E. Watson, 1923: 453; F. P. Smith, 1969: 164).

[11] For notes on early Chinese trade in ebony, see Laufer, 1919: 485-86; Wheatley, 1959: 75; Schafer, 1963: 138; I. H. Burkill, 1966, 1: 838; Hirth and Rockhill, 1966: 216 and *passim*.

[12] For details of its introduction to the United States, see Dorsett and Dorsett, 1928: 1-2.

[13] These were growing near a temple in Shantung, resembled the cultivated *D. kaki* and were called "wild persimmon" by the priests, but had green-yellow fruit, unpleasant in flavor.

[14] The persimmon is often represented in Chinese painting and other art, favored, for example, on pottery of Ch'ing times (A.D. 1644-1911) where it was ordinarily represented as red. It did not, however, seem to have any special symbolic role (Sowerby, 1940: 140) or tie with Buddhism.

[15] For a description of some of these, see Meyer, 1911: 11-14; Dorsett and Dorsett, 1928: 3-5; and "Production of fruit in Fukien Province," 1928: 1026-29.

[16] E. H. Wilson (1929: 307) noted that in Yunnan the fruit of the tree *Docynia Delavayi*, which is plentiful there, is employed in ripening persimmons. Fruits of the two plants are placed in alternating layers in large containers, then covered with rice husks and allowed to stand for ten hours, at which time the persimmons are ready for eating.

[17] This sweet seems quite similar to a persimmon pastry of the Jurchen, those twelfth-century Tungusic conquerors of North China, as described by Franke (1975: 176). The Jurchen end product was a dumpling made of a steamed paste of glutinous rice, dried persimmons, and dried jujubes to which pounded pine nuts and walnut kernels were added. Then the dumpling was shaped, and before it was consumed, it was sprinkled with honey.

[18] Chia Ming, in a fourteenth-century book, wrote that unripe red persimmons can keep for up to a year in cold, salted water, though they become a bit toxic (Mote, 1977: 232).

[19] In Japan, there was still another way of eliminating such astringency, one involving fumes of saki, rice wine. Unripe persimmons were placed in newly-empty saki casks made of the wood of *Cryptomeria*, which were then closed, made air-tight, and allowed to stand for five to fifteen days, according to the temperature. The astringency of the fruits was gone, yet their flesh remained firm and tasty, and they were readily transported (C. L. Watrous in Gore, 1911: 6-7). Following experiments with the Japanese method, researchers in the United States accomplished the same end, removal

of astringency, while maintaining firmness of the fruit, with carbon dioxide gas and without sake or alcohol. Though the researchers believed lack of oxygen to be responsible for slowing down the processes that make persimmons soft, their experiments with submerging fruit under water (or other liquids) and with simple autoasphyxiation (keeping fruit in tightly-closed containers, with the fruit depleting the oxygen therein) were either unsuccessful or did not work as well as processing with carbon dioxide (Gore, 1911). For references to recent studies of persimmon astringency and its removal by carbon dioxide, alcohol, water, and other treatments, see Itoo, 1980: 464-68.

[20] In T'ang China, too, the persimmon was said to counter drunkenness (Schafer, 1977: 94). Yet Chia Ming, in his *Yin-shih hsü-chih*, "Basic knowledge for eating and drinking" (A.D. 1368), wrote that, if eaten when drinking wine, persimmons help bring on drunkenness or heart pains (Mote, 1977: 232).

[21] The edible fruits of at least two other species of *Diospyros* are used in China. One is *D. embryopteris*, called "green persimmon" by the Chinese. Its fruit has a sticky, astringent juice believed to possess remarkable medicinal qualities, and it is a fine varnish as well (F. P. Smith, 1969: 151-52). This species, sometimes called in English the "Indian persimmon," is also found, sometimes cultivated, from India to Thailand, Malaya, and Indonesia. Its fruit is mainly used there as a dye and tannin, and for glue, strengthening fishing nets, or as a preservative or caulking on boats. It also serves as a medicinal and, when ripe, it may be eaten but is not very pleasing to the taste (Flückiger and Hanbury, 1879: 403; Watt, 1889-96, 1: 141-45; Roberts, 1931: 146-47; I. H. Burkill, 1966, 1: 843). A second species, whose fruit is eaten by the Li people of Hainan, is *Diospyros hainanensis* (G. W. Groff *et al.*, 1924: 126), which is also found in mainland Kwangtung.

[22] The ancestry of these guardian deities, known at least by the third century B.C., remains uncertain, but one argument is that their names originally meant "figurines of the spirits Shu and Yü," that "Yü" derives from *yü-chin*, or turmeric (*Curcuma longa*), that "Shu" is believed by most to be the sow thistle (*Sonchus oleraceus*), and that initially the figurines represented these plants in human form. Turmeric, according to Bodde (1975: 134-35), would have served to placate evil spirits with its pleasant aroma, but if they persisted in seeking entry to a house, the sow thistle—thorny, bitter, and unpleasant—would have served to ward them off. One may question whether the Yü figurine in question indeed took its name from turmeric, for, as Laufer (1919: 312-23) and Schafer (1963: 124-26, 185-86) have pointed out, in early times the term *yü-chin* ("*yü* gold") was applied to various species of *Curcuma* (and later to saffron as well).

[23] F. P. Smith (1969: 356) writes also of peach wood used in divination in China, observing that the Chinese character for peach consists of two parts, one meaning "omen," the other, "wood." For references to other kinds of divination in China, see footnote 8 in our chapter on "Flesh and Fowl."

[24] Of special note here is the wild peach tree *Prunus davidiana*, indigenous to Hopeh, Honan, Shansi, and Shensi. Some regard *P. davidiana* as one of the ancestors of the domesticated peach. Though most consider it a separate species, it is nevertheless closer to the domesticated peach than any other indigenous *Prunus* species. In the wild, *P. davidiana* is a shrub or small tree, but when cultivated as an ornamental it may grow thirty to forty feet tall. This tree is notable because in North China it commonly serves as rootstock for domesticated peaches, plums, and cherries. Not only do such fruits do well on *P. davidiana* stock, but they are also quite drought resistant. The fruit of *P. davidiana* itself is small, round, downy, and with scanty, tasteless flesh (Meyer, 1911: 19-20; Bois, 1927-37, 2: 192; H. L. Li, 1959: 61-62).

[25] The peach is cultivated in some native orchards in Tonkin, especially in upland regions, and at maturity the fruit is small and mostly green in color. Apparently, however, such fruits were insufficient in quality and quantity to satisfy the Chinese community there, and peaches were imported from Yunnan and North China (Crevost and Lemarié, 1917: 243). In Cochin China, according to I. H. Burkill, the peach is cultivated for its flowers not its fruit, and this is true of Thailand and Burma, too. The peach has been planted in Malaya, but either without success in fruiting or with sterility after a few years of fruiting. In Java, the peach may have been introduced by the Chinese for its flowers. In the mountains of central and western Java, the peach tree flowers but does not fruit, while in eastern Java the fruit ripens but is inferior in quality (I. H. Burkill, 1966, 2: 1842). In the Philippines, the peach tree is adapted to elevations of about 4000 feet and above, and in northern Luzon has successfully fruited at high altitudes (Wester, 1924: 144).

[26] Various writers (Hehn, 1885: 320-22; de Candolle, 1886: 221; Laufer, 1919: 539), using evidence in classical sources, have agreed that the peach arrived in Rome in the first century A.D. That the Hebrew name for peach also means "Persica" and was first mentioned in the *Mishna* (?450 B.C.-250 A.D.?) (Goor and Nurock, 1968: 202-3, 205) supports the idea of a relatively late introduction from Persia to the Mediterranean world. The possibility of a somewhat earlier arrival is raised by the peach pits found in Saite levels (664-525 B.C.) at Saqqara in Egypt (Darby *et al.*, 1977, 2: 733-35; pers. comm., L. E. Grivetti). We do not, however, know the basis for Goor and Nurock's statement (1968: 202) that it was at the time of the Eighteenth dynasty, about 1400 B.C., that peaches reached Egypt from Palestine and then spread beyond to Italy and Greece.

[27] The kernels of the wild peach *Prunus davidiana* were also used to flavor sweets and certain dishes, but they contained much prussic acid and only quite small amounts were used (Meyer, 1911: 20).

[28] For medicinal use of apricot fruit and other plant parts, see Soubeiran and Dabry de Thiersant, 1874: 270-71; Hooper, 1929: 116-17; F. P. Smith, 1969: 353-54; and Perry, 1980: 344.

[29] Moldenke and Moldenke (1952: 184-88), however, argue that the apricot, *P. armeniaca*, was probably the "apple" of the Bible, which others identify as the common apple (*Malus pumila*), citron, orange, quince, or some other fruit. If, however, Moldenke and Moldenke are right, the apricot would have reached the Mediterranean area much earlier.

[30] Bretschneider observes (1882-95, 2: 294) that in modern China the name *mei* is used for several *Prunus* species, among them some providing edible fruit, but that apparently the very best *mei* is *P. mume*.

[31] For recipes to such dips, see G. Z. Chu, 1975: 251-52, and K. Lo, 1979c: 56-57.

[32] Meyer did not further identify the species *Prunus pseudocerasus*, but it was probably *P. pseudocerasus* Koids. (= *P. cantabrigiensis* Stapf), which is called *ying-t'ao* and is cultivated in the Yangtze Valley (Uphof, 1959: 296; Zeven and de Wet, 1982: 43-44). Another possibility is that it was the related species, *P. pseudocerasus* Lindl. (= *Prunus paniculata* Edwards), which is also cultivated in China for its fruit and known as *ying-t'ao* (Bretschneider, 1882-95, 2: 299; 3: 428; Rehder, 1940: 473; Zeven and de Wet, 1982: 44).

[33] There is considerable confusion between the Chinese jujube and the Indian jujube, which most botanists place in a separate species (*Z. mauritiana* Lam. or *Z. jujuba* Lam.). Some confusion arises from the fact that Lamarck applied to the Indian jujube the same name, *Z. jujuba*, that Miller had already given to the Chinese jujube. In addition, however, some botanists have considered the two to comprise a single species, designated *Z. jujuba*, which creates further difficulties for the reader. The Indian jujube, whose fruit is used in ways similar to the Chinese jujube, can tolerate far less frost than the Chinese jujube and is cultivated in areas to its south. It is said to occur in an area stretching from Australia through Malaysia, and southwestern China (Yunnan) to India and Africa (de Candolle, 1886: 197-98; Watt, 1889-96, 6, Part 4: 368-70; S. C. Lee, 1935: 798-99; Dalziel, 1937: 299; Hayes, 1966: 400). Bois (1927-37, 2: 106) notes that it is much cultivated in tropical and subtropical parts of India and China as well as in Indochina. Zeven and de Wet (1982: 42) go further in saying that it has been grown in India, Indochina, and China since antiquity. Yet Crevost and Lemarié (1917: 217-18) describe only the Chinese jujube (*Z. vulgaris* Lam. = *Z. sativa* Gaertn.) as cultivated in gardens of Cochin China and Tonkin, with no mention of the Indian jujube. Nor does Chi Han, in his fourth century flora (H. L. Li, 1979), mention the jujube as a tree of Vietnam or China's southeast. De Candolle (1886: 198), moreover, observed that Bretschneider was not aware that the Indian jujube was found in China and that its spread beyond India may be recent. And S. C. Lee (1935: 798-99), though noting that *Z. jujuba* Mill. (the Chinese jujube) is found both wild and cultivated in China, writes only that *Z. mauritiana* Lam. (the Indian jujube) grows in Yunnan, presumably as a forest tree. Both de Candolle and Watt suggest that the Indian jujube was domesticated in India, where it is considered a poverty fruit, a situation quite different from that of the Chinese jujube in its homeland where jujubes were among the tribute sent to the royal court (Schafer and Wallacker, 1957-58: 219, 221). The difference in status may in part reflect the more careful selection the Chinese jujube has undergone through history. In the United States, where both species have been introduced, the Indian jujube is thus not as well regarded as its Chinese counterpart (Hayes, 1966: 399-400).

[34] Related species are native to that region, among them *Zizyphus spina-christi*, a thorny shrub or small tree perhaps referred to by Christ in the Sermon on the Mount, but not, as some suppose, comprising the thorns of his crown at the crucifixion. A second is *Zizyphus lotus*, the prickly lotus bush, not to be confused with the Egyptian lotus, a water lily (Moldenke and Moldenke, 1952: 247-48; Zohary and Hopf, 1988: 178).

[35] Though their shells are very hard, Pili nuts are delicious, are exported to some degree, and are considered as having encouraging prospects. Pili nuts are obtained in the Philippines principally from wild trees. They are said to be deliberately grown only on rare occasions, and then not in large plantings but in scattered ones within or in the neighborhood of hemp and coconut plantations (Wester, 1924: 148; I. H. Burkill, 1966, 1: 436; Rosengarten, 1984: 308-9).

[36] Certain other scholars (Haudricourt and Hédin, 1943: 146; Vavilov, 1949-50: 30) have also placed its origin south of China itself, or in Indochina.

[37] There are, besides *Chaenomeles sinensis*, other cultivated species of *Chaenomeles* in China. One is *C. lagenaria*, the so-called "Japanese quince," which, though grown in Japan, is of Chinese origin; another, *C. japonica*, is of Japanese origin, though grown in China. Both species, however, are cultivated in China primarily as ornamentals (Rehder, 1940: 399-400; H. L. Li, 1959: 124-25). Though the fruit of *C. japonica* may be used as medicine or fragrance in China (Meyer, 1911: 32-33; F. P. Smith, 1969: 363), it is said to be acid, astringent, and inedible (Bois, 1927-37, 2: 258).

[38] *Cydonia oblonga* is found wild in a region from the Balkans to Iran and Turkestan, and is an ancient domesticate usually placed in the Near Eastern hearth of domestication.

[39] In France, too, where the Chinese quince was introduced in 1810, it has difficulty adjusting to the north. Indeed, according to Bois (1927-37, 2: 257), it can hardly be cultivated there except in the south and southwest of the country.

[40] Schafer (1967: 184; 1977: 126-27), writing of T'ang times, observed that in the south the citron was used for decorative carving in a similar manner.

[41] For medicinal uses, see F. P. Smith, 1969: 362-63, and Perry, 1980: 341-42.

[42] There are roughly 90 species of hawthorn in temperate areas of Eurasia, and many more in North America (Rehder, 1940: 359).

[43] The leaves of at least one species, *Crataegus crenulata* Roxb., are used in China for making tea (Uphof, 1959: 112).

[44] In England the term "haw" is used for the haw tree or hawthorn *Crataegus oxyacantha*—a thorny shrub or small tree widely used as a hedge plant—and for its small red fruit. The English term has also been applied, sometimes with modifiers, to other *Crataegus* species and their fruit, among them those of China.

C. oxyacantha flowers in the month of May in England, and its blossoms, usually white, sometimes red or pink, are called "May." The tree, too, is thus sometimes called "May-tree" or "May-thorn."

[45] This term, however, is used for other species of *Crataegus* as well (Bretschneider, 1882-95, 2: 111; E. H. Wilson, 1913, 2: 29).

[46] The carambola, as mentioned previously, is also known to English-speaking people in China as "Chinese gooseberry," which has led to a certain confusion with *Actinidia*, though the two provide very different-appearing fruits. This confusion has been abetted by the Chinese name *yang-t'ao*, usually applied to species of *Actinidia*, but also, in the Canton area, to the carambola (Bretschneider, 1882-95, 2: 317-18; F. P. Smith, 1969: 14-15, 59-60).

[47] New Zealand exported 90,000 tons of fresh fruit in 1985, much of it to Europe, where it has been marketed as "kiwifruit," "kiwi," or "New Zealand gooseberry."

[48] Meyer (1911: 48-49), however, identified a third species, *Actinidia kolomikta*, as of more common occurrence in northern China, Manchuria, northern Korea, and eastern Siberia. This is a vine that from place to place occurs in thick masses in open woodlands, on occasion covering extensive areas, running over large rocks and choking out small bushes. Its fruit is enthusiastically gathered in areas where it occurs, and in eastern Siberia is dried by Russian settlers and used during winter in baking bread and pastry. At the same time, its fruit, though variable in size, is smaller than that of *A. arguta* and *A. chinensis*. This species has been cultivated since late in the last century.

[49] Another name for *Actinidia arguta* in North China is *mi-hou-t'ao*, "monkey peach," for in the mountains of Shensi, where it grows abundantly, it is prized by monkeys (F. P. Smith, 1969: 15).

[50] The fruit and other plant parts seem to be little used by the Chinese as medicine (F. P. Smith, 1969: 15; Perry, 1980: 124).

[51] There is a substance in various species of *Actinidia* that attracts cats and other felines, including tigers and lions, and leads them to damage or destroy the leaves and bark of plants in an effort to get at the attractant (Schroeder and Fletcher, 1967: 90-91). In Japan, hunters will gather vines of *Actinidia*, crush them, and throw them on the coals of a fire, and then hide in the forest and shoot wild cats that approach (Cutler, 1964: 8).

[52] Other species of *Actinidia* have been introduced to the United States and Europe as well, but, according to Schroeder and Fletcher (1967: 82-83), only the three discussed here—*A. chinensis*, *A. arguta*, and *A. kolomikta*—have shown much hope of becoming significant fruit producers. *A. arguta* was sold by catalog as a fruit plant early in this century by one nursery on Long Island, but it failed to meet with early success and was dropped. In the last few years, however, *A. arguta* has returned to the nursery trade in the United States as "the hardy kiwi," a plant capable of surviving cold winter temperatures (to -25°F.) and enjoying the advantage of being edible without peeling. Certain plant breeders in the United States are now engaged in collecting better-fruiting forms of *A. arguta*, from this country and abroad, for further improvement. In the opinion of the reviewer on whom we rely (Ruttle, 1986), finally, long after its initial introduction, the time appears right for it to gain permanent standing. He also notes that *A. kolomikta*, capable of surviving temperatures of -50°F., is being bred as well.

[53] The fruit was already called "human face" in the fourth century A.D., the name coming from its kernel which resembles a human face. At feasts, said Chi Han, it was frequently used to amuse guests (H. L. Li, 1979: 132).

[54] Most commonly mentioned in the botanical literature is *Dracontomelum mangiferum*, a tree found from India to Malaysia, Cambodia, and Vietnam, whose fruit is marketed for food in various parts of that region. In Malaysia, for example, it is consumed with fish as a pungent relish (Crevost and Lemarié, 1917: 231; Uphof, 1959: 133; I. H. Burkill, 1966, 1: 873).

[55] Chi Han called *D. dao "jên-mien-tzŭ."* Soubeiran and Dabry de Thiersant (1874: 257) and F. P. Smith (1969: 421-22) provide medical uses for a plant called *jen-mien* or *jên-mien-tzŭ*. Though they identify it as *Spondias dulcis* or *Spondias amara*, the name and Smith's description suggest that it may actually have been *Dracontomelum dao*.

[56] The name *yang-mei* is applied to the fruit of certain unrelated wild species as well, among them strawberries (*Fragaria* spp.) (E. H. Wilson, 1929: 310; F. P. Smith, 1969: 178; E. N. Anderson, 1988: 136).

[57] *Hovenia dulcis* and its products are also known by other names, for which see Bretschneider, 1882-95, 2: 314; Hooper, 1929: 74; F. P. Smith, 1969: 209; and S. Y. Hu, 1980: 13.

[58] In T'ang China, an extract of *Hovenia dulcis* also served as a sweetener, and will be mentioned again when considering honey, sugar, and other sweeteners in our "Spices and Flavorings" chapter.

[59] Though named "white mulberry" because of its fruit, in fact the fruit ranges in color from white to purple, on occasion being nearly black. Bretschneider (1882-95, 2: 328-29), moreover, noted that around Peking the fruit of *M. alba* is usually red, seldom white.

[60] There were other products of the mulberry (*M. alba*) in China as well, most notably a mulberry-bark paper which, according to Marco Polo, was used for making paper money. Bretschneider thought that Marco Polo was wrong, arguing that the paper mulberry (*Broussonetia papyrifera*), a shrub, is used in China for making paper. In fact, both species have been used for making paper, and paper from both was made into money and used in other ways as well (Laufer, 1919: 560-63; Y. S. Sung, 1966: 223; T. H. Tsien, 1985: 56-59).

[61] According to Bailey (1935, 2: 2069), the leading fruit-bearing mulberry varieties in North America were developed from *M. alba*, not *M. nigra*.

[62] Such drying of mulberries for future food use is reported (E. N. Anderson, 1988: 182) among non-Han peoples of Sinkiang.

[63] *M. alba* fruit is consumed in India (Watt, 1889-96, 5: 280-81). Though the fruit differs considerably in quality from one place to another in India, the best fruit, if observers have correctly identified the species, make fine eating. In the Dutch East Indies, *M. alba*, which was introduced long ago (probably by the Chinese for purposes of sericulture), had fruit which was regarded as a delicacy (Ochse and Bakhuizen van den Brink, 1931b, 3: 508). And in the Philippines, *M. alba* fruit is described (Wester, 1924: 132-33) as tasty, though an elevation of about 1600 feet or more is needed for the tree to yield good fruit. I. H. Burkill observed (1966, 2: 1523) that in Malaya, where *M. alba* is not common, its fruit lacks flavor and that, even where plentiful, it is little utilized. The above accounts suggest that in Southeast Asia there is a variability in fruit similar to that reported for India.

[64] Ikeda observed (1909: 39) that mulberries have never been raised for their fruit anywhere in Japan.

[65] For analyses of the food value of banana buds and flowers, see Platt, 1962: 32-33, and Leung *et al.*, 1973: 33.

[66] Records of cultivated bananas (plantains) are much earlier in India than in China, dating back to roughly 450 B.C. (*Jataka* XVI. 514, or Cowell, 1969, 5: 21; Prakash, 1961: 71, 72n). It may be, as Prakash has suggested (1961: 268), that they were cultivated long before, by people of the Indus Valley Civilization (*c.* 2500-1500 B.C.). If so, further decipherment of Indus Valley script may some day confirm this.

[67] *Ko*, as we have seen in our chapter on roots and tubers, is kudzu (*Pueraria lobata*), but the term may also be used in the sense of "coarse cloth or linen" prepared from other fiber plants (H. L. Li, 1979: 33n).

[68] Though most authorities place the earliest cultivation of abacá in the Philippines, H. L. Li (1979: 35) has noted that the plant is native to Annam; that abacá was cultivated in Indochina; and that it probably was first cultivated on the mainland, later in the Philippines.

[69] This statement is not in agreement with Perry's observation (1980: 300) that in China the carambola fruit is considered diuretic.

[70] China's best known myrobalan trees were those of the *Kuang-hsiao* temple in Canton. That temple, also known as *Ho-lin*, "myrobalan grove," had been renowned since the third century A.D. for its myrobalans, one of which was still surviving in 1960 (Schafer, 1967: 191, 303n; H. L. Li, 1979: 92).

[71] Perhaps Indians will be successful in familiarizing emblics to the American public, too, for one can sometimes find them in Indian food stores. In Berkeley, we found cans of very sweet emblic fruit labelled "Amla Murraba: Indian Gooseberry Preserve," which was claimed to be tasty, digestive, nutritious, restorative, and cooling. For other food uses of the emblic in India, see Morton, 1955.

[72] Malacca, that famed port of Southeast Asia, is believed to derive its name from that of the tree.

[73] Schafer (1963: 146) wrote initially of statements by T'ang pharmacologists that all three myrobalans (emblic, chebulic, and belleric) grew in Annam, and the emblic and belleric in southeastern China as well. Later (1967: 191), however, he wrote, apparently relying on the same pharmacologists, of the emblic and chebulic myrobalans growing in southeastern China, but the belleric only in Annam.

[74] We base this statement on data in Leung *et al.* (1973: 82, 91-92) who reported a vitamin C content for raw emblic fruit of 276 mg./100 gm. edible portion. Researchers in India, however, have reported the vitamin C content of emblic pulp to be from 540 to 747 mg./100 gm. (Hayes, 1966: 426). And Mustard (1952: 33) found whole emblic fruit from Florida to contain from 1220 to 1814 mg./100 gm, with an average of 1561 mg./100 gm.

[75] For awhile, the emblic was believed to be the world's leading natural source of vitamin C, but when testing emblics for vitamin C in 1945 it was accidentally discovered that the acerola, or West Indian cherry (*Malpighia punicifolia*), had a still higher a vitamin C content (Morton, 1955: 315-16). The latter fruit, now found in health food stores in the United States, has a vitamin C content listed by Leung *et al.* (1973: 100) as 1790 mg./100 gm., raw edible portion. Vitamin C content of the acerola, however, varies according to acidity of the fruit. In this regard, Mustard (1952: 33-34) found sweet, ripe acerola fruit from Florida to have an average vitamin C content of from 860 to 1135 mg./100 gm., but noted that other studies had found sour, ripe fruit in Florida to contain from 1579 to 2712 mg./100 gm.

[76] Emblic myrobalans were regular cargo on Indian and Persian ships journeying to early Southeast Asia and China, and one can only wonder whether they were consumed aboard to prevent scurvy among the crews.

[77] Emblic fruit (also its leaves and bark) is rich in tannin, and in India it is used in dying and tanning, providing a blackish-grey dye that, together with other substances, is an ingredient in inks and hair dyes (Watt, 1889-96, 6, Part 1: 218-21). Chi Han, writing of emblic fruit in fourth century South China, also mentions its use in such a way, to alter the color of white beards and hair (H. L. Li, 1979: 130).

[78] In the past, chebulic myrobalans were used for making drinks, as in a monastery where a tea for honored guests was made from them (H. L. Li, 1979: 92, 94). Whether these were considered more as refreshments or as medicinals is not stated.

[79] The English name "jackfruit," in its turn, derived ultimately from its Malayalam name, *chakka* or *tsjaka.*

[80] For other uses of the jackfruit in India and Southeast Asia, see I. H. Burkill, 1966, 1: 256-58, and C. A. Thomas, 1980: 158-59.

[81] Remains of wild apples have been uncovered at Neolithic levels in various archeological sites in temperate areas of Europe. Some claim that in late Neolithic times, the Lake Dwellers of Europe cultivated apples as well, but this view is disputed by others. It is clear, however, that wild crab apples were used in Europe in prehistoric times, dried for use in winter, and perhaps made into jelly or drinks, sweetened and sometimes fermented (de Candolle, 1886: 235-36; Renfrew, 1973: 136-39; Brouk, 1975: 153).

[82] In an earlier study (1914-15), Rehder had identified a dozen Chinese species of *Pyrus.*

[83] The sand pear has, however, crossed with the common pear (*P. communis*) in North America to obtain hybrids, such as the Kieffer and LeConte pears. The hybrid forms are more vigorous in growth than common pears and better in flavor than sand pears (Bailey, 1935, 3: 2869; Roach, 1985: 116).

[84] Rehder (1914-15: 231) writes that the term *pai-li* may also be used for *P. ussuriensis.*

[85] There were already a good number of pear varieties in ancient China, with twelve being recorded in the *Ch'i-min yao-shu* (sixth century A.D.) (S. H. Shih, 1962: 55).

[86] See, for example, "The crunch pears," *Sunset: The Magazine of Western Living*, September 1984, 72-75.

[87] Among the other species suggested as likely to have been significant ancestors to the cultivated apple varieties of Europe are four species indigenous to China: *M. baccata*, *M. prunifolia*, *M. spectabilis*, and *M. prattii* (Wilcox, 1958-62: 639; Watkins, 1976a: 248). All of these are also indigenous to other parts of Asia except *M. prattii*, a species of Szechwan.

[88] One such crab-apple species, whose leaves serve for making a tea called "*hai-t'ang* tea," is *M. hupehensis* (= *M. theifera*), a tree, wild but sometimes cultivated, that is widespread in northern, central, and southwestern China (H. L. Li, 1959: 124-25; Uphof, 1959: 302).

[89] *M. prunifolia* was cultivated in Japan as well, but American and European apple varieties have almost completely replaced them there (Wilcox, 1958-62: 639).

[90] A dish somewhat similar in the end product, but made with whole fruit on a stick, is "Honeyed Crab Apples" (G. B. Miller, 1966: 743).

[91] We don't know whether the same was true of Kwangtung, but in Fukien apples that fell from a tree or were rotting were sun dried and consumed mainly by the producing farmers themselves ("Production of fruit in Fukien Province," 1928: 1025).

[92] For a map of this "spontaneous" distribution in the Mediterranean and Near East, see Mathon, 1981: 37. Many believe that the domesticated pomegranate was introduced to Greece and other Mediterranean lands from Asia and that the so-called spontaneous pomegranates there are derived from cultivated ones (de Candolle, 1886: 237-39; Laufer, 1919: 276-77; I. H. Burkill, 1966, 2: 1871, 1872).

[93] Friedrich Muthmann's study of pomegranate symbolism, *Der Granatapfel: Symbol des Lebens in der Alten Welt* (1982), is such an outstanding work—careful in scholarship, excellent in well-chosen illustrations, and unusual in quality of publication—that we would call special attention to it here.

[94] There is a tradition in China, accepted by Bretschneider (1882-95, 3: 426), that the pomegranate was first brought to China from Bactria by Han traveller Chang Ch'ien (*c*. 120 B.C.). This notion is dismissed by Laufer (1919: 278-79), who observes that it is found neither in Chang Ch'ien's biography nor in the *Annals of the Han*, but first appears in the writings of post-Han writers. For further doubts about the role of Chang Ch'ien in plant introductions from the west, see Laufer, 1919: 190-91, and Y. S. Yü, 1977: 80.

[95] The time of earliest cultivation of figs in Greece was long a matter of controversy, for fig cultivation was not mentioned in the *Iliad*, and reference to it in the *Odyssey* was suspected of being of later date. Nor does the cultivated fig appear in the writings of Hesiod (Hehn, 1885: 85-86; Seymour, 1963: 339), who may have been a contemporary of Homer or lived a century or so later. Indeed, the first clear literary evidence of fig cultivation by the Greeks is found in the writings of Archilochus (*c*. 700 B.C.) (Hehn, 1885: 86). Remains of figs and representations of fig leaves or trees have, however, been found in Late Bronze Age levels on the mainland and on Crete (Vickery, 1980: 18, 32, 59). This, together with clear evidence in literature and art of much earlier fig cultivation in neighboring lands, makes it highly likely that figs were cultivated in Late Bronze Age Greece (*c*. 1500-1000 B.C.) and not just imported or collected from the wild.

[96] In Palestine in summer, the fig has thick foliage and is better than other trees at providing shelter from the sun, and it is common to see a householder sitting under the shade of a fig tree in his garden. Sometimes, too, a fig tree may overhang a house and help keep it cool. It may be that the quote from *I Kings* and similar ones elsewhere (*Micah* 4. 4; *Zechariah* 3. 10) are alluding to the shade-providing quality of the fig tree (E. W. G. Masterman in Hastings, 1910: 262).

[97] Prakash (1961: 93) writes of early Greek sources indicating that figs were regularly used in India, the final authorities for his statement being Pliny (*Natural History* XI) and Strabo. We have not been able to locate the reference given for Strabo, and the only reference to an Indian fig tree we find in Strabo (*Geography* XV. 1. 21) is to the banyan, *Ficus bengalensis*. Pliny, too, refers not to *Ficus carica* but to the banyan and its sweet-tasting fruit. Pliny drew his account of the banyan from that of Theophrastus who, with the Greeks who accompanied Alexander, called the banyan the "Indian fig" (Noehden, 1827: 121-28). In a similar vein, we found the statement in a translation of the *Jataka* (*c*.

450 B.C.) to abundant figs, ripe and sweet, but it appears immediately after mention of the banyan tree (*Jataka* XXII. 547; or Cowell, 1969, 6: 275).

98 Crevost and Lemarié (1917: 265) report that *F. carica*, though introduced to Indochina, was rare and did not grow well, though it was said to be naturalized in certain places there.

99 Though many varieties of figs will develop mature fruit without being pollinated, the fruit of others, the Smyrna figs, require pollination or immature fruit will soon drop from the tree. Pollination of Smyrna figs, whose flowers are found on the inner surface of the fruit, is accomplished by female fig-wasps (*Blastophaga psenes* or *B. grossorum*) which emerge from the flowers of the wild caprifig, usually *F. carica*, var. *silvestris*, to pollinate those of the cultivated tree. This is facilitated in Smyrna by suspending a string of caprifig fruit among the branches of a cultivated tree (Condit, 1947: 42-50). The caprifig is necessary to provide both pollen and a supply of fig wasps. This method of pollination, called caprification, is not mentioned in Chinese works (Laufer, 1919: 410) and apparently was unknown there, at least in early times.

100 Condit (1947: 105) writes of gardeners in North China protecting their fig trees by plastering them with mud and straw.

101 That its sanctity may be of great antiquity in India is suggested by the prominence of the peepul tree and peepul leaf in religious contexts on seals, seal impressions, and other remains of the Indus Valley Civilization (*c*. 2500-1500 B.C.) (Bosch, 1960: 68; Allchin and Allchin, 1982: 211, 214-15, and *passim*).

102 The famous peepul tree at Anuradhapura in Ceylon, according to Buddhist tradition, was planted in 245 B.C. If true, it would be over 2200 years old today. The tree is found in an area that had been all but abandoned for centuries, but, as an object of reverence, it continued to be tended by Buddhist priests, and to be a goal of pilgrims from around the world. The Ceylonese hold that this tree was grown from a branch of the original peepul tree at Buddh Gaya (Bodh Gaya) in India, under which the Buddha received enlightenment (Davids, 1908-26: 747-48). Grierson (1908-26: 184), writing of the tree at Buddh Gaya, insists that the peepul is a relatively short-lived tree and that it is impossible that the present tree, also an object of pilgrimage, dates from Buddha's time. Indeed he provides historical evidence that the peepul at Buddh Gaya had died more than once. He believes it possible, however, that the present tree is descended from the original peepul of Buddha, reproduced by seed from one generation to the next.

103 When Buddhism spread northwards in China into regions too cold for *Ficus religiosa*, certain linden trees (*Tilia mandshurica* and *Tilia miqueliana*)—perhaps in part because of the slight resemblance of their leaves to those of *F. religiosa*—took the latter's place as sacred trees around Buddhist temples. Not only did they assume its sacred role, but they took its Chinese name, *p'u-t'i-shu*, as well. Though Japan has its own species of lindens, *T. miqueliana*, presumably introduced from China by a Buddhist monk late in the twelfth century A.D., is the linden usually cultivated there. As in China, it is found especially around Buddhist temples and is known by the same name as that of *F. religiosa* (Japanese: *bodai*) (H. L. Li, 1963a: 133-34).

The above recalls the importance of the linden in old German folklore and literature. The linden was a holy tree under which dragons lay and where dwarfs and fairies lingered. It was also a tree under which legal proceedings took place, hence "the tree of judgment" (Skinner, 1911: 157-58).

104 Schafer (1967: 205, 307) identifies three banyan species of southeastern China and Vietnam: the "true banyan," *Ficus retusa*; "the bird banyan," *F. wightiana*; and the "large leaf banyan," *F. altissima*.

105 Aelian (second century A.D.), for example, wrote that the olive was not among the trees of India, for it is not indigenous there and doesn't do well when introduced. Pliny (A.D. 23-79), in turn, wrote that in India the olive-tree is sterile except for the wild olive (*Olea cuspidata*?) (McCrindle, 1971: 121, 141). One concludes that the early Greeks, perhaps following Alexander's invasion, or others had planted olives in India but found that they did not bear fruit there.

106 Watt (1889-96, 5: 485), however, notes that some European olive trees were planted in the Botanical Gardens at Calcutta in 1800. And Bertin *et al.* (1971: Map 17) show olives introduced in the northwest of the subcontinent in the 1800's and 1900's.

107 However, a related shrub or small tree, *Osmanthus fragrans* (sometimes *Olea fragrans*), is widely cultivated in China for its flowers, which are used as a fragrance, as in tea, alcoholic beverages, cosmetics, and medicines (Bretschneider, 1882-95, 2: 384; Ball, 1906: 688; Crevost and Lemarié, 1917: 328; Bois, 1927-37, 3: 251; F. P. Smith, 1969: 296).

[108] Many species of wild *Vitis*, however, have inedible fruit containing many needle crystals, acrid and irritating to the mouth. In Malaya, those with inedible fruit constitute a majority of the species (I. H. Burkill, 1966, 2: 2282-83).

[109] Watt (1889-96, 6, Part 4: 263) did not set a precise date, but argued that India had "a strong claim to an ancient cultivation of the vine." We have, however, found no references to the grape or grape wine in the Vedic literature (pre-800 B.C.?) or the Brahmanas (800-600 B.C.?). Panini (?350 B.C.-250 B.C.?) refers to the grape vine, its fruit, and to a type of wine imported from Afghanistan (Agrawala, 1963: 216), but the earliest convincing literary evidence of grape cultivation in India seems to occur in the writings of Kautilya (?290 B.C.-A.D. 300?) (Shamasastry, 1951: 129, 133) and later writers. Grünwedel (1901: 33-34) notes a representation from the Buddhist stupa of Sanchi, Madhya Pradesh, a human figure, mounted on a horned lion, who holds in his hand a bunch of grapes. The individual, however, is not an Indian, but a foreigner from far to the northwest. Prakash (1961: 269), in turn, writes of a bunch of grapes, including a grape leaf, held by an Indo-Greek monarch, in the Buddhist stupa at Bharhut, also in Madhya Pradesh. We have not located such a representation or mention of it in the source Prakash cited. In any case, neither stupa was built before about 250 B.C., and many representations on them date from subsequent centuries. None of the above suggests an introduction of grape cultivation to India before the fourth century B.C. at the earliest. In the seventh century A.D., Hsüan-tsang (Watters, 1904-5, 1: 177-78, 225) wrote that grape cultivation was found here and there in Kashmir and to its west, which may mean that it was uncommon elsewhere in India even at that late date.

One must leave open the possibility of much earlier cultivation of grapes in the areas adjoining Afghanistan and Iran in the far west and northwest of the Indian subcontinent. Some botanists believe the grape to be indigenous to parts of that region, and there are reports of grape pips, whatever their state of domestication, in early sites there. Large numbers of grape pips were found in the final period of Mergarh (*c.* 2500 B.C.?), located in a transitional area between the Iranian plateau and the Indus Valley (Allchin and Allchin, 1982: 148). Grapes were listed among domesticated plants found in levels dating from about 2000 B.C. and later at Pirak, a site about twenty-five miles east of Mergarh (Costantini, 1981: 277). A grape seed, identified as *Vitis vinifera*, was also found in excavations of an early agricultural community at Loebanr in Swat, in levels which may date from about 1700-1500 B.C. (Costantini, 1979a: 705, 706, 708). If the seeds in these sites are indeed from *Vitis vinifera* and not *Vitis silvestris* (see Renfrew, 1973: 127-30 for criteria distinguishing the two), it opens interesting possibilities.

[110] For a twelfth-century Chinese recipe for grape wine, see Laufer, 1919: 234.

[111] European wine, including sherry, arrived in China in the eighteenth century and after that it became a regular import (Laufer, 1919: 238-39).

[112] Distillation of liquor is usually said to have come to China during the Yüan dynasty (A.D. 1271-1368), with grape wine among the products distilled (Laufer, 1919: 237-38). This is difficult, however, to reconcile with statements in T'ang poems suggesting that distilled liquor was already made in Szechwan at that time (Schafer, 1977: 122). F. P. Smith (1969: 459) writes that the Chinese in their distillation use grape wine fermented in the regular way. Though the brandy is weak by Western standards, it brings on intoxication in Chinese far more readily than most expect.

[113] This is a substantial increase over production even thirty years ago: 48,500 tons in 1952 to 111,500 tons in 1958. The 1952 tonnage was well below those for pears (393,500 tons), citrus fruits (206,500), apples (118,000), and bananas (110,000) (N. R. Chen, 1967: 338-39). Despite the much larger present-day production of grapes, it is still dwarfed by that of apples (3.8 million tons in 1985) and pears (2 million tons) (Wittwer, 1987: 283, 284, 286).

[114] For methods of grape cultivation in China today, see H. Huang, 1980: 464-65.

[115] For identification of edible species of wild *Vitis* in China, see Bretschneider, 1882-95, 2: 316; 3: 438; Meyer, 1911: 42; Uphof, 1959: 378; F. P. Smith, 1969: 457-58; H. L. Li, 1979: 124-25.

[116] Meyer raised the possibility that *V. amurensis* be hybridized to obtain a grape more winter-hardy than *V. vinifera*. He did not, however, mention the use of *V. amurensis* as rootstock for *V. vinifera*, as is now done at a fruit research station in Manchuria (American Plant Studies Delegation, 1975: 95).

[117] The guava, *P. guajava*, is a Meso-American domesticate, whereas the strawberry guava, *P. cattleianum*, is South American. Both are found in China today. Indeed, the strawberry guava, which reached China at an early date, was once thought be native to China, which led to the species being called "Chinese guava," a term sometimes still used for it (Popenoe, 1920: 280-81).

[118] Information on all of these is included in Zee and Hui, 1981, on whose account we have drawn liberally in this paragraph.

[119] Eberhard (1952: 75-77, 96) argues that the Dragon-Boat Festival had its roots in customs of the Tai tribes of South China, with the use of boats a later accretion in coastal areas of South and Central China. Latsch (1984: 59-60), in turn, writes of reports that dragon-boat races may have begun among the Yüeh in the Spring and Autumn Period (770-476 B.C.).

[120] This name, meaning "foreign pomegranate," derives from a perceived similarity in appearance and in its many seeds.

CHAPTER EIGHT

Edible Nuts, Nut-Like Fruits, and Seeds

Nuts and Nut-Like Fruits from Species Domesticated in China
- Chinese Chestnut
- Chinese Almond
- Chinese Hazelnuts
- Ginkgo Nut
- Ling Nut

Southeast Asia's Main Contribution: The Coconut
Nuts from Central Asia and the Near East
- Persian or English Walnut
- Pistachio

The Peanut: A Gift from America
Other Edible Nuts and Seeds Consumed by the Chinese
- From Trees and Shrubs
- From Other Cultivated Plants

Oil Seeds
A Few Conclusions

The Chinese like nuts as well as other peoples do, yet nut cultivation was not well developed in traditional China, and nuts occupied an insignificant place in Chinese diet (Meyer, 1911: 51; H. L. Li, 1983: 47; E. N. Anderson, 1988: 137). As a result, our list of cultivated nuts is much shorter than that of cultivated fruits, and in general our information on them is scantier. In addition, there are questions as to the degree particular species were actually cultivated, for many of the nuts consumed in traditional China derived not from cultivated trees and shrubs but from ones that were wild or nearly so. As a result, the Western reader may face a higher percentage of unfamiliar species, domesticated and wild, in this section than in others. We have sought to make his way easier by first considering China's major, and more familiar, nuts and nut-like fruits, and then turning to other edible nuts and seeds, including oil seeds, used by the Chinese. We order the major species providing nuts and nut-like fruits by place of origin, turning first to those domesticated in China.

NUTS AND NUT-LIKE FRUITS FROM SPECIES DOMESTICATED IN CHINA

Chinese Chestnut

The chestnut genus, *Castanea*, is a small one, consisting of a dozen species native to North Africa, Eurasia, and North America (Krüssmann, 1984-86, 1: 291-92). Various of these have been cultivated, but the world's most important commercial species is the Sweet, Spanish, or European chestnut, *Castanea sativa*. Here, however, we focus on the Chinese chestnut, *C. mollissima*, the common chestnut grown in China and one of three *Castanea* species indigenous there.[1] *C. mollissima*, a deciduous tree which can reach a height of 40 to 60 feet, is native to western China, and is cultivated widely, from Peking in the north to Yunnan and Kwangtung in the south (Chun, 1921: 84; Bois, 1927-37, 2: 545; S. C. Lee, 1935: 294-95).[2] Its nuts have been eaten in China since antiquity, as indicated by chestnut remains found in a Yang-shao site that dates from the fourth or fifth millennium B.C. (P. T. Ho, 1975: 87n; K. C. Chang, 1986: 112); by the appearance of the chestnut tree among inscriptions on Shang oracle bones (H. E. Gibson, 1937: 309); and by chestnuts being included among the foods in Han tombs (Y. S. Yü, 1977: 72; Z. Wang, 1982: 53). Not only was the chestnut (*li*) a source of food, but it was one of the two important nut trees of ancient North China, highly regarded and frequently mentioned in the classical literature. The north produces the best chestnuts, but some excellent ones were also reported in the southeast (Kwangtung, Kwangsi, northern Vietnam) in T'ang times (Schafer, 1967: 183). The acceptability of the chestnut is indicated by the fact that its nuts were suitable presents for the emperor and were sent as tribute from northern Hopeh to the T'ang court, and that chestnut trees were appropriate for a nobleman to plant near his mansion and for founders of various dynasties to plant near altars dedicated to earth spirits (Bretschneider, 1882-95, 2: 306, 318-20; Schafer and Wallacker, 1957-58: 218; H. L. Li, 1983: 35).

In traditional times, chestnuts continued to be among the most common nuts used by the Chinese.[3] Since chestnut trees were raised from seed, however, there were great differences among trees in productivity and in the size and sweetness of their nuts. When harvested, the nuts might be used fresh, or, since they are very perishable, they might be dried or stored in other ways.[4] Thus in Chinese food stores, one might find chestnuts fresh, or peeled and dried, or canned in water or syrup. Early in this century, Meyer (1911: 52), whose observations centered on North China, noted that in fall and early winter large numbers of chestnuts, roasted in sand to which molasses had been added, were sold in towns. Roasted chestnuts were very popular in winter in Peking and they were said to be sold at every intersection (Tun, 1965: 83-84; K. Lo, 1976: 136-37). Chestnuts might also be boiled for use as snacks. Chestnuts of fine quality might be shipped well beyond the producing area, those of Liang-hsiang district of Hopeh being sold in Shanghai and even in Japan, where Kobe and Osaka in winter had many street stands where they were for sale (V. Y. N. Hsu and F. L. K. Hsu, 1977: 311).

As for other ways of preparing chestnuts, Schafer (1977: 98, 134) noted that the T'ang Chinese not only ate them after sun-drying or roasting, but made them into flour (as was done in Europe) and also pickled them, as by salting. For traditional China one reads also of glazed chestnuts, of minced or ground chestnuts made into "Glazed Chestnut Balls" or "Sugared Chestnut Balls" (G. B. Miller, 1966: 746), of chestnut stuffings, and of many dishes containing chestnuts, among them stir-fried ones and stews. Examples of these are the Cantonese "Braised Duck with Chestnuts and Chinese Dates" (Sakamoto, 1977: 186) and "Simmered Eel with Roast Pork and Chestnuts" (K. Lo, 1972: 120), and the Shanghai "Braised Chestnuts with Celery Cabbage" and "Stewed Chicken with Chestnuts" (W. W. Chang *et al.*, 1970: 267, 382). Chestnuts are also common in vegetarian dishes, such as the interesting "Hot Assembly of Chestnuts, Sliced Lotus Root, Ginkgo Nuts, Peanuts, Chinese Mushrooms, and Bean Curd" (K. Lo, 1974: 75). And finally one notes the well-known chestnut dessert called "Peking Dust," its name coming from the notorious dustiness of Peking, which is not far south of the Gobi Desert. This dish, which was developed not by Chinese but by European residents in the 1920's, is, in its basic form, a mound of boiled, mashed, sweetened chestnuts served with whipped cream, usually as a topping. There were variations of this, however, such as adding sherry, brandy, or rum, cream, and a garnishing of preserved kumquats or cherries, or glazed fruit or nuts (Feng, n.d.: 177-78; G. B. Miller, 1966: 747-48; K. Lo, 1971: 172; 1979c: 329; Em. Hahn *et al.*, 1973: 170; Field *et al.*, 1976: 106; C. B. T. Lee and A. E. Lee, 1980: 129).[5]

As for non-food uses, Martin Yang (1945: 37, 112) reports that in Shantung the syllables of the name for chestnuts, *li-tzŭ*, sound like the words for "favorable" and "son," and that the name for jujubes, *tsao*, is similar in sound to the word for "early." As a result, Yang notes, chestnuts symbolize "the easy birth of sons," a matter of great importance to the family, and both chestnuts and jujubes were sewn into the dress of a Shantung bride and into the bedding in the room where the married couple will live, as well as being placed in various boxes, dressers, and cabinets in the room.

Chinese chestnuts contain only a fraction of the fat found in the other major nuts of China (walnut, Chinese "almond," peanut, coconut, hazelnut), roasted chestnuts having only 2 gm./100 gm. edible portion compared to 45 to 65 gm. for those nuts in a dried form. It is also unlike them, and more like the cereal grains, in its high carbohydrate content, but it has only 3 gm./100 gm. of protein, far lower in this regard than either cereals or the major nuts. Nor is it notable in any of the usual vitamins or minerals (Leung *et al.*, 1973: 19, 23-29 *passim*).[6,7]

Chinese Almond

Domestication of the almond, *Prunus amygdalus*, is usually placed in an area stretching from the eastern Mediterranean to Central Asia, where it is found in the wild. Likely it was domesticated by the third millennium B.C. or earlier (Zohary and Hopf, 1988: 163-63). The early Chinese knew of the tree and its kernels in Persia and other lands to the west. They also imported almond kernels, as among the tribute sent from Turkestan to China in T'ang times. The tree itself was reported by the Arab merchant Soleiman to be cultivated in China in the mid-ninth century (Laufer, 1919: 407; Schafer and Wallacker, 1957-58: 219; Schafer, 1963: 141; 1977: 98), and Li Shi-chen late in the sixteenth-century wrote of it as growing everywhere "west of the Pass" (that is, in Kansu and Shensi) (Laufer, 1919: 407). Though the above would seem to leave little doubt that almonds had been cultivated at least somewhere in China or its margins, there has nevertheless been controversy among Western scholars as to whether, at least in traditional times, almonds were actually grown there and whether the kernels that foreigners in China often called "almonds" were almonds or apricot kernels. Adding to the confusion were the similarities in appearance, taste, and use between apricot and almond kernels, which sometimes led the Chinese, on their part, to call the almond kernel by the name "apricot kernel" (Laufer, 1919: 408).

Laufer (1919: 405-9), who has provided the most detailed analysis of the historical evidence, was convinced that the almond had been cultivated in China in the past. He also presented evidence from the Chinese literature suggesting that it was still cultivated there in the nineteenth century, but allowed the possibility that almond cultivation "is now extinct in China." Other authorities, among them naturalists and botanists with extensive field experience in traditional China, are more firm in their conclusion that almonds were not cultivated there.[8] To this writer, it seems reasonable to believe that the almond could not have been widely cultivated in China and been missed by so many widely-travelled, careful observers. Yet the evidence presented by Laufer seems to leave the possibility open that the almond continued to be cultivated in some places, especially in the far northwest, as identified by Li Shi-chen. It would, however, have been of quite minor importance overall, with the overwhelming share of the kernels known to foreigners as "almonds" being in fact apricot kernels. This fits with H. L. Li's statement on the matter;[9] with E. N. Anderson's observation (1988: 137) that real almonds are scarcely known; and with Meyer's conclusion (1911: 53) that the "almond cake" commonly served foreigners in traditional China was in fact made with kernels of the apricot (*hsing-jên*), *Prunus armeniaca.*[10]

There are, Meyer has noted, several varieties of *P. armeniaca* grown primarily for their seed in China. The fruit of the best variety contains a large stone with a fairly-soft shell and sweet kernel, and may be served as a snack, sometimes sugared, along with raisins and other kinds of nuts, or ground into almond flour which is made into almond cakes or cookies or into a thin porridge.[11] On occasion, such kernels may be salted, and in appearance and flavor are just like real almonds. Another variety of *P. armeniaca* has a bitter-tasting kernel that contains prussic acid and must be used in small amounts, as for flavoring sweets, or in making "almond soup" or "almond tea," a drink commonly sold along the streets (Dudgeon, 1895: 17; Meyer, 1911: 53-54; Porterfield, 1951: 31).

As for the preparation of such almond soup, Meyer noted that first rice was boiled until soft, then pounded and mixed with water until it had the consistency of milk. Then a few bitter almonds were ground up and blended in along with sugar, and the soup served hot. The soup, which was tasty and stimulating, was commonly consumed by the Chinese just before going to bed. Such "almond soup" or "almond tea" (*hsing-jên-ch'a*), well-liked as a snack not only among the people of North China but in the Ch'ing court, was also believed effective against sore throat (Davis, 1966: 227).[12] There was also a popular dish, found among those of the Ch'ing court, called "Almond Curd," a cold gelatin dessert made of water, agar-agar, ground almonds, and sugar (Davis, 1966: 144). Such almond soup and almond curd (also called almond float, almond lake, or almond junket), as made today in China, may include almond or vanilla extract, milk, and/or fruit of one sort or another (Feng, n.d.: 174; G. B. Miller, 1966: 750-52, 755, 756; W. W. Chang *et al.*, 1970: 455-56; Stallard, 1973b: 183; Field *et al.*, 1976: 107; Sakamoto, 1977: 40; K. Lo, 1979c: 324-25, 326; Y. K. Martin, 1984: 188-89). Prepared in a different, somewhat more elaborate way is the Cantonese dessert "Fried Almond Custard" (W. W. Chang *et al.*, 1970: 456). Chinese almonds are also commonly used in other ways, as in candies, cakes, and cookies, and in a broad range of main dishes, such as the Cantonese "Red and White Chicken with Almonds" and the Szechwanese "Almond Duck" (W. W. Chang *et al.*, 1970: 266, 279).[13]

Nutritionally, the kernels of *P. armeniaca* compare quite favorably with those of the true almond, being high in protein and fat, and also containing good amounts of iron, calcium, and riboflavin (Leung *et al.*, 1973: 23).

Chinese Hazelnuts

We are not dealing here with the hazelnut, or filbert, familiar to Americans, which is from a deciduous shrub or small tree, *Corylus avellana*, of European or Near Eastern origin.[14] Rather we are dealing with related species, especially *Corylus heterophylla* and *C. mandshurica*, indigenous to China.[15] These two shrubs or trees are common in the mountains of northern China, with the best hazelnuts in T'ang times reputed to come from Shensi Province. The use of hazelnuts in North China (presumably collected from the wild) goes back far in antiquity, for their remains have been uncovered at sites dating from the fifth millennium or before. That they were highly regarded as food is indicated by the hazelnut tree being included in the

Chinese classics among the leading fruit and nut trees of the ancient northern homeland, and by its nuts being regularly listed among the foods served at Han feasts (Bretschneider, 1882-95, 2: 306, 321-22; S. H. Shih, 1962: 54; F. P. Smith, 1969: 129; Schafer, 1977: 97; Y. S. Yü, 1977: 67; H. L. Li, 1983: 35; K. C. Chang, 1986: 92, 112).

Meyer (1911: 53) wrote early in this century that the Chinese hazelnut trees providing nuts were never cultivated, and, though their nuts (*chên-jên*) ultimately reached markets throughout East Asia, he insisted that they were gathered from the wild. Vavilov (1949-50: 25), however, includes *C. heterophylla* and other hazelnut species among the cultivated plants of China, though he indicates that they are mainly wild. Later authorities (Uphof, 1959: 110; Zeven and de Wet, 1982: 35) also list *C. heterophylla* and/or other species (*C. mandshurica* and *C. chinensis*) as cultivated there. To the extent that they are indeed cultivated, they represent still further Chinese offerings to the world's store of edible fruits and nuts.

The Chinese considered eating hazelnuts to be helpful in many ways; in improving the appetite, stimulating digestion, satisfying hunger, and strengthening the body (F. P. Smith, 1969: 129) and they were found in markets throughout traditional China in a dried form, as they are today in Chinese markets in the United States. They were also preserved in other ways, for one reads of sugared hazelnuts (Headland, 1914: 180) in modern China, and of hazelnuts salted or pickled and served on ceremonial occasions in T'ang times (Schafer, 1977: 134). Overall, however, we found far fewer references to hazelnuts being used in prepared dishes in China, than to chestnuts, walnuts, almonds, and peanuts. One such hazelnut dish was common at the Ch'ing imperial palace in Peking in winter, when few vegetables were to be had. Called "Stir-Fried Hazelnuts with Brown Bean Sauce," it included, among its other ingredients, pork, water chestnuts, green onions, ginger, soy sauce, and rice wine (Sakamoto, 1977: 74).

We have found no analysis of the nutritional value of the Chinese hazelnuts, but those of the European species (*C. avellana*) are like most other nuts in their high protein, high fat, and high food energy, and also contain good amounts of calcium, iron, riboflavin, and niacin (Leung *et al.*, 1973: 26).

Ginkgo Nut

The ginkgo (*Ginkgo biloba* L. = *Salisburia adiantifolia* Smith) is an interesting tree, last surviving species of the widespread ancient order Ginkgoales (originating in the Permian period, over 200 million years ago) which once included several genera and numerous species. In the Tertiary period, the ginkgo had been widespread in East Asia, Europe, and North America (Mathon, 1981: Fig. 49 on p. 59), but only in China did it survive in a natural state. It is for the above reasons that the ginkgo is often called a living fossil, remnant of a geologic time long before the appearance of man.

The ginkgo is a deciduous tree that can grow to heights of up to 120 feet, and there are male and female individuals, though some observers question whether the species is truly dioecious. Like the conifers, the ginkgo is a gymnosperm, but differs from them in its mode of fertilization and its broad, fan-shaped leaves (Figure 62) which led to one of its Chinese names, *ya-chio*, meaning "duck's foot." Some say that the ginkgo tree is unknown in a wild form. H. L. Li (1963a: 106) observes, however, that in certain parts of China it is found in what is presumed to be a wild or semi-wild state. Most often noted in this regard is an area of Chekiang Province, ten square miles in size. Some say that the Chekiang ginkgos are the remnant of an early forest, whereas others say that they are of relatively recent origin, perhaps deriving from cultivated

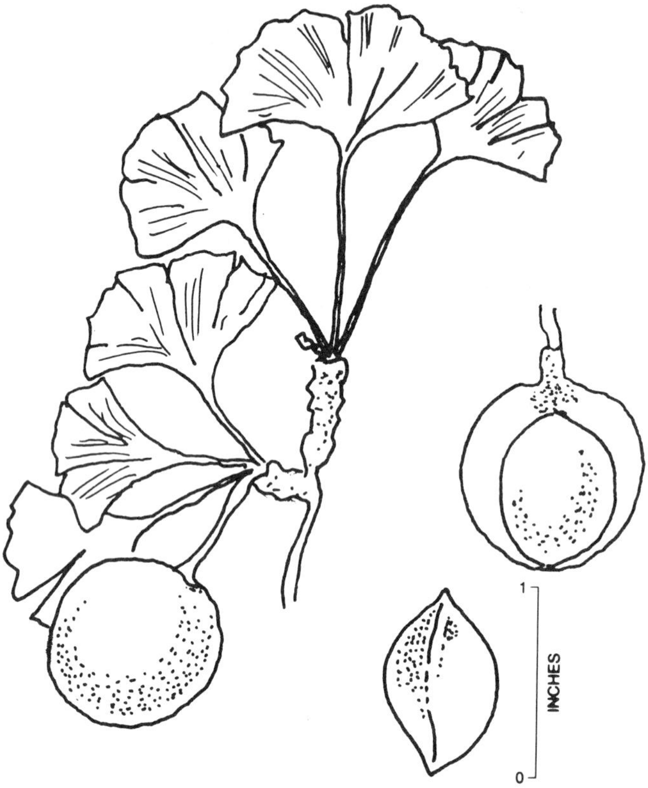

Figure 62. Leaf, stem, and nut of the ginkgo (*Ginkgo biloba*) (drawn by Shelby Hall)

ginkgo trees whose spread was encouraged by Chinese reluctance to cut them down (Detwiler, 1938: 123; H. L. Li, 1963a: 102-6; Dallimore and Jackson, 1966: 232). To this writer, it may be significant that Chekiang was the principal area of China where tree cults survived into recent times, and that associated with tree and mountain cults in China were bans on removing sacred trees and trees from sacred mountains (Eberhard, 1968: 174).

The ginkgo was not native to the Yellow River Valley and was not mentioned in early Chinese literature. Indeed the earliest clear reference to the ginkgo, as a tree found south of the Yangtze in eastern China, dates from the eleventh century A.D., during Sung times. The Sung Chinese came not only to cultivate the ginkgo, but to laud it in poetry and art (H. L. Li, 1963a: 86-92). Because of their actions and those of later Chinese, the ginkgo is now cultivated widely in China, from Manchuria to Yunnan (S. C. Lee, 1935: 6), as a graceful and appealing ornamental, long-lived, hardy, and resistant to pests and diseases. It is not only common in parks but as a bonsai plant (Zee and Hui, 1981: 23). The ginkgo tree's good qualities led to its introduction to Japan, perhaps already in Sung times; to Europe, early in the eighteenth century;

and to the United States near the end of that century. It does well in temperate regions of eastern North America as far north as central Michigan, Massachusetts, and the St. Lawrence River. It produces mature seed ("fruit") in warmer places to the south, as in Washington, D.C. where, as in other eastern cities, it is a popular street tree and where Chinese laundrymen were reported to collect it for the edible nut it contains (Bailey, 1935, 2: 1338).

It is often said that to the Chinese, Japanese, and Koreans the ginkgo is a sacred tree planted in the grounds of Buddhist temples; that certain temple trees are reputed to be more than a thousand years of age; and that Buddhist priests may in fact be responsible for the ginkgo's very survival as a species (E. H. Wilson, 1913, 2: 13, 45; Sowerby, 1940: 143; Cahen, 1943: 100, 101; Dallimore and Jackson, 1966: 232).[16] Moule (1937: 214), on the contrary, insists that the ginkgo is not a sacred tree and that it is no more associated with Buddhism than Confucianism. In China, and apparently in Japan, he writes, the ginkgo is found not just at Buddhist monasteries, but along roads, in fields, and in forests as well. H. L. Li (1963a: 93-94), drawing on Sung and Yüan accounts, concludes that though they mentioned the ginkgo as found in cultivation, they did not link it to temples or monasteries. Li acknowledges that modern temple grounds contain ginkgo trees, but he observes that trees of other species are found there as well. He sees this as part of an effort by Buddhist and Taoist priests to save part of the original forest cover of an area, to preserve indigenous species that otherwise would have been lost through agricultural activities. Though some aged trees are venerated in China, he says, they are found not only in temple grounds but elsewhere as well. Their status, moreover, is because of age, not species, and derives from an elemental reverence for nature and from concerns with geomancy.[17] Such trees, particularly in back areas of China, may be provided with incense stands and even shrines. Yet, like Moule, Li insists that the ginkgo has no special link to Buddhism as a religion; he also denies that its initial cultivation in China was for religious reasons or that it was saved from extinction by temple plantings of Buddhist or Taoist priests.

The seed dropped by a female ginkgo tree consists of a fleshy, yellow covering of unpleasant odor enclosing a white or cream-colored, soft-shelled nut with an edible kernel (Figure 62). The white color of the nut shell led to other of the ginkgo's Chinese names, *pai-kuo*, "white fruit," and *yin-hsing*, "silver apricot."[18,19]

In China, after the seed is collected its smelly covering is first removed, aided by fermentation in a container of water or by burial to hurry its decay; then, before use, the nuts are washed and sun-dried (Porterfield, 1940a: 186; Rosengarten, 1984: 287).[20,21] When raw, ginkgo nuts are not pleasant in taste, but they improve with boiling or roasting. Even so, Europeans in China seldom ate them (Fortune in Bretschneider, 1898: 503) and when they did try them, were unimpressed with their taste. One Western writer, for example, described the taste of roasted and boiled ginkgo nuts as strange, and others have called the flavor of the nuts fishy to some extent (Blasdale, 1899: 44; Detwiler, 1938: 124; F. P. Smith, 1969: 390). The raw nut is soft in texture and has a hint of bitterness. Its taste, to this writer, is unappealing and not clearly pleasant; others (Read *et al.*, 1940: 25) have described it as having an acrid, unpleasant flavor and smell. Boiled and roasted nuts are more palatable, firmer but lacking in strong taste and with a starchy quality reminiscent of a baked potato. Perhaps, we felt, they could use some added flavoring. That one can acquire a taste for the nuts, however, is clearly demonstrated by the Chinese, who consider roasted ginkgo nuts to be delicacies. They eat ginkgo nuts as appetizers, as ingredients in some dishes, among them soup and stew, or as digestives after a meal (Detwiler, 1938: 124; Porterfield, 1940a: 188; Hom and Steiman, 1981: 96; Zee and Hui,

1981: 23). They are commonly used in vegetarian fare, as with stir-fried vegetables, and in stuffings, as of chicken or duck (G. Z. Chu, 1975: 155-58, 283; Hom and Steiman, 1981: 34-35; Y. K. Martin, 1984: 24, 234). A dish with ginkgo-nut stuffing is "Eight-Jewel Duck" (Hom and Steiman, 1981: 187), and other ginkgo-nut dishes are "Eight-Treasure Vegetarian Assemblage" or "Buddhist's Vegetarian Ensemble," and, said to be a pleasure to monks or saints (lohans), "Lohan's Delight: Buddhist Vegetarian Dish" (K. Lo, 1976: 131; Y. K. Martin, 1984: 153, 218).[22] The high regard in which ginkgo nuts are held is shown in their being served at banquets and, with their shells dyed red, at weddings (F. P. Smith, 1969: 390). There is sufficient demand to make them common market items in China.[23] This demand has persisted among overseas Chinese, as shown by the ginkgo nuts they imported to Vietnam (Crevost and Lemarié, 1917: 285; Bois, 1927-37, 2: 600) and to San Francisco at the turn of this century (Blasdale, 1899: 44). Today, too, one finds them, whether in the shell, or shelled and in cans, in the markets of cities with sizeable Chinese communities.

Ginkgo nuts (whole, raw, dried) are starchy and differ strikingly from pine nuts, walnuts, peanuts, and watermelon seeds in their much lower fat (1.6 gm./100 gm.), protein (4.8 gm./100 gm.), and food energy (185 calories/100 gm.). They are also much lower in calcium (5 mg./100 gm.) and iron (1.2 mg./100 gm.), and notable mainly as a good source of vitamin C (25 mg./100 gm.) (Leung *et al.*, 1973: 19, 25, 28, 29). Like birds' nests, they seem to be a delicacy of little nutritional value. The nut, however, serves a variety of medicinal uses in China, and the seed (fruit), after lengthy soaking in rapeseed oil, is taken as a treatment for tuberculosis (Monachino, 1956: 46; F. P. Smith, 1969: 390-91; Perry, 1980: 160).

At the same time, some individuals are sensitive to contact with the fleshy seed covering, which contains acids that can inflame the skin (Perry, 1980: 160), with certain workmen in Washington, D.C. who gathered seed (fruit) from trees on the Mall having "swollen glands" for a week or so (Detwiler, 1938: 124). This toxicity, like that of poison ivy, affects some persons but not others. In some instances, too, eating the nut can be toxic. Studies of children have found that occasionally the eating of ginkgo nuts brings on restlessness, anxiety, high fever, dermatitis, vomiting, diarrhea, and convulsions, and, in some cases, death (Kao and Chen, 1944; F. P. Smith, 1969: 391; Perry, 1980: 160).[24]

We have noted previously that possible toxicity has not stopped people from growing and consuming the broad bean, *Vicia faba*, or the grass pea, *Lathyrus sativus*, and this is true, as well, of certain other foods known to pose such risks. In the case of the ginkgo tree, it is especially fortunate that such toxicity did not deter cultivation, for if it had, the species might well be extinct today.

Ling Nut

Sometimes called "water chestnut," and thus confused with the "true" or Chinese water chestnut (*Eleocharis dulcis*), is the ling nut. This is the dark, nut-like fruit of the annual aquatic plant *Trapa bicornis* (*ling*), which grows in ponds, canals, and other water bodies in China, where it was domesticated. Remains of *Trapa* have been reported in early agricultural sites near the mouth of the Yangtze in Chekiang, Kiangsu, and Shanghai: at Ho-mu-tu and in earlier levels of the Ma-chia-pang culture (*c.* 5000-3500 B.C.), as well as at Ch'ien-shan-yang (*c.* 3500-2500 B.C.) (P. T. Ho, 1975: 99; K. C. Chang, 1977a: 181, 513; 1986: 199, 254-55; T. K. Cheng, 1982: 11).

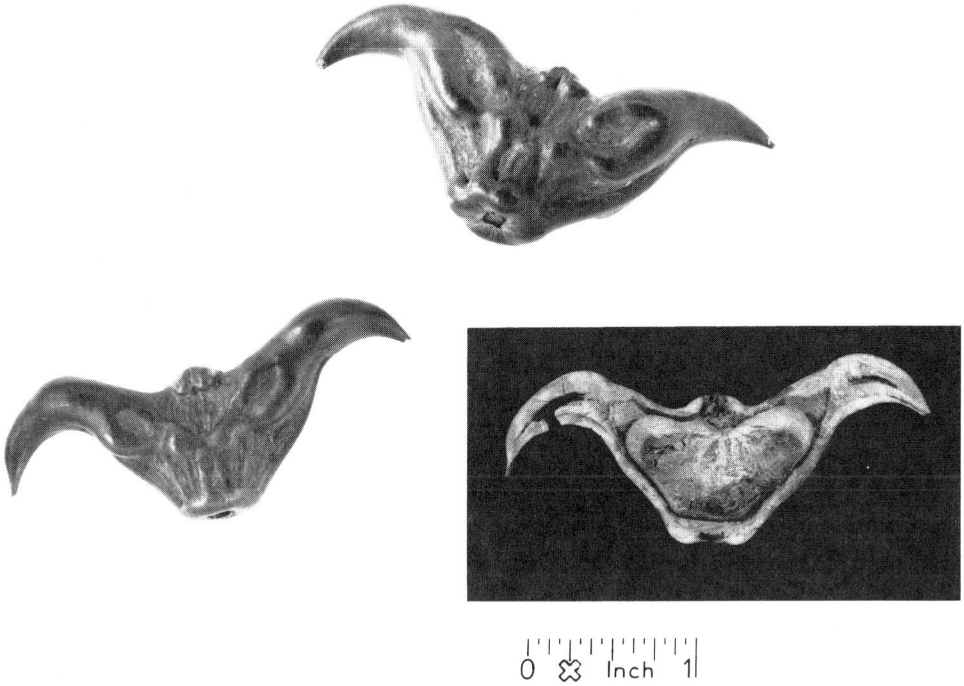

Figure 63. Ling Nut: The nut-like fruit of *Trapa bicornis*

Trapa bicornis is cultivated for food not only in China but in Japan and Korea as well. Other members of the genus *Trapa* are found in warm parts of the Old World, with one species, the European water chestnut or water caltrop (*T. natans*), eaten in Neolithic Europe and ancient Egypt, and another, the Singhara nut (*T. bispinosa*), cultivated in the Indian subcontinent and Nigeria (Watt, 1889-96, 6, Part 4: 73-75; Dalziel, 1937: 42; Uphof, 1959: 363; Darby *et al.*, 1977, 2: 681).[25] All cultivated *Trapa* species have fruit which is unusual in appearance. Ordinarily, *T. bicornis* and *T. bispinosa* fruit (Figure 63) have two opposite curved horns and resemble a bucranium or buffalo skull, one of its names in China being "buffalo-head fruit" (S. W. Williams 1895, 1: 364). The fruit of *T. natans*, for its part, has four horns, which in Europe led it to be called "water caltrop" after an ancient four-pointed metal weapon employed to defend infantry against cavalry attack by injuring the hooves of horses that step on them.

Returning to China, we note that ling nuts are well-liked as food and have been since early times. This is clearly shown by their inclusion among burial foods in one of the famous Han tombs near Ch'ang-sha (Y. S. Yü, 1977: 58), and among offerings to the gods in T'ang China (Schafer, 1977: 134). We have found few details of their cultivation in traditional China, except that in Chekiang Province, where they are widely cultivated in canals and slow-moving streams, they are planted in spring and protected by bamboo fences that are erected in the water

(Meyer, 1911: 55). Fortune (1852: 27) described a ling nut harvest along such a waterway in Chekiang which employed devices ideal for the purpose. These were round wooden tubs in which women and boys sat or kneeled. Propelling their vessels through the ling beds, they gathered and accumulated the fruit while doing little or no damage to the plants themselves.

In India, the Singhara nut, *T. bispinosa*, may be eaten raw or cooked; it may, after boiling, be prepared as porridge or ground into meal for making chapattis (Watt, 1889-96, 6, Part 4: 73-74). Raw *T. bicornis* kernels, on the other hand, should not be consumed raw, for in that state they are said to contain harmful substances, and to bring on worms, indigestion, and damage to the digestive system. Long cooking eliminates the danger (E. Watson, 1923: 338; F. P. Smith, 1969: 440; Herklots, 1972: 496; E. N. Anderson and M. L. Anderson, 1977: 330). Thus in China, the kernels of *Trapa bicornis* may be boiled or roasted as a snack, in which form they resemble a Jerusalem artichoke in taste, and are sold by vendors along the streets. They may also be fried, candied, or made into a flour that serves as a thickening agent or is used in making bread, cake, and sweets (S. W. Williams, 1895, 1: 364; Meyer, 1911: 55; E. Watson, 1923: 338; Porterfield, 1938d: 137-38; E. N. Anderson and M. L. Anderson, 1977: 330; Rosengarten, 1984: 337; E. N. Anderson, 1988: 129). Because the ling nut is associated with water, it is believed by the Chinese to be "cooling": to slake thirst, lower fever, and to be effective in treating sunstroke; the ling nut is also considered nutritious, and its flour is a food for invalids (E. Watson, 1923: 338; F. P. Smith, 1969: 440-41). In fact, if it is similar in food value to the Singhara nut (Blasdale, 1899: 41; Platt, 1962: 16-17; Leung *et al.*, 1973: 98), it is high in starch, a good source of carbohydrates, calcium, and phosphorus and, in every nutrient category considered by Leung *et al.* (1973), better than the *Eleocharis* water chestnut.

It is, however, a major culprit in transmission of the human intestinal fluke *Fasciolopsis buski*. Ova of the parasite enter the water where *Trapa bicornis* is cultivated, by way of night soil. The ova hatch in the pond; the free-swimming miricidia infect certain snails; the organisms pass through stages of growth in their snail hosts to emerge and encyst on water caltrop fruit; and then they are passed on to humans who eat the fresh fruit (Barlow, 1925: 57-59, 80-81; Robertson, 1936: 420). Only ling growing in contaminated waters are dangerous, and even growing there, only fresh fruit, for the cysts are killed when the fruit is allowed to dry or is boiled for 30 minutes or so.

SOUTHEAST ASIA'S MAIN CONTRIBUTION: THE COCONUT

Before European ships arrived on the scene, the coconut (*Cocos nucifera*; *yeh-tzŭ*) was widely dispersed on the tropical and subtropical shores of the Indian Ocean from East Africa to Southeast Asia, and on those of the Pacific as far eastward as Central America. Where the coconut originated, whether it should be regarded simply as a cultigen, and how it dispersed have been matters of controversy for over a century (de Candolle, 1886: 432-35). Some have looked to the New World as the place where the coconut originated. Guppy (1903-6, 2: 67, 413), for example, placed its origin along the Pacific coast of the Americas, with ocean currents transporting its floating nuts to Polynesia. From there the coconut was presumably carried farther westward by currents and man. O. F. Cook (1910: 304-7, 317-20, 340-41) had the coconut originating near salt deposits in an interior tropical location in northwestern South America, with some primitive American farming and maritime group carrying it westward across the Pacific in prehistoric times. Beccari (1917: 32-33, 35, 43), following certain older writers, argued that the coconut probably originated in the Asian-Pacific area. Though not rejecting the possibility that humans were agents in carrying it across the Pacific to tropical America, he insisted that it could readily have been disseminated by ocean currents. Most modern botanists (*i.e.*, Vavilov, 1949-50: 28, 30; Merrill, 1954: 194, 266-67; Uphof, 1959: 100; J. D. Sauer, 1971: 311; 1983: 218; Harlan, 1975: 75; Mathon, 1981: 70) agree with Beccari in placing the origin and domestication of the coconut palm somewhere in the Indo-Pacific area, including Southeast Asia.[26] Not only is that the region of its greatest genetic diversity and of the greatest number of coconut parasite species, but fossil remains have been found there, in Tertiary deposits in India and New Zealand and Quaternary ones in the Marianas (J. D. Sauer, 1971: 311; 1983: 218).

As for the question of human versus natural dispersal, Jonathan Sauer (1971; 1983), carefully weighing the evidence, presents what seems a quite balanced perspective of this question. *Cocos nucifera*, he observes, is best viewed as a semi-domesticated plant, with some populations completely dependent on man and others not at all. This makes it necessary to weigh the role of both natural and human factors in explaining its distribution. Natural dispersal would seem to have been readily accomplished, for the fruit is able to float at sea for several months and still remain viable. Such natural dispersal was likely responsible for much of its spread across the vast oceanic area in which it is found today. Even in pre-Columbian times, however, deliberate planting of coconuts was done in some areas of the Indian Ocean, Malaysia, and the western and central Pacific. As a result, it is reasonable to believe that they were carried by humans to certain islands of the Pacific and planted there, most notably on the Hawaiian Islands. In the absence of acceptable evidence of pre-Columbian contacts with Central America, however, Sauer thinks it more likely that the coconut arrived there by floating in the Equatorial Counter Current than by raft or canoe.

To the degree that humans were responsible for spreading the coconut across the Pacific, the Polynesians, who may have begun their eastward migration from Melanesia by 500 B.C., would deserve credit for carrying it farthest. In Melanesia they must have known the coconut in a wild form and perhaps as a cultivated plant, and in their initial migrations they may have carried coconuts with them for transplanting. To look farther back in time would be too speculative for most scholars to chance. One should, nevertheless, note that many linguists look to Indonesia and New Guinea as the area of origin of the Polynesian parental language, Proto-Austronesian, and that some have looked to the Asian mainland, including southeastern China. In any case, the place where Proto-Austronesian evolved was in or near the supposed center of coconut origin,[27] which placed those groups in a key position to carry it along, if they chose, in their earliest eastward movements. If they did so, their descendants the Polynesians may simply have been further aiding a dispersal previously carried out by ocean currents and their Proto-Austronesian ancestors.

Turning to China, one notes that references to the coconut go back to the second century B.C. (Schafer, 1967: 173), and that Chi Han in his fourth century flora provides a good description of the plant and its use. He writes of it as a tall tree with no branches but with fruit the size of watermelons (H. L. Li, 1979: 115). He mentions not only the rich and pleasant flesh of the fruit, but its milk, observing that it can make one drunk. One wonders whether in the latter comment he may be referring to a slightly-alcoholic product made by fermentation of the coconut milk, which is low in sugar; or a palm wine prepared of toddy (sap), which, when fresh, contains 10 to 12 percent sugar (Murai *et al.*, 1958: 66).[28] Tapping coconuts and other palms for toddy is found widely in Southeast Asia, India, and the Pacific;[29] in the case of the coconut, it is obtained from the inflorescence.[30]

China's neighbors the Philippines and Indonesia together account for half or more of the world's coconut acreage and production (Child, 1974: 30-31). Yet China is such a minor factor in the world scene that it is not even mentioned by Child (1974: 14-32) in his survey of areas of coconut cultivation and production. The coconut is found in China's extreme south, growing as far north in Kwangtung as latitude 21°N., which on the mainland would include only the southern part of the Lei-chou Peninsula and leave Hainan the principal area of production ("Fruit cultivation in Kuangtung," 1936: 716, 719; F. P. Smith, 1969: 121).

It is said that on Hainan the coconut tree has sixty uses,[31] being only a bit behind bamboo in that regard. Coconut logs are used in construction, of bridges for example, and coconut fiber is made into rope. Coconut shells serve as rough bowls or are carved into various useful ornamental items that sometimes are covered with beeswax and decorated with gilt letters. Coconut milk is consumed fresh or fermented, and sap is made into palm wine. Probably, as in Southeast Asia, its young buds ("cabbage") are eaten boiled, pickled, or fresh.[32] The flesh of the fruit is eaten fresh or dried, or is grated and boiled to obtain oil, or processed into copra. Coconut flesh may be "cut into strips and boiled in sugar syrup, then crystallized for use as sweetmeat," or coconuts themselves used as cooking and serving vessels as in the imperial court dish "Fillet of Chicken in Whole Coconuts with Button Mushrooms" (J. D. Keys, 1963: 4, 79).

Early in this century it was said that copra and carved shells were about the only coconut products to be exported from Hainan (Moninger, 1919: 14, 30, 39-40, 43, 44). That situation must have changed over time for in the mid-1930's over a million coconuts were sold each year in Hong Kong, Canton, and Amoy ("Fruit production in Kuangtung," 1936: 719).

In weighing the nutritional value of the coconut, one must focus on individual products. Coconut milk is mainly a safe, refreshing drink, without nutrients in significant quantities. In China it is considered "cooling" when fresh, "heating" when fermented (F. P. Smith, 1969: 121). Coconut flesh is better endowed with nutrients, with that of mature fruit high in fat (40 percent) and carbohydrate (15 percent), but low in protein (4 percent), minerals, and vitamins except for iron. Coconut toddy, on its part, is relatively high in ascorbic acid (Murai *et al.*, 1958: 50-66). Coconut milk and other products are also used medicinally in China (Soubeiran and Dabry de Thiersant, 1874: 130; F. P. Smith, 1969: 121-22), but not to the degree found in India and Southeast Asia.[33,34]

NUTS FROM CENTRAL ASIA AND THE NEAR EAST

In this section, we consider two cultivated nuts that seem to have come to China from Iranian lands to the west, one, the Persian walnut, of great importance in traditional China, and the other, the pistachio nut, unimportant.

Persian or English Walnut

Four species of the walnut genus, *Juglans*, are native to China (S. C. Lee, 1935: 233), including *Juglans regia*, the Persian walnut. *Juglans regia*, however, is also indigenous to a vast area stretching from China and the Himalayas to southeastern Europe, and its domestication is usually placed by botanists in the Near East or Central Asia (Harlan, 1975: 69; P. M. Smith, 1976: 310; Mathon, 1981: 31; Zeven and de Wet, 1982: 83; Zohary and Hopf, 1988:

163-64). By a Chinese tradition that is likely incorrect, the cultivated walnut (*ho-t'ao* or *hu-t'ao*) was introduced from the west by the noted envoy Chang Ch'ien in Han times. In any case, it seems to have come from that direction, for the earliest Chinese name for the cultivated walnut, *hu-t'ao*, means "peach of the Hu" (of Central Asian groups, especially Iranians). Probably it was not long after Han times that it was introduced, and by T'ang times it was grown in orchards in North China (Laufer, 1919: 256-65; F. P. Smith, 1969: 223; Schafer, 1977: 98; Y. S. Yü, 1977: 80).[35] Since then, the walnut has gained a significant position in Chinese agriculture and diet, and now ranks among China's most important nuts,[36] characteristic of the Wheat Region of North China where it may be cultivated in orchards or as single trees. It is also, however, grown elsewhere, being very common in Hupeh and Szechwan (Meyer, 1911: 51; Chun, 1921: 60-61; E. H. Wilson, 1929: 143, 169, 300, 309; Buck, 1956: 30).

Meyer (1911: 51) noted early in this century that the Chinese did not graft or bud their walnut trees.[37] This meant that all trees were grown from seed, which created great differences in the characteristics of walnut trees and their nuts. Some trees yielded small nuts with hard shells and poor taste, whereas others yielded large nuts of excellent flavor and with shells so thin that one could crack them with one's fingers, like a peanut.[38]

The Chinese eat walnuts in a variety of ways, some of which may seem unusual to Westerners. Walnuts may be eaten as such, raw, fried, roasted, sugared, glazed, or otherwise.[39] They may prepared in a broad range of dishes, such as Cantonese "Diced Chicken with Walnuts" (C. B. T. Lee and A. E. Lee, 1980: 180) and Peking "Stir-Fried Kidney with Walnuts" (W. W. Chang *et al.*, 1970: 331). They may be used in cakes and sweets, among them Cantonese "Walnut Cookies" (W. W. Chang *et al.*, 1970: 460-61) and "Walnut Soup" or "Walnut Tea," a sweet soup made with sugar and rice or rice flour, and sometimes with other ingredients. Such walnut soup is commonly served hot between courses at a banquet (G. B. Miller, 1966: 752-53; K. Lo, 1974: 168; 1979c: 325; G. Z. Chu, 1975: 228-29). There is also a ground walnut-sugar preparation in Szechwan that has been likened to the Central Asian halvahs made with nuts (E. N. Anderson and M. L. Anderson, 1977: 333, 354; E. N. Anderson, 1988: 167-68). And, in a parallel to European practice,[40] walnuts are sometimes ground or beaten into a paste which is then mixed with boiled water to make "walnut milk" for an infant whose mother has insufficient milk and cannot afford to hire a wet nurse (P. W. Wang, 1922: 120). Walnuts may be pressed for their oil, which has various uses, among them (like the nuts and other plant parts), medicinal ones (E. Watson, 1923: 142, 573; Hooper, 1929: 78; F. P. Smith, 1969: 223-24; Perry, 1980: 183).

Pistachio

Though one or two species of the genus *Pistacia* are found in China, *Pistacia vera*, the small tree that provides pistachio nuts (*wu-ming-tzŭ*), is indigenous to a region far to the west, and was domesticated in the Near East or Central Asia. We know the pistachio as a widely-grown nut tree in western Asia and the Mediterranean region, but it should not be overlooked that it is important, too, in Iran, Afghanistan, and Turkestan. From those lands, large amounts of pistachio nuts were exported to India in the last century where they were favorites of moneyed persons. Pistachio nuts had been exported to China from Iranian lands by T'ang times, with the Chinese believing them not only tasty but good for general well-being and sexual potency. By the mid-ninth century A.D., moreover, the tree itself was said to be cultivated in Lingnan, an unlikely part of China to plant a tree noted for its resistance to drought. Though the pistachio is still grown in the China and even is an object of interest today at a fruit breeding institute in Shensi Province, it is of quite minor importance and rarely mentioned in the literature. Thus one finds no reference to *Pistacia vera* in Bretschneider's *Botanicon sinicum* or in Meyer's survey of the fruit and nut orchards of North China, and just a single reference to it, for T'ang times, in *Food in Chinese culture* (Watt, 1889-96, 6, Part 1: 268, 273-75; Laufer, 1919: 246-53; S. C. Lee, 1935: 718-21; Schafer, 1963: 147; 1967: 192; 1977: 98; F. P. Smith, 1969: 334; American Plant Studies Delegation, 1975: 96). The walnut and pistachio are both high in protein, fat, and food energy, and contain good amounts of calcium, iron, and thiamine (Platt, 1962: 12-13; Leung *et al.*, 1973: 29).

THE PEANUT: A GIFT FROM AMERICA

The peanut (*Arachis hypogaea; hua-shêng*)[41] is a South American domesticate, its remains, in Peru, dating at least to the first millennium B.C.; its domestication, however, took place well to the southeast of Peru and likely occurred long before that date (Krapovickas, 1969: 427-29; Leppik, 1971: 188-91; Gregory and Gregory, 1976: 151-53). An overwhelming majority of scholars hold that the peanut reached China only after Columbus's discovery of America. Laufer (1907a) and Goodrich (1936-37) tried to determine its time of introduction more precisely through the use of Chinese sources, and concluded that the peanut was first introduced to Fukien from Malaysia or the Philippines. Laufer believed that date to be prior to the end of the sixteenth century, for the peanut was already grown in Chekiang Province during the period from 1573 to 1619. Subsequently Ho Ping-ti (P. T. Ho, 1955: 191-92) presented evidence from other Chinese works of the time showing that the peanut was already under cultivation south of the mouth of the Yangtze in Kiangsu and Chekiang early in the sixteenth century. Ho agreed that the peanut may have been introduced first to Fukien, and suggested that it was carried north to the Yangtze area by Fukienese cotton traders or by the Portuguese.

One cannot ignore the fact that some scholars have long held it possible that the peanut had reached China in pre-Columbian times, carried by seafarers across the Pacific Ocean from the New World.[42] The scholars in question start from a conviction that early humans, even in primitive watercraft, had the ability to voyage far across the seas, a perspective that makes plausible the idea of prehistoric plant exchanges between the Americas and Asia. For a long time, the principal evidence advanced for the idea that peanuts may have been in China in pre-Columbian times was the occurrence of a strange peanut with a hump-backed shape in southeastern China, one different from present-day American peanuts but strikingly similar to ones uncovered in prehistoric Peruvian graves (Ames, 1939: 47-48; Ed. Anderson, 1952: 167). That similarity was discounted by critics, among them Merrill (1954: 295-97) who observed that Brazil is a center of *Arachis* differentiation and that the Portuguese sea route to the Orient was by way of Brazil, the Cape of Good Hope, and Goa. Thus, in Merrill's view, when the Portuguese reached Canton in 1516 they were in a good position to introduce strange Brazilian peanuts to China, which one might also expect to find near way-stations in India and elsewhere along the Portuguese sea route to China. The case for pre-Columbian peanuts in China seemed to gain a firmer footing in the early 1960's with the surprising reports of remains identified as peanuts in two different archeological sites (Ch'ien-shan-yang and P'ao-ma-ling) in Kiangsi and Chekiang, in levels dated by the radiocarbon method to the third millennium B.C. or somewhat earlier (K. C. Chang, 1973: 526, 527; 1977a: 167, 180-81; 1977b: 27-28).

Serious doubts, however, have been raised about these discoveries: they go against quite convincing historical evidence of a post-Columbian introduction of the peanut; other plants (sesame and horse bean) found in the same levels are not known elsewhere in China until a later date; the remains may not have been of peanuts but of soybeans or some other legume; and the stratigraphy of Ch'ien-shan-yang is confused and not to be trusted (Harlan and de Wet, 1973: 54; K. C. Chang, 1977b: 27-28; 1986: 254n; H. L. Li, 1979: 126-27). Despite this, Needham and Lu (1985: 61) believe that these finds are difficult to dismiss. If, however, there had been peanuts in China at that early date, how could such an outstanding food plant have failed to spread more widely? Even if for some reason peanut cultivation had remained restricted to parts of Chekiang and Kiangsi, how could it have been overlooked in the Chinese literature until the sixteenth century A.D.? In this regard, one does note a suggestion by Wu Tê-lin in 1958 that the plant *ch'ien-sui-tzŭ* described by Chi Han in his fourth-century flora but not satisfactorily identified by botanists, was actually the peanut. This suggestion, however, is rejected by H. L. Li (1979: 124-25) who holds that the plant in question was a vine, whereas the peanut is not; that it had tendrils, which the peanut lacks; that its fruits were arranged in a manner different from those of the peanut; and that, in any case, the peanut could not have been in China in pre-Columbian times.

We leave the pre-Columbian question to return to firmer ground, the story of the peanut from the early sixteenth century onward as told by Ho (P. T. Ho, 1959: 183-89). Though it has been widely recognized that the peanut is one of America's greatest contributions to the Chinese food scene, it took centuries for peanut cultivation to spread widely there. In southeastern China, where it was introduced first, even as late as 1700 the peanut was a delicacy served at banquets, rather than a reasonably-priced everyday food. The peanut, excellent in its protein content and as a source of oil, did, however, have two special advantages: its ability to do well in sandy soils not suited to rice and other crops, and its nitrogen-fixing qualities. Because of the first advantage it was gradually adopted for use in sandy soils. Because of the second, it was incorporated into systems of crop rotation even in some areas of intensive rice cultivation, for farmers, though not aware of the role of its nitrogen-fixing root nodules, recognized the plant's value in maintaining soil fertility. Thus peanuts slowly spread until at the end of the nineteenth century they were common in cultivation and diet throughout China, south and north. In the 1929-33 village survey, they occupied 3.5 percent of all crop area in China, twelfth among all Chinese crops in acreage (Buck, 1956: 212-13). This was more than that for any other edible nut or seed other than oil seeds, the peanut's importance demonstrated by the fact that in a Peking household budget survey in 1926-27 more families bought it than any other edible nut (Gamble, 1933: 102). As noted previously, Ho included the peanut among the American food crops (together with maize, sweet potatoes, and white potatoes) that brought on a second agricultural revolution in China, and a striking increase in population over the centuries following their arrival and acceptance.

We turn now to the peanut's role as food in China, leaving the question of peanut oil until later. Over ten varieties of peanuts are found in China, but the Chinese usually classify them into two main types, small and large. The large peanuts arrived in China much later than the small ones, and were also called "foreign peanuts." In the early decades of this century, they were delicacies normally eaten roasted and salted, whereas the small ones were described as a food item consumed raw or in other ways (Meyer, 1911: 55; Skvortzow, 1920: 142-43; Read *et al.*, 1940: 26).

Roasted peanuts are regarded with special favor by the Chinese, who eat them in large quantities, and sometimes prepare them with spices as in "Spiced Roasted Peanuts" or "Fragrant Roast Peanuts," which are flavored with salt and five-spice powder (G. B. Miller, 1966: 745; W. W. Chang *et al.*, 1970: 406). In addition, the Chinese eat peanuts fried, steamed, boiled, as peanut butter, or, though not often, raw. Peanuts and peanut butter may serve in a broad range of food preparations: candy, cookies, cakes, and other sweets, including a sweet peanut soup. They may be included in other kinds of soups, too, among them the Cantonese "Pork Tail Peanut Soup" (W. W. Chang *et al.*, 1970: 113), or they may be made into sauces for dishes like the Peking "Poached Chicken Breasts with Spicy Peanut Sauce" (Kan, 1978: 96-97). They may serve in main dishes such as the Cantonese "Duck with Peanut Butter and Pineapple" (W. W. Chang *et al.*, 1970: 280) and "Pork, Bean Curd, and Peanuts with Oyster Sauce" (Sakamoto, 1977: 187-88). Peanuts are normally used, as well, in Szechwan "Kung Pao Chicken," which is commonly found on the menus of American Chinese restaurants.[43] Such extensive use of peanuts is very much to the nutritional benefit of the Chinese, for peanuts are an outstanding source of energy and protein, a good source of niacin and thiamine, and contain fair amounts of calcium, iron, and riboflavin (Leung *et al.*, 1973: 19). The thin skin surrounding the peanut seed is a far richer source of thiamine than the seed itself (Herklots, 1972: 221).

OTHER EDIBLE NUTS AND SEEDS CONSUMED BY THE CHINESE

From Trees and Shrubs

We already mentioned the jackfruit in this regard, and here we focus on edible nuts and seeds of trees and shrubs not previously considered. One seed is from *Firmiana simplex* (= *Firmiana platanifolia* or *Sterculia platanifolia*), a deciduous tree said to be native to China and Japan, and found in China from Hopeh south to Kwangtung, Kweichow, and Yunnan. This tree, known as *wu-t'ung* to the Chinese since antiquity and as the phoenix tree or Chinese parasol tree in English, is cultivated widely in China as a shade tree, often in temple grounds. Its fruit is a leaf-like follicle which bears pea-like seeds near its base. Seeds are collected and in Central China are sold on a small scale as a delicacy or used in making moon cakes (Bretschneider, 1882-95, 2: 350; Meyer, 1911: 56; E. Watson, 1923: 55-56, 577; S. C. Lee, 1935: 825-26; Rehder, 1940: 630; Bailey, 1949: 670). When dried, these seeds are high in protein, fat, potassium, and iron (Leung *et al.*, 1973: 27).

A second minor edible seed comes from *Sterculia nobilis*, an evergreen tree native to southern Asia. It was found in Annam in the fourth century A.D., and today grows in Hong Kong as well. Known as *p'in-p'o* in China and often as "pimpon" in English, it bears pods that at maturity are crimson in color and contain from one to five brown seeds with edible kernels. Ordinarily the seeds, after being shelled, are cooked or baked before they are consumed. When roasted they taste sweet, like chestnuts (H. L. Li, 1979: 128-29; Zee and Hui, 1981: 103).

Other edible seed comes from species of *Castanopsis* (from *Castanea* = "chestnut," and *opsis* = "like"), a genus of evergreen trees and shrubs native to southern and eastern Asia and western North America, some species of which are cultivated as ornamentals and may be called chinquapin (chinkapin). Various species of the genus, of which 32 are listed by S. C. Lee for China, produce edible nuts that are eaten by peasants (E. H. Wilson, 1929: 312; S. C. Lee, 1935: 296-318; I. H. Burkill, 1966, 1: 492-95 *passim*). One species, identified as *Castanopsis tibetana* by Meyer (1911: 54), was observed in small numbers in protected valleys in the mountains of Chekiang, where its chestnut-like nuts were relished by priests in temples. The paste made from nuts of another species, *C. sclerophylla* of eastern and central China, are a common dietary element among peasants (Chun, 1921: 87). According to H. L. Li (1979: 131-32), a South Vietnamese tree described by Chi Han in the fourth century was likely *Castanopsis indica*, the most common Vietnamese species and the one whose nuts are most tasty. This species is also found in Yunnan and in India, its nuts having a thinner shell than a filbert but a similar shape and taste (Watt, 1889-96, 2: 228; S. C. Lee, 1935: 311).

Two other trees sometimes cultivated for their nuts, *Torreya grandis* and *Carya cathayensis*, are rare species whose natural occurrence is limited to southeastern China, particularly Chekiang. They are relict forms likened to the ginkgo in their region of survival and in being cultivated there for their edible nuts (H. L. Li, 1963a: 105-6). *T. grandis* (= *T. nucifera*, var. *grandis*) is an evergreen tree related to the California nutmeg tree, *T. californica*, and is known to the Chinese as *fei*. Its delicious nuts (*fei-tzŭ* or *fei-shih*) were consumed in T'ang times and today are roasted and sold in Chinese markets. The tree has been under cultivation at least since Sung times, but only in northern Chekiang. That the tree has not become more widely cultivated in China is explained by its slow growth, which makes it unprofitable as a nut tree (Chun, 1921: 47; P. W. Wang, 1922: 121; S. C. Lee, 1935: 13-17; Schafer, 1977: 98; H. L. Li, 1983: 47). *Carya cathayensis* is a deciduous tree said to be the only hickory species indigenous to China, and even there it is found only in Chekiang and Kweichow. Its edible nuts (*shan-hêh-t'ao*) are found for sale in Chinese markets for use as sweetmeats; they are reported to be roasted with syrup; and oil is extracted from them for use in special pastries (Chun, 1921: 62; P. W. Wang, 1922: 121; S. C. Lee, 1935: 237-38).

Another seed-providing plant, closely related to the horse chestnuts, is *Xanthoceras sorbifolia*, a deciduous shrub or small tree indigenous only to North China (Hopeh, Shansi, Kansu) which does well under conditions of summer heat and drought. Its wood is toxic, capable of killing fish when soaked in a pond, but its fruit seems non-toxic. The fruit may resemble a small angular apple or pear in size and shape. When ripe, it splits into three sections, to give it the appearance of a cap and the tree's Chinese name *wên-kuan-kuo*, "cap fruit." *Xanthoceras sorbifolia*, which has attractive, colorful flowers, is often cultivated in China as an ornamental. It is also cultivated for the edible purple-brown seeds, each almost half an inch across, released from its ripe fruit. Though the plant is now grown in many botanical gardens in Europe and America, interest there is in its ornamental qualities not its food potential (Anonymous, 1905: 128-30; Bailey, 1935, 3: 3521; S. C. Lee, 1935: 792-93; Rehder, 1940: 592-93; Bailey, 1949: 641; H. L. Li, 1959: 155-56; F. P. Smith, 1969: 461).

Like peoples of the American Southwest, Mediterranean region, and elsewhere (Hehn, 1885: 223-27; Rosengarten, 1984: 309-15), the Chinese consumed the edible seeds (*sung-tzŭ*) of pines (*Pinus* spp.; *sung*) indigenous to their land. To appreciate the great antiquity of their use as food in China, one need note only that pine seeds have been uncovered in Yang-shao levels dating from the fourth or fifth millennia B.C. (P. T. Ho, 1975: 87; K. C. Chang, 1986: 112). Chi Han in his fourth century flora mentions excellent pine seeds produced in South Vietnam, delicious, fine smelling, and excellent on the banquet table; H. L. Li (1979: 129-30) identifies these as coming from *Pinus insularis*, a species found today in a belt stretching from the Philippines through Indochina to Assam. Li notes, however, that most edible pine seed was imported from Korea and came from the Korean pine, *P. koraiensis*, a species also indigenous to Manchuria, eastern Siberia, and northern Japan. In traditional China, pine seeds sold as delicacies in markets over the entire country came mostly from that species (whose seed are now exported by the People's Republic). Among the other species mentioned as providing edible seeds are *P. bungeana*, the lacebark pine, of Shansi, which provides a small round seed for sale; *P. armandi* of the southwest, whose seeds are consumed by local folk and sold in markets; and *P. massoniana*, a common pine of Kwangtung, Fukien, and Taiwan, but found elsewhere as well. In Central and South China, sometimes one could also obtain quite long seeds, presumably from still other species. *P. bungeana* is seldom seen in the wild, but is grown in courtyards and around temples, as in Peking where one famous temple tree is called the "Pine of the Seven Dragons" (Meyer, 1911: 53; Chun, 1921: 11-13; P. W. Wang, 1922: 121; E. H. Wilson, 1929: 312). Pine seeds were not only eaten, sometimes sugared, as snacks in traditional China, but were made into cakes and pastries, such as "Date and Pine Nut Rolls" (G. Z. Chu, 1975: 224), by those who could afford to do so. One also reads of chickens and ducks stuffed with pine needles and roasted in open air ovens at the Ch'ing palace early in this century, the needles giving the fowl an excellent taste (Der Ling, 1914: 131).

Nuts or seeds of various other shrubs and trees have also been consumed by the Chinese. For T'ang times, one reads that seed of the rice-paper plant, *Aralia papyrifera* (= *Tetrapanax papyrifera* = *Fatsia papyrifera*; *t'ung-ts'ao*), was a delicacy; that the evergreen oak *Quercus myrsinaefolia* provided edible acorns in Central China;[44] and that the seed of the Oriental arbor-vitae, *Thuja orientalis* (*po*), which came from Honan and Shansi, was well regarded as food (Schafer, 1977: 97-98). In modern China, acorns of certain oak species continue to be eaten as food or medicine (F. P. Smith, 1969: 366-68; E. N. Anderson, 1988: 137), and *T. orientalis* is often cultivated in gardens and cemeteries, its fragrant branches used at the time of New Year as symbols of longevity and happiness (Chun, 1921: 36). A wild walnut or butternut of Szechwan and Hupeh, *Juglans cathayensis* (*yeh-hêh-t'ao*), also yields nuts that are eaten, and so do wild beechnuts (*Fagus*)[45] and chestnuts of more than one species (E. H. Wilson, 1929: 309, 311-12; S. C. Lee, 1935: 233; Uphof, 1959: 200).[46]

The nut of the cashew (*Anacardium occidentale*) is prominent in American Chinese dishes and is known and used in sweets and main dishes in China. The cashew is an evergreen tree native to tropical America which the Portuguese introduced in the sixteenth century from northern Brazil to India, which in time became a leading exporter of cashew nuts. It is also found today in Southeast Asia among such neighbors of China as the Philippines, Indochina, Indonesia, Malaysia, and Burma. Temperature is the principal factor limiting the distribution of the cashew, with the tree injured or killed by freezing temperatures. This means that in India commercial production is largely restricted to the south where it does best at elevations below 1500 feet. In Indochina, Crevost and Lemarié mention it only for Cambodia, Cochin China, and Lower Annam, which suggests that it may not be suited to Tonkin. If it would do well anywhere in China, Hainan would be the place, yet we have found nothing to suggest that it is cultivated there on any scale (de Candolle, 1886: 198-200; Watt, 1889-96, 1: 232-33; Crevost and Lemarié, 1917: 237; Wester, 1924: 54; Ochse and Bakhuizen van den Brink, 1931b, 1: 33-35; I. H. Burkill, 1966, 1: 144-47; Hayes, 1966: 382-83; D. Johnson, 1973).

From Other Cultivated Plants

In China, seeds of various cultivated plants of the gourd family (*Cucurbitaceae*) are more prominent as snacks than virtually all of the nuts and nut-like fruit considered above. Prominent among these, moreover, are plants of African, Near Eastern, or American origin. One is the watermelon (*Citrullus lanatus* = *C. vulgaris*; *hsi-kua*), an African domesticate which, as we have seen, reached North China only in the twelfth century A.D. Seeds of the watermelon in China vary in color from black, brown, and red, to white. Today, they may be dried, deep-fried, parched or roasted, and are usually salted or preserved in sugar. Watermelon seeds are very popular with the Chinese. Roasted watermelon seeds are eaten as Americans do peanuts, with Wilson writing of the huge amounts of watermelon seeds consumed in China, and observing that without them a feast is incomplete, and that both scholars and coolies enjoy these tasty seeds over conversation in tea houses and restaurants. Meyer, in turn, reported that one could not enter a theater or tea house in China without being offered watermelon seeds. Watermelon seed is a significant commercial item, and it is no surprise to hear that China has forms of watermelon that have been selected, apparently over an extended period of time, for their many seeds and scanty flesh. Early in this century, the areas singled out as producing such seed melons were southern Manchuria, Chihli, Shantung, and the Yangtze Valley, but Kwangtung, with a climate like that of the American South, is also a big producer and even an exporter of watermelon seeds. Another member of the gourd family providing edible seed is the melon, *Cucumis melo*, that African or Near Eastern domesticate important in North China long before the watermelon arrived. Among the other cucurbitas providing seed are the pumpkins or squashes *Cucurbita moschata*, *Cucurbita maxima*, and *Cucurbita pepo*, which arrived from America after its discovery. Seeds of the bottle gourd, *Lagenaria vulgaris* (*hu-lu*), another common cucurbit in ancient and modern China, are also reported to be boiled in salt water and eaten in rural areas. The bottle gourd was cultivated in various parts of the Old and New Worlds in great antiquity; though its place of earliest domestication is unknown, Africa is favored by some. A non-cucurbitaceous plant of New World origin, the sunflower, *Helianthus annuus* (*hsiang-jih-k'uei* or *chao-jih-k'uei*), also provides an edible seed, which, consumed raw or roasted, is another favorite in China (Meyer, 1911: 55-56; Couling, 1917: 347; E. Watson, 1923: 457; E. H. Wilson, 1929: 339; Read, 1940: 19-20; I. H. Burkill, 1966, 1: 567; H. L. Li, 1969: 254-55; Herklots, 1972: 274, 289; E. N. Anderson and M. L. Anderson, 1977: 333; V. Y. N. Hsu and F. L. K. Hsu, 1977: 299; Zee and Hui, 1981: 40; E. N. Anderson, 1988: 132). The analyses we have found for watermelon, pumpkin, and sunflower seed (Platt, 1962: 12-13; Leung *et al.*, 1973: 28-29) show them to contain good amounts of protein, fat, food energy, iron, and calcium, and some of them are, as well, fair to good sources of certain B vitamins.

Lotus seed we have mentioned in our chapter on "Roots and Tubers." Another aquatic plant that provides edible seed is *Euryale ferox* (*ch'ien-shih*), a prickly member of the water lily family found in lakes, ponds, paddies, and ditches in southern and eastern Asia at least from India to China. It is also cultivated in those two countries and in Japan. In China, there are records of such cultivation since antiquity, and Zeven and de Wet (1982: 37) place its domestication in the Chinese/Japanese center. *Euryale ferox* bears a fruit covered with soft prickles and containing many starchy, black-coated seeds the size of peas. The seeds (foxnuts), which may be eaten fresh, fried, or roasted, are consumed medicinally[47] or for food. Because of the shape of the seed pod, the Chinese have given them the popular name *chi-t'ou*, "chicken's head" or "cock's head." In Peking early in this century, when hawkers would go down the street shouting "Old chicken heads! Just out of the river!" (Tun, 1965: 63), they were referring not to heads of fowl but to seed of *Euryale ferox*. Such seed is exported from China to overseas Chinese communities, including that of San Francisco, where, according to Blasdale, they were mainly used to add to soup, as other Americans use barley. The stalks and rhizomes of the plant, when young, are also edible, and they, too, may serve as food or medicine (Bretschneider,

1882-95, 2: 218; 3: 440-41; Watt, 1889-96, 3: 303; Blasdale, 1899: 40; Dutt, 1922: 111, 309; Hooper, 1929: 63; Bailey, 1935, 1: 1175; Uphof, 1959: 152; I. H. Burkill, 1966, 1: 999; F. P. Smith, 1969: 169-70; J. D. Keys, 1976: 141-42; Perry, 1980: 288; Regmi, 1982: 124-25; Zeven and de Wet, 1982: 37). The seeds of *Euryale ferox* are like cereals in their abundant carbohydrate (77 gm./100 gm. edible portion) and modest amounts of protein (8 gm./100 gm.) (Blasdale, 1899: 70-71; Read *et al.*, 1940: 25; Leung *et al.*, 1973: 25).

OIL SEEDS

We have already noted the importance of lard as a cooking fat in China, but in addition China had a rich variety of edible vegetable oils. Such oils have apparently been used on a considerable scale since Han times, when the wok and stir-frying seem first to have been developed. In China in the past, however, cooking oil was costly, and even in modern times stir-frying is less common than boiling and steaming, more used in restaurants than at home (E. N. Anderson, 1988: 43-44, 112-13, 151, 153).

In traditional China, the major vegetable oils consumed were soybean (most common cooking oil in North China), rapeseed (mostly forms of *Brassica campestris* or *B. juncea*),[48] peanut, cottonseed (*Gossypium herbaceum*), and sesame (*Sesamum indicum*). The plants producing these oils represent a nice balance in terms of places of domestication: some (soybean and certain of the Brassicas) from China, some (cotton, sesame,[49] some Brassicas) from elsewhere in the Old World, and one (peanut) from the New World. Of China's five major oilseeds, three were produced in large amounts in Kwangtung and Fukien: peanut, soybean, and rapeseed. Sesame and cottonseed, on the other hand, were quite minor in those southeastern provinces, their main production being in North and Central China (especially, in the 1930's, Honan, Hopeh, Kiangsu, and Hupeh). Also mentioned as providing cooking or salad oil in China are seeds of other plants, some significant, others quite minor. Among them are sunflower (*Helianthus annuus*), flax (*Linum usitatissimum*),[50] castor bean (*Ricinus communis*), hemp (*Cannabis sativa*), amaranth, opium poppy (*Papaver somniferum*), the herb *Perilla ocymoides* (= *P. frutescens*), the tea-oil shrub (*Camellia oleifera*), and the tea plant itself (*Camellia sinensis*) (A. Henry, 1893: 64; Laufer, 1919: 295; E. Watson, 1923: 46 and *passim*; Read, 1929: 433; E. H. Wilson, 1929: 343-44; T. H. Shen, 1951: 242-50, 260-71; Sealy, 1958: 1, 204-9; Y. H. Sung, 1966: 24, 215-16; F. P. Smith, 1969: 379; E. N. Anderson and M. L. Anderson, 1977: 327, 328, 343; V. Y. N. Hsu and F. L. K. Hsu, 1977: 303; Mote, 1977: 200; Spence, 1977: 270; Bray, 1984: 518-26; Wittwer *et al.*, 1987: 189, 241-50; Anderson, 1988: 67, 82).[51,52,53] They, too, represent an interesting collection in terms of their varied places of origin.

Production of sunflower seed in China has grown rapidly in recent decades, from 33,000 metric tons in 1949 to 1.7 million metric tons in 1985. This has made sunflower a major oil seed in China, and gained for the nation the rank of fourth largest sunflower producer in the world, behind only the Soviet Union, Argentina, and the United States. Because the sunflower plant is able to tolerate cold and grow in poor soils, production is concentrated in northern China, especially Manchuria (Wittwer *et al.*, 1987: 242, 244, 246-48).

Though we have not determined that its oil was consumed as food in traditional China, we feel obliged to say something about safflower (*Carthamus tinctorius*) here because it has long been cultivated in China and is now one of the world's important oil plants. Safflower is an annual plant of the thistle family, drought-resistant, spiny, and with yellow or orange-red flowers. Closely related wild plants of its genus are found from Palestine, Lebanon, and central Turkey to Central Asia and Pakistan. Present thinking favors the Near East as its primary center of origin, specifically the area of Palestine, Lebanon, Syria, Turkey, and Iraq.[54] The first evidence of cultivated safflower is for Egypt, and dates to about 1600 B.C. (Hanelt, 1961: 136; Knowles, 1976: 31-32; Mathon, 1981: 27; Zeven and de Wet, 1982: 82, 88). Safflower was also cultivated in Iran, Afghanistan, Central Asia, and India in ancient times (Watt, 1908: 277; J. C. Ray, 1933: 231; 1934: 17; Hanelt, 1961: 136-37; Prakash, 1961: 71, 93, 164; I. H. Burkill, 1966, 1: 471), and reached China about the third or fourth century A.D. (Laufer, 1919: 324; H. L. Li, 1959: 196).

In antiquity, the flowers of safflower, not the seed, were of primary importance, source of a red dye, and a coloring used in food and cosmetics (and also as medicine). Its similarities in use to saffron led to safflower being called such names as "dyer's saffron," "fake saffron," and "bastard saffron." Carthamin, the red safflower dye, had already been used on the grave-cloths of mummies in ancient Egypt (de Candolle, 1886: 164; Moldenke and Moldenke, 1952: 87), and over time it gained such importance that, with indigo, it became one of the two leading plant dyes in the world. With development of synthetic aniline dyes in the nineteenth century, however, carthamin, along with other vegetable dyes, declined in importance, and today it is not the flowers of *Carthamus tinctorius* but its seed, a source of oil, that is of paramount importance (Arnon, 1972: 380; Knowles, 1976: 31).

Turning to China, one notes there the traditional use of safflower (*hung-hua* or "red flower") as a source of dye. It is as a dyestuff that it is included among items sent as tribute to the T'ang court from Shantung, Shensi, and Szechwan (Schafer and Wallacker, 1957-58: 220).[55] A seventeenth-century account of Chinese technology by Sung Ying-hsing (1966: 76-77, 78) indicates that it continued in use as a dyestuff then, but Couling (1917: 493), early in this century, noted that though in former times safflower was widely cultivated in Szechwan for its dye, by his day it had been largely displaced by artificial dyes. Use of its flowers, as sources of dye and medicine, does, however, continue in China (Hooper, 1929: 29-30; I. H. Burkill, 1966, 1: 472; F. P. Smith, 1969: 94-95; J. D. Keys, 1976: 226; Perry, 1980: 89), their red color—following the "doctrine of signatures"—making them particularly useful in treating problems of the blood. Safflower also provides a seed from which oil is derived, and in China it was used mainly for lighting, lubricating, and making candles (E. Watson, 1923: 131). On the other hand, safflower oil is not mentioned among edible oils in our sources, the normally thorough F. P. Smith (1969: 95) writing only of its use as a lubricant and in candle-making.

A FEW CONCLUSIONS

Contrary to what we found in China's fruit scene, southeastern China, including Kwangtung, has made few if any contributions in cultivated nuts and nut-like fruit, with virtually all major Chinese domesticates (chestnut, Chinese almond, hazelnuts, ginkgo nut) seeming to derive from other parts of China. In addition, we are struck by the limited degree that China as a whole drew on Southeast Asia, India, and lands to the west for nuts and nut-like fruits. From Southeast Asia, only the coconut gained importance, and then only in China's far south. From India, there were no major contributions. And from the Near East and other lands to the west, only the walnut gained prominence, with the almond used little if at all. This may reflect the Chinese lack of interest in nut cultivation noted by Meyer (1911: 51). If so, it would account for the fact that certain of the other species yielding nuts and nut-like fruits seem to have been cared for as much (or more) for their ornamental qualities as for their food potential. We may be dealing here with an opportunity lost, something not typical of the people of China.

[1] The other two Chinese species are *C. seguinii* and *C. henryi*, both of which yield small, tasty nuts.

[2] *C. mollissima* was introduced to the United States in 1853 and again in the first decade of this century, then widely planted on an experimental basis, in large part because of its relative immunity to the Asiatic blight, which nearly exterminated the American chestnut, *C. dentata*. At present, *C. mollissima*, which is also capable of withstanding considerable heat and drought, is said to be the "most commonly planted orchard species...in the United States" (McKay and Crane, 1953: 228; Rosengarten, 1984: 57).

[3] In Peking in 1926-27, 127 of 283 families included in a household budget survey purchased chestnuts, a number exceeded only by the 271 families who purchased peanuts and the 151 who purchased melon seeds (Gamble, 1933: 102).

[4] The sixth century *Ch'i-min yao-shu* described chestnut storage in covered containers in "fine, sun-dried sand," by which means they might keep for months (S. H. Shih, 1962: 60).

[5] For a color photo of one version of Peking Dust ready for serving, see Em. Hahn *et al.*, 1973: 174.

[6] In a whole, raw form Chinese chestnuts are high in vitamin C (36 mg./100 gm., edible portion), but not when roasted (5 mg./100 gm.) (Leung *et al.*, 1973: 24).

[7] For medicinal uses of chestnuts in China, see F. P. Smith, 1969: 97-98, and Perry, 1980: 153-54.

[8] Among them are Bretschneider (in de Candolle, 1886: 219), Meyer (1911: 53-54), E. H. Wilson (1913, 2: 27; 1929: 305), Couling (1917: 12, 31-32), Porterfield (1951: 30), and F. P. Smith (1969: 40).

[9] According to H. L. Li (1959: 157), the true almond, *Prunus amygdalus*, was introduced from Persia in the medieval period but remains somewhat rare in China today and is cultivated solely in its northwestern areas.

[10] Certain other species of *Prunus* in China provide edible seed kernels as well, for example *P. japonica* (F. P. Smith, 1969: 355). In Szechwan in 1910, E. H. Wilson (1929: 305-6) discovered one such species, *P. dehiscens*, whose kernels were eaten and highly valued in the local area. He introduced the species to cultivation as having potential as a new form of almond.

[11] Almond flour, whether sweetened or unsweetened, is available in Chinese food stores in the United States. The sweetened kind is made into desserts of the type mentioned in our text; the unsweetened one, an alternative to cornstarch or flour, is used as a coating in preparing foods (Hom and Steiman, 1981: 27).

[12] This was not an isolated case. E. N. Anderson (1988: 137), for example, writes of a present-day remedy for sore throats and colds made of ground apricot kernels and congee or milk. Such a preparation is also mentioned by F. P. Smith (1969: 354), who writes further of a medicinal apricot-kernel drink, *hsing-jên-t'ang*, sold along the streets of some Chinese communities like sassafras tea was in Europe. *Hsing-jên-t'ang*, made by crushing apricot kernels in boiling water and then adding drugs and flavorings, is used in treating asthma, coughs, and discharges from the nose and eyes that may accompany a cold. For other medicinal uses of apricot kernels and other apricot products in China, see F. P. Smith, 1969: 353-54, and Perry, 1980: 344.

[13] For recipes to main dishes with almonds, see Sakamoto, 1977: 84-85, 112, 185-86, 196.

[14] For a map of the natural occurrence of the genus *Corylus*, see Krüssmann, 1984-86, 1: 377, and of *C. avellana*, see Mathon, 1981: 28.

[15] S. C. Lee (1935: 261-67) identifies six species of *Corylus* indigenous to China.

[16] Ikeda (1909: 48) observed that the ginkgo is found in warmer sections of Japan as far north as 36°N. He said nothing of the role of Buddhist priests in saving the ginkgo from extinction in Japan, but did observe that in Japan the ginkgo is a sacred tree grown mainly in the grounds of temples and shrines and not cultivated in gardens or orchards.

[17] For information on tree cults in China, see Eberhard, 1935: 248-52; 1968: 174-75.

[18] Zee and Hui (1981: 23, 107) give the Cantonese names for ginkgo nut as *pai-go* ("white nut"), *gin-go* ("silvery nut"), and *gin-hun* ("silvery plum").

[19] There has long been a dispute about the origin of the name "ginkgo." Some have suggested that the tree may have spread to Japan under the name *yin-kuo*, "silver fruit," which, through Japanese, gave rise to our name "ginkgo" (Porterfield, 1940a: 186; Dallimore and Jackson, 1966: 231; F. P. Smith, 1969: 390). Others suggest that the term comes from an error in transcription by Kaempfer (Moule, 1937: 200-201). H. L. Li (1963a: 106-11), on his part, denies that the fruit was ever known as *yin-kuo* in China or Japan, and suggests that the tree spread to Japan under the Chinese ideograph *yin-hsing*, and came, in colloquial Japanese pronunciation, to be *ginnan* or *ginkgo*, with the latter the source of the term used in the West.

[20] The Japanese prepare ginkgo nuts in a similar manner, first burying the fruit to hasten the decay of the covering, then rinsing and sun-drying them. They use the dried nuts primarily in cooking (Ikeda, 1909: 49).

[21] For the way to peel ginkgo nuts before use, see Hom and Steiman, 1981: 97.

[22] For a non-vegetarian dish with ginkgo nuts, "Chicken Cubes Stir-Fried with Ginkgo Nuts," see W. W. Chang *et al.*, 1970: 238.

[23] In Japan, there was also a considerable demand for ginkgo nuts, one second only to the demand for chestnuts (Ikeda, 1909: 49).

[24] Kao and Chen (1944) reported on seven cases of ginkgo-nut poisoning among infants and young children in Kweichow, two of which resulted in death despite "energetic treatment." Most had consumed between thirty and fifty ginkgo nuts, and all experienced a sudden onset of symptoms, usually from one to five hours afterwards. All had fever and convulsions; five of the seven were restless; four had abdominal distension; three experienced breathing difficulties, delayed light reflexes, and unconsciousness; and two had cyanosis, inequality of the pupils, stiffness of the extremities, and vomiting.

Of the two children who died, one, the youngest of the seven, was seven months old and had eaten seven nuts, whereas the other, the oldest, was four years old, and had eaten 150. According to Chinese medical books, even adults may be poisoned after consuming a thousand nuts. That people in Kweichow were aware of their special danger for children is indicated by the fact that they do not normally permit children under age ten to consume ginkgo nuts. As a result, only one of the seven afflicted children was actually native to Kweichow.

[25] *Trapa natans* has been introduced to parts of the eastern United States and by the 1930's it was established as a weed in various river systems, among them the Hudson and lower Potomac (Muenscher, 1944: 276-77), and in some places its growth was so dense as to block the water course (Porterfield, 1951: 30).

[26] For a review of the evidence on the origin of the coconut palm, see Child, 1974: 4-8.

[27] See map in Mathon, 1981: 70.

[28] That one or another of these products have long been used in what today are Kwangtung, Hainan, Kwangsi, and northern Vietnam is indicated by statements in Soubeiran and Dabry de Thiersant, 1874: 130; Schafer, 1967: 174; 1970: 44-45, 63; 1977: 121; and F. P. Smith, 1969: 121-22.

[29] For information on manufacture of palm wine from the coconut in India, see Watt, 1889-96, 2: 448-51; and in Malaysia, I. H. Burkill, 1966, 1: 618-20.

[30] Chi Han writes that a popular name for the coconut was *Yüeh-wang-t'ou*, "head of the king of Yüeh," a designation explained in terms of a feud a king of Yüeh supposedly had with a king of Champa (southern Indochina). According to this tradition, an assassin was sent to kill the king of Yüeh and hang his head on a tree, and when he did so, it immediately turned into a coconut. The king of Champa, angered, had the coconut cut open to use as a cup, as still done by southerners in Chi Han's day. The Li of modern Hainan have a story vaguely resembling this one, but it entails decapitating the heads of prisoners and planting them, with coconut trees then springing up and bearing head-shaped fruits that yield a tasty juice which ferments into a delectable wine (Schafer, 1970: 45).

[31] That this is no exaggeration is suggested by the list of useful coconut products prepared for an exhibition in India in the last century. They numbered 83 altogether and ranged from sacks, beads, toy parrots, violins, tooth brushes, and floor mats, to soap, sugar, vinegar, and distilled liquor (Watt, 1889-96, 2: 456-59).

[32] For the coconut's uses as food and otherwise in Southeast Asia, see I. H. Burkill, 1966, 1: 611-24; and in the Marshall and Caroline Islands, Murai *et al.*, 1958: 36-66.

[33] Perry (1980: 304), for example, identifies a broad range of medicinal uses of the coconut in Southeast Asia, but few for China. And Hooper (1929), in his excellent survey of drugs in Chinese pharmacies in Malaya, does not even mention the coconut.

[34] We will not consider here another of Southeast Asia's "nuts," seed of the betel-nut palm (*Areca catechu*), even though it is grown and used in southern China, and has been since antiquity. This is because the betel nut (*pin-lang*, apparently from *pinang*, one of its Malay names) (I. H. Burkill, 1966, 1: 225; S. Y. Hu, 1980: 89; E. N. Anderson, 1988: 138) is not really a food, but is chewed for other purposes, as intoxicant, stimulant, or medicinal (for example, as an anthelmintic). Betel nuts, like tea leaves, have a considerable anti-thiamine activity.

[35] "Walnut remains," not identified as to species, have been found in certain early Chinese agricultural sites dating from the fifth, sixth, or even seventh millennium B.C. (T. K. Cheng, 1982: 11; K. C. Chang, 1986: 92). The nuts may have been collected from wild trees, but, if not, the story of the walnut in China would require retelling.

[36] At the end of 1978, China was estimated to have 150 million walnut trees and a production of 118,000 tons of walnuts, some of which were exported.

[37] Grafting is now being encouraged in the People's Republic of China.

[38] In present-day Kwangsi there is a strain of walnuts called "Paper-like Walnut," whose nuts have shells no thicker than coarse paper.

[39] For the preparation of fried, sugared, or glazed walnuts, see G. B. Miller, 1966: 744; W. W. Chang *et al.*, 1970: 405; Stallard, 1973b: 185; and E. Chen, 1977: 154.

[40] Until the end of the eighteenth century, walnut milk was a common alternative to cow's milk in European homes (Rosengarten, 1984: 243).

[41] *Hua-shêng* is a contraction of *lo-hua-shêng*, meaning "propagation from the flower" or "dropping flower gives birth," which derives from the way, after the flower of a peanut plant is fertilized, that the flower stalk grows downward to the earth, where the pod develops (E. Watson, 1923: 122; Bray, 1984: 517; E. N. Anderson, 1988: 125).

[42] In pre-Linnean times, it had been widely believed that the peanut had originated in Africa, and reached the Americas as food on slave ships. Early in the nineteenth century, Robert Brown even suggested China as the peanut's place of origin, from which it was introduced to India and Africa and finally to the Americas (Ames, 1939: 44-45).

[43] For a recipe to Kung Pao Chicken, see Y. K. Martin, 1984: 117.

[44] China is rich in oaks, with fifty-four species identified by S. C. Lee (1935). Certain of them have edible acorns which people may consume as food, as mentioned briefly by Bretschneider, 1882-95, 2: 370-74 *passim*; A. Henry, 1893: 53; and F. P. Smith, 1969: 366-67.

[45] On occasion, an oil good for cooking and lighting is obtained from nuts of the wild beechnut *Fagus sinensis* (Chun, 1921: 82).

[46] The common pagoda tree (*Sophora japonica*; *huai-mi* or *huai-hua-mi*) is a common ornamental in China and Japan, and, as its name suggests, it is often grown around temples or sacred buildings. Read *et al.* (1940: 25-26, 58-59) list this tree as providing edible seed, and they and Leung *et al.* (1973: 27) provide analyses of its food value. None of our other sources, however, indicate that the seed of *Sophora japonica* is consumed as food except in time of famine, and some (I. H. Burkill, 1966, 2: 2091; Perry, 1980: 226-27) note that it is toxic and induces vomiting. The tree does provide useful plant parts (seed, pods, flowers, leaves, root, wood, gum, and bark) which serve especially as medicinals and/or sources of dye (E. Watson, 1923: 192-93, 456; Hooper, 1929: 132; E. H. Wilson, 1929: 300, 325, 372; Uphof, 1959: 341; I. H. Burkill, 1966, 2: 2091; F. P. Smith, 1969: 415-16; Perry, 1980: 226-27).

[47] For medicinal uses of *Euryale ferox* in East and Southeast Asia, see Hooper, 1929: 63; F. P. Smith, 1969: 169-70; J. D. Keys, 1976: 141; and Perry, 1980: 288. For such uses in India, where *E. ferox* is known as the "Gorgon fruit," see Watt, 1889-96, 3: 303; Dutt, 1922: 111; and Khory and Katrak, 1981: 37.

[48] Hommel (1937: 85-91) provides details, including illustrations of equipment used, of the methods of processing rapeseed oil in Chekiang and Kiangsi.

[49] There has long been controversy about the place where sesame was first domesticated, with Africa and India the leading candidates. For recent reviews of the evidence, see Bedigian and Harlan (1986) and Zohary and Hopf (1988: 126-27), who favor India as its place of origin. There is confusion as to whether sesame, flax, or some other plant is being referred to in certain early Chinese accounts, and the date of its arrival in China is uncertain. Nevertheless, most scholars (Laufer, 1919: 288-93; Schafer, 1977: 98; Y. S. Yü, 1977: 80; C. Y. Hsu, 1980: 82-83, 89; S. H. Shih, 1982: 49; Bray, 1984: 425, 525-26) believe that it arrived in Han times or somewhat afterwards. In China, as elsewhere, sesame seed is not only a source of cooking oil, but is valued for its agreeable odor (the Chinese call the oil *hsiang-yu*, or "fragrant oil") (E. Watson, 1923: 133) and taste. Considered "the king of oilseeds" by the Chinese (Bray, 1984: 525), sesame oil may be added to dishes for its fine taste, and sesame seed has long been toasted and eaten as such, or used on or in various foods such as breads, cookies, cakes, and other sweets, or even employed as hair dressing (E. Watson, 1923: 134-35; Hsu and Hsu, 1977: 199; Schafer, 1977: 98, 118; Yü, 1977: 81; Bray, 1984: 525; E. N. Anderson, 1988: 156, 168, 182).

[50] The Chinese, like the Iranians and Indians, did not make linen from flax (A. Henry, 1893: 64; Laufer, 1919: 293-96; Bray, 1984: 526).

[51] *Camellia oleifera* (considered by some to be synonymous with *C. sasanqua*) is widely cultivated in China for its seeds, from which "tea oil" is obtained. It has larger seeds than the tea plant, *Camellia sinensis*, and, by weight, its seed commonly contains more oil ("Tea oil," 1925: 13; Sealy, 1958: 1, 204, 207-8; H. L. Li, 1959: 82; I. H. Burkill, 1966, 1: 424; F. P. Smith, 1969: 81). In various places in southern Asia, the seed of other species of *Camellia*, including tea itself (*C. sinensis*), is also pressed for oil. As for China, though some (for example, E. Watson, 1923: 140, 561; E. H. Wilson, 1929: 385-86) have insisted that the tea oil of Chinese commerce is not obtained from the tea plant but from *C. oleifera*, others ("Tea oil," 1925: 13; Bray, 1984: 526) clearly indicate *C. sinensis* as one source of tea oil there. Besides its use in Chinese cooking, tea oil serves as an illuminant, hair dressing, medicine, and is also used in making soap and as an adulterant of other oils. The pulp, in turn, is processed into cakes employed as fertilizer (E. Watson, 1923: 140; Holmes, 1948: 105).

[52] For the processes of making vegetable oil in China, see C. P. Li, 1948: 142-45.

[53] For non-food uses of these and other oils in China, see particularly E. Watson, 1923: 82-148.

[54] Vavilov (1949-50: 28, 31) viewed Central Asia and India as its centers of origin, but the great diversity of safflower types in Afghanistan noted by Vavilov is now thought to reflect that land's geographic position as meeting place of Near Eastern and Indian safflower forms.

[55] It is possible that Marco Polo, in describing a dye plant of Fukien, may have been referring to safflower, but this is by no means certain (Bretschneider, 1898: 4; Cordier, 1920: 98; Yule, 1921: 225, 226n).

CHAPTER NINE

Flesh and Fowl

Importance in Diet
Pork
Chicken and Duck
Beef and Mutton
Minor Forms of Flesh and Fowl
 Goose
 Pigeon
 Muscovy Duck and Turkey
 Dog
 Cat
 Snakes
 Rats
 Insects
 Still Other Animals

IMPORTANCE IN DIET

Traditional Chinese diet was overwhelmingly vegetarian, with consumption of meat and fish far lower than in Western Europe or the United States. The 1929-33 survey (Maynard and Swen, 1956: 411) found that for villages in China as a whole, meat and fish provided only 1.7 percent of total calorie intake (pork, 1.3 percent; beef, mutton, chicken, duck, and fish, 0.4 percent), with lard another 0.4 percent. This compared with an 18 percent contribution by meat and fish for the United States, according to one estimate made at about the same time (Adolph, 1931: 41).[1] For 1931-37 it was estimated that per capita consumption of meat (including poultry) in China was 35 gm. (about one ounce) per day, and of fish 7.7 gm. (a quarter of an ounce) per day (T. H. Shen, 1951: 380-81). That amounted to 12.9 kg. (about 28 lb.) of meat per capita annually, and 2.8 kg. (6 lb.) of fish. To place those figures in perspective, one notes that in the United States at that time, per capita consumption of meat and poultry was five times as great, about 150 lb. of meat annually, and of fish, 10 lb. annually.

Within China, there were striking differences in meat and fish consumption by social class, region, and historical period.[2] The wealthy and middle classes could afford much more meat and fish than the poor, whether urban or rural. The villages of the Rice Region (including Kwangtung) had a far greater intake of meat and fish than those of the Wheat Region (Table 3). Even so, the overall picture was one of very low meat and fish consumption. As various observers have noted, in the North China Plain except among better-off families meat was an infrequent luxury usually consumed only on holidays or at wedding dinners (Cressey, 1934: 173); for the Chinese peasant, meat, including that of fowl, was a feast dish consumed on the twelve or so holidays each year or on other special occasions (Winfield, 1948: 67; E. N.

Table 3. Animal food calorie intake: 17,351 persons, 2,727 families, 136 localities, 131 hsien, 21 provinces, China, 1929-1933 (from Maynard and Swen, 1956)

Regions and Areas	Number of Localities	Daily Intake of Calories Supplied by Animal Foods per Adult- Male Unit	Percentage of Animal Calorie Intake Supplied by Different Animal Foods										
			Pork	Lard	Mutton	Eggs (chick-en)	Beef	Chick-en	Fish	Duck	Eggs (duck)	Shrimp dried	Meat, salted
CHINA	136	76	54	18	8	8	4	2	2	1	1	*	*
Wheat Region	67	32	59	7	17	10	6	*	1	-	*	-	-
Rice Region	69	122	51	31	*	5	3	3	4	1	1	1	*
Wheat Region Areas:													
Spring Wheat	13	35	51	5	36	2	6	*	*	-	-	-	-
Winter Wheat-millet	21	17	63	5	23	7	2	*	-	-	-	-	-
Winter Wheat-kaoliang	33	40	59	8	5	16	8	1	3	-	*	-	-
Rice Region Areas:													
Yangtze Rice-wheat	22	98	58	24	1	8	2	2	3	*	1	-	1
Rice-tea	19	106	54	31	*	5	1	3	4	1	1	-	-
Szechwan Rice	6	165	48	37	2	3	7	1	1	-	1	-	-
Double Cropping Rice	11	102	42	22	*	3	7	6	8	6	2	4	-
Southwestern Rice	11	196	46	44	*	2	4	2	*	1	1	-	-

* the amount is under 0.5

Anderson, 1988: 146); poorer Chinese such as peasant cultivators, it was said, consumed almost no flesh other than when celebrating the Lunar New Year (Adolph, 1931: 42; Eberhard, 1952: 39). However such low flesh consumption may have affected the nutritional state of the average Chinese, it also meant that their cholesterol intake was very low. This, according to some Chinese researchers, appears to account for the fact that the incidence of atherosclerosis in China is significantly lower and the age of onset later than in Western countries (Tung *et al.*, 1958: 600-602; Huang and Hang, 1960: 456-58).

As has often been noted, China's vegetarian diet is a concession to its shortage of land and dense human populations which can be supported only by consuming cultivated plants directly, rather than indirectly, and less efficiently, through domesticated animal intermediaries. Winfield (1948: 64) observed that though there were "more animals per crop-acre" in China than in the United States, 75 percent of them were work animals, compared to 10 percent in the U.S.; from another perspective, only 25 percent of animal units in China were kept for food, whereas in the United States, 90 percent were. To a considerable degree, moreover, the animals kept for food in China (for example the pig, chicken, duck, and carp) are self-sufficient scavengers (Buxton, 1929: 80-81, 88-89; Winfield, 1948: 66; Latourette, 1964: 491).

PORK

The traditional view among prehistorians, recently restated by Epstein and Bichard (1984: 146-51), is that the pig was first domesticated from the wild *Sus scrofa* of the Near East. The earliest remains of the domesticated form, from Çayönü in Asia Minor, date to about 7000 B.C. From the Near East, according to that view, domesticated pigs spread elsewhere, sometimes breeding with local wild boars, to reach China some millennia later. An alternative view (C. O. Sauer, 1952: 28, 31-32; Clutton-Brock, 1981: 76) is that pigs were domesticated independently in Southeast Asia or East Asia from local wild pigs, perhaps as early as, or even earlier, than in the Near East. Support for this view is claimed in the evidence of early domesticated or feral pigs in New Guinea (Bulmer, 1966; Golson and Hughes, 1976), by 3000 B.C. or before. Since wild pigs are not native to New Guinea, the suggestion is that they were introduced from Southeast Asia where they were kept in a domesticated state.[3] There are also two recent studies, one of pig cranial capacity (Hemmer, 1978) and the other of pig taxonomy (Groves, 1981: 61), which suggest that domesticated pigs of Southeast Asian and Chinese origin derive mainly from wild pigs of those regions.

The view that pigs were domesticated independently in eastern Asia has been given further credence by recent reports of remains of domesticated pigs found in excavations at the Tseng-p'i-yen caves at Kweilin, Kwangsi Province, a site dated by radiocarbon methods to roughly 9300-7000 B.C. (K. C. Chang, 1986: 100, 102-3). If these dates are accurate,[4] these may be the earliest domesticated pigs known anywhere in the world. Remains of domesticated pigs occur, as well, in early North Chinese Neolithic sites of the Tz'u-shan (Cishan) and P'ei-li-kang (Peiligang) cultures (which date roughly from 6500-5000 B.C.) (*Recent archaeological discoveries in the People's Republic of China*, 1984: 9-11; K. C. Chang, 1986: 90-91, 93), and in sites of other early cultures of North and Central China. In sites of the Yang-shao, Lungshanoid, and Lung-shan cultures (*c.* 5000 B.C.-1700 B.C.), the remains were so universal and so abundant as to lead Ho Ping-ti (P. T. Ho, 1975: 106-7) to the view that such pigs were indeed of Chinese origin. In any case, pigs were one of the most important domesticated animals of early Neolithic China, and they have continued to be important through all

Figure 64. Men carrying pigs (from Gray, 1878, 2: opp. 170)

subsequent Chinese history, as reflected in the Chinese character for "family" or "home," which consists of a character for "pig" beneath one for "roof" (C. A. S. Williams, 1941: 326).

China today has by far a larger number of pigs (*chu* or *shih*) than any other country, 331 million (1985), or roughly 40 percent of all the world's pigs; its annual pork production is over 16 million tons, which is 27 percent of all pork produced in the world (Wittwer *et al.*, 1987: 310). One explanation of the pig's important role as a meat animal in China may be sought in the claim (Adolph, 1931: 44) that it can produce twice the meat per acre of pasturage as that produced by cattle and sheep. Whether this is true or not, the pig in the traditional Chinese economy was an effective household scavenger which often survived on table scraps, chaff, and weeds, and which, in South China, was transported to market in a specially-constructed basket (Figure 64). In many places pigs were actually quartered in the family garbage pit and regularly fed human excreta and garbage. Thus they not only converted into flesh plant nutrients that the human digestive system could not absorb directly, but an important health function may have been performed, with human parasites being destroyed as they passed through the pig intestine.

It is widely held that native breeds of pigs are better suited to Chinese conditions than are European breeds (Hays, 1980: 46). They adapt better to China's range of climatic and environmental conditions. They more effectively use the high-fiber and by-product feeds that are available. They are more resistant to local diseases and parasitic infections. And their flesh is of better quality. Whether or not all of the above is so, it seems that certain Chinese breeds of pigs do bear larger litters than even the most prolific of European breeds.

Turning to the use of pork (*chu-jou*, "pig flesh"), we note in the 1929-33 village survey (Table 3) that in China as a whole, pork and lard provided over 70 percent of total human intake of animal calories, and in the Rice Region as a whole, over 80 percent. In the Double Cropping Rice Area, because of the importance of fish, chicken, and duck, pork and lard were less dominant than in any other area of the Rice Region, but even there they provided nearly two-thirds of total intake of animal calories. Pork may be eaten fresh or preserved, as by salting or smoking,[5] with the province of Yunnan renowned for its hams and other cured pork products (E. N. Anderson, 1988: 169). The dishes made of pork in China are impressive in number and variety. In part, this reflects the completeness with which the Chinese use the pig carcass for human food, as illustrated in the Chinese dishes "Baked Spareribs with Hoisin Sauce," "Pork Liver with Onions," "Kidney in Peanut Butter Sauce," "Hot Pickled Pigs' Feet," "Ginger Pigs' Feet," "Sweet-and-Sour Braised Pork Tail," "Jellied Pigs' Knuckles," and "Red-Cooked Pork Tongue." One also finds many dishes, including soups, made with ham, bacon, sausage,[6] ground pork, and pork balls. Altogether, one suspects that China's pork dishes are unmatched in their variety by those of any other people (W. W. Chang *et al.*, 1970: 288-339). Pig's blood, like the blood of other animals, is considered heating and invigorating, beneficial to the blood of the person consuming it; it is taken in various ways, sometimes in a coagulated form that may be fried or added to soup (E. N. Anderson, 1988: 144, 163, 168, 191, 194). As for lard, its role in Chinese cooking is such that in many places pigs are kept as much for lard as for pork, and are raised so as to obtain the largest amount of fat, and fat that can be separated from lean flesh as completely as possible (E. N. Anderson and M. L. Anderson, 1977: 337). Lard is used in stir-frying and deep-frying, its role being to impart a lightness and/or crispness to items being prepared, and is also said to be essential in "Chinese shrimp pastes and dim sum-type pastry doughs" (Tropp, 1982: 551). Fukien is singled out for its preference for lard over vegetable oil in cooking, a preference which may have originated because that mountainous province has abundant fodder for pigs but too little land for oilseed cultivation (E. N. Anderson and M. L. Anderson, 1977: 337; E. N. Anderson, 1988: 162-63).

Pork is commonly classified as "neutral" or "cool" in the hot-cold classification, but, often cooked with medicinals or "hot" ingredients such as ginger and pepper, is widely used as a strengthening food by women in pregnancy and after giving birth (Yeung *et al.*, 1973: 198; B. Chang, 1974: 438; Ling *et al.*, 1975: 128; M. Mathews and L. Manderson, 1981: 13; Tan and Wheeler, 1983: 43, 44; Wheeler and Tan, 1983: 53, 55).

CHICKEN AND DUCK

Poultry in China were estimated to number 1.67 billion in 1984, of which roughly 90 percent were chickens, most of the rest ducks and geese (Wittwer *et al.*, 1987: 325). The chicken (*Gallus gallus* = *Gallus domesticus*; *chi*) is descended from wild fowl of Southeast Asia and India, whether solely from the red junglefowl (*Gallus gallus*) or from other species as well (Crawford, 1984a: 298-300). The red junglefowl, it should be noted, is also found today on Hainan Island and in a narrow strip along China's border with Burma and Indochina. In early times its distribution may have extended well to the north, and as a result China cannot be ruled out as the possible place of earliest chicken domestication. The ancestor of the duck (*Anas platyrhynchos*; *ya*) is the green-headed mallard, *Anas platyrhynchos platyrhynchos*, which is found widely in the northern hemisphere. Early evidence of the duck is sparse, but it points to China and Southeast Asia as a major hearth of domestication (Zeuner, 1963: 471; Clayton, 1984a: 334-37). Domestication of both chicken and duck go back well into prehistory, and one can only speculate as to what the precise dates may have been.

In the case of the chicken, it is widely agreed that domestication was carried out not for culinary purposes, but for obtaining a ready supply of birds for divination and cock fighting, and that only later were their flesh and eggs of primary importance. Cock fighting, perhaps more popular in Southeast Asia than anywhere else in the world, is believed to have been a sophisticated offshoot of divination, one which developed when communities entered contests with each other to determine which was superior. The question was decided by divine judgment through a fight between sacred roosters (Laufer, 1927b; Crawford, 1984a: 300). Even in modern times, such cock fighting in Southeast Asia could be so important, as in Indochina in the nineteenth century (Crawfurd, 1830, 2: 269), as to lead observers to the view that chickens were kept more for that purpose than for food. Cock fighting (as well as duck fighting, pigeon fighting, and quail fighting) was found in China as well, with Canton singled out by some writers as a place where it was particularly popular.[7] Other forms of divination with fowl occurred as well. Use of chicken thighbones in divination, for example, was and is found among people of both Southeast Asia and South China, in a parallel with the inscribed oracle bones of Shang China (though these were mainly from oxen) (Odaka, 1950: 86-87; Simoons, 1961: 65-68; Eberhard, 1968: 419-32; K. C. Chang, 1977a: 210-11).[8]

Eberhard (1968: 432) suggests that the role of chickens and eggs in sport, ritual, and belief in China may derive largely from Tai peoples. If this is true, one could gain a full understanding of that role only by going beyond Han Chinese traditions regarding chickens and eggs. In a parallel with Zoroastrian belief in Iran (Hehn, 1885: 241-42), in ancient China the cock was symbolic of the sun, a yang element, light, warm, and strengthening, with a white cock preferred for exorcising the spirits of darkness, which were identified as yin, cold, and weakening. Though most people in traditional China did not understand the meaning of the custom, it was common, when a body has been left unburied for a considerable period, for a live white cock, the "soul chicken," to ride on the coffin or hearse, to avert the evil spirits of darkness or to call the soul to make certain it remains with the corpse (Cormack, 1935: 126; Burkhardt, 1953-58, 1: 98, 167-68; J. J. M. de Groot, 1964, 1: 199-201; Read, 1977a: 30). In traditional China, the cock, or white cock, also served in oath taking, and the chopping off of a cock's head was an oath "by the blood of life" that was allowed in English courts in Singapore and Hong Kong (Giles, 1876: 86; Ball, 1906: 386-87, 651; J. S. Thomson, 1909: 162).

Melanotic chickens (*wu-ku-chi* or *wu-chi*), black-fleshed and black-boned, also played a role in Chinese ritual and medicine. Such chickens are also kept by some American Indian groups, and among certain of them, especially those of Mayan language, melanotic chickens have roles in folk medicine similar to those found in China and Southeast Asia, a fact which has contributed to the view that they may have reached the New World in pre-Columbian times (Johannessen, 1981). Other evidence has also been advanced for the presence of chickens in the Americas in pre-Columbian times (Carter, 1971), though few scholars are convinced by it.[9] However the controversy is ultimately resolved, we would note that melanotic chickens originated in Asia (South China, Southeast Asia, India); that they are not a recent phenomenon (Marco Polo observed such chickens in Fukien); and that Martín de Rada, writing in the sixteenth century, considered Fukienese black-fleshed chickens to be tastier than light-fleshed ones (Boxer, 1953: 290).

Bones and clay figurines of what are said to be domesticated chickens were found in North China by about 3000 B.C., and there are claims of much earlier dates (P. T. Ho, 1975: 96, 97-98; K. C. Chang, 1977a: 157; 1986: 260).[10] There are hints that domesticated ducks, along with domesticated chickens, may have been kept in Central China at about the same time.[11] Neither chickens nor ducks, however, were as well adapted to conditions in North China as to those of Central and South China, where they have remained more important than in the north. The 1929-33 village survey, for example, found far more chickens and ducks per farm in the Rice Region than in the Wheat Region. The Double Cropping Rice Area, moreover, had more chickens and ducks per farm than any other area of the country (Buck, 1956: 246). The flesh of each of these fowl provided 6 percent of total animal calorie intake in villages of the Double Cropping Rice Area, a percentage at least double that for villages in any other area of China (Table 3). The importance of ducks in the Double Cropping Rice Area is explained by the prevalence of water and the availability of food along watercourses and in paddies after harvest.

While fowl were sometimes fed in China, mostly they ate things that otherwise would go to waste, such as insects and worms, garbage, grain remaining in the fields after harvest, and grain by-products (Dean, 1980: 52). One even reads, for southern Fukien, of maggots, which are plentiful in pits where human feces are stored, being scooped out with nets and fed to chickens and young ducks (J. P. Maxwell, 1921: 378). Fowl, in their turn, provided not only flesh and eggs, but excrement for fertilizer, and feathers and down as well. There were chicken incubators and hatcheries in certain large Chinese cities (Alsop, 1918: 188-89; Holmes, 1948: 25), and systems of husbandry appear to have been impressive and well organized, especially for ducks.

That this is not a recent development is demonstrated by the late sixteenth-century account of González de Mendoza (1853-54, 1: 153-54) who described a sophisticated system of Chinese duck husbandry, with thousands of ducks kept in cages aboard boats at night. In the morning, the ducks were permitted to leave, entering the water by means of bamboo bridges, feeding in paddies during the daylight hours, and returning when their owners, as evening approached, signaled them to return. Mendoza was struck by the fact that the ducks, when feeding in the paddies, ate snails and weeds but did not damage the rice crop. He was also struck by the unusual attention given to the hatching and survival of young birds. Late in the nineteenth century, Gray (1878, 2: 180-85) described similar systems of duck husbandry, and he, too, was impressed by the special attention given to rearing ducks: to their feed and care during the sitting period, to the number of eggs they were permitted to sit, to noise that might disturb newborn ducklings, to the food and water given them, and more. He also wrote of duck eggs hatched in institutions specializing in artificial hatching. These institutions then sold the young ducklings to duck merchants who raised them in enclosures. When sufficiently grown, such ducks were sold by the merchants to itinerant duck vendors who transported them by water, as many as two thousand to a boat. While he kept the ducks, a vendor permitted them to feed twice a day along the river or in nearby fields, thereby saving the cost of feed. So well-trained were the ducks that they returned to the boat when finally they were called. Though the itinerant sold ducks retail in communities along the way, most found their way to provision dealers who specialized in salting and drying them. Such dried ducks might be sold in various ways, as by hawkers who carried them suspended from wooden carrying sticks (Collingwood, 1868: 339).

One notes also (Clayton, 1984a: 337) what appears to be a different system of duck husbandry involving long-distance droving. Foraging in canals and paddies as a flock is driven slowly by a keeper, the ducks are gradually brought to market after travelling as much as several hundred miles over a period of six months. Still different were the non-migratory systems of duck husbandry, such as that described (C. K. Yang, 1959a: 65-66) for a village near Canton in 1948-51, which formerly supplied ducks not only to that city but shipped them by "duck junks" to Hong Kong and Macao. Only a few village families kept ducks, usually in flocks numbering 100 to 200, buying them in the city and fattening them for market over a period of two months or so. Village ducks foraged locally along streams and canals during daytime, but sometimes they were allowed into harvested fields to eat any grain they could find. Except for occasional rice chaff, however, they were not fed, but instead the principal costs of raising them involved their initial cost and those of renting land, constructing a bamboo shelter where they spent the night, and the labor needed to watch them. That ducks could play a significant role in reducing the numbers of insects and weeds in paddies, is shown by the fact that in South China it was not uncommon (though arrangements differed locally) for a rice farmer to rent ducks from a duck raiser to go through his fields (E. N. Anderson, 1988: 103).

There are said to be two main types of true ducks grown in China, Peking and Nanking. The Peking duck is white in color. The Nanking duck is more varied in color, whether brown, gray, speckled, or white. Both are commonly force-fed before slaughter. The two grow to a comparable size and have tender skin with a thick fat layer underneath it. The Peking duck, however, is renowned for roasting, whereas the Nanking duck is used for salting (T. H. Shen, 1951: 295). In China, where people like fat, the fat of ducks, which may equal 30 to 35 percent of carcass weight after gutting, is an advantage, not, as with many people in the Western world, a disadvantage.

Chicken and duck flesh are classified as "warm" or "neutral" in the hot-cold system, with chicken and chicken soup common dishes of pregnant and nursing women; duck, however, is reported by some to be "poisonous" and may be excluded from diet at such times (Koo, 1973: 141; Yeung *et al.*, 1973: 198, 200; B. Chang, 1974: 438; Ling *et al.*, 1975: 128; E. N. Anderson and M. L. Anderson, 1977: 369; M. Mathews and L. Manderson, 1981: 12, 13; Tan and Wheeler, 1983: 44; Wheeler and Tan, 1983: 53, 55). Chickens are normally eaten fresh because of their high value in that form (E. N. Anderson, 1988: 147), and ducks may be eaten fresh or preserved, as by salting and drying.[12] When eaten, moreover, the Chinese make good use of all parts of fowl,[13] including the fat and blood, the latter highly regarded as a strengthening food. The blood of a fowl may be served along with its flesh or used in other ways, with solidified chicken blood a traditional ingredient in "Hot and Sour Soup" and "Chicken Blood Soup" (K. Lo, 1979c: 195; E. N. Anderson, 1988: 163, 164, 194). Tasty snacks and other dishes are made of chicken feet,[14] wings, and liver and other giblets,[15] and whole chicken as well as other chicken parts are made into an impressive array of main dishes and soups. One of them is "Beggar's Chicken," in which, traditionally, the unplucked chicken was cleaned through its anal opening, then coated with clay and baked in a fire of wood or charcoal. When the baked clay was removed, the feathers came along with it, to reveal the juicy flesh (E. Chen, 1977: 92). Another is "Chicken's Head Soup," described as an addition to a meal in a land where very little is thrown out (Holmes, 1948: 146). Duck is as thoroughly used, with duck kidney, liver,[16] feet,[17,18] tongue, gizzard, blood, and brains considered delicacies. One nineteenth century observer reported strings of dried duck tongues for sale in a Chinese shop, and a modern food writer wrote of a recipe which begins with "two dozen ducks' tongues" (P. L. Simmonds, 1885: 172; Howe, 1971: 169; E. N. Anderson, 1988: 147). Duck cooked with taro or yams was a traditional specialty served at the autumn Festival of the Moon.[19] Overall, though China has a shorter list of main dishes made of duck than chicken, there is considerable variety among them, for example Cantonese "Pineapple Duck" (W. W. Chang *et al.*, 1970: 275; Sakamoto, 1977: 183-84), and two famous duck dishes mentioned previously, "Camphor-wood and Tea-Smoked Duck" ("Szechwan Duck") and "Peking Duck." The latter is so famous that Chinese chefs may specialize in the dish, and restaurants serve only dinners of Peking Duck (Hom and Steiman, 1981: 176). Peking ducks are raised in a special way,[20] sometimes even with a particular oven in mind,[21] and the dish is demanding to make and unusual in preparation and in the way it is eaten. Though there are different ways of preparation, varying in elaborateness, one described by Kenneth Lo (1971: 52-55; 1979c: 250-52) involves loosening the skin of the carcass from the flesh by inflating with air; then hanging the bird up to dry; and, finally, coating its skin with a sugary liquid before roasting. The roast duck has crisp, brown skin and tender flesh, and both are consumed, commonly rolled in a pancake with raw vegetables, such as spring onion and cucumber, and piquant sauces and flavorings, and then eaten with the fingers.

It is curious that in a place like Kwangtung, with such a large duck population, one should find a group of people who refused to eat it. Yet that was the case with members of a religious cult dedicated to the deity Hong-Yuen-Shuee (or Hong-Kung-Tchu-Shuee). Duck flesh was avoided by cult members because the duck was a favorite of their deity, whose life had been saved by ducks (Gray, 1878, 2: 180-81). Worship of that deity was so strong among inhabitants of one village on the island of Honam, across the Pearl River from Canton, that no ducks at all were found there.

We have found nothing to suggest a similar rejection of chicken flesh in China other than that deriving from the well-known Buddhist commitment to vegetarianism. In India, Hindus and Buddhists have had views of eating chicken that go beyond that of mere vegetarian commitment, that involve disgust with the dirty eating habits of chickens. Such attitudes are also reported among Tibetans and Mongols but, though they may occur among Chinese Buddhists, the Chinese people as a whole have been pro-chicken, and active in spreading its use in East Asia (Simoons, 1961: 69-71). The Chinese who came to the United States have retained their liking for chicken, and they also persist in their traditional preference for freshness. In Petaluma, that chicken-producing center of northern California, this means that chickens are butchered in the evening and shipped to Chinese markets in Oakland and San Francisco the next morning. The Chinese in California are also more concerned than most people about chicken flavor and prefer flesh firmer than that typical of American supermarket chickens. They also like to buy chicken with head and feet attached, for these have culinary uses and may provide signs as to how recently a chicken has been killed. Though it is against state law to sell chickens with heads and feet on, Chinese merchants in California have special permission to market them in that manner, the circumlocution being that this is a concession to Buddhism (Cost, 1987).

BEEF AND MUTTON

Since "beef" (*niu-jou*) as used in the 1929-33 village survey, refers to the flesh of both common cattle and water buffalo, and "mutton" (*yang-jou*) to that of both sheep and goats, we follow that usage and consider all four of those animals in this section.

The water buffalo seems to have been an early domesticate in Central China, possibly already at early agricultural sites dating from the fifth and fourth millennia B.C. near the mouth of the Yangtze in Chekiang, Kiangsu, and Shanghai: at Ho-mu-tu and in earlier levels of villages of the Ma-chia-pang culture (K. C. Chang, 1977a: 513; 1986: 201, 211; T. T. Chang, 1983: 75; Meacham, 1983: 158-59; Pearson, 1983: 124; Xin, 1984: 9). Such dates place these among the earliest domesticated water buffalo known in a village site anywhere, for, outside of China, the earliest remains of possibly domesticated water buffalo in such a site seem to be those at Mehrgarh in Pakistan (sixth millennium B.C.) (Meadow, 1981: 168; Allchin and Allchin, 1982: 108-9, 126). Though many uncertainties are involved, this suggests that water buffalo domestication occurred somewhere in the area from Pakistan to China, not, as some have suggested, in Mesopotamia, where the earliest domesticated water buffalo date from late in the third millennium B.C.

Common cattle, sheep, and goats were all domesticated before (some of these, well before) 6000 B.C. in the Near East or, in the case of common cattle, possibly in southeastern Europe. Though remains of common cattle and sheep, and perhaps goat, were found in sites of the Yang-shao culture (following about 5000 B.C.?) in the Yellow River area of the north, they were said not to have been in a fully domesticated state there until Lungshan times (after about 3000 B.C.?) (P. T. Ho, 1975: 93-97, 99-100, 104; 1977: 456-58; K. C. Chang, 1977a: 95, 160, 175; 1977b: 29; 1986: 113). One suspects that all three domesticates reached China from the Near East, though suggestions of possible independent domestication of sheep in China have been made (P. T. Ho, 1977: 465; E. N. Anderson, 1988: 3, 11, 144). The history of the southward spread of the sheep and goat remains to be unravelled, but on Hainan, according to Chinese sources, they were first introduced in historical times (Fenzel, 1933: 180).

In China's Wheat Region, common cattle (*niu*, "ox"; or *huang-niu*, "yellow ox") were the work animals, kept by 36 percent of all farms in the 1929-33 village survey, whereas water buffalo (*shui-niu*, "water ox") were insignificant, kept by less than 0.5 percent. In the Rice Region, by contrast, water buffalo and oxen were of equal importance as work animals, each of them kept on roughly a third of all farms. There were, however, striking differences within the Rice Region in the relative importance of the two, with the water buffalo paramount in the Double Cropping Rice Area because of its greater strength and ability to plow the heavy soils of paddies. Beef, whether from water buffalo or common cattle, accounted for 7 percent of total animal calorie intake in the Double Cropping Rice Area, a contribution matched or exceeded in few other areas of China.

Domestic sheep (*yang* is used for sheep, goats, antelope, and related species; *mien-yang*, specifically for domestic sheep) are primarily animals of China's pastoral provinces of the north and west, and very few sheep are found in the southeast. The village survey, for example, reported that only 3 percent of the Double Cropping Rice Area's farms had sheep. Domestic goats (*shan-yang*, "mountain *yang*") are found more widely in China than are sheep, but the survey reported no goats on Double Cropping Rice Area farms. This is surprising because the goat has been the far more successful of the two at adapting to the humid tropics and subtropics, and because there is a "South China goat" which is kept in Kwangtung for its flesh (Epstein, 1971: 66). In any case, the Double Cropping Rice Area's "mutton" was not a major factor in diet, providing only an insignificant share (less than 1 percent) of total animal calorie intake in the village survey. In its low mutton intake, the Double Cropping Rice Area was like other areas of China's Rice Region, but it differed from the Wheat Region where mutton contributed 17 percent of total animal calories, and in one area, 36 percent (Buck, 1956: 246; Maynard and Swen, 1956: 413). That importance partly reflected environmental factors (aridity and poor pasture, to which goats and sheep are better suited than most domesticated animals), and partly cultural ones.

China's Moslems, estimated to number 25 million in the mid-1960's, are concentrated in the west and northwest; they eat mutton and beef, but avoid pork, for they have the typical Moslem view of the pig as an unclean animal and pork as unacceptable food. As a result, when, during the Cultural Revolution, the insensitive demand was made in certain places that Moslems raise pigs, this, along with other violations of their traditional ways, led to a furious reaction (Banister, 1987: 316-17). Islamic bans on consuming pork and drinking alcoholic beverages also led Moslems to shun ordinary Chinese restaurants and to set up their own, which contributed to the establishment of a Chinese Moslem school of cooking, centered in China's northwest but also well-represented in Peking and other parts of North China.

Prominent in Chinese Moslem cuisine are lamb hot pots and barbequed lamb (K. Lo, 1971: 43-51; 1979c: 186-94), for mutton and lamb are favorite Moslem foods, as they are among Mongols and Manchus, too. In Yüan dynasty times, when the Mongols ruled China, lamb was the rulers' most important flesh, commonly prepared in traditional Mongol dishes, including ones using fat from the tail of fat-tailed sheep, a Central Asian favorite (Sinoda, 1977: 491; E. N. Anderson, 1988: 75).[22] When the Manchus, in their turn, invaded China proper in the seventeenth century and established the Ch'ing dynasty, they continued to consume their traditional mutton dishes (Davis, 1966: 104, 106, 112, 118). In addition, Peking chefs were stimulated to develop new ones, to gain for that city a reputation for having the best lamb and mutton dishes in all China, such dishes today having a special association with Moslems, Mongols, and restaurants that cater to them. At the same time, the use of lamb and mutton is prominent among other Chinese in Peking, a 1926-27 survey of household income and expenditures among 283 families there (256 Chinese, 23 Manchu, and 4 Moslem) finding that, after pork, mutton was the most common flesh (Gamble, 1933: 96).

In Han times in China, beef apparently was more in demand than pork, perhaps because of a scarcity brought on by periodic government bans on the slaughter of oxen (Y. S. Yü, 1977: 56-58, 67, 74). That it was served at Han feasts and included among foods left in the tomb of an upper-class Han lady suggests that it was fully acceptable fare. In T'ang China, however, there was a decline in beef eating attributed to Buddhism, and, though the practice continued, people felt guilty about it (E. N. Anderson, 1988: 54). In Sung times, beef was no longer included on lists of banquet or restaurant foods, which was probably a reflection of its unpopularity (Freeman, 1977: 164). In the nineteenth-century, Doolittle (1865, 2: 186-91), who wrote particularly of Foochow, reported that both common cattle and water buffalo were praised for their contributions to human life, and that people excluded beef from the meat-offerings they made to deities and did not burn candles made of bovine fat in front of the statues of deities. Strong public pressure against such slaughter was also present, and laws were instituted to protect cattle and water buffalo against slaughter for food, though these were not effectively upheld and it remained permissible to sacrifice cattle in honor of Confucius and in the emperor's sacrifices to Heaven and Earth. There were also many folk tales about butchers and others who killed cattle and ate beef, and the punishment they received for their sins. Pro-cattle tracts and posters were prepared and distributed, the severity of the offense of cattle slaughter being revealed by one poster on which it was linked with female infanticide. A tract described persons who raised cattle for slaughter or who ate beef as more evil and cruel than tigers and wolves. Butchers came in for special attention, too, being characterized as having hearts of stone. And rewards were mentioned for abstaining from beef eating, as with persons who passed their literary examinations after their families abstained from beef. Such views were found even in the Ch'ing court, where early in this century beef was banned in the neighborhood of the palace because it was sinful to slaughter and consume animals that had served as beasts of burden (Der Ling, 1914: 40).

Buddhism was clearly involved in the decline of beef eating in China, as in Japan and Southeast Asia (Simoons, 1961: 50-55), and some have suggested that the very origins of Chinese anti-beef feelings are to be found in India. Others, however, insist that such feelings preceded Buddhism in China, and that they evolved because of the usefulness of cattle as plow and draft animals. They observe that in areas where Confucian traditions were strong, the cow was venerated as a symbol of agriculture, and that many traditional Confucian scholars were as horrified at the thought of eating beef as Moslems were of eating pork (Mickey, 1947: 76-80; Löffler, 1954; Simoons and Simoons, 1968: 170-84, 190-91, 244-58; Bodde, 1975: 201-9; Palmieri, 1976: 218-36).[23] In some places, according to Eberhard (1968: 183-87), the role of cattle in cult activities may also have led to a ban on beef eating.[24]

We have not determined whether Western influence was at all effective in countering the traditional reluctance to consume beef in China, as was the case in Japan. Bans on eating meat, including beef, first appeared in Japan following the introduction of Buddhism in the sixth century A.D., and were part of the national food scene from that time onward. When, however, the Emperor Meiji ended Japan's isolation in the mid-nineteenth century, he set an example by eating beef. Beef-eating thus became symbolic of the nation's new outward-looking perspective, and soon ordinary Japanese regarded it as the sign of a civilized person. Initially people ate beef in Western-style restaurants, later in their homes (Oiso, 1976: 38-40, 47).

Returning to traditional China, we note Martin Yang's account (1945: 47) of the cattle situation in Taitou, Shantung Province, which may have been typical of that in many villages there. Oxen were regarded as members of the family, and they and cows were ordinarily kept by farmers as long as possible. People condemned the slaughter of oxen, looked down on butchers, and were sad when finally they did sell an ox or cow. In market places, Yang said, one could find "hundreds of slaughtered pigs for sale but only one or two cows"; the low price of beef, he contended, was a reflection of public aversion to slaughtering cattle. To serve beef to an honored guest would be improper and even insulting. Thus beef was not included in formal dinners, or used for festivals or religious sacrifice.

Overall, in China beef was of far less dietary importance than pork, and there were far fewer beef dishes. Like lamb and mutton, however, beef was important among Chinese Moslems, and there were also excellent beef dishes among other Chinese in various parts of the country. Perhaps best known in the United States is oyster sauce beef, a Cantonese dish which commonly includes scallions and garlic as ingredients. Scallions and garlic, as well as onions and ginger, are common in other beef dishes as well, for the Chinese believe that beef, like mutton (or lamb), has a strong flavor of its own, and should be accompanied by such strong-tasting vegetables (K. Lo, 1976: 44; 1979c: 168).

Beef and mutton are both commonly considered "hot" foods, which would appear to make them appropriate for pregnant women, who are inclined toward coldness. There are, however, other considerations, such as the previously-mentioned avoidance of mutton by pregnant women fearful that their offspring would be epileptic. In a similar vein, Chinese women in London believe that if a pregnant woman eats beef, her baby will have a dark skin; it is also considered poisonous to her after she gives birth and not advisable for other sick people to eat (Koo, 1973: 141; Yeung *et al.*, 1973: 198; Ling *et al.*, 1975: 128; M. Mathews and L. Manderson, 1981: 12, 13; Tan and Wheeler, 1983: 42, 44; Wheeler and Tan, 1983: 53, 55).

MINOR FORMS OF FLESH AND FOWL

Among the domesticated animals of minor dietary importance in traditional China are the goose, pigeon, muscovy duck, turkey, dog, cat, and silkworm, which are included among the flesh foods to be considered now. There are also scattered references, for ancient and modern China, to the eating of domesticated rabbit (as well as wild rabbit), horse (*ma*), donkey, mule, and camel (Astley, 1745-47, 4: 86; Gray, 1878, 2: 174-75; A. H. Smith, 1894: 21; Parker, 1897: 175; 1903: 271, 272; Ross, 1911: 79; Holmes, 1948: 24; Graham, 1961: 77; Simoons, 1961: 81-82; Epstein, 1971: 96; K. C. Chang, 1977b: *passim*; E. N. Anderson, 1988: 20, 42, 68, 75, 144) and, before considering the first-named animals of minor dietary importance, we would comment briefly on two of the latter, rabbit and horse.

Since establishment of the People's Republic, the raising of rabbits (*t'u*) of both Chinese and foreign breeds has expanded to the point that China now has the world's largest population of domesticated rabbits, and accounts for half of all production of rabbit flesh. China is also a significant exporter of rabbit flesh, and contributes between 30 to 40 percent of the rabbit fur entering international trade. Within China, the rabbit now ranks third as a supplier of meat, surpassed only by the pig and chicken (Wittwer *et al.*, 1987: 336).[25] In most of China rabbit flesh (*t'u-jou*) is cooked in ways similar to chicken, with F. L. Chang and G. M. Fuchs (n.d.: 20) making suggestions of specific chicken dishes well suited to rabbit.

Horse meat (*ma-jou*) was a preferred food in Han China (Y. S. Yü, 1977: 58), and one observer reported that consumption of horse meat was very widespread in the nineteenth century, too, and that horse-meat butcher shops were found in all big cities (Raveret-Wattel, 1884: 607). At the same time, horses (like donkeys, mules, and camels) are found in such small numbers in the southeastern provinces (Epstein, 1971: 94, 111, 113) that horse meat cannot have been of dietary significance there, and we will pass over it in favor of other types of flesh.

Goose

The Chinese goose, *Anser cygnoides* (*ê*), is believed to have been domesticated in China, its ancestral form being the swan goose.[26] This would give it a different origin from the domesticated goose of Europe (*Anser anser*), whose ancestral form seems to be the greylag goose. Likely the goose was domesticated in the south, perhaps by Tai peoples. Eberhard (1968: 286) notes that in early times the rearing of geese was common only in South China, and that in Sung times geese were few and costly north of the Huai, that eastward flowing river between the Yangtze and the Yellow. The time of domestication is uncertain. Bones of geese are not among animal remains listed for Yang-shao sites (fifth and fourth millennia B.C.) by Ho (P. T. Ho, 1975: 93). Clay models of geese, however, have been identified in various sites of the Ch'ü-chia-ling culture (third millennium B.C.) of the middle Yangtze Valley (K. C. Chang, 1977a: 167). Since models of what appear to be sheep, chickens, and ducks also occur in these sites, it is possible that we are dealing with domesticated animals. Whether or not this was so, what are said to be domesticated geese were found in Han archeological sites (Z. Wang, 1982: 57).

Eberhard writes (1968: 286) that in early China the rearing of geese was the responsibility of women. Gray, writing of South China in the nineteenth century, observed that geese were generally raised by farmers or others who lived adjacent to streams or rivers, where the birds could feed. They were occasionally kept in large flocks, the largest observed by Gray in nineteenth century Kwangtung consisting of nine hundred birds. In the plowing season, geese could be seen following the plowman as he turned over the soil, to get the worms he uncovered. Herders with goslings were constantly alert to the danger of hawks, and carried hollow bamboo tubes to swing and make noise to frighten the hawks away. Though far fewer than ducks, geese were numerous enough in Kwangtung to keep the daily goose market in Canton well-supplied with flesh. The daily market, with many hundreds of geese arriving by boat, was nearly as big as the celebrated annual goose festival in Nottingham (Gray, 1878, 2: 185-87).[27]

Goose flesh is mentioned quite commonly in Chinese texts from Chou times onwards, and T'ang Chinese considered goose roasted on a spit as especially fine food (K. C. Chang, 1977b: 29; Schafer, 1977: 101). Though far less important today than chicken and duck, goose and goose eggs continue to be eaten in modern South China. Goose feet, like those of duck, are prepared as delicacies, and in Kwangtung's Teochiu cuisine (a local rendering of Chaochou, now Ch'ao-an, district), a variant of Fukienese cuisine, there is a featured dish of "steamed goose, goose liver, and goose blood with garlic-vinegar dip sauce" (E. N. Anderson and M. L. Anderson, 1977: 357; E. N. Anderson, 1988: 163, 175).

Pigeon

In a live poultry market in Chinatown, Los Angeles, we were struck recently by the considerable number of caged pigeons, exceeded only by that of chickens and ducks.[28] Since this suggests an importance among the Chinese greater than implied by the scanty attention given to pigeons in most Western accounts, we give them more attention than we might otherwise have done.

The domesticated pigeon, *Columba livia*, is believed to be descended from the rock pigeon or rock dove,[29] which is found wild today from the Atlantic shores of North Africa and Europe to India (Hawes, 1984: Fig. 52.1). The earliest representation of the pigeon, in Mesopotamia, dates to about 4500 B.C. Though evidence of the bird remains scanty until classical times, some scholars believe that it may have been the earliest of domesticated birds. In ancient times, the pigeon probably enjoyed a loose relationship with man similar to that found in parts of the Near East today, with humans providing a pigeon breeding place and benefitting from the dung they collected for fertilizer and from the young birds they took for food (Brentjes, 1962: 636-38; Zeuner, 1963: 460-62; Hawes, 1984: 354). Over time the relationship would have developed from one of symbiosis to one of control and full domestication.

The domesticated pigeon appeared in India at some unknown early date, and Indians probably carried it to Malaysia, where one of its names derives from the Sanskrit (I. H. Burkill, 1966, 1: 869). From Malaysia, it may have been carried by sea to China,[30] or it may have been carried overland from India or the Near East. In either case, pigeons were being kept in China by the sixth century A.D., and may have been there much earlier (Hoose, 1938: 1, 7).

The domesticated pigeon (*ko*) has a many-sided position in traditional Chinese life which should be considered before we turn to its use as food. One should perhaps note first that Chinese farmers and others keep large numbers of them, of many breeds (Gray, 1878, 2: 180, 188-90; Levi, 1957: 58), for food, racing, or other purposes. The carrier pigeon, for example, was used by merchants to transmit information on prevailing market prices, arrival of cargoes, and such. In Hong Kong, merchants would send carrier pigeons to their partners in Canton to announce the arrival of mail packets from Europe and America. And in Peking a leading stationery shop kept carrier pigeons to communicate with its branch in Tientsin (Hoose, 1938: 2). Pigeons also play other roles far less common in the Old World than that of messenger. One is in the sport of capturing another person's pigeons by kit-pull or flock-pull, a practice which occurs in Peking and a few other places in the world (among the others, Persia and India); and another is in making music with pigeon flutes, a Chinese invention.

In capturing pigeons by flock-pull, the owner of a pigeon flock watches the sky for a pigeon flying alone and then sends up his flock to go after it and bring it to the flock's loft (Hoose, 1938: 8-14; Levi, 1957: 371; Wilford, 1962: 443). The captured pigeon, by tradition, belongs to the captor who may use it as he wishes. As for pigeon flutes, they were originated by Chinese in the Peking area, probably in early Ch'ing times. Pigeon flutes are made in great variety, of reeds, bamboo, and/or miniature gourds; they are objects of pride, carefully produced by craftsmen; and they are attached to pigeons' tails to make beautiful sounds as the birds fly (Hoose, 1938: 15-21). In the fairs of Peking there were pigeon markets at which dealers sold pigeon flutes and pigeons for sport. These were different from the markets where pigeons were sold for food. Indeed sport pigeon dealers and fanciers looked down on persons who dealt in food pigeons (Hoose, 1938: 12-13).

Though not highly regarded in China for its intelligence and lasciviousness, the pigeon is admired for the affection and care it shows to its mate and offspring. Thus to the Chinese it is symbolic of long life, faithfulness, family love, and marital fidelity (Feng, n.d.: 111; S. W. Williams, 1895, 1: 335; Howe, 1971: 171; pers. comm., Paul Wheatley),[31] which recalls its position as a sacred symbol of love and fertility in the Mediterranean, Near East, and India from ancient times onward (Watt, 1889-96, 6, Part 1: 231; Hastings, 1908-26, 1: 525; Crooke, 1926: 373-74; Zeuner, 1963: 460-62; I. H. Burkill, 1966, 1: 869; Isaac, 1970: 112; Hawes, 1984: 354). One may wonder whether such ideas were introduced to China from those lands, and even whether Buddhists, who were influential in shaping Chinese ideas in other ways, may have had a hand in this. There is, to encourage such speculation, the observation, given to Charles Darwin (1900, 1: 212) by certain informants who had lived in China, that pigeons were raised there especially by Buddhist monks. This, however, may reflect a Buddhist need for pigeons to liberate rather than any special position of sanctity, for Gray (1878, 1: 127-28) observed that it is meritorious for Buddhists to liberate sparrows or pigeons, and that at several Buddhist temples monks would purchase pigeons and keep them in cages on whose doors were written the words "preserve life." If a visitor to the temple put money in the cage, a monk would free one of the birds. This relationship thus reflects the Buddhist commitment to ahimsa, not that the pigeon was a religious symbol of the sort it was among Moslems in Asia.

The pigeon's sacred role has led various Old World peoples to reject its flesh as food (*i.e.*, many Moslem groups, Ethiopian Christians, and Orthodox Russians in Tsarist times) (Beazley, 1966: 108; Isaac, 1970: 112), and even in China, according to F. T. Cheng (1954: 59-60), northerners ordinarily do not consume pigeons, which they refer to as "holy birds."[32] Others write of pigeon or squab as popular in northern China (K. Lee and A. Branyon, 1984: 250-51), and in any case it is a common and well-liked dish in Canton. Pigeon, though not a "great dish" like duck, was also served as a delicacy at marriage banquets and on other festive occasions, and was offered daily at restaurants, prepared in a variety of ways. Examples of recipes (Feng, n.d.: 81, 122-24; Howe, 1971: 171; F. Lin, 1975: 48-49, 89; Field *et al.*, 1976: 100-101; E. Chen, 1977: 28; K. Lo, 1979a: 107-9; 1979c: 253-56; K. Lee and A. Branyon, 1984: 250-51), many identified as Cantonese, are: "Broth of Pigeon," "Fried Pigeons," "Crisp Squab," "Deep-Fried Red-Cooked Pigeon," "Minced Squab with Oyster Sauce," "Deep-Fried Aromatic Pigeons," "Lemon Squab," "Red-Cooked Lemon Pigeon," "South of the River Crispy Pigeon and Shrimp Cakes," "Casserole of Mushroom Squabs," "Braised Pigeon in Fruit Juice," and "Steamed Pigeons." Though some of these have simple names, they may in fact be gourmet dishes, with "Steamed Pigeons" a dish with salt, soy sauce, pepper, "golden needles" (tiger-lily buds), jujubes, mushrooms, scallions, smoked ham, gingerroot, sugar, "dry sherry" (rice wine), and monosodium glutamate. The braised pigeon dish is Cantonese, a savory "semi-soup" made with salt and soy sauce as well as lemon juice, orange juice, apple sauce, tomato, chutney, and chives or scallions. And "Fried Pigeons," the recipe of a Hong Kong gourmet, involves giving the birds liquor to intoxicate them before slaughter followed by preparation with onion, a green vegetable, ginger, cumin, rice wine, mushrooms, soy sauce, sugar, and salt.

Pigeon flesh is considered both "heating" and strengthening, and may be taken as a remedy for cold diseases and given to women during pregnancy and after birth (Ling *et al.*, 1975: 128; Gould-Martin, 1978: 43; Tan and Wheeler, 1983: 43). It is thus appropriate to find in the recipe "Squab Soup with Ginseng" (J. D. Keys, 1963: 88) not just squab and ginseng, but other heating foods (gingerroot and rice wine) as well. For medicinal uses, the Chinese prefer white pigeons; their excrement, for example, serves as an anthelmintic and antiscorbutic (Soubeiran and Dabry de Thiersant, 1874: 59; Hooper, 1929: 147; Hoose, 1938: 14; Read, 1977a: 47). According to Ainslie (1826, 1: 313), the Arabs and Persians consider pigeon and pigeon eggs to be aphrodisiacs. Soubeiran and Dabry de Thiersant (1874: 59), in their brief account of the pigeon in Chinese medicine, do not mention such a role; and Read (1977a: 47-48), though mentioning the above statement by Ainslie, does not indicate whether or not such views are associated with *Columba livia* in China.[33]

Muscovy Duck and Turkey

The muscovy duck (*Cairina moschata*) is descended from a large (males: 8 to 11 pounds), multi-colored (especially brown, black, and white), tree-nesting duck of the American tropics, found mainly from Central America to northern South America (J. C. Phillips, 1923, 1: 57-67). The domesticated form is larger than its wild ancestor and more variable in color, often less dark and sometimes nearly altogether white. The muscovy duck was kept by various pre-Columbian American Indian groups and in some places, such as Peru, it was the only domesticated fowl and very numerous. How the term "muscovy" came to be applied to the duck remains a matter of speculation. Some say that it is a corruption of the name of the Indian group among whom the Spanish first encountered the duck, for example the Muisca Indians of Colombia or Miskito Indians of Nicaragua. Others attribute the name to an erroneous belief that the bird produces musk or has a musky odor. And still others suggest that it derives from a European view that the duck came from Russia.

Following discovery of the Americas, the muscovy duck was introduced to Europe, Africa, and other parts of the Old World, but it has not done well in competition with the common duck, which in many places was already well-established. In part its lack of success derives from its lesser fecundity and much slower growth rate than that of the common duck; on the other hand, its flesh is leaner and with breeding and selection it may gain a more important role as a supplier of flesh. In China today the muscovy is called "foreign duck" to distinguish it from the native one. It is kept on Hainan Island and elsewhere in southeastern and eastern China, and is common, as well, in Taiwan (Gray, 1878, 2: 181; Ensminger and Ensminger, 1973: 89; Clayton, 1984b: 340-44).

We have found no detailed information on Chinese husbandry and consumption of the muscovy duck, though Pereira (1937: 160-62) wrote briefly of it for the New Territories, Hong Kong, in the mid-1930's. Fifteen years before, he noted, the muscovy duck had been found in just a single village there, but when he wrote it was quickly gaining favor, though inbreeding had left local muscovies half the size of those found in Malaya and the Philippines. Muscovy flesh, which is reddish in color, was considered inferior to that of the local white duck.

Like the muscovy duck, the turkey (*Meleagris gallopavo*) was introduced to China after discovery of the Americas. Today it is found in various parts of the country, but it is neither common nor popular. One problem is that it is inconvenient to handle, too large for a single dish. Another problem is that some view turkey flesh as coarse and unsubtle in flavor (K. Lo, 1979c: 239-40, 256). One does, however, see turkeys occasionally and the Chinese have a variety of turkey dishes. Some of these, like roast turkey, would not at first glance seem strange to Americans. The flavorings, however, may represent quite a departure, as in one "Steamed Roast Turkey," a dish for a large party, prepared with fermented black beans, scallions, gingerroot, "sherry" (rice wine), soy sauce, sugar, chili oil, and sesame oil. Also resembling American usage is "Cold White-Cut Salted Turkey," though it is Chinese in the plum sauce or hoisin sauce in which the turkey slices are dipped before eating (K. Lo, 1979c: 256-57). One sees stranger combinations or departures in the Shanghai dishes "Abalone and Asparagus and Turkey" and "Steamed Little Bao Tze with Turkey" (W. W. Chang *et al.*, 1970: 70, 144).

Dog

Though the idea of eating dogs (*Canis familiaris*; *kou* or *ch'uan*) may offend animal lovers and others in the West, the practice has been common in various parts of the world since ancient times (P. L. Simmonds, 1885: 62-67; Simoons, 1961: 91-105). There has been an overall decline in dog eating in recent centuries, but the practice has persisted even in certain places, such as Islamic North Africa (Simoons, 1981a) and Europe, where there are strong feelings against the practice. Early in this century, for example, J. S. Thomson (1909: 107) wrote of dog eating in Germany, where, he said, roughly 8000 dogs were "slaughtered for food purposes" in the previous year, 1400 of them in the cities of Kassel (Cassel) and Chemnitz. Schwabe (1979: 167-71, 173) also writes of dog eating in parts of modern Europe, and not merely in times of famine. He provides details of how the Swiss prepare "Dried Dog Meat," and presents a defense of dog eating in nutritional and other terms. Despite his convincing arguments, the strength of widely-held sentiments make it unlikely that dog flesh will soon gain a more significant dietary role in the Western world.

Present evidence suggests that the dog may have been the first animal to have been domesticated, possibly by 12,000 B.C. in the Near East; domesticated dogs, whatever their antecedents, must also have been known in very early times in northeastern Asia, for it was from there that they seem to have been introduced to North America to reach Idaho by about 8400 B.C. (Wing, 1983: 23). In North China, the dog, along with the pig, was one of the two most important domesticated animals kept by early farmers (*c.* 6500-3000 B.C.) (K. C. Chang, 1986: 93, 113).[34] Early Chinese written records reveal that dogs not only served in hunting, as guards, and in ritual and sacrifice, but as a major source of meat. Indeed the dog was considered one of the "five domesticated animals" of ancient China, and its flesh was served at ceremonial dinners and eaten even by kings (Creel, 1937: 76-77; Erkes, 1943: 204-5, 211-13; Veith, 1949: 206; K. C. Chang, 1977b: 29; E. N. Anderson, 1988: 20, 32). In recent centuries dogs of one breed or another[35] have continued to be used in hunting. In Inner Mongolia and Manchuria they serve as sledge dogs. They assist in the herding of sheep and other livestock. They serve as household pets or guards. In competition with pigs and chickens, they are village scavengers who, because of their more efficient digestive systems, are able to benefit nutritionally from eating human excrement, which is sometimes fed to them. And ultimately their flesh may be used as food for humans (Collier, 1921: *passim*; Powell, 1945: 178; Epstein, 1971: 125-45 *passim*; E. N. Anderson, 1988: 102).[36]

Many ethnic groups in China rejected dog flesh (*kou-jou*) as food, some of them, such as Moslems, because they consider dogs unclean; some, such as Buddhists, because they viewed the dog as a friend and protector of the family and opposed killing dogs and eating them; and some, non-Han minorities, because they considered dogs to be their relatives. In Kwangtung, the Yao, like other Yao groups in South China and Vietnam (the Man), believe they are descended from a dog ancestor and reject dogs as food for that reason (Lebar *et al.*, 1964: 85, 88; Eberhard, 1968: 43-50, 461-63; 1982: 81; de Beauclair, 1970: 6; Schafer, 1977: 107-8), though dogs may be sacrificed and eaten by priests and the afflicted in efforts at curing illness (C. B. Lee, 1939: 358n).[37] There seem, however, to be variations among the groups in question. The Li of Hainan are said to have stories of a dog ancestor, but Odaka (1950: 41) found no such stories among the Li he studied, and they showed no reverence toward their dogs and indeed killed dogs ritually and relished their flesh.

Among the Chinese, there has been a decline in the acceptability of dog flesh since ancient times, especially in North China. A certain Monseigneur Perny, who travelled through half of China's provinces in the nineteenth century, observed dog-meat eaten by the poor in all of them; well-to-do people, however, ate much less of it and only once or twice a year, solely in the belief that it stimulates the digestive system (Raveret-Wattel, 1884: 607). Erkes, writing in the 1940's (1943: 211), said that in North China dog flesh is eaten as food only among the poorest sorts of people, and in South China only in certain regions.

In Kwangtung, on the other hand, there is an especially strong liking for dog flesh (Figure 65), which has led to the Chinese folk saying "Loaves of steamed bread are afraid of dogs; dogs are afraid of people from Kwangtung." In the sixteenth century, Gaspar da Cruz observed a thriving market for dogs and dog flesh in Canton. Dogs were carried live in cages around the city for sale, and along the city wall were shops where dog flesh was sold raw or cooked. But, Cruz contended, it was food of the lower classes (Boxer, 1953: 134). Of the accounts of dog eating in nineteenth century Canton, that of Gray (1878, 2: 75-77), a long time resident of China, is the most detailed we have uncovered. Dogs, Gray said, were sold for their flesh on

Figure 65. Dog carcass in Canton marketplace (1985) (photo by Stephen F. Cunha)

one street every day. After butchering and cleaning, dog carcasses were displayed in the windows of restaurants specializing in dog and cat dishes. Menus, of which he provided an example, were posted on restaurant walls, with the flesh of black animals singled out because their flesh was regarded as more nutritious.[38]

That dog and cat flesh should be sold in separate restaurants raises the suspicion that, as implied by the separate system of marketing horse meat in France (Gade, 1970: 4), these were inferior meats. Yet Gray insists that the people who came to such restaurants were mainly upright merchants and craftsmen. Very few were poor. As he put it, it was more or less customary in South China for citizens to eat such food. Altogether Gray estimated that there were no less than twenty such restaurants in Canton, and in them one could see cooks stir-frying dog and cat flesh with garlic and water chestnuts. Writing in more general terms of China, Gray said that at the start of the summer, persons of all classes participate in a dog-flesh-eating ceremony which was intended to strengthen the body and counter summer illnesses and epidemics. Indeed at that time of year restaurants that regularly sold cat and dog flesh were crowded, and even street vendors who ordinarily sold other kinds of meat would display dog-flesh tidbits for sale. A special interest in dog flesh in summer does not fit with numerous observations that among Chinese dog flesh is "hot" in the hot-cold classification of foods, a strengthening food eaten in winter, but rarely in summer (Koo, 1973: 141; E. N. Anderson and M. L. Anderson, 1976: 147; H. D. R. Baker, 1983; Tan and Wheeler, 1983: 43; E. N. Anderson, 1988: 146, 191).[39]

Among other brief notes on dog eating in late-nineteenth-century Canton or South China are those of Giles (1876: 153), Ball (1906: 287-88), and Parker (1897: 175; 1903: 270), old China hands, and Collingwood (1868: 339), a naturalist who was visiting briefly. Collingwood presented a picture in keeping with Gray's: of much dog-meat eaten; of dogs being killed every morning on street corners, then skinned and prepared in suspended cauldrons, apparently out-of-doors; and of pedestrians being invited to stop and eat. Giles wrote that dog flesh was a rather expensive dish, and that dog hams were favored items in South Chinese cookery. Parker said that while in most parts of China dogs are eaten by the poor and were not usual fare, in Canton peddlers sold puppies as a dietary treat.[40] Ball, on the other hand, presented a quite different picture. He did note the Canton restaurants that served dog and cat flesh, but, he said, there were only one or two of these, and dog flesh was an uncommon food eaten only by some people.

Certainly there were pressures against dog-eating in Canton. Cantonese housewives were said to refuse to prepare dogs in their pots or at home (Collingwood, 1868: 339). Buddhist temples displayed signs opposing the eating of dogs, because they were faithful guardians of the household, and Buddhists in Canton seem to have concluded that dog-eating was contrary to divine wishes. Though they might not give up the practice, they did feel that they should not enter a temple to worship for at least three days after eating dog (Gray, 1878, 2: 76-77).

For whatever reasons, public sale of dog meat was banned in Canton in 1915, though it continued to be sold openly in suburban shops (Collier, 1921: 27). We do not know how effective the ban was or how long it continued, but H. T. Chen (1934b: 75), doing research on helminths of dogs in and near Canton in the early 1930's, reported dog flesh a common food for many Cantonese. He also wrote of regular dog-meat markets, of dog flesh costing as much or more than pork, and of villagers on Honam Island near Canton never failing to take home the carcasses of dissected dogs for eating. In present-day South China, E. N. Anderson (1988: 173, 175), while acknowledging the persistence of dog-eating among the Cantonese, observes that it is on a quite minor scale, and mainly in winter for the reasons mentioned above.

The preference in China is for the flesh of suckling pups and for puppy hams. Two Chinese methods of preparing dog ham, and two Chinese dog recipes (for "Stir-Fried Dog" and "Red-Cooked Dog") are included by Calvin Schwabe in his *Unmentionable cuisine* (1979: 172, 175). The methods of making ham differ somewhat but both involve rubbing trimmed dog legs or thighs with salt, packing them in a wooden tub, and removing and re-salting them at intervals of from five to ten days. Then they are turned, and re-packed in the tub. Altogether the curing processes continue for from two to four weeks, after which the hams are suspended for drying in a well-ventilated, shady place for five or six months. By one process, the hams are covered with sesame seeds before being hung.

It is common in China for dog and pork hams to be cured together. Indeed people say that pork hams cannot cure properly, cannot develop their special taste, unless a dog leg is among them (Schwabe, 1979: 172). One wonders whether this belief is an ancient one, perhaps from a time when dogs and pigs were the principal sources of meat in China and when it may have been customary to cure dog and pork together. Rendered dog fat is also employed to give certain Chinese dishes a special taste (Schwabe, 1979: 172).

Judging from analyses we have uncovered (Leung *et al.*, 1973: 104, 107), dog flesh has about the same amount of protein as lean pork, but less fat. Schwabe (1979: 170-71) notes the risk of trichinosis for the person eating undercooked dog flesh, but, he observes, that risk exists as well in consuming undercooked pork or undercooked flesh of other carnivorous or omnivorous animals (Dolman, 1957: 47-50; Schwabe, 1984: 546-52). There is a second risk, in various parts of the world, of contracting hydatid disease from handling dog feces or intestine (Dolman, 1957: 53-55; Schwabe, 1984: 214-18, 362-73, 419-28, 442-44, 472-84). Involved is transmission of a tapeworm (*Echinococcus granulosus* or closely-related species), a parasite that grows like a tumor in various human organs and can lead to the death of its human host. Since *E. granulosus* has been found in dogs in the Peking area (Faust, 1929: 37; H. T. Chen, 1934b: 84) and occurs in Vietnam, a strong case can be made against careless handling of dog intestine by Chinese dog eaters and against pollution of fingers or food with dog feces.

When they settled abroad, the Chinese often continued the practice of eating dog flesh. In many places, this has resulted in conflict with other peoples who disapproved of dog eating. This has been true even in Hong Kong where Chinese dog eating has led to considerable strain with the British community. One reads, for example, of a letter to the editor of a Hong Kong paper written by a British man who said that any person who eats a dog should be strung up, emasculated, and thoroughly beaten (L. Mathews, 1980). In response, a more restrained Chinese writer observed that dog flesh was an ancient and acceptable food in China and that the British attitude was still another instance of the insensitivity and brashness of Europeans in Asia. It should be noted that a ban on slaughtering and eating dogs had been in effect in Hong Kong since 1950, with severe penalties, a $1000 fine and six months in jail, for violations. However, because the law is so contrary to the tradition of the Chinese majority, it has been widely ignored. It did, however, force a change in practice. Instead of a well-organized system of dog purchase and marketing, less efficient, clandestine ones arose. Most commonly, pets were stolen or a group of friends might join in raising a puppy and, after it was six months old, serve it at a special dinner, but with only reliable and close-mouthed guests in attendance. In the winter of 1980, the controversy flared anew when the Hong Kong chapter of the Royal Society for the Prevention of Cruelty to Animals came out for repeal of the prohibition on dog slaughter and eating. The main concern was with stopping the stealing of pets, eliminating brutality in dog slaughter, and, for public health reasons, establishing control over dog butchers. Their position, however, outraged many British residents, heated exchanges occurred in the newspapers, and a special half-hour television feature examined various sides of the controversy (L. Mathews, 1980).

The situation was reminiscent of the public outcry, at about the same time, against Indochinese dog eaters in California, some of whom were also accused of stealing pets. It also recalls the campaign to ban dog-soup restaurants in Seoul, in part to make Korea look good to foreigners attending the 1988 Summer Olympics. In the Korean case, however, it was not Westerners banning dog slaughter or dog eating, but action by an Oriental dog-eating people concerned with their image in the outside world. As reported by Schiffman (1985: 1, 24), one dog-soup restaurant, though forced by Seoul officials to close, reopened after a few months in a new, inconspicuous location at the rear of a building. Though it had no signs and could be reached only by going down an alley, at noon it was full of diners, among them important government officials, enjoying a soup made of cooked dog meat and sesame seeds boiled in water.

Returning to the controversy in Hong Kong, what the RSPCA may have had in mind was a system like the one set up in Taipei, Taiwan, two decades before. In 1960, the Taipei police established regulations requiring approval by the City Health Center of all dogs for slaughter ("Regulations governing sale of dog-meat promulgated by Taipei police," 1960). A certificate of sale from a dog's owner was required; it was checked against dog registration files; and an examination was carried out to make certain that the dog was fit for human consumption. After approval, dogs had to be slaughtered within three days, and their flesh sold within five. As for marketing, only approved, registered restaurants, which were required to keep separate books on dog-meat transactions, could sell dog flesh or dog-meat dishes. It was hoped that the above would reduce the theft of dogs for sale to restaurants, and dog owners were urged to notify the police if their dogs were missing so descriptions could be entered on the "missing" list at the CHC. If an owner, after notification, did not claim his dog within seven days, an authorization for slaughter could be granted.

Meanwhile in the People's Republic itself there were campaigns of dog killing that greatly reduced the supply of dog flesh. The initial explanation, in the 1952 campaign, was that this was being done to protect the public, to thwart an American germ-warfare scheme during the Korean War involving the use of dogs to transmit disease organisms to the Chinese people (Kinmond, 1957: 162-65). The foreign reporter who wrote of this wondered, as an alternative, whether the campaign may have been a deliberate Chinese government plot to eliminate dogs because they ate too much. His suspicion was confirmed during another campaign to exterminate dogs in 1963. At that time, when army, police, and Communist Party activists were all mobilized to kill dogs, the reasons given were economic as well as hygienic. The idea was that "all animals which are not useful should be wiped out because they consume precious food" ("Chinese reds reported slaughtering dogs," 1963). Meanwhile, the newspaper report noted, those dogs not killed would be raised in cages for eating. The decrease in numbers of dogs did contribute to a striking decline in the prevalence of kala-azar (from 530,000 persons in 1951 to 10,000 at the close of 1958); this is a disease of the liver and spleen, common in North China, for which dogs were a major source of human infection (Banister, 1987: 57).

Despite the campaigns, dogs have survived in China, especially in rural areas. Even in Peking, which had an official ban on dog keeping, dogs made a comeback. By the late 1970's, one could see dogs accompanied by young boys on back streets and lanes, and there was enough of a supply of dog meat, perhaps from nearby communes, to permit a handful of restaurants to include dog meat on their menus (Munro, 1977). By 1987, China's dog population had increased greatly, with a single district of Kwangsi having more than 400,000 dogs, one for every ten people. There and in other parts of China, especially in rural areas, many thousands of people are now bitten by dogs each year, and action to reduce the numbers of dogs, especially stray and rabid ones, is again being urged ("Disease risk from more dogs," 1987: 3).

Cat

The cat (*Felis catus*) is believed to have first been domesticated in Egypt from two wild cats of the region, *Felis silvestris libyca* and *Felis chaus* (Robinson, 1984: 218-19). According to Baldwin (1971; 1975), an authority on cat domestication, semi-domesticated cats were living in farm villages in Egypt following about 5000 B.C., playing a role in destroying vermin. During the period from about 2000 to 1000 B.C., semi-domesticated cats were kept for cult purposes, and domesticated ones, in small numbers, were also found. Some domesticated cats were kept at temples; others were used by upper-class Egyptians as pets and hunting animals. Following about 1000 B.C., the cat gained enormous popularity among the Egyptian people as a whole. It was associated with and symbolic of the goddess Bast, and was even viewed as a deity in its own right. Cats could not be killed, and after death they were mourned, mummified, and buried in cat cemeteries established at various places. One cemetery at Bubastis, sacred city of that deity, was said to be several acres in size.[41] The Egyptians regarded the domesticated cat as their national animal and tried to keep it from spreading to foreign countries. At one time, they even sent envoys abroad to buy cats and bring them back to Egypt. Their efforts failed, and in the Roman period the domestic cat had diffused widely across Europe (Zeuner, 1963: 395-96; Baldwin, 1971: 77-83); it was associated with the cult of Bast/Isis; and it apparently enjoyed a high status there among her devotees.

The evidence of early domesticated cats in China is inconclusive. The bones of cats (*Felis* sp.) were found in the Shang site of An-yang. They were very few, less than ten altogether, and the cat is not among the An-yang animals K. C. Chang (1980: 139, 143) considered to have been domesticated. It is true that T. K. Cheng (1959-60, 2: 83) regarded the An-yang cat bones to have been from domesticated animals. Yet H. E. Gibson (in Sowerby, 1940: 173) found nothing in Shang oracle bone inscriptions to suggest that cats were domesticated during that period, and he and Sowerby (1940: 50) judged the cats depicted on the bones to be wild animals, likely *Felis euptilura*, the wild cat of China. There are hints that domesticated cats may have been known in China at the time of Confucius (551-479 B.C.), but, according to Baldwin (1971: 125-27), the first convincing evidence of domesticated cats in China dates from Han times. He notes that caravan traffic over the Silk Road to Persia likely started in the second century B.C., and speculates that the cat may have been introduced to China from Persia or India quite early.

Whatever the precise date of its introduction, the domesticated cat (which the Chinese call *mao*, apparently from its mewing), like the domesticated dog and rabbit, came to be a favorite of certain Chinese artists, whose paintings commonly show it playing with insects, flowers, or children (Sowerby, 1940: 50). The cat also plays a certain role in Chinese beliefs of the supernatural, though it is not accorded the sacred status it had enjoyed in ancient Egypt. More critical to China's economic well-being, the cat also assumed an important role, as it has in other parts of the world, in controlling the rodent population. In traditional China, for example, cats were always found in buildings where silkworms were kept, to guard them from rats and mice. The cat—its flesh, skull, brain, eyes, and other body parts, saliva, and excreta—also played various roles in medicine, prominent among them the treatment of rat bites and illnesses deriving from rats (Read, 1976: Sections 371-72). And among the poor, cat fur on occasion was cured for use as clothing (Baldwin, 1971: 127-28).

The idea of eating a domesticated cat raises as much emotion among many Westerners as that of eating a dog. When they have encountered the practice abroad, the reaction has frequently been one of disgust. On the other hand, as Calvin Schwabe observes (1979: 176), cat is quite like rabbit in taste; cat flesh has commonly been marketed in Europe as rabbit; and it is eaten there more than people will admit. One Spanish acquaintance from Alcántara, in Estremadura, told Schwabe of a cat stew preferred there to a comparable stew made of rabbit.[42] In a similar vein is a story my mother told me. During World War I, she lived on a farm in the Flanders, just behind the third line trenches, and her family came to lose their house cats with unusual frequency once French troops were stationed there. The family suspected that their cats (known euphemistically to the French as "roof rabbits") ended up as tasty supplements to military rations. On one occasion the French soldiers even offered the family some "rabbit stew," which offended them because they believed it to include their own cat.

Turning to modern China, we found cat dishes, such as "Steamed Cat," at Canton's Wild Animal Restaurant, and noted cages of unkempt cats being sold for eating in the large Ch'ing-p'ing market in that city. We were told that these cats were not raised especially for their flesh, that they were village cats caught and carried into the city for sale. There are also scattered references to cat-eating in former times. Freeman (1977: 154), for one, writes of "wildcats" (feral *Felis catus* and various related wild species of *Felis* occur in China) being a novelty dish of wealthy urbanites in Sung times.[43] There are also accounts, from this century and earlier, of domesticated cats being eaten, and/or of live cats being sold along the streets for eating (Astley, 1745-47, 4: 86; Gray, 1878, 2: 76; S. W. Williams, 1895, 1: 318, 777; Parker, 1897: 175; 1903: 270-71; J. S. Thomson, 1909: 107; Ross, 1911: 80). Some cat flesh for the table apparently came from stray or stolen animals, for Ball (1906: 287) writes that in Canton their house cats would disappear, and he suspected that they had been eaten. Other cat flesh, however, came from animals specially reared for food, having been fed rice, or rice and vegetables. However such cats were obtained and fed, some of them passed through the shop in Canton that sold only cat flesh (Parker, 1897: 175; 1903: 270-71) or ended up in one or another of the restaurants there that served cat flesh as well as dog flesh (Gray, 1878, 2: 75-76; Ball, 1906: 287).

As to the dietary importance of cat flesh, one cannot be certain that Europeans in China have observed and reported as fully on the matter as on that of dog flesh. It would seem, however, that, as was true of dogs, cats played a larger dietary role in the south, particularly Kwangtung, than in the north. It also seems that cat flesh was of far less dietary importance than dog flesh. S. W. Williams (1895, 1: 318, 777), for one, wrote that cat flesh was not eaten to any extent in China, and that foreign travellers exaggerated its dietary importance. Ball (1906: 287-88) reported that numerous honorable Chinese would react to eating cats with disgust, much as Europeans do, and he called cat flesh a minor food consumed mainly by the poor. The latter observation does not, however, fit with that of Gray (1878, 2: 75-76), mentioned previously, that dog and cat restaurants in Canton were frequented by respectable artisans and merchants and by only a few poor persons.

A difference of opinion among Westerners continues today as to the prevalence of cat-flesh consumption in China, and whether it is done primarily for dietary or medicinal reasons. E. N. Anderson (1988: 144, 146, 159, 173; see also E. N. Anderson and M. L. Anderson, 1977: 336), for example, writes that, contrary to the opinion of some Westerners, cat-flesh eating is more a subject of conversation than an item of diet. Indeed he observes that cats are very seldom eaten in China, and classifies cat flesh among unusual items consumed more as medicines than as foods. As we have indicated above, the Chinese do indeed regard cat flesh as having medicinal value. And sometimes they prepare it solely for medicinal purposes, as clearly demonstrated by the case of the Chinese woman who skinned a cat to give to her son as a remedy for stomach ache (Holmes, 1948: 186-87).[44] At the same time, other observers present evidence that cat flesh is also regarded as a tasty food. H. D. R. Baker (1983), for one, lists cat flesh among Cantonese delicacies, and Schwabe (1979: 177), who approaches the matter from the perspective of cuisine, has presented two Chinese recipes involving cat flesh, "Steamed Cat and Chicken" and "Dragon, Phoenix, and Tiger Soup," which would please even an exacting gourmet. He writes of the soup, moreover, as an important delicacy in China, though E. N. Anderson (1988: 146), in considering this unusual combination of snake, chicken, and cat, insists that it is consumed more as medicine than food. Also suggestive of dietary intent is Schwabe's observation that in nineteenth-century Canton cats' eyes were sold in food stores, though he was unable to find out how they were prepared.

We have found no nutritional analysis of cat flesh.

Snakes

The eating of snakes as medicine or food is reported for peoples of various parts of eastern Asia (I. H. Burkill, 1966, 2: 2076-77; Schwabe, 1979: 259-61). Though in earlier times snakes were consumed in North China, today the eating of snakes (*shê*) is considered a custom of the south, particularly of the Cantonese (B. Y. Chao, 1972: 16; Freeman, 1977: 169; Mote, 1977: 242; Schafer, 1977: 101; Summerfield, 1984: 132; E. N. Anderson, 1988: 147, 159, 173). Anderson and Anderson (E. N. Anderson and M. L. Anderson, 1977: 336; E. N. Anderson, 1988: 146, 147) write that in modern South China snakes are consumed mainly for medicinal purposes, and only once in awhile as a curiosity. In a similar vein, Lin Yutang (1935: 336), a native of Amoy (Fukien), said that he himself had never eaten snake, had not observed any of his relatives doing so, and that "snake-eating is as much a curiosity to the Chinese as it is to the foreigners." One finds other statements (Wallnöfer and von Rottauscher, 1965: 72; Schafer, 1977: 101, 129; Schwabe, 1979: 259-61; H. D. R. Baker, 1983; Summerfield, 1984: 132-34), whether referring to China or the Cantonese, that snake, at least to some, was also a delicacy, and Kenneth Lo has written (1979a: 77) of "Great Assembly of Chicken with Five Varieties of Snake" as one of the best known of Cantonese feast dishes.

The views that snake is a medicinal and delicacy are not mutually exclusive, for some of China's most prestigeous banquet dishes, such as ones with bird's nest, serve both purposes. Snake, like dog, is both a "heating" and strengthening food, and is thus typically consumed in winter, or at other times by persons in particular need of such food (Koo, 1973: 141; Ling *et al.*, 1975: 128; Tan and Wheeler, 1983: 43, 44; E. N. Anderson, 1988: 147, 175, 191, 193). The more poisonous a snake, the more valuable medicine it is (E. N. Anderson, 1988: 147), which

Figure 66. Cages of snakes for sale at a shop in Canton (1987)

explains the report (P. L. Simmonds, 1885: 241) of hundreds of dried venomous ones being imported for medicinal purposes from China by Chinese doctors in San Francisco.

In Canton and Ch'eng-tu, capital of Szechwan, we found snakes of various species[45] sold in certain marketplaces and occasionally in small shops along a street (Figure 66). Though snakes were usually kept in cages, we observed one Ch'eng-tu merchant holding aloft a live snake and extolling its virtues to potential customers (Figure 67). When a deal was struck, the snake was beheaded, then disemboweled and skinned while still writhing. Another man nearby was displaying for sale another snake, already skinned, cleaned, and tied into a neat circular bundle. In the seventeenth century, Peter Mundy observed a live four- or five-foot snake in a Kwangtung market whose mouth was sewn shut (Temple, 1907-36, 3: 190), a procedure of which we have not heard elsewhere.

In Canton today one can find snake dishes in certain restaurants, and there is at least one restaurant that specializes in them. Described (Summerfield, 1984: 132-34) as a place for adventurous tourists with strong stomachs, the restaurant window contains live snakes, includ-

Figure 67. Merchant proclaiming the virtues of a snake to potential customers (Ch'eng-tu, 1987)

ing venomous ones. The foreigner is led to a separate dining area where he can, if he wishes, start his meal with a small glass of snake bile mixed with *Mao t'ai*, a distilled liquor of very-high proof. The bile is obtained from living snakes by a deft operation performed at the dining room table. The bile drink is believed to enhance virility and to be beneficial to the heart. For his main course, the diner can choose to have the snakes killed and skinned at his table, or he can simply order from the menu. The restaurant has one menu, suitable for a group, which consists of ten dishes, but there are three specialties of the house: "Chicken Soup with Three Kinds of Snake," a soup made from the flesh of venomous snakes (with the venom removed), chicken flesh, and spices, and garnished with chrysanthemum petals; "Clear Broth of Big

Mountain Snake," a delicately spiced, somewhat sweet broth made from one or more large snakes, such as the boa constrictor; and "Struggle of the Dragon and the Tiger," a thick soup of snake and civet cat.[46] Snake wines are also made of fermented or distilled liquor in various parts of China and serve such medicinal purposes as treating malaria, paralysis, and insanity, and infestations of intestinal parasites (P. L. Simmonds, 1885: 239). Of special renown is the snake wine of the Cantonese, actually a high-proof liquor in which a venomous snake has been soaked, occasionally for years (Em. Hahn *et al.*, 1973: 146). This strong, green-colored drink is consumed as a tonic, usually in fall when the body is in need of rejuvenation. A similar drink is "Dragon-and-Phoenix Wine," made from the flesh of venomous snake and pheasant soaked in wine. Such drinks are commonly taken for reasons of health, with the exact recipe and herb ingredients varied according to the illness. A drink made of snake bile and gall bladder soaked in "wine" is believed not only to stimulate health and vigor, but to cure rheumatism (Em. Hahn *et al.*, 1973: 146). This calls to mind Schwabe's account (1979: 260) of a Chinese cafe owner he met in Montana, who aged rattlesnakes in bourbon whiskey for four to five years, then filtered the liquid and shipped it to San Francisco for sale in Chinatown as a treatment for rheumatism. Another medicine is a powder made of dried snake skin, which is imported to Malaya from China and sold in Chinese pharmacies (Hooper, 1929: 148). Relevant also is the occurrence in the *Pen-ts'ao* of snake wine along with liquors made from other animals, including dog wine, deer wine, and tortoise wine (Macgowan, 1871-72: 237-38).

Rats

As for rats, the reader should be aware that in parts of West Africa they are of considerable dietary importance, and that in certain areas of Ghana rats and mice contribute over half of the meat produced locally (Schwabe, 1979: 204). Rats were also eaten openly and in quantity in the siege of Paris during the war between France and Prussia, and they continue to be consumed today in certain parts of France (Schwabe, 1979: 204).

In China, some eating of rat (*shu*) occurred during times of hunger. Schafer (1977: 107) writes that in T'ang times rats were eaten in one city under long siege. He prefaces his account with the observation that in such situations people were forced to eat foods that were offensive ordinarily. In a similar vein, Lin and Lin (H. J. Lin and T. Lin, 1972: 145) write of the Second World War in Chungking when rats flourished, reaching such a size that cats dared not confront them. Yet people apparently were more successful in hunting and trapping them, for dishes of red-cooked flesh would appear on the tables of some Chungking people, flesh they were reluctant to identify and that some would not consume. The Chinese in the above accounts differed little from peoples of other lands facing hunger and starvation.

Other rat consumption in China, however, was an activity of normal times. In ancient China, according to K. C. Chang (1977b: 29), both the common rat and the bamboo rat (*Rhyzomys sinensis*; *chu-liu*) were hunted for food, though they seem to have been of quite minor importance. Bamboo rats were given by certain local magistrates to the imperial kitchens in T'ang times (Schafer, 1977: 100-101), so this species at least was quite acceptable food. This seems confirmed by Read (1976: Section 392), who, drawing on the *Pen-ts'ao* and modern literature on China, writes of the bamboo rat as plump, the size of a rabbit, commonly eaten as food, bland, tonic, sweet, with the flavor of duck; cooked with mutton, it makes that flesh tender.

As for other rat species in pre-modern times, Schafer (1977: 101) describes a custom found among T'ang Chinese in the south, perhaps taken from aboriginal groups: newborn rats (not otherwise identified) full of honey, crept around the dinner table making soft noises, and diners, using chopsticks, picked them up and consumed them alive. And Mote (1977: 201) writes of a Ming account that mentions rodents of several types as regular items of diet.

Accounts of nineteenth century China reveal that consumption of rats was by no means universal and that some Chinese would have been as offended at the idea as Europeans would. Indeed, much eating of rats was done for medicinal rather than culinary reasons. An old Cantonese teacher prepared and ate a rat whenever he caught one because its flesh was "warming"; and a Cantonese peasant woman ate boiled rats because she found them effective against rheumatism and chills (Parker, 1897: 175; 1903: 270). Most women who ate rats, however, seem to have done so because they were going bald, for rat flesh was viewed as effective in restoring hair. As for food uses, Parker (1903: 270) singles out Canton as China's center of rat eating, and in fact he never heard of rat eating in any province but Kwangtung. Others (Collingwood, 1868: 339; Gray, 1878, 2: 77; Ball, 1906: 288; Ross, 1911: 80) confirm its occurrence there, saying that it was not uncommon to see dried, salted rats hanging in Canton shops along with dried ducks and other dried fowl or meat. Gray writes that in winter, "when rats are in season," the windows of poultry shops along one Canton street were often filled with them. Parker (1897: 175; 1903: 270), on his part, notes that split, dried rats are usually hawked in Canton's streets "strung on a stick" like kebobs in the Near East. That consumption of rats in South China was not limited to Cantonese is clearly indicated by the observation (Odaka, 1950: 26) that the Li, tribal people of Hainan, are adept at trapping rats for dietary purposes; the rats, whether field rats or ones caught at home, are then eaten, like other flesh, after being boiled. Yet today, E. N. Anderson (1988: 175) writes that he has never observed rat eaten, and that, perhaps because of acculturation to Western ways, the Cantonese he knew were disgusted at the very idea of consuming rats.

Except for the T'ang account of newborn rats mentioned above, we have found little to suggest that rats were confined and fed or fattened for the table in China,[47] as the Romans did with another rodent, the fat or edible dormouse (*Glis glis*), which was kept in special enclosures (*glilaria*), fed on acorns and chestnuts, and stuffed and baked as a delicacy to be served at banquets (Zeuner, 1963: 415-16; C. A. Wilson, 1973: 72; Clutton-Brock, 1981: 152-53).

Insects

Insects have long been consumed in China as medicine, as food, or both.[48] Read (1940: 22-26) observed that medicinal use of insects and insect products goes far back in Chinese history; that knowledge about medicines of insect origin was exchanged with foreign lands;[49] that the *Pen-ts'ao* listed 107 kinds of "insects" used as medicine (many were insect products or were not true insects); and that thirty medicines of insect origin could still be obtained in the contemporary drug market.[50] K. C. Chang (1976: 122; 1977b: 30) lists bees, cicadas,[51] and moths among foods of animal origin in ancient China. Schafer (1977: 114) writes of a favorite pickled relish in T'ang China made of ant eggs.[52] There were, however, ethnic differences in the acceptability of insects, at least certain of them, as food, which is revealed in a Sung account referring disparagingly to the non-Han people of Hainan who "eat creatures like flies, gnats, and earthworms" after they had been steamed in sections of bamboo (Freeman, 1977: 169-70).

In traditional China, insects in some cases were raised or collected by specialists who made their livelihood in this manner on a regular basis. In other cases, people took up full-time insect collecting only when a special opportunity presented itself, as during locust infestations. And still others—children and farmers, for example—collected insects on an intermittent, part-time basis as need and opportunity arose, commonly as an interruption in their normal round of daily activities. Though insects might be consumed by the individual who captured them or by his family, they were also sold in the streets, in markets, and at stalls and shops.

The Cantonese are reputed to be China's leading insect eaters (Hoffmann, 1947: 236; Bodenheimer, 1951: 278), and it is therefore especially appropriate to approach Chinese entomophagy from the perspective of Kwangtung. In this, we rely especially on a study published in 1947 by Hoffmann, who lists six main types of insects consumed as medicine and/or food in that province: beetles, silkworm pupae, green-bottle fly larvae, giant waterbug, stinkbug, and caterpillars infected with fungus of the genus *Cordyceps*.[53]

The beetles identified by Hoffmann represent at least several different species and two families, the *Dytiscidae* and *Hydrophilidae*, and the two commonly-used beetles could be "purchased in any grocery in Canton" and "in Cantonese food shops in other large cities" such as Shanghai, Tientsin, and Peking (Bodenheimer, 1951: 277). Beetles were commonly eaten after being cooked in hot brine, which seemed to increase both their oiliness and odor, but on occasion they were fried. When consumed, their legs and certain other chitinous parts were thrown away. Though mainly an anti-diuretic medicine, they also served as a confection, eaten in Canton just like peanuts and watermelon seeds, and in Peking as restaurant fare (Bodenheimer, 1951: 273).

Kwangtung was too distant from the famed Silk Road of early times to provide silk for overland transport by caravan to the Near East and Europe. Thus its silk was produced then mainly for home use, a situation that changed only after European ships began arriving in Canton and sea trade with Europe became direct and well established ("The silk industry of Kwangtung Province," 1929: 619). By traditional times, Kwangtung had become China's second most important silk-producing region, and, from a food perspective, it enjoyed a plentiful supply of silkworm pupae (*ts'an-yung*) for human consumption. In areas where silkworms (*Bombyx mori*; *ts'an*) were cultivated in China, rural folk sometimes prepared the silk themselves, and one informant, Dr. Fen, described procedures followed in rural Shantung, Kiangsu, Chekiang, and parts of Szechwan (Bodenheimer, 1951: 275) which may have been followed in Kwangtung as well. In spring, Fen wrote, farmers would collect silkworm cocoons and then preserve them by baking or pickling with salt until they were needed.[54] The silk itself was processed mainly in summer and it was then that silkworm pupae were obtained in quantity and sun-dried for use throughout the year. Pupae from baked cocoons were preferred over those from pickled ones. Before eating such pupae, people first soaked them in water, then fried them either with onions and a sauce, or in an omelette with eggs. Such dishes were everyday fare, but also quite suitable for guests, which suggests that they were not considered an inferior food.[55] In Kwangtung, most good cocoons were sold to filatures, large factories where raw silk was processed for export. The women workers who soaked the cocoons in very hot or boiling water, unwound them, and reeled the silk were thus in an excellent position to snack on silkworm pupae as they worked, and their places of work typically were permeated with the pleasant odor of cooking food (Hoffmann, 1947: 234).

Though P. L. Simmonds (1885: 349) has written of silkworm pupae discarded in silk reeling being sold as food for the poor, other writers (Darwin, 1800: 364; Gray, 1878, 2: 224; Bodenheimer, 1951: 274-75; Kellogg, 1968: 20) say that pupae were regarded as delicacies and excellent, strengthening, and refreshing food. There are reports of silkworm pupae being "hawked about the streets" as food in China (P. L. Simmonds, 1885: 349); of silk- worms being sold at shops in cities and towns throughout the country (Bodenheimer, 1951: 274); and of roasted pupae sold at food stalls in Kwangtung (Hoffmann, 1947: 234). Hoffmann (1947: 234) was told, however, that in Central China silkworm pupae were sold, along with other wastes of silkworm rearing and processing (mulberry leaves, feces, dead worms), as feed for fish reared in ponds. In such a case, the pupae ultimately did contribute to the human food supply, but indirectly through an aquatic intermediary.[56]

In the Canton area shortly after its capture by the Japanese, Hoffmann noted fly larvae being raised commercially. The larvae apparently were principally of the species *Chrysomyia megacephala*, a green-bottle fly. Larvae were washed and sun dried, and sold principally in Canton to medicine shops but also as food. Hoffmann thought that the eating of fly larvae was a normal practice, not a matter of wartime scarcity. In support of this view is P. L. Simmonds' observation (1885: 349) in the late-nineteenth century that the Chinese raise blue-bottle fly larvae in rotting fish near the ocean and eat them flavored with spices and sauces.

The giant waterbug (*Lethocerus indicus*) was normally eaten in Canton after being dropped in boiling, salted water. As an alternative, it was placed in spiced deep fat for a short while. These bugs were expensive in prewar China and were eaten as a delicacy rather than as medicine. Typically they were displayed in large glass jars for sale in Canton, Hong Kong, and other cities where Cantonese settled. Hoffmann also notes that by eating the giant waterbug the people of Kwangtung obtained a greater fish supply, for this bug kills young fish in breeding ponds.[57]

A half dozen or more species of stinkbugs, those insects of disagreeable odor, are eaten in various parts of the world. For China, the species mentioned by Hoffmann (1947: 236) and Read (1940: 29) is *Coridius chinensis* (*Aspongopus chinensis*), a pest of beans, cucurbits, and the castor-oil plant. The Chinese have a very high regard for the aphrodisiac qualities of this insect, and consider it effective also against asthma and kidney and spleen failures. It is the main ingredient in the renowned "Black-dragon Pills" and in one form or another it is found in medicine shops everywhere in China.

Finally, according to Hoffmann, are the caterpillars (Hepialids and other types) that have become infected with fungus of the genus *Cordyceps*. Over time, the fungus, rooted in the caterpillar's neck, grows upward to a height of six to eight inches; it also grows into the body of the caterpillar until it replaces all the tissue in the insect's outer skin. Then both the caterpillar and fungus die and become dry, hard, and brown. The result is a rigid caterpillar with what appears to be a horn protruding from its neck, which serves as a drug known as "winter-worm, summer-grass" (*tung-ch'ung-hsia-ts'ao*) or "winter worm grass" (*tung-ch'ung-ts'ao*) (Ball, 1906: 72; E. Watson, 1923: 400). A dozen or so of such caterpillars, each with fungal growth longer than its body, are tied into a little bundle for shipment. Such caterpillars, according to Hoffmann, may be prepared in a broth, a delicacy or tonic, with both liquid and caterpillars consumed. Recently in a Friendship Store in Canton frequented by foreigners, we found cans (380 gm.) of "Stewed Cordyceps Sinensis with Chicken in Soup," a product manufactured by the China National Medicines and Health Products Import and Export Cor-

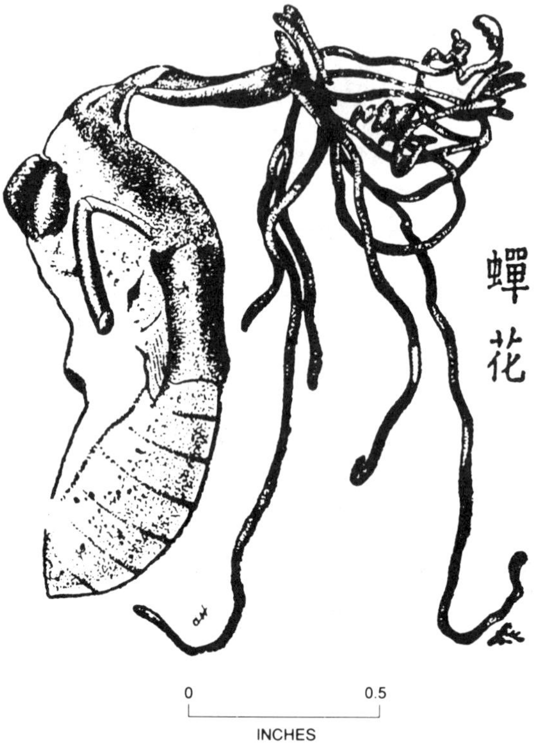

Figure 68. Cicada nymph infected with the fungus *Cordyceps sobolifera* (from Read, 1940: Figure 1)

poration, Chungking Branch (Szechwan). *Cordyceps*-infected caterpillars have a high reputation as a treatment for jaundice and tuberculosis (Read, 1940: 31). Hooper (1929: 43) observes that they rival ginseng as a strengthening agent, that they are exported, and that in Annam and Malaya they are seen as having aphrodisiac qualities as well. He speaks of them as a scarce, highly regarded medicine, among the most expensive in China, marketed for an equivalent weight of silver, and occasionally for four times that weight. As a result they were eaten only by the middle classes and well-to-do.

Other insects are also attacked by *Cordyceps* fungi and Hoffmann (1947: 234) found a few peasants near Canton who had infected cicada nymphs (Figure 68) that they sold for medicine against various ailments.

Hoffmann mentions grasshoppers in Kwangtung only in passing. He says merely (1947: 236) that specialists there made their living by collecting live grasshoppers which they sold alive as food for birds. Other observers have reported grasshoppers and locusts being used as human foods in Kwangtung or in South China in general (Read, 1940: 27, 29). Such consumption seems to have been of minor importance there as compared to North China. This is because locust plagues did not occur in the south, whereas large parts of North and Central China have been subject to frequent locust plagues over the centuries (Donovan, 1842: 26-27; Mallory, 1926: 59; Bodenheimer, 1951: 276; S. Y. Hsu, 1969). This has led to the popularity of locust cults there (S. Y. Hsu, 1969) and establishment of temples associated with locusts, which seem to have been absent in the south (Eberhard, 1968: 251).

Oliphant reported (1860: 273) that when the Tientsin area of northeast China was invaded by locust swarms in the mid-nineteenth century, hunting locusts was a popular and profitable activity for young people, who peddled fried locusts on the streets. Sowerby (1925: 237-38) also reported clouds of locusts in Tientsin and neighboring areas, and commented on the great damage they did to crops. The people, on their part, collected locusts and ate them. They also went into their fields to collect ordinary grasshoppers. These were so numerous that boatloads of grasshoppers in large baskets were carried to local markets from swampy areas nearby. Grasshoppers and/or locusts were also sold in Peking markets and grocery stores. Before use, locusts and grasshoppers were killed by boiling or steaming. Their wings and legs were pulled off, and their heads and intestines might be removed. Then they could be dry-roasted; fried in oil, or with salt and sauce; or they could be sun-dried for storage, and fried only as needed. Some locusts in grocery stores were fried and ready to be eaten, while others had only been dried, leaving the purchaser to fry them at home (Bodenheimer, 1951: 273, 274, 276).

Neither Hoffmann nor Bodenheimer report the eating of crickets (family *Gryllidae*) by the Chinese, though the practice is reported in Burma (Bodenheimer, 1951: 268), Thailand (Bristowe, 1932: 399), Indochina (Tiêu, 1928: 738), and Japan (Bodenheimer, 1951: 280). Whatever its dietary role in China, the cricket has long played other important roles there. To better enjoy cricket concerts, in T'ang times the Chinese started keeping crickets in cages, and in Sung times they created a sport of cricket fighting and a cult centering on the cricket (Laufer, 1927a: 7; Y. C. Hsu, 1929: 7; Sowerby, 1933: 159). This cult persisted into modern times. For example around the turn of this century in Peking the craze for expensive cricket cages led many nobles to become impoverished (Tun, 1965: ix, 81-83). If Y. C. Hsu (1929: 9) is correct, all species used for fighting in China were in fact crickets (family *Gryllidae*). In addition to crickets, however, katydids or long-horned grasshoppers (family *Tettigoniidae*) were held captive for their song; and in a few places, it seems, true grasshoppers (family *Acrididae*) were also kept (Laufer, 1927a: 10-11, 13; Y. C. Hsu, 1929: 16; Sowerby, 1933: 160-61). China is the only country in the world where cricket fighting has developed into a national pastime. Two places near Canton enjoyed a special reputation in this regard, and the earliest Chinese book on cricket fighting reported that the strongest and bravest crickets came from Kwangtung.[58] Cricket fighting can be an elaborate affair of pomp and ceremony. It involves the prestige of the cricket owner, is an occasion for gambling, and often arouses strong feelings. When a cricket with many victories to its credit finally dies, it may be buried with solemnity in a small coffin. Laufer, in his fascinating and definitive study *Insect-musicians and cricket champions of China* (1927a), describes the care with which owners feed their fighting crickets. But we found nothing in his or other studies to suggest than an owner would eat his crickets, even those that had been defeated, showed cowardice, or were undesirable in other ways.

Turning to the nutritional value of insects, Bodenheimer (1951: 25-34) and Taylor (1975: 54-58) have both extolled their potential as sources of protein. They also note that the insects consumed in quantity around the world are not only rich in protein, but in fat and calories, and that many insects provide significant amounts of minerals and salts, as well as vitamins. Silkworm pupae, Taylor says, have moderate amounts of vitamin A; locusts are somewhat abundant in riboflavin and niacin; and bee larvae, in vitamins A and D. The few analyses we have found of insect foods used in East Asia or elsewhere are included in Table 4. The reader will note that all are good sources of protein, and that dried caterpillars (not identified as to species) are especially rich in calcium, iron, riboflavin, and niacin. Though information presently available is quite fragmentary, one suspects that more extensive and complete analyses will demonstrate that insect contributions to Chinese diet should not be overlooked, as has so often been the case.

Still Other Animals

There are scattered references in our sources to consumption of still other animals in China's past, among them pheasant,[59] partridge, peacock, guinea fowl, deer, fox, and hare. Some of the above may have been domesticates or captives, others game animals. In traditional times, game continued to be eaten where it was available (Figure 69), and flesh of some game animals was even exported. We refer here to the tinned rice-birds from the Canton area that could be purchased in New York in the first decade of this century. These beautiful yellow birds, which were captured in nets and then drowned so that their struggles would not cause them to lose weight, were also hawked about the streets of Hong Kong (J. S. Thomson, 1909: 308).

Fish and turtles will be considered in our next chapter, and a few additional edible animals mentioned in our discussion of Chinese delicacies. Before leaving the matter of flesh and fowl consumed by the Chinese people, we feel obliged, however, to mention frogs, snails, and a few unusual creatures which often find their way into the Western literature on China.

China has some excellent habitats for frogs, yet, though southerners are noted for their taste for frogs (Schafer, 1967: 249-50; Eberhard, 1968: 202-3), we have found little information on how they were collected and marketed in traditional times, or how widespread frog cultivation may have been (frog cultivation was found in Szechwan, and today is practiced in Taiwan and the People's Republic) (Eberhard, 1968: 202-3; T. P. Chen, 1976: 129-35; Atkinson, 1977: 340). S. W. Williams (1895, 1: 778) did describe one method of capturing frogs. In a paddy known to contain mature frogs, a young live frog, little more than a tadpole, is tied to a string and bobbed up and down. Attracted by the movement, a mature frog approaches and swallows the young frog whole. It, in turn, is pulled in by the frog-fisher. The frog-fisher then extricates the young frog to be used repeatedly as long as it continues to live. The sixteenth century account of Canton by Friar Gaspar da Cruz, in turn, describes a well-organized and apparently substantial trade in frogs (probably the Chinese bullfrog, *Rana tigrina regulosa*, the common species sold for food in Hong Kong). At the city gates, frogs, kept in large tubs of water, were sold by vendors who were responsible for skinning them for the purchaser, and did so with great skill (Boxer, 1953: 131). Another report, for a city near Ning-po in Chekiang, describes frogs carried to market in baskets and tubs, with the vendor skinning and preparing them, then weighing them and selling them to a customer (Fortune, 1857: 45). Frogs have continued to be marketed in China up to the present day, for, in Kwangtung and some other places, they are considered delicacies and banquet dishes.[60] When they serve as food, frogs are called "chicken of the rice fields" by the Chinese; and frog, though more tender than chicken, does indeed resemble it in taste and in the ways it can be prepared. There are dishes in which the entire body

Table 4. Nutritional value of some insects and insect larvae used as food in East Asia or elsewhere (after Platt, 1962, and Leung *et al.*, 1973)

Food and Description	Composition of Foods, 100 grams, edible portion and as purchased																
	Food Energy (calories)	Moisture (percent)	Protein (gm.)	Fat (gm.)	Carbo-hydrate, Total (including fiber) (gm.)	Chitin (gm.)	Ash (gm.)	Calcium (mg.)	Phos-phorus (mg.)	Iron (mg.)	Sodium (mg.)	Potas-sium (mg.)	Retinol (mcg.)	β-carotene Equiva-lent (mcg.)	Thiamine (mg.)	Ribo-flavin (mg.)	Niacin (mg.)	Ascor-bic Acid (mg.)
Silkworms (*Bombyx mori*) raw, E.P.	229	60.7	23.1	14.2	0.5	-	1.5	6	252	1.2	-	-	-	-	-	-	-	-
Caterpillars, dried larvae	372	*8	55	13.8	7	9.1	-	270	-	20.0	-	-	-	-	0.35	2.0	11.2	-
Locusts, mature	134	**70	20	6.0	***	4.0	-	30	-	1.0	-	-	-	-	-	0.5	2.2	-
Termites, mature	148	75	10	12.0	***	1.2	-	12	-	1.0	-	-	-	-	-	-	-	-

* when fresh, water content 75-85 percent
** when sun-dried, water content 5 percent
*** contains too small a quantity to be significant in dietary evaluations
\- not determined

Figure 69. Bird catchers (from Gray, 1878, 1: opp. 128)

of the frog (after removing its head, guts, and skin) are used, such as the "Quick-Braised Soya Frog in Wine-Lee Sauce." There are dishes of frogs' legs, such as the Cantonese "Quick-Braised Frogs' Legs in Oyster Sauce"; the Shanghai "Frogs' Legs with Frozen Peas"; and Szechwan "Spicy Frogs' Legs," the latter prepared with hot red peppers, sweet green peppers, garlic, ginger, hoisin sauce, soy sauce, and other flavorings. There are also frog soups, such as "Tai Shih Frog Soup" made in a Cantonese restaurant with frog, chicken stock, smoked ham, bamboo shoots, cucumber, and onion and other flavorings (F. L. Chang and G. M. Fuchs, n.d.: 102, 240; W. W. Chang *et al.*, 1970: 170-72; K. Lo, 1979a: 109-11; 1979c: 257-59). Frog soup, broth, and gelatin are regarded as excellent for convalescents (Soubeiran and Dabry de Thiersant, 1874: 62).

Snails were consumed in the ancient Near East and raised in Europe for eating at least since Roman times, and it is no surprise to find that the Chinese have not overlooked such an excellent food (snails of the genus *Helix* have "more protein than T-bone steak and only 1.4 percent fat"; Schwabe, 1979: 376). Holmes (1948: 101), who was a missionary in Fukien, writes of children wading in ponds with "gourd-shaped baskets tied to their sides" and picking up snails to take home. F. H. King (1927: 155), writing of Chekiang Province, observes that several species of snails are consumed; that they are cooked in the shell and sold by vendors there; and that they are eaten as we eat popcorn. The amounts consumed were considerable, judging from the piles of shells he saw along canals outside villages.

Schwabe (1979: 376) writes that the Chinese have long known that earthworms, which are also high in protein, were "cheap, nutritious meat," and prepared them in a broth considered "a good treatment for fever." We are not certain whether these are the greenish-brown worms from paddies that, according to Ball (1906: 288), were hawked on China's streets.

Monkeys are among the more exotic creatures to be eaten by humans, but their flesh is well regarded and widely consumed around the world (P. L. Simmonds, 1885: 54-57). In China, monkey flesh is depicted along with other meats in a Manchurian tomb dating from Han times (Y. S. Yü, 1977: 60), and, though some Chinese rejected monkey flesh, other Chinese and members of minority groups, especially in southern and western China, considered it a delicacy (Graham, 1961: 19; Eberhard, 1968: 51; Read, 1976: Section 400). Monkey flesh is also considered effective against malaria and certain other ills (Read, 1976: Sections 400, 402). For the Cantonese, it is monkey brains that are sought after, a choice, expensive edible, a strengthening food, but, according to E. N. Anderson (1988: 175), more a medicine than a food. Also among exotic animals consumed, in Kwangtung among other places, are the giant salamander (*Megalobatrachus* sp.; *ni-yü*) and the pangolin, or scaly ant-eater (*Manis pentadactyla*, *M. aurita*, or *M. dalmanni*; *ch'uan-shan-chia*, "scaly hill-borer"; or *ling-li*, "hill carp," because it is viewed as a fish out of water) (S. W. Williams, 1895, 1: 328; Read, 1977b: 22-23). The giant salamander, found in rivers, streams, and marshes, averages three feet or so in length and can reach nearly six feet. Its flesh is viewed in China as "very good eating" and a delicacy (Sowerby, 1925: 74; 1940: 108), but it is also medicinal, a cure for idiocy (Read, 1977c: 76-77). The flesh of the pangolin is considered good eating (Sowerby, 1925: 40; H. D. R. Baker, 1983), and an outstanding strengthening (*pu*) food because of its remarkable appearance which is believed to reflect "great energy or unusual energy patterns" (E. N. Anderson, 1988: 147, 192, 193, 194). Pangolin scales, after roasting, are made into medicines for treating a broad range of ailments, and there is a sizeable domestic trade and import of pangolin skins and scales (Soubeiran and Dabry de Thiersant, 1874: 45-46; E. Watson, 1923: 455; Hooper, 1929: 148; I. H. Burkill, 1966, 2: 1445; Read, 1977b: 23-25).[61]

[1] This is identical with an estimate for the U.S. in 1976, when meat, poultry, and fish were judged to have contributed 18.2 percent of total calorie intake (Bywater and Baldwin, 1980: 6).

[2] For more data on consumption of meat and fish in China, see May, 1961: 36 (estimates of national per capita consumption for time periods from 1931-1957); Maynard and Swen, 1956: 401-15 (especially their regional comparisons); and Gamble, 1933: 95-102 (expenditures for fish and various kinds of meat among 283 families in Peking in 1926-27).

[3] Groves (1981: 64-66) has presented evidence that the pigs of New Guinea (*Sus papuensis*) may have a hybrid origin, deriving from two different species, *Sus scrofa* (type *vittatus*) and *Sus celebensis*, which were domesticates of Southeast Asia. The hydridization, he argues, likely occurred in Indonesia (the Moluccas) rather than in New Guinea itself.

[4] It should be noted that leading archaeologists have raised questions about the early dates reported for Tseng-p'i-yen, their suspicions being that the radiocarbon samples had been contaminated by the limestone of the cave. On the other hand, K. C. Chang (1986: 105) notes that extensive experimentation in Chinese radiocarbon laboratories has failed to demonstrate any regular tendency to vary from samples obtained in other places. In addition, he says, the Tseng-p'i-yen dates have been verified by the thermoluminescence method, which has provided dates of similar range. To us, however, it appears that the thermoluminescence dates (10,370 B.P., plus or minus 870, and 7160 B.P., plus or minus 930) are

more recent than the radiocarbon dates by one millennium at the upper end of the range and by nearly two millennia at the lower end.

[5] Legendre (1911: 583) reports that the Lolo living in the mountains of western Szechwan cut meat into strips and dry them outside or over a fire, much like the biltong of Africa. Then, when it is quite hard, they grind it, using mortar and pestle, into a powder, in which form it can keep for one to two years depending on its dryness. The Lolo consume the powder after it has been dissolved in water. The practice of drying meat strips is an ancient one among the Chinese people, too. Already in Han times, travellers, including soldiers on expeditions, carried dried provisions, mostly grain and beans but sometimes meat, and, in addition, the royal household stored such dried provisions for future use (Yü, 1977: 76-77; E. N. Anderson, 1988: 43). In the *Ch'i-min yao-shu* (sixth century A.D.), moreover, an entire chapter is devoted to methods of drying meat, with or without salting (S. H. Shih, 1962: 89).

[6] E. N. Anderson (1988: 175) writes of one Cantonese sausage prepared with a distilled liquor flavored with roses.

[7] Quail were raised by the Chinese, were trained for fighting, and were shipped across the Pacific (J. S. Thomson, 1909: 308), perhaps to Chinese communities for fighting purposes. Quail fighting was very popular in the Canton area, judging from the account in a nineteenth century gazetteer. Quail were sold at a particular place, Devil Lane, which at the peak of a sale was so crowded as to make it almost impossible to get through. Good young quail were expensive, and renowned fighting birds more so. A Cantonese quail fight, in turn, might have a large audience with considerable amounts of money exchanging hands (A. F., 1869), apparently as bets. For more information on quail fighting and pigeon fighting in China, see Gray, 1878, 1: 393-94.

[8] Divination has been prominent in Chinese life since antiquity, and modern Chinese culture is rich in its methods of divination, for which see Doolittle, 1865, 2: 331-48; Gray, 1878, 2: 1-29; Latourette, 1964: 555; and Ahern, 1981: 45-63. For the use of turtle shell in divination in China, see the section "Symbolism and Early Role in Divination" in our chapter on "Fish, Turtles, and Tortoises." For eggs in divination, see the section in our chapter on "Eggs" called "In Ritual and Belief"

[9] Crawford (1984a: 302-4), in a recent review of the evidence of pre-Columbian chickens in America, describes the zoological evidence as unsubstantial though he acknowledges that other sorts of evidence is stronger. Carter (pers. comm., G. F. Carter), in turn, writes of chicken bones in archeological sites in the New World that clearly date from pre-Columbian times, and expresses surprise that scholars refuse to acknowledge the validity of such evidence.

[10] K. C. Chang (1986: 93) writes of chicken bones, likely from a domesticated bird, in sites of the P'ei-li-kang and related cultures in North China which date from about 6500-5000 B.C. Caution is suggested before accepting such an early date, however, by the fact that in antiquity wild chickens of one species or another may have occurred in sections of the north (P. T. Ho, 1975: 98); and by the fact that chickens are not among the domesticates reported for the Yang-shao, Ma-chia-pang, or Ho-mu-tu cultures of North or Central China, which date from about 5000 to 3000 B.C. (P. T. Ho, 1975: 93; T. K. Cheng, 1982: 11; K. C. Chang, 1986: 112, 113, 201, 211). Ho Ping-ti (P. T. Ho, 1975: 95), while noting chicken bones found at the Yang-shao site of Ching-ts'un, observes that they were considered to be those of wild fowl, though one cannot determine whether they were accurately identified. There is agreement, however, that the remains of chickens found in Lung-shan and Lungshanoid sites were those of domesticated fowl, and thus use of the 3000 B.C. date seems prudent at this time.

[11] Neither Ho Ping-ti (P. T. Ho, 1975) nor K. C. Chang (1977a, 1986) report the occurrence of duck bones in sites of this antiquity in Central or North China. Chang (1977a: 167) does, however, note that clay models of ducks were found in the middle Yangtze area in certain sites of the Ch'ü-chia-ling culture, which dates from the third millennium B.C. These occur along with what seem to be models of chickens, geese, and sheep, though identifications are uncertain (K. C. Chang, 1986: 226).

[12] For details of the process of salting and drying ducks, as done in the New Territories, Hong Kong, see Pereira, 1937: 162-64. For notes on pressed cured duck, see E. N. Anderson, 1988: 175.

[13] The best-illustrated account we have found on the ways of trimming and preparing duck is that of Hom and Steiman, 1981: 165-87.

[14] E. N. Anderson (1988: 175) reports that poultry feet, which serve as stock and for munching, are imported to Hong Kong from Canada and elsewhere, and that in Hong Kong the pairs of feet used are reputed to number twice the number of fowl consumed.

[15] Hom and Steiman, 1981: 148-52 provide detailed information on the preparation of "Fried Chicken Wings," which are marinated, stuffed with ham and Chinese parsley, covered with batter, and deep-fried. For recipes to a Shanghai-style "Sweet-and-Sour Chicken Wings," Kiangsu-style "Chicken Livers and Cucumbers," and Cantonese-style "Fried Gizzards and Livers," see Gin, 1984: 134-35.

[16] For the recipe to a Szechwan "Duck Liver Soup," see Gin, 1984: 35. This soup, which can be made with chicken liver instead, also includes chicken stock, bamboo shoots, preserved Szechwan cabbage, soy sauce, rice wine, and salt and pepper.

[17] See J. D. Keys, 1963: 82-83 for a recipe to "Ducks' Feet Braised with Mushrooms, Bamboo Shoots and Water Chestnuts."

[18] Duck feet, as one would expect, are fair to good sources of calcium, with dried ones reported as having 194 mg./100 gm., edible portion, as well as being unusually high in iron (21.3 mg./100 gm) (Read *et al.*, 1940: 62; Leung *et al.*, 1973: 105).

[19] Recipes for this festival dish will be found in Feng, n.d.: 211-12 ("Yams Braised with Duck") and G. Z. Chu, 1975: 173-74 ("Taro Duck").

[20] For details of this, see Wittwer *et al.*, 1987: 334-36.

[21] E. N. Anderson (1988: 146-47) writes of a recipe for Peking Duck in one reliable Chinese cookbook which starts with the egg of a duck and instructs the cook "how to incubate, hatch, and raise the bird, so that not one second of its life is left to chance." The detail is such that it takes "several dozen pages" for the duck to arrive in the kitchen.

[22] For descriptions of the breeds of fat-tailed and fat-rumped sheep of Mongolia and China, see Epstein, 1971: 34-46, 52-53.

[23] The principal cattle involved were water buffalo, common cattle, yak, and mithan (*Bos [Bibos] frontalis*), the latter a bibovine found among tribal peoples of northeastern India and adjacent parts of Burma (Simoons and Simoons, 1968). The bull fighting involved combat between two bulls, rather than, as in Spanish bull fighting, the pitting of a man or men against a bull.

[24] T'ang pharmacologists also singled out beef for unfavorable mention. One of them, insisting that the proper use of oxen was in agriculture, said that eating flesh of the common yellow ox could bring on illness, and that of black ox should be avoided altogether (Schafer, 1977: 99).

[25] Rabbit flesh has various medicinal uses as well, and rabbit dung is highly regarded as fertilizer, in part because, according to Chinese folk tradition, it lowers the alkalinity of soils and increases the resistance of cultivated plants to insect pests and disease (Wittwer *et al.*, 1987: 337-38).

[26] See Crawford, 1984b, for a discussion of the zoological classification and ancestry of the Chinese goose; and K. C. Chang, 1977b: *passim*, for historical notes on its use as food in China.

[27] Goose dung was also sold, for good prices, in China. Wing quills were made into pens and toothpicks, and goose down and feathers were traded (Gray, 1878, 2: 186; I. H. Burkill, 1966, 1: 1118). Bird-feather clothes were made widely in early South China, and in Kwangsi these were normally of goose feathers (Eberhard, 1968: 286).

[28] That this is not a recent phenomenon is indicated in the sixteenth century account of Spanish Friar Martin de Rada, who commented on the large supply of doves and turtle-doves in Fukien (Boxer, 1953: 290).

[29] Inconsistent use of the terms "pigeon" and "dove" causes much confusion, but when employed for members of the family *Columbidae*, they are considered synonymous. The Barbary dove (usually *Streptopelia risoria*), of African origin, is the only other member of that family to be regarded as domesticated (Hawes, 1984: 351-52). Various species of the family are found wild in China, including ones belonging to the genera *Streptopelia* and *Columba* (Couling, 1917: 124; Read, 1977a: 47, 61). Y. S. Yü (1977: 56) lists remains of *Streptopelia* sp. among the foods in a Han tomb near Ch'ang-sha, Hunan, dating from the second century B.C., but whether the birds in question were cage birds or wild birds that had been hunted and killed, is not known.

[30] Crawfurd (1820, 3: 168) wrote of pigeons and doves being shipped from the Malay Archipelago to China in Ming times. This, of course, was long after domesticated pigeons were known in China, and one can only guess at the identity of the birds in question and whether they had been bred in captivity or captured from the wild.

[31] The goose, swallow, and mandarin duck (*Aix galericulata*) are also viewed in the same way (S. W. Williams, 1895, 1: 340, 790; C. S. Wang, 1934: 308-10; Eberhard, 1968: 286; Gould and Rutgers, 1969: 70; Wieger and Davrout, 1981: 503, 507). One sees this in the general custom of a bridegroom giving a goose as a wedding present to the family of his bride as an indication that he will never take another wife. Eberhard (1968: 286) writes that in Anhwei, there is an exchange of gifts, with the family of the groom presenting a male goose, and that of the bride sending back a female one. There is a benefit to the two geese as well, for, as fertility symbols, they will not be killed. In an elaboration of this custom, a pair of geese is included in the wedding procession and brought by the bridegroom to the ceremonies where the bride worships the birds as symbols of marital affection (S. W. Williams, 1895, 1: 790). Cormack (1935: 45-46) writes that in some parts of China a bridegroom's parents send two ducks and two geese to the bride's family, with one duck and one goose being returned. In the case of the mandarin duck (*yüan-yang*, from the cry of the male, *yüan*, and that of the female, *yang*), a beautiful, ornamental water fowl, the symbolism relates to the fact that the male and female are seldom apart. A pair of these ducks is commonly carried in wedding processions in China, with representations of them embroidered on bridal gowns, pillow covers, and bed coverlets, particularly on the wedding day (S. W. Williams, 1895, 1: 340; C. S. Wang, 1934: 310; Gould and Rutgers, 1969: 70).

[32] He illustrates this with the story of a government official who lived in Harbin, Manchuria, in pre-Republican times and who had a Cantonese cook who liked to gamble. Whenever the cook served pigeon for dinner the official knew that he had lost money at gambling, for few people in that city would eat pigeon and it was low in price.

[33] Pigeon dung was a famine food in Samaria in the ninth century B.C. (*II Kings* 6. 25), but we have not heard of this for China though one report mentions pigeon dung being added to a strong liquor (apparently distilled from sorghum) in the Peking area "to give it a good body" (Arlington and Lewisohn, 1935: 272). Though pigeon dung was a prized fertilizer in China, we have uncovered nothing to indicate whether the Chinese, like the Persians, build large pigeon towers. Those of Persia average 40 feet in height and are attractive in architectural form; their purpose is to facilitate gathering of pigeon dung. At one time there were more than three thousand pigeon towers just in the Isfahan area, and many of these still stand (Wilford, 1962; Beazley, 1966).

[34] Dogs, apparently domesticated, were also found in Central Chinese sites dating from 5000-3000 B.C. (K. C. Chang, 1986: 201, 211).

[35] On the breeds of dogs in China, see Epstein, 1971: 125-45.

[36] Hosie (1904: 205-6) reported that in northern Manchuria and Mongolia there were thousands of small dog and goat farms where anywhere from ten to hundreds of animals were raised each year. A bride might be given six dogs as a dowry, and from that base a large fortune might be built. Dogs were fed coarse grain and farm wastes, and were killed and skinned when six to eight months of age, and in winter when their coats were long and warm. Their flesh then was used for food and their skins sold to dealers who made them into robes, mats, and other items. Similar observations were made near Harbin in Manchuria by Powell (1945: 178), who reported that big, shaggy, black-haired Mongolian dogs were kept by Russians in a compound for their skins, which were sold in New York.

The dogs in question are not identified further by Hosie or Powell, but Epstein (1969: 129), in his book on the domesticated animals of China, considers the Hosie account in his chapter on true dogs, *Canis familiaris*. One wonders, however, whether in fact one is dealing with the raccoon dog, *Nyctereutes procyonoides* (*hao* or *he*), a different member of the family *Canidae*, but a close relative of the dog and fox. The raccoon dog, which, like the raccoon, has black patches around its eyes, is native to East Asia. Though seldom depicted in Chinese painting, the raccoon dog has been managed in captivity in that region since antiquity for its pelt (*hao-jung*; known commercially as Chinese coon, Ussurian raccoon, Japanese fox, and by other names). The keeping of raccoon dogs later spread to other parts of the world, especially to Europe (Couling, 1917: 469; Sowerby, 1940: 50; Valtonen, 1984: 215-16).

The flesh of the raccoon dog is also eaten in China, where, because of the animal's unusual fox-like appearance and perhaps because of its antics, it is among the most acclaimed *pu*, or "strengthening," foods (E. N. Anderson, 1988: 147, 183, 192, 193).

[37] The differences in perception of dogs among China's non-Han ethnic groups deserves further attention. One wonders, for example, at the similarity between the dog-ancestor views of minority groups in South China, and the observation by Powell (1945: 178) that Mongols, believing that dogs harbored human ancestral spirits, refused to kill them.

[38] See also S. W. Williams (1895, 1: 777-78) and Collier (1921: 26) on the preference for black dogs in South China. Black dogs, like whole-wheat bread in modern America, were considered most nutritious; then yellow dogs; then, least desirable, white dogs. In Chinese medicine, on the other hand, the flesh of black dogs is inferior, and the flesh and other products of white or yellow dogs are preferred (Read, 1976: Section 323; C. Y. Hsu, 1980: 217n).

[39] This is the case on Taiwan, where the body of a person who has consumed dog meat is said to have enough warmth to make it unnecessary that he cover himself, even with a quilt.

[40] This fits with the recent observation by Tan and Wheeler (1983: 44) that in Hong Kong dog flesh is "a highly valued delicacy."

[41] Around the turn of this century, excavators removed large numbers of cat mummies from Bubastis and used them as fertilizer. Nineteen tons were even shipped to England for that purpose (Clutton-Brock, 1981: 110).

[42] This "Estremaduran Cat Stew" is prepared by salting pieces of cat flesh; browning them in olive oil; adding white wine, bay leaves, and thyme; and simmering the mixture for a half hour or so. Then vegetables (peas, fava beans, carrots, turnips, potatoes, and onions) are shelled, sliced, diced, or chopped, and added, with water to cover; the stew is then simmered until the vegetables are tender (Schwabe, 1979: 176).

[43] We do not know the identity of this animal or its relationship to *Felis catus*. Lowe wrote in the 1940's (1943: 259-66) of a caged "wildcat" sold in San Francisco's Chinatown, but, since the creature weighed 55 pounds, likely it was a puma. It was purchased by his grandfather and made into wildcat soup, a powerful tonic that was consumed by his father, uncle, aunt, and grandfather, who were of Kwangtung origin.

[44] As with dogs, the flesh of black cats was usually preferred, and it commanded a higher price (Gray, 1878, 2: 76; S. W. Williams, 1895, 1: 777-78).

[45] Among the snakes commonly for sale in Hong Kong snake shops are two non-venomous species: the common rat snake (*Ptyas mucosus*) and Indo-Chinese rat snake (*Ptyas korros*), as well as three venomous ones: the many-banded krait (*Bungarus multicinctus multicinctus*), Chinese cobra (*Naja naja atra*), and king cobra (*Ophiophagus hannah*). Also identified as sold in such shops are the non-venomous copperhead rat snake (*Elaphe radiata*) and the venomous banded krait (*Bungarus fasciatus*). Specimens of the most common sea snake of Hong Kong, the venomous banded sea snake (*Hydrophis cyanocinctus*), are often captured by fishermen and sold to be eaten as delicacies (Karsen *et al.*, 1986: 73, 75, 76, 101, 102, 104, 105, 107).

[46] For more Chinese snake dishes, see Schwabe, 1979: 177, 260-61, and Samagalski and Buckley, 1984: 210.

[47] P. L. Simmonds (1885: 74), a secondary source, writes of rats and mice eaten by "the Mongol races" who fatten them like swine. We have not been able to determine his source or the accuracy of this statement.

[48] References to insect eating are scattered through the general literature on China. In addition there are articles or book chapters devoted specifically to that topic. Our principal sources on insect consumption in China are Read (1940, 1977d), who wrote on their medicinal uses; Hoffmann (1947), who described their use as human food, especially in Kwangtung; and Bodenheimer (1951: 271-79), who summarized the information on insect eating in China as part of a world survey of insects as human food. For lands to the south of China which have sizeable Chinese minorities, insect eating is described by Tiêu (1928) for Tonkin, and Bristowe (1932, 1953) for Thailand and nearby areas. In addition, there is an article by G. S. T. Chu *et al.* (1977) on beetle-eating among the Chinese of Malaya. For Chinese recipes to insect dishes, the most useful source is Schwabe, 1979: 366-75 *passim*.

[49] This may have been the case with the blister flies or blister beetles, insects from various genera of the family *Meloidae*, which provide the commercial substance cantharidin, an external irritant, blistering agent, and aphrodisiac. In Europe, the drug "Spanish fly," taken internally to increase sexual desire, derives from various species, but especially *Lytta vesicatoria* (formerly *Cantharis vesicatoria*), which is common in Spain and certain other European countries. In the last century, dried bodies of *Lytta vesicatoria* were imported to India and sold by druggists there (Watt, 1889-96, 2: 128), and one wonders when such trade in blister beetles may have begun. There were, however, species of blister beetles in India and China which were excellent substitutes, with *Mylabris cichorii* of India and China containing more cantharidin and being a more powerful vesicatory agent than *L. vesicatoria* of Europe.

In China, blister beetles had been gathered, dried, and used for medicinal purposes since ancient times (Cowan, 1865: 63; Soubeiran and Dabry de Thiersant, 1874: 68-69; Read, 1940: 23, 25; 1977d: 76-82), and Schafer (1963: 180) includes them among the essential drugs of T'ang China. Though we have not determined when such trade began, early in this century large quantities of dried Chinese blister beetles, species of *Mylabris* (*pan-mao*) (*M. cichoria* and *M.*

phalerata = *M. sidae*) from South and Central China were exported to Europe for medicinal use (E. Watson, 1923: 345-46), presumably that of aphrodisiac. Though blister beetles have been used in China to treat gonorrhea, induce abortions, and as a diuretic and for problems of the urinary tract, our sources do not reveal whether they serve as aphrodisiacs in China.

[50] Worthy of special note in terms of its usefulness is the wax-producing insect of China. This insect, commonly confused with other wax-producing species, is *Ericerus pe-la* Westwood (also designated *Coccus pe-la*, Westwood). It grows on certain evergreen trees, especially *Ligustrum lucidum*, large-leaved privet, which was planted around fields, and is known in China as *la-shu*, "wax tree"; *pai-la-shu*, "white wax tree"; or *nü-chen*, "female chastity" (because the tree symbolizes chastity) (S. W. Williams, 1895, 1: 353; E. Watson, 1923: 143). Much wax (*pai-la, ch'ung-pai-la*, or *shu-la*) comes from insects that are "domesticated," under the care of humans. They produce a fine white wax deposited on plants they frequent, as thick as a quarter of an inch on a branch. The wax is scraped from the plants, purified, made into cakes, and used for various purposes. The center of insect breeding and wax production is Szechwan, and at one time as many as ten thousand porters were said to carry wax scales from one town in the producing area across the mountains to other regions. Wax was also produced in varying amounts in other provinces, including Kwangtung, and on Hainan (Cowan, 1865: 254; Bretschneider, 1882-95, 3: 513-17; Hosie, 1897: 189-201; Sowerby, 1925: 210; E. H. Wilson, 1929: 387-92; Fenzel, 1933: 179; Graham, 1961: 20; Read, 1977d: 34-38). According to Read (1977d: 36), knowledge of insect wax dates only from Yüan times (A.D. 1271-1368). It was known to Europeans in China at least by the seventeenth century and came to be exported to Europe, being known in England as "Chinese wax," "Chinese insect wax," "insect white wax," and by other names. In China it was mainly used not for medicine but in the manufacture of candles, usually for coating candles made of animal and vegetable tallow, giving them better consistency and preventing guttering when the candles burned. This derived from the fact that insect wax melts at above 180°F., whereas animal tallow melts at about 95°F. It was also used as sizing for materials made of paper or cotton, for providing a gloss to silk, polish to furniture and stone carvings, for coating pills, and as a medicine in its own right (Hosie, 1897: 189-201; Read, 1977d: 34-38).

[51] In traditional China, boys sometimes captured male cicadas for entertainment and for their song. In Hong Kong early in this century, J. S. Thomson (1909: 38) observed young boys snaring cicadas in banyan trees, using long bamboo poles and a glue of ashes and rice paste. After capturing cicadas, the boys would tie strings around their feet to restrain them, then carry them about the streets and stimulate them to sing or encourage them to perform feats of strength (S. W. Williams, 1895, 1: 352; J. S. Thomson, 1909: 38). Cages of insects, including cicadas, were also sold in shops to amuse the wealthy (Cowan, 1865: 253).

The cicada, like the scarab in ancient Egypt, was important in early China as a symbol of immortality (Read, 1940: 31), and jade cicada amulets were placed on the tongue of a dead person (Clausen, 1954: 129). This symbolism apparently derived from the cicada's ability to emerge from the earth as a mature nymph after years of subterranean existence.

Cicadas are eaten in China as adults or as nymphs (Bodenheimer, 1951: 277). Nymphs are more common as food; they are either collected on trees in the evening or are dug from the soil; and they are generally fried before eating. Cicada nymphs infected with the fungus *Cordyceps sobolifera* (Figure 68), observes Read (1940: 31), are used for treating malaria, infantile convulsions, tetanus, and other ills. Cicada skin, in turn, is used in cases of difficult childbirth, in treating skin eruptions such as urticaria, and for removing film that grows over the eye. The latter uses, Read says (1940: 28), derive from the cicada's ability to shed its own skin. Ball (1906: 170) reported that cast off skin of the cicada was boiled in water with certain plant materials, and the liquid given to a child suffering from convulsions and fever.

[52] Read (1940: 29) provides more information on the use of certain of these insects. Pickled ant eggs, he observes, were included in sacrificial dishes eaten by ancient tribal chiefs in South China, and both bees and bee larvae were consumed by the ancient Chinese. For modern China, P. L. Simmonds (1885: 369) reports that ant larvae are a special delicacy, and that a species of red ant that nests in trees is also consumed, flavored with salt and ginger.

[53] Other insects used for food or medicine in Kwangtung or elsewhere in China are identified in other sources, especially Read (1940, 1977d) and Bodenheimer (1951: 273-78).

[54] In Fukien, where silkworm-rearing and silk-making is completely a home industry, Kellogg reports (1968: 42, 49) that pupae are killed by dropping the cocoons into hot water.

[55] Silkworm pupae are also eaten in Burma, being regarded there as delicacies and often serving as gifts (Clausen, 1954: 18-19). Donovan (1842: 6) notes, as well, references to the use of silkworms as food and medicine in Europe prior to the sixteenth century.

For China, Dr. Fen also reported (Bodenheimer, 1951: 275) that people in some hill regions of Shantung Province cultivate another silkworm, *Antherea pernyi*, on oak trees and that, though the supply is limited, they eat its pupae and regard them highly as food.

[56] For medicinal uses of silkworms and silkworm products, see Hooper, 1929: 150, and Read, 1977d: 52-64.

[57] E. N. Anderson (1988: 181) writes of a favorite sauce in northern Thailand, Laos, and presumably southern Yunnan, prepared "from fermented giant waterbug."

[58] As one might expect, there were differences of opinion in this regard, and in the matter as to which crickets had the finest song. Tun (1965: 63), writing in 1900, spoke glowingly of that produced by "little golden bell" crickets (*Homeogryllus japonicus*), which were brought to Peking for sale. "When one listens to them while lying in bed, their chirp is surpassingly clear, musical but not sad, as if they were born to be creatures of large rooms and lofty halls (*i.e.*, of noble homes). Surely their title of 'golden bell' has not been one extravagantly conferred on them."

[59] It seems that pheasant were abundant in the forests of Manchuria, and in the sixteenth or seventeenth centuries the Manchus introduced a "Pheasant Hot-Pot" to Peking. A recipe for this dish may be found in K. Lo, 1971: 71-72. For an earlier pheasant dish, one prepared by the Jurchen, that Tungusic people who conquered North China in the twelfth century, see Franke, 1975: 175.

[60] The special taste the Cantonese have for frog is illustrated in a story, whether apocryphal or not, told by F. T. Cheng (1954: 60) about Fukien farmers and the provincial Viceroy, a Cantonese. The farmers, apparently unaware of the usefulness of frogs, would throw frogs from their land onto that of their neighbors, and this commonly resulted in litigation. The Viceroy, on learning of this, ruled that the frogs be given to him, and the people, learning that he dined on them, ceased their litigation, and came to have a proper appreciation of frogs as delicacies.

[61] One reads of the tokay gecko (*Gekko gecko*) sold in Hong Kong snake shops, whether alive or pickled (Karsen *et al.*, 1986: 53), but we suspect this is for its value as a medicine (Read, 1977b: 26-27).

CHAPTER TEN

Fish, Turtles, and Tortoises

"Many people of the West poke fun at us for what may be termed our motley taste for fish food. They declare that we are omnivorous, as far as eating the products of the sea is concerned. It seems to us that if the standard of civilization is partly to be measured by the ability and ingenuity of the people to maintain life by utilizing the variety of foods placed at its disposal, then we have helped to solve one of the important problems of life."

W. C. Yen

Fish and Fishing
- Their Importance
- Fish Consumed
- Methods of Catching Fish
  - Range of Fishing Methods
  - Two Unusual Methods of Fishing
    - Fishing with Cormorants
    - Fishing with Otters
- Fish Cultivation in Ponds
- Oysters and Oyster Farming
- Fermented Fish Products
- Fish in Nutrition and Chinese Food Beliefs

Turtles and Tortoises
- Symbolism and Early Role in Divination
- Turtles as Food
  - Fresh-Water Forms
  - Sea Turtles

FISH AND FISHING

Their Importance

At the end of the last century, S. W. Williams (1895, 1: 276) wrote of China's seacoast and estuaries, its lakes and rivers, as filled with fishing vessels of contrasting sizes, well-equipped to take all forms of aquatic life. He included in this not only adult fish, but spawn which were gathered to be raised in captivity. Altogether, he noted, an enormous amount of fish, both fresh and preserved, was consumed. The fish were not only consumed locally, but were shipped throughout China and sold at low cost, within reach of all but the very poorest. Estimates of the nation's fish catch support Williams' judgment and show traditional China to have been a world leader in this regard. By one estimate (D. Riedel, 1961: 43), China's catch in 1910 amounted to 2 million metric tons, far more than that of the other leading nations of the time (Japan, 1.3 million; Russia, 1.0; United States, 1.0).[1] At the same time, China's population was very large, and national per capita consumption of fish low. Earlier we mentioned a 1931-37 estimate of 2.8 kg. (6 lb.) of fish consumed annually per capita in China as a whole. Though estimates since establishment of the People's Republic are much higher, in 1959 they were still only 6.1 kg. (13.4 lb.) per capita annually, far below those of such fish-oriented lands as Portugal (27.6 kg.) and Japan (22.5 kg.) (Borgstrom, 1962: 290).

National figures, however, are of little relevance for the Double Cropping Rice Area, for which the 1929-33 village survey found the amount of calories obtained daily from fish (including shrimp) to be eight times the national average (Maynard and Swen, 1956: 413). Assuming a similar difference in 1931-37, the annual per capita fish consumption of the province would have been comparable to that of Japan, which fits with Hoffmann's observation (1934: 43) that in Kwangtung many people eat fish every day, and some eat it at every meal.

The importance of fish in Kwangtung, as elsewhere along China's southeast coast, is based on three factors. First is a great and general liking for fish in the region. Second is a coastal location and well-developed fisheries in salt, brackish, and fresh water, using varied and sometimes unusual techniques.[2] And third is an impressive and productive system of fish farming, or fish cultivation, in ponds. In the People's Republic today, fresh-water fish ponds are said to account for 60 per cent of the nation's total fish catch (Wittwer *et al.*, 1987: 339), and Kwangtung is of special importance in that regard. Data provided by T. H. Shen (mostly for 1937) (1951: 390, 392) showed that Kwangtung not only led the nation in fresh-water fisheries, but that its annual catch of 415,000 metric tons of fresh-water fish, mostly from fish ponds, amounted to nearly three-quarters of the nation's fresh-water total.[3]

Fish Consumed

As a people both experimental in the use of food and appreciative of aquatic opportunities, the Chinese since antiquity have used a range of fish perhaps unmatched in the Western world. The picture Schafer (1977: 102-3) sketches of fish in T'ang China, for example, is one of infinite variety. One reads of marine fish, river fish, lake fish, and fish from ponds; of tasty goldfish eight-feet long; of hearty eel broth; of spicy sturgeon relish; and of carp marinated in wine sediment and sent as tribute to the royal court. As Schafer put it so well, "even the whimsical seahorse (*shui-ma*; *Hippocampus kelloggi*) could not avoid the gullets of the people of T'ang."

In modern times, Chinese diet seems to have retained fully its ancient piscine variety, with European observers amazed at the cuttlefish, jellyfish, and other species utilized, and with Ball (1906: 774) writing that to try every kind of fish consumed by the Chinese one would have to eat a different type every day for an entire year. The lengths to which people could go is revealed in Ross's observation (1911: 78-79) that even roadside ditches were emptied of water to obtain fish only as long as a finger.

In Kwangtung, the principal types of marine fish taken were shad, mackerel, sea-bream (*Sparus*), butterfish, croaker, and shark (T. H. Shen, 1951: 390).[4] Some marine fish may be eaten raw (Ta Chen, 1940: 97), though, in contrast to Japan, the practice is unusual in China.[5] And fresh, dried, salted, or dried and salted fish are employed in a great range of dishes.[6,7] Even fish heads were not wasted, for many Chinese liked to eat them, as in "Red-Cooked Fish Head and Tail, Anhwei Style" (E. Chen, 1977: 103). As Cressey has observed of China's southeast coast (1934: 338), a feast is not truly complete without a course of succulent fish and without such special treats as sea slug, shark's fin, and seaweed. One should not, however, overlook the fact that the Chinese do have preferences for fish, and that these do not necessarily match those of Western peoples. The Chinese disdain tuna and salmon, for example, along with other fish whose flesh is tough and oily, strong of taste, and unpleasant smelling; this is well illustrated in a case reported by E. N. Anderson in which Chinese, in speaking of a tuna-packing plant, referred to it as providing a fine opportunity to sell to gullible Westerners fish that was good only for fertilizer. When Anderson countered by saying that Westerners were actually fond of tuna, the Chinese were incredulous (E. N. Anderson, 1988: 141).

Methods of Catching Fish

Range of Fishing Methods

Though beyond the scope of this study to consider in detail, we would note that in traditional China fishing at sea was well-organized and regulated, and employed a variety of boats and methods of fishing. Among the methods reported for Hong Kong were various forms of trapping, seining and trawling, dip netting, gill netting, and fishing with lines (Kani, 1967: 82-90). Some fishing vessels at sea were engaged in the salt-fish trade, and might travel considerable distances from land in their search. Other vessels at sea were engaged in the fresh-fish trade, and the distances they travelled from home were smaller. One reads of large numbers of such vessels stretching to the horizon along the coast of Kwangtung. In the holds of such fresh-fish vessels were large compartments perforated so as to permit a constant exchange of water with the sea, so the fish might reach market alive and well (Gray, 1878, 2: 288-90).

Fishing in fresh water, too, was regulated and its methods quite varied (Dabry de Thiersant, 1872: 123-26, 149-73). Chinese living near P'o-yang Lake, Kiangsi Province, said that there are ninety-nine ways to catch fish. When asked "Why not a hundred?," they replied that if there were a hundred ways, no fish would be left in the lake (Hommel, 1937: 130). Among fresh water methods were luring, as by attracting fish to a boat at night with torches; driving and netting; trapping; rod fishing; trot fishing; spearing; use of the plunge basket; and netting with dip net and cast net worked from shore or a boat. In the Pearl River and elsewhere, there was also dredging for oysters, mussels, and cockles, and use of baited-baskets for catching shrimp, some of which were marketed alive (Gray, 1878, 2: 293-301; S. W. Williams, 1895, 1: 779-80; 2: 14-17; Moir, 1909; Hommel, 1937: 128-34).

The effort made at marketing live shrimp, eels, and other fish reflected a persistent preference for fresh fish among the Chinese. This is why in American Chinese restaurants that cater to a Chinese clientele one may find a tank of live fish, sometimes with crabs and lobsters, from which a diner may choose. In traditional China, live fish, including eels and water snakes, might be carried to market in water by boat or in watertight baskets. There they might be transferred to large troughs or cisterns of wood or stone in the shops of fishmongers. In one case, a fishmonger kept his fish alive in large wickerwork baskets that he suspended in a canal. On rivers and canals, in turn, one might encounter retailers of a different type: men or women in small boats containing wells of live fish for sale (Gray, 1878, 2: 293).[8] Also involved in selling live fish in Chinese cities were hawkers, each carrying a bamboo rod over his shoulder.

Suspended from one end of the rod was a tub of water containing the fish, and from the other, a block on which to chop a fish into portions as requested by customers (Gray, 1878, 2: 293).

Two Unusual Methods of Fishing

Fishing with Cormorants

Returning to methods of fishing in China, we cannot overlook two of these which, though minor in terms of total catch, demonstrate again the ingenuity of the Chinese in their food quest. One is fishing with cormorants; the second, fishing with otters.

Fishing with cormorants (*Phalacrocorax* spp.) is an ancient practice involving the taming and training of these birds, excellent divers and swimmers, to catch fish for human use. Authorities differ as to the time and place of the earliest taming of cormorants for fishing. Brentjes (1962: 642), on the basis of representations from ancient Mesopotamia, has suggested that cormorants may have been used there for fishing by about 2500 B.C. Most scholars, however, favor East Asia as the region where fishing with cormorants began. Laufer (1931: 212) writes that the earliest mention of the practice, though found in Chinese writings of the Sui dynasty (A.D. 581-618), actually referred to Japan, and Zeuner (1963: 472) suggests that it was practiced in Japan still earlier, in the fifth century A.D. Hornell (1950: 31), on the other hand, has argued that cormorant fishing began in China, and Z. G. Zhang (1984: 367) cites references to fishing with cormorants in China going back to about 221 B.C.

Whatever the time and place of its origin, cormorant fishing on streams, rivers, and lakes has been practiced on a commercial scale by the Chinese up to the present-day (Figure 70) despite opposition to the practice by Buddhists (Laufer, 1931: 225-26).[9] In China, the species ordinarily used for fishing is the common cormorant, *Phalacrocorax carbo*,[10] which is known by various names, among them *yi*, from the cormorant's cry, and *lu-tzu* (said by early Chinese sources to mean "black black," for the bird's very dark color). Certain of its other names are even more colorful: "Mr. fish-diver"; "fish-catching gentleman"; "fish crow"; and, in a different perspective, "black devil" (used in Szechwan) (Giles, 1900: 96; Laufer, 1931: 208-9; Read, 1977a: 24).

Z. G. Zhang (1984: 367) writes that in China cormorants are captured from the wild, not bred in captivity. Williams (S. W. Williams, 1895, 2: 16), however, said that cormorant eggs are often hatched under a domestic hen and the young then fed; Hornell (1950: 30) wrote that such cormorants are later used in fishing; and Zeuner (1963: 472), commenting on this practice, concluded that the cormorant is unquestionably domesticated. A cormorant's wings are first clipped to prevent its escape. Then a long string is fastened to one foot, and its other end to a stick on shore. A ring is tied around the cormorant's neck to keep it from swallowing good-sized fish it captures. When a cormorant does capture a fish, a fisherman whistles to bring the bird to him, then takes the fish and rewards the cormorant by giving it a small fish to eat. After a month of such training on shore, the cormorant's training continues for another month on the fisherman's boat or bamboo raft. Ultimately the string is removed from its foot, but the ring about its neck remains. Cormorants are in their prime when from two to ten years of age, after which their fishing activity declines (Z. G. Zhang, 1984: 367-68).

Curiously, Marco Polo did not mention cormorant fishing in China, the first European to record the practice being Friar Odoric of Pordenone, an Italian Franciscan who visited China early in the fourteenth century. More detailed descriptions of the practice were given in the sixteenth century by Father González de Mendoza (1853-54, 1: 154-56) and by Galeote Pereira,

Figure 70. Fisherman with cormorants at Kweilin (1986) (photo by Stephen F. Cunha)

a Portuguese trader. Mendoza wrote that the emperor maintained houses in all river cities where cormorants were raised for fishing, and that the birds, in the three months of the year that they were used for fishing, provided enough fish for the entire country for a year. Pereira's account, in turn, was apparently part of a description of Kweilin in Kwangsi Province. According to Pereira (Boxer, 1953: 42-43), the king kept numerous barges in rivers, and on them, cages of cormorants. The birds were given a monthly ration of rice, and lived and died in their cages. They were taken out only for royal fishing expeditions. At such times, the barges formed a circle in a shallow part of the river and the birds dived into the water and returned to their barges to deposit the fish they captured. When fish enough had been taken for their masters, the cormorants were freed to fish for themselves.

Pereira's account should not lead one to assume that Chinese cormorant-fishing was mainly royal sport. It was an activity of ordinary folk as well, in Kwangtung and elsewhere. One reads in late-nineteenth century accounts of cormorants trained in large numbers for fishing on lakes and rivers; of an individual fisherman having twelve to fifteen birds under his control; of five hundred cormorants fishing in one stretch of water only an eighth of a mile long; and of the large numbers of fish they captured for their owners who would stand on a raft or catamaran in the middle of a stream (Doolittle, 1865, 1: 55-57; Gray, 1878, 2: 297-98; S. W. Williams, 1895, 2: 16).

Cormorant fishing was taken up in England, France, and Holland in the late sixteenth or early seventeenth century, and some have suggested that Hollanders introduced it from East Asia. Training and use of the birds in Europe, however, was strongly influenced by falconry and was quite different from Chinese practice (Laufer, 1931: 206-7; Zeuner, 1963: 473).

This calls to mind the striking resemblance between the use of cormorants for taking fish and the training of falcons, other hawks, and sometimes eagles for taking small mammals, deer, and even wolves. Whereas cormorant fishing is likely of East Asian origin, falconry is believed by many to have originated in Central Asia, possibly Turkestan, Mongolia, or even Manchuria (Schafer, 1959: 293), though it was known in China at least by the third century B.C. (Schafer, 1959: 295). Its main intent, however, was to provide sport for nobility and upper classes, whereas cormorant fishing, while providing sport to some, was primarily an activity of ordinary folk seeking fish for human consumption.

Fishing with Otters

Trained otters have been used in the past by individual sports fishermen in scattered places in Europe and the Americas (Gudger, 1927: 209-19). In China, however, fishing with otters has been carried out for profit; it has been an organized affair involving more than an individual fisherman; and it is an ancient practice—known already in T'ang times—which has not died out but continued into modern times.[11]

Otter fishing has been reported in various provinces (among them Szechwan, Hupeh, Shensi, Hunan), mostly along the middle and upper Yangtze and its tributaries (Gudger, 1927: 193-205). According to certain observers of such fishing along Yangtze drainage (Hosie, 1897: 167-68; Little, 1898: 39), an otter (fresh-water otter, or *shui-t'a*) wore a leather harness around its chest. Attached to the harness was an iron chain, fastened at its far end to a boat or to a bamboo pole similar to a fishing line, set up on shore. When fishing, a fisherman first cast a circular net weighted around its edges, which settled and covered a large area of the river bottom. Then he pulled gently on the rope attached to the center of the net, bringing it to the surface, when he put an otter through a small opening in the net. The role of the otter was to root out fish from places of refuge in rocky crevices or in the mud and drive them up into the net. Then the net was pulled out of the water, bringing along both otter and, hopefully, some fish, with the otter then receiving his reward.

Other descriptions of such fishing (Swinhoe, 1870: 625; Moir, 1909: 279-81) suggest that the otter entered the water outside the net, then drove the fish under the net when it was on the bottom or still descending. One observer noted that if there were no fish in the area, an otter would even rise to the surface and inform the fisherman so that he might move to a different location. Still other accounts of Chinese fishing with otters differ in detail from the above, and some suggest different arrangements (Gudger, 1927: 200-204). One writer reported that at a word from its owner, an otter simply threw itself in the water to catch a fish and bring it back. Percival (1889: 125-26), in turn, observed use of a net and otter in quite a different way. A very large square dip-net was first lowered into the river. Then an otter was put into the water, descending to the bottom and driving any fish nearby into the area just above the net. This arrangement took full advantage of an otter's ability to herd fish, much like that of a sheep dog in herding sheep. When an otter had sufficient time to perform his task, the net was lifted from the water, carrying along both fish and otter.

Fish Cultivation in Ponds

It is curious that Buck's *Land utilization in China* fails even to mention fish ponds. Perhaps, as Drews (1951: ix-x) has argued, this reflects the ethnocentrism of the principal investigators, Westerners who, because of a non-aquatic tradition, simply overlooked its importance in the landscape and failed to recognize its vitality and significance in the Chinese economy and diet.

Documentary evidence of fish cultivation in ponds seems earlier for China than for anywhere else in the world. Already for the fifth century B.C. there is clear mention of the practice, and some have suggested that it was known long before that date. Of special interest is Drews's suggestion (1951: 23-24) that fish cultivation may not have originated among Han Chinese but spread to them from the south. That suggestion makes good sense, for aquatic agriculture was prominent in China's south since ancient times, and fish cultivation would have fit well with it, providing protein to complement the carbohydrate of taro, other starchy root crops, and, later, rice. The presence of pond fish would have been a boon to the Han Chinese as they moved southward, not only because of the ready protein they provided but because certain pond fish feed on mosquito larvae and thus reduce the danger of malaria in areas of paddy cultivation.

Whatever their origin, fish ponds became a major element in China's food scene, especially in the deltas of the Pearl and Yangtze rivers, and the country came to be one of the world's greatest centers of fish farming.[12] Fish ponds were concentrated around large cities where there was a ready market and inadequate supply of other fish. In Kwangtung, most ponds tended to be rectangular in shape and often were surrounded by mulberries or other fruit trees. Some were quite small, little more than waterholes, whereas others were two or three acres in size. Ponds ranged in depth from a few inches to as much as ten feet. Some were owned by villages, others by individuals who might have just a single pond or as many as forty or more. A pond might be rented out or worked by its owner, providing a profit of just a few dollars or a sizeable income (Hoffmann, 1934: 10-11, 18, 48-49).

Altogether, nine species of food fish were cultivated in ponds on a large scale. All but one of the nine are cyprinids, members of the carp family. Carp stand out among cultivated fish because they can live in impure water, are able to feed themselves on plant and animal life found in ponds, can convert food efficiently into growth, and provide high yields per acre (E. N. Anderson and M. L. Anderson, 1977: 335). Most important is the common carp (*Cyprinus carpio*; *li* or *li-yü*), which usually weighs from a half pound to a pound when marketed. Other fish common in ponds throughout China are grass carp (*Ctenopharyngodon idellus*; *huan* or *huan-yü*, "leisurely fish"), silver carp (*Hypophthalmichthys molitrix*; *pai-yü*, "white fish"; or *lien-yü*, "trooping fish"), bighead carp (*Aristichthys nobilis*), and black carp (*Mylopharyngodon piceus*). In Kwangtung, besides the above five, one also finds the mud carp (*Cirrhina molitorella*) and sometimes the bream (*Parabramis pekinensis*, a cyprinid and relative of the carp), and crucian carp or goldfish (*Carassius auratus*).[13] Though a valuable food fish when caught in the wild, the goldfish (*fu-yü*, "paired fish"; or *chi-yü*, "fish in filed formation") raised in ponds are ornamental. Ultimately all cyprinid species raised in ponds except the goldfish served as human food, along with other creatures that had entered the pond (for example, eels, other wild fish, and frogs).

Of cyprinid fish raised for food in ponds in China, only the common carp actually spawns in ponds. There was thus a well-developed system of capturing and rearing fry of cyprinids, with the West River (Hsi Chiang) of Kwangtung and Kwangsi provinces the main source of supply to ponds of South China, and the Yangtze to other parts of the country (Hoffmann, 1934: 12-17; S. Y. Lin, 1940a: 163, 173-76).

The one non-cyprinid cultivated species was the gray mullet (*Mugil cephalus*; *tzu-yü*), a marine and brackish water fish raised in estuarine ponds near Hong Kong, but not elsewhere in South China.[14] The gray mullet had, however, long been confined in crudely-made trap ponds along the coasts of Kwangtung and Fukien, but was not fed or cultivated. One found similar systems of confining this fish elsewhere in eastern Asia, with the gray mullet first cultivated in Japan, probably during the Tokugawa Era (A.D. 1603-1868). In Hong Kong, cultivation of gray mullet dates only to the end of World War I, and construction of ponds and methods of care reflect Western influences and not traditional ones. The capture of mullet fry in coastal waters, bays and estuaries, and brackish streams around Hong Kong was accomplished with ingenious devices (S. Y. Lin, 1940a: 177-78), and that city became a major supplier of gray mullet fry to Chinese pond operators in Southeast Asia, as well as a shipper of carp fry obtained from the mainland (Drews, 1951: 27-28, 31-33, 67-68, 137; 1961: 123-24, 138-39).

On Taiwan, fish cultivation in traditional times was not based so narrowly on carp. Instead it was like Japanese fish culture in its inclusion of non-cyprinid fish (Drews, 1951: 126-27). Though Chinese carp were cultivated on Taiwan, far more important was the milkfish (*Chanos chanos*), a seafish whose family is related to the that of the herring and is found widely in tropical and subtropical waters of the Indian Ocean and the Pacific, and which accounted for about 60 percent of all pondfish produced on Taiwan in the late 1940's. Its initial cultivation on Taiwan is credited by some to the Dutch, who occupied Taiwan from A.D. 1622 to 1662, and by others to Cheng Ch'eng-kung (Koxinga), the Ming general who took Taiwan from the Dutch. In either case, the stimulus of milkfish cultivation may have come from the Dutch East Indies where, on Java, milkfish had been reared in salt water ponds by A.D. 1400. The popularity of the milkfish was based on a supply of fry in coastal waters; on its high yields per acre (average 8700 lb.), more than other pond fish of brackish or salt water; and on the fact that ponds are built mainly along the sea in soils too saline for agriculture or for the culture of fresh water fish (Tung-pai Chen, 1952: 1-3; T. P. Chen, 1976: 1-15). In addition to carp and milkfish, other marine forms, among them prawns (*Penaeus carinatus* and *Peneopsis monoceros*), gray mullet, oysters (mainly the Japanese oyster, *Crassostrea gigas*), clams (*Meretrix lusoria*), and other species were reared on traditional Taiwan. Since the Republic of China established its headquarters on Taiwan in 1949, moreover, a wide range of additional marine species have come under cultivation or gained importance, with tilapia (*Tilapia* spp.) second, after milkfish, in amount of production in 1975, and with the eel (mainly the Japanese eel, *Anguilla japonica* and European eel, *A. anguilla*), sold locally or exported to Japan, of by far the greatest value (Tung-pai Chen, 1952: 14-15; Drews, 1961: 136-38; T. P. Chen, 1976; H. H. Chen, 1977).

Since antiquity the Chinese have held the common carp in such high regard as food as to justify further comment. One example is found in the ancient poem in which carp flesh is compared to that of other fish "as a princess of Sung is to an ordinary woman." Another example is found in a nobleman's gift of a common carp to Confucius on the birth of his first son who was then named *Li*, or "common carp." The common carp also came to symbolize good fortune, abundance, and wealth, and one of its names is "god-of-wealth fish" (Read, 1977f: 13). Its energy and ability at getting through the Dragon Gate, the noted rapids of the Yellow River, was especially admired, and when a Chinese scholar earned his final degree, he was described as having passed through the Dragon Gate. In art this was represented by a carp leaping out of the water with a dragon in the sky above. Ultimately the carp developed strong ties with Buddhism in China and elsewhere in East Asia, and even today devotees will buy and liberate young carp in honor of the Buddha's birthday. To the Chinese, notes Drews, the common carp is a status food and the king of fishes, their favorite because of its fine taste, fecundity, and ease of cultivation, for it is hardy and omnivorous and can feed on a broad range of waste products (Drews, 1951: 16-22, 33-35, 44, 164-68; Atkinson, 1977: 321-22).

In Kwangtung, fish ponds are found nearly everywhere, but Hoffmann (1934: 9, 48) identified six districts that stood out in production: Shun-te (Cantonese: Shuntak), Nan-hai (Naam-hoi), Chung-shan (Chung-shaan), P'an-yü (P'oon-ue), Hsin-hui (San-ooi), and Lo-ting (Loh-ting). In those six districts, it was estimated that 40 to 60 percent of all land was taken up with ponds. Conservatively this would amount to 150,000 acres of ponds, providing an annual harvest of fish estimated at almost 360,000 metric tons (Hoffmann, 1934: 9, 48). Five of the six districts are in the densely-settled Pearl River Delta, a low-lying area where ponds are readily built and where by-products of mulberry cultivation and silkworm rearing (silkworm pupae and waste) provided ready sources of food for pond fish. Those products were supplemented by a broad range of other foods, mainly waste products such as hog manure, night soil, grass, rice bran, and distiller's waste (Hoffmann, 1934: 21-25; S. Y. Lin, 1940a: 181-85). Sometimes, too, ducks and geese were raised close to a pond so that their excreta could serve as fish food (Hung, 1960-61: 84). Since the various species of pond carp each occupy a somewhat different feeding niche, they make an efficient use of these and other foods available in the pond.

In some areas, latrines were actually located at the edge of ponds, with fecal matter usually falling directly into the water. The practice of allowing raw feces to enter fish ponds, however, is a major element in spreading infection of the liver fluke *Clonorchis sinensis*. By way of snail and carp intermediaries, this parasite ultimately infects humans eating raw or undercooked fish, which around Canton derive especially from the so-called mulberry district and its numerous fish ponds. In a study made in Canton in the 1920's (Faust and Khaw, 1927: 124, 132-34, 137, 198-99), about a fifth of persons tested in various hospitals proved to be infected, but in the mulberry district, 42 percent of a thousand persons were infected and in some villages, 90 to 100 percent were.

Cultivators of fish took action against predators, illness and infection. Once a year at the main fish harvest, for example, it was common to remove mud from the pond (to be used as fertilizer or other purposes) or to let the mud dry in the sun, for sun-drying, like removal, was considered effective in eliminating enemies of the fish. Another sort of insurance came from the practice of alternating ponds, usually shallow, sunny ones, between fish cultivation and food crop cultivation. Among the crops commonly used in rotation were water chestnut (*Eleocharis*), arrowhead, lotus, rice, and various vegetables. Rice might also be planted along the moist edge of the pond when fish were in it, and sometimes an aquatic crop such as water chestnut would actually be raised in the pond along with the fish (Hoffmann, 1934: 18, 35-39, 65, Drews, 1951: 51-52, 141).[15]

In Kwangtung, cyprinids were harvested by net or, when a pond was drained, they were simply picked up by hand. They might be taken to market in baskets or, since live fish commanded much higher prices, they might be transported alive in water, whether in special fish boats or, using public transport, in large oval wooden tubs or tanks. Special efforts were taken, using a variety of devices, to maintain the oxygen supply so as to reduce mortality in transport. The fish boats were usually old sampan-like ones which were towed by launches or hand-propelled, and which had numerous holes in their sides to let fresh water in. Fish markets in large villages and cities were normally located on or near a canal or river to permit ready access by fish boats. Markets were usually large, open buildings, from which, after sale, most fish were taken away in wooden tubs or bamboo baskets to retail fish shops, to restaurants and hotels, to be hawked on the streets of adjacent communities, or to be sold elsewhere (Hoffmann, 1934: 39-42). In Hong Kong, where mullet were raised as well as cyprinids, marketing was handled differently, for mullet cannot readily be kept alive in a limited space like a tub and they were transported dead, sometimes iced in wooden tubs (S. Y. Lin, 1940a: 187-88; Hung, 1960-61: 85).

Oysters and Oyster Farming

A more specialized form of fish cultivation in Kwangtung was that of oysters, some raised for food and others for pearls. Cultivation of oysters for food in Chung-shan District was described by T. P. Chen in 1935, though the industry was in decline. Chung-shan District is along the estuary of the Pearl River, an area of brackish water with a tidal range of a dozen feet or more. The waters are shallow, with many islets, and the bottom is usually smooth and sandy. Cultivation of oysters for food also occurred on the opposite side of the estuary both in China proper and the New Territories of Hong Kong, and in Macao and coastal islands to the southeast.

The cultivated oyster in question is *Ostrea gigas*, a brackish-water form which can grow to be more than a foot long and weigh as much as eleven ounces. In 1935 only small numbers of these oysters remained in public beds in Chung-shan, so few that only old women and children living close at hand found it worth the effort to collect them. Thus virtually all oysters were obtained by cultivation in privately-owned beds. Some cultivators collected their own spats (young oysters), whereas others bought two-year-old oysters and raised them to marketable size in another year or so. In waters low in density and rich in organic matter there were other growers who specialized in buying and fattening mature three-year-old oysters, for well-fattened oysters are roughly three times the weight and of less watery flesh than oysters that are not fattened.

Oyster farming in the Hong Kong area is almost all located north of Yuen Long in the Deep Bay area of the New Territories. In traditional times the oysters were mostly dried or made into oyster sauce, that popular flavoring commonly used with beef and stir-fried foods. In the 1960's, however, an increasing percentage were being marketed fresh, even alive in water (Hung, 1960-61: 81-85; E. N. Anderson, 1970b: 12-13). The fishing village of Lau Fau Shan, at the end of the bus line, is the area's center of oyster production and marketing (Figure 71). Today it contains seafood restaurants and an impressive group of fish shops with tanks and metal trays full of live creatures of one sort or another. One may find fish and oysters being sold, and a string of trucks lined up nearby with tubs and tanks for transporting them to Kowloon, Hong Kong, and elsewhere. In Chung-shan District in the 1930's, on the other hand,

Figure 71. Oyster fishing vessels and oyster shells on beach at Lau Fau Shan, New Territories, Hong Kong (1987)

oysters were marketed mainly in a dried state, with some raw oysters dried directly ("dried raw oysters") and others cooked first in preparing oyster sauce, then dried ("dried cooked oysters").

In making oyster sauce in Chung-shan, a large quantity of raw oysters (over a hundred pounds altogether) was salted, washed, boiled in fresh water for twenty to thirty minutes, removed with a bamboo scoop, and drained. The drained liquid was then returned to the boiling water along with a new batch of oysters and enough water to keep the sauce from over-thickening. When, after repeated boilings the sauce was brown and of proper thickness, it was cooled in pottery tubs for a day, then decanted into pottery wine jars that were covered and sealed. In such jars, each weighing roughly 40 to 45 pounds, oyster sauce could keep for longer than a year.[16] Meanwhile the cooked oysters used in the process were dried, first for five or six hours in the sun, then for another twelve hours in the shade. Then they were coated with peanut oil to make them glossy, and marketed in large bamboo baskets. Such dried cooked oysters, which keep for about a week, were less tasty and much less expensive than the dried raw oysters, which were usually sold retail in bottles.

Except in coastal areas, most people of traditional China were unfamiliar with fresh oysters, and when they obtained oysters it was usually in a dried form. Gray (1878, 2: 299) writes that the Chinese never eat raw oysters, but consider them too cold to eat. Yet one reads of people in one coastal area, perhaps influenced by Chinese returning from Southeast Asia, considering raw oysters a great delicacy (Ta Chen, 1940: 97). The Chinese may eat salted oysters without further preparation, but usually they fry oysters alone, with meat or vegetables, or in other ways, as shown in the recipes of Kenneth Lo (1979c: 98, 293-95, 316, 319).

Dried oysters (*Ostrea* spp.) are an excellent source of protein, calcium, and phosphorus, iron, and potassium; a good source of thiamine and riboflavin; and a fair one of niacin. Oyster sauce, in turn, is an excellent source of potassium and niacin; a good source of iron; and fair one of vitamin A (Leung *et al.*, 1973: 130).

Fermented Fish Products

Fermented fish products are made in various parts of the world, including southern and eastern Asia from at least the Indian subcontinent to Japan and Korea. Within that region, however, Southeast Asia is the great center of fermented fish product manufacture, with such processing more important there than simple salting or any other method of fish preservation. Fermenting enjoys certain advantages over drying and smoking of fish in Southeast Asia. One is that fermentation by some methods can be carried out without concern for cloudiness, rain, or fuel supply; another is that fish and fish parts not normally used as human food can be made into edible substances (van Veen, 1965; Magno-Orejana, 1983).

There are two general types of fermented fish product in Southeast Asia. One is a sauce, the other a paste. The traditional ways of preparing them in Southeast Asia involved either: (1) fermentation in the presence of salt, largely by enzymes in the flesh of fish and in their guts, or (2) fermentation by a combination of fish enzymes and microbial enzymes which might "be added as starters in addition to salt" (Magno-Orejana, 1983: 273). In making such products, marine fish may be used, or freshwater fish (usually small ones), including shellfish such as oysters and shrimp.

Both pastes and sauces are normally eaten in Southeast Asia as flavoring for rice dishes, and in some places they constitute the major source of animal protein among poor people. Of the sauces, perhaps best known is the clear, brown liquid called *nuoc-mam* in Vietnam, whose use in the United States has grown with the influx of Vietnamese refugees and the opening of Vietnamese restaurants. Ordinarily in making such a sauce by traditional means, fish entrails may be removed and the fish washed. The fish are packed with salt, usually four or five measures of salt to six of fish, in large containers of pottery or wood which are then tightly sealed and buried. After fermenting for several months, a container is taken from the ground, opened, and the liquid poured off or pressed from the remaining sediment (van Veen, 1965: 230-32; Magno-Orejana, 1983). This is the sauce, a flavoring but quite a nutritious one. Though fermentation does not seem to add to the nutritional value of fish protein, it may improve its digestibility. The fermented sauce, especially if of good quality, may contain significant amounts of protein rich in essential amino acids, including lysine. Use of such fermented fish sauce thus compensates for the low lysine content of rice, and may also be a good source of calcium and certain other minerals (Leung *et al.*, 1973: 121; Magno-Orejana, 1983: 258-59). As a result, the fermented fish products of Southeast Asia cannot be dismissed, as some foreigners have done in the past, by calling them rotten, putrid, filthy, and even dangerous to health.

Along China's southeastern coast, small shrimp may be made into sauces reminiscent of *nuoc-mam*, or into pastes related to the *belachan* of Malaysia. Such products, however, are much less popular in China than in Southeast Asia, though Foochow, in Fukien, where fish sauce is used more than soy sauce, may be an exception in this regard (E. N. Anderson and M. L. Anderson, 1977: 336; E. N. Anderson, 1988: 143, 164).

E. N. Anderson (1988: 143) observes that fermented shrimp paste is used by Chinese in the far south mainly by persons who had been in Southeast Asia, and writes that the Chinese apparently learned how to make such products from ethnic groups of Southeast Asia. One should not, however, overlook the facts that use of fermented fish products in southeastern China is not a recent development, that already in T'ang times Chinese who ventured into the far south were familiar with them, and that delectable but malodorous fish dishes, unquestionably fermented, were popular among early peoples of Hainan (Schafer, 1967: 212; 1970: 64). One may speculate on why, given such long familiarity, fermented fish products are not of greater importance in China. Perhaps in part this is because as one goes north along China's coast, the weather tends to be cooler and spoilage of fish less of a problem. Perhaps, too, drying, salting, and smoking of fish, which are common in coastal China, produce products more readily shipped to interior regions. Also not to be ignored is competition from soy sauce, a flavoring quite similar to fish sauce, and a good substitute for it.

Fish in Nutrition and Chinese Food Beliefs

Fish contain abundant protein, more than milk, eggs, or beef. Fish protein, moreover, is readily digested and of good biological value. On the FAO/WHO amino-acid score, fish ranked 100, a score achieved only by eggs and beef. Fish oils are the richest known natural sources of vitamins A and D, and the flesh of carp and oily fish is a good source of vitamin D. Fish also contains various minerals, the most notable of which, because they were often deficient in Chinese diet, are iodine and fluorine (marine fish only), phosphorus, and calcium (fish, especially small ones, prepared in ways permitting the use of the calcium in their bones) (Herklots and Lin, 1940: 5-6; Borgstrom, 1962: 283-84, 302-5; Causeret, 1962; Cruickshank, 1962; Briggs and Calloway, 1979: 99, 148, 259, 288).

One also notes that recent studies published in the *New England Journal of Medicine* suggest that consumption of fish over many years may significantly reduce the risk of mortality from coronary heart disease. One study (Phillipson *et al.*, 1985) reported that a four-week diet that included fish oil significantly reduced levels of plasma cholesterol and triglycerides in persons with hypertriglyceridemia. And a second study (Kromhout *et al.*, 1985), of 852 Dutch men without coronary heart disease, found that those who consumed at least 30 gm. of fish daily over a twenty-year span had 50 percent less mortality from coronary heart disease than those who did not; the difference, moreover, seemed independent of age, blood pressure, and levels of serum cholesterol. If those results are confirmed by further studies, one might expect to find differences in mortality from coronary heart disease in China between coastal provinces where fish consumption is high and interior ones where it is low.

Further study is necessary before one can speak with confidence about the classification of fish in China as hot and cold, poisonous, or nutritious, and here we draw on a dozen or so studies, mostly of Chinese from Hong Kong or South China. Even these studies sometimes differ notably in their findings. Certain of them (Yeung *et al.*, 1973: 198, 200; E. N. Anderson and M. L. Anderson, 1976: 147; M. Mathews and L. Manderson, 1981: 12, 13; Tan and Wheeler, 1983: 44) reported that fish (or fresh-water fish) are considered "neutral" or "not hot enough to be dangerous" in terms of the hot-cold system, which, depending on the method of cooking them, makes them suitable for consumption by persons in most bodily states. Some observers (E. N. Anderson and M. L. Anderson, 1968: 26; 1976: 147; Koo, 1973: 141; Ling *et al.*, 1975: 127, 128; M. Mathews and L. Manderson, 1981: 13) also speak of fresh fish, freshwater fish, or fish soups as "nutritious" or prescribed to women during pregnancy and postpartum. On the other hand, most shellfish, including shrimp (as well as catfish or sea catfish), were reported to be hot, wet, and poisonous, and among the proscriptions for women during pregnancy and after birth (E. N. Anderson and M. L. Anderson, 1968: 24; 1976: 147; 1977: 369; Koo, 1973: 137, 141; Yeung *et al.*, 1973: 200; B. Chang, 1974: 438; Ling *et al.*, 1975: 127; M. Mathews and L. Manderson, 1981: 12, 13; Wheeler and Tan, 1983: 53, 55, 56). Some types of crabs, in turn, are "hot," whereas others are "cold" (E. N. Anderson, 1988: 142), thus filling different needs in terms of their heating or cooling qualities.

TURTLES AND TORTOISES

Symbolism and Early Role in Divination

To traditional Chinese, the turtle (most writers, following English usage, say "tortoise"),[17] along with the unicorn, phoenix, and dragon, is one of the four super-intelligent creatures that govern the four points of the compass and are believed to be prototypes for all animals with hair, feathers, scales, or shells. The ruler of the north, the turtle was the only one of the four to be found in the everyday world, a creature symbolic of strength, endurance, longevity, and, some say, immortality (Yetts, 1912: 25; Arlington and Lewisohn, 1935: 328; Burkhardt, 1953-58, 1: 138; J. J. M. de Groot, 1964, 1: 55; Read, 1977e: 1).[18] Because of the latter associations, the turtle symbol was commonly used at times of burial. A turtle figure might be found embroidered on the grave clothes of a deceased woman, or represented on a pin in her hair (J. J. M. de Groot, 1964, 1: 53, 55). A burial mound might be shaped like a turtle and even called "grave turtle." Some Chinese said that it was deliberately formed in this way because the turtle symbolized longevity. In J. J. M. de Groot's view (1964, 3: 1083, 1146), the deceased's ability to reach an advanced age was concentrated on the mound, from which it passed to his or her offspring, thereby assuring continuation of the family. Commonly, too, a carved turtle was used as the plinth, or sub-base, for memorial tablets at a place of burial. This custom is recorded already for Sui dynasty times (A.D. 581-618) when, as true also of Ming times (A.D. 1368-1644), the right to use such plinths was granted to the highest ranks of officials, but not to ones of lower rank or to ordinary people (J. J. M. de Groot, 1964, 2: 451-52; 3: 1140-41, 1146-48, 1160-61). Especially notable are the enormous turtles, each carved from a single piece of stone and supporting a memorial tablet, made for various Chinese emperors and other important persons (Creel, 1937: 186) (Figure 72).[19] That such turtle symbolism survives today is shown in a Taiwan village, where paste cakes of flour and sugar are made in the form of a turtle at festivals held in honor of the lineage patron deity. Household members may buy such turtle cakes at their lineage temple on such occasions, and take them home to assure prosperity, security, and harmony for the coming year (Diamond, 1969: 23-24).

Figure 72. Carved turtle supporting memorial tablet, Ming dynasty

Somewhat surprisingly, the turtle also bears a highly-negative image, and in the vernacular it is called *wang-pa* ("forget eight"), because it is reputed to overlook the eight rules of propriety, among them faithfulness, uprightness, and modesty. The specific violation, according to folk belief, is the unchastity of female turtles, which are held to have intercourse with snakes. The term *wang-pa* is used as an insult, one of China's worst. If one sees on a wall that term or a representation of a turtle it is a warning that if a person causes trouble he is a *wang-pa*.

To call another person a "turtle egg" is also a terrible insult meaning "miscegenated bastard" (Giles, 1900: 299; Couling, 1917: 561, 592; C. A. S. Williams, 1941: 404; E. N. Anderson, 1988: 147).

Turtle shell was widely used in divination, important in this way already in Shang times (18th-12th century B.C.). Such shell, along with bones of bovines and other animals, comprise the famed oracle bones, many of them inscribed and important sources for reconstructing the history of that period. Though such divination may have been widespread among the general public, the oracle bones at hand nearly all involve royalty. The answer to a query was obtained by interpreting the pattern of cracks that developed when hot metal rods were pressed against the shell. Special categories of persons were commonly involved in the matter of divining, but their status and roles are uncertain. In any case, the royal queries were on important matters, among them sickness and health, childbirth, weather, agriculture, travel plans, hunting and military expeditions, and sacrifices to be made (Creel, 1937: 185-96; Chi Li, 1977: 240-43; Keightley, 1978: 21-35; K. C. Chang, 1980: 31-42).[20] Several species of turtles were used in Shang divination[21] and most seem to have come not from North China, but from the Yangtze Valley and southwards (Keightley, 1978: 9; K. C. Chang, 1980: 32-33, 155). There are records of shells being brought to the court by individuals, officials, and states (K. C. Chang, 1980: 156), and some species were especially valued, as in T'ang times when an outstanding variety of turtle for divination was obtained in Nan-Yüeh (Schafer, 1967: 214).

Turtles as Food

Turtles (*kuei*) have been highly regarded sources of food in China since antiquity, and soft-shelled turtles (commonly *pieh*) and tortoises are included among the better-known *pu*, or "strengthening," foods (E. N. Anderson, 1988: 147). One reads (K. C. Chang, 1977b: 32-33) of turtle among the delicacies of ancient China; of boiled turtle, roast turtle, and turtle stew in Han times (Y. S. Yü, 1977: 67, 70); and of an extract of green turtle, a favorite of gastronomes in North China during T'ang times (Schafer, 1970: 51-52). Of special interest is the T'ang reference to what must have been green-turtle calipee, presented as tribute to the royal court by the city of Canton (Schafer and Wallacker, 1957-58: 225; Parsons, 1962: 8; Schafer, 1962: 74; 1977: 101). Calipee is a fatty, gelatinous material found immediately over the lower shell of a turtle, a delicacy used to flavor soup. Returning to Chinese associations of the turtle with longevity, one notes that Chinese in New York (1939) consumed four-fifths of all turtle sold in the Fulton Street fish market, in the expectation that they would gain its longevity (Parsons, 1962: 22).

Fresh-Water Forms

Most early Chinese references to turtles are to fresh-water forms. Schafer (1962: 73-74) identifies the ancient favorite as the soft-shelled or mud turtle *Amyda (Trionyx) sinensis* (Figure 73), which was common in fresh water throughout China.[22] This is a long-necked snapping turtle of good size which likes to bury itself in the mud or sand of streams, ponds, and lakes. Highly valued, too, by the T'ang Chinese was the giant soft-shell (*Pelochelys bibroni*; *yüan*), found in coastal South China and in deep, slow-flowing rivers. Also identified is a tasty turtle in Kwangtung that climbed trees; this was possibly *Platysternon megacephalum*, the Chinese big-headed terrapin, a nocturnal species of medium size but a good climber.[23] Fresh-water turtles continue to be eaten in modern China, with *Playsternon megacephalum*, *Chinemys*

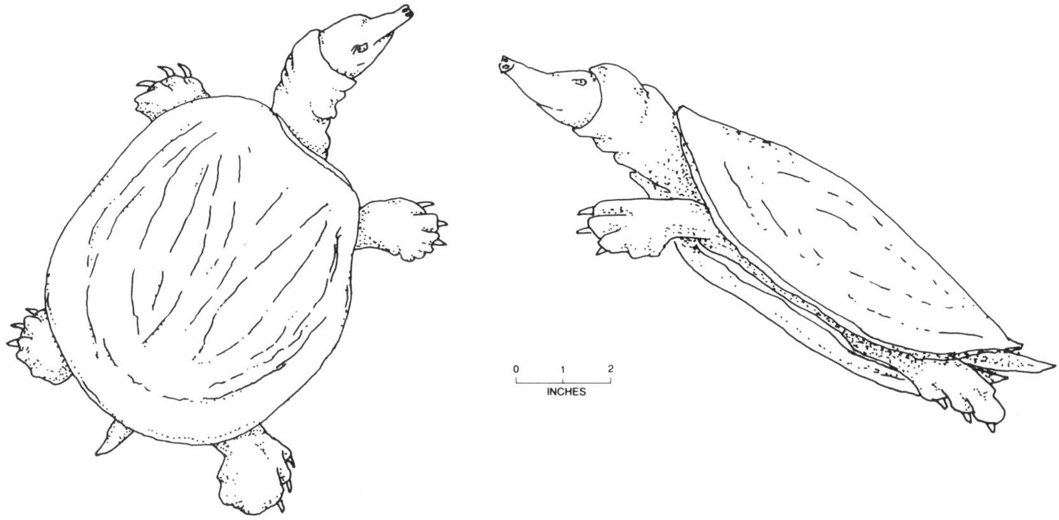

Figure 73. Soft-shelled or mud turtle (*Amyda sinensis*) (redrawn by Shelby Hall, after Sowerby, 1925)

reevesi, *Cuora trifasciata*, and *Ocadia sinensis* identified (Karsen *et al.*, 1986: 38, 39, 41, 43) as the market species of Hong Kong.

Fresh-water turtles were sometimes speared by men standing in small boats or vessels shaped like washing-tubs. In them a turtler might use a single trident, or one in each hand, to probe the muddy waters for his prey (Figure 74). Fresh-water turtles were also taken by means of a fishing line at the end of which were small hooks and a few metal weights. Commonly sitting on a stool on a river bank, the fisherman, when he spied a turtle, cast his line and, if his aim was right, the hooks, aided by the weights, penetrated the turtle's shell, permitting it to be hauled in (Gray, 1878, 2: 298-99).

E. N. Anderson (1969: 443-44) has noted that the Tan (Tonka) boat people of Hong Kong on occasion regard fresh-water turtles as "sacred fish," though this is not usually the case. Creatures that fall into the category of sacred fish are not to be captured, but if that happens inadvertently they should, if still alive, be released. If they have already died, they should be carried to a temple to be offered as a sacrifice. These observations apply only to the Tan, not

Figure 74. Man spearing fresh-water turtles (from Gray, 1878, 2: opp. 298)

to the Cantonese or other groups of the area. Yet Mao (1971: 22, 32) has made similar ones about the people of Taiwan who, he says, consider all land turtles (identified as aquatic species of the family *Testudinidae*: *Ocadia sinensis*, *Clemmys mutica*, *Chinemys reevesi*, and *Cuora flavomarginata*) as sacred, symbolic of longevity,[24] and seldom eat them or use them in other ways. Mao notes also that in China fresh-water, hard-shelled turtles (apparently excluding soft-shelled species) are sacred. Though certain of the latter are sometimes found in marketplaces, pious Buddhists, in an effort to gain merit, may purchase them to serve as scape-goats that are released in a temple pond.

Sea Turtles

The Tan place the great sea turtles more clearly in the sacred fish category than they do fresh-water turtles. This means that they should not be captured and, if that happens, they should be released.[25] Buddhists play a role in this by purchasing captured sea turtles and then freeing them in an effort to gain merit (E. N. Anderson, 1969: 443). At the same time, there is convincing evidence that Chinese eat sea turtles of one species or another, including the great sea turtles, which increase in number and species to the south. Of these, the loggerhead (*Caretta caretta*) is the only species to nest mostly outside the tropics: in East Asia on Japan, Taiwan, and mainland China from Kwangsi and Kwangtung to Hopeh (Uchida and Nishiwaki, 1981: 317; Pritchard and Trebbau, 1984: 309-10, 312). The other great sea turtles found in Chinese waters—the hawksbill (*Eretmochelys imbricata*), green (*Chelonia mydas*), leatherback (*Der-*

mochelys coriacea), and olive ridley (*Lepidochelys olivacea*)[26]—nest mainly in Southeast Asia. Though some of the latter four species nest in Japan and on Taiwan and mainland China, too, nesting is most important on the Paracel (Sisha or Xisha) and Nansha islands east of Vietnam, which politically were administered by Kwangtung Province but geographically are part of Southeast Asia. Recent data (1959-70) for the Paracel Islands show the green turtle to be the main species caught, with the loggerhead the only other species deliberately sought (C. C. Huang, 1981: 321-22).

Sea turtles may be taken in their nesting areas, or at sea by men in small sailing vessels. As in many parts of the tropics (Gudger, 1919; Hornell, 1950: 35-38; Parsons, 1962: 84-89), turtlers in Chinese waters may employ suckerfish or remora (*Echeneis* spp.) in capturing sea turtles. A ring is placed on the tail or a suckerfish, and a long line attached to the ring. Several suckerfish are taken along on a boat, and when a turtle is found asleep on the surface of the water, two or more of the fish are released to fasten themselves to the turtle's shell. This is done by means of a sucking-disk on top of the fish's head. With a sufficient number of suckerfish affixed, a turtle, despite its struggles, is unable to break loose, and is pulled into the boat (F. T. Bullen, 1904: 123-24).

The green turtle (Figure 75) is universally believed to have the best flesh of any species of great sea turtle, with the Chinese regarding it as a delicacy, the preferred turtle for great dinners (I. H. Burkill, 1966, 1: 529; Mao, 1971: 93). The olive ridley, though its flesh is not as highly regarded as that of the green, is acclaimed as food in many places (Carr, 1952: 410; Mao, 1971: 85, 93; Pritchard and Trebbau, 1984: 348). The other sea turtles, however, have flesh that is less well regarded. The loggerhead and leatherback are not good food and their flesh is even considered by some peoples to be inedible, with the loggerhead sometimes poisonous (Fitter, 1961: 117; Parsons, 1962: 6; Schafer, 1962: 73; I. H. Burkill, 1966, 1: 795; 2: 2207; Pritchard and Trebbau, 1984: 317).[27] Pritchard and Trebbau (1984: 266) insist that leatherback flesh, though tough, strong-tasting, and oily, is unquestionably edible, and on Taiwan leatherback flesh is eaten despite its bad taste (Mao, 1971: 105, 114).

Hawksbill flesh is also eaten by coastal people on Taiwan and elsewhere, but some peoples regard it as barely edible or inedible because of its reputed toxicity (Carr, 1956: 149-50; Rebel, 1974: 98-99; Carrascal de Celis, 1981: 325; Kar and Bhaskar, 1981: 367; Witzell, 1983: 48-49; Pritchard and Trebbau, 1984: 332-33).[28] The hawksbill, however, is the tortoise-shell turtle, which provides most of the tortoise shell of commerce. Tortoise shell is used in making combs, hair brushes, eyeglass frames, jewelry boxes, and in other decorative ways, as on furniture; and in economic terms the hawksbill was the most important sea turtle to the early Chinese as to other ancient peoples.[29] Tortoise shell from the South China Sea reached North China in trade or tribute from Annam, Hainan, and Kwangtung from the second century B.C. onward (Schafer, 1962: 74; 1967: 160-61, 215). Tortoise shell was so prized as to be among the items that inspired the Chinese to extend their dominion over Hainan (Schafer, 1970: 12), and as to lead one Chinese emperor in the first century A.D. to attempt to use it as currency (I. H. Burkill, 1966, 1: 951). By Sung times, the Chinese had a trade network for tortoise shell that extended not only to various parts of Southeast Asia, but to India and northeast Africa as well (Wheatley, 1959: 81-83). The green turtle—excellent of flesh but with a thin shell traditionally used mainly as veneer[30]—has now displaced the hawksbill in economic importance, for the hawksbill's numbers have been decimated by the tortoise-shell trade and it has been classified as an

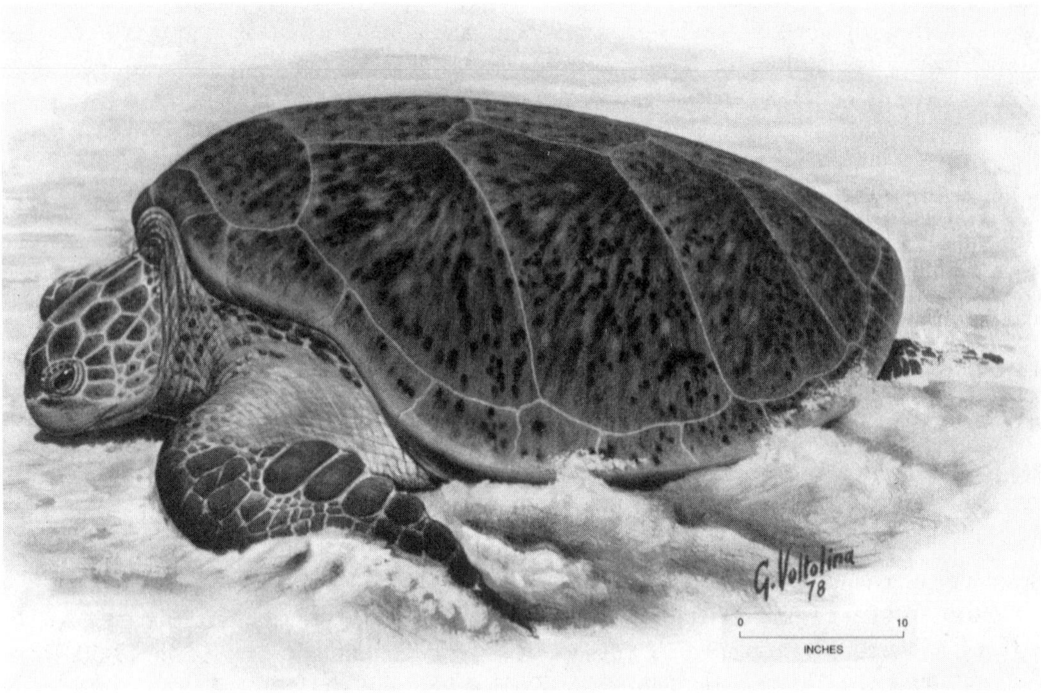

Figure 75. Green turtle (*Chelonia mydas*) (from Pritchard and Trebbau, 1984)

endangered species by the International Union for Conservation and Natural Resources (Carr and Meylan, 1979-80; Witzell, 1983: 50-51).[31]

Turning to the preparation of turtle flesh, we note F. T. Cheng's modern gourmet recipe (1954: 81-82) for the preparation of fresh-water turtle. First the flesh is cleaned, then it is cut into pieces and boiled in water and "sherry" (rice wine) for three or four hours along with chicken, mushrooms, Chinese wolfberries (*Lycium chinense*), and dried flesh of the longan. This makes a thick soup to which salt, soy sauce, and more wine are added just before serving. A soft-shelled turtle broth dating from the fifth century A.D., described by Schafer (1962: 73), seems the equal of Cheng's: turtle flesh prepared with mutton, ginger, onion, bean-relish, magnolia, wine, and rice. Schafer also observed that in ancient China or later, people were pleased to have soft-shelled turtle at a banquet; that the giant soft-shell was a princely present that was highly-regarded in soup; and that broth of sea turtle was also used and its flesh regarded as tasty.

The medicinal uses of turtle and tortoise have been many and varied, but especially prominent are their employment as tonics, as an aid in preventing and combatting fever, and as an antidote to poisons (Hooper, 1929: 148-49; I. H. Burkill, 1966, 2: 2208; Read, 1977e: 6-31). Pregnant Chinese women in ancient times, however, avoided the flesh of turtle lest their offspring be born with an unusually short neck (H. C. Hou, 1938: 348), and Chinese women in San Francisco today sometimes avoid eating turtle, along with frog, snake, shrimp, and shellfish, because of their rough or scaly exteriors or in fear that at some later time the baby will develop allergies (B. Chang, 1974: 438; Ling *et al.*, 1975: 129, 131).

[1] In 1958, nearly a decade after establishment of the Peoples Republic, China's fish catch was 6 million metric tons, somewhat higher than for second-place Japan (5.4) and far more than the Soviet Union (2.8) and the United States (2.7) (D. Riedel, 1961: 43).

[2] There is an excellent survey of Hong Kong fisheries by S. Y. Lin (1940b), which includes information on fish caught, fishing grounds and seasons, fishing vessels and the technology of fishing, as well as material on the fishermen themselves and the financing of their operations.

[3] Surprisingly, T. H. Shen (1951: 390-91) listed only 6896 metric tons of marine fish caught in Kwangtung, but 20,000 tons processed there. One suspects that his figure for fish catch was too low, perhaps much too low.

[4] For species identification of these and other marine fish used in China, see T. H. Shen, 1951: 388-91. Species identification and/or other information on marine fish and their use is also found in other sources, among them Leung *et al.*, 1973: 113-43, who provide food composition tables; S. W. Williams, 1895, 1: 346-50; Sowerby, 1925: 118-73; Herklots and Lin, 1940; Read, 1977f; and E. N. Anderson, 1988: 140-43. For identification of fresh-water fish of China, see Nichols, 1943; of Hainan, see Nichols and Pope, 1927.

[5] For more on the eating of raw fish in China, see "Background of Parasitic and Infectious Diseases" in our chapter on nutrition and health, and footnote 8 on *sushi*-eating.

[6] For generalizations on the cooking of fish and other seafood in China, and flavorings used with such dishes, see E. N. Anderson, 1988: 140-43, 157, 177; for recipes to Chinese fish and other seafood dishes, see especially W. W. Chang *et al.* (1970: 141-216) and K. Lo (1979c: 260-310).

[7] Fong and Chan (1973) found appreciable amounts of cancer-causing nitrosamines in salted fish purchased in Hong Kong markets. Such salted fish comprised virtually all the fish exported from Hong Kong to mainland China prior to the Japanese invasion (Herklots and Lin, 1940: 1).

[8] Because of serious water pollution, writes E. N. Anderson (1988: 142), this practice has ceased in China's wealthier cities.

[9] The most complete scholarly account of the history and geography of cormorant fishing is Laufer's *The domestication of the cormorant in China* (1931). For a good review of methods of cormorant fishing in ancient and modern times see Gudger's "Fishing with cormorant in China" (1926). For a brief, recent summary of information on the cormorant see Z. G. Zhang (1984), and for short accounts of cormorant fishing in Japan, see Hornell (1950: 31-33) and Joseph (1986).

[10] In Japan, though all cormorant fishing is done in rivers, sea cormorants (*Phalacrocorax capillatus*) are used exclusively in such fishing. Larger in size and more even tempered than *Phalacrocorax carbo*, these sea birds are not only better able to catch large fish but are less likely to fight with one another and with the fishermen (Joseph, 1986: 38).

[11] Fishing with otter has also been reported for parts of South and Southeast Asia (along the Indus River; along the Ganges in Bengal; in the backwaters of Cochin; and in Malaya) (Watt, 1889-96, 5: 656; Gudger, 1927: 206-8; Hornell, 1950: 33).

[12] For accounts of fish cultivation and identification of pond species in China, see Drews, 1951 and 1961, and Atkinson, 1977; for South China, especially Kwangtung, see W. E. Hoffmann, 1934; and for Hong Kong, S. Y. Lin, 1940a, and Hung, 1960-61. For brief descriptions of fish pond species and ecology in Hong Kong, Kwangtung, or South China in general, see T. P. Chen, 1934; T. Chow, 1958; E. N. Anderson, 1970b; and E. N. Anderson and M. L. Anderson, 1977: 334-35. For cultivation of fish and other aquatic creatures in Taiwan, see T. P. Chen, 1976, and H. H. Chen, 1977.

[13] Following establishment of the People's Republic, experiments were carried out aimed at cultivating various additional freshwater fish, and now *Tilapia* are widely cultivated and eels grown commercially (Atkinson, 1977: 339-40).

[14] The government of the People's Republic set up ten to fifteen marine farms, some in coastal Kwangtung, for raising mullet as well as seaweed, oysters, mussels, shrimp, crabs, scallops, sea cucumbers, and octopus (Atkinson, 1977: 341).

[15] Since the early 1950's the People's Republic has encouraged cooperatives and communes to raise carp in rice paddies, not only to obtain fish for eating but to reduce the number of mosquitoes and other undesirable insects and to improve crop yields by means of fish digging and feces (Atkinson, 1977: 337).

[16] For some Cantonese oyster-sauce dishes ("Chicken with Oyster Sauce," "Squab with Oyster Sauce," "Beef with Bean Curd and Oyster Sauce," and "Clams and Oyster Sauce") see C. B. T. Lee and A. E. Lee, 1980: 181, 194, 214, 231.

[17] The term "turtle" and "tortoise" are applied to reptiles of the order Chelonia, or Testudinata. Though popular usage of those terms differs in Britain and the United States, we follow the standard set for the United States by the American Society of Ichthyologists and Herpetologists. This entails use of "turtle" for all species of the order Chelonia, whether found on land or at sea, and "tortoise" only in a secondary sense for slow-moving terrestrial species. There are, however, problems in interpreting the literature, for writers have used "turtle" and "tortoise" in different and sometimes inconsistent ways. One notes also an inconsistency in general usage: "tortoise shell," which comes primarily from the hawksbill, a sea turtle.

[18] In Burma, like China, the turtle is venerated, and kept in large tanks near Buddhist monasteries. At one pagoda in Mandalay, it is customary for a person, after completing worship, to feed the temple turtles, and there are women who sell balls of cooked rice and other tasty bits for the sacred turtles. A similar custom was found at various places in traditional China (Read, 1977e: 8n).

[19] In a similar vein are the Chinese tales in which a giant sea turtle supports the islands where the immortals live—believed to be in the seas southeast of China (Schafer, 1967: 214). Also similar are certain traditions in India. One is in the Satapatha Brahmana, where Vishnu in the form of a turtle played an essential supportive role in the creation story, and where a live turtle—fountainhead of everything—was placed in the base of a northern altar which symbolized the universe. A second is in another early account in which India is depicted as supported on the back of a giant turtle (Stutley and Stutley, 1977: 157). In modern India, one of Vishnu's incarnations is as a turtle whose birthday is celebrated, and whose carved figure sometimes is found in the stone pavement outside of Vishnu temples (Underhill, 1921: 75-76, 138).

[20] For later times, one reads of turtle-shell divination in a matter of royal succession, and even in determining the best place for a tomb (J. J. M. de Groot, 1964, 2: 489; 3: 993).

[21] Though some writers refer to the shells used in Shang divination as "tortoise shells," only one species identified, *Testudo emys*, would, by our usage, be a tortoise, a slow-moving terrestrial form. Today this species is found in Southeast Asia. All of the four other species identified (*Ocadia sinensis, Chinemys reevesi, Mauremys mutica*, and *Cuora* sp.) were aquatic turtles which lived in streams, lakes, and other water bodies on the land. None were marine turtles.

[22] On Taiwan there are now many turtle farms where *Amyda sinensis* is raised. The turtles are fed viscera of animals, small fish, shrimp, snails, and vegetation. Such turtle cultivation is viable because the soft-shelled turtle commands a good price, the Chinese considering both its flesh and blood quite tonic and healthy. In the early 1970's, soft-shelled turtle flesh on Taiwan could cost, per pound, twice the price of roast beef (Mao, 1971: 22, 73-74; T. P. Chen, 1976: 137-44).

[23] For a photograph, and brief information, of this and other turtles of Hong Kong, see Karsen *et al.*, 1986: 38-48.

[24] Note the similarities of these species with those whose shells were used in Shang divination.

[25] This recalls the situation in Southeast Asia, where most coastal groups of Thailand and Burma as well as Moslems in the Malay Archipelago reject the flesh of sea turtles as food (Hendrickson, 1958: 457-58; Parsons, 1962: 9; Polunin and Nuitja, 1981: 353-54; Tow and Moll, 1981: 339).

[26] C. C. Huang (1981), in a recent report on the sea turtles of China, writes of the green, loggerhead, leatherback, and hawksbill, but fails to mention the olive (or Pacific) ridley, which has often been confused with the loggerhead over the years. Karsen *et al.* (1986: 45) write that little is known of the Pacific ridley, which does not nest in Hong Kong but it is often observed in the seas nearby. Mao (1971: 85) observes that the olive ridley can usually be found at the fishing port of Nan-fang-ao in Taiwan and at Hung-t'ou (Lan Yü) Island about fifty miles southeast of Taiwan. It has also been reported in various lands bordering the South China Sea: the Philippines (Carrascal de Celis, 1981), Sarawak (de Silva, 1981: 328, 334), Malaya (Tow and Moll, 1981: 339-44 *passim*), and Indonesia (Polunin and Nuitja, 1981; Suwelo *et al.*, 1981).

[27] In the United States, eating the flesh of a common box turtle has on rare occasions poisoned people, perhaps because the turtle had eaten *Amanita phalloides*, the death-cap mushroom, whose poisons are retained in the turtle's flesh (Carr, 1956: 150; Rebel, 1974: 99).

[28] Turtle poisoning may develop from eating sea turtles of other species as well. This is an occasional toxicity of unknown cause, but that probably derives from poisonous algae, invertebrates, or even sea snakes that a turtle may eat. It can develop in humans from consuming turtle flesh, fat, blood, or viscera, with symptoms commonly developing within a few hours or at most several days afterwards. Among the first symptoms are paleness; giddiness; severe intestinal pain; nausea, vomiting, and diarrhea; and sweating but coldness of the hands and feet. The mortality in reported outbreaks of such turtle poisoning is 28 percent (Halstead, 1978: 918-24).

[29] Tortoise shell from the Indian Ocean was already reaching the Mediterranean area at the time of the early Roman Empire (Parsons, 1962: 45-46), and later the Emperor Nero is reported to have had a tortoise-shell bathtub (Carr, 1952: 380).

[30] In Japan, hawksbill-turtle shell has long been imported from Southeast Asia for use by handicraft workers in making ornaments and for other decorative purposes. Recently, however, taxidermists have begun to purchase imported shells of hawksbill and green turtles for stuffing, with stuffed turtle shells marketed in Japanese vacation areas because, as in China, turtles are symbolic of good luck and longevity (Uchida and Nishiwaki, 1981: 319). In the Philippines, not only are hawksbill shells and items made of such shell exported to Japan (and elsewhere), but in certain places there is a sizeable cottage industry involved in stuffing hawksbill and green turtle shells (Carrascal de Celis, 1981: 325). We have not determined whether such stuffed shells are exported to Hong Kong or China, but, if so, one would expect them to be as popular as in Japan.

[31] We have found nothing to indicate that the penis of the green turtle (*Chelonia mydas*) served as an aphrodisiac, as it does in present-day Puerto Rico (Lowenstein, 1982: 72).

CHAPTER ELEVEN

Eggs

Eggs of Fowl
In Ritual and Belief
Types and Amounts of Fowl Eggs Used as Food
Food Uses of Fresh Eggs
Preserved Eggs: Their Processing and Uses
Nutritional Value of Eggs
Turtle Eggs

EGGS OF FOWL

In Ritual and Belief

Eggs, usually of chickens or ducks, play a major role in Chinese folk religion. One important use is in divination, a custom found widely among tribal groups in southern Asia. Eberhard (1968: 419-21) describes two forms of egg divination practiced in early and modern China. One involves tossing or throwing an egg (or eggs), with divination based on where it falls and/or whether or not it breaks. The second entails painting, blackening, or drawing on an egg, then boiling or roasting it, with divination based on patterns of cracks that appear on the shell or on examination of the yolk and/or white after cutting or dividing the egg.

Also important is the role of the egg as symbol of fertility, a symbolism also found widely elsewhere in the world, including Southeast Asia and India. In Chekiang, two red eggs (colored red for happiness and good luck)[1] may be rolled over a bride's breasts into her lap to guarantee fertility, and eggs may be used in a ceremonial washing of a bride, then presented to wedding guests (Eberhard, 1968: 421). Eggs are also presented as gifts at engagement and wedding ceremonies, and one may find red eggs on tables at marriage feasts and in the bridal room. Other persons desirous of having children may take such eggs away from the feast and eat them. Eggs, whether white or colored red, are also gifts to a newborn child or its mother,[2] and convey the wishes that the child have a long and happy life, and that the mother's fertility may continue; this is a felicitous gift at a time when she has special needs for good nutrition. Customs surrounding such gifts of eggs at the birth of a child varied from place to place in China. As described by Cormack (1935: 30-33), whose observations focussed on the Peking area, friends and relatives brought gifts to the Chinese mother and her newborn child at the time of its first bath, three days after it was born. Before the actual bathing started, an offering of incense and uncooked rice was placed in front of the ancestral tablets. Then a woman attendant brought a basin or bath and poured water in it. Next, eggs were placed in the water, one for a boy, two for a girl. A dish of lukewarm water and a spoon were then placed next to the bath, along with a plate of hard-boiled eggs, some white, some red. While the baby was bathed, each guest added a spoonful of water and an egg to the bath, while extending wishes that the child be healthy, wealthy, happy, and long-lived. A white egg conveyed the hope that the child might live to an old age and have white hair, whereas a red egg carried the wish that it have good fortune. Guests also would put jewelry or money into the bath in the hope that the child be prosperous. Ginger was put in the bath water and rubbed on the infant's navel to drive away illness; and other efforts were taken to assure the infant's well-being and future. The jewelry, money, and eggs placed in the bath, as well as the rice offered to the ancestors, were gratuities kept by the woman attendant. Later in that day, however, relatives and friends gave eggs (uncooked eggs and cooked eggs that were colored red), as well as other foods (sweetmeats, chicken), to the mother, and they, in turn, were invited to a dinner of celebration. Red eggs might also be sent to less fortunate persons by a young couple when they have their first son. They might be given to relatives and friends who visited after the birth, or they might be sent to them in the expectation of a gift for the young child. And at a party held for the baby a month after its birth, red eggs might be given to friends and relatives to take home (Gray, 1878, 2: 184; Ball, 1906: 93-94; Hutson, 1921: 2; Best, 1926: 569, 573; Waln, 1935: 84-85; Y. H. Lin, 1947: 20; Adolph, 1956: 69; Graham, 1961: 33; Eberhard, 1968: 421; Em. Hahn *et al.*, 1973: 71, 87; Yee, 1975: 72).

It was also customary at other birthdays in China for a person to eat a few boiled duck eggs, or ones kept in a red concoction. This was done in the expectation that his luck might improve. In a similar vein, he would eat a bowl of noodles, symbolic of a long life (Doolittle, 1865, 2: 222). Foreign missionaries, especially if they were newly arrived or if they were elderly and had long served in China, were commonly given gifts of eggs and noodles on their birthdays, with one missionary in Foochow receiving two hundred eggs and some noodles of great length (Holmes, 1948: 304), wishes for both good fortune and longevity.

Types and Amounts of Fowl Eggs Used as Food

Eggs of various domesticated fowl are consumed by the Chinese. Chicken eggs (*chi-tou*; *chi-luan*; *chi-tan*) are by far most important, but those of duck (*mu-luan*; *ya-tan*) and goose (*Anser cygnoides*, the Oriental tame goose considered previously) (*ê-luan*) are significant as well (Table 5). Eggs of other fowl, such as pigeon, turkey, and quail,[3] may also be eaten in China, as in Peking, where pigeon eggs were a great delicacy, a banquet food sold for much higher prices than hen eggs (Cormack, 1935: 137; Hoose, 1938: 11).[4] In overall terms, however, all eggs other than those of chicken, duck, and goose made a negligible dietary contribution.

In traditional China, eggs were much more expensive than most foods of plant origin (Guy and Yeh, 1938: Table 1), and ordinary farmers sold most of their eggs in urban areas, saving for themselves only a small percentage to be consumed as a treat at important occasions. Maynard and Swen (1956: 411-12), using 1929-33 survey data, estimated that in Chinese villages, annual egg consumption amounted to only 8 per adult-male or 41 per family (though well-to-do urban dwellers might consume many times that number). In Peking, a 1926-27 household budget survey found that, though eggs were purchased by 213 of 283 families, only half of lower-income families (with incomes less than 30 Chinese silver dollars a month) used any eggs at all, and that consumption of 1 egg per capita per week was reached only among families with incomes of $80 a month (Gamble, 1933: 100-101). Using data for 1965 and 1970, Whyte (1972: 43) estimated China's annual egg production at about nineteen per capita of human population in the People's Republic, a very low figure by Western standards. Furthermore, many eggs were exported, significantly reducing the number available for domestic

Table 5. Egg output in relation to human population in China (from Whyte, 1972)					
	Total Birds (millions)	**No. in Lay (millions)**	**Annual Lay per Bird**	**Total Eggs (millions)**	**Eggs per Head of Human Population per Year**
Chickens	265	132.5	90	11,925	15
Ducks	64	32	80	2,560	3.2
Geese	10	5	70	350	0.44
Source: Figures for total egg production from Liu and Yeh, 1965; Egg availability calculated for 1970 population.					

consumption. Egg production has increased rapidly in recent years in the People's Republic, and in 1985 it exceeded that of both the United States and Soviet Union. Despite this, per capita egg consumption is roughly a third that of the United States (Wittwer *et al.*, 1987: 325).

Restrictions on the eating of chicken eggs, or outright rejection of them as food, are found among various groups in India, Southeast Asia, and Tibet. Some of this apparently derives directly from concerns about fertility, childbirth, and sex; and some, fostered by Hinduism and Buddhism, is based on vegetarian belief or on the unclean eating habits of chickens (Simoons, 1961: 67-71; R. O. Whyte, 1974: 62). For China proper, as we noted in our section on vegetarianism, egg avoidance was found principally among Buddhists, and eggs, along with meat, fish, onions and their relatives, dairy products, and alcoholic beverages, were banned or not ordinarily consumed by monks (E. L. Liu, 1937: 134-35; Welch, 1967: 112).

Food Uses of Fresh Eggs

Fresh eggs may be steamed, boiled, fried, stir-fried, scrambled, or made into omelettes, in which forms they are commonly made or served with soy sauce. They may be prepared in soup such as egg-drop soup, or in egg fu yung, or in dishes of other sorts. They may be used in making egg-roll skins, egg noodles, or in other ways.[5] Because they are made with a product so central to Chinese cuisine, we would single out tea eggs, which have an attractive marbled surface and are a delicious, tea-flavored snack that may be eaten hot or cold, or included on a cold plate with flesh and other foods.[6] For a similar reason, we would also call attention to soy eggs, which may be made by cooking hard-boiled eggs in soy sauce for fifteen minutes or so (K. Lo. 1979c: 322).

Preserved Eggs: Their Processing and Uses

The Chinese have methods of processing eggs that enable them to reduce spoilage and lower the costs of handling. One method involves the salting of eggs, ordinarily of ducks. The usual way is to soak duck eggs in brine for 20 to 40 days, which makes the egg-white salty, and the yolk firm and bright orange in color. An alternative is to leave the eggs in brine for a week or so and then coat them completely "with a mixture of mud, salt, and chaff" (H. J. Lin and T. Lin, 1972: 166; G. Z. Chu, 1975: 293; K. Lo, 1976: 117; Schwabe, 1979: 398; Gin, 1984: 48). Salted duck eggs are a favorite of the Chinese and may be eaten in various ways, for example hard-boiled in water, sliced, and eaten with boiled rice or congee. Or they may, usually after cooking or beating, be an ingredient and flavoring in soups, vegetable dishes, or ones of seafood or flesh, as in the Cantonese "Steamed Minced Pork with Water Chestnuts and Salted Duck Egg," "Savory Steamed Lobster with Pork and Salted Duck Egg," and the custard-like "Three-Toned Eggs" (made of fresh eggs, a salted duck egg, a thousand-year-old egg, and salt) (W. W. Chang *et al.*, 1970: 193, 291, 295; H. J. Lin and T. Lin, 1972: 166; G. Z. Chu, 1975: 143, 235-36, 293; K. Lo, 1976: 117-18; Schwabe, 1979: 398; C. B. T. Lee and A. E. Lee, 1980: 208, 229; Gin, 1984: 49). Another method of preservation involves the drying and salting of duck-egg yolks (E. N. Anderson and M. L. Anderson, 1977: 336).

Most interesting of Chinese preserved eggs, however, are the so-called thousand-year-old eggs (also called hundred-year-old eggs, ancient eggs, or Ming dynasty eggs), a delicacy unusual in appearance, odor, texture, and flavor, and usually bought in the market rather than made at home. Despite their name, thousand-year-old eggs are usually cured for just a few months, and are said to be most tasty at about a hundred days. The eggs of chickens and other birds may be processed into thousand-year-old eggs, but duck eggs are the common ones used. This, and the importance of duck eggs in preservation by salting, probably derives from the fact that they are produced only in localized habitats suited to ducks and need to be shipped to market, yet spoil more quickly than chicken eggs (B. Y. Chao, 1972: 18-19; Em. Hahn *et al.*, 1973: 199; G. Z. Chu, 1975: 299; K. Lo, 1976: 116-17; 1979c: 322-23; E. N. Anderson and M. L. Anderson, 1977: 336; Schwabe, 1979: 399).

Various methods of curing, from simple to elaborate, may be followed in making thousand-year-old eggs. As detailed by Kenneth Lo (1979c: 322-23), however, salt is first dissolved in a small amount of water in a large bowl; then pine ash and lime are slowly added, and the mixture stirred until it reaches a muddy consistency. A thick layer of the mixture is applied to clean duck eggs, which are then rolled in a tray of husks, of rice or some other kind, to give them a non-adhesive coat to prevent them from sticking to anything.[7] Then they are placed in a big earthenware jar which is covered loosely and left to stand. The eggs are removed every three days and rearranged in the jar, and after fifteen days the jar is sealed and left for another month. At that time, after 45 days in total, the eggs should, according to Lo, be ready. A more lengthy procedure is mentioned by Schwabe (1979: 399), one involving an initial three months of soaking the eggs in a brine made of water, salt, lime, lye, and tea leaves. Then the eggs are dried, covered with a paste of clay, lime, ashes, and salt, and buried in the earth for further aging. By whatever process, the end product has a yolk that is green and resembles cheese, and a white that is yellow or amber and of a gelatinous consistency (G. Z. Chu, 1975: 299; Schwabe, 1979: 399).

In eating a thousand-year-old egg, one must first remove the mud and carefully clean its shell. Then one normally eats the egg, which has the smell of ammonia, uncooked. It may be eaten with hot rice for breakfast or a late night supper, or cut into pieces and served as a snack with soy sauce, sometimes garnished with gingerroot strips or slices, or with a sauce made of vinegar and shredded ginger (W. W. Chang *et al.*, 1970: 139-40; Howe, 1971: 175; G. Z. Chu, 1975: 299; F. Lin, 1975: 323; K. Lo, 1976: 116; Schwabe, 1979: 399). Such eggs may also be prepared in other ways, as in "Fried 100-Year-Old Eggs" (W. W. Chang *et al.*, 1970: 140) or "Old and Fresh Eggs," a steamed dish that includes both thousand-year-old egg yolks and fresh eggs (H. J. Lin and T. Lin, 1972: 173). The ultimate in combining types of eggs, however, is "Steamed Three Variety Eggs," which includes in a single dish thousand-year-old eggs, salted eggs, and fresh ones (K. Lo, 1979c: 320).

Nutritional Value of Eggs

The food value of fowl eggs differs somewhat by species, with those of chickens and ducks, most important of the eggs eaten by the Chinese, excellent sources of protein (which is of high digestibility and biological value); good sources of vitamin A and riboflavin; and fair ones of calcium.[8] Preservation of duck eggs by salting doesn't reduce any of these nutrients significantly, and indeed salted eggs are higher than fresh eggs in most of them. Duck eggs preserved by liming (to make thousand-year-old eggs) are slightly higher than fresh duck eggs in protein and calcium, slightly lower in iron and vitamin A, but significantly lower in riboflavin (Leung *et al.*, 1973: 111-12). Not to be overlooked in terms of nutrition are two Chinese practices: (1) that of dissolving eggshells in vinegar and giving the calcium-rich solution to sick people to drink (Holmes, 1948: 212), and (2) that of using eggshells medicinally in other ways. Though probably of quite minor importance nutritionally, it is reported that eggshells of fowl, called "Phoenix skins," are burned and ground into a powder used in treating dysuria (difficult or painful discharge of urine), and that they are imported to Malaya from China for this and other medicinal uses (Soubeiran and Dabry de Thiersant, 1874: 58; Hooper, 1929: 147). Perhaps also of nutritional relevance is the Chinese liking for fertilized eggs in which the embryo is well developed, a preference they share with certain other peoples in Southeast Asia and the Pacific region (Duval *et al.*, 1885: 302-3; Simoons, 1961: 68; Schwabe, 1979: 399). Embryonated duck eggs, judging from data provided by Leung *et al.* (1973: 111), are substantially higher in calcium than ordinary ones.

TURTLE EGGS

Eggs of fresh-water turtles have been eaten by the Chinese since antiquity. Such eggs have been highly prized and expensive, not normal fare for ordinary folk. They may be eaten fresh or preserved, as, for example, in eleventh-century China where eggs of the giant soft-shell turtle were preserved in salt (Schafer, 1962: 73). With Han conquest of the south, a greater supply of turtle eggs must have became available to them because the nesting grounds of the green and most other species of great sea turtles are more numerous on the sandy beaches of nearby tropical regions.[9]

A female green turtle apparently lays only once in several years. Then, over a period of a few months, she normally deposits from four to seven clutches of eggs (the mean number of eggs in over 8,000 clutches observed by Hendrickson [1958: 502] in Malaya and Sarawak was 104, and the maximum, 184), soft of shell and like table-tennis balls in size (Parsons, 1962: 3). A more precise comparison, in Vietnam (Richard and Nguyèn-Thi-Làu, 1961: 334), reported green turtle eggs to average nearly 31 grams, about three-quarters the average weight of a local hen's egg.

In nesting, a female green turtle moves its heavy body slowly and laboriously up a beach beyond the high-tide mark. With its flippers, it then excavates a nest in the sand, lays its eggs, covers the nest with sand, and returns to the sea. The path of the turtle, however, remains clearly imprinted in the sand for all to see, and this makes it a simple matter for humans, crabs, and other creatures to locate the nest and dig up the eggs. This may be why turtle-nesting grounds in Southeast Asia today, especially those of the green turtle, are often on islands, sometimes small, uninhabited ones, the turtle islands of such common occurrence in the region. In such places, laying turtles may have a better chance of being undisturbed by predators, and eggs of being missed by collectors; the newly-hatched turtles, as well, may be troubled by fewer of the complement of predators that so decimate their numbers.

Perhaps we should say a word here about the survival of sea turtle populations in Southeast Asia. The hawksbill, long the most important sea turtle in economic terms, is seriously threatened with extinction worldwide, for its shell has become very expensive indeed. In the Caribbean, it was said (1980) that a single hawksbill shell could return $200 or more to a turtler (Carr and Meylan, 1979-80), an unusually large amount to someone who is poor. The hawksbill, loggerhead, and leatherback, it is true, are all aided in their struggle for survival by the fact that many peoples do not regard their flesh highly. All sea turtles in Southeast Asia, moreover, continue to benefit by the rejection of their flesh as human food among certain peoples, especially Moslems. Turtle eggs, on the other hand, have continued to be popular and sought after by Southeast Asian folk everywhere.[10] On just three islands, the Turtle Islands of Sarawak, an average of nearly two million eggs a year were produced from the mid-1930's to the mid-1950's (Hendrickson, 1958: 492-93). The egg harvest has been declining there (de Silva, 1981: 334-35) and in certain other parts of Southeast Asia, but pressure on that resource continues unabated. Various of the species, including the green, are now under serious pressure.[11] It is true that Parsons (1962: 55-56, 58-59), writing a quarter of a century ago, found no convincing evidence of permanent decline in the green-turtle population of Southeast Asia. More recent writers, however, are less sanguine. The green turtle has been described as among the most exploited wild species of the Philippines; though once the most abundant sea turtle species overall, the green, like the hawksbill, has experienced a significant drop in numbers, and the overall sea turtle situation is referred to as a resource emergency (Carrascal de Celis, 1981: 325-26). Declines have been reported, as well, in other countries that border the South China Sea (de Silva, 1981: 328-29, 334-35; Polunin and Nuitja, 1981: 355-59; Tow and Moll, 1981: 339).

There have been efforts in various parts of the world to protect green turtles and their nesting grounds, and even to hatch green turtle eggs and raise the young to a point where they can be released without being exposed to the full range of predation they face in the wild. Such efforts are being made in Southeast Asia also, but are too few to stem the decline. Thus newly-hatched green turtles that have escaped the hands of egg gatherers, usually must brave a daunting array of predators—crabs, birds, snakes, and rats among them—to reach the sea where still more predators, such as sharks and other fish, await them.

We have uncovered only scattered references to Chinese trade in, and use of, sea turtle eggs. As with other foods they favor, the Chinese did develop sources of supply at home. In former times, many sea turtle eggs were also shipped from North Borneo (Sabah) to China (Parsons, 1962: 64). That trade did not continue into the late 1970's, but it was observed (de Silva, 1981: 329-30) that demand for turtle eggs is strong, for their aphrodisiac qualities, in the red-light district of Hong Kong; and that Chinese travellers take small amounts of turtle eggs as gifts when they go from Sabah to Hong Kong. We have not, however, determined whether imports of sea turtle eggs from elsewhere in Southeast Asia have ever been of much significance in China.

Sea turtle eggs are usually much more fatty than chicken eggs, and the Chinese have a special appreciation of them for this reason. Sea turtle eggs are good sources of energy, and comparable to or higher in protein than chicken and duck eggs (Penyapol, 1958: 35-36; Somadikarta and Anggorodi, 1962: 73-74). Those of the "Malaysian sea turtle" (apparently the green turtle) have been found to be a good source of vitamin A, higher in this regard than chicken and duck eggs (Chandrasekharan, 1968-69). And those of sea turtles (green, hawksbill, or both) analyzed in Thailand were also a good source of certain B vitamins and a fair one of calcium (Penyapol, 1958: 36). Indeed, Hendrickson (1958: 525-29), has presented evidence showing that using green turtles as a source of eggs rather than for flesh and fat makes more sense, yields higher profits, and is more easily managed. He argues further that the demand for green turtle soup among gourmets in foreign lands should not be allowed to bring on a net nutritional loss to people in areas where turtles live.

[1] The color red is not only a symbol of joy and good fortune in China, but is also protective against evil spirits. One example of this is the wrapping of a red cloth around the handle of an ax used to hammer the nails in the lid of a coffin to stop bad influences from the coffin reaching the person doing the work (J. J. M. de Groot, 1964, 1: 96); another is parents' placing of a bit of red cloth on the person of their young sons to keep them from being maimed by evil spirits (Doolittle, 1865, 2: 308); and still another is the use of red in the nuptial suite for protective purposes (Cormack, 1935: 85). For more on the symbolism of red in China, see Doolittle, 1865, 2: 119, 308-10, 314, and *passim*; Cormack, 1935: *passim*; Burkhardt, 1953-58, 1: *passim*; J. J. M. de Groot, 1964, 1: 28, 31, 95, 230, 326, and *passim*; Eberhard, 1968: 130, 245, 278; and A. P. Wolf, 1970: 191-94. On Chinese color symbolism in general, see Schafer, 1967: 256-61 (early China); and Giles, 1900: 55-56 and C. A. S. Williams, 1941: 76-79 (traditional China).

[2] Giles (1876: 159), writing in the mid-nineteenth century of differences in the welcome given to newborn sons and daughters in China, says that red eggs are sent at the birth of a son, but not that of a daughter.

[3] For quail egg recipes, see Howe, 1971: 176 ("Quails' Eggs with Mushrooms") and E. Chen, 1977: 54-55 (Cantonese "Fried Quail Eggs with Vegetables"). Canned quail eggs may be found in Chinatown markets in the United States.

[4] Though Read (1977a: 48) identifies various medicinal uses of pigeon eggs, we have found nothing to indicate that the high cost of pigeon eggs in Peking was related to medicinal concerns.

[5] Recipes for egg dishes of the sorts mentioned here are found in F. L. Chang and G. M. Fuchs, n.d.: 134-35, 292-93, 347-48; W. W. Chang *et al.*, 1970: especially 128-40; B. Y. Chao, 1972: 133-40, 217; G. Z. Chu, 1975: 234-35; K. Lo, 1976: 118-19; 1979b: 49; 1979c: esp. 311-21; C. B. T. Lee and A. E. Lee, 1980: 68, 195-98, 238, 253-54; Gin, 1984: 40, 48-51; K. Lee and A. Branyon, 1984: 315-320; and Y. K. Martin, 1984: 57, 58, 113-15, 148, 213.

[6] Recipes for tea eggs may be found in F. L. Chang and G. M. Fuchs, n.d.: 338; W. W. Chang *et al.*, 1970: 140; Howe, 1971: 175; Field *et al.*, 1976: 91; K. Lo, 1976: 117; 1979c: 322; C. B. T. Lee and A. E. Lee, 1980: 101-2; and F. Lin, 1975: 90. A good photograph of tea eggs, before and after their shells have been removed, may be found in K. Lo, 1979b: 49.

[7] Photos of the stages in the curing of thousand-year-old eggs may be found in Tiger and Wolf, 1985: 172-75.

[8] Though eggs also contain good amounts of iron, that iron is of low bioavailability (Bothwell *et al.*, 1979: 24).

[9] For maps of the principal areas where green turtles feed and nest in Southeast Asia, see Parsons, 1962: inside front cover.

[10] In Malaysia, many rulers in the past enjoyed exclusive rights to the eggs of river turtles, which apparently afforded some protection to turtle populations. Those rights, however, were relaxed during the colonial period, and ordinary people, always eager to obtain turtle eggs, have reduced the populations of turtle species whose eggs were collected. When controlled by the rulers of Perak, estuarine species such as *Batagur baska* were protected by the custom of allowing to be hatched one of the three clutches of eggs laid by a female. In I. H. Burkill's opinion (1966, 2: 2208), the game laws of the Federated Malay States that replaced such earlier customs have probably been less effective in preserving turtle populations. See Hendrickson (1958: 458-60) for further information on the evolution of laws affecting turtles and turtle eggs, including those of sea turtles, in various parts of Malaysia.

[11] By far the largest number of the world's leatherback turtles nest in Trengganu, Malaya, where, in 1960, females were estimated to number only between 850 and 1700. The very survival of the species was regarded as threatened because of egg collecting there (Fitter, 1961: 117, 121-25).

CHAPTER TWELVE

Spices and Flavorings

"Coming here I fended off goblins,
So it's right that I taste southern cookery—
Blended with saline and sour,
Mixed with fagara and oranges."

Han Yü (translated by Edward H. Schafer)

Ginger
Garlic and Vinegar
Salt
Honey, Sugar, and Other Sweeteners
Historical Notes
Sugar Intake in Modern China
Fagara: Chinese or Szechwan Pepper
True Pepper
Capsicum Peppers
Star Anise, Bastard Star Anise, and Anise
Cinnamon
Cinnamon in China
Ancient Sources of Cinnamon Reaching the Mediterranean and Near East
Turmeric, Saffron, and Zedoary
Cardamom
Clove and Nutmeg
Coriander
Sweet Basil
Golden Needles
Ginseng

Han Yü, quoted above, was an eighth century statesman, philosopher, and poet who was banished to what today is Kwangtung. His attitude was one of willingness to try southern cooking, to experience its different tastes. In their hearth in North China, the ancient Chinese seem to have used a quite limited range of spices and flavorings, their diet less interesting than later, when southern contributions were greater. Judging from textual evidence, the most important spices and flavorings of the Chou Period were sour plums and salt, vinegar of some sort, meat sauce, fagara (*Zanthoxylum* spp.), cinnamon, and, near the end of the period, probably soy sauce as well (K. C. Chang, 1977b: 30-31). Starting in Ch'in and Han times, conquest of the south and expansion of trade contacts with Southeast Asia and India brought a much greater range of spices and flavorings into use. For T'ang China, one reads (Schafer, 1967: 193-98; 1977: 91-92, 109-13) of an impressive assemblage of spices and flavorings, among them fagara, evodia (*Evodia officinalis*), Chinese or herbaceous cardamom (*Amomum globosum*), black or bitter cardamom (*Amomum amarum*), true cardamom (*Elettaria cardamom*), lesser galangal (*Alpinia officinarum*), Mioga ginger (*Zingiber mioga*), nutmeg, cassia or Chinese cinnamon (*Cinnamomum cassia*), coriander, cloves, sweet basil, mustard seed, licorice,[1] spring onion, dill, turmeric (*Curcuma longa*) or zedoary (*Curcuma zedoaria*), and ginseng (*Panax ginseng*).[2,3]

To fully appreciate that variety, one must note that South China, along with Malaysia and India, is one of the three great centers of spice plant domestication in eastern and southern Asia. Among the plants credited to South China by most botanists are cassia and certain types of fagara. Some botanists also suggest an origin or possible origin there of ginger (*Zingiber officinale*), star anise (*Illicium verum*), turmeric, zedoary, false cardamom (*Amomum xanthoides*), or lesser galangal (Newcomb, 1963; I. H. Burkill, 1966; Mathon, 1981; Zeven and de Wet, 1982). Certainty about origins is hindered not only by a paucity of historical records, but by the fact that some of the above spice plants occur wild in more than one center. Some may have been domesticated in a single center, others in more that one. There has, moreover, been trade by sea and land among the three centers since antiquity, and an exchange not just of spices but of spice plants as well. Monosodium glutamate, that tasteless flavor-enhancer, is ubiquitous today in Chinese cuisine at home and abroad, but it first came into large-scale production in Japan early in this century; thus in most of Chinese history, the quality of a dish could not be disguised by liberal additions of MSG, but relied on the nature and quality of its ingredients, including spices and flavorings, and the cook's skill and methods of preparation.[4]

In considering spices and flavorings individually, we focus mainly on certain of the non-soy ones, those commonly used in cooking but not, like tangerine peel and orange peel, considered previously. Other spices we consider are of historical or comparative interest.

GINGER

Ginger (*Zingiber officinale*; *chiang*) has had a long, illustrious role in Chinese cuisine. Its importance throughout history is revealed by repeated reference to it in *Food in Chinese culture*, more numerous than for vinegar, sugar, garlic, soy sauce, or any other spice or flavoring.

Ginger is a perennial herb, ordinarily grown as an annual, whose rhizome and its derivatives are not only important as spices but as medicines as well.[5] The plant is believed by botanists to have been domesticated in India, Southeast Asia, or South China. In China, ginger was already known in Chou times (K. C. Chang, 1977a: 430; 1977b: 33, 52), if not earlier. The plant was carried west from India to Africa by the Arabs, and at least by the tenth century A.D. Zanzibar was a supplier of the ginger trade. The dried rhizome reached Europe by the first century A.D., with its Greek and Latin names, like its Arabic and Persian ones, deriving ultimately from India (Watt, 1889-96, 6, Part 4: 358-60; Laufer, 1919: 545; I. H. Burkill, 1966, 2: 2338-39). In a cookbook dating from the close of the fourth century and regarded as reflecting Roman cookery, ginger was mentioned but was far overshadowed by other spices, especially pepper (ginger was in 3 percent of all dishes, pepper in 81 percent). Judging from English, French, and Italian cookbooks, ginger and saffron displaced pepper as the most frequently used spices in medieval cooking in those countries, with ginger especially popular in France. This was, however, to be just an interlude, for pepper regained its position of dominance in later times (used in 61 percent of the recipes in an eighteenth century French cookbook) (Laurioux, 1985: 45-46, 61).

China, along with India, was an important early supplier of the ginger trade, and Chinese ginger was carried far across the Old World to market, much of it for pharmaceutical purposes.[6] This is shown in the writings of Abu Mansur, tenth century Persian physician and writer on pharmacology, who considered Chinese ginger best of the three forms he identified (Laufer, 1919: 545). For long distance trade, ginger rhizomes are usually made into special products, among them dried ginger, preserved ginger, powdered ginger, and ginger oil, an essential oil used for flavoring, medicine, and perfume. Chinese preserved ginger is superior to the Indian product, and, as already reported in A.D. 1596 by van Linschoten, was much used even in India (Watt, 1889-96, 6, Part 4: 360). In the modern world, China (including Taiwan) has been the principal source of the world's preserved ginger, most of it processed in Hong Kong and destined for Western Europe. India, on the other hand, has usually been by far the principal producer and exporter of dried ginger, most of it going to the Near East, including North Africa (Edwards, 1975: 7-8, 13-14, 19; Ilyas, 1978: 256-58).

In China, ginger is grown extensively in the southern and central provinces, with some cultivation as far north as Shantung (Meyer, 1915; Porterfield, 1939a: 208, 210-11; 1951: 19; Herklots, 1972: 490). Ginger grown in the south is succulent and better for preserving than that of other regions of the country, and most preserved ginger sold and exported comes from that region (Porterfield, 1939a: 208; F. P. Smith, 1969: 465). In Kwangtung much ginger is grown in the Pearl River Delta, but it is found nearly everywhere in the province. In one district, wrote Ball early in this century (1906: 315), ginger occupied 30 percent of all level land and 70 percent of all hill cultivation.

Ginger has been consumed in China, in an impressive range of ways, since antiquity. For T'ang times, one reads of ginger as a drug, of ginger-flavored tea, ginger wine, honeyed ginger, ginger-flavored meat and, for aboriginal groups of the south, of the contents of water buffalo stomach flavored with ginger (Schafer, 1963: 141, 152, 178, 181; 1977: 100, 108, 120, 123, 131).[7] In modern China, people single out the root of young ginger (also baby ginger or spring ginger) for its delicate flavor, absence or near absence of fibers, and thinner skin than that of mature ginger. Young gingerroot may be pickled,[8] preserved in other ways, or employed as a spice, and is ordinarily used in larger amounts than mature ginger. Indeed, its tenderness and flavor are such that it may be added to dishes as much as a vegetable as a flavoring (Em. Hahn *et al.*, 1973: 64; F. Lin, 1975: 325-26; Zee and Hui, 1981: 68; Tropp, 1982: 548; K. Lee and A. Branyon, 1984: 62-63). Mature gingerroot is too strong to use in this manner, but it has many other uses. It may be dried and made into a powder, or grated and, with water, made into ginger juice (both powder and juice are used in cooking). Grated gingerroot, with sesame oil and sugar, may be made into a condiment. Gingerroot may be included in sauces and marinades, as in the Cantonese "Soft-Shelled Crabs with Ginger Sauce" and the marinated "Braised Gingered Chicken" of Peking (K. Lee and A. Branyon, 1984: 165, 205). It may be added to a soup of western Chinese type, such as Hunan "Ginger Soup with Pork and Wood Ears" (Y. K. Martin, 1984: 56-57). It may be made into a sweet soup, such as the "Ginger Soup" eaten with steamed crab and, at Chinese New Year, with dumplings (G. B. Miller, 1966: 752-53). Or it may simply be a flavoring ingredient in a dish (whether fish, flesh, noodle, vegetable, or otherwise), such as "Lobster with Ginger and Scallions" (Y. K. Martin, 1984: 89), "Fresh Ginger Chicken" (Sakamoto, 1977: 123), and "Tossed Noodles with Ginger and Scallions" (Y. K. Martin, 1984: 183). The variety in ginger storage, preparation, and use is too great to detail fully here, but one should not overlook ginger stored in sand; ginger pickled in vinegar; ginger preserved in rice wine, brine, or syrup; or crystallized ginger (F. L. Chang and G. M. Fuchs, n.d.: 359, 369; Joyce Chen, 1962: 16-18; Em. Hahn *et al.*, 1973: 61-65; G. Z. Chu, 1975: 277, 282, 290; Sakamoto, 1977: 25; C. B. T. Lee and A. E. Lee, 1980: 295-96).

Ginger is one of the *pu* or strengthening foods or medicines, and since antiquity it has enjoyed a strongly positive image for maintaining health and well being. Confucius, it is said, always had ginger when he ate (Bretschneider, 1882-95, 2: 194), a situation paralleled among people of present-day Hong Kong, who believe a little ginger should be consumed daily (Herklots, 1972: 490). By folk belief in modern China, ginger is also "hot" and may be added to dishes eaten by women during cold periods in pregnancy and post-partum; one such dish, of pigs' feet, vinegar, and other ingredients, is mentioned as important in post-partum diet, one eaten not only by the mother but by persons who visit the mother and her newborn child (Ball, 1906: 92-93; Koo, 1973: 141; Yeung *et al.*, 1973: 198, 200; B. Chang, 1974: 437, 438; Ling *et al.*, 1975: 128; Gould-Martin and Ngin, 1981: 147; M. Mathews and L. Manderson, 1981: 12, 14; Tan and Wheeler, 1983: 42; Wheeler and Tan, 1983: 51, 53, 55). Ginger is also believed to improve one's sex life and fertility, though, in the view of E. N. Anderson and M. L. Anderson, this derives from ginger's standing as a general agent in strengthening and stimulating the body. Nevertheless, they report that ginger may be tied to a large paper offering to a deity to add to ginger's fertility-enhancing qualities, and that the ginger may then be sold at auction, as to someone wishing for a son (E. N. Anderson and M. L. Anderson, 1977: 331, 369, 379). Medicinally, gingerroot is employed as an anti-wind agent, and as tonic, stimulant, and corrective for a broad range of ailments, among them headache, common cold, and influenza. It is also used to relieve the pain and discomfort of indigestion, and is thought to be useful in treating roundworm. Even its skin is used and traded abroad as a drug (Soubeiran and Dabry de Thiersant, 1874: 119; Hooper, 1929: 141; I. H. Burkill, 1966, 2: 2342-43; F. P. Smith, 1969: 465-66; B. Chang, 1974: 437; C. P. Li, 1974: 96; Wheeler and Tan, 1983: 55).

Of special interest are certain other plants of the ginger family, some of them also called ginger (*chiang*) by the Chinese. These plants, which we designate "other gingers," are of ancient use, some adopted from aboriginal groups of the south, survivors of a millennia-long search for effective medicines and good spices and flavorings. Their use reveals again the wide-ranging, experimental nature of the peoples of China. True ginger may have turned out best, but it was simply one of many plants considered.

The other gingers have caused considerable confusion among foreigners because of their similarities to true ginger. An additional problem has arisen because the Chinese have sometimes employed a single name for more than one of the other gingers. An example of this is the apparent use of the name *shan-chiang*, "mountain ginger," for three distinct plant species (H. L. Li, 1979: 38, 39).

Of the other gingers, one is actually in the same genus as true ginger. This is Mioga ginger, *Zingiber mioga*, collected or cultivated since early times as a medicinal and for its spicy roots and edible plant parts. Though characteristic of present-day Japanese cuisine, mioga ginger is seldom if ever used in Chinese dishes today (E. N. Anderson, 1988: 156). Somewhat more significant in Chinese cooking are gingers of the genus *Alpinia* (sometimes *Languas*). Great confusion was brought on when a plant specimen was sent to Kew Gardens late in the nineteenth century as the true ginger of China and was then identified as a form of *Alpinia*. This led to the belief that the common Chinese ginger was of a different genus from Indian ginger, *Zingiber officinale*. That error corrected, it is nevertheless recognized that some plants called ginger in China belong to the genus *Alpinia*, whose species occur widely in the Asian tropics; that some of these have long been collected or cultivated in China; that they are analogues of true ginger, with their rhizomes commonly used as medicine and spice; and that other plant parts also may be used as food and medicine.

There remains considerable confusion about species designations within the genus *Alpinia* and as to which species are indeed found and used in China. There are, however, two principal ones. One is greater galangal (*A. galanga*), so-called because its rhizomes are larger than those of the other widely traded species, *A. officinarum*, lesser galangal. Greater galangal is a cultivated plant of southernmost China, Southeast Asia, and India and was domesticated somewhere in the Asian tropics (Mathon, 1981: 74; Zeven and de Wet, 1982: 64). Its rhizomes, which taste like ginger mixed with pepper, are used as medicine and spice, and also to flavor and scent alcoholic beverages; its seeds, known as galangal cardamom, serve similar uses. Its rhizomes, known as *galanga* or by related names, have been imported to the Near East and Europe since early times. This product, apparently carried westward by caravan in Roman Mesopotamia on a route from Persia, was mentioned by Aëtius (sixth century A.D.) and later by Paulus Aegineta (seventh century A.D.) (I. H. Burkill, 1966, 2: 1328; J. I. Miller, 1969: 25, 51-52). Since terms similar to galanga or galangal are used for this rhizome in Arabic, Persian, and Sanskrit, Laufer (1919: 545-46) suggested that the Arabs first obtained the rhizome in India and that its name in Arabic derived from the Sanskrit *kulanja* (or *kulanjana*). However, he dismisses the claim, one that has persisted for a century, that the term "galangal" derives ultimately from the Chinese name *kao-liang-chiang* (kao-liang ginger).

One of his reasons was that *kao-liang-chiang* in China refers not to greater galangal but to lesser galangal, the most important *Alpinia* species in South China (Laufer, 1919: 546; Schafer, 1967: 194-95; F. P. Smith, 1969: 31; S. Y. Hu, 1980: 52). The name "ginger of Kao-liang" is said to refer to a district (now a town) in western Kwangtung (Bretschneider, 1882-95, 3: 117-18; E. Watson, 1923: 401). If so, this probably reflected the importance of production in that district. In T'ang China, for example, lesser galangal grew in the west of Lingnan and on Hainan, and was sent north as tribute to the royal court (Schafer and Wallacker, 1957-58: 219, Map 4; Schafer, 1967: 194-95; 1977: 111). In the 1920's, it continued to be cultivated in South China on a considerable scale, especially on Hainan, from which large amounts were shipped to other areas of China and to Europe (E. Watson, 1923: 401).

The rhizomes of lesser galangal are stronger in odor and taste than those of greater galangal, and more highly prized as spice, medicine, and scent. Its seeds (*kao-liang-chiang-tzŭ*), like those of greater galangal, are called "galangal cardamom," and serve the same uses. Early in the nineteenth century, lesser galangal was regarded as so superior to greater galangal in India, wrote Ainslie (1826, 1: 142), that it was ten times more costly. We have not determined when lesser galangal reached India, but it does antedate the Portuguese. Idrisi in the twelfth century commented on the galangal of South China, and I. H. Burkill (1966, 2: 1328-29) wonders whether the superiority of lesser galangal may have led the Arabs, ever sensitive to the quality of spices and medicines, to carry it from India to the Near East long before the Portuguese arrived. Whatever its date of arrival in Europe, it was known in European pharmacies as

"Galanga minor" to distinguish it from "Galanga major," greater galangal. Of special note is its widespread use in Russia to flavor vinegar and the liquor called *nastoica* (Dymock, 1889a: 45; Watt, 1889-96, 1: 196).[9]

We have found no analysis of the food value of lesser galangal, but the fresh rhizomes of greater galangal contain good amounts of vitamin C and fair amounts of vitamin A. Those of true ginger, in turn, are not of significant food value in any way (Leung *et al.*, 1973: 46, 47; Leverington, 1983: 316-17).

GARLIC AND VINEGAR

The importance of garlic (*suan*; *hsiao-suan*) in Chinese diet was considered in our section on "Onions and Their Relatives." Suffice it to say here that it is a regular and important flavoring in China.[10] As for vinegar (*ts'u*), it has long been of comparable significance in Chinese cooking. Its importance in Sung times is shown by its inclusion (along with firewood, rice, oil, soy sauce, salt, and tea) in a list of items that people must have every day (Freeman, 1977: 151). In Chou texts, though the word *liu* is ordinarily translated as "vinegar," nothing is known of what or how it was made (K. C. Chang 1977b: 30-31). The situation is far clearer in the sixth century A.D., when the *Ch'i-min yao-shu* described twenty ways of making vinegar. This was done in a fermentation jar over a period of a month or so, the usual ingredients being rice, water, and leaven, normally the same leaven used for making wine. Sour or damaged wine or wine lees might also be made into vinegar (C. P. Li, 1948: 208-10). In T'ang times, vinegar was manufactured from many plant sources, including wheat, rice, peaches, and grapes; sometimes, too, vinegar was flavored with flowers or leaves such as those of the kumquat (Schafer, 1977: 113).

The way of making vinegar in modern China is quite similar to that of early times. In the north, leaven, cereal, and water are fermented in a jar, the fermented mash passed through a strainer with water, and the resulting liquid concentrated out-of-doors in summer by evaporation in the sun, or in winter by freezing and then removing the ice. Sorghum and foxtail millet (*Setaria italica*) or a mixture of these are the common cereals used in making vinegar in the north, whereas rice is used in the south (C. P. Li, 1948: 210). Also mentioned as used in vinegar-making in China are wheat, grapes, wine lees, as well as peaches, jujubes ("dates"), and other fruit, the vinegar sometimes flavored with kumquat leaves or peach blossoms (E. Watson, 1923: 572; Read *et al.*, 1940: 35; C. P. Li, 1948: 210-11; Tropp, 1982: 577). There are thus various qualities of vinegar in China, and in general they are less sharp than American vinegars. Of special note among Chinese vinegars is clear rice vinegar (acid, used in cooking and pickling), red rice vinegar (milder, often serves as a dip), and thick, dark "Chinkiang" (Chen-chiang) rice vinegar from Kiangsu (of low acidity, used as a dip, in cooking, and in other ways) (J. D. Keys, 1963: 113; F. Lin, 1975: 326; Sakamoto, 1977: 32; Hom and Steiman, 1981: 41; Tropp, 1982: 577; Y. K. Martin, 1984: 27, 232).[11]

In China, vinegar is an important flavoring in dips, sauces (including sweet-and-sour sauce), dressings, and in cooking of all sorts. Sweet-and-sour sauce is common in a range of dishes, whereas other vinegar sauces and dips seem to be used especially with fish and other seafoods.[12] In cooking, vinegar may be added to dishes containing chili peppers to accentuate their hotness, as in the Szechwan "Spiced Tangerine Chicken" (H. J. Lin and T. Lin, 1972: 16, 106). And sometimes it is sufficiently prominent in a dish to be included in its name, as in the

Peking favorite "Vinegar-Tossed Chicken Cubes" (K. Lo, 1979c: 211-12) and in Szechwan and Peking "Vinegar and Pepper Soup with Sliced Fish" (K. Lo, 1979a: 206).

Mathews and Manderson (M. Mathews and L. Manderson, 1981: 12, 13) reported white vinegar among the sour foods often considered "cold" by Vietnamese and Vietnamese-Chinese women, and observed that, like sour fruit, it was proscribed for all women in the postpartum period. They note, however, that "black vinegar," used mainly by the Vietnamese-Chinese,[13] was considered tonic and was included in a postpartum pig's feet dish. Other researchers (B. Chang, 1974: 438; Ling *et al.*, 1975: 128; Yee, 1975: 69; Wheeler and Tan, 1983: 55) also list vinegar or sweet-and-sour vinegar in post-partum dishes (along with such ingredients as ginger, rice wine, eggs, pig's feet, chicken, and peanuts) described as essential in the diet of mothers after birth and during lactation. If the vinegar used in such dishes was indeed considered "cold," that quality was likely overbalanced by the "hot" ingredients used (ginger, rice wine, and peanuts).

SALT

Since antiquity salt has been produced in China by evaporation of sea water, mainly in summer months, in various coastal areas from Kwangtung to Manchuria. Salt has also been produced at salt lakes, especially in Kansu and Shansi; from brine wells, as in Szechwan and Yunnan; and from rock salt mines and other sources (E. Watson, 1923: 520-21; C. P. Li, 1948: 61; Y. H. Sung, 1966: 108-23; Schafer, 1977: 113).

The importance of salt (*yen*) in Chinese economic life is well documented. Already at the time of the Hsia dynasty (21st-18th century B.C.), salt was asked for in royal tribute; during the Shang dynasty (18th-12th century B.C.) salt was an important trade item; and during the Chou dynasty (12th century B.C.-221 B.C.) there were officials who were responsible for providing salt to the imperial household. Establishment of the earliest salt monopoly is attributed to Kuan Chung, prime minister of the country of Ch'i in the seventh century B.C., and a similar monopoly was established in Han China and continued intermittently, and with varying degrees of enforcement, into modern times. Typically, the production and sale of salt was under control of the government monopoly; salt inspectorates were set up at critical locations to deter salt smuggling; and the cost of salt could be a burden on ordinary folk, who needed considerable amounts of it for preserving vegetables (Hirth, 1887; Spencer, 1935; C. P. Li, 1948: 55; Laai *et al.*, 1962: 101-4; R. Huang, 1969: 94-99; K. C. Chang, 1980: 258; E. N. Anderson, 1988: 22, 53, 59, 61, 68, 74, 155).[14]

Schafer (1977: 113), in describing food in T'ang China, writes that pharmacologists said only good things about salt, with one writer, Ch'en Ts'ang-ch'i, asserting that salt eliminated all undesirable vapors and similar bad items that entered the human body, that it served to regulate exchanges among internal organs, and that overall it encouraged sound health.

The Chinese boil and serve rice and vegetables without adding salt. As a result, one does not see salt shakers or other containers for salt at the Chinese table, and people's salt intake comes from soy sauce and other salt-containing preparations (Hommel, 1937: 152; E. N. Anderson and M. L. Anderson, 1977: 344; E. N. Anderson, 1988: 155). Salt was extensively used, for example, in preserving fish, shrimp, and a broad range of vegetables and other foods.[15] Preservation with salt was much more common in traditional China than in the United States, because canning and refrigeration were less prevalent, making it necessary to salt a larger share of perishable foods in order to preserve them. In addition, however, the Chinese people salted foods because they taste better than fresh ones, and could thus be used in smaller amounts in facilitating the eating of rice (B. Y. Chao, 1972: 46). One example of this is the tasty salted mustard cabbage called "Red-in-snow" (*Brassica juncea*, var. *multiceps*), which may be used in soup or stir-fried with pork or other dishes.[16]

The prevalence of fresh-salting, the salting of fresh foods for a day or so, then cooking them directly, is a further example of the Chinese preference for salty dishes. In fresh-salting, salt may be used alone, or mixed with fagara (Szechwan pepper) or other spices. With fagara, the mixture is toasted, then it may be used to rub a duck inside and out, with the duck kept from several hours to two days in a cool place (or ice box or refrigerator), improving its flavor and partly curing it before further preparation.[17,18] A somewhat similar technique involves rubbing the duck with salt, and then soaking it in a brine that includes various other spices as well (H. J. Lin and T. Lin, 1972: 168). There are other fresh poultry dishes, usually chicken, in which salt is used in much larger amounts, often four to five pounds or more. In one such dish, Cantonese "Salt-Buried Chicken," salt is sprinkled lightly on a casserole, a chicken set on it and completely covered with salt, and the casserole covered and put in an oven. When baked this way, a chicken is served hot, its flesh salty, but less so than one would expect. It is also unusually fresh, and the skin is "dry and crisp" on the outside (K. Lo, 1979a: 75, 90-91).[19]

HONEY, SUGAR, AND OTHER SWEETENERS

Historical Notes

There are four species of honeybees (genus *Apis*) in the world, of which two are widely kept in a domesticated state,[20] *Apis mellifera*, the European honey bee, and *Apis cerana*, a smaller bee whose races are known as the Indian honey bee, Chinese honey bee, Japanese honey bee, or by other names. Domestication of these two species was facilitated by the fact that they construct their nests in dark cavities and are thus suited to being kept in hives. *A. mellifera* was found naturally throughout Africa, in most of Southwest Asia and Europe, and in a corridor extending eastward from the Urals into Inner Asia but not reaching China. *A. cerana*, in turn, occurred naturally from southeastern Iran and Afghanistan across India, and Southeast and East Asia as far as Japan. The two species, which cannot interbreed, overlapped in their natural distribution only in southeastern Iran (Crane, 1984: 403-5).

A. mellifera, the most numerous and widespread species and best honey-producer, is responsible for the bulk of the world's honey production today. Moreover, because beekeepers are interested in increasing honey yields, *A. mellifera* is now replacing *Apis cerana* as a domesticated species in the latter's Asian homeland ("Apis cerana at risk," 1971). In China prior to the early decades of this century,[21] however, *A. cerana* was the only domesticated honeybee species,[22] found in most parts of the country, including Kwangtung where litchi, longan, and orange were among the main honey-plants. Hives of domesticated *A. cerana* provided the bulk of traditional China's honey supply, but some was also collected from wild hives of this species, and presumably, in some places, from those of certain other wild bees.[23]

References to honey (today *mi* or *fêng-mi*) are absent in the earliest Chinese literature, according to Kellogg and Tang, and a distinct word for it appears no earlier than the third century B.C. Honey at that time seems to have been a luxury imported from the west. This may mean that beekeeping in China was a relatively late development, long after it was known in the Mediterranean area (2400 B.C.) and possibly dating only to the beginning of the Christian era or somewhat before.[24] In any case, China's domestic honey supply remained too small to meet demand, for large amounts were imported via Samarkand in the first century A.D. Chinese camel caravans bartered by silent trade at Samarkand, exchanging their silk for precious stones and honey to be carried back to China (Kellogg and Tang, 1963: 176). As a sweetener, honey was highly-regarded in early China, and one reads of honey used at Han feasts and of outstanding Tibetan honey imported to T'ang China (Schafer, 1977: 108; Y. S. Yü, 1977: 57, 67).

The first Chinese treatise on beekeeping dates from the mid-third century A.D., but even as late as the seventeenth century, according to Sung Ying-hsing (1966: 129), four-fifths of all honey came from wild hives, and only one-fifth from domestic hives. Before Western methods of beekeeping were adopted in China, hives were kept only on a small scale by farmers, mainly as a sideline, or by non-farmers largely as a hobby ("Apiculture in China," 1929: 126).[25] As a result, hives of domesticated honey bees were far fewer per square mile of farm land than in the United States or most of Europe (Crane, 1975: 144-45), with only about 500,000 in the entire country in 1949 (Chun Li, 1965: 38).[26] Because of this, because the Chinese honey bee tended to produce small quantities of honey (Kellogg, 1929: 621-23; C. G. Riedel, 1932: 277; Crane, 1975: 131), and because of China's large population, per capita honey consumption was very low, perhaps only 10 gm. annually.[27]

This fits with B. Y. Chao's observation (1972: 26) that honey is unimportant in China. One does, nevertheless, find honey in a limited number of Chinese dishes. In Canton, an entire pig may be roasted with a honey coating; the end product, called "Golden Pig," was appropriate for sacrifice because its color, symbolic of warmth and existence, was of ritual importance (E. N. Anderson, 1988: 174-75). Honey may also be brushed on marinated flesh before eating, as in "Char-siu: Cantonese Roast Pork" (Y. K. Martin, 1984: 134-35). It may be used with other ingredients in a marinade, an example being "Honey Marinade for Pork and Spare Ribs" (K. Lo, 1979c: 48). It may be applied before baking to the skin of a chicken or duck in dishes that requires a crispy skin, such as Peking Duck (C. B. T. Lee and A. E. Lee, 1980: 100; Y. K. Martin, 1984: 209). It may be found in sauces served with ham dishes such as "Ham with Lotus Seeds in Honey Sauce" (K. Lo, 1979b: 126-27; C. B. T. Lee and A. E. Lee, 1980: 151) or "Honey Pear Ham" (K. Lo, 1979c: 165). It may be used in sauces accompanying other dishes, for example "Feng Ch'eng Honey Tender Chicken," a southern dish served with a sauce made of honey, oyster sauce, soy sauce, chicken stock, rice wine, sesame oil, and MSG (Sakamoto, 1977: 176-77). Honey may also be used in sweets, such as Soochow "Honey Walnuts" (W. W. Chang *et al.*, 1970: 458); in sweetening fruit dishes or fruit preserves; as an additive to tea in making a refreshing summer beverage; or in other ways.

At the same time, because of its high cost, honey in traditional China was more a medicine than a confection (Sowerby, 1925: 265; Kellogg and Tang, 1963: 178). Though it served medicinally in various ways, for example as a cleansing agent,[28] honey was especially well-regarded for its ability to enhance longevity. Concerns with longevity, as we have seen, have been prominent in China since antiquity, and it comes as no surprise to learn that honey was an important ingredient in the elixirs of the Taoists. One even reads of an emperor granting a monthly supply of white honey, a medicine believed to confer immortality, to noted Taoist physician T'ao Hung-ching (A.D. 452-536) on his retirement (Kellogg and Tang, 1963: 177).

As noted by Schafer (1977: 108-9), honey had early competition from sweeteners little known in the West. One was tree honey, an extract of the seed, leaves, and branches of the Japanese raisin tree (*Hovenia dulcis*; *chih-chü*) (Schafer, 1977: 109). A second was commonly called "thorn honey," a manna collected from the camel thorn (*Alhagi camelorum* = *Alhagi maurorum*), one of the many mannas of western and Central Asia (Laufer, 1919: 343-50; Schafer, 1963: 187; 1977: 108-9; Donkin, 1980: 12-85).[29] Thorn honey came to China either directly or by way of intermediaries, with some from Turkestan among the tribute sent to the Chinese court in T'ang times (Schafer and Wallacker, 1957-58: 222). A third sweetener was maltose made from grains, as from rice by using barley malt (C. P. Li, 1948: 154; Y. S. Sung, 1966: 130-31, 132n; Schafer, 1977: 108). And a fourth, minor sweetener was sugar obtained from juice of the sugar palm or arenga palm (*Arenga pinnata* = *A. saccharifera*)—the leading source of palm sugar (jaggery) in Malaysia—as well as from other palm species (Schafer, 1967: 175-76; 1970: 93, 109).[30] Though now of minor importance in providing sugar in Malaysia, the sugar palm is believed to have been an early and very important source of sugar before sugar cane was cultivated (I. H. Burkill, 1966, 1: 232, 233). Sweet-stemmed sorghums, when they arrived from Africa, also came to play a role as sweeteners in northern China, with Hagerty (1940-41: 237-39) providing references from various early periods to canes of sorghum being eaten raw as a sweet, cooked and made into a syrup, and even processed into a granular sugar.[31] Yet it was not the canes of *Sorghum*, but of plants of a different genus, *Saccharum*, that were gradually to gain the position of China's premier sweetener.[32]

The sugar canes are the most important species of a genus of tropical grasses found in both the Old and New Worlds. The New Guinea area is commonly viewed at present as their place of earliest evolution, with *Saccharum robustum* probably the main wild ancestor. Some suggest a date of domestication of roughly 700-500 B.C. or even earlier. When the cultivated New Guinea forms, *Saccharum officinarum* ("the noble canes"), reached the mainland of southern Asia, it is believed that they crossed with *Saccharum spontaneum* (or other wild species and *S. spontaneum*) to give rise to hybrid clones, especially *S. sinense* ("the thin canes"). Zeven and de Wet, who believe that *S. officinarum* arrived in South China in pre-Han times, credit that region with development of the earliest *S. sinense* clones, which subsequently spread to northern India (Bertin *et al.*, 1971: Map 15; N. W. Simmonds, 1976b: 105; Zeven and de Wet, 1982: 55, 75).

I. H. Burkill describes three stages in the history of sugar cane. The first saw humans, at some remote time in the past, recognize the value of the plant and begin to breed it for its sweetness and consume it either by chewing the cane or crushing it and drinking the juice. The second stage involved learning how to boil the juice to make a solid substance something like raw sugar. And the third saw development of trade over considerable distances, which required that the product have better keeping qualities, and led to development of the techniques of sugar refining.

As for sugar cane in China, in the last century Bretschneider (1882-95, 3: 438-39) had written that it was first mentioned in various Han poems or other accounts, as *kan-chê* or *chu-chê*, a reed growing in South China having a sweet juice capable of stopping the growth of intestinal worms and countering drunkenness. H. L. Li (1979: 58) has since observed that the name *chu-chê* appeared already in Chinese accounts from the fourth century B.C., and Yu Youtai (in Wittwer *et al.*, 1987: 220) writes of the cultivation of sugar cane in China as early as the eighth century B.C. That must have been Burkill's first stage, when cane was chewed and its juice drunk, though sugar-cane syrup, as *chê-chiang*, was first mentioned in the *Elegies of Ch'u* (H. L. Li, 1979: 57), whose date is given variously as late Chou to early Han.

Sugar-cane cultivation in those times seems to have been found only in Chiao-chih (now Hanoi in Vietnam) and in China's far south (H. L. Li, 1979: 57).

A hard-sugar product called *shih-mi* ("stone honey"), small cakes made by sun-drying juice extracted from the cane, was also manufactured in Annam in the third and fourth centuries A.D., but seems to have been unknown in China (Schafer, 1963: 153; 1967: 193; H. L. Li, 1979: 55, 58). China seems to have entered Burkill's second stage, involving the manufacture of a solid sugar, from the fifth to the seventh centuries A.D. (Yu Youtai in Wittwer *et al.*, 1987: 219, 220), with clear evidence of the manufacture of granular sugar (*sha-t'ang*, "sand sugar," probably like our raw sugar, but not actually refined) in China (in Kwangtung) dating only from the beginning of the sixth century (H. L. Li, 1979: 59).

The above accounts suggest that sugar cane and techniques for making hard sugar and granular sugar were first introduced to China, via Kwangtung, from Annam. Thus we seem to be dealing with still further Nan-Yüeh contributions to China's food scene. India, however, is regarded as the ultimate place where manufacture of granular sugar was developed. Judging from Buddhist scriptures dating back roughly to 500 B.C., *sarkara*, or granular sugar (originally meaning "grit" or "gravel" in Sanskrit), was already in common use at that time, and was mentioned by various later writers, including Panini (?350-B.C.-250 B.C.?) and Kautilya (?290 B.C.-A.D. 300?) (*Jataka*. X. 442, or Cowell, 1969, 4: 10; Kautilya's *Arthasastra* II. 15. 15, or Shamasastry, 1951: 100; Prakash, 1961: 68, 251-52; Agrawala, 1963: 111, 119). From India, granular sugar was introduced to other lands. It has been observed that while the Malays have their own name for sugar cane, their names for raw sugar or brown sugar (*gula*) and sugar (*sakar* or *sagar*) come from the Sanskrit (*guda* and *sarkara*) directly or by way of Persian (Flückiger and Hanbury, 1879: 715-16; I. H. Burkill, 1966, 2: 1967). It has also been noted that certain names for sugar (or the sugar plant) in ancient South China may derive from Sanskrit (Watt, 1889-96, 6, Part 2: 32; 1908: 930; H. L. Li, 1979: 58).

There was an oft-mentioned Chinese mission sent to India in the seventh century A.D. to learn about sugar manufacture. This, however, did not mark the start of such manufacture in China but was an effort at learning how to make a better product, perhaps a better type of sand sugar, which, as indicated above, had already been manufactured in Kwangtung early in the previous century. Chinese knowledge of manufacturing refined granular sugar, *t'ang-shuang* ("sugar frost"), is credited by tradition to a monk of the T'ang period. Though he is said to have passed his skills on to others, such sugar was not well known until Sung times (Schafer, 1963: 153-54; 1967: 193; H. L. Li, 1979: 59).[33]

In early times sugar cane and cane syrup were trade items, and were also listed as tribute sent to China's royal court (Laufer, 1919: 376; Schafer and Wallacker, 1957-58: 222; Schafer, 1963: 153; H. L. Li, 1979: 55-57). In time, sugar itself played such a role, and in T'ang times it was imported from Magadha, a state in northeast India, and apparently also from Persia (Laufer, 1919: 376; Schafer, 1977: 109). The significance of the overland Persian trade is indicated by the occurrence of words for sugar in Mongol (*šikär*, *šikir*) and probably Tibetan (*ša-ka-ra*) that derive from the Persian (*šakar*) (Laufer, 1919: 376, 576, 596) and ultimately from the Sanskrit. In Sung China, there were pedlars who sold sugar cane for chewing, and ones who sold, for children, sugar-based sweets made in the form of humans, birds, other animals, flowers, and such (Gernet, 1962: 94-95).[34] Meanwhile sugar-cane cultivation and sugar processing continued to gain importance in southern China, with sugar cane a major commercial crop in sections of Szechwan and Fukien in Sung times (Freeman, 1977: 147). This led to at least a partial change in direction of sugar trade in Sung China: exports to lands bordering the South China sea (but apparently not beyond) (Wheatley, 1959: 85-87).

Southern China, for reasons of climate, has always been the country's sugar-cane producing area. In the seventeenth century, Sung Ying-hsing (1966: 124) observed that Kwangtung and Fukien together accounted for about 90 percent of all the sugar cane produced in China. In the 1929-33 village survey, the Double Cropping Rice Area had over four times the percentage of crop area devoted to sugar cane (5.8 percent) as the next most important area, immediately to its north (Buck, 1956: 212). Kwangtung was estimated to produce the greater part of the nation's sugar cane in 1935; and today it accounts for roughly 40 percent of China's acreage and production, with 90 percent coming from the five provinces of Kwangtung, Kwangsi, Szechwan, Yunnan, and Fukien (Yu Youtai in Wittwer *et al.*, 1987: 219-20).[35] Sugar cane may be chewed raw as a sweet, and one of Kwangtung's two main varieties in the mid-1930's, *jou-chê*, was used largely in this way, much of it baled and shipped to Hong Kong, Shanghai, Tientsin, or elsewhere, where it might be eaten as a snack or consumed at feasts or banquets.[36] Cane might also be processed into molasses or sugar, with Kwangtung's second main variety, *chu-chê*, used in this way ("Fruit cultivation in Kuangtung," 1936: 719).[37] Already in the seventeenth century, three kinds of sugar were made from sugar cane: white sugar, brown sugar, and rock sugar (Y. H. Sung, 1966: 125-26), and these continue in use today.

We have found little on the history of the sugar beet (*Beta vulgaris*, var. *saccharifera*; *t'ien-ts'ai*) in China, beyond the observation (Laufer, 1919: 399-400; Schafer, 1977: 108) that it was likely introduced in T'ang times, perhaps by Arabs. The possibility of using beets for making sugar was established in Europe in the mid-eighteenth century, and the beet-sugar industry there developed in the nineteenth century, along with the selection of sugar beets for increasingly higher sugar content. Hosie, who served in Manchuria until 1900 and published a thorough account of its economy in 1904, made no mention of sugar beets or sugar refineries. In 1909, however, J. S. Thomson (: 310) wrote of Manchurian sugar beets and the region's advantages of low cost of labor and high sugar content of its beets, which promised to make it a highly successful sugar-beet producer. This suggests that the commercial cultivation of sugar beets and the manufacture of beet sugar began there in the first decade of this century or shortly afterwards. In the 1920's, beets were grown for sugar mainly in the north, especially around Harbin in Manchuria, where large, modern sugar refineries had been set up (E. Watson, 1923: 542-43). Before 1937 only an insignificant amount of China's sugar came from sugar beets in Manchuria (T. H. Shen, 1951: 236), but by 1957 sugar beets were cultivated in China on nearly 400,000 acres, about 60 percent of the acreage in sugar cane. Nevertheless, sugar beet production amounted to only 14 percent of that for sugar cane (N. R. Chen, 1967: 286-87, 338-39). Since establishment of the People's Republic, small-scale sugar beet production has spread far to the south, even to Fukien, Kwangtung, and Szechwan, where sugar beets may be intercropped with sugar cane, but overall roughly 50 percent of acreage and production comes from Manchuria (Heilungkiang and Kirin), and 90 percent from Manchuria, Inner Mongolia, and Sinkiang. Production of sugar from both cane and beets has greatly increased over recent decades, to 4.45 million tons in 1985, when China was the fourth largest sugar producer in the world. The increase in sugar beet production has outstripped that in sugar cane, but 75 to 85 percent of China's table sugar still comes from sugar cane. The total production of sugar is insufficient to satisfy domestic needs, and about a million tons had to be imported in 1985 (Yu Youtai in Wittwer *et al.*, 1987: 218-24).

Though children in China are given rock sugar as candy, the Chinese are far less given to confections than are Westerners, and sweet desserts are not part of an ordinary Chinese meal at home. One encounters sweets mainly as between-meal snacks (commonly steamed or boiled buns or dumplings filled with sweet mixtures) and at banquets.[38] At a banquet, sweet dishes do not come at the end of the meal, as in the West, but between other banquet courses, often in the form of sweet soups such as "Almond Float" (K. Lee and A. Branyon, 1984: 348) and "Lotus Seed Soup" (K. Lo, 1979c: 325). Since sweet, along with sour, salty, pungent, and bitter, is one of the "five flavors" of classical Chinese cooking, its use in cooking is commonplace but always with the intent of retaining a balance among the flavors. As a result, the amounts of sugar used are ordinarily quite small. Even sweet-and-sour dishes are apt to be a bit on the tart side, with sweet-and-sour sauce commonly served separately so that the discriminating diner may use it in appropriate amounts. There are, nevertheless, regional differences in use of sugar, with eastern Chinese or Shanghai cuisine often singled out for its common use of sugar in cooking, as to counter the salty taste of soy sauce in red-cooked dishes. In other areas of China, people may prepare the same dishes but leave out the sugar (W. W. Chang *et al.*, 1970: 455; B. Y. Chao, 1972: 25; Em. Hahn *et al.*, 1973: 29; G. Z. Chu, 1975: 13-14, 55, 223; E. N. Anderson and M. L. Anderson, 1977: 362; K. Lo, 1979c: 324; C. B. T. Lee and A. E. Lee, 1980: 303; Tropp, 1982: 569-70).

Of the sugars used in cooking, white, brown, and rock are perhaps the most common forms, with one or another of these sometimes favored for dishes of particular kinds. Rock sugar, a solid, amber or clear crystallized sugar, is commonly used for braised flesh and fowl because of the dark glaze it imparts (though other sugar is also used in this way) (B. Y. Chao, 1972: 26; Sakamoto, 1977: 32; C. B. T. Lee and A. E. Lee, 1980: 303; Tropp, 1982: 570). Sugar is also used in sweet bean sauce, traditional sauce for dipping Peking duck (C. B. T. Lee and A. E. Lee, 1980: 101; Y. K. Martin, 1984: 226-27), and in fillings for sweet buns and dumplings. Coarse sugar may also be shaped into figures of humans or animals for children or for use at banquets. And sugar and other sugar-cane products serve medicinal uses as well (Soubeiran and Dabry de Thiersant, 1874: 104; F. P. Smith, 1969: 386-88; Perry, 1980: 169).

Sugar Intake in Modern China

The 1929-33 survey of Chinese villages found that for China as a whole 0.5 percent of all calories were obtained from sugar; for South China, the percentage was a bit higher (0.8 percent) (Maynard and Swen, 1956: 407, 414), probably reflecting greater cultivation of sugar cane there. In any case, consumption of sugar was only a fraction that of the United States, which during a comparable several year period averaged 10.1 percent of all calories.

Buck (1930: 357) has written in general terms that sugar is not essential in diet, and is obtained by consuming sweets in an aimless manner between meals, a practice less common in rural areas where vendors of candy are fewer. The traditional pattern in South China, as reported by E. N. Anderson and M. L. Anderson (1977: 333, 344, 349, 352-53, 355), was that people did not like very sweet foods, used very few sweets, and the poor (who consumed least sweets) seemed to have had virtually no diabetes. Modern changes, especially notable among overseas Chinese, have involved an enormous rise in sugar intake (candy, cookies, soft drinks, etc.), with white sugar supplying a group of Chinese children studied in Malaya with as much as 30 percent of their total intake of calories. The result has been a decline in nutritional standards and a striking increase in dental caries.

FAGARA: CHINESE OR SZECHWAN PEPPER

Fagara is a collective term used for shrubs and trees of the genus *Zanthoxylum* (or *Xanthoxylum*) that provide fruit used as spices and medicinals, and sometimes other useful plant parts as well. Such uses of *Zanthoxylum* are not restricted to China, but are found in India, Southeast Asia, and Japan as well (Watt, 1889-96, 6, Part 4: 323-26; Crevost and Lemarié, 1917, 313-14; Bois, 1927-37, 3: 40-42; Uphof, 1959: 388-89; I. H. Burkill, 1966, 2: 2326-27).[39] In most of that broad region, the plants utilized are wild, but a few species of *Zanthoxylum*, usually prickly shrubs, are also cultivated, mainly in East Asia. Before true pepper (*Piper* spp.) was traded to China from Southeast Asia and India, the peppery fruit of fagara served the uses to which true pepper was later put. Fagara, moreover, has continued as a spice until the present day, being commonly known in English as Chinese pepper or Szechwan pepper because it is common in Szechwan or western Chinese cuisine.

Fagara was mentioned in texts of the Chou period, and actual plant remains have been identified in a Han tomb of the second century B.C. (Bodde, 1975: 67-68; K. C. Chang, 1977b: 31; Y. S. Yü, 1977: 57). Early accounts in China reveal fagara as a seasoning in religious offerings. It was put in sacrificial wines to the gods. In ancient and medieval times, there was a fagara wine used in New Year ceremonies (Schafer, 1963: 149, 313n; Bodde, 1975: 66-68, 217). Over time, however, fagara made a transition from a spice used in religious offerings to secular condiment (Schafer, 1963: 149). It was an item of royal tribute sent from southern Shensi and Szechwan in T'ang times (Schafer and Wallacker, 1957-58: 218). Te Tsung, a T'ang emperor, is said to have added fagara and curds to his tea. Apparently the spice had gained a following among bon vivants, and perhaps ordinary folk as well, for the Buddhist monk and poet Han-shan referred with disdain to the flesh eaten by a self-centered, unthinking gourmand, among which was roast duck flavored with fagara and salt (Schafer, 1963: 149; 1977: 110). One also reads of fagara used in flavoring wine and other beverages, and in seasoning fowl and suckling-pig (Bretschneider, 1882-95, 2: 322, 323). Schafer suggests (1963: 149-50) that in T'ang times the combination of fagara and salt may have been typical of southern Chinese cooking, a forerunner of the richness of present-day Cantonese cuisine.

In early China the various species of fagara were known collectively as *chiao*, which later became a general term for fruits that are hot and pungent in taste, among them fagara, chili pepper, true pepper, and others. The *chiao* mentioned in the early literature may thus have been various species of *Zanthoxylum*, but in traditional China *Z. bungei* (= *Z. simulans*) and *Z. piperatum* were the most common ones used. *Hua-chiao* ("flower pepper") is one of the common names for *Z. bungei*, and *shu-chiao* and *ch'uan-chiao* ("Szechwan pepper") for *Z. piperatum*. *Ch'uan-chiao*, however, may also be used for *Z. bungei*. Despite problems of identification, both species are known to be cultivated in China for their small, red, pungent fruit, and *Z. piperitum* in Japan as well (Mathon, 1981: 63; Zeven and de Wet, 1982: 45).

The fruits of fagara are aromatic and mildly hot, at least initially; then, with a delay, they bring on a numbing sensation. They may be toasted and ground into a powder, sometimes with salt or salt and black peppercorns.[40] Fried foods may simply be dipped into fagara powder to give them a spicy taste (Read *et al.*, 1940: 34). Fagara, with sesame oil or sometimes with chili or chili oil, may be made into a condiment or sauce, such as "Szechwan Peppery Hot Sauce" (K. Lo, 1979b: 152, 155). It is also a common ingredient in five-spice powder, that popular Chinese condiment.[41] The general pattern is for fagara to be used with other spices and flavorings, for example in "Szechwan Peppercorn Chicken" and "Szechwan Spicy Bean Curd," where it is accompanied by chili pepper, soy sauce, ginger, garlic, scallions, and others (G. Z. Chu, 1975: 161-62, 238-39). This may derive from its oft-noted ability to enhance flavors, as in peanuts cooked with salt and fagara which are said to open the eyes of persons who have not had them prepared in this manner (B. Y. Chao, 1972: 27).

The medicinal uses served by various types of fagara in China are many, but prominent among them are those of stimulant, carminative, and diuretic (Soubeiran and Dabry de Thiersant, 1874: 258-60; Hooper, 1929: 140-41; F. P. Smith, 1969: 462-64; J. D. Keys, 1976: 181-82; Perry, 1980: 369-71). Fagara is exported from China as a medicine, much of it, one suspects, to overseas Chinese communities such as that of the Malay Peninsula (Hooper, 1929: 140-41; I. H. Burkill, 1966, 2: 2326).

Also identified as an early alternative to true pepper was the spicy seed of *Evodia officinalis*, known to the people of Fukien and Chekiang (Schafer, 1977: 110). Its name was *shih-chu-yü* or *Wu-chu-yü* ("*chu-yü* of the kingdom of Wu," an early coastal state centered on the mouths of the Yangtze but extending beyond). *E. officinalis* is a shrub or small tree native to parts of southern and central China (S. C. Lee, 1935: 682-83). F. P. Smith (1969: 462), in turn, has identified *Zanthoxylum ailanthoides* as *shih-chu-yü*, noting that its pungent fruit were not only medicinal, but eaten "as a substitute for the ordinary red-pepper." He added that it was also known as *Yüeh-chiao*, "Yüeh fagara," raising the possibility that the spice came from Wu's Yüeh neighbors to the south. Others have identified *Wu-chu-yü* as *Z. ailanthoides*, *Evodia rutaecarpa*, or some other species of these genera (Bretschneider, 1882-95, 2: 325; 3: 436).

Shrubs and trees of *Evodia* occur from the Mascarene Islands through southern and eastern Asia into Polynesia. Some species provide an aromatic oil; some provide medicine; and the leaves of a few are used as a sour flavoring (I. H. Burkill, 1966, 1: 1003-5). *E. rutaecarpa*, which is found from India to Japan, is especially well regarded as a medicinal, and the Chinese attribute many medicinal virtues to it, especially to the fruit (Hooper, 1929: 64; Uphof, 1959: 152; I. H. Burkill, 1966, 1: 1003-5; F. P. Smith, 1969: 72-73; J. D. Keys, 1976: 178-79). Though one cannot be certain, more than one species of *Evodia* may have served as pepper substitutes in China. In any case, it is clear that the Chinese have tried more than just fagara, the species of *Zanthoxylum* mentioned above, in that way.

TRUE PEPPER

Pepper (*Piper* spp.) is a sizeable genus of herbs and woody climbers native to warm, moist areas of Asia, Africa, and the Americas. Several species, mainly of India and Southeast Asia,[42] have long been domesticated for use of their fruit as spices and medicines. Best known, and most important in world trade, is *Piper nigrum*, from which our black and white pepper derive. Though some authorities place the domestication of *P. nigrum* in Southeast Asia, most place it in southwestern India, in a tropical region centering on the Malabar Coast (Newcomb, 1963: 129, 130). In India, it is only in forests of the southwest, and possibly in the Assam Hills, that *P. nigrum* is found wild, and the southwest is the region where virtually all of it is produced today (Watt, 1889-96, 6, Part 1: 261; I. H. Burkill, 1966, 2: 1776-77; Purseglove, 1968, 2: 441-42; de Waard and Zeven, 1969: 411; Ilyas, 1976: 274). Also well known is Indian long pepper, *Piper longum*, native to the eastern Himalayan foothills, to Assam and Bengal, as well as to peninsular India (Watt, 1889-96, 6, Part 1: 258; Uphof, 1959: 283; I. H. Burkill, 1966, 2: 1774). In India today, it is mostly collected from the wild, though also cultivated to a degree in Assam and Tamil Nadu (Ilyas, 1976: 277).

Though early Egyptian and Hebrew accounts are strangely silent about pepper, the spice (probably both *P. nigrum* and *P. longum*) was reaching the Mediterranean from India at least as early as the fourth century B.C. (Flückiger and Hanbury, 1879: 576; Schoff, 1912: 213; I. H. Burkill, 1966, 2: 1776; J. I. Miller, 1969: 81-82; Darby *et al.*, 1977, 2: 804).[43] Indeed, it became a staple in trade between the Roman Empire and India, with Roman ships in time coming to the Malabar Coast to pick up supplies directly. That trade became so important that pepper likely comprised over half the cargo of many Roman ships heading westward from India (Warmington, 1974: 181-84). Its value in Europe was such that in A.D. 408 when the Visigothic chief Alaric besieged Rome, he demanded "three thousand pounds of pepper" in addition to silver, gold, and other valuables (Flückiger and Hanbury, 1879: 577-78). Merchants in the pepper trade to Europe prospered, not only in Roman times but later, when the Genoese and Venetians gained prominence in the Mediterranean. Ultimately, its high cost led to a search for an alternative route direct from Europe to the pepper producing areas. It was the Portuguese, in expensive, long-continued probings down the west coast of Africa, whose efforts were finally crowned with success in this venture with the arrival of Vasco da Gama in India in A.D. 1498.

The story of the Western pepper trade, both before and after the Portuguese success, was so important economically that we could not avoid mentioning it here. Our main concern, however, is not with that romantic and well-known story, but with the Oriental pepper trade, from India and Southeast Asia to China, an eastern trade that may be of comparable antiquity. It was reported that a Han emissary was sent to the state of Nan-Yüeh, modern Kwangtung, in the second century B.C. where, at a reception, he first encountered a plant product H. L. Li has identified as black pepper (*Piper nigrum*). The emissary was told that it came from Szechwan, far to the west. That region, in turn, probably obtained it overland from India, a land also associated with pepper in other early Chinese accounts. A history of the later Han dynasty (A.D. 25-220) published in the fifth century, for example, identified black pepper as a product of India (H. L. Li, 1979: 46-53); and four centuries later a T'ang account not only associated it specifically with Magadha, that early kingdom of northeast India, but gave its Sanskrit name, *marica* (Laufer, 1919: 374).[44] That kingdom, Schafer (1963: 150) suspects, was an important pepper-producing area. It may also have been an intermediary in a pepper-trade network that originated far to the south in peninsular India.

The early Chinese knew, however, that pepper was cultivated in lands other than India, for Chi Han observed that black pepper was grown by households in what today is northern Vietnam. He also noted that some was grown near present-day Canton (H. L. Li, 1979: 46). Likely the Canton-area plants were introduced from the south, for *Piper nigrum* does not grow wild in that region today. It is said to be wild on Hainan, but attempts at domesticating it there have been described as "rather unsuccessful" and there is "no record by Europeans that it is cultivated in China" (Soubeiran and Dabry de Thiersant, 1874: 137; A. Henry, 1893: 39; F. P. Smith, 1969: 334; H. L. Li, 1979: 47-48). Thus the term the Chinese came to use for black pepper, *hu-chiao* ("foreign fagara"), was an appropriate one, and is still in use today.

Black pepper was probably first cultivated in Java by Hindu colonists who carried it there during the period from 100 B.C. to A.D. 600 (Crawfurd, 1868: 190; I. H. Burkill, 1966, 2: 1776, 1777), and by roughly A.D. 1200, Java was the main Chinese source of pepper (Wheatley, 1959: 100-101; Hirth and Rockhill, 1966: 222-23). As in early Europe, pepper seems to have remained expensive in early China, and even an object to be hoarded. In the eighth century, for example, a hundred piculs (well over a thousand pounds) of the spice was listed among the valuables seized when a Chinese official fell from grace (Schafer, 1963: 150).

When Chinese prepared Western meat dishes, according to one T'ang source, they always used black pepper (Schafer, 1963: 150), perhaps for the same reason that in the United States today it would be unthinkable to prepare a Chinese dinner without soy sauce or serve an Indian one without chutney. Pepper was not, however, restricted to Western dishes. Instead, the Chinese used it in their dishes in much the same way as fagara, and indeed they considered pepper a new, more pungent form of the indigenous spice which, over time in many places, declined in importance. In their hot-cold classification of foods, pepper is "hot," and may be added to dishes, such as "pork stomach heavily peppered," eaten by mothers after they give birth to counter the coldness that affects them at that time (Gould-Martin and Ngin, 1981: 147; M. Mathews and L. Manderson, 1981: 12, 14).

The early Chinese knew not only black pepper, but certain other types. One was Indian long pepper (*Piper longum*), which was also cultivated in Magadha. Chi Han, in his fourth century flora, wrote of *pi-po* as a product which grows in foreign lands (H. L. Li, 1979: 46). Its name *pi-po*, which remains in use, apparently derives from *pippali*, the Sanskrit name for long pepper.[45] Long pepper has the reputation of being even hotter than black pepper, and in the eleventh century was even cultivated in Lingnan (Schafer, 1963: 151). Though highly valued then, in recent times it has been described as almost "a forgotten spice," little used in cooking except in the tropics (Redgrove, 1933: 187; H. L. Li, 1979: 48). A third pepper known to the early Chinese, introduced in T'ang times, was cubeb pepper (*Piper cubeba*; *pi-ch'êng-ch'ieh*) from Malaysia (I. H. Burkill, 1966, 1: 1773-74; Schafer, 1977: 110; H. L. Li, 1979: 48). And a fourth, whose leaves were valued as a masticatory, medicinal, and additive to food and drink, was betel pepper (*Piper betle*), a plant of Southeast Asian origin that was and still is widely cultivated in southern China (I. H. Burkill, 1966, 1: 1769-72; Schafer, 1977: 110; H. L. Li, 1979: 48).[46]

From antiquity to the present day, pepper (black and white) has been used in China as spice and medicine. Medicinal usage varies from species to species, but for the four mentioned above the most common attributed qualities are stimulant, tonic for the appetite or digestion, and relief for colic, griping, and flatulence (Soubeiran and Dabry de Thiersant, 1874: 137; Hooper, 1929: 108; F. P. Smith, 1969: 103-4, 334; J. D. Keys, 1976: 146-48; Perry, 1980: 312-14). Though consumed in quantities too small to be significant in nutritional terms, black and white pepper (*Piper nigrum*) are high in calcium, iron, and certain other minerals (Farrell, 1985: 182, 185).

CAPSICUM PEPPERS

After the discovery of the New World, another great change occurred in the Chinese spice picture: the introduction of hot forms of capsicum peppers (*la-chiao*), especially *Capsicum annuum* and *Capsicum frutescens*.[47] The initial introduction of these New World domesticates was probably by way of Europe and southern Asia. Columbus himself is credited with introducing capsicum peppers to Europe. In the early 1500's, they seem to have been carried by the Portuguese to India, whether from Portugal, from a way-station in Brazil, or from elsewhere in the Americas. The Portuguese role in the introduction of such peppers to India is reflected in "Goa pepper" (for Portugal's enclave in India) being used for at least one form of capsicum pepper in Bombay (Flückiger and Hanbury, 1879: 452-53; Watt, 1889-96, 2: 135; Watt, 1908: 264-66; I. H. Burkill, 1966, 1: 449-50; Mahindru, 1982: 108). The Portuguese had occupied

Malacca in 1511, and shortly afterwards certain Portuguese merchants, sailing on Chinese junks, arrived in South China. They discovered that carrying spices to China was as profitable as carrying them to Portugal, and from that time onwards, with or without the approval of Chinese officials, the Portuguese were a factor in China's overseas trade (Boxer, 1953: xix-xxxvi). Though we have found no documentation of the matter, it seems reasonable to believe that the Portuguese, as spice merchants involved in the Indian capsicum trade, would have introduced them to South China. If not, capsicum peppers probably arrived soon after by way of the Pacific, over the galleon route established by the Spanish in 1565 from Acapulco to Manila, from which they may have been carried by junk to the Chinese mainland.

Though the story of the introduction and spread of the capsicum peppers in China remains to be told, it is an important one. Of the two principal ones, *C. annuum* plays a far greater role in South China today, with its hot forms (called variously chili, red, or cayenne peppers in English), of greater importance than its sweet ones, or bell peppers (*Capsicum annuum*, var. *grossum*) (E. N. Anderson and M. L. Anderson, 1977: 328-29; E. N. Anderson, 1988: 131).

The use of hot capsicum peppers, as noted in our section on Chinese cuisines, is conspicuous in Szechwan, Hunan, and neighboring areas. Whether green or red; fresh or dried; or whole, crushed, or powdered, such capsicum peppers are very common in western Chinese dishes, for example Szechwan "Sliced Pork with Chili Peppers" and "Diced Chicken with Chilies," and Hunan "Cured Pork Belly with Chili Peppers" (W. W. Chang *et al.*, 1970: 244; K. Lee and A. Branyon, 1984: 163, 267, 271). Hot peppers may be boiled in vegetable oil to make hot chili oil, a regular Szechwan dip and condiment; or, with oil and fagara, they may be made into hot pepper oil. They may be used, along with other ingredients, in chili sauces or pastes, which also serve as dips and condiments.[48] Chili pepper is also one of the ingredients of hoisin sauce, that soy-based dip, marinate, and flavoring used widely in Chinese cooking.

Not only did the hot capsicum peppers greatly alter the flavor of South Chinese dishes, but they significantly improved the state of nutrition. Especially critical, they can be rich sources of carotene and vitamin C (as are sweet peppers), and, in a dried form, are high in iron, calcium, and B vitamins. Moreover, as noted by Anderson and Anderson, chili peppers bear most of the year in South China, making vitamins available at seasons, such as spring, when fresh vegetables are in short supply (B. S. Platt, 1962: 28-29; Leung *et al.*, 1973: 59; E. N. Anderson and M. L. Anderson, 1977: 328-29, 347-48, 349-50; E. N. Anderson, 1988: 131).

Hot capsicum peppers, like ginger, are prominent among foods the Chinese consider "hot" (E. N. Anderson, 1970a: 210; Koo, 1973: 141; Yeung *et al.*, 1973: 198, 200; E. N. Anderson and M. L. Anderson, 1976: 147).

STAR ANISE, BASTARD STAR ANISE, AND ANISE

These three plants are often confused, and we consider them together because in one way or another they are similar. Of greatest interest in our account of Chinese food is star anise, for this spice (deriving from the fruit of *Illicium verum*, a small, evergreen tree of the Magnolia family) is believed to have been domesticated in South China or Southeast Asia (Uphof, 1959: 192; Zeven and de Wet, 1982: 39).

Today star anise is cultivated in various countries (Uphof, 1959: 192), but early in this century Kwangsi and Tonkin enjoyed a virtual world monopoly in star anise production. As mapped by Raibaudi (1931: 95), there were two major areas of cultivation and one minor one, all of them east of Nan-ning, Kwangsi. One major area embraced parts of Lung-chou District of southwestern Kwangsi and adjacent Lang Son District of northeastern Tonkin, and the other two areas, one major and one minor, were in west-central Kwangsi near its borders with Tonkin and Yunnan. Annual production about 1930 included several thousand tons of anise seed and several hundred tons of distilled anise-seed oil.[49] All of the Tonkin production of oil went to France; some of the Kwangsi anise-seed oil was exported to Tonkin and some shipped via Nan-ning and the West River to Hong Kong, and much of this exported. Seed from Tonkin and Kwangsi, in turn, was used domestically or exported. Ribaudi also referred to Hainan as an area where attempts were being made to cultivate star anise, and I. H. Burkill (1966, 2: 1245) also listed Hainan as a producing area along with Kwangsi, Kwangtung, and Tonkin.

There seem to have been changes in producing areas over time,[50] but it is likely that star anise was still another of the useful plants first cultivated by the peoples of the South China-Vietnam area. Yet it does not seem to have been of ancient cultivation there. It is not among the plants, medicinal or otherwise, found in classical Chinese sources by Bretschneider (1882-95). Chi Han did not mention star anise in his fourth century flora (H. L. Li, 1979), even though he wrote on the very region in which *Illicium verum* is cultivated today. Nor was the plant, spice, or medicine mentioned by Schafer in his accounts of T'ang China, one of which (1967) focused on what today is Kwangtung, Kwangsi, and Annam, and another (1970), on Hainan. Flückiger and Hanbury (1879: 21) report that the only early mention of star anise they discovered was for Sung times (when it was listed among tribute levied on an area of what today is Fukien), and we have found nothing earlier.

Unlike other spices originating in the Vietnam-South China region, star anise did not reach India, the Near East, or Europe in antiquity. Indeed the first to reach Europe was obtained in the Philippines by an English voyager named Candi, Candish, or Cavendish late in the sixteenth century (Flückiger and Hanbury, 1879: 21; Bretschneider, 1898: 17-18; I. H. Burkill, 1966, 2: 1245). It was only after that time, it seems, that star anise entered trade from China to India, the Near East, and Europe. Thus one reads that in India it was known only at a fairly recent date, and that late eighteenth century Persian works identify it as a new medicine (Dymock *et al.*, 1889-93, 1: 39). Star anise travelled westward from China both by sea and land. The overland route ran across the Kansu corridor. Then one branch apparently turned southwestward across Sinkiang to Yarkand before heading south to India, while another, the tea route, continued northwestward, ultimately reaching Russia.

Coming overland from the east to Europe, star anise was called, in one language or another, "Siberian anise" or "Siberian cardamom" (Flückiger and Hanbury, 1879: 21, 22; Watt, 1889-96, 4: 333; I. H. Burkill, 1966, 2: 1245). Most of that star anise was processed for its oil, which was used for the same purposes as, and commonly substituted for or mixed with, the oil of true anise (*Pimpinella anisum*). Such oil was used in medicine, in perfumes and soaps, as spice in brewing some types of beer, in making liqueurs (ones flavored with anise are especially common in Germany, France, and Italy), and for other purposes (E. Watson, 1923: 84-85). In India, the East India Company was importing sizeable amounts of star anise by the early years of the nineteenth century, and it came to be distributed widely in bazaars, mainly, it seems, as medicine, but also as spice and to obtain its aromatic oil, which was also imported in that form (Ainslie, 1826, 2: 18-20; Dymock *et al.*, 1889-93, 1: 41-42; Watt, 1889-96: 4: 330-34; Khory and Katrak, 1981: 20).[51] Today most Indian star anise is imported from China and Southeast Asia, for the tree is still not cultivated in India (Pruthi, 1976: 213; Ilyas, 1980: 258). Star anise

Figure 76. Chinese or Star anise (drawn by Shelby Hall)

is also imported from China to Malaya for flavoring curries or medicines, and for use as medicine by itself (I. H. Burkill, 1966, 2: 1245).

In English today, the names "Chinese anise" and "star anise," the common ones for *I. verum*, call attention to its source region or the shape of its fruit. Each of the long carpels, or seed-capsules, of a fruit is attached at one end to a central axis from which the carpels radiate outward, creating a striking star-like appearance (Figure 76). The Chinese, on their part, perceive the fruit differently, one common name being "eight horns" (*pa-chio*) or "eight-horned fragrance" (Bretschneider, 1898: 18; Couling, 1917: 28; E. Watson, 1923: 83; Hooper, 1929: 76; S. Y. Hu, 1980: 78; Tropp, 1982: 569) for the number of carpels it sometimes has. Picked before fully mature, the fruits are dried in the sun. As marketed, they are stiff, wood-like, rough on the outside, and brown in color. If a carpel has broken open, it somewhat resembles a boat tapered at each end. Each carpel contains but one seed, tan or brown in color but smooth and glossy, having the appearance of a small bean. Though both seed and carpel are used, it is the carpel that is aromatic and most flavorful, with a taste reminiscent of anise, fennel, and licorice.

In China, star anise is used in a variety of ways. It is one of the basic ingredients of five-spice powder. People sometimes chew star anise as a breath sweetener. They may add it to tea, sweets, and some soups and other foods as a flavoring and aromatic. It is commonly used to enhance the flavor of red-cooked dishes (slow-cooked with soy sauce), as in "Soy Sauce Chicken" (Y. K. Martin, 1984: 116), and is considered particularly good for meat and poultry ones, such as "Star-Anise Stewed Duck Tongue" (K. Lo, 1979b: 196-97) and "Fragrant Pork" (Stobart, 1970: 207).[52] Star anise oil, distilled from the fruit (and sometimes leaves as well), is used as a drug, and as a flavoring and aromatic. The fruit, too, is a medicinal, used, for example, as a carminative, diuretic, and to relieve constipation (Soubeiran and Dabry de Thiersant, 1874: 209; Hooper, 1929: 76; F. P. Smith, 1969: 214-15; Perry, 1980: 180).[53]

For three centuries Europeans thought that the star anise of commerce came from *Illicium anisatum*, a tree described by Linnaeus from Japanese plants. It wasn't until 1888 that Joseph Hooker determined that in fact the spice came from a different species in South China which he designated *I. verum*. He suggested further that the *I. anisatum* of Linnaeus might more properly be designated *I. religiosum* Sieb. and Zucc. (Watt, 1889-96, 4: 330-31; Bretschneider, 1898: 65, 694), and this is now a synonym in common usage. *I. anisatum*, however, is found in China as well as in Japan and Korea (Mathon, 1981: 58; Zeven and de Wet, 1982: 39). Indeed, some say that though it has long been cultivated in Japan, it was introduced there from China (Flückiger and Hanbury, 1879: 20; Bretschneider, 1898: 268; I. H. Burkill, 1966, 2: 1245). Buddhists are believed responsible for the introduction, and in Japan, *I. anisatum* is commonly planted around Buddhist temples, with its branches, in bundles or as garlands, sometimes placed before idols or on the tombs of friends (Ainslie, 1826, 2: 20; Flückiger and Hanbury, 1879: 20; Thunberg, 1975: 236). Its fruit, today called "Japanese star anise" or "false" or "bastard" star anise, has a smell like turpentine, and is poisonous. In Japan and China the fruit may be burned as incense, or a lamp oil extracted from them. Despite their toxicity, the fruit or leaves may be used in China as medicine, taken externally (F. P. Smith, 1969: 489; Perry, 1980: 181), or the fruit employed as an adulterant of true star anise fruits. There are, however, reports of several tragic accidents in the Orient because the former fruit has been mistaken for that of true star anise (Redgrove, 1933: 258).

Anise is the alien among the three cultivated plants considered here, the dried, ripe fruit of *Pimpinella anisum*, an annual plant domesticated in the Mediterranean or the adjacent Near East (Vavilov, 1949-50: 33; Harlan, 1975: 70; Mathon, 1981: 36; Zeven and de Wet, 1982: 102). Anise was known in Greece since antiquity, having been found in ruins in the Aegean area (on Thera-Therasia) believed to be of the late Bronze Age (K. F. Vickery, 1980: 51). It is said to be of comparable antiquity in Egypt (I. H. Burkill, 1966, 2: 1758). Some scholars, by analogy with its names in Greek (*anison*) and Arabic (*yansoon*), say that anise was the *inst* of ancient Egypt (Darby *et al.*, 1977, 2: 796). Whether or not this is so, Egypt and Crete were the principal suppliers of anise to the classical world (Flückiger and Hanbury, 1879: 310). Anise is believed to have reached India by way of Persia, but it seems not to have been known there in ancient times. There is confusion between aniseed and fennel seed in accounts of India because these two are commonly called by the same name. In any case, Watt (1889-96, 6, Part 1: 237) writes that anise was not referred to in Sanskrit accounts, and indeed we have found no clear mention of it in accounts of Indian food, spices, and medicines before Moghul times (starting in A.D. 1526). This, together with the fact that the common words for anise in many Indian languages are derived from the Arabic (I. H. Burkill, 1966, 2: 1758), fit with the suggestion (Watt, 1889-96, 6, Part 1: 237) that anise was introduced to India by Moslems. It also fits with patterns of cultivation and supply in nineteenth century India, where much anise was cultivated in the north, which bore the full force of Moslem conquest, but only small amounts elsewhere in the subcontinent. By the nineteenth century, anise had gained widespread popularity in India for making perfume (water of anise is an Indian favorite) or for use in medicine and confectioneries. It is also chewed as a breath freshener and used in some types of curry, as in vegetable and fish dishes in Bengal (Stobart, 1970: 39). Local supply, however, has remained far too small to meet demand, and the bulk of India's anise needs have been supplied by imports from Persia or Europe (Watt, 1889-96, 6, Part 1: 237-38; Ilyas, 1980: 256).

From India, anise was traded by sea to Southeast Asia, and from there to China. We found it first listed as an import by sea in Sung times, arriving in Canton from Java (Wheatley, 1959: 106; Hirth and Rockhill, 1966: 77). Yet even in this century anise seems not to have been a well-known spice in China, for some Chinese confused its seed with that of fennel (I. H. Burkill, 1966, 2: 1758). One would expect anise to have been carried overland to China from Persia or Russia in early times, for both of those anise-producing areas had regular trade contacts with China. All we have found to support this expectation is Laufer's account (1919: 196, 200) of a writer of T'ang times who compared another plant with *Pimpinella anisum*.

Today anise is cultivated in gardens in various parts of China but it does not do well in tropical lowlands. Instead, it prefers low precipitation, and in arid Kansu it grows wild. Its seeds are used both as a condiment and a medicine (F. P. Smith, 1969: 331-32; Perry, 1980: 418). Aniseed is high in calcium and iron (Farrell, 1985: 34), though it is consumed in amounts so small as to make this of little or no nutritional importance.

CINNAMON

Trees or shrubs of the genus *Cinnamomum* (family *Lauraceae*) are native to an area stretching from India and Ceylon across Southeast Asia to China, Japan, New Guinea, and northern Australia.[54] Various species of the genus have provided plant parts (especially leaves and bark) used as aromatics, seasonings, and medicines. The most important sources of bark today are *C. zeylanicum* ("Ceylon cinnamon" or "true cinnamon"), native to Ceylon, southwestern India, and the Tenasserim area of Burma, and *C. cassia* ("cassia," "cassia cinnamon," "bastard cinnamon," "Chinese cinnamon," or "Canton cassia"), native to Kwangtung, Kwangsi, and Indochina (S. C. Lee, 1935: 500; H. L. Li, 1979: 83). Other species, however, also produce bark called "cassia," among them *C. loureirii*, "Saigon cassia," and *C. burmannii*, "Batavia cassia." *Cassia*, especially from *C. cassia*, is imported in far larger amounts to the United States than true cinnamon (*Spices: A survey of the world market*, 1982, 1: 56, 59), and China is a leading producer. Stronger, more pungent than true cinnamon, cassia bark is usually marked as "cinnamon" in the United States, and true cinnamon and cassia cinnamon are used interchangeably (Parry, 1945: 35; Farrell, 1985: 67, 83-84). The same is true in present-day China (E. N. Anderson, 1988: 138).

Cinnamon in China

The earliest botanical description of *C. cassia* is that of Chi Han, who also wrote of unmixed evergreen forests of cassia at higher elevations in Ho-p'u (later a district of Kwangtung), and of gardens of cassia established in Chiao-chih (present-day Hanoi) (H. L. Li, 1979: 83). Cultivation of cassia cinnamon is thus far earlier than that of true cinnamon, which was first grown in Ceylon only late in the eighteenth century.

The word *kuei* or *kwei*, today used for cassia cinnamon, was already known to the Chinese in ancient times (Flückiger and Hanbury, 1879: 520; Bretschneider, 1882-95, 2: 383-85; 3: 443; K. C. Chang, 1977b: 31), but Laufer (1919: 543) cautions that usage of the term was too general to determine the species involved. He suggests that the Chinese first became acquainted with cassia when they began to settle the south in Han times. So impressed were they with cassia that it was included in various place names, some surviving today, as in that of Kweilin ("cassia forest"), that Kwangsi city set in a beautiful karst countryside.

Turning to nineteenth century China, one notes that a bark, the cassia lignea of pharmacists, was an important item of commerce, with an average annual export of over 3000 tons in 1880 and 1881 (W. T. Thiselton-Dyer, 1884: 19). The principal areas of production were Kwangsi, western Kwangtung, Kweichow, and Yunnan (Gray, 1878, 2: 145; Flückiger and Hanbury, 1879: 528; S. W. Williams, 1895, 2: 55; Bretschneider, 1898: 711; F. P. Smith, 1969: 107-8). Uncertainty remained about the species from which the bark derived, and in 1882 the British Colonial Office dispatched an official, Charles Ford, to visit producing areas along the West River, to report on methods of growing and harvesting, and to bring specimens back to Hong Kong. Ford found that the only species providing commercial cassia bark, leaves, or buds was *Cinnamomum cassia*, and that cultivation occurred mainly in three districts in Kwangsi and Kwangtung (W. T. Thiselton-Dyer, 1884).

One of the largest growers claimed to have more than 50,000 acres of cassia trees. Bark was first taken from trees when they were six years old or so, and at six year intervals thereafter. The bark came from small branches, each about an inch in thickness. The time of cutting was from March through late May, after which the bark was said to lose its aroma. First the branches were removed from the tree. Then they were carried to sheds or houses near the plantation, and twigs and leaves removed. Then, with a long knife, a branch was slit longitudinally and then at sixteen-inch intervals around its circumference. After it was separated from a branch, bark was scraped to remove the epidermis, then dried and tied into bundles. Twigs and leaves stripped from those branches were also marketed. Buds, on the other hand, came from older trees that had been allowed to grow among the younger, bark-producing trees of the plantation. Such older trees also provided seed for future sowing, but their bark was very thick and not normally cut.

From the plantations, bundles of bark were carried to market towns, as much as twenty-five or thirty miles away. At places where only small amounts of bark were available for shipping, it might be transported not to the market towns but carried initially by small boat, then sold to larger river boats (W. T. Thiselton-Dyer, 1884). The latter were the noted "cassia boats" which carried bark via the West River system to Canton. These were specially-constructed vessels, similar to the tea boats of the Yangtze, able to navigate the dangerous rapids they encountered (Gray, 1878, 2: 267).

In traditional China, the most common cassia products in commerce were bark (*kuei-p'i*); buds (*kuei-tzŭ*); stems, twigs, and branches; and oil (*kuei-p'i-yu*), the latter a perfume, flavoring, and drug distilled mainly from leaves along with inferior products such as twigs and broken or coarse bark.[55] Bruised, fresh leaves were also used with water in washing one's hair. Cassia buds were a flavoring similar in taste to bark. Cassia bark, in turn, was used as a flavoring, for example with pork, less often as medicine.

As flavoring, powdered cassia bark is a common ingredient in "five-spice powder." Whether as bark or powder, cassia may also be used to season sauces, meat, and poultry (G. Z. Chu, 1975: 275; F. Lin, 1975: 327; Sakamoto, 1977: 23; Hom and Steiman, 1981: 31; Y. K. Martin, 1984: 232).

As medicine, bark (with other cassia products as alternatives) served a broad range of uses. Among them, cassia was a carminative and analgesic; aid in checking nausea, vomiting, and diarrhea; prescription for post-partum problems; and a stimulant and strengthening agent whose extended use was believed to aid the complexion and give one the appearance of youthfulness (Feng, n.d.: 35; Soubeiran and Dabry de Thiersant, 1874: 119; Bretschneider, 1882-95, 2: 443-54 *passim*; Hooper, 1929: 34-35; F. P. Smith, 1969: 110; J. D. Keys, 1976: 11; Perry, 1980: 197-98). In the past, consumption of cassia was also believed to prolong life. Pao-p'u-tse, honored Taoist and alchemist of the fourth century A.D., wrote that if one took cassia along with the brains of toads for seven years, a person could walk on water and avoid aging and death; a hunchback named Chao who took that mixture regularly over a twenty-year period is reputed to have grown hair on the soles of his feet and could walk 500 li (200 miles) a day and lift a weight of a thousand chin (1,333 pounds) (F. P. Smith, 1969: 110).

Ancient Sources of Cinnamon Reaching the Mediterranean and Near East

There has been considerable controversy among scholars, reviewed recently by Casson (1984), about the sources of the cinnamon and cassia bark and cinnamon leaves (the *malabathrum* of the Romans) that reached the ancient Near East and Mediterranean, regions well beyond the closest wild cinnamon in India and Ceylon. A bark presumed by some to be cinnamon or cassia was already used by the Egyptians in the middle of the second millennium B.C., with that of Queen Hatshepsut imported from the Land of Punt, along the southern shores of the Red Sea (Flückiger and Hanbury, 1879: 520; Schoff, 1912: 82-83; Darby *et al.*, 1977: 797). The earliest certain mention of cinnamon and cassia, however, is in *Exodus* (30. 23-24) (950 B.C.-400 B.C.?) (Casson, 1984: 226), and by the second half of the first millennium B.C. cinnamon and cassia were both well-known commodities throughout the Near East and Mediterranean (Flückiger and Hanbury, 1879: 520; Moldenke and Moldenke, 1952: 75-77; J. I. Miller, 1969: 44). Various classical writers gave northeast Africa and/or Arabia as sources of cinnamon and cassia. These, however, are not wild cinnamon areas today. If they were indeed source regions, some have argued, cinnamon may have grown there in the past and since have become extinct. As an alternative, these products may have come from trees or shrubs other than cinnamon. Or cinnamon and cassia may have been shipped via Arabian and African ports, and perhaps with false stories spread by merchants there to hide their real sources of supply.

If the latter were true, what might those source regions have been? India was first mentioned as a source region by Strabo (*Geography* XV. 1. 22; H. L. Jones, 1917-33, 7: 37), who wrote about the beginning of the Christian era. In part, however, he was citing Aristobulus, who accompanied Alexander the Great on his invasion of India (327 B.C.) and reported cinnamon growing in southern India. As for cassia, Strabo wrote (XVI. 4. 25; H. L. Jones, 1917-33, 7: 365) that some people said that most of it was of Indian origin. Apuleius in the second century A.D. wrote of India's cinnamon cargoes (*Florida* 6. 2; Valette, 1960: 131), and Philostratus (*Life of Apollonius of Tyana* 3. 4; Phillimore, 1912, 1: 99-101), in the third century A.D., of cinnamon growing in the foothills of north India. All of this, however, leaves open the question of other ancient source regions, and many modern scholars have sought them in Southeast Asia and South China. That perspective has been encouraged by the Arabic and Persian words for cinnamon (*dar-sini, darcini*) which mean "wood of China" or "bark of China."

Much confusion has been brought on by differences between ancient and modern definitions of "cinnamon" and "cassia." In the ancient Near East and Mediterranean, cinnamon was bark of the finest quality, and cassia, of an inferior sort. The botanical classification of the trees from which a bark derived was not known, and did not enter into consideration (Casson, 1984: 226-30). One might, as a result, obtain both cinnamon and cassia bark from a single tree. Or cinnamon or cassia might come from several different species, and from widely scattered areas.

Despite the confusion, modern writers have sought to identify the principal species and regions from which the ancient cinnamon and cassia derived. Though differing in detail, for over a century many have had little hesitation in identifying China as a source of cinnamon products for the ancient Near East and Mediterranean. Schoff (1912: 84), for one, believed that though the Roman *malabathrum* derived from leaves of various wild cinnamon trees (*C. zeylanicum* and other species) in India, the ancients' cinnamon bark came from *C. cassia*, and probably entirely from Burma and beyond, with some, perhaps, coming by sea, through many intermediaries, from China itself. Warmington (1974: 186, 188-90, 193), in turn, contends that leaves of Chinese cinnamon reached ancient India overland, mostly by Yunnan and Burma, to be mixed with Indian cinnamon leaves and marketed with them to the west.[56] The best cinnamon bark he believes to have come from China, and the inferior cassia bark from South India. The Chinese bark he sees as carried from China on ships of Indian merchants. Casson (1984: 237-39, 245) agrees with Schoff that *malabathrum* came from species of cinnamon trees wild in India. He agrees with Schoff and Warmington that the ancient Mediterranean cinnamon bark came from mainland Southeast Asia or South China. He also sees the ancient cassia bark as coming from that region. He suggests, however, that Chinese ships were not involved in sea trade to India in those times[57] and, citing Christie (1957), implies that the cinnamon was carried on ships of Southeast Asian peoples. Christie's case, in turn, rests on the name of a ship seen in India in the first century A.D. by the Greek author of the *Periplus of the Erythraean Sea* (Par. 60; Schoff, 1912: 46) which, Christie hypothesizes, was of a type called *k'un-lun po*, "boat of the southern barbarians," by the Chinese. If he is correct, Southeast Asian seafaring peoples would be likely candidates as cinnamon merchants and early intermediaries between South China and India.

Writers who look to Southeast Asia and China as sources of cinnamon and cassia bark for the Near East and Mediterranean regions, usually see those products coming first to India, then across the Arabian Sea and up the Persian Gulf or Red Sea. J. I. Miller (1969: 145-47, 153-72), however, sees India by-passed by the earliest route, a circuitous "cinnamon route" from China and Southeast Asia that went from eastern Indonesia directly across the Indian Ocean by way of Madagascar to Rhapta, a port on the East African coast. From Rhapta, he has the route branching, with Egypt the terminus of both branches. One went by way of the Red Sea, and the other from East Africa overland to the Nile and down that river. That such trade should have taken the dangerous, difficult overland route seems highly unlikely to this writer. And in any case, a sea route by way of Madagascar has been rejected by the historian of Madagascar, Pierre Verin (1986: 30).

Both Laufer (1919: 541-44) and I. H. Burkill (1966, 1: 550-51, 555-56) have raised objections, in part because of the lack of documentary evidence, to the hypothesis that China was a supplier of cinnamon to the ancient Near East and Mediterranean. Indeed Laufer sees the first clear evidence of cinnamon as an export of China only in an account of Ibn Khordadzbeh in the ninth century A.D., though Chinese records even then remained silent on the matter. Burkill, for his part, believes that date to be too early, and denies categorically that Chinese cinnamon bark supplied the ancient Near East and Mediterranean.

This writer, in turn, is uneasy at ruling out India as an ancient bark supplier, as some have done. Several cinnamon species are native to India, their useful qualities known to peoples today (Watt, 1889-96, 2: 317-26), as presumably in ancient times. Two which are of considerable commercial importance today would also have been candidates as bark suppliers to the ancient Near East. One is *C. tamala* of North India, the other, *C. zeylanicum* of South India. *C. tamala*, whose leaves are widely used in Indian dishes today, is accepted by Casson (1984: 238) (and other authorities) as an ancient supplier of leaves, the *malabathrum* of the Romans,[58] but he does not accept it as a source of bark for the West. Yet its bark has been a commercial product in India since antiquity[59] and remains so today, the cassia cinnamon of Indian commerce (buds of *C. tamala* are marketed as well). Bark taken from its root is said to be every bit as good as the bark of the true cinnamon tree (that taken from the stem is coarser than that of *C. zeylanicum* and *C. cassia*, is marketed in larger pieces, and commonly serves as an adulterant) (Watt, 1889-96, 2: 320-21). That cinnamon bark in general was well regarded in antiquity is revealed by its Sanskrit name *tvac* or *tvaca* (also *teja*), which meant simply "bark." This usage, says Dymock (1889a: 50), reflects its position as the supreme bark. Others (Watt, 1889-96, 2: 324; Dutt, 1922: 225) give one Sanskrit name for *C. zeylanicum* as *gudatvak*, "sweet bark," a cinnamon of special quality. That the Sanskrit name *teja* was carried to Java by ancient Indian settlers and incorporated into Malay (I. H. Burkill, 1966, 1: 550), further supports the view that cinnamon was a valuable product in the Indian homeland, one perhaps even traded to Southeast Asia.

Nor can one ignore the facts that certain ancient Greek writers reported cinnamon growing in India, in the very areas in which it is found today; that they marvelled at India's cargoes of cinnamon; and that some viewed India as the principal cassia supplier.[60] Modern writers who play down India as an ancient bark supplier, focus mainly on a single account, that of the Greek author of the *Periplus*, who mentions *malabathrum* but not cinnamon or cassia among the goods supplied by India's west coast ports. That he failed to do so, however, may be explained in various ways. Most plausible, perhaps, cinnamon and cassia bark were the very items he sought as a merchant, with self interest leading him to be silent about what may have been important and low-cost sources of supply. Also used in support of China and not India as an ancient bark supplier are the Arabic and Persian words for the bark, meaning "wood of China." Yet Burkill (1966, 1: 551) cautions that "the word is capable of two meanings—Chinese wood and sweet wood." And Laufer (1919: 542n) says that the Persian term is of medieval origin, too late to use as evidence of ancient contacts. Others (Schumann, 1883: 41-42; Casson, 1984: 236-37), in turn, observe that the Arabic word is borrowed from Persian, and that it, too, came into use in medieval times. Laufer (1919: 543-44) warns further that "the word 'China' in the names of Persian and Arabic products, or the attribution of certain products to China, is not always to be understood literally. Sometimes it merely refers to a Far Eastern product, sometimes even to an Indian product."[61]

Turning to the merchants of ancient India, it seems implausible that they, as suppliers of cinnamon leaves to the Western trade, would have failed to market Indian cinnamon bark. As we have noted, they used the bark medicinally and likely in other ways as well. As middlemen in the hypothesized trade of bark arriving from China by way of Southeast Asia, why would they have overlooked domestic sources, especially the fine quality "sweet bark" of *C. zeylanicum*? After passing through the hands of intermediaries, any bark arriving from China would have been expensive, providing a special incentive for Indian merchants to seek out local alternatives. The implied financial indifference of Indian merchants is out of keeping with the spirit of enterprise that has characterized them through history.

Sea trade between India and Southeast Asia goes back very far in time, probably well into prehistory. Indian ships ventured eastward at least as far as Borneo, and, in return, those noted seafarers the Malaysians visited the shores of India (Wheatley, 1961: 184). The question of ancient cinnamon trade between the two regions remains to be investigated in detail. We would, however, note that Abu Dulaf, a tenth century Moslem traveller and first-hand observer, said that cinnamon was exported to distant lands from what may have been the Tenasserim Coast or some nearby area of Burma (Wheatley, 1961: 224, 228). True cinnamon (*C. zeylanicum*) grows wild along that coast; and additional cinnamon species, providing bark, leaves, and other products, also occur wild in Burma and other areas of Southeast Asia (I. H. Burkill, 1966, 1: 549-66). Thus it seems quite likely that this was one source region for cinnamon entering the ancient Western trade.

As for China, if it did indeed provide cinnamon bark for ancient trade to India and the West, it was likely a minor supplier. Moreover, those providing the cinnamon in earliest times would not have been ethnic Chinese for, as Laufer (1919: 541-43) has noted, until Han times the cassia cinnamon forests of South China and Indochina were controlled by other ethnic groups. Some cinnamon bark may have found its way overland, perhaps via Burma, to India. Some may have found its way by sea, through intermediaries. In such sea trade, the Nan-Yüeh, an accomplished seafaring people, may have provided the first link in the earliest chain of contacts, and other groups, such as Malays and Indians, additional ones (Wheatley, 1975: 231-32). By the seventh century A.D., however, there were many Arabs and Persians in Canton, and their ships were carrying goods without intermediaries to India and the Near East (Hirth and Rockhill, 1966: 14-15). One wonders whether it was from this direct trade of the seventh century and later that the Persian and Arabic names for cinnamon, "wood of China," came to be adopted.

Chinese-manned ships probably became involved in direct sea trade with India starting in the twelfth century (Wheatley, 1961: 61; Schafer, 1963: 12-13; Hirth and Rockhill, 1966: 18).[62] This raises another possibility, when weighed along with observations made in A.D. 1563 by Garcia da Orta, a Portuguese physician who lived in India for thirty years. According to Orta, Chinese ships carried large amounts of cinnamon bark to Ormuz, where the Persians called it *darchini*, "wood of China." But, Orta insisted, this is a mistake, since the Chinese purchase in Ceylon all the cinnamon they provide to the Persians. Can Orta have been wrong, as Bretschneider (1898: 7) believed, and the bark actually been from China? Or could the Chinese ships have carried both China and Ceylon cinnamon bark to the Persians? In any case, can such relatively late trade by Chinese have given rise to the Persian (and Arabic) names for cinnamon?

We have found nothing in our sources that actually identifies cinnamon as an early Chinese export,[63] but more on cinnamon used domestically. Presumably after the Han opened up the south, there was an increase in domestic interest and trade in cinnamon. One reads (Schafer, 1963: 180; 1967: 195) that in T'ang China, substantial amounts of cassia bark of the finest quality were sent northward each year as tribute, accompanied by a quantity of cassia buds, the unripe, dried fruit; that cassia bark and buds were among the basic drugs of the time; and that another species of cinnamon, too, provided medicinal bark.

TURMERIC, SAFFRON, AND ZEDOARY

Turmeric derives from the dried rhizomes of the perennial herb *Curcuma domestica* or *Curcuma longa* (*chiang-huang*, one reading of which is "ginger-yellow"), a member of the ginger family believed by most botanists to have been cultivated first in ancient Southeast Asia or India. The turmeric plant is unknown anywhere in a wild state. It has 62 or 64 chromosomes, and in this regard is different from other members of its genus, which have a basic chromosome number of 21. Turmeric is apparently unable to fruit, and is thus regarded by I. H. Burkill as a cultigen, likely developed from a wild, unknown ancestor through long selection by vegetative means (Sopher, 1964: 93; I. H. Burkill, 1966, 1: 716).

Its ancient history of use is best documented for India,[64] where it was known as *haridra* in Sanskrit. It may have been known in pre-Aryan times, and in any case came to be widely used as a coloring matter and cosmetic, a dye, a ritual substance, medicine, and a spice, as in curries (Ainslie, 1826, 1: 454-55; Watt, 1889-96, 2: 666-69; Prakash, 1961: 20n, 70n, 280; Sopher, 1964: 95, 99; B. Walker, 1968, 2: 216; Stutley and Stutley, 1977: 110). Those uses among Hindus are reviewed by Dymock (1891: 445-47) and Kirtikar (1912: 453-54); and among Hindus and tribal peoples by Sopher (1964: 96, 110-11). Basic to the importance of turmeric is its color, for in India yellow is a sacred color, religious and royal,[65] with erotic significance as well, conspicuous at birth, marriage, and death ceremonies, as well as in rites of cultivation. In most sections of India, a few days before the start of marriage ceremonies a Hindu bride is anointed with turmeric and oil by "five married women, or five virgins" on various parts of her body; she dons a robe dyed with turmeric and wears it until the wedding day. A gift of turmeric and oil is sent by the bride to the bridegroom for his anointment, and he returns a similar gift to her. The contract of marriage is stained or dotted with turmeric. The bridegroom ties around his bride's wrist a string with turmeric and betel nut attached, and the bride does the same for him. And finally the newly married couple smears turmeric on a section of the wall in the groom's home and worships it. A wife who commits suttee (burning herself on her husband's funeral pyre) is taken to the place of immolation in a robe dyed with turmeric, as is a married woman when she dies. A Rajput about to sacrifice himself in a desperate struggle wears a yellow garment. Yellow is the preferred color of Buddhists; yellow clothes are worn by certain Hindu ascetics; and yellow is used for certain sect marks. Formerly, it was essential not only that the color be yellow, but that it be turmeric, even though it is inferior to other yellow vegetable dyestuffs.

In classical times, turmeric was traded westward from India and perhaps Southeast Asia to the Near East and Europe, whereas saffron was carried in the opposite direction (J. I. Miller, 1969: 63). The two have often been confused with each other in southern and eastern Asia, including China (Laufer, 1919: 309 ff.; Schafer, 1963: 185-86), even though they are from quite different plants.

In Southeast Asia and Oceania, turmeric and other species of *Curcuma* have been used widely as dye, in ritual and other ways, since ancient times. Use of turmeric as a condiment, however, was found in Southeast Asia but not in native Oceania (Sopher, 1964: 95, 108-23). In China, several species of *Curcuma* are found, and *C. longa* itself is believed to be native to the southwest (Schafer, 1963: 185). Laufer (1919: 313-14) notes that though *Curcuma* species were used in China since antiquity, it is difficult to determine their species, with *Curcuma longa* not described until T'ang times. He suggests that it may have been introduced during that period, or, as an alternative, that a new, better form of turmeric arrived then. Burkill (1966, 1: 716), impressed with how well-suited turmeric is to the warm lowlands of Yunnan and Szechwan, supports the alternative view, that T'ang observers were inspired to write of turmeric by the superior quality of newly-introduced races. In either case, turmeric, other *Curcuma* species, and saffron, were referred to by the T'ang Chinese as "*yü* gold," and served as yellow dyes, medicinals, spices, and aromatics (Schafer, 1963: 124-26, 185-86), uses that have continued into modern times (E. Watson, 1923: 213-14; I. H. Burkill, 1966, 1: 716-22; F. P. Smith, 1969: 138-40; J. D. Keys, 1976: 75-76).

Saffron is a deep orange-colored aromatic powder deriving from the dried stigmas and styles of a purple-flowered crocus, *Crocus sativus*, a sterile triploid unknown in a wild state. Species closely related to cultivated *Crocus sativus* are found wild from the Mediterranean region to Iran. The ancestry of the domesticated form is uncertain but it may have originated in Greece or Italy (Brighton, 1977: 153, 154; Brian Mathew, 1977: 92-94; B. F. Mathew, 1980: 97; Basker and Negbi, 1983: 229).

Saffron is a mid-latitude species which prefers mild climates, for bulbs are killed by severe winter cold. It is, however, able to tolerate ordinary drought, and does well with only 15 to 18 inches of precipitation annually. Especially critical are conditions during flowering, with a hot, dry summer and mild, cool autumn best. Should frost or heavy rain occur during the time of flowering, severe damage can result (Bois, 1927-37, 3: 103-4; Rosengarten, 1969: 390). Ideal conditions for saffron are found in the Mediterranean region, where Spain, the leading modern saffron producer, accounts for 80 to 90 percent of world supply; the remainder comes largely from France, Italy, Greece, Turkey, India, and the People's Republic of China (Simon *et al.*, 1984: 83).

Saffron was cultivated in ancient Mesopotamia and Persia, but it is uncertain whether it was grown in early Vedic India. Indeed it will take careful historical and etymological research to determine its early history there with any degree of certainty.[66] It seems, however, that Kashmir was the first place in the subcontinent where saffron was produced commercially, and in modern times Kashmir remains India's largest producing area (about 5600 acres in the mid-1970's). Saffron is also cultivated at present in neighboring Jammu, and in Nepal, but where it has been tried elsewhere in India it has not met with much success (Randhawa and Nath, 1959: 219; Madan *et al.*, 1966: 377; Srivastava, 1968: 17; Ingram, 1969: 183; Pruthi, 1976: 202; Nauriyal *et al.*, 1976-77: 63-64, 69-70; Regmi, 1982: 150-51; Lewis, 1984: 161).

From Persia and Kashmir, saffron was traded to China, where it is known as "foreign red-flower" (*fan-hung-hua*) and "Tibetan red-flower" or "Tibetan safflower" (*tsang-hung-hua* or *hsi-tsang-hung-hua*); whether Tibet itself was able to produce saffron for this trade is uncertain, some writers saying it did but Laufer (1919: 310, 312) insisting it did not, that Tibet was simply a route by which saffron reached China. Not to be overlooked is the role of Arabs in the saffron trade and in diffusion of the plant by land and sea. One finds in China, as in many other lands, a name for saffron (*sa-fa-ang*) that derives from the Arabic word for the marketable product, *za'faran*, which in turn comes from their word for yellow (de Candolle, 1886: 166; Laufer, 1919: 311-12; Moldenke and Moldenke, 1952: 87; F. P. Smith, 1969: 132).

The time of earliest Chinese knowledge of saffron remains uncertain. Laufer (1919: 310) writes that the Chinese had some information about saffron and received it occasionally in tribute from the third to the tenth centuries A.D., but that only during the Yüan dynasty was saffron actually imported to China by Moslems, to come into common use. Schafer (1963: 124-26, 308n), on the other hand, reports clear evidence that saffron was imported and the plant itself introduced to China in the Middle Ages, and that the T'ang Chinese used it in various ways.

How widespread and common saffron cultivation may have been in traditional China is also a matter on which authorities differ. Various writers have listed China or North China as places of saffron cultivation or export, and H. L. Li (1959: 192) notes that saffron has become an essential part of the flora of China's gardens. At the same time, general accounts of Chinese agriculture and life seldom mention saffron and, when they do, one cannot be certain whether saffron or safflower is involved. Both are found in China and both are called "red flower." Li (1959: 196) cites a nineteenth-century Chinese source saying that saffron was widely grown in Szechwan during the Ming and Ch'ing dynasties. Yet Couling (1917: 493-94), possibly referring to the same situation, writes that *safflower* (*Carthamus tinctorius*) was widely grown in former times in Szechwan as a source of red dye. Laufer (1919: 311) never saw or heard of saffron being grown in China, nor was he aware of a single Chinese account to that effect. He writes that in the most comprehensive systematic botany of China, Forbes and Hemsley fail to mention saffron. We note, as well, that Bretschneider, in his *Botanicon sinicum* (1882-95), does not mention saffron, but he wrote later that though saffron is not grown in Peking, it is known to be widely grown elsewhere in China. E. Watson (1923: 210) also writes of saffron grown in China "to a certain extent" and he gives Chinese names for the plant. China is included in some but not all recent lists of countries that produce saffron, yet Hong Kong is indicated as an importer of saffron, almost all from Spain (*Spices: A survey of the world market*, 1982, 2: 129, 255; Lewis, 1984: 161; Simon *et al.*, 1984: 83).

At best, saffron would seem to have been of minor local production in traditional China, with commercial cultivation unsuited to Kwangtung and China's rainy, subtropical southeast coast. It is not mentioned in Chi Han's fourth century flora, nor have we found anything to suggest that it has ever been of significance there. It was reported (Grieve, 1931, 2: 699) as cultivated in Burma, but I. H. Burkill (1966, 1: 693) writes that it does not grow in Malaya or anywhere nearby. Nor have we found reference to saffron being cultivated elsewhere in Southeast Asia.

Since antiquity, saffron has been very expensive, and today is by far the costliest spice in the world, for it may take more than 4000 flowers to provide a single ounce of dry saffron (Madan *et al.*, 1966: 383; Basker and Negbi, 1983: 233). A product of such value presents a great temptation for adulteration, and since antiquity saffron has indeed been subject to adulteration more than any other commodity, and in quite varied ways. Certain plant adulterants were also sold in their own right, with names suggesting an affinity to saffron, as with safflower or "American saffron" (Grieve, 1931, 2: 698; Folch Andréu, 1957: 182-84; Madan *et al.*, 1966: 379-80; Rosengarten, 1969: 392-94; Pruthi, 1976: 202; Basker and Negbi, 1983: 231-33).

Saffron served among ancient Mediterranean peoples not only as a food flavoring and coloring, but as medicine. It also provided a highly-valued perfume and golden dye, saffron yellow. Saffron was used as a perfume in theaters, banquet halls, and on certain other festive occasions. In Homer's *Iliad*, the Dawn is called "saffron-robed," and no other goddess or ordinary woman wears such a robe (Seymour, 1963: 324). At other times and places in the Mediterranean, saffron was used for the clothes of royalty. Some Mediterranean peoples also viewed saffron as erotic. In the *Iliad* (XIV. 346; Butler, 1898: 233; Seymour, 1963: 324), when the deity Hera (Juno) seduces her brother-husband Zeus (Jove), the earth sprouts a couch of young grass, with dewy saffron, hyacinth, and lotus. Wedding garments among Romans and Phoenicians were dyed with saffron (I. H. Burkill, 1966, 1: 693; Sopher, 1964: 98; Basker and Negbi, 1983: 230). Some have suggested an historical link between such uses of saffron in the ancient Mediterranean and turmeric in India. Though Sopher (1964: 98-99) deems this unproven, it is impossible to ignore the striking similarities (Dymock, 1891: 441-44; Kirtikar, 1912: 443-52; Mahindru, 1982: *passim*) in the symbolism of saffron, turmeric, and other yellow plants. They occur, moreover, not just in the Mediterranean and India, but in intermediate lands such as Persia, so important in exchanges between the two.

In Southeast Asia the robes of Buddhist monks are often described by Westerners as "saffron-colored," but in reality they are dyed with turmeric or other yellow dyestuffs (Sopher, 1964: 98-99). In India, saffron is too expensive for use as a dyestuff (Watt, 1889-96, 2: 593), and in T'ang China, according to Schafer (1963: 125, 126, 186), it is uncertain whether it served as a dyestuff. Instead the Chinese seem to have used saffron mainly as an expensive medicine and perfume, and, to a lesser degree, as food coloring and flavoring (Laufer, 1919: 310, 312; E. Watson, 1923: 210; Schafer, 1963: 124-26; F. P. Smith, 1969: 132).

Since ancient times, peoples of Europe and Asia have identified a broad range of medicinal uses for saffron (Ainslie, 1826, 1: 354-57; H. Phillips, 1827, 2: 194-200; Madan *et al.*, 1966: 380; Rosengarten, 1969: 395; Pruthi, 1976: 205; Nauriyal *et al.*, 1976-77: 60; Basker and Negbi, 1983: 229-30; Simon *et al.*, 1984: 84), with the Chinese regarding it as useful for many female troubles and for other ailments as well (Soubeiran and Dabry de Thiersant, 1874: 117; E. Watson, 1923: 210; F. P. Smith, 1969: 132; Perry, 1980: 182). The medicinal claims for saffron have been discounted by practitioners of scientific medicine, and, though continuing to be well regarded in traditional systems, its overall medical use has declined in recent years (Folch Andréu, 1957; Basker and Negbi, 1983: 229-30). Nevertheless, saffron has been found to be a rich source of riboflavin, richer than any other natural source (Bhat and Broker, 1953: 544; Rajagopalan *et al.*, 1960). And in addition, rabbits fed an atherosclerosis-inducing diet and injected with a solution containing crocetin, a compound found in saffron, experienced greatly reduced levels of serum cholesterol and severity of atherosclerosis (Gainer and Chisolm, 1974). Presumably this resulted from enhanced oxygen diffusion through the blood to the tissues. It has been suggested (Grisolia, 1974) that consumption of substantial amounts of saffron daily in parts of Spain, especially Valencia Province, may be responsible for a low incidence of cardiovascular disease.

Like turmeric a perennial herb of the ginger family, zedoary (*Curcuma zedoaria*) is cultivated from Madagascar to Ceylon, India, and Southeast Asia (including Indochina), with most authorities (Vavilov, 1949-50: 27; Newcomb, 1963: 129-30; I. H. Burkill, 1966, 1: 724; Zeven and de Wet, 1982: 80) placing its domestication in east or northeast India. Its root is the main useful part, utilized in making perfume and cosmetics and as medicine, spice, and coloring (often confused with turmeric and saffron). Sometimes zedoary is used (frequently with sappan wood, from *Cæesalpinia sappan*) in making a perfumed yellow or red powder which either alone or mixed with water is thrown on participants in Holi, that major Hindu festival (Watt, 1889-96, 2: 670; Dutt, 1922: 258; Dey, 1973: 108-9). Though zedoary now has competition from aniline dyes in India, it continues to be cultivated in gardens in many parts of the country (Watt, 1889-96, 2: 670-71). Its root has a flavor reminiscent of ginger, but is milder, bitter, and slightly camphorous; in India it is less well-regarded for curries than either ginger or turmeric, and its role as spice is today quite minor (Redgrove, 1933: 56). In the Middle Ages, zedoary was among the Indian products that Arab traders carried westward, and it became one of Europe's minor spices and most highly-valued drugs. Yet today it is hardly known there, and in Britain is not even included among the officially-recognized drugs. It has also been forgotten as a spice in Europe, and is employed there today mainly in the manufacture of liqueurs and bitters (Redgrove, 1933: 56-57).

Schafer (1963: 125, 185-86; 1967: 198) has suggested that zedoary may have been among the spices of T'ang China, one of the plants collectively called "*yü* gold." Schafer also wondered whether it may have been imported from Southeast Asia and India or perhaps even been transplanted to China. Laufer (1919: 309-23, 544), who thoroughly reviewed the confusing evidence on the identity of the early *yü* gold plants, had concluded that some seem to have been forms of *Curcuma*. He did not, however, reach a firm conclusion about zedoary, writing that he knew of no Chinese name for it and remained uncertain about its presence in China. Bretschneider (1882-95, 2: 231-32), on his part, noted that certain of the *yü* gold plants of classical Chinese sources, including some with aromatic and some with non-aromatic roots, may have been turmeric or some other form of *Curcuma*. Yet he did not suggest in that account that zedoary was one of the plants involved. He later (1898: 8) cited claims by others that zedoary was native to China and that the drug was produced there. Yet it is difficult to give those claims credence. One was in a sixteenth-century account of Indian drugs and medicinal herbs by Garcia da Orta (1913: 453-58), who had lived in India for many years but had never been to China. The second was by William Roxburgh (1820-24, 1: 23) who, in an account of Indian plants, said in passing that zedoary was native not just to Bengal but to "China, and various other parts of Asia, and the Asiatic islands." Because of this and because we have not found zedoary clearly identified in sources on Chinese materia medica or botany, we remain skeptical that zedoary is native to China and uncertain whether, now or in the past, it was imported or used there on any scale.

CARDAMOM

In classical Greece and Rome, the terms *cardamomum* and *amomum* were applied to certain spices obtained by trade from the east. Those terms, however, were used in such ways as to make it uncertain what genera and species were involved (Flückiger and Hanbury, 1879: 644; I. H. Burkill, 1966, 1: 925-26). Many believe, however, that prominent among those spices were dried fruit of various species of *Elettaria* and *Amomum* from India. One, coming by sea from the Malabar Coast, may have been the "true cardamom" of today, *Elettaria cardamomum* Maton,[67,68] and others, coming almost entirely overland, seem to have been species of *Amomum* from northern India and perhaps beyond (J. I. Miller, 1969: 37-38, 67-69, 71-73).

True cardamom is a perennial herb believed to have been domesticated in South India or Southeast Asia, with India accounting for roughly three-fourths of the cardamom entering world trade in the mid-1970's (Pruthi, 1976: 63). The word "cardamom," however, is also used today for the dried fruit of species of *Amomum* used as substitutes for true cardamom, species native to India; southeastern Asia, including China; and/or the Pacific region.[69] Their dried fruits are similar in appearance to those of true cardamom, and their seeds resemble those of true cardamom in taste. Many species of *Amomum*, wild and cultivated, provide such dried fruit, and there is considerable confusion about them. I. H. Burkill (1966, 1: 132), however, identifies five which he considers major substitutes for true cardamom. These include Nepal cardamom (*Amomum subulatum* Roxb.) and Bengal cardamom (*A. aromaticum* Roxb.), both cultivated in India; round cardamom (*A. kepulaga* Sprague and Burkill) and Java cardamom (*A. maximum* Roxb.), cultivated in Malaysia and/or Indonesia; and "krervanh" (sometimes "white," "Siam," or "Cambodian" cardamom; *A. krervanh* Pierre), cultivated in Thailand, Indochina, and islands of the Malay Archipelago.

As for China, one reads that in the third century A.D. a bamboo chest filled with cardamom (identified as *A. cardamomum* L.) was sent in tribute to the imperial court from Chiao-chou (an ancient administrative unit embracing parts of present-day Kwangtung, Kwangsi, and Vietnam) (H. L. Li, 1979: 38). True cardamom and various amomum cardamoms (*A. amarum* F. P. Smith; *A. xanthioides* Wall.; *A. kepulaga* S. and B.; and *A. globosum* Lour.), principally from Indochina or Malaya, were also carried north in China in T'ang times (Schafer, 1963: 184-85; 1967: 193-94). And in Sung times two kinds of amomum cardamom, and perhaps true cardamom, were imported from Southeast Asia (especially Cambodia), and from India and Ceylon as well. In addition, Sung accounts indicate that Kwangsi provided cardamom (*A. globosum* Lour.) (Wheatley, 1959: 88-89; Hirth and Rockhill, 1966: 57, 61, 73, 77, 84, 89, 96, 221-22).

This calls attention to the fact that South China also produces cardamom from wild or cultivated plants, with several species mentioned but much confusion as to their classification and as to whether they should be considered separate species. One is Chinese, round Chinese, or herbaceous cardamom (*A. globosum* Lour.; fruit = *ts'ao-tou-k'ou* or *ts'ao-k'ou*) of Kwangsi, Kwangtung, and parts of Yunnan and Fukien, said to be a Chinese domesticate (F. P. Smith, 1969: 30; Zeven and de Wet, 1982: 47). A second is bastard cardamom (*A. xanthioides* Wall.; *sha-jên*), apparently found in Kwangtung and Kwangsi, with Canton the main place of export (Hooper, 1929: 11; I. H. Burkill, 1966, 1: 137). Another is ovoid China cardamom (*A. medium* Lour. = *A. aromaticum* Roxb.?; *ts'ao-kuo*) of Yunnan and Kwangsi, most of which is shipped by way of the West River system (Hooper, 1929: 9; F. P. Smith, 1969: 37). A fourth is black or bitter-seeded cardamom (*A. amarum* F. P. Smith; *i-chih* or *i-chih-tzŭ*), said to come from Kwangtung, Hainan, Kweichow, and Yunnan, with some shipped from Canton (Hooper, 1929: 8; F. P. Smith, 1969: 35). A fifth is hairy Chinese cardamomum (*A. villosum* Lour.; also *sha-jên*), cultivated in Kwangtung but apparently introduced from Vietnam (Hooper, 1929: 9; F. P. Smith, 1969: 38).[70] It is not surprising that Chinese cardamoms were exported to Chinese communities overseas, but it is more so that in nineteenth century India, that great center of production, the seeds (not entire seed capsules) of *A. xanthioides*, imported from China and Singapore, were found in all large markets in the south (Watt, 1889-96, 1: 223).

In Southeast Asia, cardamoms are more important as spices than in any other way. In China, on the other hand, their medicinal role is paramount,[71] a fact, according to Wheatley (1959: 89), that may have left the Chinese more ready to use forms of cardamom that were less pleasing to the taste than was true cardamom. The seeds of *A. villosum*, for one, are described as bitter, tarry, and camphor-like (F. P. Smith, 1969: 38). Yet to the degree that they were considered spices, the taste of cardamoms must have been of importance to the Chinese. This is reflected in the poetry of T'ang times where cardamom was considered a combination of godlike taste and scent with beautiful color, the red of their flowers symbolic of the land in which they grew. One might be tempted to view this merely within the context of the T'ang poets, who had an interest in colors, scents, and strong, exotic flavors that one can properly describe as unusual (Schafer, 1963: 185; 1967: 193-94). We would submit, however, that theirs may have been a perspective found in certain other parts of Chinese society as well. In any case, some cardamoms were quite pleasing in odor and taste, and in China their seed was chewed to sweeten the breath and both herb and seed used with other spices for flavoring soup and meat and fowl dishes (Sakamoto, 1977: 22; Y. K. Martin, 1984: 21, 226, 232). Even in medicinal uses, it seems, pleasantness of flavor was sometimes sought after. One reads (Schafer, 1963: 184), for example, that to fortify their minds the T'ang Chinese ate black or bitter cardamoms—known as seeds that add to one's wisdom—along with honey in millet or rice dumplings.

CLOVE AND NUTMEG

The clove and nutmeg are important in the history of spices, and remain popular in many parts of the world today. In China they do not rank with such major spices and flavorings as ginger, garlic, vinegar, chili peppers, and soy sauce. One is nevertheless justified in including them here because theirs are interesting stories that deserve to be presented from a Chinese perspective. Even more, their introduction to China involved, or seems to have involved, the Nan-Yüeh, and may represent still further Nan-Yüeh contributions to China's food scene. We consider cloves and nutmeg together because they originated in the same region and because there are similarities in their stories in China, to which they have been imported since antiquity.[72]

The clove is an evergreen tree (*Eugenia caryophyllus* (Sprengel) Bullock and Harrison = *Syzygium aromaticum* (L.) Merr. and Perry)[73] whose closest relative is found wild in forests on New Guinea, and on many of the Moluccas, or Spice Islands (Wit, 1969: 164; 1976: 216) where it is believed to have been domesticated (Vavilov, 1949-50: 30; Uphof, 1959: 149; Newcomb, 1963: 130; I. H. Burkill, 1966, 1: 976; H. L. Li, 1970: 15; Harlan, 1975: 75; Mathon, 1981: 79). Not much is known about its early history, but natives of the Moluccas do not seem to have used cloves as a spice, and only occasionally for other purposes. Because of this, Crawfurd has suggested that Indians and Chinese, who came to know and value the dried, mature, but unopened buds of the tree, are responsible for developing the clove trade, with their names for cloves coming to prevail even in Malaysia. From this perspective, one Malay name for cloves, *bunga lawang*, is said to come in part from *lavanga*, an Indian name presumably deriving from the Sanskrit, used in the trade westward to India. And a second Malay name, *chinkeh* or *chĕnkeh*, may be of Chinese origin (Crawfurd, 1820, 1: 496-98; 1868: 193-95; I. H. Burkill, 1966, 1: 976-77). Others, however, have observed that in India there is no aboriginal name for cloves, that most Indian languages have words for cloves that relate to *lavanga*, and have concluded that the Sanskrit *lavanga* more likely derived from Malay than the other way around (Watt, 1908: 527). In India, this name is said first to have occurred in the *Ramayana* (F. W. Thomas, cited in Watt, 1908: 527), but it is uncertain at what stage in the evolution of that epic (pre-350 B.C.-A.D. 250) it may have been introduced. The name *lavanga* also occurs in certain other early works,[74] and it is reasonable to conclude that the clove was known in India (as medicinal, aromatic, and spice) by the beginning of the Christian era or perhaps a few centuries before.

It is uncertain when the clove first reached the Mediterranean world. It seems not to have been known in that region in great antiquity. Though the clove was not mentioned by the Greek author of the *Periplus of the Erythraean Sea* (first century A.D.) (Schoff, 1912), that was the time that one finds the first reference in classical sources to what may have been the clove.[75] This involved the Greek name *caryophyllon*, which is believed to have come from the Arabic or Persian name for the clove, *karanful*. *Karanful*, in turn, is said by some to have come from *karampu* or similar names for cloves in South India (Watt, 1908: 527; I. H. Burkill, 1966, 1: 977). In any case, by the fourth century A.D. cloves had become well known in Europe (Flückiger and Hanbury, 1879: 281).

As for China, Nan-Yüeh seafarers are given credit for carrying cloves there from the Moluccas in the second half of the first millennium B.C. The first mention of cloves in China, in a work of the third century B.C., pictured them as a breath freshener courtiers were expected to use when in the presence of the Han emperor (Wheatley, 1959: 45). Apparently cloves were imported on a continuing basis, for they are mentioned by several Han writers (Flückiger and Hanbury, 1879: 281). Wheatley (1959: 47) writes that Yüeh voyages to the Moluccas ceased in Later Han times (A.D. 25-220) and that the clove "reappeared in China" in the seventh century, but came from the Moluccas indirectly, by way of Sumatra or Java. One notes, however, that Chi Han in his fourth century flora had identified cloves as coming from the fruit of a tree growing in Annam. Though he erred in identifying the tree from which they came (H. L. Li, 1979: 87-88, 89), his account clearly reveals that the Chinese of his day continued to be familiar with cloves. It may be that the cloves came from some forest tree of Annam,[76] or that some enterprising Nan-Yüeh traders had introduced the clove tree there,[77] with these plantings supplying the China market.

Of special interest in early China was the range of uses to which cloves were put. Schafer writes that in T'ang times it was their fragrance that seems to have been most important, and they were used both in perfume and incense and to freshen the breath. Their early name was *chi-shê-hsiang*, "chicken-tongue fragrance," which reflects the Chinese view of their shape and primary use. Their modern Chinese name, "nail fragrance" (*ting-hsiang*), is more in keeping with the perception of cloves in early Europe, where their Latin name *clavis*, "nail," led ultimately to the English "clove." Cloves also served the T'ang Chinese in medicine and magic, and to some extent in food, such as finely-cut meat "marinated in a liquor flavored with cloves." Cloves were imported to T'ang China from Indonesia, but some also came from Annam (Schafer, 1963: 159, 171-72; 1967: 198; 1977: 111) as seems to have been the case in Chi Han's day.

In modern China, cloves continue to serve as a breath freshener, aromatic, and medicinal, among other things being well regarded for their stimulant, tonic, stomachic, and carminative qualities; and for countering diarrhea and colic in children (Hooper, 1929: 61; F. P. Smith, 1969: 95-96; J. D. Keys, 1976: 140-41; Perry, 1980: 285). In cookery, their most important use is as an ingredient in "five-spice powder," though they may be used in other ways, as in various types of curry powder, and in spice mixtures used in flavor-potting sauces for meat and poultry (Y. K. Martin, 1984: 21, 226, 232).

Nutmeg (*Myristica fragrans*), also known as Banda nutmeg, is the seed of an evergreen tree which, like the clove, is believed to have been domesticated in the Moluccas or other islands of eastern Indonesia (Vavilov, 1949-50: 31; Uphof, 1959: 244; Newcomb, 1963: 130; I. H. Burkill, 1966, 2: 1550; H. L. Li, 1970: 15; Harlan, 1975: 75; Mathon, 1981: 77; Zeven and de Wet, 1982: 60), though it rarely if ever occurs there today in the wild.[78,79] The aril or outer covering of the nutmeg seed, in turn, is the source of mace. Nutmeg was an early item of trade from Indonesia to India,[80] a trade perhaps initiated by the Hindus of Java in the first centuries of the Christian era (I. H. Burkill, 1966, 2: 1550-51).[81] From India, nutmeg was carried westward by the Arabs, to reach Europe by the sixth century A.D. at least (Flückiger and Hanbury, 1879: 502-3; I. H. Burkill, 1966, 2: 1550).

Some authorities (Schafer, 1963: 185; Hirth and Rockhill, 1966, 210n) say that the earliest Chinese description of nutmeg is that of Ch'en Ts'ang-ch'i in T'ang times. The nutmeg in question came to China by merchant vessels from overseas, and it was called *jou-tou-k'ou*, one of its common names today.[82] Recently, however, H. L. Li (1979: 81-83) identified as nutmeg (*Myristica fragrans*) a plant called "*i-chih-tzŭ*," which was described by Chi Han in his flora (fourth century A.D.).[83] Chi Han reported that *i-chih-tzŭ* was produced in what today are Annam and Kwangtung,[84] and that early in the third century a governor of Chiao-chih sent a gift of *i-chih-tzŭ* preserves to Emperor Wu of Wei (H. L. Li, 1979: 81). The term *i-chih-tzŭ* in later times was used for bitter-seeded cardamom (*Amomum amarum*), but if Li is correct in his view that Chi Han's *i-chih-tzŭ* was nutmeg, the earliest Chinese knowledge of nutmeg would be four or even five centuries earlier than most scholars had suspected. And the tree must have been introduced from the Spice Islands to Annam and mainland China still earlier, perhaps by the Nan-Yüeh seafarers who first carried cloves to China.

Even if Li is wrong in his identification, the nutmeg tree seems to have been cultivated in Kwangtung or elsewhere in Lingnan by the end of the eleventh century if not sooner (Wheatley, 1959: 100; Schafer, 1963: 185; 1977: 111; Hirth and Rockhill, 1966: 210-11n). Nutmeg produced in China, however, is said to be usually dry and infested with worms (F. P. Smith, 1969: 276) and, whether for this or some other reason, it continued to be imported early in the thirteenth century (Wheatley, 1959: 100; Hirth and Rockhill, 1966: 210) as it has since.

It is often difficult to classify a product as being primarily a spice or medicinal, but that is less of a problem for nutmeg in China. Schafer (1963: 185), for example, writes that nutmeg in T'ang China apparently was a medicine but not a spice, and he mentions a broth of ground nutmeg given for digestive problems and diarrhea. F. P. Smith (1969: 276), writing of traditional China's materia medica, observes that nutmeg is mainly a medicine and rarely a spice. And Hooper (1929: 92) commented on how strange it was that the Chinese in Malaya do not make much use of nutmeg as a spice, but instead employ it as a medicine, one carried by all Chinese druggists in that country. Because its role is primarily as a drug, one can find references to nutmeg in accounts of Chinese medicine (Soubeiran and Dabry de Thiersant, 1874: 205; Hooper, 1929: 91-92; F. P. Smith, 1969: 276; J. D. Keys, 1976: 139-40; Perry, 1980: 279) far more often than in ones of Chinese food or cooking.

CORIANDER

Coriander (*Coriandrum sativum*; *hu-sui*) is a hardy annual herb which provides two products highly regarded around the world. One is its dried mature fruit, coriander seed, a valued spice. The second is its leaves, known in the United States by the plant's Spanish name, cilantro, or by an incorrect one, Chinese parsley.

Coriander is believed by most botanists to be of Mediterranean or Near Eastern origin (Vavilov, 1949-50: 33;[85] Harlan, 1975: 70; Mathon, 1981: 28; Zeven and de Wet, 1982: 119), and even today is a weed in Egypt and common, too, in fields of grain in Palestine (Moldenke and Moldenke, 1952: 86). What seems to be coriander is mentioned in the Ebers Papyrus, that Egyptian medical treatise of the mid-sixteenth century B.C. Coriander remains, in turn, have been identified in the tomb of Tutankhamen (mid-fourteenth century B.C.) and in a tomb at Deir el-Bahari dating from the Twenty-first Dynasty (*c.* 1085-950 B.C.) (Darby *et al.*, 1977, 2: 798). Coriander was known to the early Hebrews, for its seed is compared to manna in both *Exodus* (16. 31) and *Numbers* (11. 7), and the plant is frequently mentioned in the Talmud (Moldenke and Moldenke, 1952: 86). The ancients used both coriander leaves and seed as medicine as well as flavoring or spice.

The evidence on its spread eastward needs to be weighed again in the light of recent archeological excavations. It was, however, mentioned in the *Bundahishn* (West, from SBE, cited in Laufer, 1919: 192), a Zoroastrian text based on a lost section of the original *Avesta*, which attests to its presence in ancient Persia. Indeed Laufer (1919: 297-99) has argued that Persia should be included, along with Egypt and Palestine, as part of the ancient homeland of coriander; he also believes that coriander growing was highly developed in Persia, and that its coriander spread in all directions, reaching both India and China. Though coriander has been identified in early Sanskrit writings, Laufer (1919: 284, 298) notes that there is no real Sanskrit name for coriander, that some names for it simply mean "grains" and that another, *kustumburu*, is of Persian origin.[86] He therefore concludes that coriander was introduced to India from Persia. Unlike saffron, which was also introduced by way of Persia, coriander adapted well to Indian conditions. Today the plant is found in all parts of the country, its leaves eaten as a vegetable, and its seeds, a seasoning and curry ingredient, used also in sweets, liquor, and as the source of an essential oil (Watt, 1889-96, 2: 568-69).

Turning to China, Laufer observes (1919: 297-99) that only in the sixth century A.D. does one find unquestionable literary evidence of the coriander plant, but with nothing said about from whence it had arrived.[87] Though he searched further, Laufer failed to turn up other information on the introduction of coriander, forcing him to rely on linguistic data in seeking an answer. He concludes that *hu-sui*, one common Chinese name for coriander, seems to be a transcription of a Persian name for it, one form of which he gives as **koswi*.

One should not overlook the possibility that the coriander plant reached China, whether overland or by sea, by way of India. We would note, however, that neither the plant nor seed were mentioned in Chi Han's fourth century flora (H. L. Li, 1979), Schafer's accounts (1967, 1970) of South China and Hainan in T'ang times, or monographs by Wheatley (1959) and Hirth and Rockhill (1966) on Sung maritime trade. There would have been problems in establishing coriander for seed production in much of Southeast Asia. Coriander likes places with a warm somewhat dry summer, and the lowland tropics of that region are not suited for seed production. If one is interested in seed there, coriander must be grown at higher elevations. Ochse and Bakhuizen van den Brink (1931b, 3: 705-7) have noted that most coriander seed used in the

Dutch East Indies was imported, and I. H. Burkill (1966, 1: 673), noted that in Malayan bazaars it is Indian coriander that is sold. At the same time, coriander can be and is cultivated for its leaves in both lowland and highland regions of both countries. Presumably because of this, coriander was cultivated only on a modest scale, with the Chinese singled out in the Dutch East Indies as especially given to consumption of its leaves as a vegetable. The pattern of cultivating coriander for its leaves may also prevail in Annam, for Crevost and Lemarié (1917: 165) write only of coriander leaves, used to provide a pleasant, spicy odor to certain dishes.

After its introduction to China, coriander, like garlic, was included among the five vegetables of strong odor forbidden to geomancers and Taoist monks (F. P. Smith, 1969: 28). That proscription seems to have had little or no influence on the general public who use both seed and leaves with relish. Coriander leaves are widely consumed there today, used by cooks as we would parsley. It serves as a garnish and flavoring, as in fish and fowl dishes such as the Cantonese "Duck with Coriander" (C. B. T. Lee and A. E. Lee, 1980: 191-92) and the Soochow "Steamed Floured Carp" (G. Z. Chu, 1975: 184-85), or in soups such as "Fish Soup with Coriander" (K. Lo. 1979b: 36). Use of coriander leaves is said to be especially common in North China (E. N. Anderson and M. L. Anderson, 1977: 331; Y. K. Martin, 1984: 232).

The fruit, leaves, and root of coriander all are used in medicine, with carminative, stomachic, and corrective qualities most often mentioned for the seed (Hooper, 1929: 43; F. P. Smith, 1969: 127; J. D. Keys, 1976: 198-99; Perry, 1980: 415). The leaves do in fact contain good or very good amounts of calcium, iron, and vitamins A and C, and the seed is very high in riboflavin, calcium, and iron (Farrell, 1985: 95).

SWEET BASIL

Sweet basil, *Ocimum basilicum* (*lo-lê*; *hsiang-ts'ai*), is an annual herb indigenous to the tropics of the Old World, now widely cultivated for its leaves, a valued aromatic flavoring. It is one of several Old World species of herbs and shrubs of its genus to have been domesticated (Watt, 1889-96, 5: 440-45; Crevost and Lemarié, 1917: 155-56; Bois, 1927-37, 3: 178-182; Ochse and Bakhuizen van den Brink, 1931b, 2: 355-57; Dalziel, 1937: 462-63; Uphof, 1959: 251; I. H. Burkill, 1966, 2: 1597-1603). Botanists have differed about the place of sweet basil domestication, but have generally focussed on an area extending from Persia to Southeast Asia. Authors of two recent works (Zeven and Zhukovsky, 1975: 49; Mathon, 1981: 77) place domestication of *O. basilicum* in Southeast Asia, though apparently without much conviction. Laufer (1919: 586-90) favored Persia, though he believed that the plant reached India in antiquity. And Vavilov (1949-50: 32) placed its origin in his Central Asiatic Center, in which he includes Afghanistan and areas to its north as well as northwest India (Punjab, Kashmir, North-West Frontier Province). Though the regions favored by Laufer and Vavilov are almost entirely outside the tropics usually given as the plant's original home, it should be noted that *O. basilicum* is reported to be indigenous both to Persia and Sind (northwest India) (Watt, 1889-96, 5: 441; Laufer, 1919: 587-88).

In India, the sweet basil plant is not only indigenous and widely cultivated, but a related species, also a flavoring or spice, enjoys an unusual role in Hindu society that must have attracted attention to others of its genus as well. We refer here to *tulasi* or *tulsi*, holy basil (*Ocimum sanctum*), one of Hinduism's most sacred plants, linked in legend to the deities Krishna and Vishnu, among others. *Tulasi* is cultivated in or near Hindu temples and homes everywhere. In most Brahmin homes it is grown in a pedestal of brick or plaster which bears the name *Vrindavana*, and women of the household worship it each evening, circumambulating it, watering it with care, and putting a lamp nearby for the night. Even though Hindus of all castes may not worship *tulasi* , it is an object of veneration to all, one that no gardener would injure or destroy (Underhill, 1921: 129). The plant should be collected only for ritual or medical purposes (Dymock, 1889b: 83). *Tulasi* is said to kill mosquitoes and demons, to cure disease and cleanse the air; it is a common belief that evil spirits would not risk approaching a place where *tulasi* grows, that the plant bestows offspring on those without children and riches on the poor (Dymock, 1889b: 83; Gonda, 1970: 111). Hindu poets have written that those who cultivate *tulasi* gain sanctity, that the plant directs them to heaven and guards them from calamity (Watt, 1889-96, 5: 444). Its sanctity, moreover, has led to the use of its woody stems (from plants of large size) for making beads word singly or in necklaces (Watt, 1889-96, 5: 444; Gonda, 1970: 111, 113).

The special status of *tulasi* relates to the belief that it represents the hair of a beautiful, virtuous woman named Vrinda who ultimately became Tulasi, wife of Vishnu.[88,89] It is said that if the plant, which Tulasi's spirit is believed to enter each evening, is injured, Vishnu is enraged (Dymock, 1889b: 83). The plant figures in Hindu ceremony in various ways. Especially prominent is its role in the fall celebration of Tulasi's marriage to Vishnu, an event which marks the start of the Hindu marriage season. At that time, the plant is ceremonially wed to Vishnu, who may be represented by an image or a stone[90] in which the deity lives (Underhill, 1921: 130; B. Walker, 1968, 2: 585; Gonda, 1970: 111; Stutley and Stutley, 1977: 306; P. Thomas, 1980: 94).[91] Also prominent is the use of leaves of the variety *villosum* in taking oaths at legal proceedings. It is said that this custom was started by the British during the colonial period, that they sought an equivalent of the Bible as used in oath-taking in the Western world, and decided that *tulasi* was a good alternative for use with Hindus. A Brahmin first places *tulasi* leaves in a person's hand, then swears him in, and finally the leaves are chewed and swallowed (Watt, 1889-96, 5: 445).[92] Water containing *tulasi* leaves may be given to worshippers in a temple, or poured into the mouth of a dying man. The leaves, in turn, may be scattered over the head of a dead man, or branches placed on his chest (Dymock, 1889b: 83; Mahindru, 1982: 74). And in the Puranas one reads of bodies being cremated using *tulasi* wood, with absolution of sins and other great benefits for the deceased (Mani, 1975: 798).

Turning to sweet basil (*O. basilicum*), one notes in the ancient and modern Near East and Mediterranean certain similarities in its role with that of *tulasi* in India. In ancient Greece, for example, basil was associated with mourning, and in ancient Persia, it was planted on graves. It may be that such knowledge led John Keats to portray sweet basil in a mourning context in his poem "Isabella or, the pot of basil." When Isabella exhumed the head of her murdered lover, Lorenzo, she wrapped it up, placed it in a garden pot, covered it with mold, and planted above it sweet basil, which she watered with her tears (Garrod, 1939: 215-35).[93] An even more striking parallel is found between *tulasi* in India and sweet basil in modern Greece. In Indian myth and folk belief, *O. sanctum* grew on Tulasi's grave. In modern Greek belief, basil grew on Christ's grave. Because of association with Tulasi, *O. sanctum* was worshipped in India. Apparently because of its association with Christ, *O. basilicum* is venerated in the Eastern Church. Sweet basil sprigs are taken to church by Greek women to be blessed on St. Basil's Day. Returning home, they scatter basil on the floor to assure good luck for the following year. They and their families eat a small amount of sweet basil in the belief that this will keep them free of sickness. Basil placed in a cupboard, Greek women are convinced, will keep their clothing free of moths and rats for the year (Kincaid, 1908: 2, quoting P. Bent in "The Cyclades"). Also of note is the association of basil with fertility, love, and sexual stimulation in ancient Rome, and in modern Italy where a girl with a date might wear a bit of basil in her bosom or behind an ear (Maple, 1980: 44; D'Andrea, 1982: 32; Farrell, 1985: 37-38).[94] There are striking inconsistencies as well, with basil also considered a plant of the devil in ancient Greece, symbolic of hate and misfortune (D'Andrea, 1982: 32). Overall, however, the parallels are striking enough as to justify careful investigation, not just in the Mediterranean world but in Persia as well.

Pruthi has written (1976: 29) also of a widespread belief in India that growing sweet basil near homes and temples guarantees happiness. We have not found such a statement in other sources, and wonder whether Pruthi may have meant *tulasi* instead of sweet basil. If not, it could be that just as the name *tulasi* was sometimes applied in India to other species of *Ocimum*, that its attributes, to a degree, were bestowed on them. When writing of the early Hindu conquest of Java, I. H. Burkill (1966, 2: 1598) writes not that Hindus there considered *tulasi* sacred, but that they viewed the *Ocimum* species so. He observed further that early Hindus in Java used some common name for the various species of *Ocimum*, a name related to the Sanskrit *surasi* and *tulasi*, and that its Malay derivative, *sĕlaseh*, also a common name for *Ocimum* species, now extends as far as the Philippine Islands. As to survivals among modern Malays of Hindu practices regarding *Ocimum*, it has been suggested that their association of *sĕlaseh* with love may represent a survival of Hindu marriage practices. Also noted (T. F. Thiselton-Dyer, 1889: 158) among Malays is the planting of a type of basil on graves.

Though Burkill did not suggest that early Hindus actually introduced sweet basil or related plants to Malaysia, Laufer (1919: 588-90) wrote that *O. basilicum* and perhaps other species, including *O. sanctum*, had been introduced from India along with their Sanskrit names. He presented a strong linguistic case that *O. basilicum* was also introduced to China from India. This was based on the fact that at the time of the first description of the plant in Chinese literature (sixth century A.D.), its Chinese name, *lo-lê* (**la-lak,* **ra lak*), was almost identical with one of its Sanskrit names, *karalaka*.

Lo-lê, one Chinese name for sweet basil, was also mentioned in earlier accounts, from the fourth and fifth centuries A.D., and in one account its popular designation was "vegetable of the goddess Hsi Wang Mu," Queen Mother of the West (Laufer, 1919: 588; H. L. Li 1979: 72, 76-77). This raises the interesting question of whether, if sweet basil had indeed been introduced from India at about that time, the idea of *tulasi* sanctity was known to the Chinese and perhaps even came along, to be applied to sweet basil as well. Our sources, unfortunately, are silent on this question. H. L. Li (1969: 258), drawing on the *Ch'i-min yao-shu*, presents it as one of the valuable and commonly-used culinary herbs of northern China in the sixth century A.D. Schafer (1977: 112) includes it among the flavorings of T'ang times, and elsewhere (1963: 158, 162) he identifies the plant as a provider of perfume in those days. Our sources on traditional China are equally silent about any possible ritual role for sweet basil and related species. It provides perfume, and is sown in gardens to mask the bad odor of fertilizer. It is used in various medicinal ways (F. P. Smith, 1969: 289; Perry, 1980: 189) and it serves as a flavoring, though it is seldom mentioned in Chinese cookbooks. The leaves of sweet basil contain abundant calcium, and good or very good amounts of iron, riboflavin, and vitamins A and C (Leung *et al.*, 1973: 33).

GOLDEN NEEDLES

Another widely-used flavoring in traditional China, especially in northern dishes, were "lily" buds, usually dried. They are known as "golden needles" (*chin-chên*), "golden needle vegetable" (*chin-chên-ts'ai*), "yellow-flower vegetable" (*huang-hua-ts'ai*), and by similar names. Golden needles, most poetic of their names, derives from their yellow or yellowish-brown color, and the fact that they are thin, usually about three inches long. They have a pungent, pleasant odor, and a sweetish moldy taste that adds an unusual flavor to dishes. They are also valued for their texture, and may serve as an inexpensive condiment, vegetable, or medicine.[95] Before use in cooking, dried golden needles are soaked in water, usually for twenty minutes or more. They may be added to soups, particularly hot and sour ones such as "Peking Sour and Peppery Soup" (G. Z. Chu, 1975: 93-94; Tropp, 1982: 552; Y. K. Martin, 1984: 202). They are common in vegetarian dishes, such as "Buddhist's Vegetable Ensemble for the Gods" (K. Lo, 1979b: 39). They may be added to noodles, or dishes of fish or meat, such as the classic northern one, "Mu Shu Pork" or "Yellow Flower Pork" (Couling, 1917: 308; Stout, 1933: 97-99; 1934: 95-96; H. L. Li, 1959: 111; G. Z. Chu, 1975: 280; Sakamoto, 1977: 25, 114; K. Lo, 1979b: 39; C. B. T. Lee and A. E. Lee, 1980: 296; Tropp, 1982: 552; K. Lee and A. Branyon, 1984: 79, 260-61; Y. K. Martin, 1984: 17, 139, 229).[96]

Golden needles may derive from the tiger lily or other species of *Lilium* ("Lily flowers and bulbs used as food," 1889: 116-17; Davy, 1898; E. H. Wilson, 1913, 2: 63; Stout, 1933: 99; F. P. Smith, 1969: 204; Harrington, 1984: 174-75) whose bulbs were mentioned in our chapter on "Other Vegetables." They may also come from the day-lilies, which are not lilies in the narrowest sense but species of *Hemerocallis*. The original home of the day-lilies was in temperate Asia from the Caucasus eastward, but mainly in China, Japan, and Siberia (Stout, 1935: 206). Altogether, fifteen or so species of day-lilies have been identified, and a few have been cultivated in China since ancient times for their edible flowers (H. L. Li, 1959: 109-13; 1979: 66-67).[97] Such cultivation is found also among the Japanese and, to a degree, among the Vietnamese as well (Blasdale, 1899: 20; Crevost and Lemarié, 1917: 172, 174; Bois, 1927-37, 1: 493; 3: 245-46; Uphof, 1959: 183; I. H. Burkill, 1966, 1: 1155). Best known and most widely-cultivated in traditional China was *H. flava*, the Lemon Day-lily, but also cultivated were *H. fulva*, the Fulvous Day-lily, and *H. minor* (= *H. graminea*), the Dwarf Yellow Day-lily (Bretschneider, 1882-95, 2: 241-42; Blasdale, 1899: 44; H. L. Li, 1959: 112; 1979: 66-67; I. H. Burkill, 1966, 1: 1155).[98]

In traditional China, day-lilies were cultivated as food crops in various regions, but particularly in Szechwan (whose cuisine makes liberal use of golden needles) and the Yangtze Valley (Stout, 1933: 97; E. N. Anderson, 1988: 167). Couling (1917: 308) writes of flowers from *H. flava* as a regular export from the old port city of Chen-chiang at the junction of the Yangtze River and the Grand Canal in Central China. The lilies were commonly grown between plantings of potatoes and cabbage in poor, sandy soils in an east-west belt in the provinces of Kiangsu, Anhwei, and Honan. Flowers were picked in the spring, sun-dried, pressed, packed, and carried to market. There they were dried again and packed in large bales for further shipment. The lilies were reproduced not from seed but by replanting young bulbs.

Dried day-lily flowers, commonly in packages weighing a half pound or pound, are found in food stores everywhere in China, even in remote areas. One can also find them in herb stores in China. Though medicinal uses of day-lilies in China are of less importance than food or even ornamental ones (H. L. Li, 1959: 111), flowers (and shoots and roots) do serve a range of medicinal uses (Soubeiran and Dabry de Thiersant, 1874: 110; McClure and Hwang, 1934: 11; F. P. Smith, 1969: 204; Perry, 1980: 237). Of special note here are two ancient traditions associated with day-lilies. One is that if a pregnant woman wears flowers of *H. minor* in her girdle, she will give birth to a male rather than a female child (H. L. Li, 1959: 110; 1979: 66-67; F. P. Smith, 1969: 204). A second is that day-lilies, as drug or talisman, are able to dispel human sorrow, giving rise to the name "the plant of forgetfulness" (Bretschneider, 1882-95, 2: 241; H. L. Li, 1959: 110). The latter may be the origin of the modern belief that day-lily flowers, when consumed, deaden pain, a quality which leads them to be prescribed for women at childbirth. An extract of day-lily flowers is also believed to cleanse the blood, to remove its poisons and waste products, and to act against anemia. In fact, day-lily flowers in a dried form contain good to excellent amounts of iron, and, fresh, of vitamins A and C (Leung *et al.*, 1973: 44).

Day-lilies were grown as medicinals and ornamentals in eastern Europe at least as early as the first century A.D., and it has been speculated (G. P. Baker, 1937: 400-401) that they were carried along with silk in the trade that developed with China in the first few centuries before Christ. The historical evidence, however, has not been given the careful attention it deserves, and one can only say that the story of the early spread of day-lilies from China to the West is yet to be told.

GINSENG

Ginseng (*shên, jên-shên, shên-ts'ao*) is more a medicinal than a spice or flavoring, and is not mentioned in most Chinese cookbooks published in the United States. Nevertheless in China it may be added to food and drink to prevent illness, for it belongs among the strengthening foods, like bird's nest and ginger, which add to one's *ch'i*, vital essence. Ginseng is also "hot" and a cure for cold ailments; thus it is prescribed as a tonic to strengthen pregnant women and to help counter the "cold" state resulting from childbirth. One study in San Francisco found it also given to infants (Koo, 1973: 141; B. Chang, 1974: 438; Ling *et al.*, 1975: 128, 130; Gould-Martin, 1978: 43; Wheeler and Tan, 1983: 53, 55). Thus we include ginseng here not because it is fully at home among spices and flavorings but because it seems to fit best in a chapter that includes ginger. Wherever one put it, ginseng has to be considered somewhere in this book, for it is a highly-esteemed product in China, rich in its cultural associations[99] and important in Chinese views of health and well-being.

The most-prized ginseng of China, hereafter referred to simply as "ginseng," is the root of a perennial herb, *Panax ginseng*, native to North China, Manchuria, Korea, and southern parts of the Soviet East Asia. It is of the same family (*Araliaceae*) as sarsaparilla and ivy, and is collected wild or is cultivated for its curative, strengthening, stimulating, and rejuvenating powers. The views surrounding ginseng in China are somewhat similar to those in early Europe and the Near East about mandrake (*Mandragora officinarum*), that perennial herb of the Nightshade family (*Solanaceae*) native to the Mediterranean region and western Asia.[100] The roots of both plants were believed to derive their power from supernatural forces, thus falling into the category of magical drugs. Both roots were seen as resembling the human body in form, and as having the ability to act in some fashion (Figure 77). Collecting them entailed risks that were minimized by ritual and other special observances. And both were seen as rejuvenating agents that served a broad range of medicinal uses and were considered aphrodisiacs as well.[101]

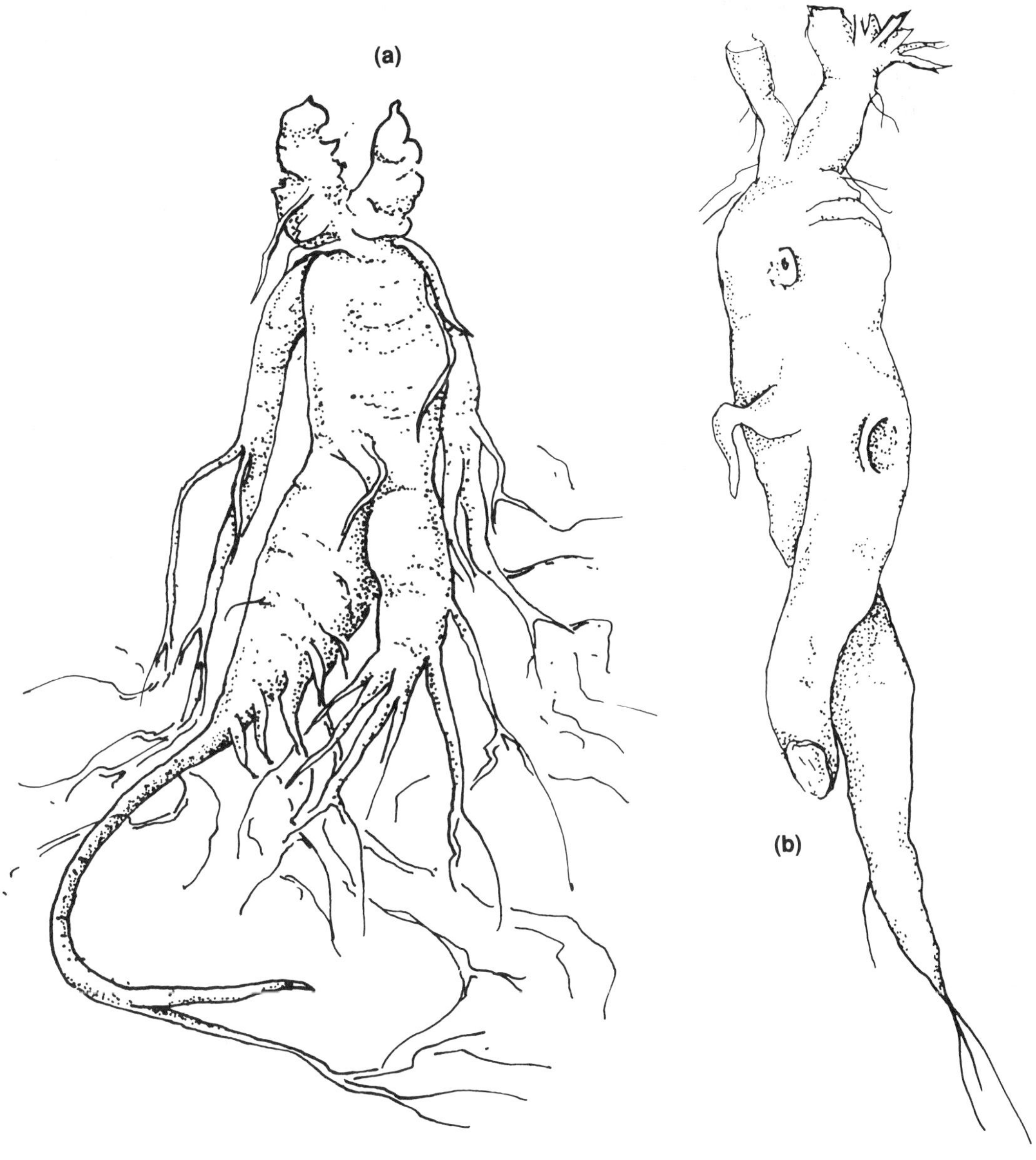

Figure 77. (a) Ginseng (*Panax ginseng*) and (b) Mandrake (*Mandragora officinarum*) (drawn by Shelby Hall)

Ginseng has been collected in China since great antiquity, but it has always been very expensive and used mainly by the wealthy. It was an early item of royal tribute, with the T'ang court receiving man-shaped ginseng roots from eastern Shansi, Hopeh, and Manchuria (Schafer and Wallacker, 1957-58: 219, Map 5). Ginseng was also a precious gift that might be bestowed on someone dear, as when the emperor dispensed it as a special favor to a valued high official who had suffered a major decline in health (Couling, 1917: 206; Schafer, 1963: 190). In at least one case, a special search was initiated at great cost to obtain ginseng of special effectiveness. This was in 1882, when an empress was nearly dead and a few special friends set out on a search for some of the most precious ginseng, which they found only after many great risks and horrors (O'Gorman, 1893: 262).

At various times in Chinese history, it is said, ginseng root has been worth as much, by weight, as silver, and roots of unusually-high quality or at times of shortage, as much or more than gold (Bretschneider, 1882-95, 3: 24; O'Gorman, 1893: 262; Ball, 1906: 316; Wallnöfer and von Rottauscher, 1965: 38; J. P. Hou, 1978: 44). It is thus understandable that heavy pressure developed on the supplies of wild ginseng. Because of overexploitation, as well as forest clearing associated with the spread of farming, the supply of wild ginseng in China greatly declined. Even as early as T'ang times most ginseng came from Korea (Bretschneider, 1882-95, 3: 20; Schafer, 1977: 112).

By Ch'ing times, Manchuria was the major remaining domestic source of supply. A monopoly over ginseng had been established immediately after the Ch'ing dynasty was founded, and the government, using both coercion and remuneration, sought to control all collecting, transporting, and marketing of ginseng (Symons, 1981: 3, 75-81).[102] The story of Ch'ing management of the ginseng monopoly has been examined in detail in a fine study by Symons (1981), and here we would note only that it was a many-sided and interesting affair, but one that would not have come about had ginseng not been such an expensive commodity and important source of imperial revenue.

Ch'ing rulers had compelling reasons which, if they alone had been the basis of action, might have saved Manchuria's wild ginseng resource from destruction. For one, a substantial portion of the revenues of the imperial household derived from the ginseng monopoly. For another, the Manchus, at least before their acculturation to Han ways, wanted to preserve their homeland, to avoid having it overrun by Chinese settlers from the south. At the same time, there were population pressures elsewhere in China and a need for settlers in Manchuria to counter the Russian danger from the north. As a result, Ch'ing policy vacillated, but the migration of Chinese settlers continued, and Manchuria's population increased along with the threat to its wild ginseng resource. Unable to stop ginseng destruction by agricultural expansion or illicit ginseng collection, the Ch'ing court then sought to obtain as much ginseng for itself before it was all gone (Symons, 1981: v-vi, 75-81).

Ginseng cultivation is referred to in Chinese poems of T'ang and Sung times (J. P. Hou, 1978: 44), but the first certain evidence of cultivation that we have is in the sixteenth-century *Pen-ts'ao*, in which Li Shi-chen writes of ginseng grown just like other vegetables in Korea (Bretschneider, 1882-95, 3: 22). Large-scale cultivation in Manchuria began only in the eighteenth century, and today its Fu-sung area alone accounts for 60 percent of China's ginseng production (J. P. Hou, 1978: 44).[103] The cultivated product, however, is considered much less potent than the wild, now quite rare, one.

Neither the ginseng plant nor its products seem to have found their way to India, the Near East, or Europe in ancient times. There has been speculation that a plant mentioned in *Ezekiel* (27. 17) may have been ginseng, but Moldenke and Moldenke (1952: 166-67) view this as highly unlikely. Nor have we found reference to ginseng in accounts of ancient India or of ancient trade among India, China, and the Western world. Ibn Khordadzbeh, ninth-century Arab geographer, did mention ginseng as exported by Moslems from "Sila," which was somewhere in East Asia, perhaps in Korea. Watt (1889-96), in his exhaustive work on Indian economic products, does mention ginseng (*Panax ginseng*) as used in China and Japan, but says nothing to suggest it was used in India. Earlier in the nineteenth century, Ainslie (1826, 1: 154-55) had reported that only on rare occasions was ginseng brought to India from China. As for Europe, it may have reached there only in the seventeenth century A.D. (Bae, 1978: 19). The Chinese and Koreans may have been unable to convince foreign merchants of ginseng's merits, or, if they did, the merchants may have been unable to develop a market for it in India, the Near East, and Europe. If so, one wonders whether ginseng failed to be accepted in those areas because of its high cost and because mandrake already filled the niche ginseng might have occupied.[104]

Also to be noted here are other species of *Panax* used in ways sometimes similar to *P. ginseng*. One is *P. pseudo-ginseng*, whose subspecies are found wild or cultivated from northeastern India and southern Tibet through northern Burma, Thailand, Indochina, and China to Japan.[105] A second is *P. quinquefolium*, American ginseng, which is native to, and cultivated in, the United States and Canada.[106] China, as the largest consumer of ginseng, has imported both American ginseng and some sort of ginseng from Japan,[107] but these are viewed as inferior to the *Panax ginseng* of Manchuria and Korea.

Ginseng gatherers in Manchuria, according to an account early in this century, considered their work both hard and risky, for it entailed seeking the root in mountain forests, with the collector almost on his knees. When a search was successful, the ginseng had to be dug up with great care, and with watchfulness for panthers and tigers, which also hunted ginseng and would attack a man who possessed it. Above all, gatherers feared the small red-eyed devil who protected the ginseng, and who sometimes assumed its form. Such a devil might blind a hunter, or suck his blood (Ossendowski, 1924: 118-20). Because of the honor in which ginseng was held, and perhaps in fear of the risks involved, certain men before starting their collecting activities would, for a month preceding, follow a vegetarian diet and abstain from sexual relations and strong alcohol. On the day he left home, a gatherer, in a further effort at assuring success and safety, made himself clean by bathing in cold water, and also made offerings to the gods. When he found a ginseng plant, he feared that it might vanish if he were unworthy. He did not permit it out of his sight, but prostrated himself before it while saying aloud "Don't flee. I am a pure man, my soul is free from having been soiled, my heart is open and free of any bad intentions" (Symons, 1981: 7). The methods of ginseng cultivation and harvesting in modern times are much more prosaic.

In modern China, ginseng may be taken as a medicine in various ways. It may also be added to various dishes out of conviction that this will help prevent illness. There is, for example, a soup made of ginseng and squab, both of which are strengthening (J. D. Keys, 1963: 88; Em. Hahn *et al.*, 1973: 70). And for seventeenth century Thailand (whether among Thais or Chinese is not clear) one reads of a soup of ginseng, bird's nest, and chicken (Tachard, 1981: 271). The latter soup contained three of China's best-known strengthening foods, two of them very costly indeed. The Chinese also use ginseng brandy or ginseng wine made by soaking the root in strong vodka (E. N. Anderson and M. L. Anderson, 1977: 342; V. Y. N. Hsu and F. L. K. Hsu, 1977: 313). Though manufactured in North China or Korea, it is also widely used in modern South China.

Mandrake, though its use continued in the Near East and India, gradually fell from favor as a medicinal in Europe from the sixteenth century onward (C. J. S. Thompson, 1934: 237). This was largely because of the rise of scientific medicine, whose adherents failed to confirm claims made for mandrake. Ginseng has continued to be important in Chinese medicine. The meaning of *jên-shên*, common Chinese name for the root, is given in English as "man root" or "man-shaped root," and other early names for ginseng meant "divine herb" or "terrestrial essence." Traditional Chinese belief holds it to be the earth's power living in a root, a man-shaped form of the spirit of the earth (S. Y. Hu, 1976: 11). Ginseng, according to S. Y. Hu (1976: 13), is not, as some Westerners say, a panacea in Chinese medicine, but a drug given to those who have signs of weakness that need to be cured by strengthening medicines. As such it may be employed both for specific ailments and to sustain the weak, treat the ill, revitalize the aged, and restore the dying.[108]

Practitioners of traditional medicine in China and neighboring countries remain convinced of its value, and there are many testimonials to this, including that of the widely-travelled Lin Yutang (1935: 335-36) who described it as being the most constant and strongest tonic known to humanity. With the emphasis on traditional medicine following establishment of the People's Republic, there has been increasing research there on its effectiveness. Scientists in other countries, especially Japan, Korea, and the Soviet Union, have been active as well.[109] In the 1960's, Soviet pharmacologists (N. Brekhman, 1965: 148; I. I. Brekhman and I. V. Dardymov, 1969: 46) cited experiments with humans that support the traditional Chinese view. Ginseng was found to be a strengthening agent or tonic, increasing the efficiency of both the physical and mental processes, improving one's ability to concentrate, and preventing overfatigue. Animal experiments, in turn, found ginseng to have a remedial effect on such illnesses as radiation sickness, experimental neurosis, cancer, and alloxan diabetes, and to bring on a striking increase in resistance to a wide range of biological, chemical, and physical factors. The medicinal value of ginseng, nevertheless, remains in dispute, with many persons trained in scientific medicine doubting its effectiveness, and with American researchers in repeated efforts unable to identify for ginseng any major medicinal virtues (S. Y. Hu, 1976: 23; J. D. Keys, 1976: 86). There is also the risk, for some persons, that consumption of large amounts of ginseng over a long period of time will bring on a ginseng-abuse syndrome with the common symptoms of diarrhea, skin eruptions, sleeplessness, nervousness, hypertension, and edema (Siegel, 1979). There has nevertheless been a continuing interest in ginseng among certain Western researchers,[110] and it has gained acceptance among some enthusiasts of natural foods and drugs. The United States is estimated to have five to six million users of ginseng (Siegel, 1979: 1614), and it is readily obtained in a broad range of forms in herb and health food stores, drug stores, and markets in Western countries. One can sometimes even find soft drinks containing ginseng in ordinary supermarkets, our favorite being a ginger and ginseng soda.

[1] Various species of licorice (*Glycyrrhiza*), mostly from northern China, Mongolia, and Manchuria, are used as medicinals. Licorice is second in importance as a drug only to ginseng, and is believed to have many useful properties, among them tonic and rejuvenating ones. Schafer mentions that while licorice might be added to dishes in T'ang China, likely it was more significant in medicine than in cooking (Schafer, 1977: 92, 112). For other notices of licorice in China, its occurrence and medicinal uses, see Bretschneider, 1882-95, 3: 13-15; Hooper, 1929: 71; F. P. Smith, 1969: 196; C. P. Li, 1974: 67; J. D. Keys, 1976: 120-21, 314; Perry, 1980: 217-18, 392; and E. N. Anderson, 1988: 196, 198.

[2] Fenugreek (*Trigonella foenum-graecum*), a Near Eastern domesticate (its seeds have been identified in Egyptian and Southwest Asian sites as early as 3000 or 4000 B.C., according to Renfrew, 1973: 188, and Zohary and Hopf, 1988: 111), was introduced to China in Sung times. Its Chinese name (*hu-lu-pa*), like its Malay one, derives from the Arabic ***hulba*** or ***helba***. The plant, mentioned in Sung times as grown in Kwangtung and Kweichow, is now cultivated more widely in China. Fenugreek seeds are often mentioned as medicinal in the literature on China, but seldom identified as a flavoring, though they are an ingredient in some of the curry powders that have come into use in recent decades. The

dietary role of fenugreek in other parts of the Old World is better documented. In India, fenugreek seeds are common in curry; and in Ethiopia, in the tasty hot sauce that is added to stew (*wot*), a dietary mainstay (Bretschneider, 1882-95, 1: 65; Watt, 1885-96, 6, Part 4: 86-88; Laufer, 1919: 446-47; I. H. Burkill, 1966, 2: 2221-22; F. P. Smith, 1969: 442-43; Darby *et al.*, 1977, 2: 801-2; Perry, 1980: 228; Zeven and de Wet, 1982: 97; E. N. Anderson, 1988: 137). In ancient Rome, fenugreek seed was cooked and eaten as a food; and in modern Egypt and India, fenugreek greens are prepared as salad or cooked vegetable. In Egypt, its germinated seed serves as flavoring in bread, and overall fenugreek is so well regarded medicinally as to lead to the hawker's cry "Healing from God, Helba!" (Watt, 1889-96, 6, Part 4: 87, 88; Moldenke and Moldenke, 1952: 34; Renfrew, 1973: 188; Darby *et al.*, 1977, 2: 801-2). The Egyptians of Rosetta toast fenugreek seed and make of it a sort of coffee, to which lemon juice is added (Ainslie, 1826, 1: 131); in India, fenugreek seed serves in making a beverage, an alternative to true coffee, or it is used as an adulterant of coffee (Watt, 1889-96, 6, Part 4: 88); and in northwest Ethiopia the Amhara make a beverage of ground fenugreek seed, honey, and water (Simoons, 1960: 156-57). Our sources are silent, however, on whether or not fenugreek serves such uses in China.

[3] Mint (most often mentioned is *Mentha arvensis, po-ho*) is found in China, both wild and cultivated in gardens, and has long been used in cooking, commonly to flavor vegetables, and in making a sort of tea. Mint leaves also serve medicinal purposes, and are exported from Canton to Malaya where they may be found in Chinese pharmacies (Laufer, 1919: 198; E. Watson, 1923: 121-22; Hooper, 1929: 89-90; I. H. Burkill, 1966, 2: 1477-79; F. P. Smith, 1969: 263; Zeven and de Wet, 1982: 39). Peppermint oil was also distilled for export and domestic uses, and as a source of menthol, with Canton described as the leading center of menthol trade in China (E. Watson, 1923: 121-22; Hooper, 1929: 89-90). Because dietary use of mint as a flavoring is on quite a small scale and seldom mentioned in works on Chinese food, however, we will not consider it further.

[4] For critical views on the use of MSG in Chinese cooking, see Em. Hahn *et al.*, 1973: 77-78, and E. N. Anderson, 1988: 173, 205-6.

[5] For a recent review of ginger, its forms, composition, cultivation, and processing, see Leverington, 1983.

[6] This was a two-way affair, with China importing as well as exporting ginger products. In Sung times, for example, ginger peel was imported from Cambodia, presumably for medicinal use (Wheatley, 1959: 106-7).

[7] Ginger, in addition, was, for prospectors, an indicator plant believed to grow where copper or tin occurs (Schafer, 1963: 251).

[8] A recipe for pickled spring ginger will be found in K. Lee and A. Branyon, 1984: 303.

[9] Its medicinal uses by the Chinese are identified by Soubeiran and Dabry de Thiersant (1874: 125-26); Hooper (1929: 81-82); F. P. Smith (1969: 31-33); J. D. Keys (1976: 73); and Perry (1980: 436-37).

[10] In the fifth and sixth centuries A.D., according to H. L. Li (1969: 258), another plant, called "water garlic" (probably *Allium nipponicum*), was used in both wild and cultivated forms. Today it is no longer cultivated.

[11] For details of the process of making vinegar in ancient and modern China, see C. P. Li, 1948: 208-11.

[12] For recipes to seafood sauces and dips, see F. L. Chang and G. M. Fuchs, n.d.: 362-63, and K. Lo, 1979c: 40-41, 57-58.

[13] This vinegar is not identified further, but we wonder whether it may have been the thick, dark brown Chinkiang vinegar mentioned above.

[14] See Hirth (1887) for the early history of the salt monopoly in China; Spencer (1935: 356) for a map of China's salt production areas and consumption districts in the 1930's; C. P. Li (1948: 55-65) for processes of making salt from the sea, lakes, and wells; and S. Y. Lin (1940-41) for methods of salt manufacture in Hong Kong.

[15] See Schafer, 1977: 115, for T'ang China; and E. N. Anderson and M. L. Anderson, 1977: 340, for modern South China.

[16] For photographs of Red-in-snow, details of its salting, and recipes in which it is used, see Y. K. Martin, 1984: 25, 140-41, 213, 234. For further information on Red-in-snow and its use, see W. W. Chang *et al.*, 1970: 92, 93, 159, 398; B. Y. Chao, 1972: 154, 154-55, 160, 198; and Herklots, 1972: 220-22.

[17] When finally cooked, the duck in this and similar dishes is normally allowed to cool before eating. For full details on preparation of duck in this way, see F. L. Chang and G. M. Fuchs, n.d.: 296; W. W. Chang *et al.*, 1970: 273; B. Y. Chao, 1972: 104-5; E. Chen, 1977: 97; and C. B. T. Lee and A. E. Lee, 1980: 45-46. For fresh-salted meat and fish recipes, see B. Y. Chao, 1972: 71-72, 115.

[18] Recipes for fresh-salted chicken dishes may be found in W. W. Chang *et al.*, 1970: 219-20, and C. B. T. Lee and A. E. Lee, 1980: 94.

[19] Other recipes of this sort may be found in W. W. Chang *et al.*, 1970: 220; K. Lo, 1976: 76; 1979c: 200-201; and E. Chen, 1977: 17. Good illustrations of the procedures involved, along with a recipe, are presented in Hom and Steiman, 1981: 115-17.

[20] *Apis florea*, a small, primitive honeybee found wild in much of southern Asia as far west as southeastern Iran and Arabia, is also reported to be domesticated, but only in Oman (Crane, 1984: 405).

[21] It is uncertain what the exact date of earliest *A. mellifera* introduction to China may have been. Oschmann wrote that the first European honeybees were introduced to China in 1925 (Crane, 1975: 414). On the other hand, Y. M. Evsyukov said that Italian and Russian races of bees had been regularly introduced to China since 1913 (Crane, 1960: 5). And C. G. Riedel wrote in the early 1930's (1932: 278) that over the preceding twenty years various missionaries had introduced hives of European bees. He also wrote that about 1928 many Chinese, expecting to make a great deal of money, began to keep European honey bees (of the Italian variety). He described the episode as a craze, with over 200,000 colonies of bees imported, mainly from Japan, via Tientsin alone. The keeping of European bees received a further boost following establishment of the People's Republic, and by 1960 China had roughly a million colonies of *A. cerana* and more than 350,000 of *A. mellifera*, 84 percent of the latter of Italian race and 16 percent, north Russian (Crane, 1960: 5). The decline of *A. cerana* continued, and by the mid-1970's, Crane (1975: 414) wrote that while *A. cerana* was still kept in some parts of China, elsewhere it had been replaced by *A. mellifera.*

[22] The other Asian honeybee species, *A. dorsata* and *A. florea*, build their nests in the open and lack the protection from winter cold afforded to *A. cerana* by its nesting in cavities where bees remain virtually inert in clusters during the winter. Certain races of *A. cerana*, as a result, were able to establish themselves in eastern Asia as far north as Hokkaido, the Soviet East Asia, and Manchuria, whereas *A. dorsata* and *A. florea* were restricted to tropical Asia ("Apis cerana at risk," 1971: 141; Crane, 1975: 412, 420).

[23] One notes (Crane, 1983: 15; 1984: 404), in this regard, that Hainan and the adjacent mainland are far enough south to be within the range of certain tropical species: the tropical Asian honeybee, *A. dorsata*, and bees of the stingless group, *Meliponini.*

[24] For a discussion of the earliest representations of beekeeping in the Near East and Mediterranean, see Crane, 1983: 35-39. This work, *The archaeology of beekeeping*, is a study of the relationship between bees and humans through history, as revealed by material objects that have been uncovered. Unfortunately it contains only a few brief notices of China. Earlier, however, Crane (1960: 4) had written that beekeeping was known in China for over three millennia, but provides no specific documentation for this claim.

[25] Kellogg (1937: 141), writing of traditional Fukien, notes that superstitions about bees are common. It was considered good fortune when a swarm of bees alighted on a house, but bad fortune when the swarm abandoned it. Thus when the owner of a bee colony dies, his family talks to the bees so that they will not leave. Some people, moreover, were reluctant to keep bees out of such concerns; one farmer, asked by Kellogg why he did not keep bees, answered that he was afraid that they might leave and bring ill fortune to his household.

[26] For general descriptions of beekeeping in traditional China, see Gray, 1878, 2: 199-201; "Apiculture in China," 1929; and C. G. Riedel, 1932. For beekeeping in Fukien Province, see Kellogg, 1937.

[27] It is difficult to estimate per capita honey consumption in China today. In 1984, China was estimated to have 5.7 million bee colonies. China's bee colonies were more numerous than those in the United States (4.3 million in 1984), though exceeded by those in the Soviet Union (8.2 million). Annual honey production per colony in China (1984) averaged 20.2 kg., somewhat below that for the United States (21.8) and the Soviet Union (24.1), but above that for certain other countries ("World honey crop reports," 1985). By the 1980's, China had also become the largest honey exporter and second largest honey producer (exceeded only by the Soviet Union) in the world ("World honey crop reports," 1984, 1985). A large portion of China's exports have gone to Japan, for which it is the leading foreign supplier. And in the first ten months of 1986, China provided 29 million pounds of the honey imported by the United States, more than any other country, including the traditional leading suppliers, Mexico, Argentina, and Canada (*Honey Market News*, Vol. 57, No. 12 [January 13, 1987], p. 8).

Per capita consumption of honey in China, however, remains quite low, for China's population is huge and its supply, after exports, is greatly reduced. We estimate the 1983 per capita Chinese honey consumption at only 36 grams (honey production estimate, 100,000 metric tons, less exports, 64,000 metric tons, left a supply for domestic consumption of 36,000 metric tons for a human population of one billion). Our estimate for China is well below that provided for recent years (1985, 1986) by the *Honey Market News*: 100 grams per capita annually. Our calculation, however, would have been exactly 100 grams had we ignored the fact that nearly two-thirds of China's production is exported. With either estimate, China's annual per capita honey consumption would be well below that of the United States and U.S.S.R. (500 to 600 gm. in 1985 and 1986) and still further below that of West Germany (1200 to 1300 gm.) (*Honey Market News*, Vol. 57, No. 12 [January 13, 1987], p. 5).

[28] Gould-Martin (1978: 40) observed that honey, like brown sugar and the juice of sugar cane, was especially effective in cleansing the body of poisons. For other medicinal uses of honey and wax in China, see Read, 1977d: 10-17.

[29] For a map of the distribution of *Alhagi* species and areas of collection of alhagi manna, see Donkin, 1980: Fig. 3 and Map 2. For uses of *Alhagi*, see Donkin, 1980: 17-26.

[30] A sweet white beet seems to have been introduced to China from the west in T'ang times, but it is uncertain whether it was used in any way other than as a vegetable (Laufer, 1919: 399-400; Schafer, 1977: 108).

[31] Sweet-stemmed sorghum has long been cultivated in India for making sugar-candy, for which the cities of Ajmer and Bikaner, in India's arid west, were renowned (Watt, 1908: 1041). As for China, however, few of our modern sources mention sugar being made from sorghum, and some say it was not so used. De Candolle (1886: 382), for example, wrote that the Chinese make a distilled liquor from sorghum but not sugar. V. D. Collins (1865: 91), who was sent to China in the last century to study methods of cultivating sorghum and processing its juice into sugar, reported its canes chewed for their sweetness but found nothing to suggest that sugar was made from them. F. P. Smith (1969: 386), in turn, lists sweet-stemmed sorghum with various forms of sugar cane, but does not specify whether it is made into sugar. E. Watson (1923: 547), who does give a Chinese name for sorghum sugar (along with names for various other sugars, such as maple sugar, maize sugar, and milk sugar), and who says that it is made in a similar way to ordinary cane sugar, fails to make it clear whether he is referring to a product made in China or an import. That syrup and sugar are obtained from sweet sorghum in China today is confirmed by Yu Youtai (Wittwer *et al.*, 1987: 206), who describes these products, which are particularly important in Kiangsu and Chekiang, as of major importance in the food industry. He also reports development of a hybrid between sugar cane and sorghum which yields more sugar than either parent.

[32] At least in some places, honey long continued to rival sugar as a sweetener, as in seventeenth-century northwestern China, where Sung Ying-hsing (1966: 130) reported honey to equal cane sugar in importance.

[33] For details of the manufacture of cane sugar in early China, see C. P. Li, 1948: 155-58.

[34] Detailed information on how to make animal-form hollow candy through the use of molds was given by Sung Ying-hsing in the seventeenth century (Y. H. Sung, 1966: 129).

[35] For a map of areas of sugar-cane and sugar-beet production in China, see Wittwer *et al.*, 1987: 221.

[36] In Chekiang shops where sugar cane was sold, water was commonly poured on the cane to keep it fresh, which, since the water was not boiled, presented a danger to the person consuming it (Holmes, 1948: 36, citing Mrs. T. D. Holmes).

[37] For brief accounts of sugar making in traditional China, see E. Watson, 1923: 542, and Hommel, 1937: 111-14. For sugar making in seventeenth-century China, see Y. H. Sung, 1966: 126-28.

[38] Not to be overlooked among Chinese sweets is "Sweetened Water," made by adding sugar to boiling water, and permitting it to dissolve. Such sweetened water may be consumed after cooling, sometimes with additional flavoring; or it may, as by a Ch'ing empress early in this century, be consumed hot (Der Ling, 1914: 92-93; G. B. Miller, 1966: 755-56).

[39] Medicinal use of certain *Zanthoxylum* species is also reported in the New World and in tropical Africa (Uphof, 1959: 388-89).

[40] The ways of doing this are described in F. Lin, 1975: 208, and Hom and Steiman, 1981: 69-70.

[41] A recipe for five-spice powder is given in K. Lee and A. Branyon, 1984: 338.

[42] Also of note is West African black pepper, Ashanti pepper, or Benin pepper, the dried fruit of *Piper guineense*. In West Africa this pepper is usually collected from the wild, though sometimes it is cultivated or permitted to grow in forest clearings and even trained to grow on trees remaining uncut. Though it serves similar uses, West African pepper has never been regarded by Europeans as a substitute for the black pepper of Asia (*Piper nigrum*) (Flückiger and Hanbury, 1879: 651-54; Dalziel, 1937: 16-17, 471-72). *Piper clusii*, also of West Africa, is little different from *P. guineense*, and Bois (1927-37, 3: 24) regards it as a variety of the latter.

[43] Though pepper may have been in Greece long before, Hippocrates, Aristotle, and Antiphanes (all of the fourth or fifth centuries B.C.) are among the earliest classical writers to have written of pepper (J. I. Miller, 1969: 82).

[44] Certain authorities (Hirth and Rockhill, 1966: 223n; H. L. Li, 1979: 49) have interpreted early Chinese sources as saying that pepper was imported to China from India by way of Persia and the Silk Road. Laufer (1919: 374n), however, reads those sources as merely identifying pepper as a product of Persia, not that it was exported from there to China.

[45] Sometimes confused with Indian long pepper is Javanese long pepper or Dutch long pepper (*Piper retrofractum*), a Southeast Asian species more pungent even than Indian long pepper. This pepper has also been exported from Java to China for a long time (I. H. Burkill, 1966, 2: 1781-82).

[46] Betel pepper is best known for the use of its fresh leaves, along with a slice from the nut (*pin-lang*) of the areca palm (*Areca catechu*) and sometimes lime and other ingredients, in betel chewing. Betel chewing is a widespread practice in South and Southeast Asia, and is also rather common in the southern provinces of China, including Kwangtung. In North China, on the other hand, it is nearly unknown, perhaps because fresh betel-pepper leaves have been difficult to obtain (Soubeiran and Dabry de Thiersant, 1874: 138; I. H. Burkill, 1966, 1: 224). Those leaves contain good amounts of calcium and seem to be unusually high in vitamin A (Leung *et al.*, 1973: 35).

[47] In former times botanists often identified just one or two species of domesticated capsicum peppers, usually *Capsicum annuum* and *Capsicum frutescens* (de Candolle, 1886: 288-90; I. H. Burkill, 1966, 1: 450). Today, however, two or three additional species are commonly recognized. One species is *C. baccatum*, cultivated mainly in sections of South America. Another is *C. chinense*, more widely cultivated in South America, as well as in Mexico and the Caribbean. It is, however, closely related to *C. frutescens* and likely should be combined with it. A third species is *C. pubescens*, a highland form cultivated in the Andes and certain parts of upland Mexico and Central America (Heiser, 1976: 265-67; Zeven and de Wet, 1982: 179-180). One or another of these species is also reported in certain other parts of the world, including China (F. P. Smith, 1969: 92).

[48] Recipes for hot chili oil may be found in F. L. Chang and G. M. Fuchs, n.d.: 371; Sakamoto, 1977: 27; Hom and Steiman, 1981: 30; K. Lee and A. Branyon, 1984: 337; and Y. K. Martin, 1984: 225-26. For recipes to hot pepper oil, see F. Lin, 1975: 208-9; and for chili in sauces or pastes, F. L. Chang and G. M. Fuchs, n.d.: 367, 368; Hom and Steiman, 1981: 30; K. Lee and A. Branyon, 1984: 336; and Y. K. Martin, 1984: 226.

[49] For a description of the distillation of star anise oil by the Chinese, see Dymock *et al.*, 1889-93, 1: 40.

[50] The history of star anise production and cultivation in China remains to be written, but it seems that areas of production have changed over time. In Sung times, a tribute of star anise was levied on one area of Fukien (Flückiger and Hanbury, 1879: 21), and in the middle of the nineteenth century Rondot (1848: 11-12) reported that the best star anise arrived in Canton by junk from Fukien, and only smaller amounts from Kiangsi, Yunnan, and some places in Kwangtung. At least one other observer mentioned Yunnan as an area of production in the last century (Flückiger and Hanbury, 1879: 21). Yet writers early in this century (Couling, 1917: 28; Hooper, 1929: 76; Raibaudi, 1931: 95) write only of Kwangsi, or Kwangsi and Kwangtung, as areas of star-anise production or cultivation.

[51] Sometimes in Indian markets one also finds the dark brown fruit of an Indian species, *Illicium griffithii*. Though slightly aromatic in taste initially, it is bitter and astringent afterwards (Dymock *et al.*, 1889-93, 1: 40-41).

[52] For other recipes using star anise see Em. Hahn *et al.* (1973: 46-47), Field *et al.* (1976: 79), and K. Lo (1979a: 106) for "Braised Star Anise Beef"; and Hom and Steiman (1981: 217-19) for "Chinese Beef Stew." The Chinese have observed, too, that one or two carpels of star anise added to chicken before roasting greatly improves its flavor.

[53] Though certain of these accounts (Soubeiran and Dabry de Thiersant, Smith) refer to star anise as *Illicium anisatum*, they wrote at a time before it was discovered that the spice derived from a different species, *I. verum*.

[54] For a map of their distribution, see Mathon, 1981: 57.

[55] For details of the distillation of cassia oil, see C. P. Li, 1948: 148-51.

[56] The Greek author of the *Periplus* (first century A.D.) did report caravans reaching India from China via the Silk Road. He mentioned no cinnamon products among their goods, however, only silk in various forms (Schoff, 1912: 48).

[57] Chinese accounts do mention voyages, with the help of the Nan-Yüeh, to the South Seas starting in the second century B.C. Some have placed Huang-chih, most distant of the places visited, along the Bay of Bengal, but this is far from certain (Wheatley, 1959: 18-19). In any case, there is nothing to suggest that cinnamon was carried as a trade item in those times.

[58] The linguistic case is strong, with the Sanskrit *tamala-pattra* (leaves of the tamala, *C. tamala*) becoming the Greek *tamalabathra* and *malabathron* (Casson, 1984: 238, 245) and the Roman *malabathrum.*

[59] One reason Casson (1984: 237-38) gives for eliminating India and Ceylon as possible cinnamon bark suppliers to the western trade is that they failed to utilize their cinnamon trees before the thirteenth century A.D. or afterwards. Curiously, he seems not to have considered early Indian sources. Such an effort would require a careful search of the Sanskrit literature, a task beyond the scope of this book. We did, however, make a brief effort in that direction, and it clearly shows that cinnamon bark and leaves were both used in ancient India. Kautilya's *Arthasastra* (?290 B.C.-A.D. 300?) mentions powdered cinnamon bark (*choca* or *coca*), from *C. zeylanicum* if Prakash is correct, used to flavor a liquor (*Arthasastra* II. 25. 121; Shamasastry, 1951: 133; Prakash, 1961: 279, 282). Bark was also mentioned in the second century A.D. medical treatise by Charaka (*Charaka samhita*), a court physician in Kushan, a state in northwest India. This was well beyond the occurrence of *C. zeylanicum.* Yet, if translators are correct in their identification of species, both bark and leaves of *C. tamala* (*patra*) were in medicinal use, and the bark of *C. zeylanicum* (*tvac*) as well (P. Ray and H. N. Gupta, 1965: 69, 75). In a similar vein, one considers the *Ashtanga-hridaya-samhita*, a medical treatise by Vagbhata II (?A.D. 700-870?). The bark and root of *C. zeylanicum* (*tvac*, *coca*, or *varanga*) were medicinals, as were the leaves of *C. tamala* (*patra* or *tamala*) (Hilgenberg and Kirfel, 1941: 42, 115, 120, 317, 319, and *passim*; Godbole *et al.*, 1966: 55-56). Also frequently mentioned was a medicinal compound of cinnamon bark (*tvac*), *C. tamala* leaves, and cardamom, known, as it is today (Dutt, 1922: 225), by the name *trijataka.*

[60] Curiously, Ceylon is not mentioned as an ancient cinnamon marketer. The earliest references we have uncovered to what seems to be Ceylon as a supplier of cinnamon to commerce are in the writings of al-Kazwini (d. A.D. 1283), who drew liberally on earlier accounts including that of Abu Dulaf (tenth century A.D.) (Ethé, 1868: 229; Schumann, 1883: 47; Yule, 1921, 2: 315), and in a letter from Malabar by Dominican Friar Menentillus of Spoleto in A.D. 1292 or 1293. A large quantity of cinnamon, according to Menentillus, was shipped from Ceylon (Yule, 1866, 1: 213). This fits with Marco Polo's observation at about the same time that the Kingdom of Malabar itself had a considerable amount of cinnamon as well as ginger and pepper (Yule, 1921, 2: 389). Further evidence of cinnamon products in Ceylon is found in writings of Ibn Batuta (mid-fourteenth century) (Gibb, 1969: 254-55), Hieronimo di Santo Stefano (R. H. Major, 1857a: 5), and Nicolo Conti (fifteenth century) (R. H. Major, 1857b: 7-8). According to the latter, there were many cinnamon trees in Ceylon, and both bark and fruit were used. One can only guess how much earlier Ceylon cinnamon was exploited.

[61] In this regard, one recalls that the muscovy duck came not from Moscow, but from South America; that American saffron (safflower) was domesticated in the Near East, not the Americas; and that the Italian *granturco* (maize), did not originate in Turkey, but in Meso-America.

[62] Schoff has suggested (1912: 84) that from the third to the sixth centuries A.D., Chinese ships were active in sea trade from China to Persia, and that they carried cinnamon in that trade. Though his assertion is accepted by certain other students of ancient trade in the Indian Ocean (Warmington, 1974: 190; Casson, 1984: 260), Laufer (1919: 541n) has questioned it, noting that no reference was cited. Others (Yule, 1866, 1: lxxvii-lxxxii; Heyd, 1879, 1: 33-34) did cite Arab and Persian sources that report Chinese ships trading in the Persian gulf in the fifth century A.D., and Chinese sources of the seventh and eighth centuries to that effect. Schafer (1963: 12-13), as a specialist on T'ang China, is unconvinced, noting that the "Persian Argosies" in Chinese accounts often referred merely to ships trading with the Persian Gulf, commonly with Malay or Indian crews. He also notes that when Arab ninth and tenth century accounts refer to "Chinese" ships in Persian Gulf ports, they actually mean vessels in the China trade, in the same way that we refer to "East Indiamen" and "China clippers."

[63] A careful weighing of Chinese documentary evidence by a sinologist is needed to cast further light on China as an early source of cinnamon traded to the west. Our brief search of historical sources on China (among them Bretschneider, 1882-95; 1898; Yule, 1921; Boxer, 1953; Wheatley, 1959; Schafer, 1963; Hirth and Rockhill, 1966) has uncovered nothing about early exports of cinnamon from China.

[64] For information on the cultivation and processing of turmeric in India, see Ilyas, 1978: 259-63.

[65] In traditional China, yellow was also the royal color, worn by the emperor and his lineal descendants, prominent in imperial weddings and funerals, and widely used in temples and other buildings that were the property of the imperial family or sustained by the imperial government (Doolittle, 1865, 2: 452, 455, 456, 457, 459; Giles, 1900: 55; Cormack, 1935: 67-70 *passim*, 91; C. A. S. Williams, 1941: 76-77).

[66] An initial problem is the route by which the saffron plant reached India. The word *kunkuma*, one of the Sanskrit names for saffron, is remarkably like its name in Assyrian (*karkuma*), Arabic (*kurkum*), Hebrew (*karkom*), and Persian (*kurkum* or *karkam*). Laufer (1919: 320-21) thinks it likely that the saffron plant was introduced to Kashmir from Persia, and, in support, cites another Sanskrit name for saffron, *vahlika*, meaning "originating from the Pahlava."

Also at issue is the time the saffron plant was first introduced to India. Saffron is not mentioned by Macdonell and Keith in their exhaustive *Vedic index of names and subjects* (1912), and may not have been known to the earliest Sanskrit writers (prior to about 600 B.C.). It has been suggested that cultivation of saffron at Padmapura (modern Pampur), in Kashmir, first came about when Vagbhata II, author of an eighth century A.D. Indian medical treatise, received saffron bulbs from a patient and planted them (Mahindru, 1982: 51). That this was the earliest saffron cultivation in Kashmir and in India, however, does not fit with the Buddhist legend that Madhyantika, earliest Buddhist apostle in Kashmir (third century B.C.), planted saffron there (Watters, 1904-5, 1: 262-63; Laufer, 1919: 320-21); or with references to the use of saffron in the *Ramayana* (*c.* 350 B.C.-A.D. 250), and its cultivation in Uttar Pradesh (*Ramayana* 2. 94. 9; 5. 7. 9; 7. 91. 19). Nor does it fit with mention of saffron (*kunkuma* or *kumkuma*) by Charaka (?A.D. 80-180?) in his medical treatise (*Charaka Samhita* 23. 117, etc., or P. Ray and H. N. Gupta, 1965: 64); by Susruta (*c.* A.D. 350) in his treatise (*Susruta Samhita* 6. 4a; 7. 6; 8. 27-36; 38. 22; 46. 100, or Bhishagratna, 1963, 1: 346, 521; 2: 732, 740, 748); or with the seventh century A.D. account by Chinese Buddhist pilgrim Hsüan-tsang of what seems to be saffron cultivation in Kashmir (Watters, 1904-5, 1: 261, 262-63). Also to be weighed is the statement (Mahindru, 1982: 20), without specific citation, that saffron was used at the time of the *Taittiriya Samhita*, a handbook for priests dating to pre-Christian times (700-300 B.C.?); it was poured into the sacrificial fire at a wedding, and added as an aphrodisiac to sweets eaten by the newly-weds.

We have not determined the pattern of trade in Indian saffron over time, but note that in A.D. 1632 Peter Mundy found saffron from both Kashmir and Kishtwar, then separated politically, far down the Ganges Valley in Patna (Temple, 1907-36, 2: 154).

[67] The Vikings, in their turn, obtained cardamom in Constantinople in early times. Today it is still very popular in Scandinavia, as well as in Germany and Russia, with Sweden usually the world's leading importer of cardamom. In that area of Europe cardamom is used for flavoring various liqueurs as well as cake, pastry (including Danish pastry), bread, and other baked goods. It is also added to certain meat dishes and to prepared meats. The special taste of Swedish meatballs derives in part from the cardamom used in their preparation. Cardamom is also sometimes used in preparing German *Sauerbraten*, and as one of the spices used with pickles and pickled herring (Stobart, 1970: 61; Loewenfeld and Back, 1978: 86-87; Lewis, 1984: 55; Farrell, 1985: 60-63).

[68] In India, especially in the north, cardamom is an essential ingredient of curry, and in the Near East it is used in pilau. It serves in India as medicine, in betel chewing, as a masticatory, and in other ways. And it is widely used in the Near East, along with sugar, to flavor coffee, a usage responsible for that region importing more cardamom than any other. Saudi Arabia alone nearly equalled Sweden as the largest importer of Indian cardamom in 1952-53, the two together taking more than a third of India's exports (Flückiger and Hanbury, 1879: 648; Indian Council of Agricultural Research, 1953: 49-50; Stobart, 1970: 61; Pruthi, 1976: 68-69; Lewis, 1984: 54-55).

[69] We would say a word here about a related spice of West Africa, Melegueta pepper or "Grains of paradise." Though called "pepper," the plant is more closely related to the amomum cardamoms of India and southeastern Asia than to true pepper. Indeed botanists formerly classified it as *Amomum melegueta*, part of the same genus. Now, however, along with over fifty related African species (among them the korarima cardamom of East Africa), it is in a separate genus, *Aframomum* (I. H. Burkill, 1966, 1: 61, 927-28). Reaching medieval Europe by trans-Saharan trade from West Africa, the spicy seeds of *Aframomum melegueta* were commonly used to spice beer, wine, and other liquors. When European ships finally reached West Africa by sea in the fourteenth and fifteenth centuries, they were able to do away with intermediaries and gain direct access to producing areas of coastal Liberia, which they called, after the spice, the Grain Coast. One doesn't know how it got there, but *A. melegueta* is now cultivated to a small degree in gardens in India (Watt, 1889-96, 1: 222; Pruthi, 1976: 55). *A. melegueta* was also introduced to Singapore on at least two occasions in the past, but apparently it did not succeed (I. H. Burkill, 1966, 1: 61). As for China, we find no reliable information that it is found there.

[70] In this paragraph we generally follow Bretschneider, 1882-95, 3: 120-24; Hooper, 1929: 8-11; I. H. Burkill, 1966, 1: 132-38; and F. P. Smith, 1969: 30-31, 35-39. Our list, however, may not be complete, for Bretschneider (1882-95, 3: 122) and Schafer (1963: 185) reported that the round cardamom of Java (*Amomum kepulaga*) was cultivated in Kwangtung and Kwangsi by the eleventh century A.D., though regarded as inferior to the imported product. We have found no mention of it in contemporary accounts.

[71] For information on the medicinal uses of cardamoms in China, see Soubeiran and Dabry de Thiersant, 1874: 121-25; Hooper, 1929: 8-11; F. P. Smith, 1969: 30, 35-38; J. D. Keys, 1976: 74-75, 76; and Perry, 1980: 437-38.

[72] Chinese trade in cloves and nutmeg has continued to the present day. Though we have found no data on imports to China proper, we note that from 1976 to 1980, Hong Kong imported significant amounts of cloves, nutmeg, and mace and that some of these were re-exported, with one destination presumably being China. In 1980, on the other hand, China, along with Singapore, became a major supplier of cloves to Hong Kong (*Spices: A survey of the world market*, 1982, 2: 129, 131).

[73] These two are considered by Wit (1969: 164) to be acceptable scientific names for cloves. Other common synonyms are *Caryophyllus aromaticus* L., *Eugenia caryophyllata* Thunb., and *Eugenia aromatica* (L.) Baill.

[74] *Lavanga* is mentioned in the *Pinda Niryukti* (54), a Jain work dating from the third century B.C.; in Kautilya's *Arthasastra* (II. 15. 21) (?290 B.C.-A.D. 300?); and by Charaka (?A.D. 80-180?) (*Samhita* 26. 210, etc.) and Susruta (*Samhita* 46. 202) (*c.* A.D. 350) (Prakash, 1961: 70, 92, 281; P. Ray and H. N. Gupta, 1965: 65).

[75] Though some scholars think reference was to another plant, the Greek word later used for cloves, *caryophyllon*, was already mentioned in Pliny's *Natural history* (XII. 15) in the first century A.D. Pliny's description is of an Indian "grain resembling that of pepper, but larger and more brittle...reported to grow on the Indian lotus-tree...imported here for the sake of its scent."

[76] According to Crevost and Lemarié (1917: 309), Loureiro had described the clove tree, called by the natives "wild clove," in the forests of one province of Annam. They themselves, however, were unable to find such trees there or to get information about them from residents. Nevertheless they couldn't eliminate the belief that a wild clove tree does indeed occur in Indochina, for in the market of one community one found cloves in small numbers that were purchased by Annamese and Chinese.

[77] Chi Han was not the only early writer to say that cloves were produced in Annam. The T'ang pharmacologist Su Kung said the same thing, which led Schafer (1963: 171) to believe that the clove tree had been introduced to Annam.

[78] Some authorities (Flückiger and Hanbury, 1879: 502; Watt, 1889-96, 5: 311) have said that nutmeg is found wild, especially in the Banda Islands in the Moluccas. I. H. Burkill (1966, 2: 1550) says that it was, and is, rare in the wild. Others (Flach and Cruickshank, 1969: 330; Zeven and de Wet, 1982: 60) say categorically that it no longer occurs in the Moluccas in a wild condition. The problem derives in part from an effort by the seventeenth-century Dutch to preserve their monopoly of nutmeg production by limiting its production to Banda and Amboina and destroying all nutmeg trees on other islands (Flückiger and Hanbury, 1879: 504; I. H. Burkill, 1966, 2: 1551).

[79] The origin of the genus *Myristica* is placed in New Guinea and adjacent islands, where certain wild species sometimes provide useful products and where one, Papuan nutmeg (*Myristica argentea*), found both wild and cultivated (Flach and Cruickshank, 1969: 329-30), provides seed and aril used as spices.

[80] There are species of *Myristica* native to the Indian subcontinent, but none of these are of any value in producing spices. One of them, *M. malabarica*, does provide seed, sometimes called Bombay nutmeg, used in India as a medicinal, and aril, Bombay mace. Both seed and aril are inferior products said to be lacking in flavor and aroma. They may, however, be used as substitutes for, or adulterants of, nutmeg and mace (Watt, 1889-96, 5: 314-15; Redgrove, 1933: 296; I. H. Burkill, 1966, 2: 1556).

The nutmeg tree, *M. fragrans*, was introduced to India and late in the last century was cultivated in the Nilgiri Hills, without much commercial success, and on a small scale in other parts of the country (Watt, 1889-96, 5: 312). It continues to be grown in India today, mostly in Tamil Nadu and Kerala in South India, but production has not been able to meet domestic needs (Pruthi, 1976: 167-68).

[81] There was a Sanskrit name for nutmeg, *jatiphala*, which appeared in the *Kasyapa Samhita* (second century B.C.-*c.* 400 A.D.) (Prakash, 1961: 164, 280). It has been suggested that the name *pala*, used for nutmeg in most of Malaysia, came from the second part of its Sanskrit name. If so, this would support the idea that Indians initiated the early trade

in nutmeg, as does the fact that on some of the Molucca Islands one finds such Sanskrit names (I. H. Burkill, 1966, 2: 1550-51).

[82] Schafer (1963: 185, 322n) gives the meaning of *jou-tou-k'ou* as "fleshy cardamom," cardamom being a spice with which nutmeg was sometimes confused in Chinese literature. Others (Hirth and Rockhill, 1966: 210n; H. L. Li, 1979: 82) note that Ch'en Ts'ang-ch'i said that the *jou-tou-k'ou* is called by foreigners or in foreign lands *chia-chü-lê*, which sounds like the Arabic name for cardamom, *qaqulah*. That name was also one of a country, apparently on the western side of the Malay Peninsula (Schafer, 1963: 185). J. I. Miller (1969: 59), in turn, has suggested that the Sanskrit names for nutmeg and mace, which he gives as *jati-phala* and *jati-kosa*, gave rise to the Chinese *jou-tou-k'ou*. The question of names for nutmeg in early China, Southeast Asia, and India, their affinities and origins, is clearly one that needs careful attention from a linguist.

[83] One notes a similarity between *i-chih-tzŭ* and *nikuzuku*, the Japanese word given by Crevost and Lemarié (1917: 309) for nutmeg.

[84] Crevost and Lemarié (1917: 310) were told by some people in Indochina that nutmeg was found in southern Annam, but they were not given precise information. In addition, certain Annamese retail merchants sold very small amounts of nutmeg which, they claimed, came from Annam, but without being more explicit. However, despite requests for specimens and official inquiries in Annam's provinces, as well as personal observation at shows, Crevost and Lemarié were unable to confirm that nutmeg was indeed found there.

[85] Vavilov (1949-50: 37) also lists Abyssinia (Ethiopia) as a center of origin of certain forms of coriander, which others have considered a subspecies.

[86] Prakash (1961: 281), on the other hand, says that the word *kustumburu* appears to be Dravidian.

[87] Laufer, who dismisses various claims for an earlier introduction of coriander, accepts it as one of the culinary vegetables of the Yellow River area as listed in the *Ch'i-min yao-shu*, the sixth-century agricultural encyclopedia. For that list, see S. H. Shih, 1962: 49-50.

[88] According to one version of this Vishnu legend as told by P. Thomas (1980: 93-94), Vrinda was renowned for her fidelity to her husband Jalandhar, an Asura (a divine being, or "demon") and he, in turn, was "invincible so long as his wife remained faithful to him." Jalandhar became involved in a struggle with the deity Indra, who solicited aid from Vishnu and Shiva who determined to destroy Jalandhar's invincibility by seducing his wife. Shiva, confident of himself, simply urged Vrinda to leave her husband and go away with him, only to be rebuked and driven off. Vishnu, following a more devious route, took the form of Jalandhar and, in that guise, succeeded in raping Vrinda. Discovering his identity, Vrinda cursed Vishnu to turn him into a stone, and he, in turn, cursed her and turned her into the *tulasi* plant. Over time, however, the episode was forgotten and Vrinda, now called Tulasi, became Vishnu's dearly beloved wife.

In a somewhat different version of this legend, told by B. Walker (1968, 2: 584), Vrinda was the wife of a man of low caste who gained divine power because of her outstanding virtue. So powerful was he that even Shiva could not beat him, and Shiva then led Vishnu to adopt the form of Vrinda's husband and to seduce her, after which Shiva killed her husband. Learning what had happened, Vrinda cast herself onto her husband's burning funeral pyre, whereupon Vishnu, eager that her attractions not be lost, made her body into the Gandaki River of Nepal, and her hair into the *tulasi* plant. Occasionally, it is believed, Vishnu visits that river, now a sacred place, to lie in the water while clasping Vrinda in his arms. The black stone found on the river bed is considered the sperm of Vishnu and it, too, has a symbolism, value, and ritual role among Hindus.

For other versions of the above accounts and other stories of Tulasi see Kincaid, 1908: 3-8; and Mani, 1975: 797-98.

[89] For Vrinda in Krishna lore, see Watt (1889-96, 5: 444-45); Stutley and Stutley (1977: 306); and P. Thomas (1980: 94).

[90] This is one of the black stones mentioned in Footnote 88. Known as *salagrama*, such stones are considered to be inhabited by Vishnu, and thus sacred. All homes of orthodox persons will have such a stone for purposes of worship, when it may substitute for an image of the deity (Kincaid, 1908: 8; Underhill, 1921: 131).

[91] A similar marriage ceremony between Tulasi and Vishnu may be performed at the time a well or tank is first put into use, to assure its power to fertilize garden plants and the suitability of its water for drinking (Crooke, 1926: 64).

[92] The goodness and purity of *tulasi* are revealed also in folk sayings. An undeserving son of an illustrious father may, for example, be designated "marijuana growing on a *tulasi* plant." In a similar vein is a Marathi saying meaning that on occasion one must do evil in order to accomplish good: "it is necessary to put an onion in the root of the *tulasi* plant" (Kincaid, 1908: 8). For the Hindu perception of onions as impure, see our section on "Onions and Their Relatives" in Chapter 5.

[93] In this, Keats follows Boccaccio, whose Lisabetta also watered with her tears a pot where her lover's head was buried.

[94] T. F. Thiselton-Dyer (1889: 101) reported use of sweet basil in "Voigtland" (Vogtland district in Germany?) in testing a woman's chastity. If the basil withered in her hands, a woman was judged impure.

[95] We have found no thorough scholarly account of the use of flowers in Chinese cuisine. The subject, however, deserves attention from some culinary scholar, for, as G. Z. Chu (1975: 80) has noted, flowers are an essential element not only in Chinese tea, but in cooking, preserving, and ornamenting. The use of flowers in tea, to be mentioned again in our section on beverages, is well known in the West, but there are other uses as well. Rose petals, for example, are not only added to black tea to make "rosebud tea," but they are made into a scented liquid, "Dew of Roses," which serves for making a cooling summer beverage. They are also made into a sweet paste that serves to sweeten and flavor food. Jasmine flowers, in turn, may be added as a scent to food and wine and other beverages (H. L. Li, 1959: 95-96, 129), and cassia blossoms are salted or sugared and employed in seasoning sweet soups and other sweets (F. Lin, 1975: 328). One also reads of candied jasmine and violets, of pickled banana blossom (Wright, 1913: 141), and of chrysanthemum wine (C. S. Wong, 1967: 80) and "Chrysanthemum Fish Hot Pot" (K. Lo, 1979c: 74), indicating that one is indeed dealing with a rich and varied phenomenon deserving further investigation. That dietary use of flowers is not a recent phenomenon is shown by early references, as to peony sauce as a seasoning in Han times (Yü, 1977: 67) (for the touching story of the peony, "king of flowers," in China, see H. L. Li, 1959: 22-36).

[96] For recipes involving use of golden needles with pork, see Stout, 1933: 98, and B. Y. Chao, 1972: 55.

[97] The *hüan-ts'ao* ("plant of forgetfulness") referred to in the *Shih-ching* has been identified as a species of *Hemerocallis* (Bretschneider, 1882-95, 2: 241; Keng, 1974: 401), which may indicate that golden needles were already used in early Chou times.

[98] Popular names used for these species are quite varied. *H. flava* is often called "Tall Yellow Day-lily" or "Custard Lily"; *H. fulva*, "Orange Day-lily," "Tawny Day-lily," or "Europa Day-lily"; and *H. minor*, "Grass-leaved Day-lily" or "Dwarf Day-lily."

[99] See O'Gorman, 1893, for information on cultural views of ginseng in China and Korea. For a recent collection of tales about ginseng from Manchuria, Korea, and other countries, see Kimmens, 1975.

[100] The story of the rise and fall of mandrake is a fascinating one of ancient beginnings; rich cultural associations, superstitions, and folk beliefs; and modern survivals. One finds the plant in both the original Hebrew version and the King James version of *Genesis* (30. 14-24), though it was dropped from some later accounts. Jacob, married to the sisters Leah and Rachel, had fathered four children by Leah but none by Rachel though he loved her most. Eager to conceive, Rachel obtained a mandrake root, known in Hebrew as "love-apple," found in a field by her sister's son. It was not the mandrake, however, but God, who much later enabled Rachel to conceive and give birth to a son. His name was Joseph, of fame because he was sold into slavery by his half-brothers, yet rose to a high position in Egypt and later became the ancestor of two of ancient Israel's tribes (Frazer, 1918, 2: 372-75; Moldenke and Moldenke, 1952: 137-39, 283).

The ancient Greeks respected mandrake for its ability to arouse sexual passion and, by implication, to enhance conception in women. So closely was mandrake associated with love that Aphrodite, Goddess of Love, was called Mandragoritis, "She of the mandrake." Just as the Chinese take precautions to assure the safety of the ginseng collector, the ancient Greeks took such measures when collecting mandrake. In obtaining the first mandrake root, the collector was advised to draw a circle three times around the plant with a sword, then face to the west when cutting it. In obtaining the second root, the collector should dance around the plant while speaking constantly of love matters. As a further precaution, it was recommended that he remain windward of the plant when digging it up, lest the odor fell him.

Certain uses of mandrake were also similar: the modern Chinese steep ginseng in liquor to make an alcoholic drink, whereas the ancient Greeks put mandrake in wine or vinegar for use in love charms (Frazer, 1918, 2: 375-76). Mandrake and mandrake wine were narcotics that induced such stupor as to make a person insensitive to pain, and, as a result, was used as an anesthetic before surgery. In the Roman world, mandrake wine or "death wine" was given to persons being crucified or to other persons experiencing torture or an extended or painful death (C. J. S. Thompson, 1934: 221-33). The root itself was also kept as a charm or carved into a puppet which, among other things, was expected to make a barren woman fertile and its possessor wealthy. These came to be associated with witchcraft and, during the Inquisition, possession of a mandrake puppet led many women to be burned at the stake (Frazer, 1918, 2: 376-84). As is true for ginseng, there is a sizeable literature on mandrake, and we refer the reader particularly to Frazer, 1918, 2: 372-97; C. J. S. Thompson, 1934; and Moldenke and Moldenke, 1952: 137-39, 283.

[101] That ginseng is used as an aphrodisiac or treatment for impotence, in East Asia and elsewhere, is attested to by Collyer, 1903: 29; S. Y. Hu, 1976: 23; J. D. Keys, 1976: 86; Siegel, 1979: 1614, 1615; Perry, 1980: 43; and Carlson, 1986: 245, 246. As in the case of ginger, mentioned previously, in China the aphrodisiac qualities of ginseng are secondary to and derive from its role as a general strengthening and repairing agent (E. N. Anderson and M. L. Anderson, 1977: 369; E. N. Anderson, 1988: 192-93).

[102] Ginseng was not the only commodity subject to state monopoly in pre-modern China. China's first monopolies, of salt and iron, were already set up in 119 B.C., during Han dynasty times. In T'ang and Sung times, salt monopolies were also in operation and, though they were not as effective, tea and liquor monopolies as well (Symons, 1981: 69).

[103] In the People's Republic, the world's second largest producer in 1980, ginseng has been successfully grown in Shansi and Hopeh, regions that supplied wild ginseng in ancient times, and also in Yunnan, far south of its occurrence in the wild (J. Y. P. Chen, 1962: 15). Since World War II, new centers of commercial ginseng cultivation have also developed in South Korea and the Soviet Union (Brekhman, 1965: 149; Baranov, 1966: 404, 405; C. P. Li and R. C. Li, 1973: 250), with South Korea the world's largest producer of ginseng in 1980 (A. W. Carlson, 1986: 243).

[104] Even India seems to have been mandrake country, for it is reported there in modern times and there are names for it in many Indian languages. Though we have not reconstructed its history there, we note that it was listed among the medicinal plants in the medical treatise of Charaka (?A.D. 80-180?). The plant's name was *laksmana*, meaning "possessed of lucky signs or marks" (P. Ray and H. N. Gupta, 1965: 65; Khory and Katrak, 1981: 478). Ainslie (1826, 1: 206-8) included mandrake in his *Materia indica*, but without giving details of its role there. And others (Watt, 1889-96, 5: 143-44; Khory and Katrak, 1981: 478-79) note that it is sold in the markets of northern India, is part of India's materia medica that, among other things, is considered an aphrodisiac. The root is also seen as resembling the human body, and is worn as a charm.

A related species, *Mandragora caulescens*, is found in Sikkim at elevations of 12,000 to 13,000 feet above sea level, but Watt (1889-96, 5: 144) writes that it is unknown whether its root has the medicinal qualities ascribed to mandrake.

[105] There is considerable confusion about the classification of Asian *Panax* species. Here we follow Hara's classification (1970) and draw on his other works (1966, 1971) and on H. L. Li's oft-cited study (1942) for additional data.

[106] For earlier accounts, some of them popular, of the history of American ginseng, its collection and cultivation, see Kains, 1903; L. O. Williams, 1957: 344-46; Hardache, 1974: 9-83; Dixon, 1976: 31-47; and J. P. Hou, 1978: 81-89. A. W. Carlson, in a recent article (1986), provides the best up-to-date survey we have found of United States ginseng collection, cultivation, and export; and McMahan (1981), the best maps of contemporary wild ginseng occurrence and harvesting. Before the beginning of this century most ginseng exported from the United States was harvested from the wild. Since that time, cultivated ginseng has dominated the export market, and as a producer in 1980 the United States was behind only South Korea and the People's Republic of China. Ninety percent of American cultivated ginseng was then produced in Marathon County, Wisconsin. American ginseng was of such high value that it is shipped to the Orient by air, much of it going via Hong Kong to southern China (A. W. Carlson, 1986).

[107] *Panax pseudo-ginseng*, subspecies *japonicus* Hara (= *P. japonicus* C. A. Meyer) is a native wild plant of Japan. On occasion, its root is utilized in Japan as a drug. *Panax ginseng*, though not indigenous to Japan, was introduced as a cultivated plant early in the eighteenth century, and ginseng production subsequently developed in Nagano, Shimane, and Fukushima prefectures on Honshu (Shibata, 1974: 69; Bae, 1978: 36-37). One often reads in the literature of the past two centuries that the Chinese consider Japanese ginseng, like American ginseng, inferior to that of Korea. Authors differ, however, on whether the Japanese ginseng involved was *Panax pseudo-ginseng*, subspecies *japonicus* or *Panax ginseng*. Since the latter was a cultivated plant in Japan, we suspect that in recent centuries it was marketed in larger quantities, and therefore the main object of comparison. If this is true, it would fit with the long-standing Chinese tradition of ranking ginseng (*Panax ginseng*) from different places in terms of quality. In early China, the best ginseng came from North China. Later, with exhaustion of the North Chinese supply, Manchurian ginseng was regarded as superior to the Korean product. And in modern times, Japanese ginseng, most recent *Panax ginseng* to reach the Chinese market is, along with American ginseng, held in lowest regard (Soubeiran and Dabry de Thiersant, 1874: 200; Bretschneider, 1882-95, 3: 18-25; O'Gorman, 1893: 262, 263; F. P. Smith, 1969: 302).

[108] For information on ginseng and its various commercial products, their preparation and medicinal uses, see especially S. Y. Hu, 1976, but also Soubeiran and Dabry de Thiersant, 1874: 200-202; Hooper, 1929: 100-101; Wallnöfer and von Rottauscher, 1965: 42; F. P. Smith, 1969: 301-4; C. P. Li, 1974: 36; J. D. Keys, 1976: 85-86; J. P. Hou, 1978: 52-58; and Perry, 1980: 42-43.

[109] Soviet interest in ginseng led not only to the expanded cultivation mentioned earlier, but to extensive pharmacological and clinical research. Already in 1949, a special Ginseng Committee, responsible for co-ordinating ginseng research and development, was organized as part of the Siberian Department of the USSR Academy of Sciences. Appropriately, the Committee was under the Siberian Department's Far East branch, in a region where extensive planting of ginseng was taking place. From 1949 until 1965, the Committee held twenty-three sessions attended by Soviets and foreigners, and published several volumes as well as Committee minutes (Brekhman, 1965: 149).

[110] For reviews of the results of medical and pharmacological research on ginseng, see C. P. Li and R. C. Li, 1973; Karzel, 1974; and Bae, 1978: 59-218. For ginseng from the perspective of bodily harmony, as a bridge between scientific and traditional Oriental medicine, see Fulder, 1980.

CHAPTER THIRTEEN

Four Chinese Delicacies

"Fish is what I like, so are bear's paws;
but if I cannot have both, I will forgo the fish and choose the bear's paws."

Mencius (translated by James Legge)

Bird's Nest
Shark's Fin
Bear's Paw
Bêche-de-Mer

In his *Musings of a Chinese gourmet*, F. T. Cheng (1954) singles out for attention ten of China's most renowned delicacies. We have considered certain of these already, and here we focus on four of the remaining ones: bird's nest, shark's fin, bear's paw, and bêche-de-mer. We choose these four because, unlike some of Cheng's delicacies, they fall clearly into the category of foods that highlight important occasions, costly foods that add to the status of those who eat them (E. N. Anderson and M. L. Anderson, 1977: 373).

BIRD'S NEST

One of the most distinctive Chinese delicacies is edible bird's nest (*yen-wo*), the white, translucent, gelatinous nest of various species of small swifts called swiftlets. The principal commercial species are the "White-nest" or "Edible-nest Swiftlet" (*Collocalia fuciphaga*, now sometimes *Aerodramus fuciphagus*) and the "Black-nest Swiftlet" (*Collocalia maxima* = *Aerodramus maximus*).[1] Though related species and genera of swiftlets occur from the Seychelles and Mascarenes through South and Southeast Asia to New Guinea, northern Queensland, and islands of the western and southwestern Pacific, *C. fuciphaga* and *C. maxima* have a much more limited distribution. *C. fuciphaga* is found almost entirely in Southeast Asia, from the islands of the Bay of Bengal to Indonesia, Vietnam, and the Philippines; and *C. maxima* from the eastern Himalayas to Borneo and nearby Palawan in the Philippines (Medway, 1963: 37, 38, 43-47; 1969: 57; Smythies, 1981: 187-89; Ali and Ripley, 1983: 29; Cranbrook, 1984: 145-48; Tikader, 1984: 98-99).

The highest quality nests are those of *C. fuciphaga*, called "white" because they consist entirely of nest-cement, a salivary secretion, occasionally with a few feathers of the nesting birds and bits of green vegetable matter. The nests of *C. maxima*, in turn, are called "black" because of the black feathers mixed with the nest-cement (Medway, 1963: 37; 1969: 58; Smythies, 1981: 189; Cranbrook, 1984: 147-48). White nests, which require far less cleaning before use, command the highest prices by far, perhaps twenty-five times that of black nests (Smythies, 1981: 186).

Repeated in the European literature is the notion that edible birds' nests consist of saliva and seaweed, or seaweed and half-digested food regurgitated by the swiftlets (Marshall and Folley, 1956: 383; Smythies, 1981: 189). Though small amounts of seaweed may indeed find its way into edible birds' nests (and certain nests of one minor commercial species, *Collocalia unicolor* [*Aerodramus unicolor*] of India, may be largely of seaweed), vegetable matter of any sort is a quite minor part of the nests of the two principal commercial species (Marshall and Folley, 1956: 386; Medway, 1963: 37; 1969: 58; Smythies, 1981: 189; Cranbrook, 1984: 147-48).

Those two swiftlet species are cave swiftlets, customarily nesting in crevices on coastal cliffs, in sea caves, inland caves, or similar man-made situations in dim light or total darkness (Medway, 1963: 41; 1969: 60; Smythies, 1981: 187, 188-89; Cranbrook, 1984: 145, 147, 148). They may live in cave colonies, as at Niah Cave in Sarawak, which number a million individuals or more. They find their way to nests in the darkness deep within a cave by echolocation, an ability possessed only by swiftlets of a proposed new genus, *Aerodramus*.[2] Their echolocation capabilities are less than those of bats, sufficient for the swiftlets to find their way to their nests on the roof or walls of a cave but not to locate, in the darkness of night, the insects on which they feed. Thus they feed in the hours of daylight, taking insects on the wing (Medway, 1969: 60-62).

Birds' nests seem to be a food only of the Vietnamese (in Tonkin and Cochin China) and Chinese (Medway, 1963: 38, 40, 44). Japanese do not use such nests (Watt, 1889-96, 2: 508) and we have found nothing to suggest that they were part of the traditional diet of other peoples in Southeast Asia. The Thai, for example, collect birds' nests for sale to Chinese or for feasts given in honor of important Chinese guests, but despite long association with Chinese the Thai have not incorporated them into their cuisine (Landon, 1973: 220; R. A. de Groot, 1983: 70-72).[3] Similarly, the Land Dayaks of Sarawak, though active for centuries in collecting and selling birds' nests for the China market, nevertheless do not eat them (Medway, 1958: 465).

As for China, Medway writes that none of the four swiftlet species producing edible nests are found there; that swiftlets of other species are rare; and that the Chinese must always have relied on overseas sources for significant amounts of birds' nests. One should, however, note that such swiftlets are found in Vietnam, an early Chinese domain, and on the Paracel Islands (Medway, 1963: 37, 38, 41), long visited by Chinese ships; and that *Collocalia fuciphaga*, the White-nest Swiftlet, has recently been reported on Hainan itself (Xian and Zhang, 1983: 125). That nests of Hainan area swiftlets are utilized is suggested by Swinhoe (1871-72: 68), a nineteenth-century traveller who reported that Tychow, Tyfar, and other small islands just south of Hainan are reputed to provide birds' nests of outstanding quality, superior to all others in taste, though they are rare. Swinhoe noted further that they were collected from caves in the fall. In addition, the naturalist Sowerby (1931: 136), though without further documentation, wrote of birds' nests collected in coastal China and Taiwan.

Harrisson (1959: 7) found tools of the types used in taking edible birds' nests in the Niah caves in Borneo at levels that suggested a sizeable traffic in birds' nests with China in T'ang times, a view supported by T. K. Cheng (T. K. Cheng, 1982: 43, 50). There is, however, no mention of edible birds' nests in Chinese sources prior to late in the Ming dynasty. Ming accounts of Malacca in the first half of the fifteenth century fail to mention edible birds' nests. Nor are they mentioned in writings of the Arab travellers and merchants who preceded the Portuguese in Southeast Asia, nor in any European account through the sixteenth century. Yet in the early seventeenth century, European observers described birds' nests as an unusual but choice food in Indochina, and they reported a well-developed trade in birds' nests, most of them intended for China, at Batavia (now Djakarta) in the Dutch East Indies. This evidence led Medway (1963: 39-43, 47) to the view that the use of birds' nests in China first became widespread at some time from the mid-fifteenth to the early seventeenth century.

One suspects that if the eating of "white birds'-nests" originated in areas controlled politically by China, it may first have occurred along its southernmost margins, with Hainan and Vietnam the most likely places. An alternative possibility is that long before the Southeast Asian sources of white nests were developed, the Chinese, or some aboriginal group in South China, had consumed nests of some inferior-nest swiftlet. One candidate is *Collocalia innominata* or *C. brevirostris innominata* (= *A. brevirostris innominata*), a subspecies found, it seems, eastward from the Andaman Islands and eastern Himalayas to Malaya and Sumatra, and northward to Central China (Etchécopar and Hüe, 1978: 500-501; Ali and Ripley, 1983: 28; de Schauensee, 1984: 277; Tikader, 1984: 98). Though nests of this subspecies are of little value because of the abundant grass and feathers they contain, Hopwood (1919: 859) reported them collected and sold to the poor in Burma early in this century, and I. H. Burkill (1966, 1: 646), that they were collected in Malaya.[4] There is, as well, a reference to edible nests, though inferior ones, from a cave in South China (Medway, 1963: 38), with no mention of the swiftlet species from which they came. One wonders, nevertheless, whether certain inferior swiftlet nests were eaten on a small scale in South China long before their first mention in the Chinese literature. If so, expansion of trade with Southeast Asia would merely have made available nests of far better quality, which then gained high status and general popularity.

Why such an unusual food should have come into use at all remains a mystery, but one should note that in China birds' nests are not only a delicacy but a strengthening food and medicine (Marshall and Folley, 1956: 383; Medway, 1969: 58; Read, 1977a: 54; Smythies, 1981: 186; R. A. de Groot, 1983: 68).[5] One reads (R. A. de Groot, 1983: 69-70) of the Chinese belief that the nest is made by a swiftlet from windblown sea foam, that it is rich in minerals and other valuable elements from the sea, a distillation of the bird's qualities of vigor, strength, perseverance, potency, and fidelity. In traditional and modern China, bird's nest is considered nutritious and purifying, its soup suitable for women during pregnancy as well as for young children (Koo, 1973: 141; Ling *et al.*, 1975: 128; Tan and Wheeler, 1983: 43). A wealthy opium smoker, in his morning-after depression, might also consume bird's nest in soup or as a sweet (steamed or cooked in water with sugar). This is believed to refresh, stimulate, and revive his bodily strength and balance. Bird's nest is also held to contribute to manly vigor, youthful skin, wisdom, and longevity. The sweet is prescribed by Chinese doctors as tonic and invigorating, for tuberculars and persons with other long, debilitating sicknesses (P. L. Simmonds, 1885: 142; Hooper, 1929: 146-47).

Whatever the origin of bird's-nest eating, once the practice developed among the Chinese, they and their middlemen in Southeast Asia became active in seeking out sources of supply. They stimulated the opening of production at new caves there, many small and low, others large and lofty. In the latter, collecting was a risky undertaking, in which men, sometimes in teams, climbed poles or ladders as much as 200 to 300 feet high, to gain access to nests. At the big Niah Cave, such men were professionals who used cloths to wrap their feet when climbing a pole, with ropes, candles, and lanterns to light their way, specialized tools to pry the nests loose, nets to catch the nests, and sacks to carry them (Medway, 1957, 1958). Such operations were owned or controlled in various ways involving private individuals, officials, and governments. In Siam, one clever eighteenth-century Chinese entrepreneur became wealthy by obtaining from the Crown exclusive rights to exploit the caves on two tiny islands for birds' nests, until finally the king, made aware of their value, took away the concession (Medway, 1963: 44; R. A. de Groot, 1983: 70-71). Other Chinese individuals or firms, however, continued to be granted concessions for birds' nests in that land, selling their birds' nests to local Chinese or exporting them to China. In 1939, however, a law was enacted that was intended to eliminate the Chinese from the business: it reserved concessions for development by the Thai government from that time onward (Landon, 1973: 219-20). On Java in the mid-nineteenth century the Dutch East Indies government purchased the rights to birds' nest collecting at the renowned caves of Karangbolong, and they continued under government control with the independence of Indonesia (Medway, 1963: 46; 1969: 58). In Sarawak, the government permitted local ownership to continue, but since the mid-1930's issued regulations affecting the main areas of nest collecting (Medway, 1969: 58-59). And in Burma, where the taking of birds' nests was a government monopoly, rights were rented or sold to individuals at auction (Crawfurd, 1829: 426; Hopwood, 1919: 858-59).

Already in the seventeenth century, birds' nests were valuable commodities shipped from various parts of Southeast Asia to Tonkin and China. Batavia became the important entrepôt in the bird's-nest trade of the time, with an estimated four million nests passing through that port each year in the eighteenth century (Medway, 1963: 39-40). In the nineteenth-century, Singapore joined Batavia as an important bird's-nest emporium, and supplies reached China from Java, Sumatra, Borneo, Thailand, Vietnam, and even as far as Fiji (P. L. Simmonds, 1885: 142-43; Parker, 1897: 177; 1903: 272; Hooper, 1929: 146-47; I. H. Burkill, 1966, 1: 645-46). The trade in birds' nests spread eastward to Burma and India, as well, with the Mergui Archipelago, Tenasserim and Arakan coasts, the Andaman and Nicobar islands, and the west coast of India among the sources of supply. The prime movers in this western trade, according to Watt (1889-96, 2: 505-8; 1908: 138), were Chinese merchants living in Rangoon who exported the birds' nests to China.

The price of birds' nests has been subject to considerable fluctuation, but in early 1987 in Hong Kong ordinary nests sold for about U.S. $375 a pound, and nests of top quality up to $1000 a pound, and even more (Kristof, 1987). At one time in the early 1980's the very best nests were said (R. A. de Groot, 1983: 72) to be worth nearly $300 an ounce in Hong Kong, compared to about $400 for an ounce of gold. From a collector's perspective, sale of a single "white" nest in Java in 1961 provided three times as much money as earned by a full day's labor (Medway, 1969: 58). It is thus understandable that widespread theft, poaching, and smuggling of such nests have developed, and that great care is taken to thwart them. Also understandable are the worries about maintaining the bird's-nest supply, involving efforts at conservation and at giving useful swiftlets an advantage over competitors. In Sarawak, for example, some birds' nesters have killed and destroyed nests of bats and birds of no economic value to make more room in a cave for swiftlets producing edible nests (Medway, 1957: 260). There has also been legislation seeking to minimize the loss of life among young birds as a result of collection practices. And now there is concern about the impact agricultural pesticides may have on swiftlet well-being and survival (Cranbrook, 1984: 160). Though a historical geography of birds' nests has yet to be written, evidence at hand makes it clear that we are dealing with demand for a food that has had a major economic impact, and an ecological one as well, over a considerable geographic area.

Late in the nineteenth century it was estimated that eight and a half million nests reached Canton annually (E. Balfour, 1884, 1: 366). When they arrived there or at other Chinese ports, if it had not already been done they were carefully cleaned, dried, and packed. Prices were lower in Canton and higher in regions more distant from the source of supply. They ranked after ginseng in cost and were consumed especially by the well-to-do, with many of the finest quality birds' nests shipped to the imperial court in Peking (P. L. Simmonds, 1885: 140-43; S. W. Williams, 1895, 1: 780; 2: 397-98; Parker, 1903: 272; Hooper, 1929: 146-47). Because of the expense of birds' nests, agar-agar may be added to small amounts of bird's nest in making dishes (H. J. Lin and T. Lin, 1972: 84). In addition, the Japanese make imitation nests of agar-agar or other materials for export to China, which not only resemble real nests in shape, but even bear small "traces of feathers" (E. Balfour, 1884, 1: 366; E. Watson, 1923: 329).

Swiftlet nests, which come in different grades,[6] lack a strong flavor, and thus the taste of foods made with bird's nests, among them soup, poultry stuffing, dumplings, and sweets,[7] derives to a large extent from other ingredients with which they are prepared. In one of Canton's better restaurants, the many-floored Dai Tung Restaurant, two such recipes we found on the menu were "Stewed Bird's Nest with Crab Cream" (about U.S. $11) and "Stewed Bird's Nest with Shredded Chicken" (about U.S. $8.50). To appreciate the full range of possibilities, however, one might consider the recipes for bird's nest soup that appear in cookbooks and elsewhere (F. T. Cheng, 1954: 78-79; J. D. Keys, 1963: 107-8; Davis, 1966: 168; Howe, 1971: 157-58; B. Y. Chao, 1972: 169-70; H. J. Lin and T. Lin, 1972: 83-84, 182-83; Em. Hahn *et al.*, 1973: 191; E. Chen, 1977: 118; K. Lo, 1979c: 75; and R. A. de Groot, 1983: 73).

Bird's-nest cement is fairly rich in protein which, however, is deficient in lysine, methionine, and tryptophan as compared to beef, pork, and tuna protein; its protein is also far more expensive in cost per pound (Y. O. Chang and M. S. Dunn, 1957). The nests, even when soaked, also contain significant amounts of calcium, iron, and riboflavin (Read *et al.*, 1940: 66-67; Leung *et al.*, 1973: 110).

SHARK'S FIN

Shark's fin (*yü-ch'ih* or *sha-yü-ch'ih*) is another great delicacy in China, served at banquets and special dinners and, where found on restaurant menus, quite expensive. In the Dai Tung Restaurant in 1987 we found shark's fin dishes to be two to three times as expensive as bird's nest ones: about U.S. $25 for "Premier Shark's Fin in Clear Soup" and U.S. $23 for "Premier Shark's Fin with Brown Sauce." Shark's fin seems to have first gained favor in Sung times (Freeman, 1977: 154-55). Domestic supply has long been too small to meet demand, and the Chinese have been forced to obtain sharks' fins from abroad. Late in the last century, the Indian ports of Bombay and Madras were identified as large suppliers of "sharks'" fins to China, mainly from the tiger shark (*Galeocerdo cuvieri* or *G. tigrinus*) and shark-like guitar fishes (*Rhynchobatus* spp.). Other suppliers mentioned were the Philippines, Dutch East Indies, Burma, and the Hawaiian Islands (P. L. Simmonds, 1885: 318). In the 1920's, most of the shark's fin imported to China came from India, where fins (from at least fourteen species of sharks, rays, and certain other fish) were obtained in the Persian Gulf, Arabian Sea, and Indian Ocean, with Bombay and Karachi identified as very important in the shark's fin trade (E. Watson, 1923: 527). The trade network may have been even more extensive in the 1950's, for

F. T. Cheng (1954: 73-74) listed as suppliers not only India and the Philippines, but also Korea, Japan, Mexico, Venezuela, and Norway. There are different qualities of shark's fin,[8] but the finest quality and most expensive type, according to Cheng, is "Manila yellow," from the Philippines. Yellow in color, it is tastier, more tender than other types of shark's fin, and more of the fin is edible.

When fresh, shark's fin has little flavor, but when cured its flavor is more pronounced. Curing for the Chinese market as practiced in Malaya involves removing the fin, coating it with lime dust, rubbing the flesh with salt, and then sun-drying it (I. H. Burkill, 1966, 2: 2035). Ways of preparing shark's fin for eating have been described by various authors.[9] The most detailed account we have found, however, is that of F. T. Cheng (1954: 74-78), who notes that preparation differs according to the quality of the fin, with more elaborate preparation for compact shark's fin of the finest quality. Whatever the quality, he observes, it is best to soften the fin by soaking in cold water for three days, though one can skip this stage if there is insufficient time. The fin is then prepared for cleaning by simmering for four to five hours or until skin and bone can readily be removed. In simmering, the water is changed frequently to prevent the unpleasant odor of decayed flesh and bone from affecting the fin. Then the fin is cleaned, with careful attention to the removal of decayed flesh. Fin of low quality may lack sufficient gelatin to hold together and may be loose or even separate in the process of cleaning. Cooks make a special effort to avoid separation, however, for a dish of compact shark's fin is three times or more as expensive in restaurants as one of loose fin. In Canton, the cook uses a bamboo net in later stages of preparation to keep the fin compact or, since a bamboo taste may be imparted to the fin, he may instead employ a net of fine silver wire. With the fin clean, it is again simmered for forty-five minutes with ginger, onion, and wine, the liquid being changed twice to remove what fish taste remains in the fin.

Then the preparation differs according to whether the fin is of lesser quality (loose or separated) or better quality (compact). Lesser quality fin is usually cooked in a sort of soup with chicken or crab flesh and served with them. Better quality fin, on the other hand, is cooked with ham and pork in a double boiler with water and wine until tender, for about three hours. Then it is carefully removed from the pot, separate from the other ingredients, and is put in another pan with specially-prepared chicken and ham bouillon. The fin is cooked for ten minutes to permit the bouillon flavor to penetrate it. Soy sauce and, if needed, salt is added, and the dish is served immediately.

Kenneth Lo (1979c: 299) writes that shark's fin can almost be regarded as "a culinary joke" because it takes three days to prepare but just three minutes to consume. Long preparation is necessary, he adds, because shark's fin beforehand is as hard to cut or chew as rhinoceros' horn. That the Chinese should develop such a dish is eloquent testimony to their unusual commitment to culinary experimentation as well as to the high regard in which they hold shark's fin, which, like bird's nest, is considered strengthening by the Chinese. In terms of food value, shark's fin (dried, soaked, and drained) is very high in calcium and high in iron (Leung *et al.*, 1973: 137). Though it is also high in protein, that protein appears to be only of fair quality (Holmes, 1948: 210).

BEAR'S PAW

The third of China's well-known delicacies is bear's paw (*hsiung-chang* or *hsiung-fan*), which may come from any of the several bear species (*Ursus* spp.) found in China.[10] Like bird's nest and shark's fin, it is a strengthening food (Soubeiran and Dabry de Thiersant, 1874: 44-45; Read, 1976: Section 361) served as a special dish, for example at expensive feasts. Pollard (1921: 193) was told by a Chinese informant in Szechwan that during their winter hibernation, black bears are able to ward off hunger by licking their paws, which is why the Chinese value them so highly. Apparently people's reasoning was that if licking their paws can sustain creatures as large, powerful, and dangerous as adult bears over an entire winter, the paws must be very strengthening indeed.

Bear's paw, unlike bird's nest and shark's fin, is also considered a delicacy in the Western world. Many Europeans and North Americans have long recognized that bear's paw, like pig's feet, is very tasty indeed. In Germany, salted, smoked bear's paw was set aside for royalty (P. L. Simmonds, 1885: 52, 60-61). And in Britain, even though overhunting led to the extinction of bears before the Norman conquest, the excellence of bear's paw was not forgotten, for bears were brought from the continent for the sports of bear-bating and bear-hunting. In his sixteenth century *Booke of hunting*, Turbervile wrote disparagingly of bear's flesh, yet acknowledged that "their feet are the best morsel of them, for they be delicate meat" (C. A. Wilson, 1973: 77, 82).

The story of bear's paw as a delicacy in China is yet to be told, but it may well be the earliest used of the first three delicacies we have considered in this chapter. Bird's nests, after all, are first mentioned in Chinese accounts only late in the Ming times (though trade may go back to T'ang times) and shark's fin appears first to have become popular in Sung times, whereas bear's paw was already a delicacy in ancient China. This is demonstrated in the quote by Mencius (372-289 B.C.) which introduced this chapter, and is further confirmed by bear's paw being among foods served at a Han feast (Y. S. Yü, 1977: 67).

In those times, however, bears were far more numerous in China than in the present day. One suspects that over the centuries since, a growing shortage of bear's paws was accompanied by a steady increase in their price and in restrictions on their use. In any case, Chinese merchants residing in Nagasaki in Ch'ing times included bear's paw, along with bird's nest, shark's fin and other delicacies, among the dishes served on first-class occasions, but for second-class ones it was omitted though the other two delicacies were not (Spence, 1977: 277). Today, according to F. T. Cheng (1954: 92), the supply of bear's paw is so small that it is turning into a delicacy of historical significance only. It was found, he said, mainly in North China, and was satisfactorily prepared solely by chefs of the culinary schools of Shantung and Honan. Where bear's paw is still used, it is a scarce and costly dish, and commonly other items, such as buffalo's foot or beef tongue, are substituted for it.

Bear's paw entered into domestic trade in China, sometimes being carried considerable distances to market. We have found nothing to indicate whether Chinese merchants sought out foreign sources for bear's paw as they did bird's nest and shark's fin, though this would certainly have been in keeping with their entrepreneurial spirit. In California today, wildlife authorities each year find bodies of illegally-killed black bears with their paws removed, as well as the gallbladder, and, in some cases, the head, too. That Chinese and Korean demand is behind these killings is indicated by the fact that one can find such gallbladders in traditional drug stores, as in Korea Town in Los Angeles (pers. comm., Donald Koch). It remains unknown, however, in whose pot the paws end up. Bear's gallbladder contains a concentration of gall (*hsiung-tan*), or bile, which, like bear's paw, is highly-regarded for its strengthening qualities, and is a very expensive item in Chinese commerce, used especially for problems of the liver and abdomen.[11]

F. T. Cheng (1954: 92-93) provides a recipe for the preparation of bear's paw as a delicacy. The paw is first coated with clean mud and baked in an oven. When the mud dries and hardens, the paw is removed and permitted to cool. Then the mud is broken loose, with the hairy skin of the paw adhering and coming off with it. The flesh that remains is then simmered to make it soft, with the water changed regularly to eliminate its gamy odor and flavor. When quite soft and without taste, the paw is simmered with chicken flesh, lean ham, wine, and a small amount of water until a rich, thick sauce is obtained. For serving, the paw, like ham, is cut into slices and served with the sauce. A second recipe, called "Braised Bear's Paw," is provided by Sakamoto (1977: 231-33), but it is actually made with beef tongue, a substitute. Nevertheless, it is prepared in the style of Szechwan, a region, she observes, that is noted for its bear's paw dishes. This dish, which sounds quite elegant, includes, in addition to beef tongue, shrimp, pork, chicken, green onions, chili peppers, ginger, white pepper, salt, oil, and a seasoning mixture consisting of soy sauce, rice wine, sugar, and MSG. Bear's paw is described as having a special flavor, similar to that of fatty ham of outstanding quality, but far superior because it lacks greasiness. "It is so smooth and delicious," says Cheng, "that it simply melts in one's mouth."

Perhaps because of its rarity, Leung *et al.* (1973) did not include bear's paw among the East Asian foods they analyzed in nutritional terms. We do, however, have the word of gourmet F. T. Cheng (1954: 92) that "it must be highly rich in nutritious properties," a view held also by J. D. Keys (1963: 73).

BÊCHE-DE-MER

The bêche-de-mer or trepang of commerce is a dried product that comes from many species (and several genera) of sea cucumbers or sea slugs, invertebrates of the family *Holothuroidea*, which live on the sea floor, especially on coral reefs. Bêche-de-mer vary in color from white to gray, brown, and black. The creature may be taken in shallow water simply by picking it up, or obtained by spearing or diving. While native peoples in the source areas are active in the taking of bêche-de-mer, I. H. Burkill writes (1966, 1: 1201), the actual processing, which varies in detail, is in the hands of Chinese. By the method he describes, the bêche-de-mer is first thrown into boiling water to kill it. Then it is removed from the water, split, gutted, and dried. Then it is commonly smoked, not to improve the taste but to preserve the flesh. In traditional times, bêche-de-mer were imported to China in gunny sacks or boxes and, when sold, were ordinarily four to eight or nine inches in length.[12]

English-speaking people commonly regard the bêche-de-mer as unsightly, and its taste, uninspiring. One such person, for example, described it as being dirty light or dark brown in color, stiff and loathsome, and possessed of a strong fishy smell (P. L. Simmonds, 1885: 445), and another (Parker, 1903: 272), as tough and virtually without flavor, something "between...French *escargots* and cooked seaweed." Its tastelessness, however, is an asset to the Chinese, for, as we shall see below, it takes on and accentuates the flavors of other items cooked with it. The high regard that the Chinese have for bêche-de-mer leads them to consume more of it than any other people,[13] with their effort in the fishery for bêche-de-mer perhaps comparable to that of the Portuguese in anchovies. The Chinese consider bêche-de-mer a delicacy, and it is served at important dinners and banquets. It is also highly regarded as a strengthening food, one believed to have the ability to rejuvenate and to improve a man's virility and sexual capabilities. It is thus no accident that the Chinese do not call the bêche-de-mer "sea worm" (*bicho de ma* or *becho de mar*), like the Portuguese, or "sea cucumber," like the English, but instead, according to E. Watson (1923: 326), honor it with the name "sea ginseng" (*hai-shên*).

The story of Chinese trade in bêche-de-mer remains to be told, but it goes far back in history and, as with bird's nest and shark's fin, extended into seas quite distant from the homeland. It is said that a thousand years ago Chinese fishermen already obtained bêche-de-mer from the waters of Ceylon, and that five centuries ago they had numerous boats in that trade (I. H. Burkill, 1966, 1: 1200). Early in the nineteenth century, bêche-de-mer was described as a more important item of commerce from island Southeast Asia to China, in amount and value, than perhaps all others except pepper. The largest market dealing in bêche-de-mer was Macassar, where no less than thirty-six varieties were sold, the best of them for unbelievable prices. The Chinese network, however, extended far beyond, from India and Ceylon on the west to New Guinea and Australia on the east. In Queensland, Chinese immigrants fished for sea cucumber, and at one time a regular junk visited ports there to buy dried bêche-de-mer for export. In most of the region, however, the fishing itself was done by other ethnic groups, and there were many English and American ships catching sea cucumbers for the Chinese market, with Americans getting as far west as the Caroline Islands in their quest. The trade to China, however, was in the hands of Chinese who were responsible for annual imports from 1865 to 1875 averaging about 1200 tons (Crawfurd, 1820, 3: 440-44; P. L. Simmonds, 1885: 444-48). In the 1930's, I. H. Burkill wrote (1966, 1: 1201) of bêche-de-mer activities extending from Japan in the north, and Fiji in the east, to Arabia in the west.

Bêche-de-mer requires extended soaking before cleaning and use, and has a jelly-like texture, a fishy odor, and little taste. Though it may serve as the main ingredient in a Chinese soup or stew, its role is that of imparting and harmonizing the flavors of the other ingredients (K. Lo, 1979c: 284). Recipes for its preparation are legion, but we would note especially "Bêche-de-mer Gourmet," a dish prepared with prawns, Chinese mushrooms, tree fungus, onions, ginger, soy sauce, and other flavorings; "Bêche-de-mer in Cream Stock," the cream stock in this case not a dairy product but prepared with duck, chicken, pork, scallions, and ginger (H. J. Lin and T. Lin, 1972: 63, 183); "Red-cooked Bêche-de-mer with Pork, Mushrooms, and Bamboo Shoots"; "Butterfly Bêche-de-mer," a dish including smoked ham, chicken, pigs' knuckles, eggs, bean sprouts, scallions, ginger, and other ingredients (K. Lo, 1979c: 284-85); and "Happy Family," called "a gastronomic Ark" because of the variety of fish, fowl, and meat it contains (H. J. Lin and T. Lin, 1972: 150; Schwabe, 1979: 361).

In leaving the matter of bêche-de-mer, we would note that holothurin, a toxic substance obtained from sea cucumbers, has been demonstrated to have a significant antitumor effect (C. P. Li *et al.*, 1974: 110-11). We would also note the view of zoologist Sidney Townsley of the University of Hawaii that sea cucumbers are an underexploited source of protein which could provide large yields under cultivation (Schwabe, 1979: 360). Indeed, dried muscle of the sea cucumber *Stichopus japonica* is reported to contain more than 80 percent protein, and even soaked, nearly 19 percent, though the protein appears to be of only fair quality (Holmes, 1948:

210; Leung *et al.*, 1973: 136, 245). The soaked muscle is also rich in iron and contains fair amounts of calcium.

To our knowledge, no research has been conducted on the ecological disturbances that may be brought on by the Chinese demand for delicacies such as those considered above. That serious problems can develop is revealed in a story recently released in Peking ("Mice infestation," 1985) about a catastrophic increase in the population of mice in southern Kwangsi, with hordes of hungry mice doing major damage to crops of rice, maize, and other cereals. According to experts, the problem was brought on by the large-scale killing of owls, weasels, and snakes, predators of rodents but also human delicacies in South China. As a solution to the problem, some Chinese economists passionately urged the populace to start eating mice, and asked restaurants to include mice and rat dishes on their menus. An alternative, of course, is to reduce the killing of the predator species regarded as delicacies. That this alternative is being followed is shown by a report dated May, 1988 (*Insight*, **4**, No. 20, May 16, 1988: 38) that a Peking restaurant was facing criminal charges because it was serving the flesh of animals that are protected species. Authorities had determined that, since the beginning of the year, the restaurant had served 41 owls, 30 giant salamanders, and 70 boa constrictors, and, in addition, refrigerators in the restaurant yielded the remains of a pangolin, a scarce South China tiger, 28 rhesus monkeys, and other creatures.

[1] The additional species of swiftlets producing inferior edible nests are identified by Medway as *Collocalia troglodytes*, restricted to the Philippines, and *C. unicolor* (*Aerodramus unicolor*), to India, as in the Nilgiri Hills along the southwest coast, and in Ceylon (Medway, 1963: 37-38; G. M. Henry, 1971: 155; Ali and Ripley, 1983: 29). Their nests include not only feathers, but other substances, mainly vegetable matter of one sort or another.

[2] Medway and Pye (1977: 226-27), on the basis of ability to echolocate, included most swiftlets formerly classified as *Collocalia* in the new genus *Aerodramus*. The "White-bellied Swiftlet" (*C. esculenta*) and "Giant Swiftlet" (*Hydrochous gigas*), neither of which can echolocate, and the "Pygmy Swiftlet" (*C. troglodytes*), whose echolocation abilities were unknown, were left in their former genera. Thus the genus *Aerodramus* includes at least four swiftlet species or subspecies reported to make edible nests (*A. fuciphagus*, *A. maximus*, *A. unicolor*, and *A. brevirostris innominata*).

[3] In the seventeenth century, the French Jesuit Tachard (1981: 271) observed birds nests' in Thailand, and wrote that "these people" used them. His statement followed a paragraph dealing with the Chinese, and one cannot be certain whether by "these people" he meant Chinese in Thailand or Thais. There was, however, at least one effort by a Thai company to popularize birds' nests among their people. That was in 1939, when the company advertised that nests were available in leading shops, and argued that though such nests had previously been used only by Chinese, they actually come from Thailand and that Thais should eat them (Landon, 1973: 220).

[4] Identification of swiftlet species is difficult at best, and continual taxonomic revisions (see Medway, 1965-66) have contributed further confusion. If, however, the swiftlets mentioned by Hopwood and Burkill were indeed *Collocalia innominata*, we are dealing with a swiftlet species that at least in Burma and Malaya, provides edible nests, however inferior.

[5] Various writers (for example, Koo, 1973: 126, and Smythies, 1981: 186) also attest to the reputation of bird's nest as an aphrodisiac.

[6] For information on the grades of bird's nests, see E. Watson, 1923: 328-29.

[7] One sweet dish, of bird's nest and rock sugar, is commonly served at wedding feasts (G. B. Miller, 1966: 847).

[8] For information on this, see E. Watson, 1923: 527-28.

[9] Among them are Ball, 1906: 291-92; J. D. Keys, 1963: 103-4; Howe, 1971: 159; K. Lo, 1971: 99-100; 1979a: 26, 128, 246; 1979c: 299-301; B. Y. Chao, 1972: 131-32; H. J. Lin and T. Lin, 1972: 87-88; and E. Chen, 1977: 116-17.

[10] For identification of the bear species of China, see E. H. Wilson, 1913, 2: 184-88; Couling, 1917: 586; Sowerby, 1940: 51-52; and Read, 1976: Section 361.

[11] For more information on bear's gall in China, see E. Watson, 1923: 322. Watson wrote that bear's gall was obtained mainly from Tibet, Mongolia, and China's "three eastern provinces," and was shipped from ports of North and Central China to other parts of the country.

[12] There are different qualities of bêche-de-mer, but they are ordinarily divided into two classes, "black" and "white," with black the highest quality. For more information on this, see E. Watson, 1923: 327.

[13] Among neighboring peoples who consume bêche-de-mer are the Japanese, who also eat the intestine and, when it is present, even the roe (Hornell, 1950: 202).

CHAPTER FOURTEEN

Beverages, Including Dairy Products

"Tea is of a cold nature...
If one is generally moderate but is feeling hot or warm, given to melancholia,
suffering from aching of the brain, smarting of the eyes
troubled in the four limbs, or afflicted in the hundred joints,
he may take tea four or five times.
Its liquor is like the sweetest dew of Heaven."

Lu Yü, The classic of tea (T'ang work translated by Francis Ross Carpenter)

Introduction
Tea
- Its History in China
- Its Use in Traditional Times

Alcoholic Beverages
- Problems with Terminology
- Alcoholic Beverages in Antiquity
  - Grain Wines
  - Grape Wine, Mead, Toddy, and Koumiss
  - Distilled Liquors
- Alcoholic Beverages in Traditional China
  - Principal Kinds
  - Use of Alcoholic Beverages

Milk and Dairy Products
- Origin and Spread of Dairying in the Old World
- Hypotheses Advanced for the Failure of Dairying and Use of Animal Milk to be Accepted Universally
- History of Dairying in China
- Dairying and Dairy Products in Modern China
- Nutritional and Medical Perspectives

INTRODUCTION

In considering beverages, we turn first to the traditional world's most-important non-alcoholic beverages—tea, coffee, and cocoa—and the position they gained or failed to gain in China. Today we associate coffee drinking with Arabia, yet that practice seems to have been known there only in the thirteenth century A.D. or later. Coffee (*Coffea arabica*) is an Ethiopian domesticate of uncertain antiquity which may not have been known in Arabia at the time of Mohammed, who was thus not able to ban coffee as he had wine and other intoxicants. Though its use was initially condemned by Moslem clergy, coffee fit a dietary niche in the alcohol-free environment of Islamic Arabia, and it came to be accepted there and later in other parts of the Islamic world as well. The Arabic name for coffee, *kahwa*, which found its way into languages of Persia, India, and Southeast Asia along its eastern route of dispersal, is also the source, whether directly or indirectly, of its Chinese names *k'a-fei* or *chia-fei* (Watt, 1889-96, 2: 460-63; E. Watson, 1923: 379; I. H. Burkill, 1966, 1: 627-29; S. Y. Hu, 1980: 9). It would seem likely that Arab Moslem merchants, who seem to have been in Canton at least by the eighth century, would have imported coffee after it came into use in Arabia. Whether or not they did so, coffee came under cultivation in the Dutch East Indies at the close of the seventeenth century and coffee beans likely reached China not long after. In the People's Republic, coffee is now being cultivated in Yunnan Province, at least some of it among minority groups (Tregear, 1980: 169; E. N. Anderson, 1988: 139). Cocoa (*k'ou-k'ou* or *ko-ko*) is made from the beans of cacao (*Theobroma cacao*), an American domesticate which did not reach Asia until after the discovery of the New World. Spanish ships carried the plant across the Pacific, possibly introducing it to Indonesia (Celebes) in the mid-sixteenth century and certainly to the Philippines by the end of the seventeenth century (de Candolle, 1886: 314, 415-16; Watt, 1908: 364-66; Merrill, 1954: 314n; I. H. Burkill, 1966, 1: 627-29; 2: 2187-88; Ferwerda, 1976: 258).

When coffee and cacao first reached China, they found a formidable competitor, tea, firmly established there, and they have never successfully challenged its supremacy.[1] S. W. Williams (1895, 1: 776), writing late in the nineteenth century, said that coffee and cocoa were unknown in China. More recently, Anderson and Anderson mention coffee as gaining popularity among modern overseas Chinese, and among recent imports to South China. Fruit drinks have long been used in China, and sodas and fruit squashes are now widely used, but they are recent, having become popular with students only in the 1930's. Soybean milk, long used as a beverage, has also gained favor in recent times (Dudgeon, 1895: 17; E. N. Anderson and M. L. Anderson, 1977: 342, 371; V. Y. N. Hsu and F. L. K. Hsu, 1977: 307). Since none of the above beverages were of great importance in traditional China, we will say no more of them, but turn instead to tea, China's leading beverage.

TEA

Its History in China

Tea (*Camellia sinensis* = *Thea sinensis*) is an evergreen shrub or tree whose center of genetic diversity is in an area stretching from northeastern India to Central China. The species is divided into two varieties by Sealy (others, however, give them specific rank): "China tea," *C. sinensis*, var. *sinensis*, and "Assam tea," *C. sinensis*, var. *assamica*. China tea is the hardier of the two. Typically it is a shrub with small leaves, and may be indigenous to China, with Sealy accepting the claim that tea is wild in southern and western Yunnan. Assam tea is typically a tree with larger, usually thinner leaves, and seems to be indigenous to warm areas from Assam through Burma and Indochina to southern China, including Hainan. There is, however, considerable overlapping between the two, and differences of opinion as to whether or not they had separate origins. The place of earliest tea domestication is also uncertain, whether Southeast Asia, southwestern China, Tibet, or elsewhere, and so is the identity of its domesticators

(Kingdon-Ward, 1950; Sealy, 1958: 112-23; H. L. Li, 1959: 83-85; Visser, 1969: 460-61; 1976: 18-19; Zeven and de Wet, 1982: 46).

In ancient China, water seems to have been the everyday drink of ordinary folk. Fermented alcoholic beverages, most of them from grains, were also used at times of feast and ritual and at many meals, as were soups of one sort or another (Dudgeon, 1895: 1, 4; K. C. Chang, 1977b: 30, 37, 40-41, 42; Y. S. Yü, 1977: 68-69, 73; Pirazzoli-t'Serstevens, 1982: 50, 72). Perhaps the ordinary people of ancient China, like people of traditional times, preferred their water after boiling, or, like the poor in certain interior districts of modern China where tea was too expensive or not obtainable, they drank the water left after boiling grain (Cormack, 1935: 135).[2] The ancient Chinese must have also used herbal teas of various sorts, perhaps including some of the thirty-five substitutes for true tea listed by a Chinese author of Yüan times (Dudgeon, 1895: 13, 64-65). Such teas continued to be used in traditional China, sometimes as substitutes for true tea, by peasants and other poor persons (E. H. Wilson, 1929: 385). Some herbal teas in ancient China, like true tea, likely had a bitter taste. In any case, when first coming into use in North China, tea was commonly confused with other bitter plants which were designated by the same written character. One of the ancient meanings of the character was "sow thistle, bitter herbs, bitter," and another, "leaf from which a bitter drink is made, tea plant, tea" (Bodde, 1942: 75-76, 76n).

Many scholars believe that tea was initially used as a medicine, and Watt (1908: 237) suggests a second stage in which tea was used as a vegetable (see paragraph on *leppett* below). Sealy (1958: 124-25), in a similar vein, observes that in Assam, Burma, Indochina, and southernmost China, the leaves of Assam tea (*C. sinensis*, var. *assamica*) have long been a stimulant, whether chewed (see *mieng* below) or made into a drink, but that tea was not a beverage until rather recent times. He suggests that use of *assamica* tea spread north from that area to the Chinese, and that initially they used it a medicinal drink. Then they discovered that var. *sinensis* provided a pleasant and stimulating beverage, and began using it on an everyday basis. If he is right, tribal peoples, many of them Tai-speaking, played a critical role in the sequence of events that led the Chinese to become the world's leading tea-drinking people.

The earliest clear literary reference to tea drinking in China, or anywhere else, is for Szechwan during Han times, but some place tea cultivation there still earlier. Initially it seems to have been a medicine, but it then became an everyday beverage whose use spread until, by late T'ang times, it had become general in China. Szechwan continued to be the nation's foremost tea producer until the eleventh century, but T'ang tribute lists show that tea and tea leaf-buds (used for the best teas) came to the royal court at Ch'ang-an (Hsi-an) or Lo-yang from eight provinces that stretched from Szechwan in the west to Fukien and Anhwei in the east (Figure 78).[3] T'ang was a time when tea sellers were very popular, and when there arose a class of wholesalers who purchased tea from peasants and stockpiled and shipped it. T'ang was when the first treatises were written on tea and its cultivation. It was when the state made its first efforts to establish a tea monopoly for the revenue it could provide. That the status of tea was increasing, too, is revealed by the fact that it became customary in T'ang times to offer tea to a guest, as in England today; that experts would argue about the ranking of teas, where one could obtain the best water for making tea, and such; and that the finest tea moved northward by barge over the Grand Canal to the capital. Tea retained its position as the highest-status Chinese beverage even in Sung times after it had become a necessity for even the most lowly of people. Tea houses, which have played an important role in Chinese social life since, were

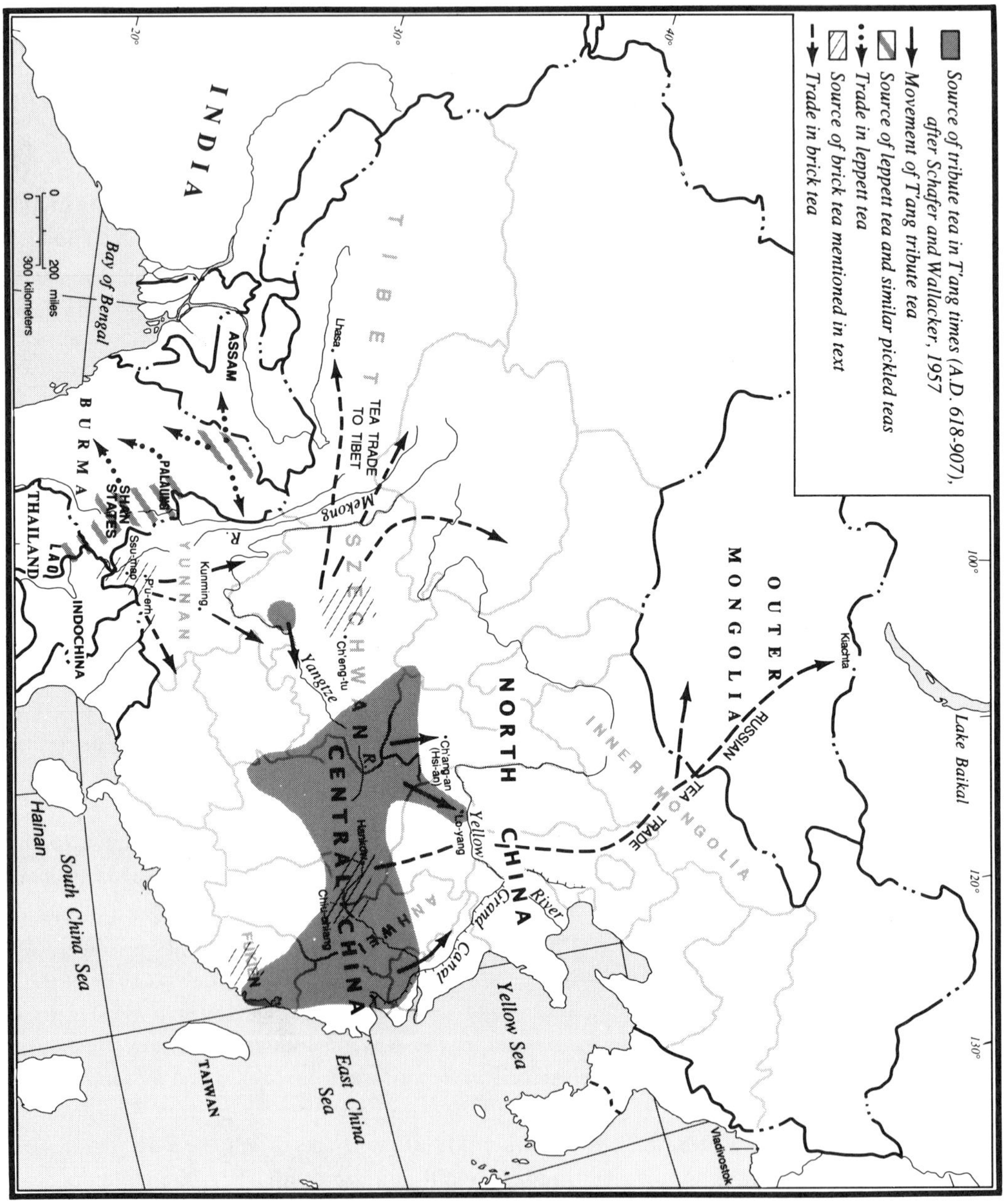

Figure 78. Tea sources and trade mentioned in the text

already known in T'ang and Sung times, and some them were large, handsomely-decorated places (Bretschneider, 1882-95, 2: 130-31; Dudgeon, 1895: 1; Watt, 1908: 211; E. H. Wilson, 1929: 375; Bodde, 1942; Goodrich and Wilbur, 1942: 195-96; Schafer and Wallacker, 1957-58: 222, Map 1; Needham and Lu, 1962: 452; Eberhard, 1969: 200; F. R. Carpenter in Lu Yü, 1974: 12-17; Freeman, 1977: 156, 159, 160; Schafer, 1977: 122-23, 136; Y. S. Yü, 1977: 70; H. L. Li, 1983: 49-50; Bray, 1984: 75, 428; Blofeld, 1985: 48).

In early Szechwan, tea was prepared with flour, salt, and ginger, but in later China it was commonly made without other ingredients, or mixed with flowers, tangerine peel, or other scents and flavorings.[4] The pastoral peoples of China's western and northern borders, however, prepared their tea with salt, milk, and/or butter, and during Yüan times butter tea made inroads in China proper (E. N. Anderson, 1988: 76). Among the border peoples were the Mongols who, according to Laufer, may have spread the habit of tea drinking to western Asia in the thirteenth century, along with their name for tea, *ch'ai*, which, in turn, came from the North Chinese *ch'a*.[5] The first European publication to mention tea dates from the mid-sixteenth century, and the first tea shipments arrived in Amsterdam and Paris early in the seventeenth century. The Dutch East India Company, a major factor in this trade, sent its first embassy to the Emperor of China in 1655. Interestingly, since this was during the Ch'ing (Manchu) dynasty, tea in the imperial court was drunk not in the Chinese way, but in the way of the border pastoralists, with warm milk and salt added. The Chinese teas shipped to Western Europe came from China's southeastern ports, and the names for tea in that part of Europe (Spanish: *té*; French: *thé*; Dutch: *thee*; etc.) derive from the Fukienese *tay*.[6] China dominated the tea trade by sea to Western Europe until the nineteenth century, when plantation tea produced in India and Ceylon came to challenge the Chinese teas (H. Phillips, 1827, 2: 285-87; de Candolle, 1886: 119; Ball, 1906: 681-82; Couling, 1917: 551; Laufer, 1919: 553-54; H. L. Li, 1959: 83-84; I. H. Burkill, 1966, 1: 421-22; Eberhard, 1969: 200; Schafer, 1977: 123; Flaumenhaft and Flaumenhaft, 1982: 154-55; Bray, 1984: 429).[7]

Its Use in Traditional Times

In contrast to India and Ceylon, traditional China derived little of its tea from plantations, and instead, tea usually came from small plots that could not be used for other crops. The tea plant cannot take drought but it does well under hot conditions if there is sufficient precipitation. As a result, 95 percent of all tea was produced in Central and South China between latitudes of 24° and 32°N. Though the region of tea cultivation is somewhat similar to that of rice, the latter is usually an irrigated crop of lowland areas, and tea of unirrigated hill-slopes. Altogether 66 percent of all China's tea in the 1930's came from the hills south of the middle and lower Yangtze Valley, a region of hot, rainy summers, and typically with precipitation of from 60 to 80 inches annually (S. W. Williams, 1895, 2: 40-41; Ball, 1906: 682-83; F. H. King, 1927: 287; Buxton, 1929: 110; Cressey, 1934: 330-31).

In China, teas are said to number "ten-thousand," indicating that there are many of them, that they are of indeterminate number.[8] This is a testimony to the long-time selection and care devoted to tea in China, like the mulberry but in contrast to the disinterest shown for tree crops of most other sorts (E. N. Anderson, 1988: 110). Some Chinese teas are of indifferent taste and low price, others outstanding in taste and very costly, and often known by unusual names such as "Old Man's Eyebrows," "Hairy Crab," "Sparrow's Tongue," "Water Nymph," and "Clear Distance." The essential forms of tea, however, are (a) black, or fermented; (b) oolong, semifermented; (c) green, or unfermented; (d) scented (as with flowers of jasmine, gardenia, chrysanthemum, narcissus, rose, orange, lychee, osmanthus, or other plants);[9] and (e) "brick," tea which is steamed and compressed into a solid or semi-solid product of some sort.[10]

Though tea is China's national drink and most people consume it in amounts far greater than in the Western world, those amounts vary considerably from person to person. At one end of the spectrum were people who drank many cups every day, such as the old monk of T'ang times who attributed his longevity to drinking 40 to 50 cups of tea on a normal day and often more than a hundred cups. At the opposite extreme were poverty-stricken people who could not afford tea regularly.

Not everyone observed the rules, but there were proper ways of making and serving tea[11] and an etiquette of tea drinking in traditional China. An oft-noted custom in pre-Republican times was the honoring a formal guest by an official, who, holding a tea cup in both hands, presented it to the guest, who was then expected to rise, receive the cup in the same manner, and sit down only when the host had done so.[12] Giles writes of tea used in such formal meetings as "guest tea," not intended for drinking but for signaling the end of the visit. When a host raised his teacup, this indicated that the visit was at an end. Should an official guest overstay his welcome or the host have another appointment, the host might, as an alternative, give a gentle hint by touching his cup with his fingers. An official guest, on his part, might conclude the meeting by raising his cup or inviting the host to drink his tea, which, when he did so, marked the end of the visit. "A guest," writes Giles, "no sooner raises the cup to his lips than a dozen voices shout to his chair-coolies" (Giles, 1876: 16-17; Ball, 1906: 259-60; Waln, 1935: 45, 48; C. J. H. Halcombe, cited in Holmes, 1948: 285; F. T. Cheng, 1954: 150).

Tea may be prepared in pots or cups with boiling water, and is consumed hot and normally without sugar or other additives. Today the Chinese often keep hot water in thermos bottles for use in making tea during the day. Before the time of thermos bottles, they commonly kept tea warm by placing a pot of hot tea over glowing charcoal, or in a covered basket or wooden container insulated on all sides by closely-packed cotton, except where its spout emerged. At meals, tea was customarily served beforehand and afterwards, but not with the meal itself.[13] Tea was also served at a broad range of other social and business occasions, as to guests visiting a home, to customers at better shops, and to participants in official conferences. When a Chinese writes to a friend to invite him to visit at home, according to K. Ch'en, the proper expression is "the tea is ready," and indeed a cup of tea is the first thing such a visitor is offered when he arrives (Holmes, 1948: 284). Tea was such a universal symbol of welcome that on one occasion a bandit, after threatening to use force and insisting on a search of his intended victim's possessions, was offered a cup of tea by the quick-thinking victim. The bandit's manner changed at once and he later departed on the mere assurance by the victim that he had neither money nor firearms. The bandit also forgot to take his hat, perhaps because he was so flustered by the unexpected hospitality.

China had no elaborate tea ceremony like that of Japan but tea was used regularly in rituals and on other important occasions. It was customary, for example, for a potential bridegroom to send a gift of tea, symbolic of unity, loyalty, and steadfastness, to the family of his fiancée. It was also the practice for people to offer tea at rites and to present it to the image of a deity being carried in a procession through a neighborhood (Doolittle, 1865, 2: 127-28; Ball, 1906: 259-60; F. H. King, 1927: 77; Hommel, 1937: 159, 311; Dudgeon, 1895: 6-7; F. T. Cheng, 1954:

149-50; Needham and Lu, 1962: 452; Osgood, 1963: 167; W. W. Chang *et al.*, 1970: 500; B. Y. Chao, 1972: 230; E. N. Anderson and M. L. Anderson, 1977: 377-78).

Tea houses in traditional China were found in villages, towns, cities, and also in the countryside, especially at scenic spots such as springs or ancient shrines. They did, however, differ notably in number from region to region.[14] They also differed in size and quality of construction, ordinary village tea houses sometimes having mud floors and thatch roofs, and the finest tea houses of urban areas occupying fine buildings in attractive surroundings. Tea houses were noisy, happy places of social exchange, with patrons talking with friends and others about all manner of things, business and social, sometimes for many hours. In Szechwan, where Spencer considered village tea houses to be community social centers, time spent in them was so great as to lead Chinese from other provinces to refer to Szechwanese as "tea-house loafers." Tea houses commonly served not only tea but snacks and perhaps light foods of other sorts, such as sweets, soup, congee, and noodles. In some places, in fact, they were mostly restaurants, and in others they sold more wine than tea (Spencer, 1942; K. Lo, 1972: 49-68). Their usual patrons were men, who might be entertained by singing, story-telling, recitations of poetry, and even sexual pleasures (Gray, 1878, 2: 72-75; E. N. Anderson and M. L. Anderson, 1977: 371; Blofeld, 1985: 47-59).

Besides being a beverage in China, tea is a flavoring. One tea-flavored food, mentioned previously, is tea eggs. Examples of tea-flavored main dishes are "Tea Smoked Chicken" (C. B. T. Lee and A. E. Lee, 1970: 95-96) and "Stir-Fried Shrimp with Dragon-Well Tea Leaves," the tea in question being one of China's best green teas (W. W. Chang *et al.*, 1970: 182).

The processing of tea leaves into a pickled tea or vegetable preserve called *leppett* or *letpet*, is found in Burma, especially in the Shan States, and in certain hill areas between Burma and Assam. The tea leaves in question come from plantings of Assam tea, *C. sinensis*, var. *assamica*. Early in this century such pickled tea, transported in large amounts by coolies and pack animals, was the most valuable product traded from the Shan hill country to lowland Burma.[15] The product was made by softening tea leaves through steaming or boiling; rolling them and allowing them to cool; compressing them into pits, or into bamboo tubes that are drained and stoppered; and, finally, burying and aging them. The Burmese regard the end product as a delicacy and traditional food essential for ceremonial needs (they commonly prepare it with sesame, oil, and garlic, in which form it has a taste reminiscent of olives). Small gifts of leppett would accompany Burmese invitations to feasts and ceremonies, being a polite suggestion that the invitee make a small monetary contribution for the feast. In the last century, leppett tea was sold not just in Burma and parts of India, but in many areas of China's southwest. We have not determined whether the Shan living in China made the product, but they did obtain pickled tea in trade from the Palaung, a Mon-Khmer people. Burmese Shans, Burmese, and Chinese and Indians living in Burma are also buyers of Palaung pickled tea. A preserved tea used by the Lao of northern Thailand, called *mieng* or *miang*, is made in a similar way, but for chewing. Chewing of *mieng* is widespread among the Lao and almost essential for men involved in heavy labor such as rowing or poling (Crawfurd, 1829: 115, 450; Watt, 1889-96, 2: 74; 6, Part 3: 449; 1908: 235-37; "Lao tea," 1892; "Leppett tea," 1896; Ferrars and Ferrars, 1901: 65, 70, 72; Nisbet, 1901, 1: 175, 446; 2: 191; Milne, 1924: 234-37; Ukers, 1935, 1: 237, 451; 2: 398; Sealy, 1958: 1, 122-23). Tea leaves contain an anti-thiamine factor, and chewing 12 gm. of tea leaves daily can destroy 11 mg. of thiamine daily under some conditions.

In northern and northeastern Thailand, the continuous chewing of such tea leaves may be related to the high incidence of beriberi there (Vimokesant *et al.*, 1974: 373-74).

Brick tea deserves special mention here, for it has long been used in China and exported, though not common in Western Europe or the United States. The practice of compressing tea into compact forms apparently developed because of a need for products that could stand rough transport to distant markets. This is why Szechwan is prominent in brick tea manufacture, for it supplies the Tibetan market by way of rough mountain trails. Some Szechwan brick tea for the Tibetan market is made from fine quality leaves which, after drying and fermenting, are steamed in a cloth over a boiler, placed with boiled glutinous rice flour into a mould, and, with hand pressure, made into a brick which is then dried and wrapped. In early times, blood was commonly used as the cementing agent, but, because of Buddhist objections, that practice has ceased. Szechwan's finest-quality brick tea was intended for distant areas of Tibet such as Lhasa, whereas brick teas of lower quality were made for sale in eastern Tibet. These were prepared of coarse tea leaves, sweepings, and leafy twigs, some an inch or more in circumference, to get a product described by one British observer as "crows' nests pressed into cakes." If twigs comprised a large enough part of the material, a strong enough brick could be made just through steaming and compression, without a cement of rice flour.[16] Whatever their manufacture, bricks were normally used in Tibet for butter tea. This is a combination of hot tea, butter, and salt mixed in a churn, then re-heated to provide a dietary staple that resembles soup more than tea as we know it. Butter tea may have *tsamba*, parched barley flour, added to it, and sometimes the mixture was kneaded by hand and consumed as dough pellets (S. W. Williams, 1895, 1: 241; 2: 50; Hosie, 1897: 93-95; H. R. Davies, 1909: 279; E. H. Wilson, 1929: 377-84).

A second type of brick tea, which many consider to be a separate form, tablet tea, was developed for the Russian tea trade. It was made using tea dust, often in later times with steam or hydraulic presses, to form hard, sharply-defined small bricks (tablets) bearing attractive designs. For superior bricks, tea dust was carefully selected as to quality. A second, coarser brick tea which included leaves and stalks was made for the Russians to trade to Mongols and other Central Asian peoples along the way, who commonly prepared it with salt and milk or butter.

Tea had been traded regularly from China to Russia since the beginning of the eighteenth century. In the initial period, all tea was handled by officials and agents designated by China and Russia, and all passed through the commercial settlement of Kiachta, south of Lake Baikal and along the border of Outer Mongolia and the Russian Empire.[17] A caravan from China to Russia by this route covered a distance of thousands of miles and took a period of six months, with caravans across Mongolia consisting of 200 to 300 camels. The tea that reached Russia was used in some places as a sort of currency, and was consumed in various ways, including that of placing a lump of sugar between the lips and sipping tea through it. The tea trade by caravan reached its height from 1860 to 1880, and the Russians came to operate factories for making bricks in China, especially at Hankow (Han-k'ou) and Chiu-chiang (Kiukiang) on the Yangtze River, in China's most important tea-producing area. The caravan trade via Kiachta was hurt by the opening of the Suez Canal, with sea shipments to Odessa via the canal becoming important in the 1880's. Completion of the Trans-Siberian Railroad to Vladivostok in 1900 led to the final decline of the caravan trade via Kiachta. Russia continued, however, to be the major importer of Chinese tea, taking two-thirds of China's tea exports in the years before the first World War. After the war, however, there was a sharp drop in that trade (Gray, 1878, 2: 215-16;

Moncrieff, 1890; S. W. Williams, 1895, 1: 206-8; Couling, 1917: 551; Ukers, 1935, 1: 294, 305-6; 2: 96-100, 337; F. P. Smith, 1969: 84; Klein, 1987: 240).

Brick teas are also made in various of China's provinces for marketing at home or abroad for medicinal purposes. In Fukien, for example, there is a brick tea, made of coarse tea leaves (aged for over a decade) together with other ingredients, that is commonly used to reduce fevers (Gray, 1878, 2: 216). Perhaps most famous of the medicinal brick teas are those called P'u-erh tea (*P'u-erh ch'a*), a name deriving from that of a prefecture in southern Yunnan. These teas are said to have gained recognition beyond that province only during the Yüan dynasty, when Mongol troops of the Kublai Khan and the Chinese troops they were fighting learned of their virtues. Today the term P'u-erh tea embraces a broad range of teas of distinctive taste from Yunnan which may be marketed in a loose-leaf form, as bricks, or, on occasion, as paste. Late in the last century, however, the term was used for a form of tea, thought to be Assam tea (*C. sinensis*, var. *assamica*), that was produced along a tributary of the Mekong in the Shan States, part of what today is southernmost Yunnan. It was shipped north, especially to Ssu-mao and Yunnanfu (Kunming), for further processing and shipment to other parts of China. Some of the finest P'u-erh tea was made into balls as large as a man's head for the Chinese royal court, which led to its other name, "man-head tea." There is also another, more common, compressed form of P'u-erh tea made of good quality tea: a flat, circular cake several inches across which can be found in Chinese grocery stores in Oakland and San Francisco. And there are still other forms of P'u-erh tea, including lower qualities of various types, some marketed just in Yunnan or intended for Tibet. Whatever its place of origin in later times (when some came from Szechwan), P'u-erh tea enjoys an outstanding reputation as a medicinal, and is found in large drug stores throughout China as well as in wealthy monasteries in Tibet. Because it is strong and bitter, it is taken in small amounts, seldom more than a cup or two at a time, as a digestive after a large meal. It is highly regarded as a stimulant and an aid to digestion, for its ability to dissolve fats, neutralize poisons, remove obstructions, and stimulate secretion. Indeed the Chinese have told what appear to be tall tales about its powers as a solvent. One is that if P'u-erh tea is added to a pot of meat or chicken, even the bones are turned into broth, and that the pot cannot be of iron lest too much metal be dissolved, making the dish inedible ("P'u-êrh tea," 1889; H. R. Davies, 1909: 96-97, 293; E. H. Wilson, 1929: 384-85; Sealy, 1958: 123-25; F. P. Smith, 1969: 85; Blofeld, 1985: 75-77).

The Chinese believe ordinary teas, as well, to be effective in promoting health and well being, and specifically green and semifermented teas rather than black ones.[18] A long list of the claimed beneficial effects of tea drinking has been drawn from the Chinese literature by Blofeld (1985: 155-60). Tea drinking is said to lessen thirst, eliminate fatigue and depression, create a sense of well-being, encourage alertness and clarity of thought, and contribute to longevity. It is also said to have more specific physiological benefits, among them aiding digestion, increasing the flow of blood in the body, hastening the removal of alcohol and other harmful substances from the body, acting to relieve pain in the limbs and joints, countering anemia, enhancing disease resistance, helping to keep skin youthful, impeding tooth decay, and giving zest to the eyes. Blofeld writes that most of the claims (except those about longevity and tooth decay) have been sustained by medical research, and he provides brief summaries of the evidence, as well as noting that certain vitamins and minerals are found in tea. Like others, he cautions about the caffeine in tea, for the caffeine in five cups of tea daily can bring on nervousness, sleeplessness, headache, and other symptoms. Not mentioned, however, are the facts that the tannin of tea may hinder digestion and that tea, whether drunk as a beverage or chewed in a fermented form, contains a thiamine inhibitor and can cause thiamine deficiency in persons with low thiamine intake (Vimokesant *et al.*, 1974; J. D. Keys, 1976: 189-90; Nutrition Reviews, 1976: 145-46; Briggs and Calloway, 1979: 460).

ALCOHOLIC BEVERAGES

Problems with Terminology

In China, alcoholic beverages of all kinds are called *chiu* (Cantonese: *chau*), a term with an original meaning of "beer" or "ale," but today used in the general sense of "alcoholic beverage." Though *chiu* is commonly translated into English as "wine," today and in the past most Chinese alcoholic beverages are not made from the fermented juice of grapes or other fruit, and are therefore not wines in that sense, but beers, ales, or distilled drinks made of cereals. The word *chiu* may be accompanied by qualifiers to identify a beverage more precisely, such as *shao-chiu* ("burnt wine") or *huo-chiu* ("fire wine") for distilled liquors; *huang-chiu* ("yellow wine")[19] for rice wine; and, in modern times, "beer *chiu*," "whiskey *chiu*," or "brandy *chiu*" for beverages of sorts asked for by foreigners. Despite this, there can be considerable confusion in the matter, as in the case of a Westerner at a Chinese affair who, after agreeing that he would like some wine, is given beer, whiskey, or brandy, perhaps because these are foreign "wines" (F. T. Cheng, 1954: 121). Confusion may also be caused by foreign use of the term "samshu" (from *san-shao*, "thrice fired") not just for a distilled sorghum liquor known by that name, but as a general one that embraced fermented liquor of all sorts (Dudgeon, 1895: 20; Giles, 1900: 245; Couling, 1917: 497, 603). We would, therefore, urge caution on the reader of English-language accounts of Chinese alcoholic beverages. In what follows, we attempt to minimize confusion by using the term "wine" not in the restricted sense for fermented beverages made from grapes or other fruit, but for fermented beverages of all sorts, including ones made of rice (rice wine) and other grains (*i.e.*, millet wine). In what follows, we do not, however, include distilled alcoholic beverages among the wines, but place them in a separate category.

Alcoholic Beverages in Antiquity

Grain Wines

"Wines" (fermented beverages) have been made in China from fruit juices, cereals, and other substances since antiquity. In breweries uncovered in Shang sites in Honan and Hopeh there were remains of plums, peaches, and jujubes along with an assortment of vessels and even a funnel (Xing and Tang, 1984: 56). Despite this, grain wines, especially those made with millets, were far more important in ancient North China (K. C. Chang, 1977b: 30). Flavorings, aromatics, and medicines were commonly added to such wines (an initial intent may have been to keep the wines from becoming sour), which were used not only at meals and feasts, but at sacrifices and other ritual occasions (K. C. Chang, 1977b: 30; Chi Li, 1977: 254; E. N. Anderson, 1988: 43). The sixth-century *Ch'i-min yao-shu*, which pictures life in North China, focuses on grain wines made from millet (common millet and foxtail millet) and rice, with ferment cakes of wheat and rye (T. C. Huang and Y. T. Chao, 1945-47).[20] Not much later, in T'ang times, rice wines, made especially with glutinous rice, were produced and shipped northward in such amounts as to make millet wine (mainly from glutinous millets) an oddity of China's far north (Schafer, 1977: 119).

Grape Wine, Mead, Toddy, and Koumiss

The reader will recall, from our section on the grape, that grape wine has been made in China since Han times. Of particular importance was an elongated, purple wine grape, known as "Mare Teat," which was introduced in the seventh century A.D. and seems to have given a new impetus to grape-wine manufacture in China (Sampson, 1869a: 53; Laufer, 1919: 232; Schafer, 1963: 142-43; 1977: 121-22). The area around T'ai-yüan, capital of Shansi, was celebrated during T'ang times for its grape wines, with such wines sent to the T'ang royal court in tribute from both northern Shansi and Turkestan (Schafer and Wallacker, 1957-58: 231). Shansi continued to be a center of grape-wine manufacture in Mongol times, but the founder of the Ming dynasty, after accepting grape wine from T'ai-yüan the first time it was presented in tribute, ruled that it should not be presented again. This seems to have brought on a decline of grape-wine making, but grapes, later described as exceptionally large and tasty, were still grown in Shansi when the early Jesuits arrived, and provided wine for them and for those who followed (Cordier, 1920: 75-76; Yule, 1921, 2: 13, 15-16).

As shown previously, honey was not abundant in China and the available supply was used as a sweetener and in medicine. This meant that little was left for the making of mead. Thus, though mead has been made by the Mongols and Chinese since early times, it must have been of quite minor importance for there are relatively few clear references to it in accounts of Chinese alcoholic beverages (Schafer, 1963: 141; 1977: 121; F. P. Smith, 1969: 417; Mote, 1977: 207; Needham *et al.*, 1980: 139, 142n).

Fermented toddy, or palm wine, mentioned briefly in our consideration of sago, was a product of the far south, one made from the juice of the sugar palm (*Arenga saccharifera*), coconut palm, and other species (Schafer, 1967: 174-76; 1970: 121).

Koumiss, fermented mare's milk, is associated with the peoples of Inner Asia (Mongols, Turkic groups, Tibetans), though, as will be indicated in our section on milk and dairy products, it gained a certain following in China proper at one historical period or another, especially when nomadic influences were powerful.

Distilled Liquors

Li Shih-chen, in the *Pen-ts'ao* (late sixteenth century), identified as distilled liquors: (1) one called *shao-chiu*, or "burnt *chiu*," a term he considered synonymous with *huo-chiu*, "fire *chiu*"; and (2) *araki* or *arrack*, terms widely used in the Old World for "distilled liquor," which he rendered as *a-la-chi*. He then wrote that the manufacture of burnt *chiu* was not known in ancient China, but had first begun there in Yüan times (A.D. 1271-1368). Li's statements have been used by many Chinese and Western scholars to support the view that distillation was unfamiliar to the ancient Chinese, that it was a procedure developed in the West, and that the Chinese first distilled liquor in Yüan times (Dudgeon, 1895: 18, 23; Laufer, 1919: 237-38; F. P. Smith, 1969: 417, 419). This view has long been questioned by other Chinese and Western scholars who have found burnt *chiu* and/or the distillation of alcohol mentioned in Chinese literature prior to the Yüan dynasty (Schafer, 1967: 190; 1977: 122, 139n; Needham *et al.*, 1980: 133, 133n; Xing and Tang, 1984: 56-57). Forbes (1948: 10-11), for example, notes references to such distillation by two Sung Chinese chemists. One of them, in a treatise on alcoholic beverages, briefly mentions famous distillers; and the other discusses distilled liquors and their manufacture. Forbes believed, however, that there was no evidence of such distillation in China before Sung times, when distillation apparatus was very likely introduced from the West.

This view is challenged by Needham *et al.* (1980: 132-58), who, in a lengthy weighing of the evidence, present striking new insights into the history of distillation and distilled liquor in China. They cite a Chinese account from late in the third century A.D. that tells of a grape *chiu* in "the Western regions" that remained drinkable for years, even as long as a decade, a *chiu* that could make one drunk for days. Other Chinese accounts suggest that such a beverage may have already been known by the second century B.C. or even before. Needham *et al.* doubt that the *chiu* in question was distilled, but think that it was frozen-out wine, in which alcohol levels are raised by freezing fermented beverages, then separating the ice from the strong liquid remaining. If so, these are likely the earliest references to frozen-out wine anywhere in the world. Such frozen-out wine, in the Needham view, was the predecessor of all "strong liquor," a drink that laid the groundwork for development of alcohol distillation in China by whetting people's appetite for such products. Even more, the argument goes, the Chinese were the first to distill alcoholic beverages, perhaps by the sixth or seventh centuries A.D., long before such beverages were made in the Western world.[21] Needham *et al.* suggest the possibility that there was still another step in the sequence of invention, one deriving from the Taoist perception of yin and yang. By such thinking, frozen-out wine, a strong liquor obtained by subjecting wine to extreme cold, would have been regarded as pure yin. This, in turn, may have led people to subject wine to the opposite, extreme heat, to bring on something that was pure yang. The result was burnt *chiu*, the first distilled liquor. The relative ease of the distillation process for making strong liquor led it to prevail over that of freezing wine. It was applied in China both to cereal wines and grape wines, and much later it may have spread to the Western world, perhaps by way of Central Asia.[22] It may even have been that the *chiu* the Chinese sold in Hainan, Annam, and widely in Southeast Asia in Sung times was actually distilled liquor and not fermented wine as many have believed. If so, this would be an early example of a form of trade, from an advanced people to less advanced ones, that could greatly facilitate the spread of more-advanced cultures, as it did in the case of European overseas expansion.

Alcoholic Beverages in Traditional China

Principal Kinds

Fermented alcoholic beverages of great variety were marketed in traditional China, but the most common ones were made of rice, with glutinous rice yielding the best wine, followed by common millet and foxtail millet (F. P. Smith, 1969: 417). Sorghum wine is also common, and grain wines of other sorts. The best known rice wine is *Shao-hsing chiu*, named for a city and district in Chekiang famous in China since early times for its rice wines. It comes in many brands and varies in length of aging and quality of taste. *Shao-hsing chiu*, which also received highly favorable notices from various early European travellers, was marketed throughout traditional China and also exported, for example to California and Australia. *Shao-hsing chiu* resembles sherry in strength and taste, and like sherry it may serve as a drink or be used in cooking. In English-language cookbooks, sherry is often substituted for Shao-hsing wine. At dinners and ceremonies, Shao-hsing is normally served from a small wine pot and drunk in tiny cups, the wine not cold or at room temperature but warmed. To be good, according to gourmet

F. T. Cheng (1954: 122), it should have been aged seven years or more, though really good *Shao-hsing chiu* is hard to come by. Cheng observes, however, that when fully mature, after about fifteen years, "it is exquisite, having the bouquet of that famous French wine *Château-Yquem*, minus the sweetness." This must be the sort of rice wine the Chinese are referring to in the common saying, "*Shao-hsing* is the polished scholar; *san-shao* is a rowdy" (F. P. Smith, 1969: 420).

Most common among China's distilled liquors[23] are those made of sorghum (*kao-liang*), which are known as *kao-liang chiu*. Their production and consumption is concentrated in North China where most sorghum is produced, but they are made and consumed elsewhere as well. In Szechwan and Yunnan, for example, as much as nine-tenths of the sorghum harvest goes into making distilled liquor. *Kao-liang chiu* is reminiscent of vodka in its clarity, strength,[24] and flavor. It comes in various forms, one of which, *mei-kwei-lu* ("rose dew"), is flavored with the petals of a particular species of rose (*Rosa rugosa*). Like rice wines, distilled sorghum liquors are served warm,[25] and may be used in cooking, and as a replacement for rice wine when the latter is weak.

Distilled liquor may also be made from rice, lesser millets, other cereals, and a broad range of products, among them potatoes, sweet potatoes, grapes, sugar cane, and, among Inner Asian peoples, from koumiss.[26] In modern China, two of the best-known distilled liquors, named for their places of manufacture, are *Fen-chiu* (for a district or a nearby river in Shansi), a strong product first made in T'ang times; and *Mao-t'ai* (for a town in northwestern Kweichow), also strong, first made in Sung times (but marketed beyond the local area starting only in the eighteenth century) (Doolittle, 1865, 2: 230; Couling, 1917: 603; E. Watson, 1923: 460, 577; Arlington and Lewisohn, 1935: 272; F. P. Smith, 1969: 459; W. W. Chang *et al.*, 1970: 502-3; V. Y. N. Hsu and F. L. K. Hsu, 1977: 308; K. Lo, 1979a: 265; Needham *et al.*, 1980: 149-50; Bray, 1984: 451; Y. K. Martin, 1984: 232; Needham and Lu, 1986: 427; Zheng, 1987). *Mao-t'ai* is made from sorghum and wheat and is 106-proof, yet is described as "mellow and sweet" (Zheng, 1977: 40). It has won various international awards, and gained special press attention when Premier Chou En-lai served it to President Nixon in 1972.

Many early Europeans reaching China found domestic grape wines, which they called "Mandarin wine," unpalatable. This, however, presented an opportunity for trade, and in the eighteenth century East India vessels made it a practice to carry wine to China. Sherry was preferred because of its strength and ability to take heat, and at one time it sold in Canton for more than double its price in Cadiz. Spanish vessels also carried wine to Macao by way of the Philippines, much of this wine purchased by Chinese and finding its way to the royal household in Peking (Laufer, 1919: 238-39). Perhaps because of familiarity with European wines, certain Chinese ultimately came to cultivate European grapes for wine making. We refer here to the Singapore Chinese merchants who in 1895 established a company, mentioned previously, called the Chang Yü Pioneer Wine Company at Yen-t'ai (Chefoo) in Shantung. Their initial purpose was to manufacture wine from the native grapes of that province. Unfortunately, the wine turned out to be expensive and of low quality, so then American and Austrian vines were imported, with the first wines from these grapes marketed in 1914 (Couling, 1917: 603). Production of grape wines obtainable in Peking the 1930's was said to come from Shantung and Shansi (Arlington and Lewisohn, 1935: 272), and North China has continued to produce such wines until the present day. With establishment of the People's Republic, there has been a striking increase in the production of wine grapes (from roughly 100 metric tons before to about

30,000 tons in 1980) and wine (H. Huang, 1980: 465-66). Though production remains small by Western standards, one can now find Chinese grape wines readily in Hong Kong, which, as a sophisticated, international city, is also well-stocked with grape wines from Europe, Australia, and other parts of the world. In the South Chinese cities we visited, by contrast, one has to seek out grape wines, often in government Friendship Stores which cater to foreign visitors. Even these stores normally sell only one or two kinds of domestic grape wine, the best brands we tasted being "Great Wall" and "Dynasty." Both are medium-dry white wines (alcohol content about 12 percent) from North China and are quite acceptable in taste; indeed, the former has won awards in international competitions in London and Madrid.

Also deserving brief mention are the alcoholic beverages, whether fermented or distilled, made with fruit, flowers, or other flavorings, aromatics, or medicinals. Among these are some we have mentioned previously, and others as well, including some medicinal ones that can properly be described as unusual.[27]

As late as 1895, S. W. Williams (1895, 1: 776) was able to write that beer, apparently Western-type beer, was unknown in China. It must have been introduced soon after, for C. P. Li (1948: 182) reports that such beer was introduced to China in the final decades of the Ch'ing period (C. P. Li, 1948: 182). Today one finds many brands of Western-type beer in China, with two-thirds of the nation's beer in 1984 produced in northern and northeast China, where most barley is produced (Shao Qiquan, Lu Wei, Li Ansheng, and Wei Rongxuan in Wittwer *et al.*, 1987: 210). That is the region, as well, of China's most famous Western-type beers, such as the internationally renowned Tsingtao Beer (first made by Germans, who occupied Tsingtao from 1897 to 1914), but such beers made elsewhere, among them the Pearl River Beer marketed in Canton, are quite acceptable.

Use of Alcoholic Beverages

Europeans have often commented on the fact that the Chinese are a temperate people, that few of them are heavy drinkers. This has not, however, been true of all groups in China throughout history. In Shang and Chou times there seems to have been some quite heavy drinking, especially among the nobility. In Yüan times, the Mongols were great drinkers, with Mongol emperors, several of whom died of alcoholism, holding great drinking parties whose extremes have been amply documented by Chinese and Western observers. There are also accounts from other historical periods of excessive drinking and its effects among the Chinese: of drinking parties, drunkenness, and drunken brawls (Soothill, 1923: 196, 198; C. P. Li, 1948: 182-83; Schafer, 1954: 5, 23-27, 52, 81-82; Osgood, 1963: 167-68; C. S. Wong, 1967: 86-90; Mote, 1977: 207; Spence, 1977: 278; K. C. Chang, 1980: 210-11; E. N. Anderson, 1988: 48).

There were, however, countervailing forces against consumption or excessive consumption of alcoholic beverages. Some derived from Buddhism, Islam, and certain Christian sects. Taoists did not ban alcoholic beverages, but said that they should be taken at proper times and in moderation. They did, however, speak out against drunkenness as damaging to one's spirit and reducing longevity (Maspero, 1981: 498-99). According to the *Confucian analects*, that sage set no limit on the wine he drank, but did not permit himself to become bewildered by it (Legge, 1893: 232). The Taoist and Confucian view of alcoholic beverages and drunkenness was held by most Chinese, who did not avoid liquor but, out of concern with virtue, morality, and regard for family and society, considered drunkenness bad. This led to moderation and created strong social pressure against excessive drinking, which entailed a loss of standing.

There are many expressions of this view in Chinese writings which, in one way or another, have acted to reinforce popular thinking and behavior. One finds among the classical sayings memorized by Chinese school boys in traditional China, the following one: "When you are still young, do not drink wine; to get drunk is a hideous thing." A more complete expression appeared in a popular tract on the rules of propriety: "The crimes committed on this earth are nearly all caused by wine, lust, money, and anger...It is true that the weddings, the funerals, the sacrifices, the rejoicings and so forth, cannot go off without drinking. The proverb says, 'Without wine, there is no feast.' But, when drinking, moderation is to be observed; you must avoid, for sake of drinking a few more glasses, to get intoxicated so that you will say but foolish things, beat people and swear against them. The proverb says, 'When one is drunk, one does nothing good'" (Wieger and Davrout, 1981: 143, 228).

Alcoholic beverages were also commonly used in ritual offerings, and consumed at meals and on festive occasions. Normally a host would determine the drink to be offered, with the guests, contrary to American custom, not given a choice of drinks. The drink was usually served in small cups and sipped rather than drunk. The number of drinks that guests at a dinner could consume was also determined in large part by the host, and a dinner guest would not think of asking for more. A host was usually generous with his drinks, and sometimes even sought to get some guests tipsy. A guest, in turn, might take action to prevent this. He might, for one, refuse drinks, though he would usually offer explanations so as not to offend his host. As an alternative he might leave his cup unemptied, or, according to Giles, he might hire one of the young men singers at a party to carry out by proxy whatever heavy drinking is called for (Doolittle, 1865, 2: 230-31; Giles, 1876: 154; Dudgeon, 1895: 53-63; Kulp, 1925: 325; F. T. Cheng, 1954: 117-21; C. S. Wong, 1967: 86; V. Y. N. Hsu and F. L. K. Hsu, 1977: 308).[28] Western observers have long suspected, however, that any explanation of low Chinese liquor consumption must go beyond the social realm. One suggestion is that tea is so ubiquitous and well established as to exclude liquor from many social affairs. Another suggestion, made late in the last century, is that opium, whose use had increased notably in towns and cities, came to displace liquor in some situations and led to a notable decline in liquor consumption and drunkenness (Dudgeon, 1895: 43-44). A third suggestion is that the Chinese custom of drinking wine not cold but warm speeds its effect on a person and tends to reduce the amount he consumes. And a fourth is the fact, long noted by foreigners, that after drinking even small quantities of liquor the faces of the Chinese may flush, and they may become confused and talk loudly. The Chinese recognize this too, for they have an expression, "to get red faced," for being slightly intoxicated (E. N. Anderson and M. L. Anderson, 1977: 343; E. N. Anderson, 1988: 121). Scientific research has confirmed this phenomenon, finding that Orientals (Taiwanese, Japanese, and Koreans) may flush and exhibit mild to moderate symptoms of drunkenness after consuming amounts of alcohol that have no discernible effect on whites (Wolff, 1972, 1973).[29] Whether this may have been a factor inhibiting drunkenness or not, and whatever the relative contribution of social pressures and competition from tea and opium to Chinese moderation in liquor consumption, sobriety has long been a characteristic of the Chinese. Among the Chinese on Taiwan, for example, Chafetz (1964: 298) was told that there were at most ten instances of alcoholism over a seventeen year period.

Besides serving as drinks, wines are common ingredients in Chinese cooking, as in "Drunken Chicken," a dish which may be made with chicken, chicken stock, salt, onion, gingerroot, and rice wine (Field *et al.*, 1976: 89). Also used are fermented rice-wine lees, which may serve in making sweets, such as the Shanghai or Eastern "Wine-Sediment with Orange Segments" (E. Chen, 1977: 153-54). They may, as an alternative, be used as a flavoring in other dishes, for example the South Chinese "Quick-Fried Sliced Chicken Breast with Sea Clams in Wine-Sediment Paste," and the Peking recipe "Sliced White Fish in Wine-Lee Sauce" (K. Lo, 1979a: 52, 134). Such dishes have been prepared since early times, as shown by the "white fish in wine lees" sent in tribute to the royal court from Anhwei in T'ang times. The white fish in question is the common carp, *Cyprinus carpio* (Schafer and Wallacker, 1957-58: 226). As for grape wine, it is used in only a few Chinese dishes (Sakamoto, 1977: 33). There are also various medicinal uses of alcoholic beverages (F. P. Smith, 1969: 417-21, 459), which we have mentioned previously.

Wine and other strong alcoholic beverages are considered "hot" in the hot-cold system of traditional China. Strong alcoholic beverages were thus prescribed as postpartum food and added to postpartum dishes. M. Mathews and L. Manderson (1981: 12, 14) found that Vietnamese-Chinese women liked champagne best and that in the week after giving birth they consumed from "three to ten bottles" of it whereas ethnic Vietnamese women preferred rice wine, Dubonnet, or brandy. Wheeler and Tan (1983: 55), in turn, report that Chinese women in London consume boiled chicken "with rice wine, brandy, or benedictine," and much ginger. Beer, because it is classified as "cold," is not included among the postpartum alcoholic beverages, and among San Francisco Chinese women is forbidden during pregnancy (Ling *et al.*, 1975: 129). One explanation given by Chinese in Hong Kong as to why beer, unlike other alcoholic beverages, is "cold," is quite interesting. They say that Westerners, who eat very large amounts of "hot" foods, would die without consuming a lot of something cold, and beer is the only food or drink they use in such amounts (E. N. Anderson, 1970a: 209-10; E. N. Anderson and M. L. Anderson, 1976: 148).

MILK AND DAIRY PRODUCTS

Origin and Spread of Dairying in the Old World

In considering the use of dairy products as food, one is dealing with an interesting historical and geographic problem which is best presented in its broadest perspective before turning to the Chinese scene. The milking of domesticated animals and consumption of their milk by humans are culture traits, learned by people.[30] These traits were not found among humans in the hunting and gathering stage which characterized almost all of human history, for early hunters and gatherers possessed no domesticated herd animals (many did have dogs), and thus lacked that prerequisite of dairying. As a result, early hunters and gatherers drank milk, that of their mothers or nursemaids, as infants, but after weaning they normally consumed no milk or other dairy products for the rest of their lives.

Dairying became possible for humans only with the domestication of herd animals suitable for milking. Such domestication first occurred, so far as is now known, at about 9000 B.C., the date of the earliest remains of domesticated sheep found in the Near East. Over the millennia that followed, goats, common cattle, and other animals joined Near Eastern herds. Our search of the archeological record, however, has failed to uncover convincing evidence that these animals were actually milked by humans in those earliest days of domesticated herd animals.

When and where dairying first began remains uncertain, but there are reports of a milking scene on a vase uncovered at an Anatolian site that may date roughly from 6200 to 5600 B.C., and there is clear evidence of dairying in Mesopotamia and the Sahara Desert of North Africa dating from about 4000 B. C. Following that time, the practices of dairying and human consumption of animal milk gradually spread to other parts of Eurasia and Africa. Even as late as A.D. 1500, however, there still remained large areas of the world in which dairying was absent. It was not found, of course, among any of the surviving hunting and gathering groups, wherever they were found. Dairying was absent among American Indian agriculturalists, though some of them, the peoples of the Andes, did have llamas and alpacas that might have served dairy purposes. It was also absent among the agriculturalists of almost a third of Africa, a region centered on the Congo Basin.[31] And it was absent, or at least not the usual practice, among the cultivators of much of Southeast and East Asia (Simoons, 1970), as well as those of New Guinea and Oceania.

The situation in traditional Southeast and East Asia is especially interesting from a geographic and historical perspective, for dairy products are prominent in the cuisines of peoples to the west (India, Tibet, and the Inner Asian frontiers of China), but little used in most of Southeast Asia, China, or in Japan or Korea. In this regard, conditions in the Malay Peninsula, as we saw it nearly three decades ago, are quite revealing, for that region is a meeting place of two immigrant groups, Indians and Chinese, who differ strikingly in their traditional use of dairy products. Malaya is tropical, and presents serious problems of animal feeding, nutrition, and disease to setting up successful dairying operations, and there were few private-sector Chinese engaged in dairying. There were, however, a considerable number of Indian dairymen who somehow had managed to identify and exploit situations, as along roads, where their dairy cattle and water buffalo could feed. Indian dairymen might also harvest wild grasses, or buy them from others; or they might buy certain edible wastes of agricultural processing, in one case pineapple scraps at a cannery. The success of Indian dairy enterprises rested firmly on a persistent demand for dairy products in the Indian community, for few if any Chinese, Malays, or Europeans would buy milk or other dairy products from an Indian dairyman. Indian demand, however, was far greater than Chinese. As shown in a household budget survey conducted in Malaya in 1957-58 (Malaya [Federation], Department of Statistics, 1959), the average monthly consumption of fresh milk and clarified butter by the Indians dwarfed that of the Chinese in every income category. For all income groups, average consumption in urban areas was six imperial pints of fresh milk per month for Indians, and less than a pint for Chinese. In terms of value of milk and milk products consumed, Indians also ranked higher than Chinese, a contrast most pronounced in rural areas and at the lowest income levels. Poor Indians would thus spend sizeable portions of their income on dairy products, whereas poor Chinese favored other foods. Even when Chinese in Malaya, largely through Western influence, came to consume dairy products (most conspicuously, ice cream) in good amounts, those products did not gain a significant role in the preparation of Chinese dishes. Striking proof of this is found in the results of a survey conducted in Singapore about 1940, an ambitious effort in which the surveyors attempted to identify the ingredients of all dishes sold in restaurants and along the streets of that city (Canton *et al.*, 1940). Of the 494 Chinese dishes found by the surveyors, dairy products of some sort were ingredients of only two, and both involved use of milk products in what seemed to be completely Western dishes. One was Quaker Oats to which beaten eggs were added, with the mixture then flavored with tinned condensed milk and sugar. The second was cookies made of imported flour, presumably wheat, and sugar, eggs, and butter. Of the 167 Indian dishes, by contrast, 41 were prepared with dairy products of one sort or another, usually imported ghi.

Hypotheses Advanced for the Failure of Dairying and Use of Animal Milk to be Accepted Universally

Returning to the general question of the milking and non-milking traditions, one notes striking differences in attitude. Many milking peoples looked on the practice of milking as simply one of their daily chores, necessary to obtain food they regarded as essential. Persons of non-milking tradition, on the other hand, viewed the matter differently. Some were simply indifferent to the possibility of milking animals. Some, however, considered the practice of milking to be unnatural and even immoral, as with Buddhists who held the taking of milk from a cow to be robbing its calf.[32] As for milk and milk products, some persons of non-milking tradition were just not interested in them. Others regarded them with pronounced negative feelings, as in the view that milk is a disgusting bodily secretion, unclean, like urine.[33] They might, as with certain Chinese, say that Europeans have a butter smell about them (Headland, 1914: 174), or, like present-day Japanese, say that people influenced by foreigners have a butter odor about them (Browning, 1986: 1). When Europeans did get persons of non-milking tradition to drink milk, moreover, some of them developed intestinal symptoms which in former times were presumed to be psychosomatic in origin. The non-milking views identified above, which we have designated "the non-milking attitude," is in large part a reflection of the cultural conservatism that Eduard Hahn (1896: 77-82) and Carl Sauer (1952: 87) hold largely responsible for the failure of dairying to spread throughout East Asia. In Sauer's words, people "did not take on the habit of milking because it was strange to their ways and they did not like it." This fits with Berthold Laufer's statements (1914-15: 168) that the situation is a matter of history, reflecting differences in the evolution of Chinese economic and social life from that of peoples to the west; and with K. C. Chang's view (1977: 7) that the failure of dairying to gain a major role in diet is essentially a reflection of Chinese culture, which accepted or rejected things from abroad according to whether they were congenial to existing patterns.

Other factors have also been invoked for the failure of dairying to spread. In East Asia, environmental and ecological hindrances have been noted by many. Their role, however, has been overemphasized by some, most notably the anthropologist Marvin Harris (1968: 368) who has argued that dairying was simply unsuited to Chinese conditions of intensive agriculture and dense populations. There was not enough land, he argues, both to grow crops for humans and to provide pasture for dairy animals. To do so would have placed an intolerable burden on the Chinese subsistence economy. Largely ignored by persons who hold such views, however, is the fact that in much of the Indian subcontinent, in places with similar densities of human population, dairying is well-established, and milk and milk products are highly valued foods. Whatever the problems of feeding dairy animals in India, people somehow have been able to find ecological niches for them, to integrate dairying into their agricultural systems. One might properly ask why the Chinese, an equally ingenious people, could not have followed the Indian example, as they did in seeking out India's methods of sugar refining.[34]

Closely related to the environmental and ecological hypotheses is the argument that dairy products were simply too expensive to produce in China in competition with alternative foods, such as soybean and pork products, and that this was probably the main reason why dairy products were avoided. Though high cost would indeed be likely to reduce the use of dairy products, like other expensive foods, by ordinary folk, one might ask whether it is likely to have excluded such products from diet to the degree found in much of China. One notes that dairy products are costly in India as well, and that the poor can afford only small amounts of them, yet they continue to use them. One might also observe that the Chinese are noted for their willingness to spend money for food, and that expense, though it reduces consumption, has not excluded such costly items as ginseng and bird's nests.

Anderson and Anderson (E. N. Anderson and M. L. Anderson, 1977: 326, 341; E. N. Anderson, 1988: 145-46) have also noted the traditional Chinese argument that they rejected dairy products partly to distinguish themselves from nomads living on their frontiers, peoples they regarded as inferior in culture, and partly to avoid becoming dependent on them for a supply of dairy products, as they already were for horses. One wonders, however, why the Chinese, if they felt concerns about dependence, did not, like the Indians, find niches suitable for dairying in China itself and enjoy what milk products they themselves could produce without reliance on the nomads. More plausible as an explanation is the prejudice the Chinese display for nomadic groups on their frontiers. At the same time, views of cultural superiority have not stopped the Chinese from taking over and sometimes even seeking out food plants and animals from various peoples they regarded as barbarians, for, as K. C. Chang (1977b: 7) has noted, foods have been taken over from foreigners since earliest historical times, and the Chinese are not "nationalistic to the point of resisting imports."

Another possible explanation for the failure of the Chinese to take up dairying is that a high percentage of them (probably more than 90 percent) at some time after weaning experience a sharp drop in the enzyme lactase in the small intestine. Such a drop, which is under genetic control, reduces an individual's ability to hydrolyze lactose, milk sugar, into monosaccharides (glucose and galactose) that can be absorbed, and may lead a person to become lactase deficient, a lactose malabsorber. Some lactose malabsorbers are also lactose intolerant, developing symptoms such as intestinal gas, distension, flatulence, cramps, and diarrhea when they consume, at a single sitting, lactose-rich dairy products in large amounts (containing lactose equivalent to or more than that found in eight ounces of milk) (J. D. Johnson *et al.*, 1974).[35] One may ask the question whether the Chinese associated milk drinking with the symptoms of lactose intolerance and, as a result, came to reject all dairy products as food. In weighing the matter, one must recognize first that some persons classified as lactose malabsorbers are able to consume considerable amounts of milk and other lactose-rich dairy products, even when first trying them, without developing symptoms; that other malabsorbers, though symptomatic initially, over time develop a tolerance to lactose which permits them to consume lactose-rich dairy products in quantity without ill effect; and that if a person experiences such symptoms he may or may not associate them with consumption of dairy products. Even if he does make such an association, an intolerant person may simply reduce the amount of lactose-rich dairy products consumed at one time to the point that he does not experience symptoms. As an alternative, he may remain symptom-free by consuming dairy products (such as fermented milk or hard cheese) in which processing has reduced or eliminated the lactose.

There may, of course, be a relationship between a group's history of milk use and its prevalence of adult lactose malabsorption. Where such prevalences are low, what appears to have happened is that a genetic condition, persistence of high lactase activity through life, became widespread because dairy products were critical dietary elements, and because people consumed fresh milk or other lactose-rich dairy products in good quantities over several millennia. However high the prevalence of lactose malabsorption, an entire population was not prevented from taking up dairying because of the genetic condition (Simoons, 1981b), as demonstrated by the high percentages of lactose malabsorbers among apparently healthy adolescents and adults of two milk-using groups of pastoral tradition in China: 87.9 percent of a group of 198 Mongols from Inner Mongolia, and 76.4 percent of a group of 195 Kazakhs from Sinkiang (Y. Wang *et al.*, 1984: 103-5).[36]

To understand the failure of dairying and milk use to become central in the Chinese scene, in our opinion, one must view the matter in the broadest terms and be receptive to the possibility that several deterrents may have been involved. The Chinese situation, after all, is not unique, for as of A.D. 1500 there were many non-dairying peoples in the world, including other peoples in East Asia. Since such peoples were especially numerous in places distant from the Near Eastern hearth where dairying originated, one might question whether recency of exposure to dairying and milk use may have been a factor in some cases. In addition, one cannot ignore the fact that some peoples simply view the milking of herd animals as a bizarre practice, and milk an offensive animal secretion, unacceptable as food. Certain individuals among non-milking, as well as among milking groups, may have been encouraged in their rejection of milk by experiencing the symptoms of lactose intolerance,[37] though this is unlikely to have deterred an entire group from taking up the use of dairy products. We think it important, as well, not to overlook ethnic relationships that may have encouraged or discouraged dairying and milk use. In India, the Indo-European Aryans, a semi-pastoral people who may have invaded the subcontinent at about 1500 B.C., not only had an enormous impact on Indian culture in general, but they seem to have been vigorous in spreading dairying in India. Continuing that tradition are the modern Hindus who, through influence on or acculturation of India's tribal peoples, have fostered the spread of dairying and milk use. Indians, it should be noted, have since antiquity also furthered the cause of dairying in Southeast Asia. Paul Wheatley (1965) has demonstrated that dairying spread from India to various parts of Southeast Asia in the first millennium A.D. along with Hinduism and Mahayana Buddhism. It was taken up there largely by the court circles to provide milk and milk products needed for ritual purposes. When those Indian religions declined in many parts of Southeast Asia following A.D. 1000, there was no further need for ritual offerings and the practice of dairying ceased.

History of Dairying in China

Returning to China, one notes that the ancient Chinese seem not to have been a dairying, milk-using people (Creel, 1937: 81-83; P. T. Ho, 1975: 113-14; K. C. Chang, 1977b: 29). There were, however, two early outside influences that have long acted to encourage dairying. As in Southeast Asia, the spread of Buddhism from India encouraged the ritual use of dairy products, and the view that dairy products are clean and healthful food and some of them medicinal (Eberhard, 1952: 10; P. T. Ho, 1975: 114n; Schafer, 1977: 106, 107, 134; Sabban, 1986b: 40-42, 44-45).[38] Another influence came from neighboring peoples, such as Tibetans and Mongols, with Wolfram Eberhard suggesting (pers. comm.) that such encouragement of dairying in China occurred particularly during two periods of history: between A.D. 250 and 1000, when alien cattle-breeding societies ruled North China; and during the Mongol period (A.D. 1271-1368) and immediately afterwards. This is in agreement with Sabban's finding (1986b: 36-55), based on culinary texts dating from the sixth to the nineteenth centuries, that there were historical periods in which the Chinese displayed sophistication in the preparation and use of a variety of dairy products. It also agrees with Schafer's observation (1977: 105-6) that the prejudice against dairy products in China broke down after Han times, and that by T'ang times they were significant dietary elements among the upper classes in part because of the close ties established between North China's aristocracy and noble families of nomads on China's borderlands, and in part because of a slow mixing of peoples. This fits with the fact that in T'ang China fermented milks, including koumiss, fermented mare's milk, were far more in demand than fresh milk (Schafer, 1963: 153; 1977: 105-6; E. N. Anderson, 1988: 55).

The milk-using period from Han to T'ang has been described as an "interval," and T'ang, as the apparent peak in Chinese use of milk products (E. N. Anderson, 1988: 55). At the same time, the use of koumiss remained common in Sung times (A.D. 960-1279), and it was not looked upon as an especially alien, barbarian, or nomadic product; there was an imperial office responsible for making koumiss, a common banquet food; and there were restaurants that specialized in its sale (Freeman, 1977: 156, 160, 167; E. N. Anderson, 1988: 66).

The Mongol conquest in the thirteenth century might have been expected to stimulate milk use in China, for koumiss was important among the Mongols both as the principal everyday beverage and ceremonial drink. Among the palace buildings was one for preparing koumiss, which was so much in demand at Mongol banquets that the most prestigeful of them were popularly known as "koumiss feasts" (Mote, 1977: 206-7). Marco Polo mentioned a herd of over 10,000 white horses kept by the Khan; the mares provided milk for him, his family, and members of one privileged tribe (Palladius, 1876: 27). In the biography of T'u-t'u-ha, a Kipchak prince and general in the army of Kublai Khan, there is reference to the Kipchak sending black mares' milk to the Mongol court in China (Bretschneider, 1876: 249). That a gift should be shipped such enormous distances, from what is now the Ukraine to China, is a further testimony to the regard in which such milk was held by the Mongols.

At the same time, if E. N. Anderson (1988: 66) is correct, the Mongol example served to reaffirm the Chinese perception of dairy products as alien, associated with enemies of the nation. In any case, Eberhard notes that shortly after Mongol times, milk and milk products were gradually dropped where they had come into use by Chinese. This came about, he says, principally because of a significant increase in population following A.D. 1500, a growing need to concentrate on food crop production, and consequent shortage of grazing land for all but plow animals, which had little if any milk to spare. E. N. Anderson (1988: 66) writes of a rise in Chinese national feeling in Ming times strengthening the resolve not to consume dairy products. Milk use did not cease, however, and for Ming dynasty times (A.D. 1368-1644), there are references to butter made of cow's milk and to milk sacrifices being offered (M. Hermanns, 1949: 128). There are also reports for the fifteenth century of butter used in a Taoist sacrifice, and its use by Buddhist monks in China proper late in the seventeenth century (pers. comm., W. Eberhard).

In their homeland the Manchus are said not to have practiced dairying in the past (Shirokogoroff, 1924: 130-31), yet there are references to an office for tea and milk in the Ch'ing imperial household; to a Ch'ing emperor having milk for breakfast (Spence, 1977: 281); to milk tea or milk tea with salt drunk at the Ch'ing court and served at official functions there and elsewhere (Nieuhoff, 1673: 41, 74, 119; Parker, 1897: 176; 1903: 271); and to a distilled milk liquor that came from the emperor's cellar (Nieuhoff, 1673: 123). During the Ch'ing period, there are also references to the use of dairy products in east Central China: in Chekiang (Spence, 1977: 264) and at the city of Soochow, in what today is Kiangsu, where much butter and milk were consumed (Montanus, 1671: 366). The conclusion seems inescapable that use of dairy products continued in some places and in some circles. That they were not major dietary elements among the general populace is probably due to the fact that neither the Mongols nor any other nomadic group had been able to exert an influence on Chinese culture as profound as that of the Aryans on Indian culture.

Dairying and Dairy Products in Modern China

Turning to traditional China of the nineteenth and twentieth centuries, one notes the inclusion of dairy products among the foods banned or not normally consumed by Buddhists (Welch, 1967: 112), a situation strangely at odds with the evidence that Buddhism fostered their use. An opposite influence, by some of China's border peoples, led certain of their Chinese neighbors to consume dairy products. It is well documented that dairying occurred traditionally along China's western borders (Hermanns, 1948: 70-71, 127, 128; Simoons, 1970: 575-77; E. N. Anderson and M. L. Anderson, 1977: 340-41). Some Chinese settlers in Inner Mongolia consumed dairy products made by the Mongols, and some along the Tibetan border used milk and milk tea in imitation of the Tibetans (Lattimore, 1932: 305; Frick and Eichinger, 1952: 132). One even reads of Mongols carrying large amounts of butter, packed in sheep stomachs, to Peking early in this century (Headland, 1914: 173-74), though this was probably for sale by Mongol and other residents of pastoral tradition. Chinese there used little, if any, butter, and many border Chinese, too, remained hesitant or unwilling to consume butter and other dairy products, according to frequent reports by observers. This is illustrated by the story of a Chinese man who had long lived on the edges of Mongol country. Coming to Peking, the Chinese pretended to be a Mongol prince. Though he knew Mongol speech and customs and was successful for a while in his deception, some local Chinese doubted him. As a test, they arranged a banquet. Rather than the Chinese dishes he had been given previously, they provided Mongol foods, including milk, butter, and cheese, to observe his reaction. Confronted with those foods, the imposter left precipitously, for he was one of those unable to eat dairy products (Creel, 1937: 81).[39]

Also active in dairying were certain Indians of the Hong Kong area who, in about 1920, maintained herds of water buffalo, of the Delhi breed, for dairy purposes (Levine, 1922: 8). It is likely, however, that these herds, as is the case with Indian herds in Southeast Asia, were established to provide milk to Hong Kong's Indian community. A decade before, the Western community in Hong Kong was served in part by tinned products obtained from ships in port, and in part by a small herd of American cows which were kept in stalls in a cool upland area and fed with grass that was cut and carried to them (J. S. Thomson, 1909: 54). The effort, which entailed Western cattle, was likely quite marginal in economic terms, as compared to that of the Indians, yet Western influences were by far more important in encouraging milk use among the Chinese, especially by infants and children. Change was most marked among groups influenced by Western thought, such as converts to Christianity, hospital patients, persons educated in Western-oriented schools, and the middle and upper classes, especially in urban areas. At Canton Christian College about 1920 there was a college dairy in which studies of the dairy potential of Chinese water buffalo were carried out, and which provided at least some milk to teachers and students. It is said that the first water buffalo dairy had been established in Canton in the first decade of the twentieth century, but that water buffalo had been milked in certain Kwangtung villages long before (Levine, 1922: 4-5, 8, 10-11, 26).[40] Whatever its source, milk was hawked in the streets of Canton late in the last century, even late at night (S. W. Williams, 1895, 1: 776). The Cantonese are said to have drunk not fresh but curdled milk of cows and buffaloes, which they regarded highly. Such curdled milk, made with sugar and vinegar, they consumed particularly on summer evenings, whether at home or in restaurants called "cow-milk saloons" (Gray, 1878, 2: 154).

Yunnan may be the one place in China proper where dairy products became an integral part of regional diet in traditional times. According to FitzGerald (1941: 40), the Minchia, a tribal group of Yunnan whose linguistic affiliations are unclear, milked their cattle and made two kinds of cheese, one of which was sold as a delicacy in Yunnanfu (Kunming) to Yunnan Chinese. Milk was also sold in Yunnanfu (H. R. Davies, 1909: 312; Osgood, 1963: 161), and yogurt, milk curd, cheese, and koumiss were made or used by Yunnan Chinese (E. N. Anderson and M. L. Anderson, 1977: 340-41, 355; E. N. Anderson, 1988: 74, 169).[41] Knowledge of how to process dairy products may have come to the Yunnan Chinese in part from Tibetans, for early in this century cows were said to have been regularly milked only in the Tibetan part of the province (H. R. Davies, 1909: 312). Some observers, however, suggest that milking among Yunnan Chinese derived ultimately from the Mongols and their armies. In support of this, it is observed that today such knowledge is found mainly among the province's Moslem residents and that they claim to be descended from troops who arrived in Yunnan in the thirteenth century as part of the forces of Kublai Khan (Paul Buell, cited by E. N. Anderson and M. L. Anderson, 1977: 341; E. N. Anderson, 1988: 74, 169).

On a national scale, however, the consumption of dairy products in traditional China remained negligible, and they were virtually absent in rural diet (Adolph, 1929: 39; Lindstedt, 1939: 378E-79E). The situation was somewhat different in large cities, where middle-class families, through Western influence, came to consume a certain amount of milk (fresh and tinned) and butter. In one survey of thirty North Chinese middle-class families in the summer of 1922, milk was used by three (Adolph, 1925: 4). In Peking, a household survey of 283 families in 1926-27 found that milk, mostly imported, was purchased by 49, mostly upper-income families (Gamble, 1933: 103). In Peking in the 1930's dairy products remained scarce, somewhat expensive, and were consumed in very small amounts (Lindstedt, 1939: 378E; Lang, 1946: 94-95). And for China as a whole, Chiao (1944: 22) was still able to write in the mid-1940's that there was very little if any dairying, and that in the wartime Szechwan cities of Chungking and Ch'eng-tu, it was "difficult to buy butter or cheese at any price." Since establishment of the People's Republic, efforts have been made to improve and expand dairying, and in 1972 China had roughly 900,000 milk cows, described as ordinary holsteins (American Plant Studies Delegation, 1975: 102). In northwestern Heilungkiang Province cows were milked for the first time starting in 1945, and in Kwangtung at certain buffalo breeding stations local buffalo cows were crossed with bulls of the Murrah breed, one of India's best milking buffaloes, in an effort to raise milk yields (Epstein, 1971: 5, 28).[42] One reads of a dairy farm near Peking with eight hundred cows (May, 1961: 13); of a Shanghai Dairy Company which has eleven sizeable herds able to supply most of that city's fresh milk needs (American Plant Studies Delegation, 1975: 103); and of milk also obtainable in most Chinese cities (A. W. Chung, 1979: 82). It seems that at least some government experts, probably unaware of Marvin Harris's view that dairying is incompatible with the intensive, subsistence agriculture of China, have found or developed suitable niches for dairying even near its great cities. This success, it should be emphasized, occurred at a time when China's population densities are greater, and competition for land more intense, than ever before.

The success of dairying in the People's Republic rests, to a considerable degree, on the use of wastes for feeding dairy animals. One reads of cows fed a protein concentrate made in part from food processing remains such as oil meal, grain millings, and brewers wastes. One effort that would gain an approving nod from Indian dairymen in Southeast Asia, is that of a commune bordering Shanghai, a large supplier of vegetables to that city. The commune reduces the concentrate fed to its 106 milk cows so as better to use other feeds available: vegetable wastes such as cull vegetables and plant parts not sold, and weedy grasses removed from cultivated and uncultivated fields and from roadsides. In addition, Chinese observers have noted that the grasslands, concentrated in the semiarid lands of the north and west, cover nearly 40 percent of China, an area three times that under cultivation, and that there are great possibilities of improving them for grazing purposes (American Plant Studies Delegation, 1975: 101-2; Yu Youtai in Wittwer *et al.*, 1987: 65-81). A contrary view has recently been taken by Wang Qing, Director of the Beijing Food Research Institute, who contends that throughout China's history cow's milk was both an elite food and expensive to produce, that it should be produced for infants and the aged but not for the general population, that the overwhelming majority of Chinese are lactose intolerant, and that increasing production of soybean milk should be emphasized instead of dairying. It is said that Wang Qing succeeded in turning the Chinese away from dairy development and toward expansion of soybean milk production (Simonds, 1988). It will be interesting to see what happens to the effort at dairying, but even before any retrenchment, consumption of dairy products remained neither general nor large, with Mellander (1973: 253) observing that, in a dry or fresh form, milk serves as a weaning food, but is little consumed by adults, and Newman *et al.* (1981: 43), that milk is estimated to comprise less than 1 percent of the Chinese diet.

Nutritional and Medical Perspectives

The nutritional merits of milk and dairy products, particularly as sources of protein, calcium, and vitamin D, are too well known for us to dwell on here. We would, however, note that the near absence of dairy products from the diets of most Chinese meant that some critical nutrients had to be obtained from other foods. This matter will be taken up in our section on nutrition. Here we would note only that one benefit to the Chinese of substituting soybeans for dairy products is a reduced consumption of animal fat and, probably, a lowered incidence of heart disease. One might also wonder whether low milk consumption may also have reduced to some degree the incidence of senile cataracts among the Chinese. We have in mind the hypothesis (Simoons, 1982) that some types of senile cataracts may be brought on by decades-long consumption of milk and milk products, especially in persons who use large amounts of milk, who have high lactase activity, lactose hydrolysis, and galactose absorption. Such persons, according to the hypothesis, are likely to suffer repeated small galactose challenges, accumulation of galactitol in the lens, and to have a greater likelihood of developing senile cataracts. Though the hypothesis remains highly speculative, an initial study in Naples (Rinaldi *et al.*, 1984) supports the idea that persons with high lactase activity are particularly at risk of developing senile and presenile cataracts.

One cannot leave the question of dairying and milk use in China without observing that the milk of various animals (among them dog and pig) (Read, 1976: *passim*), and of humans as well (Cooper and Sivin, 1973: 227-34; Read, 1976: Section 422), has served medicinal uses. With respect to human milk, we note especially the curious custom in China of old persons consuming mother's milk in the hope that it will rejuvenate them and prolong their lives.[43] This custom may be linked to Taoist desires for longevity. In any case, it goes back at least to Han times, when a certain prime minister, Chang Ts'ang, is reputed to have reached the age of 180 by drinking the milk of his wives (Ware, 1966: 105). Even more impressive is the Sung account of a man who reached the age of 240 by consuming the milk of the wives of his great grandchildren (Read, 1976: Section 422). The custom continued to be found in traditional China, with reports of women selling their milk especially for the use of motherless infants or aged persons in various Chinese cities, among them Shanghai (S. W. Williams, 1895, 1: 776), Amoy, and cities of Szechwan (Dudgeon, 1895: 18; I. L. B. Bishop, 1900, 2: 6). In some cases, such milk was sold in a small cup or other container, with Amoy said to be the only Chinese city known to Europeans where such milk was actually hawked, for sale to old men (Parker, 1897: 176; 1903: 271). In other cases, the old person would actually nurse like an infant, as with an empress dowager late in the last century who had a wet nurse to provide her with milk (Dudgeon, 1895: 18). There is also a Chinese account of a woman so dedicated to her mother-in-law, who was very old and toothless, that she withheld milk from her child so as to permit the old woman to suckle her breasts. That account included an illustration of the act of suckling, a scene that Gray (1878, 2: 154) had also seen occasionally on painted porcelain cups.

[1] For a geographical accounting of coffee and tea preferences around the world, see Berdichevsky, 1976.

[2] Various observers report that the Chinese consider cold liquids, including water, as unhealthy, injurious to the intestines. Most Chinese at Foochow, says Doolittle, would endure thirst for a long time before drinking cold water, and would much prefer warm or hot water. That this is not a modern phenomenon is shown by twelfth-century author Chuang Chho, who wrote that even when travelling ordinary Chinese are careful to drink only boiled water. In traditional times, particularly in North China, hot water was a usual roadside beverage, and cold water rarely consumed (Doolittle, 1865, 1: 46; Dudgeon, 1895: 16-17; S. W. Williams, 1895, 1: 776; Hosie, 1897: 137; Parker, 1897: 183; 1903: 279-80; Needham and Lu, 1962: 451-52). At the same time, one must be careful about making generalizations, for one seems to be dealing with preferences that may or may not always be followed in practice. In his twelfth century travels up the Yangtze River, Lu Yü repeatedly mentions cold water and its taste; once he writes of Taoist priests along a difficult road serving hot drinks to the travellers and of a pond not far away whose water tasted "sweet and cold" (C. S. Chang and J. Smythe, 1981: 85). In modern times at Kao Yao, a Chinese village in Yunnan, boiled hot water was the common drink of a good number of middle class families and few men would drink water cold and unboiled. At the same time, women and young children drank cold, unboiled water, and certain poor families drank such water exclusively (Osgood, 1963: 164-65).

For an interesting account of waters of various sorts, their qualities, and uses, see T. T. Chang (1933). Chang's account is a summary of a volume written by Chia Ming (A.D. 1268-1374?), who in turn drew his material principally from Chinese medical writings.

[3] Schafer (1977: 123) writes that in T'ang times the teas most celebrated in North China seem to have been those of Szechwan and Chekiang, with the excellent teas of Fukien, Kwangtung, and Kwangsi barely touched on in the *Ch'a-ching*, a definitive eighth-century work on tea.

[4] The practice of adding salt and ginger to tea did not cease, and in the sixteenth century a Chinese writer noted that rural people add to their tea some ginger and salt—"heating" agents to counter the "cooling" effect of the tea. There are also reports of such tea in traditional China (S. W. Williams, 1895, 2: 51-52; Cormack, 1935: 136), and of people in China's southwestern provinces preparing a soup of brick tea along with such ingredients as sesame, ginger, and citrus peel (Blofeld, 1985: 140).

[5] That the Mongol name for tea, *ch'ai*, is found in Persian, Indian, Turkish, Neo-Greek, and Russian (Laufer, 1919: 553) is an indication of the importance of the overland route from North China.

[6] Italy has two words for tea, *tè* coming from southeastern China, and *cià* from North China.

[7] Soon after, they came to dominate the world's tea trade. Whereas China contributed 42 percent of world tea exports in 1896, that figure dropped to only 11 percent in 1925. During that same period, the contribution of India and Ceylon jumped from 48 to 69 percent (Cressey, 1934: 331).

[8] For an account of the many forms of Chinese tea, see Blofeld, 1985: 60-90.

[9] For details of the processing of green, black, and scented tea in traditional China, see Fortune, 1847: 203-24; 1857: 197-212, 206-12; Gray, 1878, 2: 206-13; S. W. Williams, 1895, 2: 42-49; and Ukers, 1935, 1: 302-5.

[10] "Brick tea" is not always used in the same sense. Here we use the term to refer to such compressed teas in general, regardless of type of tea used in their preparation (shoots, leaves, dust, stalks, twigs), cementing agent, method of compression, or shape of the finished product (whether a plate, tablet, cone, ball, or whatever).

[11] See F. T. Cheng, 1954: 152-56, and Blofeld, 1985: 120-40.

[12] Insensitivity to this custom created a serious problem for an American who visited a Chinese official on a matter of business. He was received with great courtesy, and when a servant brought a cup of tea, the official, taking the cup in both hands, presented it with ceremony to his American visitor. The latter, who was thirsty and unaware of Chinese customs, immediately drained the cup. At that, his host, offended, refused to consider their business and indeed the unfortunate American was gently forced to leave (K. Bard, cited in Holmes, 1948: 285-86).

[13] Fukien and parts of Kwangtung are exceptions to this, where tea may be served with a meal (B. Y. Chao, 1972: 230; K. Lo, 1976: 147). Kenneth Lo (1976: 17-18) also writes, apparently about China in general, that he has often seen tea drunk with meals, particularly when, during working hours, meals are rushed and eaten without soup.

[14] Spencer (1942: 56) makes a case for Szechwan being especially well-endowed with tea houses, and in one village, less than a decade old, nearly half of the thirty shops on its single street were tea houses. By contrast, Spencer found no tea houses in Kweichow, and at some Kweichow fairs had difficulty "even in getting a cup of tea to drink."

[15] Late in the last century European tea planters in Chittagong also started making pickled tea for the Burmese market, and apparently met with a degree of success (Watt, 1889-96, 6, Part 3: 449).

[16] For more complete descriptions of the way in which brick tea is made, see Watt, 1908: 235; E. H. Wilson, 1929: 378-84; and Ukers, 1935, 1: 305-7.

[17] This northern route across Siberia was developed by Russian fur traders, who by the close of the seventeenth century had reached the Pacific Ocean. Besides tea, the Chinese brought silk, porcelain, and gems to Kiachta (Franck and Brownstone, 1984: 107-8).

[18] In the two Chinese communities studied by Anderson and Anderson (in the New Territories of Hong Kong, and in Malaya) (E. N. Anderson and M. L. Anderson, 1976: 147), herbal teas were considered very "cold," to be purchased from herbalists and taken in small amounts for hot ailments. The coldness of herbal teas was the reason they were prohibited in the "cold" postpartum period to the Vietnamese-Chinese women studied by M. Mathews and L. Manderson (1981: 12, 13). Ordinary tea, by contrast, was considered "hot" or "neutral" by these women and was prescribed during that period. Tan and Wheeler (1983: 44; Wheeler and Tan, 1983: 53) found that Chinese women in London classified tea, along with rice, oil, and salt, among the "basic and essential foods," though neither "beneficial" nor "repairing." They also noted, however, that Chinese teas were banned to pregnant women, though the women could not explain this proscription. As indicated in the quote at the beginning of this chapter, however, Lu Yü (1974: 60) clearly regarded tea as cooling. Cormack (1935: 136), in turn, quotes a seventeenth-century Chinese account that describes tea as having a cooling nature, and Blofeld (1985: 154) writes that in traditional China green tea was considered a cooling drink. Moreover, among Chinese in Canada and San Francisco tea (not identified further) was considered "cold," and, at least in San Francisco, it was banned to pregnant women for that reason (Yeung *et al.*, 1973: 198; Ling *et al.*, 1975: 129). Most forms of medicinal tea (*Thea sinensis*) considered by F. P. Smith (1969: 84-85) are also described as "cooling" or used in treating fever or other "hot" ailments.

[19] The term *huang-chiu* is also used for a millet wine in northeastern China (Hommel, 1937: 145).

[20] For accounts of the preparation of such rice and/or millet wine in early and modern China, see E. Watson, 1923: 576; T. C. Huang and Y. T. Chao, 1945-47; and C. P. Li, 1948: 184-206.

[21] Hellenistic-Byzantine stills were not capable of distilling alcohol in the seventh century because at that time they did not have the head- or condenser-cooling necessary for such distillation (Needham *et al.*, 1980: 140, 157).

[22] The consensus of scholars is that distillation of liquor in the New World began in post-Columbian times. On the other hand, Needham and Lu (1985: 56-57), noting that traditional ways of distilling liquor in Mexico were Chinese and Mongolian rather than Hellenistic and that certain vessels and tubes found in ancient Mexican sites could have served for distillation of liquor, raise the possibility that knowledge of distillation had spread across the Pacific to the New World in pre-Columbian times.

[23] For descriptions of Chinese stills and/or the manufacture of distilled liquor, see Guppy, 1884; Hosie, 1904: 226-35; Hommel, 1937: 141-46; C. P. Li, 1948: 198, 206-8; and Needham *et al.*, 1980: 62-68, 103-58 *passim.*

[24] *Kao-liang chiu* in North China usually has an alcohol content of 65 percent (Needham *et al.*, 1980: 68). For details on its manufacture, see Guppy, 1884, and Hosie, 1904: 226-35.

[25] In Kwangtung in the seventeenth century, Peter Mundy wrote of warm arrak ("rack") served from an unusual container ("bottle") with an opening through which hot coals were inserted and with a grate through which the ashes might drop (Temple, 1907-36, 3, Part 1: 194).

[26] For example, the Monguors of the Kansu-Tibetan borderlands, who speak a Mongolian language, distill liquor from wheat and barley (Schram, 1954: 117-18), and the Li tribal people of Hainan make a distilled liquor of sweet potatoes (Odaka, 1950: 28).

[27] See, for example, F. P. Smith, 1969: 418-19.

[28] His efforts at remaining sober might be thwarted by participation in various drinking games, among them finger-guessing games. There are various versions of the finger-guessing games, but all seem to involve competition between two persons. In one such game described by Dudgeon, contestants would each throw out one of their fists with certain fingers extended, and then immediately guess the total number of fingers involved, both their own and their opponent's. Should one contestant guess right and the other guess wrong, the latter loses and is required to drink a cup of wine. It was not uncommon for all persons sitting at a table to be playing a finger-guessing game at once, usually with considerable boisterousness. Such finger-guessing games are said no longer to be popular in China, but one does encounter them from time to time at Chinese dinners and other affairs in Malaysia (Gray, 1878, 2: 65-66; Dudgeon, 1895: 36; C. S. Wong, 1967: 91; V. Y. N. Hsu and F. L. K. Hsu, 1977: 313).

[29] Edwin M. Lemert (1982: 86-93), writing about alcohol consumption among American Indians (who are also reported to flush more than whites after alcohol consumption, and also to differ in their ability to metabolize alcohol), cautions against giving too much weight to such evidence in explaining drinking behavior. Flushing, he insists is "not an insuperable obstacle to drinking." He also notes that though American Indians, Chinese, and Japanese are all Mongoloids, Indians have serious alcohol problems and Chinese and Japanese do not. Moreover, he observes, mixed-blood Indians seem to be similar to pure-bred ones in alcohol consumption and alcohol-related problems. In Lemert's view, present evidence gives little support to the view that there is a racially-based "vulnerability to the effects of alcohol," and he favors a sociocultural model of some sort to explain Indian drinking patterns.

[30] One recalls, in this regard, a similar culture trait involving the breast-feeding of other animals by women. This practice, however, has a distribution and aims quite different from those of dairying. In some cases it is simply a reflection of a woman's affection for a pet. In other cases, though affection may still be involved, economic concerns are present as well, as among certain pig-complex peoples of Melanesia. In still other cases, breast-feeding of animals is linked to ritual needs, as among the Ainu of Japan who nursed cubs in connection with their bear cult. And, in still others, animals are nursed out of concern with the well-being of the human mother or child. Though this practice was widespread around the world, when we wrote on the matter several years ago we found no evidence of it among the Chinese or any other East Asian people apart from the Ainu (Simoons and Baldwin, 1982: 432). Since then, we were given an unidentified piece from a syndicated column in an American newspaper which claimed that a Ch'ing queen who considered her Pekinese dogs sacred had furnished each dog with a nursemaid.

[31] That region of non-milking was in the African tropics, centered on the Congo Basin but extending westward in a belt along the Guinea Coast of West Africa, and eastward across Mozambique to the shores of the Indian Ocean. That was a region largely infested with tsetse flies that brought on sleeping sickness in domesticated animals, some of them more seriously affected than others. In any case, many groups within that region of non-milking had domesticated animals (goats, sometimes sheep and cattle) capable of being milked, yet they did not milk them (Simoons, 1954).

[32] This view is reported for China (Laufer, 1914-15: 168; Creel, 1937: 80-81) and also for Southeast Asia.

[33] For an excellent review of the factors—biological, cultural, and individual (psychological)—involved in the development of food likes and dislikes, see Rozin and Vollmecke, 1986.

[34] Considering the rejection of milk products by most Southeast Asian peoples, E. N. Anderson (1988: 146) suggests that the region is not suited to cattle herding; that preparing milk products was too laborious, costly, and risky; and that conditions for keeping milk were poor, even when it was preserved. One cannot, however, ignore (1) the success of immigrants from Bangladesh, India, and Pakistan in developing a viable dairy economy in various parts of Southeast Asia, as on Singapore Island near sea level and virtually at the equator; (2) that much of the Indian subcontinent, like Southeast Asia, is hot and humid, and presents a similar danger of milk spoilage and contamination with disease organisms, yet people have ways of dealing with such problems; and (3) that certain dairy products prepared by Asian peoples have excellent keeping qualities (I have a decades-old, dried milk ball [*kurut*] from Afghanistan that I regularly show to my students and which remains quite edible).

[35] Though, to our knowledge, it has never been advanced as an explanation for the general Chinese failure to consume animal milk, one notes the ancient South Chinese belief that a toxin in dairy products brought on beriberi in Mongolia and Manchuria; this, writes T. Lee (1940: 319), led the Chinese to forbid milk to beriberi patients.

[36] These percentages are little different from the 92.3 percent prevalence of adult lactose malabsorption among a group of 248 Han Chinese from northeastern China included in the study. The failure of the Mongols and Kazakhs to have lower prevalences of such lactose malabsorption is explained by the investigators in terms of the insignificance of fresh milk in the diet of older children and adults, and their use of fermented milk products, commonly diluted, only seasonally. This, in the investigators' view, was insufficient to have created the selective pressures needed to favor the gene for persistence of high lactase activity throughout life (Y. Wang *et al.*, 1984: 104-5).

[37] Symptoms of lactose intolerance in humans tend to produce a "danger-based rejection" of milk, rather than a distaste (taste aversion) for it (Rozin and Vollmecke, 1986: 442).

[38] Devadatta, the Buddha's cousin and follower, was incensed by Buddhist monks' lax behavior, and sought to incorporate a ban on milk and curd into the rules to which they were subject (Rockhill, 1884: 87-88). Devadatta's demand was refused and he and his followers formed a rival sect which persisted from the Buddha's time (*c.* 500 B.C.) until at least the end of the fourth century A.D. and which presumably required monks to give up dairy products. Certain Mahayana Buddhists in India, however, were also reported (Watters, 1904-5, 1: 57) to have refrained from milk and milk products. This rejection, rooted in the concept of non-violence, was apparently observed only by a handful of devout Buddhists, especially in Mahayana monasteries.

[39] Anyone who has observed the enthusiasm of Chinese in Hong Kong and Southeast Asia for ice cream may wonder about this man's objections. One should note, however, that Chinese attitudes toward milk and dairy products may cover a broad spectrum. This, in turn, may have contributed to differences in the ways Western observers, among them reputable sinologists, have characterized general Chinese attitudes toward milk and dairy products: from mere indifference to strong feelings of rejection. Laufer (1914-15: 167) and Creel (1937: 81) refer to a profound aversion to consuming animal milk (or dairy products in general) among the Chinese. Adolph (1956: 68) speaks of the taste of milk and its products being disgusting to great numbers of present-day Chinese. Whyte (1972: 32) writes of a "traditional distaste" for milk; and May (1961: 13), of "an ingrained distaste for milk which will be difficult to overcome." Parker (1897: 176; 1903: 271) adds that the Chinese hold cheese "in abomination" and consider it as a sort of "rotten milk," a view echoed by Lin Yutang, who said: "the only thing we have not discovered and will not eat is cheese. The Mongols could not persuade us to eat cheese, and the Europeans do not have a greater chance of doing so." Snapper (1965: 377-78), in turn, says that the North Chinese "is unwilling to drink milk, designates butter as 'cow oil' and is unable to understand how educated persons can eat the smelly thing we call cheese." Anderson and Anderson (E. N. Anderson and M. L. Anderson, 1977: 326, 341), in the mildest statement of all, say that the Chinese are disinterested in dairy products, but do not harbor strong aversion to drinking milk, a practice which has readily been accepted throughout East Asia today. They do note, however, an objection to smelly cheese, which is regarded by some Chinese as "a putrefied mucous discharge of an animal's guts." Whatever their differences in describing Chinese attitudes, Western observers agree that there is no religious taboo in China against milk use.

[40] Levine, writing in 1922, suggested that such dairying may have begun a century before, but dairying of some sort goes back much further in the Canton area. One notes, for example, that tea made with salt and warm milk was offered as a toast of welcome by the Ch'ing viceroys in Canton at a dinner given in 1655 for two Dutch ambassadors (Nieuhoff, 1673: 41), which implies a local source of milk of some kind. Schafer, moreover, has noted specifically (1977: 92-93) that water buffalo milk was used in the Canton area in T'ang times.

[41] The Haw, Chinese from Yunnan who live in a few villages of northern Thailand, make a considerable effort at dairying, milking common cattle and water buffalo as well as goats (pers. comm., O. Gordon Young).

[42] For data on milk yields and composition of Chinese water buffalo, common cattle, yak, and camel, see Epstein, 1971: 7, 8-9, 10, 12, 17, 18, 22, 28, 30, 119-20.

[43] This custom is not restricted to China, having been reported among the Miao of Thailand by Bernatzik (1947, 1: 226-27).

CHAPTER FIFTEEN

Notes on Food in Nutrition and Health in Traditional China

> "Partly due to lack of knowledge of Chinese eating habits, evaluations by Western experts of the nutritional status of China's population often fall short of reality."
>
> *T.-H. Cheng*

Parasitic and Infectious Diseases
Foods, Nutrient Intake, and Nutritional Deficiency
- Calories: Adequacy and Sources
- Protein and Protein Deficiency
- Vitamin A and Deficiency States
- Thiamine Deficiency and Beriberi
- Riboflavin Intake and Deficiency
- Niacin Deficiency and Pellagra
- Vitamin C Deficiency and Scurvy
- Calcium, Vitamin D, Rickets, and Osteomalacia
- Iron and Iron Deficiency
- Iodine Deficiency and Goiter
- Selenium Deficiency in Keshan Disease

Food, Nutrition, and Care of Those at Greatest Risk: Mothers and Children
- Pregnancy and Abortion
- Birth and Birth Weights
- Disposal of Unwanted Infants
- Confinement and the Feeding of Infants and Nursing Mothers
- Weaning
- Mortality among Infants and Mothers
- Growth Rates of Chinese Infants and Children
- Birth Rates, Death Rates, and Population Increase

Chinese Customs, Good Health, and Nutritional Balance

Significant changes in matters of health and disease have taken place in China since establishment of the People's Republic. However deserving that matter may be of full presentation to Western readers, to do so would be a major undertaking best done by medical and nutritional researchers, and well beyond the scope of our study. Our intent here is a more limited one: to describe the state of health and nutrition of the Chinese people in their homeland in the nineteenth and early twentieth centuries, with a special focus on the role of foods and food habits as they contributed to the health and well being of the Chinese people.

PARASITIC AND INFECTIOUS DISEASES

Justus von Liebig (1878: 453), the father of organic chemistry, extolled the sophisticated techniques of the Chinese farmer that led to high crop yields, dense human populations, and an unusual equilibrium between man and nature. A critical element in that equilibrium was the practice of returning human excrement, night soil, to the cultivated fields.[1] F. H. King (1927: 172-73) estimated that each year in East Asia as a whole, over 182 million tons of night soil were returned to the fields, containing more than a million tons of nitrogen, 376,000 tons of potassium, and 150,000 tons of phosphorus. Such night soil was collected at homes throughout China, whether in rural or urban areas, for it was viewed as an important economic resource. In southern Fukien, for example, entrepreneurs would build latrines for public use, and they did this not out of concern for public health or comfort but in order to obtain valuable night soil for sale (J. P. Maxwell, 1921: 378). In a large urban area the value of such night soil was considerable, and it could bring in substantial revenues. For example, the International Concessions in Shanghai received $31,000 in gold from a Chinese contractor in 1908 for rights to collect and remove 78,000 tons of night soil. The Shanghai night soil, which was sold to farmers in surrounding rural areas, was transported by several flotillas of boats operating on a daily basis throughout the year (F. H. King, 1927: 171-72).

Our interest here, however, is not in the economics of night soil or in its critical role in agriculture, but in the health implications of its use. Much night soil was collected in cities by coolies who then sold it to farmers in the countryside, typically carrying it at night in pails suspended from carrying poles (hence the term "night soil"). Farmers, in turn, would accumulate both purchased and household excreta, storing it mainly in large stoneware vats having a capacity of from five hundred to a thousand pounds, and ultimately carrying it to a garden or field, where, usually after being diluted with water, it was applied to crops with a long-handled dipper (F. H. King, 1927: 67-68, 175-76). If night-soil storage is sufficiently long, the numbers of pathogenic organisms can be greatly reduced or eliminated altogether. Reports differ considerably, however, on the customary length of storage, with four to eight months the highest estimate we found, for southern Fukien (J. P. Maxwell, 1921: 378). In a study in the Canton area, on the other hand, Oldt (1924-25: 98-99) found the ordinary length of night-soil storage to be from ten days to two weeks before it was applied to crops, whereas hookworm eggs remained viable in storage for three to six weeks. After fertilization, moreover, viable hookworm larvae were found in the soil for at least seven to ten days (in one case for six weeks), ample time to infect humans working in the fields.[2]

Survival and transmission of fecal-borne pathogenic organisms in South China were favored, as well, by other factors, including environmental and demographic conditions and cultural views and practices: high temperatures and abundant moisture; high densities of human population; abundance of animals such as dogs, cats, and rats that serve as reservoirs for certain parasites that infect humans; the common practice among young children of defecating at will around household and village; general indifference to the presence of human excrement in the household and environment; the custom of going barefoot or in grass sandals when cultivating; generally poor hygienic conditions; and widespread pollution of water and food (H. T. Chen, 1933; 1934a; 1934b; Lamson, 1935: 278-81; Winfield, 1948: 258-59). J. P. Maxwell (1921:

378-80) provided insights into the problem as he saw it in southern Fukien. He noted that village latrines might be located close to wells and in at least one case a latrine was cracked and in a position to pollute the household water supply. In addition to village latrines, moreover, jars and buckets were found at regular intervals along a road for people to urinate in. This and other urine, often including a certain amount of fresh fecal material, was then diluted and used for watering vegetables, especially garlic, leeks, and cabbage. Though it has often been observed that eating raw foods is rare among the modern Han Chinese, Maxwell noted that these three vegetables were consumed raw or partially cooked, so that when children were old enough to eat vegetables, infection with the intestinal round worm *Ascaris lumbricoides* was nearly universal,[3] occurring in 95 to 100 percent of young people.[4,5]

Robertson's account (1936) of conditions in the Yangtze Delta gives a somewhat different perspective on the problem. It was not unusual in that region, he wrote, for night soil from towns and cities to be carried to rural areas in the holds of sampans which, on the return journey, were loaded with fresh vegetables. Even if those vegetables were not contaminated by night soil in cultivation or transmission, they were commonly washed in the contaminated water of streams or ponds before they were sold by merchants or used at home. Proper cooking, of course, removed the danger, but certain types of vegetables were eaten raw or partially cooked, not sufficient for eliminating helminthic ova. Among the most common plant foods consumed raw were onions, ginger, lettuce, radish, cucumber, coriander, and a form of melon, all of which were commonly washed and then, without cooking, eaten with flavorings such as soy sauce, sesame paste, or vinegar. In addition, there were many other vegetables that were cooked for just a few minutes. Of special interest in this regard are plants that serve as intermediate hosts for helminths, such as the previously mentioned role of water caltrop (as well as water chestnut and water bamboo) in transmitting the parasitic organism *Fasciolopsis buski* (Robertson, 1936: 418).[6]

Irrigation systems in southern Fukien, according to Maxwell (J. P. Maxwell, 1921: 378), were built so that water would flow from one paddy field to another, enabling the disease organisms, which enter a paddy when night soil is applied, to enter neighboring paddies and, ultimately, a stream or river whose water is drunk by people living along the shores. We have noted the reluctance of Chinese to drink unboiled water, but Maxwell wrote that sometimes people scooped up water from a paddy with their hands and drank it. Other observers have noted that in Chinese homes water was sometimes not boiled if fuel was in short supply or if time was pressing; the result was that enough unboiled water was consumed to constitute an important means of spreading disease (Winfield, 1948: 259; Whyte, 1972: 24).

The fecal-borne diseases causing highest mortality in rural areas of traditional China were typhoid fever, dysentery, and cholera, and all were among the nation's five leading diseases in that regard (Banister, 1987: 51). Also serious, however, were a considerable number of fecal-borne parasitic diseases, among them ancylostomiasis (hookworm infection), ascariasis, clonorchiasis, and schistosomiasis, which commonly affected adults in a debilitating way, bringing on illness, contributing to malnutrition, and aggravating other medical problems. Typhoid fever was quite common in Canton where, besides being spread by contaminated food and water, typhoid bacilli were found on public eating utensils such as chopsticks, tea cups, and rice bowls (Kwauk, 1929). Cholera is a disease of great antiquity in China, and was among the most important of China's epidemic diseases (Jefferys and Maxwell, 1910: 86-87; Lamson, 1935: 278). It also occurred in an endemic form, and was reported to be highly endemic in a region of South China centered on Kwangtung Province (one of only two such highly endemic regions in the world) (Simmons *et al.*, 1944: 471). Cholera might bring on mild illness, as with a carrier of the disease, or it might result in death, sometimes within a few hours of the initial attack.

The broad range of associations between fecal-borne diseases and environmental and socio-cultural factors in traditional South China remains imperfectly understood.[7] Oldt (1926: 1062-63) reported, however, that in Kwangtung hookworm was transmitted largely in fields fertilized by night soil, and was found in farmers (average 60 percent of all farmers) but not in urban people (use of urban night-soil would thus tend to reduce the likelihood of soil infestation). Oldt also reported differences in the incidences of hookworm infection according to the types of crops cultivated, with rates of 65 percent and 70 percent among persons engaged in vegetable gardening and sericulture, but only 20 to 40 percent among rice growers. He explained this difference in terms of the application of night soil in paddies when they are under water, which obstructs the hatching of eggs and growth of larvae. This explanation fits with the finding that the larger the amount of land farmers had under rice, the lower the incidence of hookworm infection among them.

The case of hookworm infection is just one of many separate stories, but it is clear that other fecal-borne diseases have similar associations with human activities. One example is the heavy infections of liver fluke (*Clonorchis sinensis*), common in parts of Kwangtung Province (in Canton in the mid-1920's, more than 20 percent of the public were infected) but very rare in an endemic form in Central China and completely absent in North China. The high rate of infection in Kwangtung derived from eating a favorite regional dish of raw or partially-cooked fish (Oldt, 1926: 1061; Faust and Khaw, 1927: 122-37; McKinley, 1935: 426-32; Sung, 1940). Though in the past (*i.e.*, T'ang and Sung times) consumption of raw fish had been popular elsewhere in China, including North China (Sinoda, 1977: 490; E. N. Anderson, 1988: 55, 142),[8] in modern times fish are usually cooked thoroughly and humans rarely contract clonorchiasis. A second example, from Chekiang and elsewhere, is the eating of raw fresh-water crabs which transmit to humans the lung fluke *Paragonimus westermani* (Jefferys and Maxwell, 1910: 140-42; J. L. Maxwell, 1931). A third is that between schistosomiasis and water contact by farmers along irrigation ditches and streams and rivers that harbor infected snails (McKinley, 1935; T. H. Cheng, 1971; "Report of the American schistosomiasis delegation to the People's Republic of China," 1977; Banister, 1987: 57).

The overall result was that China was perhaps the world's richest country in its intestinal fauna (Lamson, 1935: 279-80), that sanitation constituted the major health problem, and that fecal-borne diseases caused a quarter of all deaths, or a total of more than 4 million annually (Winfield, 1948: 255). During the Japanese invasion (1937 to 1945) from 5 to 10 million Chinese died in the conflict, but during the same period fecal-borne diseases brought on 30 to 35 million deaths (Winfield, 1948: 112). Individual and national burdens of intestinal parasites were enormous. On a national level, when China had a population estimated at 500 million, Winfield (1948: 128) judged that ascarids (round worms) living in Chinese intestines totalled eight billion and weighed 130,000 tons, or about the weight of 2 million Chinese adults. That estimate was based on an assumption that rural people averaged twenty ascarids each. Around Canton, however, one survey found 98 percent of rural folk infected with ascarids and they averaged thirty five per person (Winfield, 1948: 128). In the silk-cultivating areas of the lower Yangtze Valley it was not unusual for an individual to have hundreds or thousands of hookworms, a serious matter in view of the fact that burdens of only three to four hundred hookworms bring on anemia in which the patient's hemoglobin is reduced 80 percent (Winfield, 1948: 115-16). Such parasitic burdens may well have occurred in China for thousands of years,[9] acting not only to reduce nutrition and retard growth ("Ascariasis, giardiasis and growth," 1983), but to slow human population increase and favor the survival of individuals who could tolerate such burdens.[10] Despite major campaigns against them,[11] such parasitic infections continue to be serious in the People's Republic, where, by 1979-81 estimates, 20 million people still had clonorchiasis, 2 to 3 million had paragonimiasis, and ascariasis and ancylostomiasis remained common (Banister, 1987: 66).

In addition to the fecal-borne diseases, many other diseases were serious problems in rural China. Among those causing highest mortality were smallpox, tuberculosis, measles, diphtheria, pneumonia, skin disease, and malaria, but others, such as tetanus, scarlet fever, typhus, plague, and leprosy,[12,13] were significant as well (Banister, 1987: 51). Though most of those diseases are not food related, we will comment briefly on several of them: malaria because it was a critical factor in the settlement of Central and South China by the Han Chinese; plague because it was a socially-disruptive disease; smallpox, tetanus, and venereal diseases because of their effect on infants and young children, groups of special concern to be dealt with later in more detail; and tuberculosis because it has continued to be a serious problem among American Chinese.

Malaria, which probably originated in Africa, seems to have been known in China well before the beginning of the Christian era, possibly several centuries B.C. (Hoeppli, 1959: 47-48, 49; Ackerknecht, 1965: 88-89; Lu and Needham, 1976: 5-6). In traditional times it was found in many places, and was endemic in the Yangtze Valley and in South China (Jefferys and Maxwell, 1910: 47-48; McKinley, 1935: 309; Simmons *et al.*, 1944: 476) where prevailing conditions (heat, humidity, and stagnant bodies of water) favored the survival of mosquitoes. In earliest times, more than two thousand years ago, the Han Chinese, like Europeans, lived in a temperate, mid-latitude environment, and regarded Ling-nan (the hot, sub-tropical south) as unhealthy, malarial, and to be avoided, a place suitable for banished persons, escaped criminals, and others seeking refuge (Wiens, 1954: 131-32). The conquest and settlement of the south was a long process, and the Han Chinese, who were mainly valley farmers, settled in the very places where malaria was most serious and, as a result, they suffered especially high mortality. Over time, however, clearing and settlement seem to have reduced the conditions favorable to mosquitoes. In addition, natural selection, and perhaps interbreeding with tribal peoples native to the south, may have given the Han Chinese greater resistance to malaria. In any case, a relatively healthful situation ultimately came to prevail there (Wiens, 1954: 337).

The Chinese medical literature contains various explanations of malaria, especially: (1) that it was brought on by spirits which possessed the human body; (2) that it came about because of eating improperly, whether drinking bad water or eating certain fruits, such as bananas; (3) that it developed because of an imbalance between yin and yang; and (4) that it derived from vapors or stagnant waters found in particular places (Hoeppli, 1959: 273-77). Particularly common was the conviction that miasmas, such as mist from ponds or pools in woodlands and mountains or from swamps in early morning, were the cause of malaria (Wiens, 1954: 338). This and other fears led frontier Han Chinese to follow "four dont's" in an effort to avoid the disease: (1) don't get up early in the morning (to avoid early mists), (2) don't eat to fullness (to avoid gastritis), (3) don't have excessive sexual relations (to maintain health and vigor), and (4) don't bathe (to prevent colds). In addition, almost all Chinese, regardless of their education or virtuousness, partook of the one critical element, opium, which was smoked in the belief that it warded off malaria (Wiens, 1954: 338-39). Should these efforts fail, a broad range of other treatments were available to the afflicted (Hoeppli, 1959: 279-89).

Plague epidemics have struck various regions of Eurasia since antiquity, with the best known of them, the Black Death which appeared in the fourteenth century, decimating the populations of the Near East and Mediterranean region. The disease is caused by the bacillus *Pasteurella pestis*, and is usually transmitted from diseased rats to humans by bites of the rat flea. Certain historians have looked to Central Asia as the place in which the great plagues of history originated. Whether or not that is true in every case, there is strong evidence that the Black Death did originate in Central Asia and spread westward to the Near East and Europe and eastward to China. Important in the spread of the Black Death, it has been argued, was the great trade route that extended from North China across Inner Asia to the Black Sea (Dols, 1977: 15-16, 35-38, 42-43), which, if true, means that what was a great boon to China in trade turned out to be very destructive in terms of disease transmission. In any case, plague epidemics had been known in ancient China, and they continued into modern times, several of them striking China in the nineteenth and early twentieth centuries. Plague was also endemic for extended periods in some places, among them Hong Kong and Yunnan. The accounts we have uncovered (J. S. Thomson, 1909: 326-29; Jefferys and Maxwell, 1910: 45, 67-78; McKinley, 1935: 309, 312; Ta Chen, 1940: 177-78) suggest that an average of a hundred cases of plague were reported to hospitals in South China annually; that the overwhelming number of cases went unreported; that when epidemics struck, cities and villages alike were affected; that a single village could suffer thousands of dead and great disruption; and that in such an epidemic healthy persons might leave their homes to live in hastily-erected shelters in the hills. It was Chinese immigrants who introduced plague to North America by way of San Francisco, with the earliest cases recorded there in 1900, and with 187 cases (87 of them fatal) reported in 1907, after the earthquake (Ackerknecht, 1965: 8, 12, 20-21).

Smallpox was described in China by Ko Hung in the fourth century A.D., and it has been suggested that it was introduced there in the first century A.D. (Cartwright, 1972: 116). The Chinese developed an inoculation against severe smallpox by the early eleventh century, long before inoculations against smallpox were used in Europe. The Chinese inoculation, called variolation, used smallpox itself (obtained from persons with a mild case of the disease), in contrast to vaccination, in which cowpox was used. However, many parents were negligent about giving such inoculations to their children, and epidemics were said to sweep through South China periodically (Kulp, 1925: 56; Ackerknecht, 1965: 62; Cartwright, 1972: 123). In the Peking area late in the last century, half of all adults bore the scars of smallpox, and virtually all of them, at one time or another, had the affliction (Parker, 1897: 181), contracted by variolation or other means. Giles (1876: 5) wrote that the common form of smallpox in Peking was mild, but even there smallpox could kill, with the Emperor T'ung Chih, who was eighteen years old, dying of the disease in 1875. That this danger was considerable is shown by the case of the Chinese man who, asked the number of his children, named all of them except one. When asked about this, he explained that the unnamed child, his youngest, had not yet had smallpox (Headland, 1914: 238). Smallpox was said to be responsible for most blindness, a common affliction in China. Mongols who came to Peking had a special fear of catching smallpox there and of introducing it to their homeland, where they were said to be free of the disease. Even in the mid-1940's smallpox killed tens of thousands of people annually in China. (Winfield, 1948: 108).

Also important among Chinese health problems, but little documented, were infections such as tetanus, which caused very high mortality among newborn children and their mothers. By one estimate, tetanus killed 30 to 80 percent of newborn children in certain villages in Hopeh Province.

Venereal diseases were common, especially in cities where widespread prostitution facilitated their spread. Though rural folk had far lower rates of venereal infection (4 percent of male farmers who were patients in Canton hospitals), during times of war or revolution, soldiers—who had unusually high rates of infection (45 percent of soldier-patients in Canton hospitals)—spread these diseases widely in the countryside as well. As elsewhere in the world, venereal diseases, especially syphilis, were responsible for sterility, miscarriages, defective children, insanity, and other serious health problems and death (Lamson, 1935: 348-57).

Tuberculosis was common and general in the south, one of the principal killing diseases in traditional China, and, by one estimate in the 1930's, it led to two million deaths annually (Jefferys and Maxwell, 1910: 43-44; Lamson, 1935: 328; McKinley, 1935: 311). Despite major efforts at vaccination and treatment in the People's Republic, tuberculosis continues to be a serious problem in China, with a 1979-80 survey finding 6.6 million people with active tuberculosis, and another 1.6 million under observation because their sputum contained tuberculosis bacilli (Banister, 1987: 63). Chinese migrating abroad were often tubercular, and for many years, tuberculosis was the leading cause of mortality among overseas Chinese. Mortality rates from tuberculosis among American Chinese (526 per 100,000) in 1921 were six times those of American whites (Winslow and Koh, 1924: 348). In San Francisco in 1930, more Chinese died from tuberculosis (261 per 100,000) than from any other cause, and their tuberculosis mortality rates were three times those of the city population as a whole. There was a decline in rates of mortality from tuberculosis among San Francisco Chinese in the 1930's, and heart disease and cancer became as great or greater causes of mortality by the early 1940's (Geiger *et al.*, 1939: 7, 24, 28; 1945: 4, 12, 14). Nevertheless, American Chinese rates of mortality from tuberculosis continued to be far above those of American whites, and even in the early 1970's the people of San Francisco's Chinatown had the highest incidence of tuberculosis in the United States (Carp and Kataoka, 1976: 33; H. King, 1976: 496-97).

FOODS, NUTRIENT INTAKE, AND NUTRITIONAL DEFICIENCY

Leaving our sketch of parasitic and infectious diseases, we turn to matters of food and nutrition: to individual nutrients, the adequacy of their intake, important food sources, and health implications. Then we will turn to the question of food, nutrition, and care of mothers and children for, of all Chinese, they were by far at greatest risk.

Calories: Adequacy and Sources

The Committee on Nutrition of the Chinese Medical Association (1939: 304-5)[14] set the recommended calorie requirement for an adult Chinese in a temperate climate and living not by manual labor but in an ordinary way, at 2400 per day. This was the same amount considered adequate for Europeans. The committee recognized that Chinese on average weigh less and thus need fewer calories than Europeans, but set the same requirement nonetheless because Chinese

Table 6. Comparison of sources and amounts of protein in the average rural Chinese diet with diet in the United States (after Adolph, 1946b)

Foods	Average Amount of Protein in Rural Chinese Diet (gm. per capita per day)	Average Amount of Protein in U.S. Diet (gm. per capita per day)
Cereal and legumes	71	29
Vegetables and fruits	4	5
Meats and eggs	5	31
Milk	0	18
Total	80	83

foods, which are mainly vegetarian, have a lower coefficient of digestibility. Higher minima were set for persons engaging in manual labor and for nursing women.

Summarizing thirty dietary surveys conducted in China from 1925 to 1943, J. H. Liu and C. K. Chu (1943: 96-97) reported an average daily per capita calorie intake of about 2800 prior to the war (1937) and 2600 during the war. The range in the survey averages was from 2045 (wartime college students) to 3453 (wartime armies) calories daily per capita.[15] Overall, average intakes reported by Liu and Chu seem quite adequate, but, as Maynard and Swen noted (1956: 409), in many localities the intake of calories fell below the average, as low as 1400 per day. Moreover, since the Chinese peasant lived so little above the poverty line, a poor harvest could quickly lead to an inadequate intake of calories. In addition, pregnant and nursing women and workers engaged in heavy labor could have inadequate intakes.

Table 1, prepared by Adolph in the mid-1940's, shows the sources of calories in the average rural Chinese diet and the average diet in the United States. In rural China, cereals and legumes make an overwhelming contribution, 88 percent of total calorie intake compared to 38 percent in the United States. In Chinese diet, in contrast to that of the United States, very small percentages of calories are contributed by vegetables (5 percent), fats (4 percent), and meat and eggs (3 percent), and no calories significant enough to include, by fruit, sugar, or milk. Judging from the 1929-33 village survey, the only root crops making a significant contribution to calorie intake in China are sweet potatoes (2.6 percent of total calories supplied by staple crops) and white potatoes (0.9 percent), with white potatoes making a greater calorie contribution in the Wheat Region; sweet potatoes, about the same in both the Wheat Region and Rice Region.

Protein and Protein Deficiency

For China as a whole, the average daily intake of protein in rural areas, by Adolph's estimate (Table 6), was 80 gm. per capita, not much below the 83 gm. average in U.S. diet. Within China, however, there were considerable variations by income, state of war or peace, and region. In 30 pre-war and wartime dietary surveys considered by J. H. Liu and C. K. Chu (1943: 96-97), the range in average daily per capita protein intakes was from 47 to 112 gm. Indications are that South China in general had lower protein intake (67 gm. per man daily, according to Adolph, 1929: 39) than North China (86 gm. daily). Within the south, moreover, protein intake seems to have been lower in the Double Cropping Rice Area than in any other area (Maynard and Swen, 1956: 419) (Table 7).

Table 7. Daily intakes of protein in China per adult-male unit and percentage of protein supplied by different classes of food: 17,351 persons, 2,727 families, 136 localities, 131 hsien, 21 provinces, China, 1929-1933 (from Maynard and Swen, 1956)

Regions and Areas	Number of Localities	Number of Adult-Male Units	Intake of Protein (gm.)	Percent of Calories as Protein	Percentage Distribution of Protein							
					Seeds and Their Products				Vegetables		Animal Products	Fruits
					Grains	Soybeans	Other Legumes	Total	Potatoes	Others		
CHINA	136	13,341	100	12	76	9	7	92	2	2	4	*
Wheat Region	67	6,966	108	14	73	11	10	94	2	2	2	*
Rice Region	69	6,375	91	11	78	6	5	89	2	3	6	*
Wheat Region Areas:												
Spring Wheat	13	1,554	107	13	78	5	10	93	4	1	2	*
Winter Wheat-millet	21	2,030	95	13	78	8	11	97	1	1	1	*
Winter Wheat-kaoliang	33	3,382	117	14	67	17	9	93	3	2	2	*
Rice Region Areas:												
Yangtze Rice-wheat	22	2,086	98	11	81	6	6	93	1	2	4	*
Rice-tea	19	1,687	92	11	80	8	2	90	2	3	5	*
Szechwan Rice	6	571	85	12	71	6	10	87	2	4	7	*
Double Cropping Rice	11	974	82	10	78	4	3	85	6	1	8	*
Southwestern Rice	11	1,057	91	11	75	8	5	88	1	4	7	*

* the amount is under 0.5

More significant than total protein intake in China is the quality of protein consumed. Adolph (1944) noted that 85 to 90 percent of the Chinese population is rural; that rural Chinese diets are largely vegetarian, with 95 percent of protein deriving from vegetable sources; and that vegetable proteins in general have a lower biological value and coefficient of digestibility than do animal proteins. Soybeans have rightly been extolled as a rich source of protein of high biological value, and one of China's most important contributions to the world's store of domesticated plants.[16] It should be noted, however, that three quarters of all dietary protein in China derived from cereals, and less than 10 percent from soybeans. The result, Adolph observed, is that the average Chinese diet may be quite lacking in protein. Thus, in 1939 the Committee on Nutrition of the Chinese Medical Association argued that because Chinese diet is essentially vegetarian, minimum protein allowances should be 50 percent higher than those of the West, 80 gm. per day for an adult weighing 55 kg., or about 1.5 gm. per kg. Though it has been argued that this figure is too high (Adolph, 1946a: 966), the basic concern about protein quality remains valid.

Severe protein deficiency disease (kwashiorkor) in young children was probably rare in normal times in China. At the same time, China has not been free of the problem of protein deficiency (Ji Chen, 1942). Snapper (1965: 26-27) writes of North China that people live on the borderline of protein adequacy, with a decrease in income at once reducing the intake of better quality protein even though intake of calories remains adequate. The problem was especially great for vulnerable groups such as pregnant and lactating women and young children, with the latter having protein needs, in terms of grams per unit of body weight, of two or more times that of an adult male (Passmore *et al.*, 1974: 16-21; Whyte, 1974: 186-87).

Vitamin A and Deficiency States

Many writers on modern China have observed, whether in general terms or for specific regions, that vitamin A intake is low or inadequate, or that symptoms of vitamin A deficiency are common among children, other groups, or in the general population (Agnew, 1926-29: 59; Frazier and Hu, 1931: 507-8; Reiss, 1936; Guy and Yeh, 1938: 212; Chiao, 1944: 18; Simmons *et al.*, 1944: 68; Adolph, 1946b: 869; 1951: 129; Winfield, 1948: 73; T. H. Shen, 1951: 168; Chou and Marlatt, 1953: 305; Jolliffe and Tung, 1956: 316-17; May, 1961: 28; Snapper, 1965: 23-26; Whyte, 1972: 26, 82; R. L. Freedman, 1978: 2).[17] Surveys conducted in 1929-33 and 1937 found that villages of the Rice Region of China in general had markedly lower vitamin A intake than those of the Wheat Region, one survey reporting it at roughly half that of the Wheat Region. At the same time, the north is singled out by E. F. Yang (1958: 89) as the region with the greatest vitamin A deficiency problem because of the previously mentioned limited availability of vegetables in winter. This fits with the fact that in the 1920's and 1930's reports of xerophthalmia and keratomalacia were more common in northern than in southern China (J. H. Liu and C. K. Chu, 1943: 100, 103).[18]

Since traditional Chinese diet contained little or no milk, butter, or cheese, and usually only small amounts of liver, fish liver oils, eggs, or other animal foods with significant quantities of vitamin A_1 (retinol), this vitamin was derived mainly from the carotene in plant foods. In South China, the richest plant sources of carotene were chili peppers and vegetables such as carrots; spinach, water spinach (*Ipomoea aquatica*), and other dark leafy vegetables; certain cabbages; green peppers; and sweet potatoes (yellow varieties) (Guy and Yeh, 1938; Kwoh *et al.*, 1941; R. S. Harris *et al.*, 1949; Fan and Hou, 1953; Platt, 1962; Leung *et al.*, 1973; E. N. Anderson and M. L. Anderson, 1977: 329-30, 346-47; R. L. Freedman, 1978: 2; E. N. Anderson, 1988: 131). Certain other fruits and vegetables contained lesser but important amounts of carotene, among them persimmons, leeks, tomatoes, and certain legumes such as the soybean and field pea. Since green leafy vegetables were available throughout the year in South China, it was relatively simple, according to E. F. Yang (1958: 89), to meet vitamin A requirements, but this was not so in North China, where celery cabbage, the principal winter vegetable, was low in carotene.

Within South China, the Double Cropping Rice Area seems to have been characterized by a higher intake of vitamin A than most areas of the Rice Region (Maynard and Swen, 1956: 426-27). That vitamin A deficiency is still a problem in Kwangtung is suggested by Fry's findings (1965: 101-3) in an orphanage in Hong Kong of an average daily vitamin A intake (952.5 I.U.) well below minimum safe levels, providing only a third of the physiological needs of a thirteen to fifteen year old boy (2,884 I.U.) as set down by INCAP. It is suggested further by Fehily's report (1945: 469) that in Hong Kong on occasion vitamin A deficiency was found in infants who were given rice paste or pre-masticated boiled rice, and in unwanted and neglected female infants. Fehily did not, however, know of vitamin A deficiency among adults there, likely, she said, because of widespread consumption of fish which were commonly eaten whole.

Various researchers, among them nutritionists and epidemiologists, have long been interested in the possible protective role of vitamin A (used here to include beta carotene, retinol, and synthetic forms of vitamin A) against cancer. In a recent review, Mettlin (1984) found considerable support from both animal and epidemiological studies for the hypothesis that vitamin A intake is protective, especially against lung cancer. Two studies cited by Mettlin involved consumption of green or green and yellow vegetables among East Asian populations. One was a prospective study by Hirayama, published in 1977, of 256,000 Japanese adults over a ten-year span. In Japan, green and yellow vegetables contribute an average of 44 percent of total vitamin A intake. Males and females, smokers and non-smokers, experienced significantly lower lung cancer mortality if they ate such vegetables daily as compared to those who did not. The second was a retrospective study by MacLennan *et al.* (1977: 854, 857, 859, 860) of Singapore Chinese, which reported a significant increase in risk of lung cancer among persons who consumed green leafy vegetables (such as kale or mustard greens) less than once a week. The Singapore Chinese in question were Cantonese or from other groups originating in China's southeast. Though we have found no such studies for Kwangtung itself or for other parts of China, if, as appears likely, there were significant regional differences in vitamin A intake, one might, if the hypothesis is correct, expect them to be reflected in contrasting incidences of lung cancer.

Thiamine Deficiency and Beriberi[19]

Judging from the writings of Sun-szu-mo (A.D. 581-673), beriberi was rare in the ancient Han homeland of North China. There the common cereals were wheat and millet, which were rich in thiamine and eaten whole or lightly milled. Beriberi seems to have become common only in the fourth century A.D. as the Han Chinese migrated southward into areas where rice was the dominant food crop. Beriberi was considered a special risk of life in Nan-Yüeh in T'ang times and before, and a broad range of cures were prescribed for it (Schafer, 1967: 133).[20] Chinese medical writers did not associate beriberi with the consumption of rice, but suggested, instead, other causes, such as alcoholism, sexual indulgence, and miasmas. Nevertheless, their remedies did include some items (among them bran, black soybean, liver, and kidney) that contain significant amounts of thiamine (T. Lee, 1940: 316-20; E. F. Yang and B. E. Read, 1940; Lu and Needham, 1951: 15).[21]

The early regional difference between North and South China in the incidence of beriberi has persisted into modern times. In the first half of the twentieth century, thiamine deficiency and beriberi were not usually serious in the wheat-millet-sorghum areas of North China,[22,23] but remained so in the parts of the south where rice was the dietary staple (J. H. Liu and C. K. Chu, 1943: 99, 103; Chiao, 1944: 19; Simmons *et al.*, 1944: 68-69; Wilkinson, 1944: 655-56; T. H. Shen, 1951: 168).[24] This came about because people preferred milled rice, especially highly polished forms that in milling lose most of their thiamine. Milled white rice is preferred over brown rice in China in part because it is more readily stored, cooked, and digested; in part because it is bland in taste and can better take on the flavorings of other ingredients in a dish; and in part because, like white flour in various countries, it is higher in status (E. N. Anderson and M. L. Anderson, 1977: 345; E. N. Anderson, 1988: 115-16). The result was that, assuming a minimum daily requirement of vitamin B_1 for an adult male of 300 to 350 I.U., people in South China had a daily intake (250-322 I.U.) near the lower margin of their needs or below, whereas people in North China were significantly above (450-690 I.U.) (Lu and Needham, 1951: 19).

The situation was exacerbated in South China by changes that took place late in the nineteenth century. Prior to that time, people ground their rice at home daily, as needed, and much vitamin B_1 remained on the grain (Lu and Needham, 1951: 18-19). This changed when modern milling machinery came into use, and consumption of highly polished rice, which is far lower in vitamin B_1, became significant (E. N. Anderson and M. L. Anderson, 1977: 345; E. N. Anderson, 1988: 116, 135). Storing of polished rice after milling, moreover, facilitated the growth of bacteria and molds, and required that the grain be washed before use, a process that removed still more of its vitamin B_1 (Lu and Needham, 1951: 19). The problem was more severe in towns and cities, where highly polished rice was the norm, and less severe in rural areas, where much rice continued to be lightly-milled. The result was that thiamine deficiency and beriberi were more common in towns and cities, especially in South China (Lindstedt, 1939: 371E).[25] The problem became worse in crisis situations, as shown by data for Hong Kong to which many people fled from the mainland following the Japanese invasion of China. From 1937 to 1940, Hong Kong, whose rice was imported, highly milled, and stored for an extended period of time, experienced a steady increase in deaths from beriberi, which in the latter year numbered over 5700 and was one of the city's leading causes of death (Wilkinson, 1944: 655-56; Fehily, 1945: 468; 1947: 220).[26]

The thiamine content of rice in traditional China depended on more than the method of milling and washing the grain. It decreased, as well, with length of storage, degree of fermentation, and with cooking (E. F. Yang, 1938). As shown on Table 8, polishing of "first grade" raw rice by the traditional village process of milling in the Shanghai area led to a 72 percent loss in thiamine content. The total loss increased to 84 percent after washing and to 91 percent after cooking by the customary means.

There is one estimate that a Shanghai factory worker obtained 153 I.U. of thiamine from non-cereal foods per day.[27] If he ate rice that was polished, washed, and cooked in the customary way, he obtained another 47 I.U. for a total of 200 I.U. (E. F. Yang, 1938: 62-63),

Table 8. Loss of thiamine in preparation of rice by traditional means (after E. F. Yang, 1938)		
	Thiamine Content (Micrograms per 100 gm.)	Total Thiamine Loss after Each Processing Stage (%)
Whole rice	313	—
"First grade" raw rice (polished, whole kernels)	87	72
Washed	50	84
Cooked (thirty minutes)	29	91

which was below the lower limit of 300 I.U. mentioned above. If, however, half-milled, washed, cooked rice was substituted, the total daily thiamine intake rose well above that point to 263 I.U. Use of parboiled, washed, cooked rice would have raised the total much higher, to 469 I.U., and it is unfortunate in nutritional terms that the process of parboiling rice, which has been used since antiquity in South India, has not become common practice in China (E. N. Anderson, 1988: 116).

Recognition of the thiamine deficiency problem led Chinese authorities to introduce unpolished government rice into the diets of soldiers and school children regardless of their preferences, with good results (J. H. Liu and C. K. Chu, 1943: 103; Chiao, 1944: 19). Sometimes, too, broader events in society inadvertently led to declines in consumption of highly-polished rice. One reads that during the Japanese invasion and afterwards a rice shortage led to broader use of lightly-milled rice and a great reduction in the incidence of beriberi (T. H. Shen, 1951: 168), with British medical officers returning to Hong Kong after defeat of the Japanese amazed to find that beriberi was not widespread (Fehily, 1947: 220).

The degree to which tea consumption may have contributed to the occurrence of beriberi remains to be investigated. It should be noted, however, that in Thailand Vimokesant *et al.* (1974: 372-74) found that when schoolboys fifteen to seventeen years old consumed a liter of tea daily, 2.1 mg. of thiamine were destroyed. In view of the generally low thiamine intake in South China,[28] tea consumption may have been a significant contributing factor to the beriberi problem. Raw fish and other seafood also contain an anti-thiamine factor (Briggs and Calloway, 1979: 173), but one can only speculate on what effect, if any, fish consumption had on thiamine availability in China.

Riboflavin Intake and Deficiency

Riboflavin was named and first synthesized in 1935, and riboflavin deficiency was first studied experimentally in humans in 1938. A few years later, H. C. Hou (1940b; 1941) described cases of riboflavin deficiency in hospital patients and refugee camp inhabitants in Shanghai. One symptom common in riboflavin-deficient persons, cheilosis (reddening of the lips, cracks at the corners of the mouth), occurred in 44 percent of the children examined (eighty-two of 186) in the refugee camp. Of the six refugee children who were treated with riboflavin doses, all were healed of cheilosis, or partially so, by the eighth day. Though cheilosis is not limited to riboflavin-deficient persons, its relatively common occurrence among poor Chinese in Shanghai (H. C. Hou, 1941: 314) suggests that riboflavin deficiency may have been common as well in the general population. For South China, Fehily (1947: 221) found that of 1,215 children examined in a child welfare clinic in Hong Kong following the Japanese occupation, 113 (9 percent) showed signs of riboflavin deficiency either alone or combined with thiamine deficiency; and Jolliffe and Tung (1956: 312-15) found various clinical signs of riboflavin deficiency in many sixth grade children (eleven to thirteen years old) on Taiwan, and estimated, on these grounds, that 60 to 70 percent of them were deficient in riboflavin. This fits with the observation that on Taiwan the daily per capita intake of riboflavin (1953 est.) was 0.51 mg., well below the 1.4-1.6 mg. recommended by WHO (1974) for children ten to twelve years old. Judging from a study by Chen and Huang in 1956-57 (J. S. Chen and C. S. Huang, 1960: 594, 598-99), riboflavin deficiency may be even more serious among Taiwan aborigines. These researchers found many clinical signs of riboflavin deficiency among the children they studied; they judged that 55 to 65 percent of children in fourth and fifth grades suffered from riboflavin deficiency; and they found that adults of the Paiwan aboriginal group had an average daily riboflavin intake of only 0.14 mg.

In a study of 205 orphanage children in Hong Kong (ages five to eighteen years) in August, 1961, Fry (1965) found serious deficiencies of riboflavin, and estimated that the average boy thirteen to fifteen years of age had a daily intake of 0.67 mg. of riboflavin, which was less than half of the 1.7 mg. daily riboflavin intake recommended by the WHO (1974) for a boy of that age. Similar results were obtained in a one-year study by Fry and Baykan (1972: 102) of 410 orphanage children (ages six to seventeen years).

One notes that in China, certain sources of riboflavin that are very important in the West (milk, cheese, eggs, meat) are consumed in very small amounts. The Chinese do make more regular use of organ meats which are rich in riboflavin (especially liver, kidneys, and heart), but these are not major dietary elements, and most Chinese probably derive much of their riboflavin from sources that are not rich in the vitamin. Roots and tubers and whole grain cereals, for example, have only modest amounts of riboflavin (.07 mg./100 gm. for brown rice), and the milling and cooking of cereals greatly reduces those amounts (to .01/100 gm. in milled, cooked rice). Because they are consumed in volume, however, cereals can make a significant contribution in this regard.[29] Among the roots and tubers, the sweet potato deserves special mention, for though its tubers contain little of the vitamin (.05 mg./100 gm., edible portion), they may be a major dietary item, and in addition they provide greens rich in riboflavin (0.16-0.35 mg./100 gm.) (Leung *et al.*, 1973: 14, 60; Caldwell and Enoch, 1972: 311). Also deserving mention are pulses such as soybeans[30] and mung beans, as well as various green vegetables, all of which are rich in riboflavin.

Niacin Deficiency and Pellagra

Pellagra is a nutritional-deficiency disease related to low absorption of niacin and has commonly occurred in certain parts of the world where maize is a dominant food. Maize does contain niacin, but in a bound form, and it is relatively low in tryptophan, the amino acid from which the body itself can produce niacin. According to Briggs and Calloway (1979: 188), pellagra is brought on in maize-consuming populations by the simultaneous presence of three conditions: (1) limited quantities of available niacin, (2) limited quantities of tryptophan, and (3) an imbalance brought on by the fact that maize contains sizeable amounts of amino acids of other sorts as compared to tryptophan. Techniques of cooking maize with alkalis, which are common among American Indians, increase the availability of niacin, and, as a result, pellagra is uncommon in areas of Meso-America where maize is important. Even lacking such alkali processing, any deficiencies of maize with respect to niacin can be remedied by consuming foods, such as brewer's yeast and soybeans, that are rich in niacin or tryptophan-containing protein. However, where maize dependence is high and other dietary sources of tryptophan, niacin, or both, were lacking (such as the American South during the Great Depression, present day India, and South Africa), pellagra could become a major disease (Katz, Hediger, and Valleroy, 1974: 766).

In the Mediterranean region and Europe, there is no description of pellagra until the introduction of maize after discovery of the New World (Davidson and Passmore, 1969: 453). A similar situation seems to have existed in China, where T. Lee (1940: 315) reported difficulty in uncovering a detailed account of pellagra in the historical record. This may have been because in traditional China, despite the absence of alkali processing, maize was seldom of such dietary importance as to create a serious risk of pellagra. In the 1929-33 village survey, for example, maize was found to occupy less than 10 percent of China's total crop area (13.3 percent in the Wheat Region, 5.5 percent in the Rice Region), and it was the leading crop in only 3 percent of a total of 164 localities (Buck, 1956: 210-11). Lindstedt (1939: 371E) was thus able to write in traditional times that pellagra is rarely encountered in China. Jolliffe and Tung (1956: 315-16), though they judged that about 10 percent of sixth-grade children studied on Taiwan were niacin deficient, did not find classical pellagra or "skin lesions suggestive of niacin deficiency."[31] And Fry (1965: 102-3), after a survey of niacin intake among orphanage children in Hong Kong, concluded that an average thirteen to fifteen year old boy consumed 15.5 mg. of niacin daily, enough to meet more than 100 percent of his physiological needs (as set by INCAP, 1954),[32] and a percentage higher than that for any of the nine nutrients considered. Though we have found no such studies in Kwangtung, the 1929-33 village survey found the Double Cropping Rice Area to have the lowest percentage of cropland (only 1.4 percent) devoted to maize of any of China's agricultural areas (Buck, 1956: 211), and a map presented by Liu and Chu (J. H. Liu and C. K. Chu, 1943: Chart 1) suggests that Kwangtung may have fewer cases of pellagra than almost any other part of China.

It is true that reports of pellagra occur for various parts of China (C. S. Yang and K. K. Huang, 1934; Morris *et al.*, 1940; J. H. Liu and C. K. Chu, 1943; Snapper, 1965: 26), but usually the disease was associated with other vitamin deficiencies, and occurred among such deprived groups as war refugees and very poor people. Fehily (1945), for example, did not include pellagra on a list of deficiency diseases of Hong Kong prior to the Japanese invasion, and Wilkinson (1944: 656) observed that the disease did not occur in its standard form in Hong Kong before 1940.

Vitamin C Deficiency and Scurvy

One finds scattered statements that Chinese diet, in general or for certain regions, is likely to be below optimum in vitamin C intake. Nevertheless, whatever standard of recommended intake of vitamin C one uses,[33] the minimum needed to prevent or cure scurvy is only 10 mg. per day and a small serving (50 gm.) of greens can provide three times that amount (Passmore *et al.*, 1974: 35). Perhaps because of this, one finds no mention of scurvy in early Chinese literature (T. Lee, 1940: 314), and in modern China the disease appears to be relatively uncommon (Lindstedt, 1939: 372E; C. F. Wang, 1943: 80; Holmes, 1948: 217, 220-21). Scurvy could become common at times of flood or other crisis (J. Morgan and A. S. Gault, 1941; Fehily, 1947), but it was rare in Hong Kong and in 1937-41 no adult case was reported there, probably because people eat green vegetables every day (Wilkinson, 1944: 657-58). Scurvy was said to occur in North China in normal times but only among poor families, particularly in late winter and spring; more common there and in other parts of China is subclinical vitamin C deficiency, which may occur in springtime when the supply of fruit is low and the price expensive (Chiao, 1944: 19). According to Maynard and Swen (1956: 428-29), the Double Cropping Rice Area exceeds all other areas of China in the amount of vitamin C available. Fry's study of Hong Kong orphanage children (1965) found an average intake of 32 mg. daily, which was over 100 percent of the 30 mg. recommended by the WHO (1974).

Some of the richest common sources of vitamin C, each containing 20 mg. or more of ascorbic acid per 100 gm. of edible portion, as purchased, are listed in Table 9. Many of these sources, however, are quite minor in Chinese diet, with sweet potatoes the only major source of vitamin C among foods listed by Maynard and Swen (1956: 413) as providing the bulk of dietary calories in the Double Cropping Rice Area.[34] In addition, significant amounts of vitamin C are obtained from green leafy vegetables, especially cabbages and mustards; radishes (greens and roots); tomatoes; peppers, both sweet and red; certain root, stem, or aquatic plants; and, to a lesser degree, fruits, among them citrus, longan, litchi, jujube, haws, and others.[35] Also not to be overlooked in this regard are bean sprouts and white potatoes.[36]

One cannot, of course, estimate vitamin C intake by the quantity of ascorbic acid in a limited sample of fresh, raw vegetables. For one thing, ascorbic acid may vary greatly from one week, month, or season to another (H. C. Hou, 1936; T. F. Yü, 1940). It may also be reduced greatly by length of storage and by preparation and cooking (C. F. Wang, 1942; Chiao, 1944: 24; Snapper, 1965: 22; Passmore *et al.*, 1974: 33-34; Sood and Bhat, 1974). Thus poor people who buy cheap vegetables that are no longer fresh would obtain much less vitamin C from them than otherwise. Stir-frying in hot oil is one way of cooking meat and vegetables, after cutting or chopping them into small pieces. T. F. Yü (1940) found that cooking by sautéing, which might be expected to have results similar to stir-frying, results in significant vitamin C losses in vegetables, most vegetables losing 50 percent or more of their vitamin C after two to three minutes (Table 10). He also found that the loss of vitamin C increases sharply with the duration of sautéing, with two-thirds or more lost in cabbages and green pepper overcooked ten minutes (Table 11), as may occur when food preparation is on a large scale in institutions. Use of copper vessels and adding of alkali in cooking (to help maintain the fresh green color) also increase the loss of ascorbic acid through facilitating oxidation (T. F. Yü, 1940: 524, 533).

There is also a significant loss of vitamin C in boiling vegetables in water for five minutes. However, much of the vitamin remains in the water (Table 12) and such vegetable water can be a good source of vitamin C. Water from Chinese cabbage, for example, has a vitamin C content a fifth that of orange juice. Even boiling as long as thirty minutes, the minimum time required to destroy intestinal parasite ova, does not materially alter the vitamin C content of vegetable water (Su, 1938: 354-56). Thus intake of vitamin C is significantly enhanced in China by frequent use of vegetables in soups and by the widespread use of vegetable soup and vegetable water by pediatricians (to provide infants with a cheap source of vitamin C) (Su, 1938: 351; T. F. Yü, 1940: 528-29). It is also known that whereas pickled vegetables themselves contain little vitamin C, the brine normally contains much more (Table 13).

Table 9. Some good plant sources of vitamin C in China	
Roots and Tubers (raw, edible portion)	mg. per 100 gm.
Chinese radish (*Raphanus sativus*)	42
Lotus root (*Nelumbium speciosum*)	38
Sweet potato (*Ipomoea batatas*) (yellow variety)	30
(pale variety)	21
Turnip (*Brassica rapa*)	22
Green Leafy Vegetables	
(raw, edible portion of leaves or leaves and stems)	
Bitter cucumber (*Momordica charantia*) (leaves)	170
Chinese kale (*Brassica alboglabra*)	137
Ceylon spinach (*Basella rubra*)	86
Mustard cabbage (*Brassica juncea*)	73
Taro (*Colocasia antiquorum*) (leaves)	63
Amaranth (*Amaranthus gangeticus*)	62
Spinach (*Spinacea oleracea*)	56
Chinese white cabbage (*Brassica chinensis*)	53-62
Turnip greens (*Brassica rapa*)	47
Water spinach (*Ipomoea aquatica*)	47
European cabbage (*Brassica oleracea*)	46
Water cress (*Nasturtium officinale*)	46
Chinese rapeseed (*Brassica campestris*)	45
Celery cabbage (*Brassica pekinensis*)	38
Other Vegetables (raw, edible portion)	
Broad bean (*Vicia faba*) (sprouts)	110
(bean)	26
Bitter cucumber (*Momordica charantia*) (fruit)	57
Tomato (*Lycopersicon esculentum*) (ripe fruit)	23
Wax gourd or winter melon (*Benincasa hispida*) (fruit)	20
Chinese celery (*Apium graveolens*) (unbleached)	20
Fruits (raw, edible portion)	
Emblic myrobalan (*Phyllanthus emblica*)	276
Wampee (*Clausena lansium*)	148 (?)
Guava (*Psidium guajava*)	132
Papaya (*Carica papaya*) (ripe)	73
Longan (*Euphoria longana*)	56
Pummelo (*Citrus grandis*)	53
Litchi (*Litchi chinensis*)	50
Chinese jujube (*Zizyphus jujuba* Mill.)	46
Sweet orange (*Citrus sinensis*)	43
Tangerine (*Citrus reticulata*)	42
Kumquat (*Fortunella margarita*)	40
Carambola (*Averrhoa carambola*)	38
Mango (*Mangifera indica*) (ripe)	36
King orange (*Citrus nobilis*)	32
Pineapple (*Ananas comosus*)	22
Oriental persimmon (*Diospyros kaki*) (ripe)	20
Chinese olive (*Canarium album*)	20
Nuts and Seeds (whole, raw, edible portion)	
Ginkgo (*Ginkgo biloba*) (dried)	25

Table 9. (Continued) Some good plant sources of vitamin C in China	
Spices and Flavorings (raw, edible portion)	mg. per 100 gm.
Sweet pepper (*Capsicum annuum*) (green fruit)	103
Red pepper (*Capsicum frutescens*) (fruit)	96
Lemon daylily (*Hemerocallis flava*) (leaves)	88
Coriander (*Coriandrum sativum*) (leaves and stems)	78
Chinese leek (*Allium odorum*)	36
Sweet basil (*Ocimum basilicum*) (leaves)	26

(after Leung *et al.*, 1973, and M. J. Caldwell, 1972)

Table 10. Loss of vitamin C: Comparison between boiling vegetables in water and cooking in sizzling hot oil (after T. F. Yü, 1940)

Vegetables	Vitamin C (mg. per 100 gm.)			Vitamin C % of loss	
	Raw	Boiled in water*	Sautéed **	Boiled in water*	Sautéed **
Kohlrabi	21.6	6.7	8.6	69.0	60.0
Mung bean sprouts	8.7	1.7	2.5	80.5	71.3
Asparagus	9.1	5.6	5.1	38.5	44.0
European cabbage	25.0	9.4	9.7	62.4	61.2
Cauliflower	31.5	23.3	20.3	26.0	35.6
Small cabbage	44.9	16.2	18.3	63.9	59.3
Spinach	10.7	2.3	4.4	78.5	58.9
Tomato	10.0	8.6	6.9	24.0	31.0

* 5 minutes
** 2-3 minutes

Table 11. Effect of duration of sautéing in boiling oil on the vitamin C content of vegetables (after T. F. Yü, 1940)

Vegetables	Wt. of Sample (gm.)	Duration of Sautéing (minutes)	Vitamin C Content (mg. per 100 gm.)
European cabbage	1000	0 (raw)	22.6
		13 (just done)	9.9
		18 (5 min. overcooked)	8.5
		23 (10 min. overcooked)	4.3
Small cabbage	1000	0 (raw)	18.6
		12 (just done)	8.8
		17 (5 min. overcooked)	5.8
		22 (10 min. overcooked)	2.9
Green pepper	400	0 (raw)	42.8
		8 (just done)	25.2
		13 (5 min. overcooked)	19.0
		18 (10 min. overcooked)	14.4

Table 12. Vitamin C content of four vegetable waters (after T. F. Yü, 1940)

	Raw Vegetable (mg. per 100 gm.)	Vegetable Water (mg. per 100 gm.)
Chinese cabbage	13.4	8.0
Small cabbage	18.4	13.1
European cabbage	39.7	26.5
Turnip	25.1	11.6

Table 13. Vitamin C content of pickled vegetables (after T. F. Yü, 1940)

	Room Temperature	Vitamin C (mg. per 100 gm. of vegetables)		
		Raw Vegetables	Pickled Vegetables	Brine
Cabbage	25°C	15.6	1.7	12.0
	30°C	26.9	5.3	9.2
Turnip	25°C	18.3	1.8	16.9
	30°C	13.6	2.1	7.9
Carrot	25°C	3.4	1.3	1.9
	30°C	3.3	negligible	negligible

Calcium, Vitamin D, Rickets, and Osteomalacia

Over recent decades the question has repeatedly been raised as to what minimum intake standards for calcium are appropriate to non-Western populations, and whether low calcium intake constitutes a health problem. Against this background one must approach cautiously the repeated statements (Maynard, 1936: 426; Adolph, 1938: 133; Winfield, 1948: 72; T. H. Shen, 1951: 167; Maynard and Swen, 1956: 425) that calcium deficiency is among the most consistent and critical nutritional problems in China, for such statements have often been based on comparisons with minimum intake standards for the Western world that now seem unreasonably high.

It should, however, be noted that Chinese calcium intakes are often low even by comparison with the new, lower standards, such as that set down in 1961 by the FAO/WHO Expert Group on Calcium Requirements. Whereas that standard sets a daily recommended intake of 400 to 500 mg. per adult man, in some sections of China calcium intakes were below those levels. This seems to have been most pronounced in the Double Cropping Rice Area, where daily calcium intake averaged only 329 mg. per adult male in the 1929-33 survey, well below the averages of 417 to 579 mg. in the wheat areas of North China (Figure 79). Within the Double Cropping Rice Area, furthermore, some localities had daily calcium intakes far below average, as low as 170 mg. per adult male (Maynard and Swen, 1956: 423).[37]

It is difficult to determine, however, what the nutritional impact of such low calcium intakes *per se* might have been for adult men or for other sex and age groups. Fry's study of Hong Kong orphanage children, for example, found an average daily intake of merely 260 mg. (Fry *et al.*, 1967: 958). A later, year-long study (Fry *et al.*, 1967; Fry and Baykan, 1972), in which some Hong Kong orphans received 375 mg. supplements of calcium daily, found no significant differences in height or bone density between them and children who did not receive

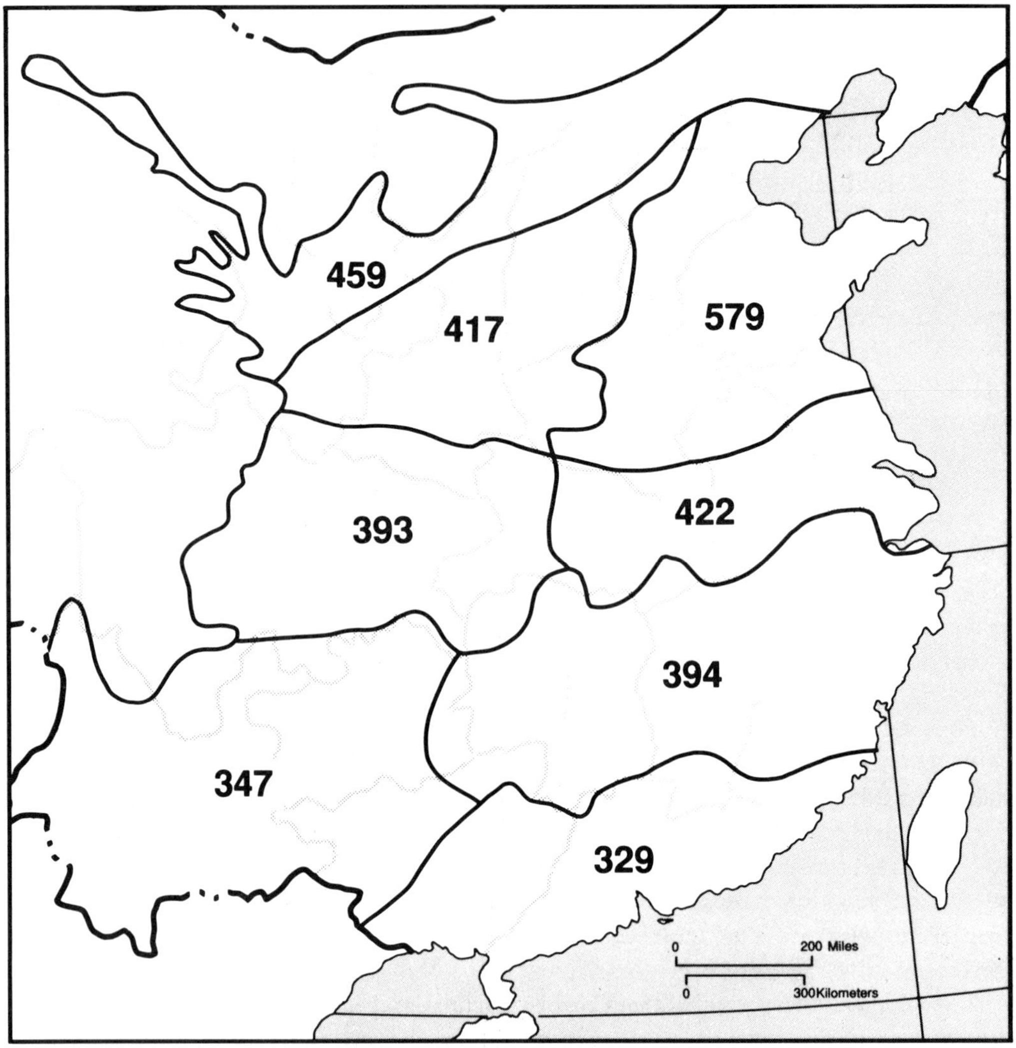

Figure 79. Calcium intake per adult-male unit per day (in milligrams) (redrawn from Maynard and Swen, 1956: 422)

the calcium supplements. The children with low calcium intake seem to have adapted, without apparent ill-effect, to the situation.

Traditional Chinese diet was overwhelmingly vegetarian, with little meat or eggs and little or no animal milk or other dairy products. After weaning, virtually all a person's dietary calcium derived from plant sources, with cereals and legumes, as major dietary elements, contributing the bulk, about four-fifths, of the total calcium intake (Adolph, 1939). Although leafy vegetables, together with soybeans, soybean curd, and leeks, were the most economical plant sources of calcium, leafy vegetables comprised only about 5 percent of the entire diet and a small share of the total calcium intake (Adolph, 1939: 330-32; Winfield, 1948: 72). Studies were made of various Chinese leafy vegetables and all except spinach were found to have high coefficients for calcium retention (Kung, Yeh, and Adolph, 1938). This led to suggestions that consumption of leafy vegetables be encouraged, along with bone meal and other calcium additives.

Of special note here are the customs that, whether or not people perceive of them in those terms, increase calcium intake and absorption in China. One is the custom of giving sesame seed, an outstanding source of calcium when whole, for consumption by a woman who has lost her hair after giving birth to a child (Holmes, 1948: 212). A second is the practice, in Hong Kong and presumably elsewhere in south coastal China, of consuming both the heads of fish and their little bones (Fry and Baykan, 1972: 103). In addition, the Chinese may utilize the calcium in eggshells by dissolving them in vinegar and consuming the resulting solution, and enhance the absorption of bone calcium by the sweet-and-sour method of preparation. Of special relevance is the custom, found in some parts of China, of giving pregnant women ribs of pork that have been cooked in sweet-and-sour sauce, the vinegar of which dissolves the calcium in the bones and permits it to be digested (Winfield, 1948: 72). The ribs are typically cut into pieces an inch and a half in size, which are then boiled for an hour in a solution of vinegar, soy sauce, salt, sugar, and rice. A single serving of this dish was found to contain 580 mg. of phosphorus and 550 mg. of calcium (deriving mostly from the bone), about half of the daily calcium intake recommended by the WHO for women in the latter half of pregnancy (Clinton, 1933: 871-72; Hoh *et al.*, 1933; Passmore *et al.*, 1974: Table 1). For South China, pickled pig's feet prepared with vinegar and sugar are a favored dish (A. W. Chung, 1979: 82), and it is a great credit if one gives two pig's feet to a pregnant Cantonese woman. The pig's feet are chopped into small pieces and cooked in rice vinegar for as much as a dozen hours, then sealed in containers. A mother anticipates having a dozen or more pairs of pig's feet before she gives birth, after which she eats them four or five times daily (Clinton, 1933: 871-72). One also reads of what seems to be the same or a similar dish, a broth made of dark vinegar, ginger, and pig's feet given to the mother of a newborn child in the Canton area to assist her in producing additional milk for her infant (Yee, 1975: 69); and of a Madame Wang, who was so skilled at preparing such broths (and ones of chicken stewed with Chinese gin) that the mother of a newborn child she visited would speedily and invariably return to her normal condition of health (L. Shaw, cited in Holmes, 1948: 183).

Chinese peasants spend much time exposed to sunshine out-of-doors. Thus one would expect ultraviolet radiation to produce for them sufficient calciferol (vitamin D_2) for adequate calcium metabolism. There are, however, literary allusions, possibly dating back to the seventh century B.C., to rickets and osteomalacia (sometimes called adult rickets), diseases commonly involving both vitamin D and calcium metabolism (T. Lee, 1940: 320-23; Lu and Needham, 1976: 6, 8, 9).[38] There are also modern clinical references to the diseases (S. H. Liu, 1940; Snapper, 1965: 14-22) as not unusual in China. Yet Lindstedt (1939: 372E), while acknowledging that vitamin D intake is low in China, observes that rickets is common to some degree mainly in Peking where a mild form is found in certain infants. This fits with a recent survey in the Peking area of North China which reported mild rickets in 20 percent of children under seven years of age (R. L. Freedman, 1978: 13). In sunny South China, on the other hand, vitamin D deficiency does not present a serious problem, and rickets is uncommon (E. F. Yang, 1958: 88). Wilkinson (1944: 658), for example, wrote that fully-developed rickets seems to be unknown in Hong Kong, and Fehily (1945: 469) found no symptoms of the condition in the five thousand or so infants and children she observed in that city.

road crews working on the famous Burma Road, 50 percent; and among certain aboriginal people in the mountains, goiter was nearly universal.[43]

One native explanation of goiter, reported for Kansu, is that it is brought on by drinking the water of springs or of mountain streams that flow over the exposed roots of trees such as willows (Kelly and Snedden, 1960: 175). Other explanations, used in the Sino-Tibetan border country (Liljestrand, 1933-34: 199-202), are that "monkey pouch," as the condition is called, derives from consuming certain local or interior salts; from drinking certain waters, as from stagnant hill reservoirs; from eating wild thyme; or, some said, it simply occurs naturally in certain districts. As for the salt explanation, one notes that some, but not all, interior salts are indeed quite low in iodine (Kelly and Snedden, 1960: 175, 177).[44] As for water as a possible goitrogen, we note first that similar beliefs occur widely in Eurasia. In the one area of endemic goiter in Korea, for example, people say that the disease will develop in persons who drink water draining from "the decaying roots of the edible pine" (Kelly and Snedden, 1960: 180). In a village on Bali where endemic goiter is prevalent, people drink water from a goitrogenic spring cursed by two goiter goddesses (Kelly and Snedden, 1960: 170). In one goitrous region in northwestern Punjab, there is a stream that since antiquity has been know as Gilhari Khud ("goiter stream") (Kelly and Snedden, 1960: 157). And in Europe, many bodies of water are believed to be goitrogenic, perhaps the most interesting being the "conscripts' springs" of Savoy, where military recruits would go to drink water so as to have their necks swell, thus permitting them to avoid military service (Kelly and Snedden, 1960: 113).

If there is truth in the above tales, the waters in question may have been unusually low in iodine, high in lime, or both. With respect to lime, there is the oft-noted correlation, made in India and elsewhere, between the continued drinking of hard water rich in calcium salts, as in limestone areas, and high incidence of goiter among humans. That calcium may indeed play a role in bringing on goiter has been confirmed by studies with laboratory rats. If the diet of rats is relatively low in iodine, goiter can develop if they consume sizeable amounts of calcium. The calcium is believed to act indirectly in causing iodine deficiency through lowering the metabolic effects of the thyroid hormone thyroxine and thus increasing iodine needs (Langer, 1960: 13, 20-21; Roche and Lissitzky, 1960: 362).

In a similar vein is the suggestion, made with reference to one region in Szechwan, that consumption of large amounts of cabbage may "tip the scales to the side of iodine insufficiency" (Kelly and Snedden, 1960: 175). This is in keeping with the results of studies of humans and laboratory animals elsewhere, which support the view that consuming large amounts of cabbage, kale, turnips, and other Brassicas over a long time has a goitrogenic effect (Kelly and Snedden, 1960: 66, 90, 91, 105; Milcu, 1960: 281; Roche and Lissitzky, 1960: 359). Similar effects have been suggested for other foods, among them soybeans, peanuts, and pearl or bulrush millet, *Pennisetum americanum* (Roche and Lissitzky, 1960: 359-62; "Millet - a possibly goitrogenic cereal," 1983).

Kwangtung and other coastal areas of China, in contrast to the inland regions mentioned above, are nearly free of endemic goiter (Kelly and Snedden, 1960: 149, 174-79).[45] This relates not only to the importance there of food from the sea, but to ways of producing salt and methods of food preparation and consumption. With respect to food preparation, an average of 20 percent of a fish's iodine disappears when it is fried, 23 percent when grilled, and 58 percent is lost to the water in boiling (Harrison *et al.*, 1965: 76). Thus, the Chinese pattern of brief cooking, commonly by stir-frying, would minimize the loss of iodine, as would the practices of using cooking water in soup and reusing oil from deep-frying.

Selenium Deficiency in Keshan Disease

Selenium is an element which occurs in strikingly different amounts from place to place on the earth's surface (Kubota *et al.*, 1967; Shamberger, 1981b). Since the 1930's, it has been known that if selenium is plentiful in the environment it can be toxic and cause disease in animals (Underwood, 1971: 323). Since the 1950's, it has been established that selenium deficiency is more widespread around the world than selenium excess; that sizeable geographical areas, for example New Zealand and Scandinavia, are characterized by selenium deficiency; that such deficiency can contribute to certain diseases in animals; and that these can be prevented by selenium supplements. What role, if any, selenium has in human health and disease, remained unknown. Underwood noted in 1971 (: 324) that neither selenium poisoning nor selenium deficiency had convincingly been demonstrated in humans, even among people living in places where these are found "naturally in livestock." And in 1976, R. F. Burk (: 105) observed that it has not been clearly established that selenium was linked causally to any human disease condition.

Subsequent research led the Food and Nutrition Board of the National Research Council in 1980 to classify selenium as an essential nutrient (Levander, 1982: 70). Though its contribution to human health and disease remains a matter of investigation, it has been suggested that selenium may have a role in bringing on or preventing heart disease, cancer, arthritis, dental caries, and other such conditions.[46] Of most significance for our study, however, is the possible link between selenium deficiency and Keshan disease (X. Chen *et al.*, 1980; 1981; G. Yang *et al.*, 1984), an endemic cardiomyopathy named for the area of Manchuria where it was first described in 1935. The disease, believed by people of affected areas to derive from a soil condition, is now known to occur in a long northeast-southwest belt extending from Heilungkiang Province, where Keshan is located, to Yunnan Province along China's southern border. Incidence rates in endemic areas can be as high as 11 percent, and account for more cases than all other cardiac diseases (W. H. Yu, 1982: 1203-4). In the south, children from two to seven years of age are likely to contract the disease, whereas in the north, it is women of child-bearing age and children less than fifteen years old (Keshan Disease Research Group *et al.*, 1979: 471).

Keshan disease is found principally among peasants in rural areas. It is more common in hill and mountain regions, less so in the plains; and it is absent in coastal locations. The selenium link has been demonstrated by various surveys. In one, Chinese researchers collected samples of human hair in Shantung Province, from a hilly area, where Keshan disease was common, to the seacoast, where it was absent. The hair of hill residents had an average selenium content of only 0.097 to 0.123 p.p.m., in contrast with 0.234 and even 0.331 along the coast (Keshan Disease Research Group, 1979: 478). Soybeans and sweet potatoes, as well, were found to have much lower selenium content in areas where Keshan disease occurred than in those where it was absent. Other studies have shown that the selenium content of blood in persons from affected areas commonly dropped below 0.01 p.p.m., whereas in unaffected ones the minimum blood selenium content was estimated to be about 0.04 p.p.m. (Keshan Disease Research Group, 1979: 479). The most convincing evidence for the selenium/Keshan disease link is found in a study carried out from 1974 to 1977 among children (one to nine years of age) in an area of Szechwan Province where Keshan disease was very common. Altogether 8,495 children were involved in 1974 and 12,749 in 1977, the latter being 99 percent of all children of ages one to nine years in four communes. Prior to the experiment, Keshan disease, whether acute, subacute, or chronic, was found in about ten children per 1000 among local one to nine year olds. Starting in 1974 one group of children received supplements of selenium (0.5 or 1 mg. of sodium selenite weekly), and the incidence of Keshan disease was 2.2 per 1000 in 1974 and 1.0 per 1000 in 1975. In contrast, the control group, which received no selenium supplements, had an incidence of 13.5 per 1000 in 1974 and 9.5 per 1000 in 1975. Starting in 1976, both groups were given selenium supplements and the disease incidence in the combined group dropped to 0.32 per 1000 in that year and to zero in 1977 (Keshan Disease Research Group...*et al.*, 1979: 471).

The Chinese researchers do not believe that selenium deficiency alone brings on Keshan disease, for certain epidemiological factors cannot be explained in those terms. Instead, they suggest, selenium deficiency may be a primary cause of the myocardial lesions typical of the disease, supplemented by other factors, perhaps dietary or environmental (Keshan Disease Research Group, 1979: 479-82; Keshan Disease Research Group *et al.*, 1979: 475; X. Chen *et al.*, 1980; 1981; W. H. Yu, 1982; G. Yang *et al.*, 1984: 224-29). Considered against evidence derived from studies of animals and humans in other parts of the world (see, for example, Shamberger and Willis, 1976; Underwood, 1977: 314-24; "Selenium in the heart of China," 1979; Passwater, 1980; Shamberger, 1981a; 1981b; 1983), the question of selenium deficiency in humans is deserving of more consideration.[47]

FOOD, NUTRITION, AND CARE OF THOSE AT GREATEST RISK: MOTHERS AND CHILDREN

Pregnancy and Abortion

By tradition in China, a good mother sees that others in the family have the best and most food, first her husband, then male children, then female children. She herself was last, eating the "least and poorest" (S. S. Chang, 1973: 153). This changed, however, during pregnancy and lactation, when she was normally given special consideration. This was especially so at her first pregnancy and if the child turned out to be a boy, for her family standing was greatly enhanced because she had contributed to its perpetuation. Failing to provide sons, a woman's standing in the household suffered, for she was held accountable for that failure, and her dealings with her husband and mother-in-law, protector of the family, were likely to become strained (Levy, 1963: 114).[48]

The most complete account we have uncovered of precautions taken by pregnant Chinese women is that of Manderson and Mathews (1981) (Table 14). Their account is based on interviews in Australia with forty refugee women from Vietnam, of whom twenty five were ethnic Chinese (nine Cantonese, nine Teochiu (Ch'ao-an) Fukienese from eastern Kwangtung, and seven of other dialects). Certain behavioral precautions reported by Manderson and Mathews derive from concern that the delivery be normal. Very heavy work, it is held, may bring on miscarriage or premature delivery, and idleness may lead to prolonged labor. The recommended path is moderate activity, as much as a woman would engage in if not pregnant. Other behavioral precautions relate to wishes for the mother and fetus to remain healthy and for the newborn child to be normal. If a pregnant woman sleeps in the afternoon, her child may be excessively large. If she stares at a deformed individual or if the father hammers a nail, her

Table 14. Foods prescribed and proscribed throughout pregnancy among Chinese women in Australia (from Manderson and Mathews, 1981)

Prescribed or Permissible Foods		Proscribed or Restricted Foods
Affecting 100%	Affecting ≥ 80%	Affecting > 20%
Sweet fruit	Sour fruit	Pineapple
Orange	Lemon	Melon
Sugar cane	Grapes	Green coconut
Raw vegetables	Ripe papaya	Green papaya
Cooked vegetables	Mango	Long banana
Peanut	Short bananas	Mung beans
Potato	Spinach	Chili
Boiled rice	Cabbage	Buffalo
Glutinous rice	Sweet potato	Black and white duck
Onion	Ginger	Salt water fish
Garlic	Sauces	Shellfish
Pork	Pork offal	Vinegar
Fresh water fish	Chicken	Coffee
Eggs	Duck	Herbal infusion
Cow's milk	Soy milk	Salt
	Tea (Chinese)	Pepper
	Tea (Ceylon)	Alcohol
	Carbonated drinks	*Tan kuei*
	Water	
	Honey	
	Vitamins	

child may be misshapen. Of special danger to mother and child is spirit possession, though such danger is minimized by carrying out rituals and by limiting social exchanges, especially attendance at funerals, weddings, and shrines.[49]

Receiving special treatment by Manderson and Mathews are the dietary precautions of pregnancy. These are followed so that mother and child will be in a state of equilibrium, thereby avoiding sickness and miscarriage, and guaranteeing the offspring an ideal personality and good physical traits. Dietary precautions are based in part on the hot-cold system of belief described previously. They are also based on the belief that some foods are tonic (enhancing health and energy and serving to increase the blood supply) or anti-tonic; that some foods and illnesses have "wind"; that a handful of foods have magical properties; and that others are neutral or not part of any of the above systems (Manderson and Mathews, 1981: 3). The ultimate goals were to avoid illnesses brought on by eating too much food of a particular category and, when imbalances did develop in the pregnant woman's body and that of the fetus, to provide foods that return them to a balanced state.

There were considerable differences of opinion among the women interviewed as to how foods should be classified under the above categories, and thus on when and in what quantities they should be consumed. Moreover, the acceptability of some foods remained unchanged during an entire pregnancy, whereas that of others varied from one period of the pregnancy to another.

During the first three months of pregnancy, when weakness and nausea may occur, a woman is believed to be feeble, cold, and lacking in vigor. In the following three months, she is in a neutral condition. And in the final three months, she is hot and tonic. In the first three months of pregnancy, hot foods are prescribed (among them ginger, black pepper, alcohol, and ripe mangoes) as are tonic ones (pork, steamed rice, *tan-kuei* [an angelica herbal tonic], and other medicines). Avoided are "cold" foods (among them melon, green coconut, spinach, and mung beans) and anti-tonic ones (for example, lemon, pineapple, and vinegar). Also avoided by a few pregnant women, out of concern with bringing on a miscarriage, are foods that are "too hot," such as chili and alcohol.

During the second three months of pregnancy, a neutral time, the woman starts eating "cold" foods, and these become important in the final three months, to counter the hot condition that prevails then. At the same time, few hot foods are eaten, or they are avoided altogether, to spare the mother heat, discomfort, indigestion, and constipation, and to assure that the newborn child will not have skin problems. Tonic foods are taken only in limited amounts in the last three months to reduce the danger of having a large baby and difficulties in birth.

Although Manderson and Mathews acknowledge that in certain other parts of the world traditional beliefs about hot and cold foods have a negative impact on human health and nutrition, that is not the case among the women they studied. They write that such beliefs may guarantee dietary variety as well as nutritional balance both in a Western sense and in traditional ones (Manderson and Mathews, 1981: 7). In any case, the nutritionist should be aware that, whatever individual variations exist, none of the women questioned the basic concepts of the traditional system of belief.

The literature surveyed to date fails to provide much information on induced abortion in traditional China, its prevalence and context. Orleans (1973: 207) writes that abortion was never subject to the moral or legal barriers so common in the Western world. Other accounts, in turn, suggest that abortion served as a means of birth control (H. D. R. Baker, 1979: 5). Such abortions seem to have become more common in the People's Republic, especially since the 1950's (Orleans, 1973: 207; see also Bowers, 1970: 9). Orleans writes of many articles in medical journals in the mid-1960's describing abortion procedures. He also writes of efforts to develop simple methods and equipment to be used by medical staff in carrying out abortions in the countryside. Orleans says (1973: 207) that the increase in abortions was not so much a matter of government pressure, as one of increased facilities and personnel. Mosher (1983a: 225-29, 254-61, and *passim*), writing in 1983 after field work in Kwangtung Province, describes a very different situation, with strong government pressure to abort over-quota infants.

Though Orleans does not estimate the prevalence of induced abortion in the People's Republic, Yeh and Chow (1973: 219) do: from 4.5 to 6 million induced abortions in 1970 out of a total number of pregnancies of from 30.7 to 33.9 million.

Birth and Birth Weights

To assist at birth, which normally occurred at home, it was customary to have present a traditional midwife and female relatives.[50] Since traditional midwives received no training in modern medical beliefs and practices,[51] unhygienic conditions prevailed, and both mother and child were at great risk of infection. It was not unusual for a Chinese woman to have given birth to from five to twelve, or even more, children with only one or two surviving, or none at all. For the infant, the greatest risk was tetanus neonatorum, brought on by the use of dirty scissors or by tying the umbilical cord with unsterilized cloth and applying herbal medicines and even cow dung or road dust. Death of infants from tetanus often occurred, with convulsions, eight to ten days after birth. Such deaths were so commonplace that the Chinese designated tetanus

Table 15. Regional differences in birth weight of Chinese babies (from C.-C. Lee, 1948)

Author	Locality	No. of Cases	Weight in gm.
Lee	Peking	1437	3052
King and Tang	Shantung	1000	3090
Eno	Shanghai	2260	3128
Oppenheim	Fukien	942	3159
Uttley	Kwangtung	5437	3018
Lee	Ch'eng-tu	3550	3038
	Average of Chinese Newborn		3081
	Average of American Newborn		3400

neonatorum the "nine-day disease." As for mothers, the greatest danger was puerperal, or childbed fever (Lamson, 1935: 285-86; Ta Chen, 1940: 179-80; Winfield, 1948: 109). Also causing many deaths among Chinese mothers were complications, such as the osteomalacia common in Shansi Province, that blocked delivery. Altogether, wrote Winfield in 1948 (: 109), "6 to 8 percent of all deaths in China occur as a result of childbirth."

Ta Chen, in 1940 (: 179-80), said that modern midwifery was gaining ground in China, and had become popular in middle and upper class families, leading to a drop in the prevalence of tetanus neonatorum. Even before that time, women of the upper classes enjoyed the advantages of better nutrition and health (Levy, 1963: 114), which favored survival of mothers and children.

As of 1948, at least six studies presented data on birth weights of infants in China. Summarizing that data, involving altogether over 14,000 infants, C. C. Lee (1948: 155) reported an average birth weight of 3081 gm. as compared to 3400 for American newborn. Boys averaged a hundred grams more than girls. Birth weights also correlated with season of birth (heaviest babies were born in winter and spring), and with the mother's age (birth weights increased with maternal age up to forty-one to forty-five years) and parity (birth weights increased up to the seventh pregnancy). Lee found little difference in birth weights during the war years despite the fact that mothers had less nutritious food during that time. At the same time, he identified regional differences in birth weight (Table 15) which may in part reflect differences in the mothers' nutritional standing. Kwangtung babies were lighter, 3018 gm. on average, than babies from any other region.

Millis (1953a) carried the matter of Chinese birth weights into a broader framework, comparing newborn weights of homeland Chinese with Singapore Chinese, Singapore Indians, and certain other ethnic groups around the world. Birth weights of Singapore Chinese infants averaged 6.9 lb. (3130 gm.) for boys and 6.7 lb. (3039 gm.) for girls. Singapore Chinese newborn, like those of the homeland, weighed less than English or American newborn; first-born babies were significantly lighter than those of any other pregnancy; mean birth weights tended to increase with parity; and Chinese and Indian women of better economic status on average bore larger infants.[52]

Disposal of Unwanted Infants

Though it was only one way of getting rid of unwanted infants in traditional China, infanticide, because it shocked so many foreign observers, is unusually prominent in the Western literature. Yet at a conference held late in the nineteenth century to consider the prevalence of infanticide in China, seventeen Western authorities divided nearly evenly on the question of whether the practice was more common there than in other countries of the world ("The prevalence of infanticide in China," 1886: 25). Neither they nor anyone else could be certain, of course, for the act had long been looked on by the Chinese as immoral and illegal, and was normally done in secrecy and seldom revealed later. It is clear, however, that many Chinese considered it all too common (P. T. Ho, 1959: 58-62).

Infanticide in China was mostly of two types. One involved the killing of illegitimate or sickly or abnormal infants of whatever sex; the latter were commonly disposed of by a woman's mother-in-law because of the heavy expense necessary to care for the child ("The prevalence of infanticide in China," 1886: 28; Topley, 1974: 248). The second, and more common, type of infanticide involved female infants, for a male infant was expected to grow into a man and to support his parents in their old age, whereas a female, when married, would leave her family and help maintain her husband's aged parents and not her own. Infanticide of both types was more prevalent among the poor, and more common during times of war and famine when powerful economic motives, involving the family and its preservation and well being, were more likely to prevail over the bonds of affection a mother and family normally have for an infant. A well-off family that already had several sons would usually be pleased to have a daughter, but some wealthy families in places such as Fukien were reported to practice female infanticide for various reasons, such as dislike for the burden of caring for large families or concern about the payment of dowries in the distant future ("The prevalence of infanticide in China," 1886: 26; Headland, 1901: 34; P. T. Ho, 1959: 60). The midwife was commonly involved in disposing of an infant, or the parents might assume that responsibility. One way was simply to neglect it. Drowning, whether in a tub or some natural body of water, was also common;[53] and there are reports, as well, of infants being suffocated, strangled, buried alive, or killed in other ways ("The prevalence of infanticide in China," 1886: *passim*; Couling, 1917: 249).

As background to the question of female infanticide, one should note the striking differences in greeting accorded a newborn child in traditional China. If it is a boy, everyone congratulates the parents for the "great happiness" that has arrived, but a girl may be greeted with scant or no congratulations, for it is merely "a small happiness" (Headland, 1901: 35-37). The differences in attitude are further reflected in traditional Chinese sayings that it is "bad business raising girls for other people" (Headland, 1901: 35) and that daughters are merchandise "on which one loses" (Mosher, 1983a: 195). There is one report of a series of five or six female infants born to a mother being killed one after another, in the hope that ultimately she would give birth to a boy ("The prevalence of infanticide in China," 1886: 38).[54] In recognition of the economic hardship of raising a daughter in nineteenth-century Ning-po, a Chinese organization called "Society for Saving Infants" gave cash awards to poor parents who had just had a female child ("The prevalence of infanticide in China," 1886: 36-37; Couling, 1917: 249).

Even if economic pressures led to a decision to get rid of a female infant, many families preferred to sell it, give it away, or abandon it rather than letting it die of neglect at home or killing it outright. Thus in certain districts of Kwangtung and Fukien, as reported in the last century, a poor woman might sell a young daughter to a neighbor or friend who would raise it to be the wife of a son (Doolittle, 1865, 2: 206; Gray, 1878, 2: 50). There was also a market for young girls to be used as servants, mistresses, and such, but they had to be old enough to carry out such roles, and not all families could support a daughter until that age was reached (H. D.

R. Baker, 1979: 7). If not, a female infant might be sent to a foundling home. Such institutions are not new in China, for already in Sung times the royal court established foundling homes ("The prevalence of infanticide in China," 1886: 39; Gernet, 1962: 149). Most nineteenth-century cities, in turn, had officially-supported foundling homes where female infants could be taken ("The prevalence of infanticide in China," 1886: 26, 27), and there were also homes supported by Chinese officials and other wealthy or upper-class persons, as well as ones supported by Christian churches or missions.[55] That foundling homes could be quite effective in saving lives is indicated by the case of one such institution (called "orphanage") in nineteenth-century Hupeh, which was reputed "to have saved the lives of some ten thousand female babies within the first three years of its founding" (P. T. Ho, 1959: 61). At one time, four hundred infant girls were said to have been maintained in the Catholic home in Foochow. Canton, in turn, had a Foundling Hospital, supported by the salt tax, where there were five hundred infants, mostly girls. After they were eight or ten months old, the girls were sold, ostensibly to childless married couples or to families who wanted daughters for their sons (Gray, 1878, 2: 50), for some parents preferred to raise a girl to be their son's wife rather than raise their own daughter (Headland, 1901: 35). Unfortunately, some young girls who were sold or disposed of in other ways by foundling homes ultimately became slaves or prostitutes.

Though the existence of such foundling homes tended to reduce the amount of infanticide, the practice continued to attract public attention and concern, as shown by repeated mention of it in the literature, and by the moral tracts and government edicts directed against it (Doolittle, 1865, 2: 205-9; Gray, 1878, 1: 232-33; 2: 49-51; "The prevalence of infanticide in China," 1886: *passim*; S. W. Williams, 1895, 2: 239-42; Ball, 1906: 347-50; Ross, 1911: 98, 104, 193, 197, 241; Couling, 1917: 249; Cormack, 1935: 28; Lamson, 1935: 560-61; Lang, 1946: 46n; P. T. Ho, 1959: 58-62; Gernet, 1962: 148-50).[56] Inquiries by an American missionary in 57 localities in two regions of nineteenth-century Fukien, led to the conclusion that in one region 25 percent of all infant females were killed, and in the other, 40 percent (S. W. Williams, 1895, 2: 240). Those regions of Fukien were reputed to have the highest known incidences of infanticide in China, though Kwangtung, Chekiang, and Anhwei were also repeatedly singled out by nineteenth-century observers as places where infanticide was common. Other observers of that period noted that infanticide was most common in South China (or South and Central China) but practically unknown in North China ("The prevalence of infanticide in China," 1886: 31-32, 34, 39, 40, 41, 44-45). Yet Li Tsung-hsi, a governor-general of Kiangsu and Anhwei in the mid-nineteenth century, reported hearing that in certain parts of his home province of Shansi, all female infants, except perhaps the first to be born, were ordinarily drowned (P. T. Ho, 1959: 60). A nineteenth-century county history, in turn, said that in rural Hupeh and parts of Hunan it was customary for families to raise two sons and one daughter, and to dispose of subsequent children (P. T. Ho, 1959: 58). A general estimate for China, by a Chinese observer or observers early in this century, was less shocking: that from 5 to 10 percent of female babies "were abandoned or done away with for economic reasons" (Ross, 1911: 193, 241).

As for Kwangtung, Couling (1917: 249) wrote early in this century of missionary estimates that in some districts only a third of all female infants were permitted to live; this ultimately led to a shortage of marriageable young women there, and parents had to go elsewhere to obtain wives for their sons. There were campaigns against the practice in Kwangtung, and penalties for persons found guilty (Couling, 1917: 249; Lamson, 1935: 561). Lang (1946: 46, 150-52,

253), on the basis of research conducted from 1935-37 in several parts of China, concluded that infanticide was not uncommon, that it primarily, if not entirely, involved female infants, but that much more frequently young daughters were not killed outright but were given less food than sons, thereby weakening them and reducing their ability to survive epidemics.[57] Levy (1963: 290), who was in China in 1943-44, wrote of changes in attitudes toward infant daughters that resulted from increasing employment of women in factories. While this led to the placing of a higher value on female life, he noted, female infanticide and the selling of daughters still occurred in many places when times were difficult. C. K. Yang (1959a: 19) found no infanticide in a Kwangtung village studied from 1948 to 1951, though he reported that babies were sometimes abandoned.[58] The one case he describes is of a female infant left by a poor family on the doorstep of a city house that seemed prosperous. This practice is also reported by others, and H. D. R. Baker (1979: 7) writes that in Hong Kong in recent years banks are preferred places for abandoning infants. Female infanticide persisted in Kwangtung and elsewhere in China even after the People's Republic was established. Steven W. Mosher (1983a: 252; 1983b), for one, heard reports of female infanticide in a Pearl River Delta commune after he left it in 1979-80.[59] Mosher observed that infanticide in China had seemed to be dying out twenty or thirty years ago, but is now making a comeback not for the reasons given by the government ("feudal attitudes of backward villagers") but because of official birth-control policies which permit a couple to have only one child. This policy, and the lack of a social security system to care for parents in their old age, creates tremendous pressure on a married couple, as was true in traditional China, to have male and not female offspring. Mosher's observation that female infanticide survives in the People's Republic has been confirmed by reports in the Chinese press (Banister, 1987: 221-21, 427) and by Premier Zhao Ziyang's condemnation of the practice at a political meeting in late 1982. It is also reflected in the 1982 census by the changing sex ratios among cohorts of children born from mid-1978 onwards (Banister, 1987: 119, 220-21). As for infanticide involving sick and defective infants, Banister (1987: 224-26) thinks it likely that since establishment of the People's Republic a larger share of infants with birth defects have been permitted to live. She expresses the view that, as defective individuals reach their reproductive years, more hereditary diseases are being passed on to future generations, but she also reports governmental efforts, involving physical examinations, testing, and counseling, to minimize the likelihood of birth defects.

Confinement and the Feeding of Infants and Nursing Mothers

By Chinese tradition, if the family can afford it a mother remains at home to recover for an entire month after birth of a child. During that month she is subject to restrictions known collectively as "doing the month." As described by Pillsbury (1978), following interviews on Taiwan and the mainland, the restrictions, altogether a dozen, relate to three principles: (1) avoiding actions that bring on disease or bodily ailments, (2) avoiding offense to gods or other humans through pollution with fetal blood, and (3) following a dietary regime to counter the cold imbalance brought on by giving birth. Seven restrictions relate to hopes of avoiding disease and illness: (a) that the mother remain inside her own house for the entire month, (b) avoid exposure to wind, (c) avoid washing her body or hair, and (d) remain in bed as much as possible and not walk or move around. She is also (e) not to read or cry (for fear of affecting her eyesight), (f) to avoid raw and "cold" foods, and (g) should avoid getting ill (since confinement illnesses may linger throughout life). In addition to (a) above, there are four restrictions based on concern that fetal blood will anger the gods or pollute humans: (a) do not enter someone else's home, (b) do not have sexual intercourse, (c) do not eat at the family table, and (d) do not burn incense to the gods (which would be disrespectful). Efforts to return the mother's body to a state of balance involve avoiding raw and "cold" foods, and eating chicken and various other dishes viewed as "heating," or beneficial in other ways. As reported for a Kwangtung village (Yee, 1975: 69), a mother, during confinement, avoids cold drinks and fresh fruit and vegetables, though she is permitted to eat salted or cured vegetables and meats. Special dishes given her included a thick broth of vinegar, ginger, and pig's feet, and another mixture called "mother's brew," a soup made of carefully-selected ingredients. Basically it is a chicken soup, a food highly-regarded for its strengthening qualities. In addition, however, it contains gingerroot, which is both "heating" and an aid in dispelling wind from the mother's body. Also among the soup's ingredients are other "hot" foods (peanuts, dried jujubes, and rice wine or whiskey) and ones (dried wood fungus and lily buds) that invigorate and/or kill pain and cleanse the blood.[60]

Pillsbury records three quite different views of the usefulness of doing the month, as held by physicians. First, she notes (1978: 16), are American physicians who know little of Chinese ways, who find the custom strange but admit that it may provide some benefit to the mother, such as rest and good food. Second are those Chinese physicians trained in Western medicine who generally discount traditional Chinese beliefs and who criticize especially the restrictions on washing and certain of those relating to diet. Third are the Chinese physicians trained in Western medicine who nevertheless see value in some aspects of "doing the month," for example increasing the mother's protein intake, providing her with rest, and keeping her from drafts. Pillsbury herself suggests (1978: 19) that in terms of Western medical knowledge some of the practices are effective, but others appear not to be.

Chinese confinement practices must be reckoned with since they are still followed in mainland China, Taiwan, and among overseas Chinese. Pillsbury (1978: 19) reports that every one of more than one hundred Chinese and Chinese-Americans interviewed considered "doing the month" as valuable in terms of physical and mental health. Refugee women who gave birth in an Australian hospital were distressed when, a day or two after delivery, hospital staff urged them to get out of bed, walk, and exercise (M. Mathews and L. Manderson, 1981: 14). This, they believed, would lead to various ills such as varicose veins and pains in the back and legs. They were also told to shower and wash their hair, but most tried not to do so in the belief that showering would permit cold and wind to enter their bodies and render them susceptible to arthritis. Failure to observe behavioral restrictions, moreover, was believed responsible for the headaches, backaches, and skin eruptions that developed, and women believed more future illness would also result. Where women were served hospital foods that were restricted or proscribed by tradition, some tried to compensate by taking traditional medicines or large amounts of ginger, pepper, or salt. In any case, most women were bothered because they could not follow traditional dietary codes, though out of politeness they hesitated to request changes in hospital ways.

Platt and Gin (1938: 344) provide a good overall picture of traditional Chinese infant-feeding procedures following birth.[61,62] Right after birth, a Chinese baby was given a solution containing a native purgative, an act repeated a few hours later. The intent was to force from the infant's body the "hot" substances characteristic of the late fetal period. Breast feeding started the second day. By tradition, breast milk was considered the best, or even the only way to feed a newborn child (A. W. Chung, 1979: 81). A mother, moreover, regarded it as a duty to nurse her own child (Kao, 1948: 234). The result was that in China, except for gentry and a few well-educated mothers who were disinterested, breast feeding by the mother was the rule. In two surveys, 90 and 95 percent of Chinese infants were found to be nursed by their mothers (Platt and Gin, 1938: 344; Kao, 1948: 234).

It was recognized that a nursing woman must be provided with good food for satisfactory lactation.[63] This belief was behind the Chinese custom (Kao, 1948: 233-34) of presenting gifts, usually food, to the mother of a newborn child. Ordinarily a mother giving birth for the first time received a greater number of gifts. The total varied also according to the wealth of the family, but usually numbered from five to forty gifts, including such nutritionally valuable foods as eggs, chickens, fish, and pork.[64] Guy (1936: 437-38) notes that in the Peking area a mother is fed many eggs in the twenty or thirty days after she gives birth. The average number of eggs consumed in that time is about a hundred. Poor women might be able to obtain only twenty eggs or so, but women from families able to afford it commonly ate four or six eggs every day during that time. Hutson (1921: 2) writes that in Southwest China a mother measures her happiness by the quantity of eggs she consumes, and that whereas in most cases gifts are of twenty to thirty eggs, among the wealthy the gift may be five hundred or even a thousand eggs. The abundance was such that a Chinese mother usually had enough food to serve her nutritional needs for a month or two, and a mother from a well-to-do family would also benefit by gifts of clothing, bedding, and jewelry.

If a mother were unable or unwilling to breast feed her child, she might feed it artificially. In this regard, it is interesting that though a nursing woman would herself sometimes consume soybean milk, this drink was not ordinarily fed to infants in traditional China (Guy, 1936: 440; Platt, 1936: 418; Platt and Gin, 1938: 345, 351; Jelliffe, 1968: 58).[65] Similarly, in most of traditional China the diet of infants did not include animal milk; though in the People's Republic cow's milk is now available in most cities, it is less well regarded than mother's milk (A. W. Chung, 1979: 82).[66]

Another alternative in feeding an infant was to employ a wet nurse. It was common in pre-Communist times, for example, for the children of gentry to be nursed by a servant (Levy, 1963: 115). In those times there was also a well-established system of wet nursing in China (Platt and Gin, 1938: 345-46; Kao, 1948: 235-36),[67] though, in further testimony to the high regard in which boys were held, Kao (1948: 236) found that two-thirds of all wet nurses were used for male infants. A wet nurse might be hired on the basis of recommendations by relatives or acquaintances of the baby's parents, or located through a servant agency in a city. In the latter case, the wet nurse was commonly a poor woman from the countryside; such a woman would normally have nursed her own infant for two months, then left it with others, traveled to a city, and become affiliated with an agency. The agency owner would provide an infant for her to nurse, to continue her milk secretion. He also provided, for payment, room and board, and when the wet nurse obtained employment he received a share of her pay. There were well-defined qualities expected of desired in a wet nurse (she should be a youthful woman with good health, good looks, and a pleasant manner who has borne her first child; and her milk should be thick and white, and liked by the child). Most wet-nursed children had a single nurse, but about a fourth of such children had more than one, in sequence. Of special concern was the danger that the infant would contract diseases of the eye or skin or even syphilis or tuberculosis, and it became commonplace, before hiring a wet nurse, to have her take a complete physical examination.

Most Chinese women (70 percent in one study) nursed their infants irregularly, when the baby cried (Platt and Gin, 1938: 344; F. L. K. Hsu, 1948: 202; Kao, 1948: 234). Other women—especially in larger communities—followed a schedule, nursing every three or four hours. Nursing was also done at night (85 percent of women in one survey did so); ordinarily there were fewer feedings at night, and these were facilitated by the usual practice of a mother and her infant sleeping in one bed (Platt and Gin, 1938: 344; Kao, 1948: 34). There was no accepted length of feeding; instead a mother ordinarily nursed her infant until it fell asleep (Platt and Gin, 1938: 344; F. L. K. Hsu, 1948: 202). In any case, a child's needs were generally met whether or not a mother was at home, for most Chinese women do not hesitate to expose their breasts in public when nursing their infants (A. W. Chung, 1979: 82-83).

Mixed feeding began quite early in the lives of most Chinese infants, often in the first month and usually in first six months. Such mixed feeding was commonly started because of an insufficient supply of mother's milk, but this was not universally so. One reads that if a family could afford it, an infant was fed "a thin paste of rice flour or of starch cake...slightly sweetened" once or twice a day starting on the second day after birth, at the commencement of breast feeding (Platt and Gin, 1938: 344). A few days later, if the mother's milk was ample, the infant might refuse this supplementary food but in most cases infants continued to take rice flour in small amounts.

As an infant grew older, the amounts and types of additional foods were gradually increased. In the earliest stages, rice products in liquid or paste form were customary, but later soft sugared rice, soup, eggs, biscuits, oranges, and other types of food were added (Platt and Gin, 1938: 344-45). Throughout China one found the custom of premasticating foods before feeding them to an infant, which increased the danger of infection but provided the infant with meat and vegetables it might not otherwise consume (Platt and Gin, 1938: 352; Jenner, 1968: 42). Of special interest was the feeding, primarily at weaning, of jujubes, whether cooked, mashed, or as an extract. Chinese medical belief holds that eating jujubes (a good source of vitamin C) increases the volume of blood and helps circulation (Platt and Gin, 1938: 345).

Weaning

In their survey, Platt and Gin found that breast feeding stopped by the end of the first year in about a third of all cases. Whether or not it stopped, the child was regarded as virtually weaned at that time. However, breast feeding sometimes continued long afterward, into the second, third, or fourth years and, in a few cases, even to the end of the fifth year (Platt and Gin, 1938: 350). Breast feeding was prolonged partly because it was believed that it would prevent a new pregnancy and partly because it was held to benefit the infant (Millis, 1955a: 157), yet Millis (1954: 324-26) found no correlation in Singapore Chinese breast-fed babies between duration of breast feeding and weight gain. The cessation of breast feeding usually occurred because the mother became pregnant again, had too little breast milk, or because the child had grown large.[68]

Mortality among Infants and Mothers

Estimates of infant mortality in China from 1919 to 1934, as collected by Ta Chen (1946), ranged from 110 to 555 per 1000 live births. Orleans (1973: 200-201) estimated for pre-Communist China as a whole a median infant mortality of 275 per 1000 births. This was higher by far than that for any of the forty-two countries (1927-40) included in a table published in 1941 by the School of Public Affairs, Princeton University (Table 16). In Hong Kong in 1909, the number of children who died at less than one year of age was 87 percent of the number of births reported that year. In Chungking, a Western physician who had long practiced there estimated at about the same time that 75 to 85 percent of infants die before they were three years of age (Ross, 1911: 102-3). By far the greatest contributors to this high infant mortality rate were China's poor families (Lang, 1946: 140, 150; C. K. Yang, 1959a: 18), among whom rates of infant mortality could be staggering. Among cases that readily came to informants' minds were those: of one woman who gave birth to eleven children, all of whom died; of another woman who had seven children, all of whom died young; and of a third who had eleven children, of whom two survived (Ross, 1911: 102-3). With establishment of the People's Republic, infant mortality rates dropped sharply. On one commune, for example, infant mortality was only 11.5 per 1000 live births by 1971, and lower still, 5.7 per 1000, in 1972. In Shanghai, a similar drop occurred. We have not found data for more recent years, but a striking drop in infant mortality also occurred in Hong Kong, from 327 per 1000 in 1940 (before the Japanese occupation) to 100 in 1946 (after the occupation had ended) (Fehily, 1947: 222).

Though we have uncovered no precise data, there is one estimate that the maternal death rate in pre-Communist China was at least three times that of Western countries (Lamson, 1935: 271), which might make it 200 per 10,000 live births, or more.

Growth Rates of Chinese Infants and Children

The largest collection of anthropometric data that we have uncovered on the growth rate of Chinese in traditional times are those of C. H. Wu and S. C. Soong (1941), in a compilation of published data and data of their own entailing altogether about 40,000 individuals.[69] Unfortunately those authors failed to break their data down by region, and we turn instead to an earlier study by Stevenson (1925); included in the latter study were 10,863 persons (ages two to seventy years) from all provinces of China. In the north, more subjects were from Chihli than from any other province; in the center, from Kiangsu; and in the south, from Kwangtung. Perhaps his most notable finding was that South Chinese have a shorter period of growth, with its phases, especially pubescent acceleration, coming sooner than among other Chinese.[70] Even though their growth phases are delayed, however, Central and North Chinese develop into taller and heavier adults. Comparing the height of Chinese adults with a worldwide standard published in Germany, Stevenson judged North Chinese males and females above average in height; Central Chinese, average; and South Chinese, below average (Table 17). In South China, however, there were greater individual variations in height than in the other regions, because, says Stevenson, southerners are a more heterogeneous population, with less mixing among ethnic groups.

For the post-World War II period, we have uncovered various studies dealing with growth and the skeletal maturation of children in South China. S. T. Chan *et al.* (1961) studied 287 healthy, normal children (ages three months to twenty years) in Hong Kong, most of whom were born in Kwangtung Province and were from poor families. Heights and weights of boys

Table 16. Infant mortality in various countries (from Ta Chen, 1946)

Country	Year	Infant Mortality(per 1,000 live births)
China	1927	275
Chile	1939	225
Egypt	1939	198
Rumania	1939	176
India, British Provinces	1938	167
Ceylon (Br.)	1939	166
Yugoslavia	1938	144
Poland	1938	140
Portugal	1938	139
Bulgaria	1939	138
Philippine Is. (U.S.)	1937	137
Hungary	1940	134
Jamaica (Br.)	1938	129
Mexico	1940	126
Greece	1937	122
Lithuania	1939	122
Czechoslavakia	1938	121
Spain	1938	119
Salvador	1938	117
Japan	1938	114
Puerto Rico (U.S.)	1939	112
Guatemala	1938	101
Italy	1939	97
Argentina	1939	92
Uruguay	1939	82
Estonia	1938	77
Belgium	1938	72
Latvia	1939	70
Finland	1938	68
France	1938	66
Germany	1940	65
Canada	1939	61
Denmark	1939	58
England and Wales	1940	55
Union of South Africa	1939	50
New Zealand	1938	50
U.S. (continental)	1940	48
Switzerland	1939	43
Norway	1937	42
Sweden	1939	39
Netherlands	1940	39
Australia	1939	38

Table 17. Average height of Chinese adults in the 1920's (from Stevenson, 1925)

	Males	Females
North China	5 ft. 6 ½ in. (above average)	5 ft. 2 in. (above average)
Central China	5 ft. 5 in. (average)	5 ft. 0 in. (average)
South China	5 ft. 4 in. (below average)	4 ft. 11 ½ in. (below average)

and girls were similar to those in previous studies of South Chinese children; but heights, weights, and weight-height indices were lower at every age than those in the Brush Foundation Study in Cleveland. In a second study, of several thousand Hong Kong children (almost all were of Kwangtung origin), K. S. F. Chang *et al.* (1963) found that the principal effect of differences in socio-economic background was on the rate of growth, especially in height, with favored children reaching maturity one to two years sooner. By age eighteen, however, there were no significant differences in height or weight among socio-economic groups. Though S. T. Chan *et al.* (1961) had included some information on skeletal maturation in their article, that matter was taken up more fully in the second Hong Kong study (reported in Low *et al.*, 1964). By the Greulich-Pyle standards, the Chinese children were behind in skeletal development when young (before twelve years for girls, fifteen years for boys) but ahead when older. A significant difference in rate of skeletal maturation was found to exist among socio-economic groups, with more rapid maturation in the high group than in the low one.

For Chinese in Singapore, there are several studies by Millis. In one (Millis, 1953b), she focused on gains in weight and length of Chinese infants from lower income families; weight gain was comparable to those of white American and Australian infants in the first twenty weeks of life, but slowed significantly after six months of age. At one year of age a Singapore Chinese baby weighed four to five pounds less. Another study (Millis, 1954) compared breast-fed and artificially-fed infants of lower income Chinese families in growth and health (at twenty-four weeks of age breast-fed infants were heavier and less likely to be sick). Still another study (1955b) weighed the influence of family economic background on growth in the first year of life (infants from wealthier families increased more rapidly in weight).[71,72]

Birth Rates, Death Rates, and Population Increase

According to Orleans, most experts would probably agree that in pre-Communist China the birth rate was between 40 and 45 per 1000 population. This would have been among the highest birth rates of any country in the world. At the same time, death rates were very high, with the median, from year to year, probably 34 per 1000 (Orleans, 1973: 200-201). The resulting annual rate of natural increase, of roughly six to eleven per thousand, was among the lower ones of the world, comparable to those of the United States and most countries of Western Europe.[73] Given the nation's enormous population base, such an increase was nevertheless sufficient to raise China's population to 600 million by 1950, just after the People's Republic came to power.

As a result of serious efforts at population control in the People's Republic in recent decades, the crude birth rate dropped to about 30-31 per 1000 by 1971. However, because of equally-striking declines in the death rate, to about seven to 7.5 per 1000, the rate of natural increase, about 22 to 24 per 1000, was much greater than in pre-Communist times. Though that rate of increase dropped in subsequent years, to an estimated 11.5 in 1983, even this seems as high or higher than it was prior to 1949 (Banister, 1987: 46, 228). In the last few years, moreover, the rate of population increase has started to rise again, from eleven per thousand in 1985 to 14.4 in 1987 (compared to about nine per thousand in the U.S., and twenty per thousand in India). At such a rate, it is estimated that China's total population, which was roughly a billion people in 1982, will grow to 1.5 billion by 2010, posing a serious threat to the nation's goal of raising standards of living (Ignatius, 1988: 1).[74]

CHINESE CUSTOMS, GOOD HEALTH, AND NUTRITIONAL BALANCE

In seeking to assess the adequacy of Chinese diet and nutrition, one is struck by the tendency for observers to fall into two camps. In one are the "China enthusiasts" who stress the ingenuity of the Chinese people over several millennia in developing a system of agriculture capable of supporting dense populations, in coming to use a rich assemblage of plants and animals, and in learning to prepare them in ways that meet their basic nutritional needs. It is the latter, the adequacy of Chinese diet, which the enthusiasts usually stress. At the opposite pole are the "realists," some of whom, it is true, are Westerners burdened by ethnocentrism. The realists are struck by the poverty of traditional China, its widespread disease, the succession of famines in China's history, and the low life expectancy of her people. Recognizing the enthusiast-realist dichotomy, one is less dismayed by the range of general statements one reads for the 1940's and 1950's: that China enjoys decent nutrition, but with the common East Asian lack of balance (M. C. Balfour *et al.*, 1950: 78); that nutritionally its diet is not very well balanced (M. Yang, 1945: 233); that its diet is unsatisfactory in nutritional terms (T. H. Shen, 1951: 165); and that the bulk of the Chinese are at the mere subsistence level or below it (Rowe and Kendall, 1954, 2: 520).

The China enthusiasts laud the clever and effective ways, given the restraints of a low energy economy, in which the Chinese had shaped their environment so as to support dense populations with less ecological damage than typical of most Western countries. They admire the broad range of foods the Chinese incorporated into their diet; their methods of food processing and preparation that are sometimes unusual and effective in terms of nutrition; their great concern with creating flavorful dishes; and their development of cuisines that rank among the world's best.

All who have interests in food and nutrition must grant the validity of most of the China enthusiasts' arguments. At the same time, it is hard not to acknowledge that traditional China had very serious problems indeed. It was a land that periodically suffered flood, drought, political upheaval, war, and famine, with great loss of life and a burden of nutritional deficiency on many survivors. The succession of epidemics which swept China, the endemic diseases, and the great burden of intestinal parasites all contributed further to malnutrition, illness, and death. Women typically went through many more pregnancies than would otherwise have been necessary; and many more lost their lives in pregnancy and childbirth than was reasonable. All of the above, together with the high prevailing infant mortality, so reduced life expectancy that many persons never contributed significantly to the nation's well being. In one sense, one might view this as a waste of a nation's nutrients in a never-ending struggle to sustain a large population, and it calls to mind Adolph's observation (1929: 40) that the nutritionist's concern should not be merely with helping people survive, but also with optimal metabolism and how this might be achieved.

[1] Animal dung is also used as fertilizer by the Chinese, and, as F. H. King observed (1927: 176-78), people would seek out animal droppings, whether along city streets or country roads, and bring them home for use. On one farm where cows were powering an irrigation pump, a boy was observed with a large wooden dipper catching animal excreta before they fell to the ground and then putting them in a container set aside for that purpose.

[2] Untreated night soil continues to be collected and used in the People's Republic, with collection in villages and communes, as always, in the pre-dawn hours, and transport by men with buckets or by canal boat. In larger cities, however, night soil is now being processed to destroy disease-producing organisms, and the treated product is shipped in tanks by truck or boat to the countryside (Ensminger and Ensminger, 1973: 44).

[3] That such infections were known in antiquity is demonstrated by the fact that *A. lumbricoides* was the first human helminth to be mentioned in the Chinese medical literature, appearing in the *Huang-ti su-wen, The Yellow Emperor's classic of internal medicine* (Hoeppli, 1959: 20-21). Though this classic work has been modified and added to over time by commentators, a large part of the text existed in Han times, and some of it is earlier, possibly much earlier (Veith, 1949: 4-9; Hoeppli, 1959: 20).

[4] Opium smokers had much lower infections of *Ascaris lumbricoides*, which, according to J. P. Maxwell (1921: 380), derives from the fact that they are lazy and that few of them work in the fields. In a similar vein, Taiwan children from urban Taipei in 1956 were found to have incidences of intestinal parasites half those of children from rural areas (Jolliffe and Tung, 1956: 319, 320).

[5] Interestingly, *Ascaris lumbricoides* was used therapeutically in Chinese medicine, as in treating eye diseases (Hoeppli, 1959: 168-71).

[6] Writing of South China, Holmes (1948: 226), while observing that people cooked nearly everything and did not eat [raw] salads, identified the water chestnut as one item that did not have to be cooked.

[7] As it did to the earlier Chinese medical writers whose works were reviewed by Hoeppli and Ch'iang (1938). Some of those writers were far off the mark, but several, in emphasizing the need to avoid raw and cold foods, came quite close to the understanding we have today. Among these is Chang Chieh-pin (early eighteenth century A.D.), who wrote that among the foods that produce worms, those consumed raw and cold are most influential, and that substances that have had contact with water or been stored under damp, dark conditions are likely to be wormy.

[8] Sinoda (1977: 490) writes about ten different ways of preparing *sushi*, principally of carp, in North China in T'ang times, but he described Sung as China's greatest period of *sushi*-making, and notes historical accounts of K'ai-feng and Hangchow (Hang-chou) that mention many kinds of *sushi*: of fish, prawns, clams, other marine products, pork, mutton, duck, goose, as well as other kinds. A special delicacy of literary men, he notes, was a *sushi* made of sparrows.

[9] Of interest here is the suggestion that geographic patterns of hookworm distribution in East Asia and among American Indians may provide evidence of early trans-Pacific contacts (Needham and Lu, 1985: 63-64).

[10] For Chinese anthelmintic prescriptions, see Hoeppli, 1959: 308-20.

[11] Included in the fight against schistosomiasis was a mass effort to eliminate snails, intermediate hosts of the disease organism, by draining canals and other waterways, turning over the soil, and killing every snail uncovered before letting water back in (Banister, 1987: 57).

[12] The first evidence of leprosy in China dates from about the sixth century B.C. (Lu and Needham, 1976: 6), and some believe that it was first introduced to China via early sea routes from India. In any case, the disease seems to have found excellent conditions in the region, for it became widespread in China and endemic and commonplace in Kwangtung (Jefferys and Maxwell, 1910: 46-47, 92-95; J. L. Maxwell, 1929; Lamson, 1935: 301; McKinley, 1935: 178, 310). By one estimate (Lamson, 1935: 301), China as a whole had half the lepers of the entire world, a total of 1.2 million in 1951, and, despite efforts to fight it, the disease is still found in the People's Republic (Ackerknecht, 1965: 117; Banister, 1987: 55).

[13] For a Kwangtung folktale of leprosy being spread by a virgin to the first man having intercourse with her, see Eberhard, 1968: 134. For Chinese views of eliminating one's leprosy by passing it on to someone else through sexual intercourse, see Lamson, 1935: 309.

[14] In pre-Communist times, the most cited nutritional standards for China were those set by the Committee on Nutrition, Chinese Medical Association (1939) which, in turn, were based on those of the Health Organization of the League of Nations (1936). Since the mid-1940's, Western writers on nutrition in China have tended to use (1) the 1939 standards or revisions of them; (2) standards set by the National Research Council of the United States; (3) the revision of NRC standards by INCAP (1954) for peoples of Central America; or (4) those set by WHO/FAO. Because they are relatively recent and have a world orientation, we generally follow the latter recommendations, as published by the WHO in 1974.

[15] For a discussion of intake of calories in the People's Republic as of 1960, see W. K., 1961: 125-26.

[16] For a brief review of soybean research in China and elsewhere, see Adolph, 1946a: 965.

[17] That vitamin A deficiency already occurred in early China is demonstrated by numerous references to night blindness. Many Chinese plant and animal remedies for night blindness were high in vitamin A or provitamin A, with some having ten to twenty times the total vitamin A activity of commercial cod liver oils (Mar and Read, 1936; T. Lee, 1940: 314-16).

[18] One suspects that these data were skewed by conditions of the Sino-Japanese war. In Shanghai refugee camps, for example, vitamin A deficiency was second only to that of vitamin B_1 (thiamine), and xerophthalmia and other deficiency states were commonplace (H. C. Hou, 1940a: 3, 11).

[19] Our search of the literature on traditional China has provided information on thiamine, riboflavin, and niacin, but little or nothing on other B vitamins.

[20] That period was also one when the use of tea was spreading in China, and, because it contains an anti-thiamine factor, it is possible that it, too, contributed to the increase in incidence of beriberi.

[21] See E. F. Yang and B. E. Read, 1940, for the thiamine content of plant beriberi remedies in China.

[22] Note, however, the findings of Guy and Yeh (1938); these researchers described two basic diets of the Peking area in North China: (a) a whole cereal (maize/millet)-soybean diet, almost entirely vegetarian, which was typical of the overwhelming majority of rural people in North China; and (b) a milled cereal (wheat/rice)-meat diet, followed by the well-to-do. Whereas the whole cereal-soybean diet contained abundant thiamine, the milled cereal-meat diet was judged markedly low in this vitamin.

[23] In times of crisis, beriberi could become serious in any part of China, as during the Sino-Japanese conflict among Chinese refugees in Shanghai. In one refugee hospital there in late 1937, beriberi, the most common deficiency disease, accounted for 18 to 48 percent of all monthly admissions (H. C. Hou, 1940a: 3, 11).

[24] Newman *et al.*, writing in 1981 of the People's Republic, indicate that thiamine deficiency is still of consequence in mainland South China. Subclinical thiamine deficiency was reported to be common among Chinese children on Taiwan in the mid-1950's, but beriberi was absent, though it occurred among adults (Jolliffe and Tung, 1956: 310, 317, 325). Such symptoms, on the other hand, were rare and mild among Paiwan and Taiyal aboriginal children studied on Taiwan in 1956-57, for millet and sweet potatoes and not rice were the staples of both tribal groups. Among Paiwan adults, daily per capita thiamine intake amounted to 1.59 mg. (J. S. Chen and C. S. Huang, 1960: 594), well above the WHO recommended intakes of 0.9 and 1.2 mg. for adult men and women.

[25] This was also true in Japan after highly-polished rice came to be popular and widely-used in large cities such as Osaka and Edo in the eighteenth and nineteenth centuries. The rural-urban difference in Japan is well illustrated in a story told by Oiso (1976: 47): a Japanese warrior from a rural area, when on duty at Edo, developed an illness (beriberi) that could not be diagnosed, but on returning to the countryside, he recovered. This led many Japanese to call beriberi "the disease of Edo," which even early in the Meiji era (beginning 1867) was thought to derive from poisoning or contamination.

[26] Of special note in this regard is the report (Fehily, 1941) that 18 percent of infants brought to a welfare center in Hong Kong had infantile beriberi, a condition brought on by a toxic substance in the milk of thiamine-deficient mothers.

[27] For the thiamine content of common Chinese foods, see R. S. Harris *et al.*, 1949; Leung *et al.*, 1973; and Guy and Yeh, 1938 [Peking foods].

[28] We found one survey (Fry, 1965) that provides specific data on amounts of thiamine consumed per day by children in mainland southeast China. The survey involved Chinese children in a Hong Kong orphanage who were somewhat better off with respect to food than most children, yet the daily thiamine intake of a representative fourteen-year-old boy was estimated to be .76 mg., below the 1.2 mg. WHO recommended intake for a boy thirteen to fifteen years of age.

[29] Fry (1965: 102) provides the diet of a typical boy of fourteen in a Hong Kong orphanage. Though this cannot be taken as typical of Hong Kong or China's southeast, it does demonstrate the critical importance of contributions from foods (rice, taro, wax gourd, tofu, and onions) that are low in riboflavin. If we read her data correctly, such low-riboflavin foods contributed just over half (51 percent) of the total daily intake of .41 mg.: eggs, 32 percent; rice, 22 percent; pork, 17 percent; taro and wax gourd, 10 percent each; tofu, 7 percent; and onions, 2 percent.

[30] Processing of soybeans into tofu, however, removes most riboflavin.

[31] Clinical symptoms of niacin deficiency were far more common among the Paiwan and Taiyal aboriginal children studied by J. S. Chen and C. S. Huang (1960: 597, 599) in 1956-57. Per capita daily intake of niacin per Paiwan adult, moreover, was only 2.85 mg., far below the WHO recommended intake of 19.8 mg. for an adult man.

[32] That figure, however, was a bit below the WHO recommended niacin intake of 19.1 mg. for a boy of that age.

[33] The WHO, following the Report of a Joint FAO/WHO Expert Group in 1970, recommended 20 mg. of ascorbic acid per day for infants and children less then thirteen years old, and 30 mg. for adolescents and moderately active adults (Passmore *et al.*, 1974: 35-36, Table 1).

[34] Of special interest is the finding by J. S. Chen and C. S. Huang (1960: 597) that adult Paiwan aborigines on Taiwan had an estimated daily per capita intake of ascorbic acid of 128 mg., four times the 30 mg. WHO recommended intake for adults, and that this came mainly from the sweet potato and its leaves.

[35] For the vitamin C content of Chinese food, see T. J. Chu and B. E. Read, 1938, and Su and Liu, 1938.

[36] White potatoes (raw, edible portion) have 18 mg. of ascorbic acid per 100 gm. as purchased. Soybean sprouts have 10 mg./100 gm.; and mung bean sprouts, also important in South Chinese diet, have 18 mg./100 gm. (Leung *et al.*, 1973).

[37] Taiwan was governed by Japan at the time of the 1929-33 survey, but per capita daily calcium intake seems to have been even lower there than on the mainland, with one estimate of an average of 225 gm. for 1935-39 (excluding the calcium obtained from calcium carbonate added to rice to whiten it) (Jolliffe and Tung, 1956: 310).

[38] Used in the treatment of rickets in China are tiger bones, antelope horn, and rhinoceros horn, though the basis for their reputed effectiveness remains unknown (T. Lee, 1940: 321).

[39] Excess dietary phosphorus or a low calcium/phosphorus ratio may lead to the development of bone diseases such as osteoporosis and periodontal disease (Schaafsma and Visser, 1980). Though we have uncovered no studies for China that attempt to link bone disease to phosphorus intake or calcium/phosphorus ratios, we would note various studies that contain information on the phosphorus content of Chinese food, dietary phosphorus intake, and calcium/phosphorus ratios. See especially Hawks, 1936; Kung and Yeh, 1938; C. C. Y. Chang, 1939; P. S. Tang and L. H. Chang, 1939: 503; R. S. Harris *et al.*, 1949: 33; and Maynard and Swen, 1956: 421-25; note also Adolph, 1929: 39; 1946a: 966; and 1946b: 868.

[40] The WHO daily recommended intakes (Passmore *et al.*, 1974: Table 1) of iron were 5-9 mg. for moderately-active adult men, 5-10 mg. for young children, 5-18 mg. for male adolescents, and 5-10 mg. for female adolescents below age thirteen. For female adolescents, however, recommended intakes rose to 12-28 mg., and for adult women, to 14-28 mg. According to the authors of the WHO study, for pregnant and lactating women whose iron levels have been at these levels throughout life, iron intake should be the same as for other adult women. For women with low iron stores when pregnancy starts, however, the requirement is increased.

[41] On Taiwan, low per capita daily iron intake, possible deficiencies in other nutrients, and high levels of infection with intestinal parasites led to a significant amount of anemia (13 percent) in one group of sixth-grade school children studied (Jolliffe and Tung, 1956: 310, 317-18, 325).

[42] One survey of two aboriginal groups (Taiyal and Paiwan) in Taiwan found that they consumed no sea food, little table salt, and that 35 percent of a group of fourth and fifth grade children had quite clear signs of endemic goiter (J. S. Chen and C. S. Huang, 1960: 600).

[43] With establishment of the People's Republic, major efforts were made to prevent and treat endemic goiter around the country. The main preventive was iodized salt, which was used with quite good results (Kelly and Snedden, 1960: 179-80).

[44] Some common salts produced in interior regions are quite high in iodine; other are not. The same is true of salt produced in coastal regions. In one survey (H. Wang and F. W. Cheng, 1935), involving 132 salt samples produced in various parts of China, eight of the ten districts having salt with highest average iodine content were in inland provinces, Szechwan and Shensi. The mean iodine content of all Szechwan salt samples was 753 micrograms per kilogram. Another study (F. W. Cheng and P. N. Liu, 1936), involving seventeen common salt samples produced in Kwangtung, reported a mean iodine content for all samples of 85 micrograms per kilogram, which, combined with results of the first study, would rank Kwangtung third among China's provinces in the iodine in its salts. The samples of sea salt from Fukien, by contrast, contained only 6 micrograms per kilogram, lowest average for all provinces (H. Wang and F. W. Cheng, 1935). This brings to mind McClendon's observation (1939: 15) that sea salt is not necessarily a sufficient iodine

source. Whatever the source of Chinese salt, moreover, refining reduces the iodine content (Adolph and Whang, 1932: 348, 351; H. Wang and F. W. Cheng, 1935: 353).

[45] Adolph and Whang (1932: 350), writing of the coastal Shanghai-Soochow area, found that farm families had twice the daily iodine intake of city families. They explain this in terms of the greater consumption of green and root vegetables by farm families and the tendency of city families to consume delicacies low in iodine. Even so, city families had sufficient iodine intake.

[46] For a review of relevant evidence in this matter, see Spallholz *et al.*, 1981: *passim*, and Levander, 1982: 75-77.

[47] We have uncovered only brief mention of other elements in Chinese diet and health: a study of the mineral content of Chinese foods (C. Y. Chang, 1939); one of the copper content of Chinese foods (Adolph and Chou, 1933); reference to barium poisoning leading to an endemic disease called "Kiating paralysis" (Chiao, 1944: 20-21); reference to excessive fluorine intake (Chiao, 1944: 20); and analyses of food, water, and salt for fluorine content (F. W. Cheng and C. N. Tsao, 1936; Reid, 1936; L. T. Cheng and T. P. Chao, 1940a; 1940b). Chinese teas in general are high in fluorine content (Reid, 1936; L. T. Cheng and T. P. Chou, 1940b).

[48] It is no surprise, therefore, to read (Cormack, 1935: 28-29) of a woman going to a child-giving goddess to pray for a son, promising the goddess gifts if the prayer is answered.

[49] Cormack (1935: 29) wrote that in some places in China an expectant mother may keep a "long, sharp knife on her bed" so that evil spirits will be intimidated and stay away, and that she puts other protective devices, such as figures of wild animals made of paper, around the room, by the windows, or outside the door.

[50] Details of birth procedures among ethnic Chinese from Vietnam are given in M. Mathews and L. Manderson, 1981: 11.

[51] It was estimated in 1930 that China had more than 100,000 traditional midwives who lacked an understanding of the anatomy and biology of pregnancy, of hygiene, and of how to care for a newborn infant (Lamson, 1935: 286).

[52] We found no mention of macrosomia in the literature of China, and only Millis's studies in Singapore (1953a: 170; 1959) provide an idea of how common it may be. Among about 48,000 infants from low income Chinese families included in one study (1959: 166), 405 had birth weights of 9 lb. or more. This amounted to 0.8 percent of all Chinese newborn, as compared to 0.3 percent of Indian newborn.

[53] In the middle of the last century, Doolittle (1865, 2: 206) wrote that the most common way of killing a female infant was to drown it in a tub of water. Other ways of killing such an infant, he said, were to cast it into a stream or, as reported for a few rural families, to bury it alive out of conviction that their next infant will then be a boy.

[54] There is also a report of another mother bearing one female child after another and of all of them being killed. Then, when she gave birth to still another girl, it was burned to deliver a harsh "lesson to the soul which had chosen to clothe itself so frequently in female bodies" ("The prevalence of female infanticide in China," 1886: 27).

[55] The first Catholic orphanage was opened in Shansi in 1632, and between 1844 and 1872, Catholic missionaries set up 101 orphanages in various areas of China ("The prevalence of infanticide in China," 1886: 27).

[56] A contrary view is that of the scholar Wang Shih-to (1802-1889) who argued that large scale female infanticide was one way of stopping population increase (P. T. Ho, 1959: 61).

[57] In support of this conclusion, Lang cites statistics showing that among children under age five (in a rural district in Hopeh and among poor families in Peking) there were as many females as males, but that in all subsequent age groups, until roughly age fifty, males were more numerous. This, along with the selling of women to be prostitutes or concubines, created a scarcity of women among the poor. Also of relevance is Lang's finding that in Catholic orphanages in China there were eight times as many girls as boys (Lang, 1946: 78n, 150-51).

[58] In some places in the past, there were more formal arrangements for abandoning infants so as to maximize their chances of being adopted by some charitable person. In a city near Shan-t'ou (Swatow) in Kwangtung, a basket was suspended against a wall, with matting to protect it from sun and rain, in which parents might leave such an infant in the hope that someone might adopt it (Ball, 1906: 349-50). In Amoy in Fukien, there was a hole in the city wall where, by custom, such an infant was deposited.

[59] Mosher received national publicity ("Stanford morality," 1983; Wessells, 1983; Sun, 1985) when he was dropped from the Stanford Ph.D. program in anthropology, ostensibly not because his results were embarrassing to the People's Republic but because of his procedures in collecting field data in China.

[60] More detailed information on proscribed and prescribed post-partum foods may be found in Pillsbury, 1978: 13-14; M. Mathews and L. Manderson, 1981; and Wheeler and Tan, 1983: 54-56.

[61] Most of the subjects studied by Platt and Gin were from East-Central provinces of China; the authors suggest (1938: 351), however, that similar customs generally prevail in the rice areas of the south.

[62] There are various studies on the feeding of infants and nursing women. Guy (1936) dealt mainly with diets of nursing mothers and young children in Peking. Platt and Gin (1938) discussed methods of nursing and infant feeding among 387 subjects from various provinces of China, especially Kiangsu (200 subjects), Chekiang (111), Kwangtung (17), and Hupeh (17). Kao (1948) described Chinese breast-feeding practices in general terms. A. W. Chung (1979) wrote especially of changes in breast-feeding practices in the People's Republic. Millis (1954; 1955a; 1956), writing of Singapore Chinese, was concerned with artificial feeding, breast feeding, weaning, and the duration of breast feeding as related to weight gain and illness.

[63] Analysis of the milk of eighty-seven Chinese mothers by Tso in 1924 found no appreciable difference from milk of Western mothers (Kao, 1948: 234).

[64] For other foods provided the nursing mother, see Guy, 1936 (North China); Platt and Gin, 1938: 345 (mostly East-Central China); and A. W. Chung, 1979: 82 (People's Republic in general). For eggs as gifts to the mother of a newborn child, see our chapter on "Eggs."

[65] See the section on use of soybean products in our chapter on "Cereals and Pulses." In an article by the Chinese Medical Association of the People's Republic (1974: 151), it was observed that both cow's milk and soybean milk were being produced as part of an effort to encourage "scientific child rearing among the masses."

[66] See also Mellander (1973: 253) for use of animal milk in feeding infants and children in the People's Republic.

[67] In the People's Republic, hiring of wet nurses is viewed as unacceptable exploitation; however, it is acceptable for a neighbor or relative with abundant breast milk to volunteer to wet nurse an infant besides her own (A. W. Chung, 1979: 81).

[68] Millis (1955a: 157-58) found that half of all cases of early weaning (before the twelfth week) among Chinese mothers in Singapore came about because of failure of lactation; and nearly all other cases were based on the mother's employment needs, indifference, or illness. Early weaning was far more common among middle and upper income Chinese mothers than among those of lower income.

[69] By the mid-1920's, various studies of the growth of children had been carried out in China proper, most notably those of Shirokogoroff and Appleton (1924); Appleton (1925; 1926); Stevenson (1925); and N. Keys and W. W. Cadbury (1926). These studies, which cast light on traditional patterns in China, apparently were not followed up until the 1960's when researchers at the University of Hong Kong published significant new data.

[70] Low *et al.* (1964: 1333) in a study of Hong Kong children reported a similar short period of growth and rapid "rate of maturation in [the] early teens."

[71] Additional articles (Millis, 1956; 1958) compared Singapore Chinese and Indians in the influence of breast feeding on first-year growth of infants, and in growth of preschool children.

[72] For the United States, we note the studies by Appleton (1927; 1928) contrasting growth of Kwangtung Chinese children (ages six to twenty years) in Hawaii with children in China; and one by Lasker (1946) comparing a broad range of physical characteristics of American-born with immigrant Chinese.

[73] Banister (1987: 228-29) cites Chinese studies published in 1980-81 that give China's pre-1949 crude birth rate as only 35 or 38 per 1000. She also cites these or other recent studies that give China's crude death rate for that period as only 25 or 28 per 1000. The result, a rate of natural increase of between 7 and 13 per 1000, is in line with that suggested by Orleans.

[74] For a wealth of information on changes in fertility in China since establishment of the People's Republic, see Banister, 1987: 227-96 and *passim*.

CHAPTER SIXTEEN

Conclusion

"Once in five days I see the flesh of spotted pig,
Once in ten days I find yellow chicken porridge.
Men of this land—meal after meal—eat yam and taro;
They present me with 'odorous rats' and boiled bats."

Su Shih (translated by Edward H. Schafer)

The above statement was made by Su Shih, statesman and poet, during his exile on Hainan in the eleventh century A.D. His reaction to aboriginal food on the island, at best one of boredom, was probably quite similar to the reactions of other Han Chinese when, at still earlier dates, they were exposed to the cooking of Canton, or its predecessor, P'an-yü. Certainly there is nothing in the historical record to suggest that Chinese from the north looked on early Cantonese cooking with much favor. Yet in China today, people from many parts of the country regard Cantonese as one of the nation's best regional cuisines, and even as its premier cuisine. One is dealing here with a striking change, over time, in people's views of Cantonese cookery, a change which almost certainly reflected notable improvements in its quality.

Some observers, as we have seen in our chapter on cuisine, explain Canton's rise to culinary greatness partly in terms of the arrival there of chefs from the north following the fall of the Ming dynasty. Those chefs, according to this view, brought with them a store of knowledge that, encouraged by culinary rivalry among Canton's monied merchant families, came to be applied to the rich variety of foodstuffs available locally. One must concede the plausibility of that view. Canton is located in a subtropical area that enjoys an abundance and great diversity of local foodstuffs coming both from the land and sea, many of them foodstuffs not found in the lower Yangtze area or North China. The arrival of chefs from the north, a region of early culinary renown, would have had a quite salutary effect on Cantonese cookery. And the culinary rivalry of wealthy merchant families would have been an equally positive, and perhaps decisive, factor. Our study of China's foods from the perspective of the southeast, however, suggests that additional factors must have been involved as well. Perhaps most critical was the oft-noted culinary adventuresomeness of the Cantonese, which likely had roots in attitudes of the pre-Han peoples of the area. One also needs to consider Canton's unusual geographic situation in weighing its rise to culinary greatness. That situation was not only critical in providing Canton with wealthy merchant families in the first place, but it enabled the city to obtain plants, animals, and culinary inspiration from a variety of sources. One was North China, an early agricultural hearth whose store of edibles was regularly enriched by plants and animals introduced from Central Asia and lands to the west. A second was Southeast Asia, an even more important agricultural hearth on whose margins Canton was located. Southeast Asia is where the chicken, duck, and water buffalo, all of great importance in China, are believed to have been domesticated. The wild swine of Southeast Asia were likely the ancestors of China's pigs. Southeast Asia was also a major contributor to the world's store of food plants, among them roots and tubers, such as yams and taro; cereals, such as rice; and a broad range of other food plants as well. Of special note among Southeast Asia's culinary contributions to China were fruits and spices and flavorings, their number so great as to make "Fruits" and "Spices and Flavorings" our longest chapters by far.

Canton was admirably situated with respect to the sea routes across Southeast Asia and on to India. Moreover, since the coastal peoples of southeastern China, ancient and modern, had a strong orientation to the sea, they were able to probe far across that region in their search for edible substances and other items of value that were not found, or were not abundant, in their homeland. There were few, if any, parts of the vast region of Southeast Asia that merchants from Canton did not enter. In the quest for delicacies, such as bird's nest, shark's fin, and bêche-de-mer, they developed trade networks that reached still farther, to Australia, India, and beyond. The belt stretching from India across Southeast Asia to China is also the world's richest source of spices, and this must be reckoned with in explaining why that belt gave birth not to one, but to several of the world's great cuisines. They emerged not just in China and India, the high civilizations at either end of the belt, but also in the spice lands along the way, in Indochina, Malaysia, Thailand, and Burma.

We would argue that their geographic position gave the Cantonese an unusual advantage in the quest for culinary excellence, an advantage perhaps even greater than that enjoyed by the people of North China. Sinologist Berthold Laufer has convincingly demonstrated, in his magisterial work *Sino-Iranica* (1919), that North China's excellent geographic position enabled it to benefit from overland exchanges, via the Silk Road, with Iranian peoples and, through them, with India, the Near East, Europe, and Africa. Canton and China's southeast enjoyed a similar advantage, one relying, however, mainly on sea rather than land routes. Yet the story of its foreign trade and exchanges as they have contributed to China's culinary evolution, while not ignored, still remains to be told with the thoroughness of Laufer. Studies do exist, among them those of Wheatley (1959) and Hirth and Rockhill (1966), that cast light on limited aspects of those exchanges, but much more is required. Scholars need to sift and weigh a broad range of material, from archeology, history, linguistics, botany, zoology, and other disciplines, before we have a coherent, reasonably-complete understanding of South China's ties with Southeast Asia and India. We would suggest, in keeping with Laufer's term "Sino-Iranica," that "Sino-Indica" would be an appropriate designation for such an investigation, one which should provide new insights into factors that have shaped China's food scene.

In a sense, our effort has helped in setting the stage for a study of Sino-Indica, for we have collected historical material on a broad range of domesticated food plants and animals that were involved in exchanges with Southeast Asia and India. In such exchanges, a prominent place must be given to the Nan-Yüeh, or South-Yüeh, those early coastal fishing and seafaring peoples of southeastern China and Vietnam. Whatever the linguistic and cultural affiliations of the Nan-Yüeh may have been, those peoples are likely responsible for a significant number of early introductions of food plants, flavorings, and spices from Southeast Asia. Nan-Yüeh seafarers, for example, are given credit for carrying cloves to China from the Moluccas late in the first millennium B.C. They probably first introduced, from what today is Vietnam, sugar cane and methods of making hard sugar and granular sugar, the Chinese olive, and perhaps the fruit banana. They themselves, the Cantonese, or other peoples of southeastern China or

Vietnam, may have been the first to cultivate star anise, cassia cinnamon, certain forms of cardamom, true water chestnut, padi-straw mushroom, and litchi and longan. Though this is a tentative and partial list, it is sufficient to confirm the region's importance, and to demonstrate a need for further investigation of the role of Canton and China's southeast in the nation's culinary evolution.

Needed now are more complete studies, by specialists, of the regions in question. Why such studies of southeastern contributions have not been more conspicuous in China may, in part, derive from "Great Han chauvinism," in which Chinese tend to look on non-Han minorities as inferior and backward and their customs distasteful (Banister, 1987: 314-15). Such attitudes, in turn, have often led to an ignoring or down-playing of minority contributions. Related to the above is the prevailing custom of viewing Chinese history from a North Chinese perspective, one focussing on the Han Chinese conquest and settlement of the south as bringing civilization to that region. We would suggest that a proper study of Sino-Indica requires that the history of South China be viewed from a truly southern perspective, one which pays attention to all factors that have contributed to culinary evolution, one not unduly burdened by concerns with the story of the Han Chinese political penetration of the south.

REFERENCES

A. F. (1869). Quail fighting in Canton, *Notes and Queries on China and Japan.* **3**: 22.
Abe, Shigenobu, and Takashi Kaneda (1972). The effect of edible seaweeds on cholesterol metabolism in rats, pp. 562-65 **In:** *Proceedings of the Seventh International Seaweed Symposium.* Sapporo, Japan, August 8-12, 1971. Tokyo: University of Tokyo Press.
Abrahams, I. (1908-26). Symbolism (Jewish), pp. 143-45 **In:** Vol. 12 of James Hastings (ed.), *Encyclopaedia of religion and ethics.* Edinburgh: T. & T. Clark. 13 vols.
Ackerknecht, Erwin H. (1965). *History and geography of the most important diseases.* New York and London: Hafner Publishing Company.
Adolph, William H. (1923). Diet studies in Shantung, *China Medical Journal.* **37**: 1013-19.
_______ (1925). A study of North China dietaries, *Journal of Home Economics.* **17**: 1-7.
_______ (1929). Aspects of nutrition and metabolism in China, *Scientific Monthly.* **29**: 39-43.
_______ (1931). A 4,000-year food experiment, *Australian Journal of Dentistry.* **35**: 40-47.
_______ (1938). Vegetarian China, *Scientific American.* **159**: 133-35.
_______ (1939). Calcium in the Chinese dietary, *Proceedings of the 6th Pacific Science Congress.* **6**: 329-32.
_______ (1944). The protein problem of China, *Science.* **100**: 1-4.
_______ (1946a). Nutrition research in China, *Journal of the American Dietetic Association.* **22**: 964-70.
_______ (1946b). Prewar nutrition in rural China, *Journal of the American Dietetic Association.* **22**: 867-69.
_______ (1951). Nutrition under the Chinese Communist government, *Scientific Monthly.* **73**: 128-30.
_______ (1956). What early man discovered about food, *Harper's Magazine.* **1**: 67-70.
Adolph, William H., and Hsi-chen Liu (1939). The value of the sweet potato in human nutrition, *Chinese Medical Journal.* **55**: 337-42.
Adolph, William H., and Pao-chun Whang (1932). Iodine in nutrition in coastal mid-China, *Chinese Journal of Physiology.* **6**: 345-52.
Adolph, William H., and T'ung-pi Chou (1933). Copper in Chinese food materials, *Chinese Journal of Physiology.* **7**: 185-88.
Agnew, Mary C. (1926-29). Dietary studies in Szechwan, *Journal of the West China Border Research Society.* **3**: 55-60.
Agrawala, V. S. (1963). *India as known to Panini.* Second Edition. Varanasi: Prithvi Kumar.
Agricultural Research Service of the United States Department of Agriculture (1971). *Common weeds of the United States.* New York: Dover Publications.
Ahern, Emily M. (1973). *The cult of the dead in a Chinese village.* Stanford: Stanford University Press.
_______ (1974). Affines and the rituals of kinship, pp. 279-307 **In:** Arthur P. Wolf (ed.), *Religion and ritual in Chinese society.* Stanford: Stanford University Press.
_______ (1978). Sacred and secular medicine in a Taiwan village: A study of cosmological disorders, pp. 17-39 **In** Arthur Kleinman, Peter Kunstadter, E. Russell Alexander, and James L. Gate (eds.), *Culture and healing in Asian societies.* Boston: G. K. Hall & Co.; Cambridge, Mass.: Shenkman Publishing Company.
_______ (1981). *Chinese ritual and politics.* Cambridge: Cambridge University Press.
Ainslie, Whitelaw (1826). *Materia indica.* London: Longman, Rees, Orme, Brown, and Green. 2 vols.
Ainsworth-Davis, J. R. (1931). *Cooking through the centuries.* London: J. M. Dent and Sons; New York: E. P. Dutton.
Ali, Salim, and S. Dillon Ripley (1983). *Handbook of the birds of India and Pakistan. Vol. 4.* Delhi, London, and New York: Oxford University Press.
Allchin, Bridget, and Raymond Allchin (1982). *The rise of civilization in India and Pakistan.* Cambridge, London, New York: Cambridge University Press.
Alsop, Gulielma F. (1918). *My Chinese days.* Boston: Little, Brown, and Company.
American Plant Studies Delegation (1975). *Plant studies in the People's Republic of China: A trip report of the American Plant Studies Delegation.* Submitted to the Committee on Scholarly Communication with the People's Republic of China. Washington, D.C.: National Academy of Sciences.
Ames, Oakes (1939). *Economic annuals and human cultures.* Cambridge: Botanical Museum of Harvard University.
Anderson, Edgar (1943). A variety of maize from the Rio Loa, *Annals of the Missouri Botanical Garden.* **30**: 469-75.
_______ (1945). What is Zea Mays? - A report of progress, *Chronica Botanica.* **9**: 88-92.
_______ (1952). *Plants, man and life.* Boston: Little, Brown and Company.
Anderson, Eugene N., Jr. (1969). Sacred fish, *Man.* **4**: 443-49.
_______ (1970a). *The floating world of Castle Peak Bay.* Anthropological Studies, No. 4. Washington, D.C.: American Anthropological Association.

_______ (1970b). Traditional aquaculture in Hong Kong, *Journal of Tropical Geography*. **30**: 11-16.
_______ (1972). *Essays on South China's boat people*. Taipei: Orient Cultural Service.
_______ (1977). Edible boundary lines: Foods as ethnic markers among Chinese, Paper for the Annual Meeting of the Association of Asian Studies, 1977. 16 pp.
_______ (1980). "Heating" and "cooling" foods in Hong Kong and Taiwan, *Social Science Information*. **19**: 237-68.
_______ (1984). "Heating and cooling" foods re-examined, *Social Science Information*. **23**: 755-73.
_______ (1987). Why is humoral medicine so popular?, *Social Science and Medicine*. **25**: 331-37.
_______ (1988). *The food of China*. New Haven and London: Yale University Press.
Anderson, Eugene N., Jr., and Marja L. Anderson (1968). Folk medicine in rural Hong Kong, *Ethnoiatria*. **2**: 22-28.
_______ (1973). *Mountains and water: Essays on the cultural ecology of south coastal China*. Asian Folklore and Social Life Monographs, Vol. 54. Taipei: Chinese Association for Folklore.
_______ (1976). Folk dietetics in two Chinese communities, and its implications for the study of Chinese medicine, pp. 143-75 **In**: Arthur Kleinman, Peter Kunstadter, E. Russell Alexander, and James L. Gate (eds.), *Medicine in Chinese cultures*. DHEW Publication No.(NIH) 75-653. Washington, D.C.: U.S. Government Printing Office.
_______ (1977). Modern China: South, pp. 319-82 **In**: K. C. Chang (ed.), *Food in Chinese culture*. New Haven: Yale University Press.
André, Jacques (1981). *L'alimentation et la cuisine à Rome*. New Edition. Paris: Les Belles Lettres.
Andrews, Alfred C. (1961). Acclimatization of citrus fruits in the Mediterranean region, *Agricultural History*. **35**: 35-46.
Anonymous (1905). The Chinese chestnut (*Xanthoceras sorbifolia*), *Flora and Sylva*. **3**: 128-30.
"Apiculture in China" (1929). *Chinese Economic Journal*. **4**: 126-34.
"Apis cerana at risk" (1971). *Bee World*. **52**: 141-42.
Appleton, V. B. (1925). Growth of Chinese, *American Journal of Diseases of Children*. **30**: 43-49.
_______ (1926). Further study of the growth of Chinese, *China Medical Journal*. **40**: 259-64.
_______ (1927). Growth of Chinese children in Hawaii and in China, *American Journal of Physical Anthropology*. **10**: 237-52.
_______ (1928). Growth of Kwangtung Chinese in Hawaii, *American Journal of Physical Anthropology*. **11**: 473-500.
Arasaki, S., and T. Arasaki (1983). *Vegetables from the sea*. Tokyo: Japan Publications.
Arena, Jay M. (1974). Nutritional status of China's children: An overview, *Nutrition Reviews*. **32**: 289-95.
Arlington, L. C., and William Lewisohn (1935). *In search of Old Peking*. Peking: H. Vetch.
Arnon, I. (1972). *Crop production in dry regions. Vol. II: Systematic treatment of the principal crops*. London: Leonard Hill Books.
Arora, R. K. (1977). Job's tears (*Coix lacryma-jobi*)—a minor food and fodder crop of northeastern India, *Economic Botany*. **31**: 358-66.
"Ascariasis, giardiasis and growth" (1983). *Nutrition Reviews*. **41**: 149-51.
Astley, Thomas (ed.) (1745-47). *A new general collection of voyages and travels*. London: Thomas Astley. 4 vols.
Atkinson, Clinton E. (1977). People's Republic of China, pp. 321-44 **In**: E. Evan Brown (ed.), *World fish farming: Cultivation and economics*. Westport, Conn.: AVI Publishing Company.
Aubert, M., J. Aubert, and M. Gauthier (1979). Antibiotic substances from marine flora, pp. 267-91 **In**: H. A. Hoppe, T. Levring, and Y. Tanaka (eds.), *Marine algae in pharmaceutical science*. Berlin and New York: Walter de Gruyter.
Bae, Hyo-won (ed.) (1978). *Korean ginseng*. [Seoul]: Korea Ginseng Research Institute, Republic of Korea.
Bailey, L. H. (1894). *Some recent Chinese vegetables*. Cornell University Agricultural Experiment Station, Horticultural Division, Bulletin No. 67. Ithaca, N.Y: Cornell University.
_______ (1935). *The standard cyclopedia of horticulture*. New York: Macmillan. 3 vols.
_______ (1949). *Manual of cultivated plants most commonly grown in the United States and Canada*. Revised Edition. New York: Macmillan.
Baker, G. P. (1937). Hemerocallis: The day lily, *Journal of the Royal Horticultural Society*. **62**: 399-411.
Baker, Hugh D. R. (1968). *A Chinese lineage village: Sheung Shui*. Stanford: Stanford University Press.
_______ (1979). *Chinese family and kinship*. New York: Columbia University Press.
_______ (1983). The rituals of the table, *Times Literary Supplement (London)*, June 24, p. 6726.
Balasingam, E. (1967). The ecology and conservation of the leathery turtle *Dermochelys coriacea* (Linn.) in Malaya, *Micronesica*. **3**: 37-43.
Baldwin, James A. (1971). *The origin and diffusion of the domestic cat: A study in the geography of domestication*. Unpublished M. A. Thesis in Geography, University of Texas, Austin.
_______ (1975). Notes and speculations on the domestication of the cat in Egypt, *Anthropos*. **70**: 428-48.
Balfour, Edward (1884). *Cyclopaedia of India and of eastern and southern Asia*. Third Edition. London: Bernard Quaritch. 9 vols.

Balfour, Marshall C., Roger F. Evans, Frank W. Notestein, and Irene B. Taeuber (1950). *Public health and demography in the Far East: Report of a survey trip, September 13-December 13, 1948*. New York: Rockefeller Foundation.
Ball, J. Dyer (1906). *Things Chinese*. Fourth Edition. New York: Charles Scribner's Sons.
Banerjee, Anukul Chandra (1977). *Studies in Chinese Buddhism*. Calcutta: Firma KLM (P) Ltd.
Banga, O. (1957). Origin of the European cultivated carrot, *Euphytica*. **6**: 54-63.
_______ (1963). *Main types of the Western carotene carrot and their origin*. Zwolle: Tjeenk Willink.
_______ (1976). Radish: *Raphanus sativus* (Cruciferae), pp. 60-62 **In:** N. W. Simmonds (ed.), *Evolution of crop plants*. London: Longman.
Banister, Judith (1987). *China's changing population*. Stanford: Stanford University Press.
Baranov, A. (1966). Recent advances in our knowledge of the morphology, cultivation, and uses of ginseng (*Panax ginseng* C. A. Meyer), *Economic Botany*. **20**: 403-6.
Bardach, J. E., J. H. Ryther, and W. O. McLarney (1972). *Aquaculture: The farming and husbandry of freshwater and marine organisms*. New York: John Wiley & Sons.
Barlow, Claude Heman (1925). *The life cycle of the human intestinal fluke Fasciolopsis buski (Lankaster)*. American Journal of Hygiene, Monographic Series, No. 4. Baltimore: American Journal of Hygiene.
Barrau, Jacques (1956). *L'agriculture vivrière autochtone de la Nouvelle-Calédonie*. Noumea: Commission du Pacifique Sud.
_______ (1965a). L'humide et le sec, an essay on ethnobotanical adaptation to contrastive environments in the Indo-Pacific area, *Journal of the Polynesian Society*. **74**: 329-46.
_______ (1965b). Witnesses of the past: Notes on some food plants of Oceania, *Ethnology*. **4**: 282-94.
_______ (1970). La région Indo-Pacifique comme centre de mise en culture et de domestication des végétaux, *Journal d'Agriculture Tropicale et de Botanique Appliquée*. **17**: 487-503.
Basker, D., and M. Negbi (1983). Uses of saffron, *Economic Botany*. **37**: 228-36.
Baslow, Morris H. (1969). *Marine pharmacology*. Baltimore: Williams and Wilkins Co.
Batu, A.-T., T.-Than, and T.-Toe (1976). Iron absorption from Southeast Asian rice-based meals, *American Journal of Clinical Nutrition*. **29**: 219-25.
Beadle, G. W. (1972). The mystery of maize, *Field Museum of Natural History Bulletin*. **43**: 2-11.
Beazley, E. (1966). The pigeon towers of Isfahan. *Iran*. **4**: 105-9.
Beccari, O. (1917). The origin and dispersal of *Cocos nucifera*, *The Philippine Journal of Science, C. Botany*. **12**: 27-43.
Beckers, Christian, and Francois Delange (1980). Iodine deficiency, pp. 199-217 **In:** John B. Stanbury and Basil S. Hetzel (eds.), *Endemic goiter and endemic cretinism: Iodine nutrition in health and disease*. New York: John Wiley & Sons.
Bedigian, Dorothea, and Jack R. Harlan (1986). Evidence of cultivation of sesame in the ancient world, *Economic Botany*. **40**: 137-54.
Bender, A. E., D. S. Miller, and E. J. Tunnah (1953). Biological value of algal proteins, *Chemistry and Industry*. 1953: 1340-1.
Benedict, Paul K. (1942). Thai, Kadai, and Indonesian: A new alignment in southeastern Asia, *American Anthropologist (N.S.)*. **44**: 576-601.
Benemerito, A. N. (1935-36). Mushroom culture in Canton, *Philippine Agriculturalist*. **24**: 624-34.
Bennett, Amanda (1984). China winces as cost of state subsidies soars, *Wall Street Journal (Western Edition)*, Tuesday, July 3, p. 24.
Berdichevsky, Norman (1976). A cultural geography of coffee and tea preferences, *Proceedings of the Association of American Geographers*. **8**: 24-29.
Bernatzik, Hugo Adolf (1947). Akha und Meau: *Probleme der angewandten Völkerkunde in Hinterindien*. Innsbruck: Wagner'sche Univ.-Buchdruckerei. 2 vols.
Bertin, J., J. J. Hémardinquer, M. Keul, and W. G. L. Randles (1971). *Atlas of food crops*. Paris and The Hague: Mouton.
Best, A. E. (1926). Chinese folklore. Relating to conception and maternity, *China Medical Journal*. **40**: 564-74.
Bezuneh, Taye, and Asrat Feleke (1966). The production and utilization of the genus *Ensete* in Ethiopia, *Economic Botany*. **20**: 65-70.
Bhat, J. V., and Rajul Broker (1953). Riboflavine and thiamine contents of saffron, *Crocus sativus* Linn., *Nature*. **172**: 544.
Bhishagratna, Kaviraj Kunjalal (1963). *An English translation of the Sushruta Samhita based on original Sanskrit text*. Chowkhamba Sanskrit Series, Vol. 30. Varanasi: Chowkhamba Sanskrit Series Office. 3 vols.
Bishop, Carl Whiting (1922). The geographical factor in the development of Chinese civilization, *Geographical Review*. **12**: 19-41.
_______ (1938). Long-houses and dragon-boats, *Antiquity*. **12**: 411-24.
Bishop, Isabella L. Bird (1900). *The Yangtze Valley and beyond*. New York: G. P. Putnam's Sons. 2 vols.
Black, W. A. P. (1953). Seaweeds and their value in foodstuffs, *Proceedings of the Nutrition Society*. **12**: 32-39.

Blasdale, W. C. (1899). *A description of some Chinese vegetable food materials and their nutritive and economic value.* U.S. Department of Agriculture, Office of Experiment Stations, Bulletin No. 68. Washington, D.C.: U.S. Government Printing Office.

Blofeld, John (1985). *The Chinese art of tea.* London, Boston, and Sydney: George Allen & Unwin.

Bodde, Derk (1942). Early references to tea drinking in China, *Journal of the American Oriental Society.* **62**: 74-76.

_______ (1975). *Festivals in classical China: New Year and other annual observances during the Han Dynasty, 206 B.C.-A.D. 220.* Princeton: Princeton University Press.

Bodenheimer, F. S. (1951). *Insects as human food.* The Hague: W. Junk.

Bois, D. (1927-37). *Les plantes alimentaires chez tous les peuples et à travers les âges.* Paris: Paul Lechavalier. 4 vols.

Bonavia, E. (1888). *The cultivated oranges and lemons etc. of India and Ceylon.* London: W. H. Allen and Co. 2 vols.

Bond, D. A. (1976). Field bean: *Vicia faba* (Leguminosae-Papilionatae), pp. 179-82 **In:** N. W. Simmonds (ed.), *Evolution of crop plants.* London and New York: Longman.

Borgstrom, Georg (1962). Fish in world nutrition, pp. 267-360 **In:** Vol. 2 of Georg Borgstrom (ed.), *Fish as food.* New York and London: Academic Press.

Bosch, F. D. K. (1960). *The golden germ: An introduction to Indian symbolism.* Indo-Iranian Monographs, Vol. 2. 'S-Gravenhage: Mouton and Company.

Bothwell, Thomas H., Robert W. Charlton, James D. Cook, and Clement A. Finch (1979). *Iron metabolism in man.* Oxford, London, Edinburgh, Melbourne: Blackwell Scientific Publications.

Bowers, John Z. (1970). Medicine in mainland China: Red and rural, *Current Scene.* **8**: No. 12 (June 15), 1-11.

Boxer, C. R. (1953). *South China in the sixteenth century.* Hakluyt Society Works, Second Series, No. 106. London: Hakluyt Society.

Bradshaw, Wilda Hockenberry (1934). Experiences of a dietician in South China, *Trained Nurse and Hospital Review.* **92**: 171-75.

Brand, Donald D. (1971). The sweet potato: An exercise in methodology, pp. 343-65 **In:** Carroll L. Riley, J. Charles Kelley, Campbell W. Pennington, and Robert L. Rands (eds.), *Man across the sea.* Austin and London: University of Texas Press.

Bray, Francesca (1984). *Agriculture,* Part 2 of Vol. 6 (Biology and Biological Technology) of Joseph Needham, *Science and civilisation in China.* Cambridge: Cambridge University Press.

Brekhman, I. I., and I. V. Dardymov (1969). Pharmacological investigation of glycosides from ginseng and *Eleutherococcus, Lloydia.* **32**: 46-51.

Brekhman, N. (1965). Ginseng in the USSR, *Indian Journal of Public Health.* **9**: 148-49.

Brentjes, Burchard (1962). Nutz und Hausvögel im Alten Orient, *Wissenschaftliche Zeitschrift der Martin-Luther-Universität Halle-Wittenberg.* **11**: 635-702.

Bressani, Ricardo, Ramiro Paz y Paz, and Nevin S. Scrimshaw (1958). Chemical changes in corn during preparation of tortillas, *Journal of Agricultural and Food Chemistry.* **6**: 770-74.

Bretschneider, E. (1876). Notes of the mediaeval geography and history of central and western Asia, *Journal of the North China Branch, Royal Asiatic Society (N.S.).* **10**: 75-307.

_______ (1882-95). *Botanicon sinicum: Notes on Chinese botany from native and Western sources.* London: Trübner and Co; Shanghai, Hong Kong, Yokohama, and Singapore: Kelly and Walsh. 3 vols.

_______ (1898). *History of European botanical discoveries in China.* London: Sampson Low, Marston and Company. 2 vols.

Bridgman, Robert F. (1974). Traditional Chinese medicine, pp. 1-21 **In:** John Z. Bowers and Elizabeth F. Purcell (eds.), *Medicine and society in China.* New York: Josiah Macy, Jr. Foundation.

Briggs, George M., and Doris Howes Calloway (1979). *Bogert's nutrition and physical fitness.* Tenth Edition. Philadelphia, London, Toronto: W. B. Saunders Company.

Brighton, Christine A. (1977). Cytology of *Crocus sativus* and its allies (*Iridaceae*), *Plant Systematics and Evolution.* **128**: 137-57.

Bristowe, W. S. (1932). Insects and other invertebrates for human consumption in Siam, *Transactions of the Royal Entomological Society of London.* **80**: 387-404.

_______ (1953). Insects as food, *Proceedings of the Nutrition Society.* **12**: 44-48.

Brouk, B. (1975). *Plants consumed by man.* London and New York: Academic Press.

Browning, E. S. (1986). Unhappy returns: After living abroad, Japanese find it hard to adjust back home, *Wall Street Journal (Western Edition),* May 6, pp. 1, 24.

Brunken, Jere, J. M. J. de Wet, and J. R. Harlan (1977). The morphology and domestication of pearl millet, *Economic Botany.* **31**: 163-74.

Buck, John Lossing (1930). *Chinese farm economy: A study of 2866 farms in seventeen localities and seven provinces in China.* Chicago: University of Chicago Press.

_______ (ed.) (1956). *Land utilization in China.* Facsimile reprint of Vol. 1 of the original 1937 edition. New York: Council of Economic and Cultural Affairs.

Buck, John Lossing, Owen L. Dawson, and Yuan-li Wu (1966). *Food and agriculture in Communist China*. Stanford: Hoover Institution on War, Revolution, and Peace.

Bühler, George (trans.) (1882). *The sacred laws of the Âryas, as taught in the schools of Âpastamba, Gautama, Vâsishtha, and Baudhâyana*. Part 2, Vol. 14 of *The sacred books of the East*, edited by F. Max Müller. Oxford: Clarendon Press.

______ (trans.) (1886). *The laws of Manu*. Vol. 25 of *The sacred books of the East*, edited by F. Max Müller. Oxford: Clarendon Press.

Bullen, Frank T. (1904). *Denizens of the deep*. New York: Fleming H. Revell Company.

Bulmer, S. (1966). Pig bone from two archaeological sites in the New Guinea highlands, *Journal of the Polynesian Society*. **75**: 504-5.

Burk, Raymond F. (1976). Selenium in man, pp. 105-33 **In:** Vol. 2 of Ananda S. Prasad and Donald Oberleas (eds.), *Trace elements in human health and disease*. New York: Academic Press. 2 vols.

Burke, J. Henry (1967). The commercial citrus regions of the world, pp. 40-189 **In:** Vol. 1 of Walter Reuther, Herbert John Webber, and Leon Dexter Batchelor (eds.), *The citrus industry*. Revised Edition. Berkeley: Division of Agricultural Sciences, University of California.

Burkhardt, V. R. (1953-58). *Chinese creeds and customs*. Hong Kong: The South China Morning Post, Ltd. 3 vols.

Burkholder, Paul R., and Gurdial M. Sharma (1969). Antimicrobial agents from the sea, *Lloydia*. **32**: 466-83.

Burkill, H. M. (1985). *The useful plants of west tropical Africa. Volume 1*. Supplement to the Second Edition, edited by R. W. J. Keay and F. N. Hepper. Kew: Royal Botanic Gardens.

Burkill, I. H. (1951). The rise and decline of the greater yam in the service of man, *Advancement of Science*. **7**: 443-48.

______ (1966). *A dictionary of the economic products of the Malay Peninsula*. Kuala Lumpur: Ministry of Agriculture and Co-operatives. 2 vols.

Bushnell, G. H. S. (1976). The beginning and growth of agriculture in Mexico, *Philosophical Transactions of the Royal Society of London, Ser. B*. **275**: 117-20, 127-28.

Butler, Samuel (1898). *The Iliad of Homer*. London, New York, and Bombay: Longmans, Green, and Company.

Buxton, Leonard Halford Dudley (1929). *China, the land and the people: A human geography*. Oxford: Clarendon Press.

Bywater, A. C., and R. L. Baldwin (1980). Alternative strategies in food-animal production, pp. 1-30 **In:** R. L. Baldwin (ed.), *Animals, feed, food and people*. American Association for the Advancement of Science, Selected Symposium, 42. Boulder: Westview Press.

Cable, W. J. (1984). The spread of taro (*Colocasia* sp.) in the Pacific, pp. 28-33 **In:** S. Chandra (ed.), *Edible aroids*. Oxford: Clarendon Press.

Cahen, Edward (1943). The ginkgo through the ages, *Journal of the Royal Horticultural Society*. **68**: 99-105.

Caldwell, M. J. (1972). Ascorbic acid content of Malaysian leaf vegetables, *Ecology of Food and Nutrition*. **1**: 313-17.

Caldwell, M. J., and I. C. Enoch (1972). Riboflavin content of Malaysian leaf vegetables, *Ecology of Food and Nutrition*. **1**: 309-12.

Callen, E. O. (1967). The first New World cereal, *American Antiquity*. **32**: 535-38.

Campbell, C. G. (1976). Buckwheat: *Fagopyrum* (Polygonaceae), pp. 235-37 **In:** N. W. Simmonds (ed.), *Evolution of crop plants*. London and New York: Longman.

Canton, N. A., J. L. Rosedale, and J. P. Morris (1940). *Chemical analyses of the foods of Singapore*. Singapore: Government Publications Bureau.

Carlone, Raffaele (1957). Le specie del genere *Malus* (Miller) esistenti in Cina e loro utilizzazione in frutticoltura, *Genetica Agraria*. **7**: 278-304.

Carlson, Alvar W. (1986). Ginseng: America's botanical drug connection to the Orient, *Economic Botany*. **40**: 233-49.

Carp, Frances M., and Eunice Kataoka (1976). Health care problems of the elderly of San Francisco's Chinatown, *Gerontologist*. **16**: 30-38.

Carr, Archie (1952). *Handbook of turtles*. Ithaca, N.Y.: Comstock Publishing Associates.

______ (1956). *The windward road*. New York: A. A. Knopf.

Carr, Archie, and Anne Barkau Meylan (1979-80). Extinction or rescue for the hawksbill?, *Oryx*. **15**: 449-50.

Carrascal de Celis, Nelma (1981). The status of research, exploitation, and conservation of marine turtles in the Philippines, pp. 323-26 **In:** Karen A. Bjorndal (ed.), *Biology and conservation of sea turtles. Proceedings of the World Conference on Sea Turtle Conservation, Washington, D.C., 26-30 November 1979*. Washington, D.C.: Smithsonian Institution Press.

Carter, George F. (1963). Maize to Africa, *Anthropological Journal of Canada*. **1**: No. 2, 3-8.

______ (1964). Archaeological maize in West Africa: A discussion of Stanton and Willett, *Man*. **64**: 85-86.

______ (1971). Pre-Columbian chickens in America, pp. 178-218 **In:** Carroll L. Riley, J. Charles Kelley, Campbell W. Pennington, and Robert L. Rands (eds.), *Man across the sea: Problems of pre-Columbian contacts*. Austin and London: University of Texas Press.

Cartwright, Frederick F. (1972). *Disease and history*. London: Rupert Hart-Davis.

Casson, Lionel (1984). *Ancient trade and society*. Detroit: Wayne State University Press.

Causeret, Jean (1962). Fish as a source of mineral nutrition, pp. 205-34 **In:** Vol. 2 of Georg Borgstrom (ed.), *Fish as food*. New York and London: Academic Press.

Cavaletto, Catherine G. (1980). Lychee, pp. 469-78 **In:** Steven Nagy and Philip E. Shaw (eds.), *Tropical and subtropical fruits*. Westport, Connecticut: AVI Publishing Company.

Chafetz, Morris E. (1964). Consumption of alcohol in the Far and Middle East, *New England Journal of Medicine*. **271**: 297-301.

Chan, S. T., K. S. F. Chang, and F. K. Hsu (1961). Growth and skeletal maturation of Chinese children in Hong Kong, *American Journal of Physical Anthropology*. **19**: 289-300.

Chan, Sue S. (1973). The Chinese diet and pregnancy, *Papers of the Kroeber Anthropological Society*. **47-48**: 148-64.

Chandra, S. (1984). Preface, pp. ix-x **In:** S. Chandra (ed.), *Edible aroids*. Oxford: Clarendon Press.

Chandrasekharan, N. (1968-69). The Malaysian sea turtle eggs as a source of vitamin A, *The Medical Journal of Malaya*. **23**: 214-15.

Chang, Betty (1974). Some dietary beliefs in Chinese folk culture, *Journal of the American Dietetic Association*. **65**: 436-38.

Chang, Chang-ying (1939). Mineral contents of Chinese foods, *Chinese Journal of Physiology*. **14**: 133-46.

Chang, Chang-ying, and Hsien Wu (1940). The occurrence of vitamin D in fresh leafy vegetables, *Chinese Journal of Physiology*. **15**: 253-62.

Chang, Chun-shu, and Joan Smythe (trans.) (1981). *South China in the twelfth century; A translation of Lu Yu's travel diaries, July 3-December 6, 1170*. Chinese University of Hong Kong, Institute of Chinese Studies, Monograph Series, No. 4. Hong Kong: Chinese University Press.

Chang, Flora L., and Gaynell M. Fuchs (n.d.). *Five treasures of Chinese cuisine*. Honolulu: Oriental Publishing Co.

Chang, Isabelle C. (1970). *What's cooking at Changs': The key to cooking Chinese*. New Revised Edition. New York: Liveright.

Chang, K. S. F., Marjorie M. C. Lee, W. D. Low, and E. Kvan (1963). Height and weight of Southern Chinese children, *American Journal of Physical Anthropology*. **21**: 497-509.

Chang, Kwang-chih (1970). The beginnings of agriculture in the Far East, *Antiquity*. **44**: 175-85.

_______ (1973). Radiocarbon dates from China: Some initial interpretations, *Current Anthropology*. **14**: 525-28.

_______ (1976). *Early Chinese civilization: Anthropological perspectives*. Harvard-Yenching Institute, Monograph Series, Vol. 23. Cambridge and London: Harvard University Press.

_______ (1977a). *The archaeology of ancient China*. Third Edition. New Haven and London: Yale University Press.

_______ (ed.) (1977b). *Food in Chinese culture: Anthropological and historical perspectives*. New Haven and London: Yale University Press.

_______ (1980). *Shang civilization*. New Haven and London: Yale University Press.

_______ (1986). *The archaeology of ancient China*. Fourth Edition. New Haven and London: Yale University Press.

Chang, S. T. (1965). Cultivation of the straw mushroom in S. E. China, *World Crops*. **17**: No. 3 (September), 47-49.

Chang, Shu-ting (1977). The origin and early development of straw mushroom cultivation, *Economic Botany*. **31**: 374-76.

Chang, T. T. (1933). Chia Ming's Elements of Dietetics, *Isis*. **20**: 324-34.

Chang, Te-tzu (1976a). Rice: *Oryza sativa* and *Oryza glaberrima* (Gramineae-Oryzeae), pp. 98-104 **In:** N. W. Simmonds (ed.), *Evolution of crop plants*. London and New York: Longman.

_______ (1976b). The origin, evolution, cultivation, dissemination, and diversification of Asian and African rices, *Euphytica*. **25**: 425-41.

_______ (1976c). The rice cultures, *Philosophical Transactions of the Royal Society of London, Ser. B*. **275**: 143-57.

_______ (1983). The origins and early cultures of the cereal grains and food legumes, pp. 65-94 **In:** David N. Keightley (ed.), *The origins of Chinese civilization*. Berkeley: University of California Press.

Chang, Wonona W., Irving B. Chang, Helene W. Kutscher, and Austin H. Kutscher (1970). *An encyclopedia of Chinese food and cooking*. New York: Crown Publishers.

Chang, Yet-oy, and Max S. Dunn (1957). Amino acid composition of bird's nest protein, *Food Research*. **22**: 182-83.

Chang, Yü T'ien, and H. Bruce Collier (1936). The vitamin C content of Szechwan fruits and vegetables, *Chinese Journal of Physiology*. **10**: 435-44.

Chao, Buwei Yang (1972). *How to cook and eat in Chinese*. Third Edition. New York: Vintage Books.

Chao, Yuen Ren (1976). *Aspects of Chinese sociolinguistics*. Stanford: Stanford University Press.

Chapman, V. J. (1970). *Seaweeds and their uses*. Second Edition. London: Methuen.

Chapman, V. J., and D. J. Chapman (1980). *Seaweeds and their uses*. Third Edition. London and New York: Chapman and Hall.

Char, Tin-yuke (1969). Origin of the Hakka, pp. 1-25 **In:** Tin-yuke Char and C. H. Kwock, *The Hakka Chinese: Their origin and folk songs*. San Francisco: Jade Mountain Press.

Chen, Cheng-siang (1970). *The agricultural regions of China*. Edited and with an appendix by M. Y. Nuttonson. Silver Spring, Maryland: American Institute of Crop Ecology.

Chen, Esther (1977). *Traditional Chinese southern and eastern cuisine*. Seattle: By the author.

Chen, H. T. (1933). A preliminary report on a survey of animal parasites of Canton, China, rats, *Lingnan Science Journal*. **12**: 65-74.

_______ (1934a). Helminths of cats in Fukien and Kwangtung provinces with a list of those recorded from China, *Lingnan Science Journal*. **13**: 261-73.

_______ (1934b). Helminths of dogs in Canton, with a list of those occurring in China, *Lingnan Science Journal*. **13**: 75-87.

Chen, Hsi-huang (1977). Taiwan, pp. 345-58 **In:** E. Evan Brown (ed.), *World fish farming: Cultivation and economics*. Westport, Connecticut: AVI Publishing Company.

Chen, James Y. P. (1962). Highlights in pharmacology in mainland China, *Annual Review of Pharmacology*. **2**: 11-16.

_______ (1973). Chinese health foods and herb tonics, *American Journal of Chinese Medicine*. **1**: 225-47.

Chen, Ji (1942). Nutritional edema in children, *American Journal of Diseases of Children*. **63**: 552-80.

Chen, Joseph S., and C. S. Huang (1960). Dietary and clinico-nutritional survey of aborigines in Taiwan, *Metabolism*. **9**: 594-603.

Chen, Joyce (1962). *Joyce Chen cook book*. Cambridge, Mass.: Joyce Chen Associates.

Chen, K. C., T. F. Yü, S. H. Liu, and H. I. Chu (1940). Studies on the vitamin C requirement of Chinese patients with scurvy, *Chinese Journal of Physiology*. **15**: 119-42.

Chen, Nai-ruenn (1967). *Chinese economic statistics: A handbook for mainland China*. Chicago: Aldine Publishing Company.

Chen, T. P. (1934). A preliminary study of association of species in Kwangtung fish ponds, *Lingnan Science Journal*. **13**: 275-83.

_______ (1935). The oyster industry of Chung-shan, *Lingnan Science Journal*. **14**: 63-72.

_______ (1976). *Aquaculture practices in Taiwan*. Farnham, Surrey: Fishing News Books.

Chen, Ta (1940). *Emigrant communities in South China*. New York: Institute of Pacific Relations.

_______ (1946). *Population in modern China*. Chicago: University of Chicago Press.

Chen, Tung-pai (1952). *Milkfish culture in Taiwan*. Chinese-American Joint Commission on Rural Reconstruction, Fisheries Series, No. 1. Taipei: Joint Commission on Rural Reconstruction.

Chen, Tung-tou, Hsien Wu, and Chang-ying Chang (1941). Further experiments with soybean diets, *Chinese Journal of Physiology*. **16**: 91-96.

Chen, Wenhua (1984). Agricultural science and technology in ancient China, pp. 49-55 **In:** *Recent discoveries in Chinese archaeology*. Beijing: Foreign Languages Press.

Chen, Xiaoshu, Guangqi Yang, Junshi Chen, Xuecun Chen, Zhimei Wen, and Keyou Ge (1980). Studies on the relations of selenium and keshan disease, *Biological Trace Element Research*. **2**: 91-107.

Chen, Xiaoshu, Xuecun Chen, Guangqui Yang, Zhimei Wen, Junshi Chen, and Keyou Ge (1981). Relation of selenium deficiency to the occurrence of Keshan disease, pp. 171-75 **In:** Julian E. Spallholz, John L. Martin, and Howard E. Ganther (eds.), *Selenium in biology and medicine*. Westport, Connecticut: AVI Publishing Company.

Cheng, F. T. (1954). *Musings of a Chinese gourmet*. London: Hutchinson.

Cheng, F. W., and C. N. Tsao (1936). Fluorine content of Chinese common salt, *Journal of the Chinese Chemical Society*. **4**: 388-92.

Cheng, F. W., and P. N. Liu (1936). Additional note on the survey of iodine content in Chinese common salt, *Journal of the Chinese Chemical Society*. **4**: 49-51.

Cheng, Libin T., and T. P. Chou (1940a). The fluorine content of Chengtu well water, *Journal of the Chinese Chemical Society*. **7**: 36-39.

_______ (1940b). The fluorine content of foodstuffs in Szechuan, *Chinese Journal of Physiology*. **15**: 263-68.

Cheng, T. H. (1969). Production of kelp - A major aspect of China's exploitation of the sea, *Economic Botany*. **23**: 215-36.

Cheng, Te-k'un (1959-60). *Archaeology in China*. Cambridge: W. Heffer and Sons.

_______ (1982). *Studies in Chinese archaeology*. Institute of Chinese Studies, Centre for Chinese Archaeology and Art, Studies Series, No. 3. Hong Kong: Chinese University Press.

Cheng, Tien-hsi (1971). Schistosomiasis in mainland China: A review of research and control programs since 1949, *American Journal of Tropical Medicine and Hygiene*. **20**: 26-53.

Chiao, Tsai (1944). Problems of nutrition in present-day China, pp. 16-25 **In:** Harley Farnsworth MacNair (ed.), *Voices from unoccupied China*. Chicago: University of Chicago Press.

Child, Reginald (1974). *Coconuts*. Second Edition. London: Longman.

China: Development of olive production (1980). Report of a Study Tour in the People's Republic of China, 25 September - 19 October 1979. FAO Plant Production and Protection Paper, No. 23. Rome: Food and Agriculture Organization of the United Nations.

Chinese Medical Association (1974). Child health care in New China, *American Journal of Chinese Medicine*. **2**: 149-58.

"Chinese reds reported slaughtering dogs" (1963). *Wisconsin State Journal*, July 7, Section 4, p. 3.

Chiu, B. T. (1954). Notes on the marine algae of Hong Kong. I. The Chlorophyceae, *Hong Kong University Fisheries Journal*. **1**: 39-48.

_______ (1956). Seaweeds of economic importance in South China, pp. 171-72 **In:** T. Braarud and N. A. Sørensen (eds.), *Second International Seaweed Symposium*. London and New York: Pergamon Press.

_______ (1958). Seaweeds of economic importance in Hong Kong, *Hong Kong University Fisheries Journal*. **2**: 132-33.

Chiu, Milton M. (1984). *The tao of Chinese religion*. Lanham, New York, London: University Press of America.

Chiu, Y. T. (1931). Analyses of Chinese foods. I. Chinese wines, *Lingnan Science Journal*. **10**: 391-98.

_______ (1932). Analyses of Chinese foods, III. Chinese teas, *Lingnan Science Journal*. **11**: 409-17.

_______ (1938). Analyses of Chinese foods, *Chinese Medical Journal*. **54**: 435-41.

Chopra, R. N., R. L. Badhwar, and S. Ghosh (1965). *Poisonous plants of India*. Revised Edition. New Delhi: Indian Council of Agricultural Research. 2 vols.

Chou, Te-ch'in, and Abby L. Marlatt (1953). Some factors in the Chinese diet affecting carotene utilization, *Journal of Nutrition*. **51**: 305-15.

Choudhury, B. (1976). *Vegetables*. Fourth Revised Edition. New Delhi: National Book Trust, India.

Chow, T. (1958). The bionomics of pondfish culture in the New Territories, Hong Kong, *Hong Kong University Fisheries Journal*. **2**: 130-32.

Christie, Anthony (1957). An obscure passage from the Periplus, *Bulletin of the School of Oriental and African Studies, University of London*. **19**: 345-53.

Chu, G. S. T., J. R. Palmieri, and J. T. Sullivan (1977). Beetle-eating: A Malaysia folk medical practice and its public health implications, *Tropical and Geographical Medicine*. **29**: 422-27.

Chu, Grace Zia (1975). *Madame Chu's Chinese cooking school*. New York: Simon and Schuster.

Chu, T. J., and B. E. Read (1938). The vitamin C content of Chinese foods. Part II, *Chinese Journal of Physiology*. **13**: 247-56.

Chun, Woon Young (1921). *Chinese economic trees*. [Shanghai]: Printed by the Commercial Press, Ltd.

Chung, Arthur W. (1979). Breastfeeding in a developing country: The People's Republic of China, pp. 81-86 **In:** Dana Raphael (ed.), *Breastfeeding and food policy in a hungry world*. New York: Academic Press.

Chung, H. L., and J. C. Ripperton (1929). *Utilization and composition of oriental vegetables in Hawaii*. Hawaii Agricultural Experiment Station, Bulletin 60. Washington, D.C.: U.S. Government Printing Office.

Chung, Shih-fan (1965). Studies on favism in Kwangtung Province, *Paediatrica Indonesiana*. **5**, Supplement: 880-85.

Claiborne, Craig, and Virginia Lee (1972). *The Chinese cookbook*. Philadelphia: Lippincott.

Clausen, Lucy W. (1954). *Insect fact and folklore*. New York: Macmillan.

Clayton, G. A. (1984a). Common duck, pp. 334-39 **In:** Ian L. Mason (ed.), *Evolution of domesticated animals*. London and New York: Longman.

_______ (1984b). Muscovy duck, pp. 340-44 **In:** Ian L. Mason (ed.), *Evolution of domesticated animals*. London and New York: Longman.

Clinton, Frances (1933). Notes on Chinese diets, *Journal of Home Economics*. **25**: 871-72.

Clutton-Brock, Juliet (1981). *Domesticated animals from early times*. Austin: University of Texas Press; London: British Museum (Natural History).

Cole, J. (1979). *Amaranth from the past - for the future*. Emmaus, Pa.: Rodale Press.

Collier, V. W. F. (1921). *Dogs of China and Japan in nature and art*. New York: Frederick A. Stokes Company.

Collingwood, C. (1868). *Rambles of a naturalist*. London: John Murray.

Collins, G. N. (1909). A new type of Indian corn from China, *U.S. Department of Agriculture, Bureau of Plant Industry, Bulletins*. **161**: 1-30.

Collins, Varnum D. (1865). Sorgo, or northern Chinese sugar cane, *Journal of the North China Branch, Royal Asiatic Society (N.S.)*. **2**: 85-98.

Collyer, C. T. (1903). The culture and preparation of ginseng in Korea, *Transactions of the Korea Branch, Royal Asiatic Society*. **3**: 18-30.

Committee on Nutrition, Council of Public Health, Chinese Medical Association (1939). Minimum nutritional requirement for China, *Chinese Medical Journal*. **55**: 301-23.

Condit, Ira J. (1947). *The fig*. Waltham, Mass.: Chronica Botanica Co.

Condit, Ira J., A. N. Benemerito, and Wang Hao Chen (1937-38). Citrus fruits and their culture in South China, *California Citrograph*. **23**: No. 3, 38-39, 69, 104.

Conklin, Harold C. (1963). The Oceanian-African hypotheses and the sweet potato, pp. 129-36 **In:** Jacques Barrau (ed.), *Plants and the migrations of Pacific peoples: A symposium*. Honolulu: Bishop Museum Press.

Cook, O. F. (1910). History of the coconut palm in America, *Contributions from the United States National Herbarium*. **14**: 271-342.

Cooley, J. S. (1948). Sweetpotatoes—World production and food value, *Economic Botany*. **2**: 83-87.

Cooper, William C., and Nathan Sivin (1973). Man as a medicine: Pharmacological and ritual aspects of traditional therapy using drugs derived from the human body, pp. 203-72 **In:** Shigeru Nakayama and Nathan Sivin (eds.), *Chinese science: Explorations of an ancient tradition*. Cambridge, Mass., and London, England: MIT Press.

Cordier, Henri (1920). *Ser Marco Polo: Notes and addenda to Sir Henry Yule's edition, containing the results of recent research and discovery*. London: John Murray.

Cormack, Mrs J. G. [Annie] (1935). *Everyday customs in China*. Edinburgh and London: Moray Press.

Cost, Bruce (1987). Chinatown chickens heads and feet above the rest, *San Francisco Chronicle*, July 1, Section FF, p. 4.

_______ (1988). China's mysterious vegetable, *San Francisco Chronicle*, August 22, Section ZZ-7, p. 3.

Costantini, Lorenzo (1979a). Notes on the palaeoethnobotany of protohistorical Swat, pp. 703-8 **In:** Vol. 2 of Maurizio Taddei (ed.), *South Asian Archaeology 1977*. Papers from the Fourth International Conference of the Association of South Asian Archaeologists in Western Europe, held in the Istituto Universitario Orientale, Naples. Naples: Seminario di Studi Asiatici, Istituto Universitario Orientale.

_______ (1979b). Wood remains from Shahr-i Sokhta: A source of information for the ancient environment and technology in protohistoric Sistan, pp. 87-121 **In:** Vol. 1 of Maurizio Taddei (ed.), *South Asian Archaeology 1977*. Papers from the Fourth International Conference of the Association of South Asian Archaeologists in Western Europe, held in the Istituto Universitario Orientale, Naples. Naples: Seminario di Studi Asiatici, Istituto Universitario Orientale.

_______ (1981). Palaeoethnobotany at Pirak: A contribution to the 2nd millennium B.C. agriculture of the Sibi-Kacchi Plain, Pakistan, pp. 271-77 **In:** Herbert Härtel (ed.), *South Asian Archaeology 1979*. Papers from the Fifth International Conference of the Association of South Asian Archaeologists in Western Europe, held in the Museum für Indische Kunst der Staatlichen Museen Preussischer Kulturbesitz Berlin. Berlin: Dietrich Reimer Verlag.

Cotterell, Yong Yap, and Arthur Cotterell (1975). *The early civilization of China*. New York: G. P. Putnam's Sons.

Couling, Samuel (1917). *The encyclopaedia sinica*. Shanghai: Kelly and Walsh.

Coursey, D. G. (1967). *Yams*. London: Longmans.

_______ (1976). Yams, pp. 70-74 **In:** N. W. Simmonds (ed.), *Evolution of crop plants*. London and New York: Longman.

_______ (1983). Yams: *Dioscorea* spp. (Dioscoreaceae), pp. 555-601 **In:** Harvey T. Chan, Jr. (ed.), *Handbook of tropical foods*. New York and Basel: Marcel Dekker.

Cowan, F. (1865). *Curious facts in the history of insects*. Philadelphia: J. B. Lippincott.

Cowell, E. B. (ed.) (1969). *The Jataka, or stories of the Buddha's former births*. First published by Cambridge University Press, 1895-1907. London: Luzac & Company, Ltd. 6 vols.

Cranbrook, Earl of (1984) (see also Medway, Lord). Report on the birds' nest industry in the Baram District and at Niah, Sarawak, *Sarawak Museum Journal (N.S.)*. **33**: No. 54, 145-70.

Crane, Eva (1960). Beekeeping in the People's Republic of China, *Bee World*. **41**: 4-8.

_______ (1975). *Honey: A comprehensive survey*. London: Heinemann.

_______ (1983). *The archaeology of beekeeping*. London: Gerald Duckworth & Co.

_______ (1984). Honeybees, pp. 403-15 **In:** Ian L. Mason (ed.), *Evolution of domesticated animals*. London and New York: Longman.

Crawford, R. D. (1984a). Domestic fowl, pp. 298-311 **In:** Ian L. Mason (ed.), *Evolution of domesticated animals*. London and New York: Longman.

_______ (1984b). Goose, pp. 345-49 **In:** Ian L. Mason (ed.), *Evolution of domesticated animals*. London and New York: Longman.

Crawfurd, John (1820). *History of the Indian Archipelago*. Edinburgh: Archibald Constable and Company. 3 vols.

_______ (1829). *Journal of an embassy from the Governor-General of India to the Court of Ava, in the year 1827*. London: Henry Colburn.

_______ (1830). *Journal of an embassy from the Governor-General of India to the courts of Siam and Cochin China*. London: Henry Colburn and Richard Bentley. 2 vols.

_______ (1868). On the history and migration of cultivated plants used as condiments, *Transactions of the Ethnological Society of London*. **6**: 188-206.

Creel, Herrlee Glessner (1937). *Studies in early Chinese culture*. Baltimore: Waverly Press.

Cressey, George Babcock (1934). *China's geographic foundations: A survey of the land and its people*. New York: McGraw-Hill.

_______ (1955). *Land of the 500 million: A geography of China*. New York: McGraw-Hill.

Crevost, Ch., and Ch. Lemarié (1917). *Catalogue des produits de l'Indochine. Produits alimentaires, Tome 1*. Hanoi: Gouvernement Général de l'Indochine.

Crooke, William (1885). Egg-Plant—Potatoes—Onions—Unlucky, *Panjab Notes and Queries*. **3**: No. 27 (December), 41.

_______ (1896). *The popular religion and folk-lore of northern India.* London: Archibald Constable. 2 vols.

_______ (1926). *Religion and folklore of northern India.* Oxford: Oxford University Press.

Crozier, Ralph C. (1968). *Traditional medicine in modern China.* Cambridge: Harvard University Press.

Cruickshank, Ethel M. (1962). Fat soluble vitamins, pp. 175-203 **In:** Vol. 2 of Georg Borgstrom (ed.), *Fish as food.* New York and London: Academic Press.

Cutler, Hugh (1964). Chinese gooseberry, New Zealand gooseberry, or kiwi fruit, *Missouri Botanical Garden Bulletin.* **52**: No. 9, 7-9.

Dabry de Thiersant, P. (1872). *La pisciculture et la pèche en Chine.* Preceded by an introduction on fish culture by J. L. Soubeiran. Paris: Librairie de G. Masson.

Dallimore, W., and A. Bruce Jackson (1966). *A handbook of Coniferae and Ginkgoaceae.* Fourth Edition, edited by S. G. Harrison. London: Edward Arnold (Publishers) Ltd.

Dalziel, J. M. (1937). *The useful plants of west tropical Africa.* London: Crown Agents for Oversea Governments and Administrations.

D'Andrea, Jeanne (1982). *Ancient herbs in the J. Paul Getty Museum gardens.* Malibu, Cal.: The Museum.

Danford, Darla E. (1986). The nutritional and other health perils and benefits of geophagia, *1986 AAAS Annual Meeting Program, 25-30 May, Philadelphia,* p. 82.

Darby, William J., Paul Ghalioungui, and Louis Grivetti (1977). *Food: The gift of Osiris.* London, New York, San Francisco: Academic Press. 2 vols.

Darwin, C. (1900). *The variation of animals and plants under domestication.* New York: D. Appleton. 2 vols.

Darwin, E. (1800). *Phytologia, or the philosophy of agriculture and gardening.* London: J. Johnson.

Davids, T. W. Rhys (1908-26). Wisdom tree, pp. 747-49 **In:** Vol. 12 of James Hastings (ed.), *Encyclopaedia of religion and ethics.* Edinburgh: T. & T. Clark. 13 vols.

Davids, Thomas W. Rhys, and Hermann Oldenberg (trans.) (1885). *Vinaya texts.* Part 3, Vol. 20 of *The sacred books of the East,* edited by F. Max Müller. Oxford: Clarendon Press.

Davidson, Stanley, and R. Passmore (1969). *Human nutrition and dietetics.* Edinburgh and London: E. & S. Livingstone.

Davies, D. Roy (1976). Peas: *Pisum sativum* (Leguminosae-Papilionatae), pp. 172-74 **In:** N. W. Simmonds (ed.), *Evolution of crop plants.* London and New York: Longman.

Davies, H. R. (1909). *Yün-nan: The link between India and the Yangtze.* Cambridge: University Press.

Davis, Lucille (1966). *Court dishes of China: The cuisine of the Ch'ing dynasty.* Rutland, Vt.; Tokyo, Japan: Charles E. Tuttle Company.

Davy, J. Burtt (1898). Lily bulbs and flowers as food, *Erythea.* **6**: 26.

Daweewarn, Dawee (1982). *Brahmanism in South-East Asia.* New Delhi: Sterling Publishers Private Limited.

de Beauclair, Inez (1961). "Miao" on Hainan Island, *Current Anthropology.* 2: 394.

_______ (1970). *Tribal cultures of southwest China.* Asian Folklore and Social Life Monographs, Vol. 2. Taipei: Orient Cultural Service.

de Candolle, Alphonse (1886). *Origin of cultivated plants.* Second Edition. New York: D. Appleton and Company.

de Groot, J. J. M. (1896). *Le code du Mahâyâna en Chine.* Verhandelingen der Koninklijke Akademie van Wetenschappen te Amsterdam, Afdeeling Letterkunde, New Series, Vol. 1, No. 2. Amsterdam: Johannes Müller.

_______ (1964). *The religious system of China.* First published at Leyden by E. J. Brill, 1892-1910. Taipei: Literature House, Ltd. 6 vols.

de Groot, Roy Andries (1983). On the trail of bird's nest soup: Caves, climbs and high stakes, *Smithsonian.* **14**: No. 6 (September), 66-75.

de Langhe, E. (1969). Bananas (*Musa* spp.), pp. 53-78 **In:** F. P. Ferwerda, and F. Wit (eds.), *Outlines of perennial crop breeding in the tropics.* Miscellaneous Papers, No. 4, Agricultural University Wageningen. Wageningen: H. Veenman and Zonen.

de Schauensee, Rudolphe Meyer (1984). *The birds of China.* Washington, D.C.: Smithsonian Institution Press.

de Silva, G. S. (1981). The status of sea turtle populations in East Malaysia and the South China Sea, pp. 327-37 **In:** Karen A. Bjorndal (ed.), *Biology and conservation of sea turtles. Proceedings of the World Conference on Sea Turtle Conservation, Washington, D.C., 26-30 November 1979.* Washington, D.C.: Smithsonian Institution Press.

de Waard, P. W. F., and A. C. Zeven (1969). Pepper (*Piper nigrum L.*), pp. 409-26 **In:** F. P. Ferwerda, and F. Wit (eds.), *Outlines of perennial crop breeding in the tropics.* Agricultural University Wageningen, Miscellaneous Papers, No. 4. Wageningen: H. Veenman and Zonen.

de Wet, J. M. J. (1977). Domestication of African cereals, *African Economic History.* **3**: 15-32.

de Wet, J. M. J., and J. R. Harlan (1971). The origin and domestication of Sorghum bicolor, *Economic Botany.* **25**: 128-35.

de Wet, J. M. J., K. E. Prasada Rao, M. H. Mengesha, and D. E. Brink (1983). Domestication of *sawa* millet (*Echinochloa colona*), *Economic Botany.* **37**: 283-91.

Dean, William F. (1980). Poultry production, pp. 52-64 **In:** J. A. Hoefer and P. J. Tsuchitani (eds.), *Animal agriculture in China.* Committee on Scholarly Communication with the People's Republic of China, Report No. 11. Washington, D.C.: National Academy Press.

Dennis, John V., and Charles R. Gunn (1971). Case against trans-Pacific dispersal of the coconut by ocean currents, *Economic Botany.* **25**: 407-13.

Der Ling, Princess (1914). *Two years in the Forbidden City.* New York: Moffat, Yard, and Company.

Der Marderosian, Ara, John Beutler, Walter Pfendner, Jeffrey Chambers, Robert Yoder, Eileen Weinsteiger, and Joseph Senft (1980). Nitrate and oxalate content of vegetable amaranth, pp. 31-41 **In:** *Proceedings of the Second Amaranth Conference.* Emmaus, Pa.: Rodale Press.

Detwiler, S. B. (1938). The ginkgo, *Soil Conservation.* **4**: 122-24.

Devadas, Rajammal P., and S. Saroja (1980). Availability of iron and B-carotene from amaranth to children, pp. 15-21 **In:** *Proceedings of the Second Amaranth Conference.* Emmaus, Pa.: Rodale Press.

Dey, Kanny Lall (1973). *The indigenous drugs of India.* Second Edition. New Delhi: Pama Primlane, The Chronica Botanica.

Diamond, Norma (1969). K'un Shen; A Taiwan village. *Case Studies in Cultural Anthropology,* edited by George and Louise Spindler. New York: Holt, Rinehart and Winston.

Diao, Richard K. (1967). The national minorities of China and their relations with the Chinese communist regime, pp. 169-201 **In:** Vol. 1 of Peter Kunstadter (ed.), *Southeast Asian tribes, minorities, and nations.* Princeton: Princeton University Press. 2 vols.

"Disease risk from more dogs" (1987). *China Daily (Beijing),* **Vol. 6**, No. 1793, Saturday, April 25, p. 3.

Dixon, Pamela (1976). *Ginseng.* London: Duckworth.

Doggett, H. (1976). Sorghum: *Sorghum bicolor* (Gramineae - Andropogoneae), pp. 112-17 **In:** N. W. Simmonds (ed.), *Evolution of crop plants.* London and New York: Longman.

Dolman, C. E. (1957). The epidemiology of meat-borne diseases, pp. 11-108 **In:** *Meat Hygiene.* WHO Monograph Series, No. 33. Geneva: World Health Organization.

Dols, Michael W. (1977). *The Black Death in the Middle East.* Princeton: Princeton University Press.

Donkin, R. A. (1980). *Manna: An historical geography.* Biogeographica, Vol. 17. The Hague: W. Junk.

Donovan, E. (1842). *Natural history of the insects of China.* A new edition by J. O. Westwood. London: Henry G. Bohn.

Doolittle, Justus (1865). *Social life of the Chinese.* New York: Harper & Brothers. 2 vols.

Dorsett, P. H., and J. H. Dorsett (1928). *Culture and outdoor winter storage of persimmons in the vicinity of Peking, China.* Circular No. 49. Washington, D.C.: U.S. Department of Agriculture.

Drews, Robin A. (1951). *The cultivation of food fish in China and Japan.* Unpublished Ph.D. Dissertation, University of Michigan, Ann Arbor.

_______ (1961). Raising fish for food in Southeast Asia, pp. 121-43 **In:** Vol. 1 of Georg Borgstrom (ed.), *Fish as food.* New York and London: Academic Press.

Du, Swun-deh (1952). Favism in West China, *Chinese Medical Journal.* **70**: 17-26.

Dudgeon, John (1895). *The beverages of the Chinese.* Tientsin: Tientsin Press.

Dupuis, Jacques (1975). Un probème de géographie medievale: Diffusion du mais et traversée de l'Atlantique à l'epoque précolumbienne, *Les Cahiers d'Outre-Mer.* **28**: 5-13.

Dutt, Uday Chand (1922). *The materia medica of the Hindus.* With a glossary of Indian plants by George King. Calcutta: Adi-Ayurveda Machine Press.

Duval, Mathias and others (1885). Sur les oeufs pourris comme aliment en Chine, *Bulletins de la Société d'-Anthropologie de Paris.* **8 (Sér. 3)**: 299-303.

Dymock, W. (1889a). On the narcotics and spices of the East, *Journal of the Anthropological Society of Bombay.* **2**: 36-54.

_______ (1889b). The flowers of Hindu poets, *Journal of the Anthropological Society of Bombay.* **2**: 80-91.

_______ (1891). On the use of turmeric in Hindoo ceremonial, *Journal of the Anthropological Society of Bombay.* **2**: 441-48.

Dymock, William C., J. H. Warden, and David Hooper (1889-93). *Pharmacographia indica: A history of the principal drugs of vegetable origin met with in British India.* London: Trübner & Co.; Byculla, Bombay: Education Society's Press. 6 vols.

Eberhard, Wolfram (1935). Zur Volkskunde von Chêkiang, *Zeitschrift für Ethnologie.* **67**: 248-65.

_______ (1952). *Chinese festivals.* New York: Henry Schuman.

_______ (1968). *The local cultures of South and East China.* Translated from the German by Alide Eberhard. Leiden: E. J. Brill.

_______ (1969). *A history of China.* Berkeley and Los Angeles: University of California Press.

_______ (1982). *China's minorities: Yesterday and today.* Belmont, Cal.: Wadsworth.

Edie, Harry H., and Bess W. C. Ho (1969). *Ipomoea aquatica* as a vegetable crop in Hong Kong, *Economic Botany.* **23**: 32-37.

Edkins, Joseph (1893). *Chinese Buddhism: A volume of sketches, historical, descriptive, and critical.* London: Kegan Paul, Trench, Trübner, and Co.

Edwards, D. (1975). *The market for dried, preserved and fresh ginger.* London: Tropical Products Institute, Ministry of Overseas Development.

Embrey, Hartley Clarke (1921). The investigation of some Chinese foods, *China Medical Journal.* **35**: 420-47.

_______ (1922). The American Red Cross diet in the famine districts of China, *American Journal of Public Health.* **12**: 514-15.

Ensminger, M. E., and Audrey Ensminger (1973). *China - The impossible dream.* Clovis, Cal.: Agriservice Foundation.

Epstein, H. (1971). *Domestic animals of China.* New York: Africana Publishing Corporation.

Epstein, H., and M. Bichard (1984). Pig, pp. 145-62 **In:** Ian L. Mason (ed.), *Evolution of domesticated animals.* London and New York: Longman.

Erkes, Eduard (1943). Der Hund im alten China, *T'oung Pao.* **37**: 186-225.

Espino, R. B. (1935-36). Construction of mushroom beds in the Canton region, *Philippine Agriculturalist.* **24**: 752-62.

Etchécopar, R. D., and François Hüe (1978). *Les oiseaux de Chine.* Papeete, Tahiti: Les éditions du pacifique.

Ethé, Hermann (1868). *Zakarija ben Muhammed ben Mahmûd el-Kazwîni's Kosmographie.* Leipzig: Fues's Verlag (R. Reisland).

Evans, G. M. (1976). Rye: *Secale cereale* (Gramineae-Triticinae), pp. 108-11 **In:** N. W. Simmonds (ed.), *Evolution of crop plants.* London and New York: Longman.

Evreinoff, V. A. (1954). Les ancêtres de nos abricotiers, *Journal d'Agriculture Tropicale et de Botanique Appliquée.* **1**: 431-40.

Fan, W. H., and H. C. Hou (1953). Carotene contents of certain Nanking vegetables and fruits, *Chinese Medical Journal.* **71**: 127-34.

Farrell, Kenneth T. (1985). *Spices, condiments, and seasonings.* Westport, Conn.: AVI Publishing Company.

Faust, Ernest Carroll (1929). The animal parasites of the dog and cat in China, *Lingnan Science Journal.* **8**: 27-44.

Faust, Ernest Carroll, and Oo-keh Khaw (1927). *Studies on Clonorchis sinensis (Cobbold).* American Journal of Hygiene, Monographic Series, No. 8. Baltimore: American Journal of Hygiene.

Fehily, Lydia (1941). Infantile beri-beri in Hong Kong, *Journal of Tropical Medicine and Hygiene.* **44**: 21-26.

_______ (1945). Nutritional deficiencies in Hong Kong before the Japanese invasion, *British Medical Journal.* **2**: 468-70.

_______ (1947). Deficiency diseases in reoccupied Hong Kong, *British Medical Journal.* **1**: 220-22.

Fei, Hsiao-tung (1946). *Peasant life in China: A field study of country life in the Yangtze Valley.* New York: Oxford University Press.

Fei, Hsiao-tung, and Chih-i Chang (1945). *Earthbound China: A study of rural economy in Yunnan.* Chicago: University of Chicago Press.

Feine-Dudley, Laurie B. (1981). A provisional key to some edible species of the family Amaranthaceae, pp. 35-47 (Appendix II) **In:** G. J. H. Grubben and D. H. van Sloten (eds.), *Genetic resources of amaranths.* Rome: International Board for Plant Genetic Resources.

Feng, Doreen Yen Hung (n.d.). *The joy of Chinese cooking.* New York: Grosset and Dunlap.

Fenzel, Gottlieb (1933). Die Insel Hainan; Eine landeskundliche Skizze, dargestellt auf Grund eigener Reisebeobachtungen und des vorhandenen Schrifttums, *Mitteilungen der Geographischen Gesellschaft in München.* **26**: 69-221.

Ferguson, A. R. (1984). Kiwifruit: A botanical review, pp. 1-64 **In:** Jules Janick (ed.), *Horticultural Reviews. Vol. 6.* Westport, Conn.: AVI Publishing Company.

Ferrars, Max, and Bertha Ferrars (1901). *Burma.* London: Sampson Low, Marston and Company; New York: E. P. Dutton and Co.

Ferwerda, F. P. (1976). Coffees: *Coffea* spp. (Rubiaceae), pp. 257-60 **In:** N. W. Simmonds (ed.), *Evolution of crop plants.* London and New York: Longman.

Field, Michael, and the editors of Time-Life Books (1976). *Recipes: Chinese cooking.* New York: Time-Life Books.

Fitter, Richard S. R. (1961). The leathery turtle or luth, *Oryx.* **6**: 116-25.

FitzGerald, C. P. (1940). The religion of a non-Chinese tribe of Yunnan, *Man.* **40**: 178-82.

_______ (1941). *The tower of five glories: A study of the Min Chia of Ta Li, Yunnan.* London: Cresset Press.

_______ (1972). *The southern expansion of the Chinese people.* London: Barrie and Jenkins.

Flach, M., and A. M. Cruickshank (1969). Nutmeg (*Myristica fragrans* Houtt. and *Myristica argentea* Warb.), pp. 329-38 **In:** F. P. Ferwerda and F. Wit (eds.), *Outlines of perennial crop breeding in the tropics.* Agricultural University Wageningen, Miscellaneous Papers, No. 4. Wageningen: H. Veenman and Zonen.

Flaumenhaft, Eugene, and Mrs Eugene Flaumenhaft (1982). Asian medicinal plants in seventeenth century French literature, *Economic Botany.* **36**: 147-62.

Flückiger, Friedrich A., and Daniel Hanbury (1879). *Pharmacographia: A history of the principal drugs of vegetable origin, met with in Great Britain and British India.* Second Edition. London: Macmillan and Company.

Fogg, Wayne H. (1983). Swidden cultivation of foxtail millet by Taiwan aborigines: A cultural analogue of the domestication of *Setaria italica* in China, pp. 95-115 **In:** David N. Keightley (ed.), *The origins of Chinese civilization.* Berkeley: University of California Press.

Folch Andréu, Rafael (1957). Una droga que tiende a desaparecer del tesoro medicinal: El azafrán (estudio histórico), *Farmacognosia.* **17**: 145-224.

Fong, Y. Y., and W. C. Chan (1973). Bacterial production of di-methyl nitrosamine in salted fish, *Nature.* **243**: 421-22.

Food and Agriculture Organization of the United Nations (1984). *FAO production yearbook. Vol. 37, 1983.* Rome: FAO.

Food Policy and Food Science Service, Nutrition Division, FAO (1970). *Amino-acid content of foods and biological data on proteins.* FAO Nutritional Studies, No. 24. Rome: Food and Agriculture Organization of the United Nations.

Forbes, R. J. (1948). *Short history of the art of distillation.* Leiden: E. J. Brill.

_______ (1965). *Studies in ancient technology. Vol 3.* Leiden: E. J. Brill.

Forrest, R. A. D. (1965). *The Chinese language.* London: Faber and Faber.

Forsyth, William G. C. (1980). Banana and plantain, pp. 258-78 **In:** Steven Nagy and Philip E. Shaw (eds.), *Tropical and subtropical fruits: Composition, properties and uses.* Westport, Connecticut: AVI Publishing Company.

Fortune, Robert (1847). *Three years' wanderings in the northern provinces of China.* London: John Murray.

_______ (1852). *A journey to the tea countries of China.* London: John Murray.

_______ (1857). *A residence among the Chinese.* London: John Murray.

Franck, Irene M., and David M. Brownstone (1984). *To the ends of the earth: The great travel and trade routes in human history.* New York and Oxford: Facts on File Publications.

Franke, Herbert (1975). Chinese texts on the Jurchen: A translation of the Jurchen monograph in the San-ch'ao pei-meng hui-pien, *Zentralasiatische Studien.* **9**: 119-86.

Frazer, James George (1918). *Folk-lore in the Old Testament.* London: Macmillan and Company. 3 vols.

Frazier, Chester N., and Ch'uan-k'uei Hu (1931). Cutaneous lesions associated with a deficiency of vitamin A in man, *Archives of Internal Medicine.* **48**: 507-14.

Freedman, Maurice (1970). Ritual aspects of Chinese kinship and marriage, pp. 163-87 **In:** Maurice Freedman (ed.), *Family and kinship in Chinese society.* Stanford: Stanford University Press.

Freedman, Robert L. (1978). Review of recently published work dealing with nutrition research in Mainland China, *World Review of Nutrition and Dietetics.* **30**: 1-22.

Freeman, Michael (1977). Sung, pp. 141-76 **In:** K. C. Chang (ed.), *Food in Chinese culture.* New Haven and London: Yale University Press.

Frick, Johann, and Franz Eichinger (1952). Tiere im Volksleben, pp. 125-47 **In:** *Ethnographische Beiträge aus der Ch'inghai Provinz (China).* Folklore Studies, Supplement 1. Peking.

Fried, Morton Herbert (1953). *Fabric of Chinese society: A study of the social life of a Chinese county seat.* New York: Praeger.

"Fruit cultivation in Kuangtung" (1936). *Chinese Economic Journal and Bulletin.* **18**: 716-25.

Fry, Peggy Crooke (1965). The nutritive value of the diets of Hong Kong children, *Journal of Tropical Pediatrics and African Child Health.* **10**: 100-104.

Fry, Peggy Crooke, and Suna Goksu Baykan (1972). Quantities of food and nutrients consumed at different ages and seasons by Hong Kong Chinese children, *Nutritional Reports International.* **6**: 99-106.

Fry, Peggy Crooke, Ruth M. Leverton, and Suna Goksu (1967). Growth of Hong Kong children: On diets containing rice or rice and wheat with and without nutrient supplements, *American Journal of Clinical Nutrition.* **20**: 954-59.

Fulder, Stephen (1980). *The root of being: Ginseng and the pharmacology of harmony.* London: Hutchinson.

Fuson, C. G. (1929). The peoples of Kwangtung: Their origin, migrations, and present distribution, *Lingnan Science Journal.* **7**: 5-21.

Gade, Daniel W. (1970). Horsemeat as human food in France, *Ecology of Food and Nutrition.* **5**: 35-39.

Gainer, John L., and G. M. Chisolm III (1974). Oxygen diffusion and atherosclerosis, *Atherosclerosis.* **19**: 135-38.

Galinat, W. C. (1971). The origin of maize, *Annual Review of Genetics.* **5**: 447-78.

Gamble, Sidney D. (1933). *How Chinese families live in Peiping.* New York and London: Funk & Wagnalls Company.

Gargan, Edward A. (1988). China tries to skewer a sacred cow, *New York Times (National Edition),* Thursday, February 10, p. 4.

Garrod, H. W. (ed.) (1939). *The poetical works of John Keats.* Oxford: Clarendon Press.

Geiger, J. C., Emmett E. Sappington, Roslyn C. Miller, and Hilda F. Welke (1939). *The health of the Chinese in an American city. San Francisco.* San Francisco: Department of Public Health.

Geiger, J. C., Roslyn C. Miller, Hilda F. Welke, and Eunice Gibson (1945). *The health of the Chinese in an American city. San Francisco.* San Francisco: Department of Public Health.

Gerber, Paul, James D. Dutcher, Eugene V. Adams, and John H. Sherman (1958). Protective effect of seaweed extracts for chicken embryos infected with influenza B or mumps virus, *Proceedings of the Society for Experimental Biology and Medicine.* **99**: 590-93.

Gernet, Jacques (1962). *Daily life in China on the eve of the Mongol invasion 1250-1276.* Translated by H. M. Wright. New York: Macmillan.

Getty, Alice (1962). *The gods of Northern Buddhism.* Rutland, Vt.: C. E. Tuttle Co.

_______ (1971). *Ganesa: A monograph on the elephant-faced god.* Second Edition. New Delhi: Munshiram Manoharlal.

Gibb, H. A. R. (trans.) (1969). *Ibn Battúta: Travels in Asia and Africa, 1325-1354.* New York: Augustus M. Kelley Publishers.

Gibson, C. (1964). *The Aztecs under Spanish rule: A history of the Indians of the Valley of Mexico 1519-1810.* Stanford: Stanford University Press.

Giles, Herbert A. (1876). *Chinese sketches.* London: Trübner and Co.; Shanghai: Kelly and Co.

_______ (1900). *A glossary of reference on subjects connected with the Far East.* Third Edition. Shanghai: Kelly and Walsh.

Gin, Maggie (1984). *Regional cooking of China.* San Francisco: 101 Productions.

Glover, I. C. (1979). Prehistoric plant remains from Southeast Asia, with special reference to rice, pp. 7-37 **In:** Vol. 1 of Maurizio Taddei (ed.), *South Asian Archaeology 1977.* Papers from the Fourth International Conference of the Association of South Asian Archaeologists in Western Europe, held in the Istituto Universitario Orientale, Naples. Naples: Seminario di Studi Asiatici, Istituto Universitario Orientale.

Godbole, S. R., G. S. Pendse, and V. A. Bedekar (eds.) (1966). *Glossary of vegetable drugs in Vagbhata.* Poona: Indian Drug Research Association.

Gode, P. K. (1941-42). Some notes on the history of the fig (*Ficus carica*) from foreign and Indian sources, *New Indian Antiquary.* **4**: 125-36.

_______ (1943). Some notes on the history of the fig - Does the word "phalga," used by Caraka and Susruta mean "añjira," *Indian Historical Quarterly.* **19**: 62-65.

Golson, J., and P. J. Hughes (1976). The appearance of plant and animal domestication in New Guinea, pp. 88-100 **In:** J. Garanger (ed.), *La préhistoire Océanienne.* Paris: Centre National de la Recherche Scientifique.

Gonda, J. (1970). *Visnuism and Sivaism: A comparison.* London: Athlone Press, University of London.

González de Mendoza, Juan (1853-54). *History of the great and mighty kingdom of China and the situation thereof.* Edited by George T. Staunton. Hakluyt Society Works, Nos. 14 and 15. London: Hakluyt Society. 2 vols.

Goodrich, L. Carrington (1936-37). Early notices of the peanut in China, *Monumenta Serica.* **2**: 405-9.

Goodrich, L. Carrington, and C. Martin Wilbur (1942). Additional notes on tea, *Journal of the American Oriental Society.* **62**: 195-97.

Goor, Asaph, and Max Nurock (1968). *The fruits of the Holy Land.* Jerusalem: Israel Universities Press.

Gore, H. C. (1911). *Experiments on the processing of persimmons to render them nonastringent.* U.S. Department of Agriculture, Bureau of Chemistry, Bulletin 141. Washington, D.C.: U.S. Government Printing Office.

Gould, J., and A. Rutgers (1969). *Birds of Asia.* New York: Taplinger Publishing Company.

Gould-Martin, Katherine (1978). Hot cold clean poison and dirt: Chinese folk medical categories, *Social Science and Medicine.* **12**: 39-46.

Gould-Martin, Katherine, and Chorswang Ngin (1981). Chinese Americans, pp. 130-71 **In:** Alan Harwood (ed.), *Ethnicity and medical care.* Cambridge: Harvard University Press.

Graham, David Crockett (1961). *Folk religion in southwest China.* Smithsonian Miscellaneous Collections, Vol. 142, No. 2. Washington, D.C.: Smithsonian Institution.

Grant, A. J. (1897). *Herodotus.* London: John Murray. 2 vols.

Gray, John Henry (1878). *China: A history of the laws, manners, and customs of the people.* London: Macmillan and Company. 2 vols.

Gregory, W. C., and M. P. Gregory (1976). Groundnut: *Arachis hypogaea* (Leguminosae-Papilionatae), pp. 151-54 **In:** N. W. Simmonds (ed.), *Evolution of crop plants.* London and New York: Longman.

Grierson, G. A. (1908-26). Gaya, pp. 181-87 **In:** James Hastings (ed.), *Encyclopaedia of religion and ethics.* Edinburgh: T. & T. Clark. 13 vols.

Grieve, M. (1931). *A modern herbal.* London: Jonathan Cape. 2 vols.

Grimm, M. R. (1952). Iodine content of some marine algae, *Pacific Science.* **6**: 318-23.

Grisolia, Santiago (1974). Hypoxia, saffron, and cardiovascular disease, *Lancet.* **2**: 41-42.

Grist, D. H. (1953). *Rice.* London, New York, Toronto: Longmans, Green and Company.

Groff, Elizabeth H. (1919). Soy-sauce manufacturing in Kwangtung, China, *Philippine Journal of Science.* **15**: 307-16.

Groff, George Weidman (1921). *The lychee and lungan.* New York: Orange Judd Company; Canton: Canton Christian College.

_______ (1929-30). Culture and varieties of Siamese pummelos, *California Citrograph.* **15**: 442, 459, 462.

Groff, George Weidman, and C. W. Howard (1924-25). The cultured citrus ant of South China, *Lingnaam Agricultural Review.* **2**: 108-14.

Groff, George Weidman, Edward Ding, and Elizabeth H. Groff (1924). An enumeration of the McClure Collection of Hainan plants, *Lingnaam Agricultural Review.* **2**: 9-44, 115-39.

Groves, C. (1981). *Ancestors for the pigs: Taxonomy and phylogeny of the genus Sus.* Technical Bulletin No. 3. Canberra: Department of Prehistory, Research School of Pacific Studies, Australian National University.
Grünwedel, Albert (1901). *Buddhist art in India.* Translated by A. C. Gibson. Revised and enlarged by Jas. Burgess. London: Bernard Quaritch Ltd.
Gudger, E. W. (1919). On the use of the sucking-fish for catching fish and turtles: Studies in *Echeneis* or *Remora*, *American Naturalist.* **53**: 289-311, 446-67, 515-25.
_______ (1926). Fishing with the cormorant in China, *American Naturalist.* **60**: 5-41.
_______ (1927). Fishing with the otter, *American Naturalist.* **61**: 193-225.
Günther, Albert C. L. G. (1864). *The reptiles of British India.* London: Ray Society.
Guppy, H. B. (1884). Samshu-brewing in North China, *Journal of the North China Branch, Royal Asiatic Society (N.S.).* **18**: 163-64.
_______ (1903-6). *Observations of a naturalist in the Pacific between 1896 and 1899.* London: Macmillan and Company. 2 vols.
Guy, R. A. (1936). The diets of nursing mothers and young children in Peiping, *Chinese Medical Journal.* **50**: 434-42.
Guy, R. A., and K. S. Yeh (1938). Peking diets, *Chinese Medical Journal.* **54**: 201-22.
Hagerty, Michael J. (trans.) (1923). *Han Yen-Chih's Chü lu (Monograph on the oranges of Wên-chou, Chekiang).* Reprint from T'oung Pao, Ser. 2, Vol. 23. Leiden: E. J. Brill.
_______ (1940-41). Comments on writings concerning Chinese sorghums, *Harvard Journal of Asiatic Studies.* **5**: 234-60.
Hahn, Eduard (1896). *Die Haustiere und ihre Beziehungen zur Wirtschaft des Menschen.* Leipzig: Duncker und Humblot.
Hahn, Emily, and the editors of Time-Life Books (1973). *The cooking of China.* New York: Time-Life Books.
Halstead, Bruce W. (1978). *Poisonous and venomous marine animals of the world.* Revised Edition. Princeton: Darwin Press.
Hammond, Norman (1976). The early history of American agriculture: Recent research and current controversy, *Philosophical Transactions of the Royal Society of London B.* **275**: 120-28.
Handbook of nonprescription drugs (1982). Seventh Edition. Washington, D.C.: American Pharmaceutical Association.
Hanelt, Peter (1961). Zur Kenntnis von *Carthamus tinctorius* L., *Die Kulturpflanze.* **9**: 114-45.
_______ (1972). Zur Geschichte des Anbaues von *Vicia faba* L. und ihrer verschiedenen Formen, *Die Kulturpflanze.* **20**: 209-23.
Hanelt, Peter, Helga Schäfer, and Jürgen Schultze-Motel (1972). Die Stellung von *Vicia faba* L. in der Gattung *Vicia* L. und Betrachtungen zur Entstehung dieser Kulturart, *Die Kulturpflanze.* **20**: 263-75.
Hara, Hiroshi (1966). *The flora of eastern Himalaya.* Tokyo: University of Tokyo Press.
_______ (1970). On the Asiatic species of the genus *Panax*, *Journal of Japanese Botany.* 45: 197-212.
_______ (1971). *The flora of eastern Himalaya: Second report.* Tokyo: University of Tokyo Press.
Hardache, Val (1974). *Woodland nuggets of gold: The story of American ginseng cultivation.* Second Edition. Northville, Michigan: Holland House Press.
Harlan, Jack R. (1975). *Crops and man.* Madison, Wisconsin: American Society of Agronomy, Crop Science Society of America.
_______ (1976). Barley: *Hordeum vulgare* (Gramineae - Triticinae), pp. 93-98 **In:** N. W. Simmonds (ed.), *Evolution of crop plants.* London and New York: Longman.
_______ (1977). The origins of cereal agriculture in the Old World, pp. 357-83 **In:** Charles A. Reed (ed.), *Origins of agriculture.* The Hague, Paris: Mouton Publishers.
_______ (1986). Lettuce and the sycomore: Sex and romance in ancient Egypt, *Economic Botany.* **40**: 4-15.
Harlan, Jack R., and A. Stemler (1976). The races of sorghum in Africa, pp. 465-78 **In:** J. R. Harlan, J. M. J. de Wet, and A. B. L. Stemler (eds.), *Origins of African plant domestication.* The Hague, Paris: Mouton.
Harlan, Jack R., and J. M. J. de Wet (1973). On the quality of evidence for origin and dispersal of cultivated plants, *Current Anthropology.* **14:** 51-62.
Harries, H. C. (1978). The evolution, dissemination and classification of *Cocos nucifera* L., *Botanical Review.* **44**: 265-320.
Harrington, Geri (1984). *Grow your own Chinese vegetables.* Pownal, Vt.: Garden Way Publishing.
Harris, David R. (1981). The prehistory of human subsistence: A speculative outline, pp. 15-35 **In:** Dwain N. Walcher and Norman Kretchmer (eds.), *Food, nutrition and evolution.* New York: Masson Publishing USA.
Harris, Marvin (1968). *The rise of anthropological theory.* New York: Thomas Y. Crowell Company.
Harris, Robert S., Florence K. C. Wang, Ying H. Wu, Chi-hsuen S. Tsao, and Lenore Y. S. Loe (1949). The composition of Chinese foods, *Journal of the American Dietetic Association.* **25**: 28-38.
Harrison, Michael T., Sheena McFarlane, Ronald M. Harden, and Edward Wayne (1965). Nature and availability of iodine in fish, *American Journal of Clinical Nutrition.* **17**: 73-77.
Harrisson, Tom (1959). New archaeological and ethnological results from Niah Caves, *Man.* **59**: 1-8.

Hart, Henry H. (1954). *Poems of the hundred names.* Third Edition. Stanford: Stanford University Press.

Hashimoto, Mantaro J. (1973). *The Hakka dialect.* Princeton-Cambridge Studies in Chinese Linguistics, No. 5. Cambridge: Cambridge University Press.

Hashimoto, Oi-kan Yue (1972). *Phonology of Cantonese.* Princeton-Cambridge Studies in Chinese Linguistics, No. 3. Cambridge: Cambridge University Press.

Hashmi, Z. A. (1945). Organized duck breeding in India, *Indian Farming.* **6**: 306-8.

Hastings, James (ed.) (1908-26). *Encyclopaedia of religion and ethics.* Edinburgh: T. & T. Clark. 13 vols.

_______ (ed.) (1910). *Dictionary of the Bible.* London: T. & T. Clark.

Haudricourt, A. G. (1962). Domestication des animaux, culture des plantes et traitement d'autrui, *L'Homme.* **2**: 40-50.

Haudricourt, A. G., and L. Hédin (1943). *L'Homme et les plantes cultivées.* Paris: Gallimard.

Hawes, R. O. (1984). Pigeons, pp. 351-56 **In:** Ian L. Mason (ed.), *Evolution of domesticated animals.* London and New York: Longman.

Hawks, Jean Elizabeth (1936). Preparation and composition of foods served in Chinese homes, *Journal of the American Dietetic Association.* **12**: 136-40.

Hayes, W. B. (1966). *Fruit growing in India.* Third Revised Edition. Allahabad: Kitabistan.

Hays, Virgil W. (1980). Swine production, pp. 41-51 **In:** J. A. Hoefer and P. J. Tsuchitani (eds.), *Animal agriculture in China.* Committee on Scholarly Communication with the People's Republic of China, Report No. 11. Washington, D.C.: National Academy Press.

Headland, Isaac Taylor (1901). *The Chinese boy and girl.* New York, Chicago, Toronto: Fleming H. Revell Company.

_______ (1914). *Home life in China.* New York: Macmillan Company.

Hehn, Victor (1885). *The wanderings of plants and animals from their first home.* Edited by James Steven Stallybrass. London: Swan Sonnenschein and Company.

Heiser, C. B., Jr. (1976). Peppers: *Capsicum* (Solanaceae), pp. 265-68 **In:** N. W. Simmonds (ed.), *Evolution of crop plants.* London and New York: Longman.

Helbaek, Hans (1971). The origin and migration of rye, *Secale cereale L.*; A palaeo-ethnobotanical study, pp. 265-80 **In:** Peter H. Davis, Peter C. Harper, and Ian C. Hedge (eds.), *Plant life of south-west Asia.* Edinburgh: Botanical Society of Edinburgh.

Helm, Johannes (1954). *Lactuca sativa* L. in morphologisch-systematischer Sicht, *Die Kulturpflanze.* **2**: 72-129.

Hemmer, H. (1978). Geographische Variation der Hirngrösse in *Sus scrofa* und *Sus verrucosus* - Kreis (Beitrag zur Problem der Schweindomestikation), *Spixiana.* **1**: 309-20.

Hendrickson, John (1958). The green sea turtle, *Chelonia mydas* (Linn.), in Malaya and Sarawak, *Proceedings of the Zoological Society of London, Ser. A.* **130**: 445-535.

Henry, Augustine (1893). *Notes on the economic botany of China.* Shanghai: Presbyterian Mission Press.

Henry, G. M. (1971). *A guide to the birds of Ceylon.* Second Edition. London, New York, Melbourne: Oxford University Press.

Herklots, G. A. C. (1972). *Vegetables in South-East Asia.* London: George Allen & Unwin.

Herklots, G. A. C., and S. Y. Lin (1940). *Common marine food-fishes of Hong Kong.* Second Enlarged Edition. Hong Kong: G. A. C. Herklots.

Hermanns, Matthias (1948). *Die A Mdo Pa - Grosstibeter: Die Sozial-Wirtschaftlichen Grundlagen der Hirtenkulturen Innerasiens.* Freiburg: Philosophische Fakultät der Universität Freiburg in der Schweiz.

_______ (1949). *Die Nomaden von Tibet.* Vienna: Herold.

Herrmann, Albert (1966). *An historical atlas of China.* New Edition, edited by Norton Ginsburg. Chicago: Aldine Publishing Company.

Hesseltine, C. W., and Hwa L. Wang (1967). Traditional fermented foods, *Biotechnology and Bioengineering.* **9**: 275-88.

Heyd, Wilhelm (1879). *Geschichte des Levante-handels im Mittelalter.* Stuttgart: Cotta'schen Buchhandlung. 2 vols.

Higham, C. F. W. (1984). Prehistoric rice cultivation in Southeast Asia, *Scientific American.* **250**: 138-46.

Hilgenberg, Luise, and Willibald Kirfel (1941). *Vagbhata's Astangahrdayasamhita: Ein altindischer Lehrbuch der Heilkunde. Leiden: E. J. Brill.*

Hilu, Khidir W., and J. M. J. de Wet (1976). Domestication of *Eleusine coracana, Economic Botany.* **30**: 199-208.

Hirth, Friedrich (1887). Notes on the early history of the salt monopoly in China, *Journal of the North China Branch, Royal Asiatic Society (N.S.).* **12**: 53-66.

Hirth, Friedrich, and W. W. Rockhill (trans.) (1966). *Chau Ju-kua: His work on the Chinese and Arab trade in the twelfth and thirteenth centuries, entitled Chu-fan-chi. Vol. 1* [Reprint of the 1911 book published in St. Petersburg]. New York: Paragon Book Reprint Corp.

Ho, Kê-ên (1967). The Tanka or boat people of South China, pp. 120-23 **In:** F. S. Drake (ed.), *Symposium on historical, archaeological and linguistic studies on southern China, South-East Asia and the Hong Kong region.* Hong Kong: Hong Kong University Press.

Ho, Ping-ti (1955). The introduction of American food plants into China, *American Anthropologist.* **57**: 191-201.

_______ (1956-57). Early-ripening rice in Chinese history, *Economic History Review (Ser. 2)*. **9**: 200-218.
_______ (1959). *Studies on the population of China, 1368-1953*. Cambridge, Massachusetts: Harvard University Press.
_______ (1969). The loess and the origin of Chinese agriculture, *American Historical Review*. **75**: 1-36.
_______ (1975). *The cradle of the East: An inquiry into the indigenous origins of techniques and ideas of Neolithic and early historic China, 5000-1000 B.C.* Chicago: University of Chicago Press.
_______ (1977). The indigenous origins of Chinese agriculture, pp. 413-484 **In:** Charles A. Reed (ed.), *Origins of agriculture*. The Hague, Paris: Mouton Publishers.
Hodge, W. H. (1956). Chinese water chestnut or matai - A paddy crop of China, *Economic Botany*. **10**: 49-65.
Hodgson, Robert W. (1917). *The pomegranate*. Bulletin No. 276, Agricultural Experiment Station, College of Agriculture. Berkeley: University of California Press.
Hodgson, Robert Willard (1967). Horticultural varieties of citrus, pp. 431-591 **In:** Vol. 1 of Walter Reuther, Herbert John Webber, and Leon Dexter Batchelor (eds.), *The citrus industry*. Revised Edition. Berkeley: Division of Agricultural Sciences, University of California.
Hodous, Lewis (1929). *Folkways in China*. London: A. Probsthain.
Hoeppli, R. (1959). *Parasites and parasitic infections in early medicine and science*. Singapore: University of Malaya Press.
Hoeppli, R., and I-hung Ch'iang (1938). The origin of human helminths according to the old Chinese medical literature, *Monumenta Serica*. **3**: 579-601.
Hoernle, A. F. R. (ed.) (1893-1912). *The Bower manuscript*. Archaeological Survey of India, New Imperial Series, Vol. 22. Calcutta: Superintendent, Government Printing.
Hoffmann, William E. (1934). Preliminary notes on the fresh-water fish industry of South China, especially Kwangtung Province, *Lingnan University Science Bulletin*. **5**: 1-70.
_______ (1947). Insects as human food, *Proceedings of the Entomological Society of Washington*. **49**: 233-37.
Hoh, Pik Wan, Jessamine Chapman Williams, and Charles S. Pease (1933). Possible sources of calcium and phosphorus in the Chinese diet: Determinations of calcium and phosphorus and in a typical Chinese dish containing meat and bone, *Journal of Home Economics*. **25**: 728.
Holden, J. H. W. (1976). Oats: *Avena spp.* (Gramineae - Aveneae), pp. 86-90 **In:** N. W. Simmonds (ed.), *Evolution of crop plants*. London and New York: Longman.
Hole, Frank, Kent Flannery, and James Neely (1965). Early agriculture and animal husbandry in Deh Luran, Iran, *Current Anthropology*. **6**: 105-6.
Holmes, Maybel Marion (1948). *A source book of Chinese food habits*. Unpublished Ph.D. Dissertation in Home Economics, Cornell University.
Hom, Ken, and Harvey Steiman (1981). *Chinese technique*. New York: Simon and Schuster.
Hommel, Rudolf P. (1937). *China at work*. New York: John Day.
Hooper, David (1929). On Chinese medicine: Drugs of Chinese pharmacies in Malaya, *The Gardens' Bulletin, Straits Settlements*. **6**: 1-163.
Hoose, Harned Pettus (1938). *Peking pigeons and pigeon-flutes*. Miscellaneous Publications, Vol. 2. Peking: College of Chinese Studies, California College in China.
Hopf, Maria (1969). Plant remains and early farming in Jericho, pp. 355-59 **In:** Peter J. Ucko and G. W. Dimbleby (eds.), *The domestication and exploitation of plants and animals*. Chicago: Aldine Publishing Company.
Hopwood, Cyril (1919). Notes on some nests recently found in South Tenasserim, *Journal of the Bombay Natural History Society*. **26**: 853-59.
Hornell, James (1950). *Fishing in many waters*. Cambridge: Cambridge University Press.
Hosie, Alexander (1897). *Three years in western China*. Second Edition. London: George Philip & Son.
_______ (1904). *Manchuria: Its people, resources and recent history*. Second Edition. London: Methuen and Company.
Hou, H. C. (1936). The variation of vitamin C content in certain fruits and vegetables, *Chinese Journal of Physiology*. **10**: 221-36.
_______ (1937). Distribution of vitamin C in Chinese foods. I. Vegetables, *Chinese Journal of Physiology*. **12**: 249-62.
_______ (1938). Dietary principles in ancient Chinese medicine, *Chinese Medical Journal*. **53**: 347-52.
_______ (1940a). Diet and nutritional deficiencies among refugees, pp. 1-13 **In:** H. C. Hou, P. G. Mar, T. G. Ni, and B. E. Read, *Nutritional studies in Shanghai*. Chinese Medical Association, Special Report Series, No. 12. Shanghai: Division of Physiological Sciences, Henry Lister Institute of Medical Research.
_______ (1940b). Riboflavin deficiency among Chinese. 1. Ocular manifestations, *Chinese Medical Journal*. **58**: 616-28.
_______ (1941). Riboflavin deficiency among Chinese. 2. Cheilosis and seborrheic dermatitis, *Chinese Medical Journal*. **59**: 314-25.
Hou, Joseph P. (1978). *The myth and truth about ginseng*. South Brunswick: A. S. Barnes and Company; London: Thomas Yoseloff Ltd.

Howard, C. W. (1923). *The sericulture industry of South China.* Canton: College of Agriculture, Canton Christian College.

Howe, Robin (1971). *Far Eastern cookery.* New York: Drake Publishers.

Howells, W. W. (1983). Origins of the Chinese people: Interpretations of the recent evidence, pp. 297-319 **In:** David N. Keightley (ed.), *The origins of Chinese civilization.* Berkeley: University of California Press.

Hsia, Chih-tsing, and others (1954). *China: An area manual.* Edited by David Nelson Rowe and Willmoore Kendall. Chevy Chase, Md.: Operations Research Office, Johns Hopkins University. 2 vols.

Hsieh, Chiao-min (1973). *Atlas of China.* Edited by Christopher L. Salter. New York: McGraw-Hill.

Hsu, Cho-yun (1980). *Han agriculture: The formation of early Chinese agrarian economy (206 B.C.-A.D. 220).* Edited by Jack L. Dull. Seattle and London: University of Washington Press.

Hsu, Francis L. K. (1948). *Under the ancestors' shadow: Chinese culture and personality.* New York: Columbia University Press.

_______ (1983). *Exorcising the trouble makers: Magic, science, and culture.* Westport, Conn. and London: Greenwood Press.

Hsu, L. C. (1974). Nutrition - from China to the West; Art-science quality of nutrition, *Ecology of Food and Nutrition.* **3**: 303-14.

Hsu, Peng-cheng, and William H. Adolph (1940). Chinese celery cabbage as supplement to a cereal diet. 1. Growth, digestibility, roughage effect, biological value, and calcifying potency, *Chinese Journal of Physiology.* **15**: 275-84.

Hsü, Princeton Sung-shi (1967). The Chuang people of South China, pp. 115-19 **In:** F. S. Drake (ed.), *Symposium on historical, archaeological and linguistic studies on southern China, South-East Asia and the Hong Kong region.* Hong Kong: Hong Kong University Press.

Hsu, Shin-Yi (1969). The cultural ecology of the locust cult in traditional China, *Annals of the Association of American Geographers.* **59**: 731-52.

Hsu, Vera Y. N., and Francis L. K. Hsu (1977). Modern China: North, pp. 295-316 **In:** K. C. Chang (ed.), *Food in Chinese culture.* New Haven and London: Yale University Press.

Hsu, Yin-ch'i (1929). Crickets in China, *Peking Society of Natural History Bulletin.* **3**: 3, 5-41.

Hu, C. C. (1930-31). Citrus culture in China, *California Citrograph.* **16**: 502, 543-44.

_______ (1934). *The history and distribution of citrus fruit in China.* Bulletin No. 31, New Series. Nanking: College of Agriculture and Forestry, University of Nanking.

Hu, Hsien-chin (1948). *The common descent group in China and its functions.* Viking Fund Publications in Anthropology, No.10. New York: Viking Fund.

Hu, Shiu-ying (1948). Shien—some noteworthy edible herbs of China, *Herbarist.* **14**: 30-36.

_______ (1956). Malva—a herb of high nutritive value, *Herbarist.* **22**: 22-30.

_______ (1976). The genus *Panax* (Ginseng) in Chinese medicine, *Economic Botany.* **30**: 11-28.

_______ (1980). *An enumeration of Chinese materia medica.* Hong Kong: Chinese University Press.

Huang, Chu-chien (1981). Distribution of sea turtles in China seas, pp. 321-22 **In:** Karen A. Bjorndal (ed.), *Biology and conservation of sea turtles. Proceedings of the World Conference on Sea Turtle Conservation, Washington, D.C., 26-30 November 1979.* Washington, D.C.: Smithsonian Institution Press.

Huang, Hsing-tsung (1986). Plants and insects in man's service, pp. 471-553 **In:** Part 1 of Vol. 6 (Biology and biological technology) of Joseph Needham, *Science and civilisation in China.* Cambridge: Cambridge University Press.

Huang, Huibai (1980). Viticulture in China, *Hortscience.* **15**: 461-66.

Huang, K'o-wei, and Chen-piao Hang (1960). The low incidence of atherosclerosis and its relation to certain dietary features among Chinese, *Chinese Medical Journal.* **80**: 455-58.

Huang, Ray (1969). Fiscal administration during the Ming Dynasty, pp. 73-128 **In:** Charles O. Hucker (ed.), *Chinese government in Ming times.* New York and London: Columbia University Press.

Huang, Su-huei (1972). *Chinese cuisine: Wei-Chuan cooking book.* Taipei: Wei-Chuan Foods Corp.

Huang, Tzû ch'ing, and Yün-ts'ung Chao (trans.) (1945-47). The preparation of ferments and wines by Chia Ssû -hsieh of the Later Wei Dynasty, *Harvard Journal of Asiatic Studies.* **9**: 24-44.

Hubbard, George D. (1929). The Pearl River Delta, *Lingnan Science Journal.* **7**: 23-35.

Hughes, Richard (1972). A toast to monkey head, *Far Eastern Economic Review.* **76**: 27-28.

Hulsewé, A. F. P. (1979). *China in Central Asia: The early stage: 125 B.C.-A.D. 23.* Sinica Leidensia, Vol. 14. Leiden: E. J. Brill.

Hume, H. Harold (1957). *Citrus fruits.* New York: Macmillan.

Humm, H. J. (1947). Agar - A pre-war Japanese monopoly, *Economic Botany.* **1**: 317- 29.

Hung, Mak Shiu (1960-61). The fish ponds and oyster beds of the Wang Chau area, *Journal of the Geographical, Geological, and Archaeological Society, University of Hong Kong.* **1**: 79-86.

Hurlbut, Floy (1939). *The Fukienese: A study in human geography.* Muncie, Indiana: Published by the author.

Hush, Joanne, and Peter Wong (1976). *The Chinese menu cookbook.* New York: Holt, Rinehart and Winston.

Hutson, James (1921). *Chinese life in the Tibetan foothills.* Shanghai: Far Eastern Geographical Establishment.

Hutton, J. H. (1968). *The Sema Nagas.* Second Edition. Bombay: Oxford University Press.

_______ (1969). *The Angami Nagas.* Second Edition. Bombay: Oxford University Press.

Hymowitz, T. (1970). On the domestication of the soybean, *Economic Botany.* **24**: 408-21.

_______ (1976). Soybeans: *Glycine max* (Leguminosae - Papilionatae), pp. 159-62 **In:** N. W. Simmonds (ed.), *Evolution of crop plants.* London and New York: Longman.

Hymowitz, T., and C. A. Newell (1981). Taxonomy of the genus *Glycine,* domestication and uses of soybeans, *Economic Botany.* **35**: 272-88.

Ignatius, Adi (1988). China's birthrate rises again despite a policy of one-child families, *Wall Street Journal (Western Edition),* Thursday, July 14, pp. 1, 14.

Ikeda, T. (1909). *The fruit culture in Japan.* Tokyo: Seibido.

Ilyas, M. (1976). Spices in India, *Economic Botany.* **30**: 273-80.

_______ (1978). The spices of India - II, *Economic Botany.* **32**: 238-63.

_______ (1980). Spices in India - III, *Economic Botany.* **34**: 236-59.

Indian Council of Agricultural Research (1953). *Report of the Spices Enquiry Committee.* New Delhi: Indian Council of Agricultural Research.

Ingram, Jean S. (1969). Saffron (*Crocus sativus* L.), *Tropical Science.* **11**: 177-84.

International Board for Plant Genetic Resources (1982). *Working group on the genetic resources of Citrus (Tsukuba, Japan, 4-6 November 1981), Report.* Rome: IBPGR Secretariat.

Isaac, Erich (1959). Influence of religion on the spread of citrus, *Science.* 129: 179-86.

_______ (1970). *Geography of domestication.* Englewood Cliffs, N.J.: Prentice-Hall.

Ishibashi, Masayoshi, and Toshio Yamamoto (1960). Inorganic constituents in seaweeds, *Records of Oceanographic Works in Japan.* **5**: 55-62.

Itoo, Saburo (1980). Persimmon, pp. 442-68 **In:** Steven Nagy and Philip E. Shaw (ed.), *Tropical and subtropical fruits.* Westport, Conn.: AVI Publishing Company.

Jacobi, Hermann (trans.) (1884). *Gaina Sûtras. Part 1.* Volume 22 of *The sacred books of the East,* edited by F. Max Müller. Oxford: Clarendon Press.

Jain, S. K., and Deb Kumar Banerjee (1974). Preliminary observations on the ethnobotany of the genus *Coix, Economic Botany.* **28**: 38-42.

Jao, Tsung-i (1967). The Shê settlements in the Han River Basin, Kwangtung, pp. 101-9 **In:** F. S. Drake (ed.), *Symposium on historical, archaeological and linguistic studies on southern China, South-East Asia and the Hong Kong region.* Hong Kong: Hong Kong University Press.

Jarrige, C., and M. Tosi (1981). The natural resources of Mundigak, pp. 115-42 **In:** Herbert Hårtel (ed.), *South Asian Archaeology 1979.* Papers from the Fifth International Conference of the Association of South Asian Archaeologists in Western Europe, held in the Museum für Indische Kunst der Staatlichen Museen Preussischer Kulturbesitz Berlin. Berlin: Dietrich Reimer Verlag.

Jefferys, W. Hamilton, and James L. Maxwell (1910). *The diseases of China.* Philadelphia: P. Blakiston's Son and Co.

Jeffreys, M. D. W. (1957). The origin of the Portuguese word zaburro as their name for maize, *Bulletin de l'Institut Francais d'Afrique Noire.* **19**: 111-36.

_______ (1967). Pre-Columbian maize in southern Africa, *Nature.* **215**: 695-97.

_______ (1971). Pre-Columbian maize in Asia, pp. 376-400 **In:** Carroll L. Riley, J. Charles Kelley, Campbell W. Pennington, and Robert L. Rands (eds.), *Man across the sea: Problems of pre-Columbian contacts.* Austin and London: University of Texas Press.

Jelliffe, Derrick B. (1968). *Infant nutrition in the subtropics and tropics.* Second Edition. Geneva: World Health Organization.

Jelliffe, Derrick B., and F. J. Bennett (1960). Indigenous medical systems and child health, *Journal of Pediatrics.* **57**: 248-61.

Jenner, Alice (1968). Social, emotional and cultural influences as related to eating patterns and malnutrition, *Canadian Nutrition Notes.* **24**: 37-43.

Johannessen, Carl L. (1981). Folk medicine uses of melanotic Asiatic chickens as evidence of diffusion to the New World, *Social Science and Medicine.* **15D**: 427-34.

Johnson, Dennis (1973). The botany, origin, and spread of the cashew *Anacardium occidentale* L., *Journal of Plantation Crops.* **1**: 1-7.

Johnson, John D., Norman Kretchmer, and Frederick J. Simoons (1974). Lactose malabsorption: Its biology and history, *Advances in Pediatrics.* **21**: 197-237.

Johnston, H. W. (1966). The biological and economic importance of algae, Part 2, *Tuatara.* **14**: 30-63.

_______ (1972). A detailed chemical analysis of some edible Japanese seaweeds, pp. 429-35 **In:** *Proceedings of the Seventh International Seaweed Symposium.* Sapporo, Japan, August 8-12, 1971. Tokyo: University of Tokyo Press.

Jolliffe, Norman, and Ta-cheng Tung (1956). Nutrition status survey of the civilian population of Formosa, *Metabolism* **5**: 309-27.

Jones, Henry A., and Louis K. Mann (1963). *Onions and their allies: Botany, cultivation, and utilization*. London: Leonard Hill Books; New York: Interscience Publishers.

Jones, Horace Leonard (trans.) (1917-33). *The geography of Strabo*. London: William Heinemann; New York: G. P. Putnam's Sons. 8 vols.

Joseph, Lawrence E. (1986). Man and cormorant, *Audubon*. **88**: 2 (March), 38-41.

Joshi, A. B., and M. W. Hardas (1976). Okra: *Abelmoschus esculentus* (Malvaceae), pp. 194-95 **In:** N. W. Simmonds (ed.), *Evolution of crop plants*. London and New York: Longman.

Kains, Maurice Grenville (1903). *Ginseng; its cultivation, harvesting, marketing, and market value, with a short account of its history and botany*. New Edition. New York: O. Judd Company.

Kale, F. S. (1937). *Soya bean; its value in dietetics, cultivation and uses*. Second Edition. Baroda, India: Baroda State Press.

Kan, Lilah (1978). *Introducing Chinese casserole cookery*. New York: Workman Publishing.

Kani, Hiroaki (1967). *A general survey of the boat people in Hong Kong*. New Asia Research Institute, Southeast Asia Studies Section, Monograph Series, No. 5. Hong Kong: Chinese University of Hong Kong.

Kano, Tadao, and Kokichi Segawa (1956). *An illustrated ethnography of Formosan aborigines. Vol. 1, The Yami*. Tokyo: Maruzen.

Kao, Yung-en (1948). Breast feeding in China, *Acta Paediatrica Scandinavica*. **36**: 233-37.

Kao, Yung-en, and Shui-chao Chen (1944). A clinical study of ginkgo-nut poisoning among children in Kweichow, *Chinese Medical Journal (Chengtu Edition)*. **62A**: 64-66.

Kar, C. S., and S. Bhaskar (1981). Status of sea turtles in the eastern Indian Ocean, pp. 365-72 **In:** K. A. Bjorndal (ed.), *Biology and conservation of sea turtles. Proceedings of the World Conference on Sea Turtle Conservation, Washington, D.C., November 26-30, 1979*. Washington, D.C.: Smithsonian Institution Press.

Karsen, Stephen J., Michael Wai-neng Lau, and Anthony Bogadek (1986). *Hong Kong amphibians and reptiles*. Hong Kong: Urban Council.

Karzel, K. (1974). Pharmacological aspects of ginseng. A review, pp. 49-55 **In:** *Proceedings of International Ginseng Symposium*. Seoul: Central Research Institute, Office of Monopoly, Republic of Korea.

Katz, Michael (1975). Parasitic infections, *Journal of Pediatrics*. **87**: 165-78.

Katz, S. H., M. L. Hediger, and L. A. Valleroy (1974). Traditional maize processing techniques in the New World, *Science*. **184**: 765-73.

Keightley, David N. (1978). *Sources of Shang history: The oracle-bone inscriptions of Bronze Age China*. Berkeley: University of California Press.

Kellogg, Claude R. (1929). Studies on the Chinese honey-bee, *Lingnan Science Journal*. **7**: 613-23.

_______ (1937). Bee-keeping in Fukien Province, *The China Journal*. **26**: 139-42.

_______ (1968). *Entomological excerpts from southeastern China (Fukien Province)*. Claremont, Cal.: Published by the author.

Kellogg, Claude R., and Chung-chang Tang (1963). Honey and its uses in China, *American Bee Journal*. **103**: 176-78.

Kelly, F. C., and W. W. Snedden (1960). Prevalence and geographical distribution of endemic goitre, pp. 27-233 **In:** *Endemic goitre*. WHO Monograph Series, No. 44. Geneva: World Health Organization.

Keng, Hsuan (1974). Economic plants of ancient North China as mentioned in *Shih Ching* (Book of Poetry), *Economic Botany*. **28**: 391-410.

Kennard, William C., and Harold F. Winters (1960). *Some fruits and nuts for the tropics*. U.S. Department of Agriculture, Agricultural Research Service, Miscellaneous Publication No. 801. Washington, D.C.: U.S. Government Printing Office.

Keshan Disease Research Group of the Chinese Academy of Medical Sciences (Beijing) (1979). Epidemiological studies on the etiologic relationship of selenium and Keshan disease, *Chinese Medical Journal*. **92**: 477-82.

Keshan Disease Research Group of the Chinese Academy of Medical Sciences (Beijing), Antiepidemic Station of Xichang District (Sichuan), and Antiepidemic Station of Mianning County (Sichuan) (1979). Observations on effect of sodium selenite in prevention of Keshan disease, *Chinese Medical Journal*. **92**: 471-76.

Keys, John D. (1963). *Food for the emperor: Recipes of Imperial China with a dictionary of Chinese cuisine*. [Los Angeles]: Ward Ritchie Press.

_______ (1976). *Chinese herbs: Their botany, chemistry, and pharmacodynamics*. Rutland, Vt.: Charles E. Tuttle Company.

Keys, N., and W. W. Cadbury (1926). An age-height-weight study of Cantonese school boys, *China Medical Journal*. **40**: 14-24.

Khory, Rustomjee Naserwanjee, and Nanabhai Navrosji Katrak (1981). *Materia medica of India and their therapeutics*. Delhi: Neeraj Publishing House.

Kimmens, Andrew C. (ed.) (1975). *Tales of the ginseng*. New York: William Morrow and Company.

Kimura, Tsuneo (1952-53). On digestibility and absorption rate of certain sea weeds, *Eiyo-Shokuryo Gakkai (Kyoto).* **5**: 176-79 (in Japanese).

Kincaid, C. A. (1908). *The tale of the tulsi plant and other studies.* Bombay and London: The Times of India Office.

King, F. H. (1927). *Farmers of forty centuries.* New York: Harcourt, Brace, and Company.

King, Haitung (1976). Selected epidemiological aspects of major diseases and causes of death among Chinese in the United States and Asia, pp. 487-550 **In:** Arthur Kleinman, Peter Kunstadter, E. Russell Alexander, and James L. Gate (eds.), *Medicine in Chinese cultures.* Washington, D.C.: U.S. Government Printing Office.

Kingdon-Ward, F. (1950). Does wild tea exist?, *Nature.* **165**: 297-99.

Kinmond, William (1957). *No dogs in China: A report on China today.* Toronto: University of Toronto Press.

Kinsley, David (1986). *Hindu goddesses.* Berkeley: University of California Press.

Kirby, R. H. (1950). Seaweeds in commerce, *Colonial Plant and Animal Products.* **1**: 183-216, 284-93.

Kirtikar, K. R. (1912). The use of saffron and turmeric in Hindu marriage ceremonies, *Journal of the Anthropological Society of Bombay.* **9**: 439-54.

Kitagawa, Evelyn M., and Philip M. Hauser (1973). *Differential mortality in the United States: A study of socio-economic epidemiology.* Cambridge: Harvard University Press.

Klein, Richard M. (1987). *The green world: An introduction to plants and people.* Second Edition. New York: Harper & Row, Publishers.

Knowles, P. F. (1976). Safflower: *Carthamus tinctorius* (Compositae), pp. 31-33 **In:** N. W. Simmonds (ed.), *Evolution of crop plants.* London and New York: Longman.

Koo, Linda C. (1973). Traditional Chinese diet and its relationship to health, *Papers of the Kroeber Anthropological Society.* **48**: 116-47.

_______ (1987). Concepts of disease causation, treatment and prevention among Hong Kong Chinese: Diversity and eclecticism, *Social Science and Medicine.* **25**: 405-17.

Koul, A. K. (1974). Job's tears, pp. 63-66 **In:** Joseph Hutchinson (ed.), *Evolutionary studies in world crops.* London: Cambridge University Press.

Koutras, Demetrios A., Josip Matovinovic, and Robert Vought (1980). The ecology of iodine, pp. 185-95 **In:** John B. Stanbury and Basil S. Hetzel (eds.), *Endemic goiter and endemic cretinism: Iodine nutrition in health and disease.* New York: John Wiley & Sons.

Krapovickas, A. (1969). The origin, variability and spread of the groundnut (*Arachis hypogaea*), pp. 427-41 **In:** Peter J. Ucko and G. W. Dimbleby (eds.), *The domestication and exploitation of plants and animals.* Chicago: Aldine Publishing Company.

Kratochvil, Paul (1968). The Chinese language today. London: Hutchinson.

Kriger, Sara Finn, and William H. Kroes (1972). Child-rearing attitudes of Chinese, Jewish, and Protestant mothers, *Journal of Social Psychology.* **86**: 205-10.

Kristof, Nicholas D. (1987). Quest for nests makes soup prices soar, *Sacramento Bee*, Wednesday, January 28, p. F5

Kromhout, Daan, Edward B. Bosschieter, and Cor de Lezenne Coulander (1985). The inverse relation between fish consumption and 20-year mortality from coronary heart disease, *New England Journal of Medicine.* **312**: 1205-9.

Krüssmann, Gerd (1984-86). *Manual of cultivated broad-leaved trees and shrubs.* Translated by Michael E. Epp. Portland, Oregon: Timber Press. 3 vols.

Kubota, J., W. H. Allaway, D. L. Carter, E. E. Cary, and V. A. Lazar (1967). Selenium in crops in the United States in relation to selenium-responsive diseases of animals, *Journal of Agricultural and Food Chemistry.* **15**: 448-53.

Kulp, Daniel Harrison, II (1925). *Country life in South China.* New York: Bureau of Publications, Teachers College, Columbia University.

Kung, Lan-chen, and Hui-lan Yeh (1938). Nitrogen, calcium, and phosphorus balances of rural adolescent boys on low cost diets, *Chinese Journal of Physiology.* **13**: 285-306.

Kung, Lan-chen, Hui-lan Yeh, and William H. Adolph (1938). The availability of calcium in vegetable food materials, *Chinese Journal of Physiology.* **13**: 307-16.

Kunstadter, Peter (1967). China: Introduction, pp. 149-67 **In:** Vol. 1 of Peter Kunstadter (ed.), *Southeast Asian tribes, minorities, and nations.* Princeton: Princeton University Press. 2 vols.

Kwauk, Z. Y. (1929). Isolation of typhoid bacilli from restaurant utensils in Canton, *Lingnan Science Journal.* **8**: 533-34.

Kwoh, Tsuin-ho, Chung-hua Lee, and Tson-pen Sun (1941). A note on the carotene content of some Chinese fruits and vegetables, *Journal of the Chinese Chemical Society.* **8**: 54-59.

Laai, Yi-faai, Franz Michael, and John C. Sherman (1962). The use of maps in social research: A case study in South China, *Geographical Review.* **52**: 92-111.

Laderman, Carol (1983). *Wives and midwives: Childbirth and nutrition in rural Malaysia.* Berkeley: University of California Press.

_______ (1987). Destructive heat and cooling prayer: Malay humoralism in pregnancy, childbirth and the postpartum period, *Social Science and Medicine.* **25**: 357-65.

Lamson, Herbert Day (1935). *Social pathology in China.* Shanghai: Commercial Press.

Landon, Kenneth Perry (1973). *The Chinese in Thailand.* Reprint of the 1941 edition. New York: Russell & Russell.

Lang, Olga (1946). *Chinese family and society.* New Haven: Yale University Press.

Langer, P. (1960). History of goitre, pp. 9-25 **In:** *Endemic goitre.* WHO Monograph Series, No. 44. Geneva: World Health Organization.

Lao, Bomin (1984). A 7,000-year-old oar, p. 11 **In:** *Recent discoveries in Chinese archaeology.* Beijing: Foreign Languages Press.

"Lao tea" (1892). *Bulletin of Miscellaneous Information, Royal Gardens, Kew.* 1892: 219-22.

Larsen, B. A., and W. W. Hawkins (1961). Nutritional value as protein of some of the nitrogenous constituents of two marine algae, *Chondrus crispus* and *Laminaria digitata, Journal of the Science of Food and Agriculture.* **12**: 523-29.

Lasker, G. W. (1946). Migration and physical differentiation: A comparison of immigrant with American-born Chinese, *American Journal of Physical Anthropology.* **4**: 273-300.

Latourette, Kenneth Scott (1964). *The Chinese, their history and culture.* Fourth Edition. New York: Macmillan; London: Collier-Macmillan.

Latsch, Marie-Luise (1984). *Chinese traditional festivals.* Beijing: New World Press.

Lattimore, Owen (1932). Chinese colonization in Inner Mongolia: Its history and present development, pp. 288-312 **In:** *Pioneer settlement: Cooperative studies.* Special Publications, No. 14. New York: American Geographical Society.

Laufer, Berthold (1907a). Note on the introduction of the ground-nut into China, *Proceedings of the International Congress of Americanists, Fifteenth Session.* **1**: 259-62.

_______ (1907b). The introduction of maize into eastern Asia, *Proceedings of the International Congress of Americanists, Fifteenth Session.* **1**: 223-57.

_______ (1914-15). Some fundamental ideas of Chinese culture, *Journal of Race Development.* **5**: 160-74.

_______ (1919). Sino-Iranica: Chinese contributions to the history of civilization in ancient Iran, *Field Museum of Natural History, Anthropological Series.* **15**: 185-630.

_______ (1927a). *Insect-musicians and cricket champions of China.* Anthropology Leaflets, No. 22. Chicago: Field Museum of Natural History.

_______ (1927b). Methods in the study of domestications, *Scientific Monthly.* **25**: 251-55.

_______ (1929). The American plant migration, *Scientific Monthly.* **28**: 239-51.

_______ (1930). Geophagy, *Field Museum of Natural History, Anthropological Series.* **17**: 2, 97-198.

_______ (1931). The domestication of the cormorant in China and Japan, *Field Museum of Natural History, Anthropological Series.* **18**: 201-63.

_______ (1934). The lemon in China and elsewhere, *Journal of the American Oriental Society.* **34**: 143-60.

_______ (1938). The American plant migration. Part 1: The potato, *Field Museum of Natural History, Anthropological Series.* **28**: 1-132.

Laurioux, Bruno (1985). Spices in the medieval diet: A new approach, *Food and Foodways.* **1**: 43-76.

Lebar, Frank M., Gerald C. Hickey, and John K. Musgrave (1964). *Ethnic groups of mainland Southeast Asia.* New Haven: Human Relations Area Files Press.

Leber, Bärbel H. S. (1977). *Marine algae as human food in the Pacific region.* Unpublished M. A. Thesis in Geography, University of California, Davis.

Lee, C. B. (1939). Local history, social organization and warfare, *Lingnan Science Journal.* **18**: 357-69.

Lee, Calvin B. T., and Audrey Evans Lee (1980). *The gourmet Chinese regional cookbook.* Secaucus, N.J.: Castle Books.

Lee, Chen-chia (1948). Birth weights of full term infants in West China, *Chinese Medical Journal.* **66**: 153-55.

Lee, Chung-hua, Tsuin-ho Kwoh, and Tson-pen Sun (1941). Ascorbic acid contents of Chungking fruits and vegetables, *Journal of the Chinese Chemical Society.* **8**: 67-73.

Lee, Karen, and Alaxandra Branyon (1984). *Chinese cooking secrets.* Garden City, N.Y.: Doubleday and Company.

Lee, Shun-ching (1935). *Forest botany of China.* Shanghai: Commercial Press, Limited.

Lee, T'ao (1940). Historical notes on some vitamin deficiency diseases in China, *Chinese Medical Journal.* **58**: 314-23.

Legendre, A. F. (1911). The Lolos of Kientchang, western China, *Annual Report, Smithsonian Institution.* 569-86.

Legge, James (trans.) (1893). *Confucius: Confucian analects, the great learning and the doctrine of the mean.* Second Edition. Chinese Classics, Vol. 1. Oxford: Clarendon Press.

_______ (trans.) (1895). *The works of Mencius.* Second Edition. Chinese Classics, Vol. 2. Oxford: Clarendon Press.

Lemert, Edwin M. (1982). Drinking among American Indians, pp. 80-95 **In:** Edith Lisansky Gombert, Helene Raskin White, and John A. Carpenter (eds.), *Alcohol, science and society revisited.* Ann Arbor: University of Michigan Press; New Brunswick: Rutgers Center of Alcohol Studies.

"Leppett tea" (1896). *Bulletin of Miscellaneous Information, Royal Gardens, Kew.* 1896: 10-16.

Leppik, E. E. (1971). Assumed gene centers of peanuts and soybeans, *Economic Botany.* **25**: 188-94.

Lessing, Ferdinand D. (1976). Über die Symbolsprache in der chinesischen Kunst, pp. 95-184 **In**: Vol. 91 of *Asian Folklore and Social Life Monographs*. Taipei: Orient Cultural Service.

Leung, W. T. W., R. R. Butrum, and F. H. Chang (1973). *Food composition table for use in East Asia, Part 1*. DHEW Publication No. (NIH) 73-465. Bethesda, Md.: U.S. Department of Health, Education and Welfare, and Food and Agriculture Organization of the United Nations.

Levander, O. A. (1982). Clinical consequences of selenium intake and its relationship to vitamin E, *Annals of the New York Academy of Sciences*. **393**: 70-82.

Leverington, Rowland E. (1983). Ginger, pp. 297-350 **In:** Harvey T. Chan, Jr. (ed.), *Handbook of tropical foods*. New York and Basel: Marcel Dekker.

Levi, W. M. (1957). *The pigeon*. Sumter, S. C.: Levi Publishing Co.

Levine, C. O. (1922). The water buffalo for dairy purposes, *Lingnaam Agricultural Review*. **1**: 1, 1-30.

Levring, T., H. A. Hoppe, and O. J. Schmid (1969). *Marine algae: A survey of research and utilization*. Hamburg: Cram, de Gruyter, and Co.

Levy, Marion J., Jr. (1963). *The family revolution in modern China*. New York: Octagon Books.

Lewicki, T. (1974). *West African food in the middle ages*. London: Cambridge University Press.

Lewis, Y. S. (1984). *Spices and herbs for the food industry*. Orpington, England: Food Trade Press.

Li, C. P. (1974). *Chinese herbal medicine*. DHEW Publication No. (N.I.H.) 75-732. Bethesda, Md.: John E. Fogarty International Center for Advanced Study in the Health Sciences, National Institutes of Health.

Li, C. P., A. Goldin, and J. L. Hartwell (1974). Antineoplastic substances from the sea: A review, *Cancer Chemotherapy*. **4**: 97-129.

Li, C. P., and Ruth Cheng Li (1973). An introductory note to ginseng, *American Journal of Chinese Medicine*. **1**: 249-61.

Li, Ch'iao-p'ing (1948). *The chemical arts of old China*. Easton, Pa.: Journal of Chemical Education.

Li, Chun (1965). Beekeeping in China, *Gleanings in Bee Culture*. **93**: 38-39.

Li, Chi (1977). *Anyang*. Seattle: University of Washington Press.

Li, Hui-lin (1942). The Araliaceae of China, *Sargentia*. **2**: 1-134.

_______ (1952). A taxonomic review of the genus Actinidia, *Journal of the Arnold Arboretum*. **33**: 1-61.

_______ (1959). *The garden flowers of China*. Chronica Botanica, Vol. 19. New York: Ronald Press.

_______ (1963a). *The origin and cultivation of shade and ornamental trees*. Philadelphia: University of Pennsylvania Press.

_______ (1963b). *Woody flora of Taiwan*. Philadelphia: Morris Arboretum; Narberth, Pa.: Livingston Publishing Company.

_______ (1969). The vegetables of ancient China, *Economic Botany*. **23**: 253-60.

_______ (1970). The origin of cultivated plants in Southeast Asia, *Economic Botany*. **24**: 3-19.

_______ (1979). *Nan-fang ts'ao-mu chuang: A fourth century flora of southeast China*. Hong Kong: Chinese University Press.

_______ (1983). The domestication of plants in China: Ecogeographical considerations, pp. 21-63 **In:** David N. Keightley (ed.), *The origins of Chinese civilization*. Berkeley, Los Angeles, London: University of California Press.

Li, Shu-fan (1964). *Hong Kong surgeon*. New York: E. P. Dutton & Co.

Liebig, Justus von (1878). *Chemische Briefe*. Sixth Edition. Leipzig and Heidelberg: Carl Winters.

Liener, Irvin E., and Madhusudan L. Kakade (1980). Protease inhibitors, pp. 7-71 **In:** Irvin E. Liener (ed.), *Toxic constituents of plant foodstuffs*. Second Edition. New York: Academic Press.

Liljestrand, S. H. (1933-34). A survey of endemic goitre, with special reference to the Gold Rivers District in the Chino-Tibetan border marches, *Journal of the West China Border Research Society*. **6**: 196-207.

"Lily flowers and bulbs used as food" (1889). *Bulletin of Miscellaneous Information, Royal Gardens, Kew*. 1889: 116-18.

Lin, Florence (1975). *Florence Lin's Chinese regional cookbook*. New York: Hawthorn Books.

Lin, Hsiang Ju, and Tsuifeng Lin (1972). *Chinese gastronomy*. New York: Pyramid Publications.

Lin, K. Y. (1939). The economics of Yao life, *Lingnan Science Journal*. **18**: 409-23.

Lin, S. Y. (1940a). Fish culture in ponds in the New Territories of Hong Kong, *Journal of the Hong Kong Fisheries Research Station*. **1**: 161-93.

_______ (1940b). The fishing industries of Hong Kong, *Journal of the Hong Kong Fisheries Research Station*. **1**: 5-160.

_______ (1940-41). Salt manufacture in Hong Kong, *Hong Kong Naturalist*. **10**: 34-38.

Lin, Yueh-hwa (1947). *The golden wing: A sociological study of Chinese familism*. London: Kegan Paul, Trench, Trübner & Co.

Lin, Yutang (1935). *My country and my people*. New York: Reynal and Hitchcock.

Lindqvist, K. (1960). On the origin of cultivated lettuce, *Hereditas*. **46**: 319-50.

Lindstedt, H. (1939). Food consumption habits in China, *International Review of Agriculture*. **30**: 363E-89E.

Ling, Stella, Janet King, and Virginia Leung (1975). Diet, growth, and cultural food habits in Chinese-American infants, *American Journal of Chinese Medicine.* **31**: 125-32.

Little, Archibald John (1898). *Through the Yang-tse gorges: Or trade and travel in western China.* Third and Revised Edition. London: Sampson Low, Marston and Company.

Liu, Bo, and Yun-sun Bau (1980). *Fungi pharmacopoeia* (Sinica). Oakland, Cal.: Kinoko Company.

Liu, En-lan (1937). Pootu, a lost island, *Economic Geography.* **13**: 132-38.

Liu, J. Heng, and C. K. Chu (1943). Problems of nutrition and dietary requirements in China, *Chinese Medical Journal.* **61**: 95-109.

Liu, S. H. (1940). The role of vitamin D in the calcium metabolism in osteomalacia, *Chinese Medical Journal.* **57**: 101-18.

Lo, Hsiang-lin (1967a). The southward expansion of Chinese civilization and the advancement of learning in Kwangtung Province, pp. 139-49 **In:** F. S. Drake (ed.), *Symposium on historical, archaeological, and linguistic studies on southern China, South-East Asia and the Hong Kong region.* Hong Kong: Hong Kong University Press.

_______ (1967b). The Yüeh bronze drums and their manufacture and use, pp. 110-14 **In:** F. S. Drake (ed.), *Symposium on historical, archaeological, and linguistic studies on southern China, South-East Asia and the Hong Kong region.* Hong Kong: Hong Kong University Press.

Lo, Kenneth H. C. (1971). *Peking cooking.* New York: Pantheon Books.

_______ (1972). *Chinese food.* Harmondsworth, Middlesex: Penguin Books.

_______ (1974). *Chinese vegetarian cooking.* New York: Pantheon Books.

_______ (1976). *Food the Chinese way.* Secaucus, N.J.: Chartwell Books.

_______ (1979a). *Chinese provincial cooking.* London: Elm Tree Books.

_______ (1979b). *The complete encyclopedia of Chinese cooking.* London: Octopus Books.

_______ (1979c). *The encyclopedia of Chinese cooking.* New York: A & W Publishers.

Löffler, Lorenz G. (1954). *Zur sakralen Bedeutung des Büffels und Gajals für Ahnen-kult, Ernteriten und soziale Feste im Gebiet Südostasiens.* Unpublished Ph.D. Dissertation, Johannes-Gutenberg-Universität, Mainz.

Loewenfeld, Claire, and Philippa Back (1978). *The complete book of herbs and spices.* Second, Revised Edition. Newton Abbot and London: David and Charles.

Loh, Bei-fang (1942). Aménagement et soins à donner aux oiseaux de cage en Chine, *Bulletin de l'Université l'Aurore.* **3 (Série 3)**: 643-84.

Lontoc, Aurea V., Evelyn O. Ladines, Julita G. Gonzales, and Concepcion A. Ayroso (1982). Nutritive value of some Philippine seaweeds. Part 1: Essential mineral elements, *Philippine Journal of Nutrition.* **35**: 220-30.

Low, W. D., S. T. Chan, K. S. F. Chang, and M. M. C. Lee (1964). Skeletal maturation of southern Chinese children, *Child Development.* **35**: 1313-36.

Lowe, Pardee (1943). *Father and glorious descendant.* Boston: Little, Brown and Company.

Lowenstein, Jerold M. (1982). Sea turtles, aphrodisiacs, and machismo, *Oceans.* **15**: 72.

Lu, Gwei-djen, and Joseph Needham (1951). A contribution to the history of Chinese dietetics, *Isis.* **42**: 13-20.

_______ (1976). Records of diseases in ancient China, *American Journal of Chinese Medicine.* **4**: 3-16.

Lu, Yü (1974). *The classic of tea.* Translated and introduced by Francis Ross Carpenter. Boston, Toronto: Little, Brown and Company.

Lynch, S. R., J. L. Beard, S. A. Dassenko, and J. D. Cook (1984). Iron absorption from legumes in humans, *American Journal of Clinical Nutrition.* **40**: 42-47.

McClendon, J. F. (1933). Iodine and goiter with especial reference to the Far East, *Journal of Biological Chemistry.* **102**: 91-99.

_______ (1939). *Iodine and the incidence of goiter.* Minneapolis: University of Minnesota Press.

McClure, F. A. (1933). Methods and materials of Chinese table plant culture, *Lingnan Science Journal.* **12**, Supplement: 119-49.

_______ (1938). Notes on bamboo culture, *Hong Kong Naturalist.* **9**: 4-18.

_______ (1956). Bamboo in the economy of oriental peoples, *Economic Botany.* **10**: 335-61.

McClure, F. A., and Ts'ui-mae Hwang (1934). The flora of a Canton herb shop, *Lingnan University Science Bulletin.* **6**: 1-32.

McCollum, G. D. (1976). Onion and allies: *Allium* (Liliaceae), pp. 186-190 **In:** N. W. Simmonds (ed.), *Evolution of crop plants.* London and New York: Longman.

McCrindle, John W. (1971). *Ancient India as described in classical literature.* Reprint of the 1901 edition. St. Leonards: Ad Orientem Ltd.; Amsterdam: Philo Press.

Macdonell, Arthur Anthony, and Arthur Berriedale Keith (1912). *Vedic Index of Names and Subjects.* London: John Murray. 2 vols.

Macgowan (1871-72). On the "mutton wine" of the Mongols and analogous preparations of the Chinese, *Journal of the North China Branch, Royal Asiatic Society (N.S.).* **7**: 237-40.

McIntosh, Susan Keech, and Roderick J. McIntosh (1981). West African prehistory, *American Scientist.* **69**: 602-13.

McKay, John W., and H. L. Crane (1953). Chinese chestnut - A promising new orchard crop, *Economic Botany.* **7**: 228-42.

McKinley, Earl Baldwin (1935). *A geography of disease.* Washington, D.C.: George Washington University Press.

MacLennan, R., J. Dalosta, N. E. Day, C. H. Law, Y. K. Ng, and Shanmugaratnam (1977). Risk factors for lung cancer in Singapore Chinese, a population with high female incidence rates, *International Journal of Cancer.* **20**: 854-60.

McMahan, Linda (1981). A summary of the biology and management of American ginseng, pp. 3-16 **In:** *Proceedings of the Third National Ginseng Conference.* Asheville, North Carolina. [Raleigh]: North Carolina Department of Agriculture.

Madan, C. L., B. M. Kapur, and U. S. Gupta (1966). Saffron, *Economic Botany.* **20**: 377-85.

Madlener, J. C. (1977). *The sea vegetable book.* New York: Clarkson N. Potter.

Mager, J., M. Chevion, and G. Glaser (1980). Favism, pp. 265-94 **In:** Irvin E. Liener (ed.), *Toxic constituents of plant foodstuffs.* Second Edition. New York: Academic Press.

Magno-Orejana, Florian (1983). Fermented fish products, pp. 255-95 **In:** Harvey T. Chan, Jr. (ed.), *Handbook of tropical foods.* New York and Basel: Marcel Dekker.

Maharidge, Dale (1985). Alabama blacks contend U.S. now the villain on voting rights, *Sacramento Bee*, Sunday, September 29, pp. A 1, 15, 16.

Mahindru, S. N. (1982). *Spices in Indian life.* New Delhi: Sultan Chand & Sons.

Major, R. H. (trans.) (1857a). Account of the journey of Hieronimo di Santo Stefano, in his *India in the fifteenth century.* Hakluyt Society Works, No. 22. London: Hakluyt Society.

_______ (trans.) (1857b). The travels of Nicolò Conti, in the East, in the early part of the fifteenth century, in his *India in the fifteenth century.* Hakluyt Society Works, No. 22. London: Hakluyt Society.

Malaya (Federation), Department of Statistics (1959). *Household budget survey of the Federation of Malaya, 1957-1958.* Kuala Lumpur: Department of Statistics, Federation of Malaya.

Mallory, Walter Hampton (1926). *China: Land of famine.* New York: American Geographical Society.

Manderson, Lenore (1987). Hot-cold food and medical theories: Overview and introduction, *Social Science and Medicine.* **25**: 329-30.

Manderson, Lenore, and Megan Mathews (1981). Vietnamese behavioral and dietary precautions during pregnancy, *Ecology of Food and Nutrition.* **11**: 1-8.

Mani, Vettam (1975). *Puranic encyclopaedia: A comprehensive dictionary with special reference to the epic and Puranic literature.* Delhi: Motilal Benarsidass.

Mao, Shou H. (1971). *Turtles of Taiwan.* Taipei: Commercial Press.

Maple, Eric (1980). *The secret lore of plants and flowers.* London: Robert Hale Limited.

Mar, Peter G., and B. E. Read (1936). Chemical examination of Chinese remedies for nightblindness, *Chinese Journal of Physiology.* **10**: 273-84.

Marriott, John, and Pamela A. Lancaster (1983). Bananas and plantains, pp. 85-143 **In**: Harvey T. Chan, Jr. (ed.), *Handbook of tropical foods.* New York and Basel: Marcel Dekker.

Marshall, A. J., and S. J. Folley (1956). The origin of nest-cement in edible-nest swiftlets (*Collocalia* spp.), *Proceedings of the Zoological Society of London.* **126**: 383-89.

Martin, Robert Montgomery (1838). *The history, antiquities, topography and statistics of eastern India.* London: Wm. H. Allen and Co. 3 vols.

Martin, Yan-kit (1984). *Chinese cooking: Step by step techniques.* New York: Random House.

Maspero, Henri (1981). *Taoism and Chinese religion.* Translated by Frank A. Kierman, Jr. Amherst: University of Massachusetts Press.

Massal, Emile, and Jacques Barrau (1956). *Food plants of the South Sea Islands.* Technical Paper No. 94. Noumea, New Caledonia: South Pacific Commission.

Matheson, Pamela (1951). Ascorbic acid in Chinese gooseberries, *Food Preservation Quarterly.* **11**: 34.

Mathew, B. F. (1980). *Crocus* L., pp. 92-99 in Vol. 5 of T. G. Tutin, V. H. Heywood, N. A. Burges, D. M. Moore, D. H. Valentine, S. M. Walters, and D. A. Webb (eds.), *Flora Europaea.* Cambridge: Cambridge University Press.

Mathew, Brian (1977). *Crocus sativus* and its allies (*Iridaceae*), *Plant Systematics and Evolution.* **128**: 89-103.

Mathews, L. (1980). Hong Kong is talking about a dogmeat reversal, *San Francisco Examiner & Chronicle*, February 3, Sunday Punch Section, p. 1.

Mathews, Megan, and Lenore Manderson (1981). Vietnamese behavioral and dietary precautions during confinement, *Ecology of Food and Nutrition.* **11**: 9-16.

Mathon, Claude-Charles (1981). *L'Origine des plantes cultivées.* Collection Écologie appliquée et sciences de l'environnement, No. 5. Paris: Masson.

Maxwell, J. Preston (1921). Intestinal parasitism in south Fukien, *China Medical Journal.* **35**: 377-82.

Maxwell, James L. (1929). Leprosy: The problem of its eradication from China, *Lingnan Science Journal.* **8**: 535-36.

_______ (1931). Paragonimiasis in China, *China Medical Journal.* **45**: 43-49.

May, Jacques M. (1961). *The ecology of malnutrition in the Far and Near East.* New York: Hafner Publishing Company.

Mayers, William Frederick (1874). *The Chinese reader's manual.* Shanghai: American Presbyterian Mission Press; London: Trubner and Co.

Maynard, L. A. (1936). The improvement of the diet of the Chinese farm family, *Chinese Medical Journal.* **50**: 425-33.

Maynard, Leonard A., and Wen-yuh Swen (1956). Nutrition, pp. 400-436 **In**: John Lossing Buck (ed.), *Land utilization in China.* Facsimile reprint of the 1937 edition published by the University of Nanking. New York: Council on Economic and Cultural Affairs.

Meacham, William (1983). Origins and development of the Yüeh Coastal Neolithic: A microcosm of culture change on the mainland of East Asia, pp. 147-175 **In**: David N. Keightley (ed.), *The origins of Chinese civilization.* Berkeley: University of California Press.

Meadow, R. H. (1981). Early animal domestication in South Asia: A first report on the faunal remains from Mehrgarh, Pakistan, pp. 143-69 **In**: Herbert Härtel (ed.), *South Asian Archaeology 1979.* Papers from the Fifth International Conference of the Association of South Asian Archaeologists in Western Europe, held in the Museum für Indische Kunst der Preussischer Kulturbesitz Berlin. Berlin: Dietrich Reimer Verlag.

Medway, Lord (1957) (see also Cranbrook, Earl of). Birds' nest collecting, *Sarawak Museum Journal (N.S.).* **8**: No. 10, 252-60.

_______ (1958). Birds' nesting among the Land Dayaks, *Sarawak Museum Journal (N.S.).* **8**: No. 11, 465-69.

_______ (1963). The antiquity of trade in edible birds'-nests, *Federation Museums Journal (N.S.).* **8**: 36-47.

_______ (1965-66). Field characters as a guide to the specific relations of swiftlets, *Proceedings of the Linnean Society of London.* **177**: 151-72.

_______ (1969). Studies on the biology of the edible-nest swiftlets of South-east Asia, *Malayan Nature Journal.* **22**: 57-63.

Medway, Lord, and J. D. Pye (1977). Echolocation and the systematics of swiftlets, pp. 225-38 **In**: Bernard Stonehouse and Christopher Perrins (eds.), *Evolutionary ecology.* Baltimore, London, Tokyo: University Park Press.

Mellander, Olaf (1973). Nutrition and health care in preschool children, *Journal of Tropical Pediatrics and Environmental Child Health.* **19**: 253-57.

Merrill, Elmer Drew (1954). *The botany of Cook's voyages.* Chronica Botanica, Vol. 14, No. 5-6. Waltham, Mass.: Chronica Botanica Company.

Mertz, Walter (1981). The essential trace elements, *Science.* **213**: 1332-38.

Messer, Ellen (1987). The hot and cold in Mesoamerican indigenous and Hispanicized thought, *Social Science and Medicine.* **25**: 339-46.

Mettlin, Curtis (1984). Epidemiologic studies on vitamin A and cancer, *Advances in Nutritional Research.* **6**: 47-65.

Meyer, Frank N. (1911). *Agricultural explorations in the fruit and nut orchards of China.* U.S. Department of Agriculture, Bureau of Plant Industry, Bulletin No. 204. Washington, D.C.: U.S. Government Printing Office.

_______ (1915). A field of ginger in Shantung Province, China, opp. p. 896 **In**: *Plant Immigrants.* U.S. Department of Agriculture, Bureau of Plant Industry, Office of Foreign Seed and Plant Introduction, Bulletin 110.

"Mice infestation" (1985). *Sacramento Bee*, Thursday, December 12, p. A 21.

Michanek, G. (1979). Seaweed resources for pharmaceutical uses, pp. 203-35 **In**: H. A. Hoppe, T. Levring, and Y. Tanaka (eds.), *Marine algae in pharmaceutical science.* Berlin and New York: Walter de Gruyter.

Mickey, Margaret Portia (1968). *The Cowrie Shell Miao of Kweichow.* Harvard University, Papers of the Peabody Museum of American Archaeology and Ethnology, Vol. 32, No. 1. Cambridge: Peabody Museum.

Milcu, Stephen (1960). The study of experimental goitre, pp. 279-313 **In**: *Endemic goitre.* WHO Monograph Series, No. 44. Geneva: World Health Organization.

Miller, Gloria Bley (1966). *The thousand recipe Chinese cookbook.* New York: Atheneum.

Miller, J. Innes (1969). *The spice trade of the Roman Empire.* Oxford: Clarendon Press.

"Millet - A possibly goitrogenic cereal" (1983). *Nutrition Reviews.* **41**: 113-16.

Millis, Jean (1953a). Factors influencing the birth weight of offspring, *Medical Journal of Malaya.* **7**: 168-78.

_______ (1953b). Gain in weight and length in the first year of life of Chinese infants born in Singapore in 1951, *Medical Journal of Australia.* **1**: 283-85.

_______ (1953c). Growth curves of Singapore-born Chinese infants during the first year, *Medical Journal of Malaya.* **7**: 296-301.

_______ (1954). The influence of nutrition on the growth rate in the first year of life of Chinese infants born in Singapore in 1951, *Medical Journal of Australia.* **1**: 322-26.

_______ (1955a). Some aspects of breast feeding in Singapore, *Medical Journal of Malaya.* **10**: 157-61.

_______ (1955b). The effect of economic background on body size of Chinese infants in the first year, *Journal of Tropical Pediatrics.* **1**: 156-60.

_______ (1956). The influence of breast feeding on weight gain in infants in the first year, *Journal of Pediatrics.* **48**: 770-75.

_______ (1958). Growth of preschool Chinese and southern Indian children in Singapore, *Medical Journal of Malaya.* **12**: 531-39.

_______ (1959). Distribution of birth weights of Chinese and Indian infants born in Singapore: Birth weight as an index of maturity, *Annals of Human Genetics*. **23**: 164-70.

Milne, Mrs Leslie (1924). *The home of an Eastern clan: A study of the Palaungs of the Shan States*. Oxford: Clarendon Press.

Mitchell, Robert D. (1971). The grass pea: Distribution, diet, and disease, *Yearbook of the Association of Pacific Coast Geographers*. **33**: 29-46.

Moir, Gordon (1909). Some methods of fishing in China, *Badminton Magazine of Sports and Pastimes*. **29**: 279-89.

Moldenke, Harold N., and Alma L. Moldenke (1952). *Plants of the Bible*. Waltham, Mass: Chronica Botanica Company.

Monachino, Joseph (1956). Chinese herbal medicine - Recent studies, *Economic Botany*. **10**: 42-48.

Moncrieff, Alexander (1890). Compressed or tablet tea, *Bulletin of Miscellaneous Information, Royal Gardens, Kew*. 1890: 109-12.

Moninger, M. M. (ed.) (1919). *The isle of palms: Sketches of Hainan*. Shanghai: Commercial Press, Ltd.

_______ (1932). The Hainanese Miao and their food supply, *Lingnan Science Journal*. **11**: 521-25.

Montanus, Arnoldus (1671). *Atlas Chinensis*. London: Printed by Tho. Johnson for the Author.

Morgan, Harry Titterton (1942). *Chinese symbols and superstitions*. South Pasadena, Cal.: P. D. and Ione Perkins.

Morgan, Julia, and A. S. Gault (1941). Study of an outbreak of scurvy. I. Clinical features, *Chinese Medical Journal*. **60**: 141-55.

Morris, H. H., M. S. Hwang, and P. T. Kuo (1940). Pellagra among the war refugees in Shanghai. Its associated deficiencies and nicotinic acid treatment, *Chinese Medical Journal*. **57**: 427-41.

Morse, W. J. (1947). The versatile soybean, *Economic Botany*. **1**: 137-47.

Morse, William R. (1926-29). The practices and principles of Chinese medicine, *Journal of the West China Border Research Society*. **3**: 82-104.

_______ (1934). *Chinese medicine*. New York: P. B. Hoeber.

Morton, Julia F. (1955). The emblic (*Phyllanthus emblica* L.): A rich but neglected source of vitamin C, *Proceedings of the Florida State Horticultural Society*. **68**: 315-21.

Moseley, George V. H., III (1973). *The consolidation of the South China border*. Berkeley, Los Angeles, London: University of California Press.

Mosher, Steven W. (1983a). *Broken earth: The rural Chinese*. New York: Free Press; London: Collier-Macmillan Publishers.

_______ (1983b). Why are baby girls being killed in China?, *Wall Street Journal (Western Edition)*, Monday, July 25, p. 13.

Mote, Frederick W. (1977). Yüan and Ming, pp. 193-257 **In:** K. C. Chang (ed.), *Food in Chinese culture*. New Haven and London: Yale University Press.

Moule, A. C. (1937). The name *Ginkgo biloba* and other names of the tree, *T'oung Pao*. **33**: 193-219.

Muenscher, Walter Conrad (1944). *Aquatic plants of the United States*. Ithaca, New York: Comstock Publishing Company.

Mukherjee, Sunil Kumar (1953). The mango - Its botany, cultivation, uses and future improvement, especially as observed in India, *Economic Botany*. **7**: 130-62.

_______ (1972). Origin of mango (*Mangifera indica*), *Economic Botany*. **26**: 260-64.

Mulinos, Michael G., and George B. Jerzy Glass (1953). The treatment of constipation with a new hydrasorbent material derived from kelp, *Gastroenterology*. **24**: 385-93.

Munro, R. H. (1977). Comeback of the Chinese pet dog, *San Francisco Examiner & Chronicle*, April 10, Sunday Punch Section, p. 5.

Murai, Mary, Florence Pen, and Carey D. Miller (1958). *Some tropical South Pacific foods*. Honolulu: University of Hawaii Press.

Murdock, George P. (1959). *Africa: Its peoples and their culture history*. New York, Toronto, London: McGraw-Hill.

Mustard, Margaret J. (1952). Ascorbic acid content of some miscellaneous tropical and subtropical plants and plant products, *Food Research*. **17**: 31-35.

Muthmann, Friedrich (1982). *Der Granatapfel: Symbol des Lebens in der Alten Welt*. Schriften der Abegg-Stiftung Bern, Band 6. [Fribourg]: Office du livre.

Nakayama, T. O. M., and C. O. Chichester (1963). Astringency of persimmons (*Diospyros kaki*, L.), *Nature*. **199**: 72-73.

National Research Council (1984). *Amaranth: Modern prospects for an ancient crop*. Washington, D.C.: National Academy Press.

Nauriyal, J. P., Rajendra Gupta, and C. K. George (1976-77). Saffron in India, *Arecanut and Spices Bulletin*. **8**: 59-72.

Nayar, N. M. (1973). Origin and cytogenetics of rice, *Advances in Genetics*. **17**: 153-292.

Nayar, N. M., and K. L. Mehra (1970). Sesame: Its uses, botany, cytogenetics, and origin, *Economic Botany*. **24**: 20-31.

Needham, Joseph, and Gwei-djen Lu (1962). Hygiene and preventive medicine in ancient China, *Journal of the History of Medicine and Allied Sciences*. **17**: 429-78.

_______ (1985). *Trans-Pacific echoes and resonances; Listening once again.* Singapore and Philadelphia: World Scientific.

_______ (1986). *Botany*, Part 1 of Vol. 6 (Biology and biological technology) of Joseph Needham, *Science and civilisation in China.* Cambridge: Cambridge University Press.

Needham, Joseph, and Wang Ling (1956). *History of scientific thought*, Vol. 2 of Joseph Needham, *Science and civilisation in China.* Cambridge: Cambridge University Press.

Needham, Joseph, Ping-yü Ho and Gwei-djen Lu (1980). *Spagyrical discovery and invention: Apparatus, theories and gifts*, Part 4 of Vol. 5 (Chemistry and chemical technology) of Joseph Needham, *Science and civilisation in China.* Cambridge: Cambridge University Press.

Nelson, H. G. H. (1974). Ancestor worship and burial practices, pp. 251-77 **In:** Arthur P. Wolf (ed.), *Religion and ritual in Chinese society.* Stanford: Stanford University Press.

Newcomb, Robert M. (1963). Botanical source-areas for some oriental spices, *Economic Botany.* **17**: 127-32.

Newman, Jacqueline M., Isobel Contento, and Elaine Kris (1981). Perspectives on food and nutrition in the People's Republic of China, *Journal of Nutrition Education.* **13:** 43-45.

Nichols, John Treadwell (1943). *The fresh-water fishes of China.* Vol. 9 of Ruth Tyler (ed.), *Natural history of Central Asia.* New York: American Museum of Natural History.

Nichols, John Treadwell, and Clifford H. Pope (1927). The fishes of Hainan, *Bulletin of the American Museum of Natural History.* **54**: 321-94.

Niehoff, Arthur, and Natalie Meister (1972). The cultural characteristics of breast-feeding: A survey, *Journal of Tropical Pediatrics and Environmental Child Health.* **18**: 16-20.

Nieuhoff, John (1673). *An embassy from the East-India Company of the United Provinces, to the Grand Tatar Cham Emperor of China, delivered by their excellencies Peter de Goyer and Jacob de Keyzer, at his imperial city of Peking.* Second Edition. London: Printed for the Author at his house in White-Friers.

Nisbet, John (1901). *Burma under British rule - and before.* Westminster: Archibald Constable and Co. 2 vols.

Nisizawa, Kazutosi (1979). Pharmaceutical studies on marine algae in Japan, pp. 243-64 **In:** H. A. Hoppe, T. Levring, and Y. Tanaka (eds.), *Marine algae in pharmaceutical science.* Berlin and New York: Walter de Gruyter.

Noehden, George Henry (1827). Account of the banyan-tree, or Ficus indica, as found in the ancient Greek and Roman authors, *Transactions of the Royal Asiatic Society of Great Britain and Ireland.* **1**: 119-32.

Norman, B. (1972). *Tales of the table: A history of Western cuisine.* Englewood Cliffs, N.J.: Prentice-Hall.

Norris, E. R., M. K. Simeon, and H. B. Williams (1937). The vitamin B and vitamin C content of marine algae, *Journal of Nutrition.* **13**: 425-33.

Nutrition Reviews (1976). *Present knowledge in nutrition.* Fourth Edition. New York and Washington, D.C.: The Nutrition Foundation.

Ochse, J. J., and R. C. Bakhuizen van den Brink (1931a). *Fruits and fruitculture in the Dutch East Indies.* Batavia: G. Kolff and Co.

_______ (1931b). *Vegetables of the Dutch East Indies.* The Hague: Martinus Nijhoff. 4 vols.

Ochse, J. J., M. J. Soule, Jr., M. J. Dijkman, and C. Wehlburg (1961). *Tropical and subtropical agriculture.* New York: Macmillan. 2 vols.

Odaka, Kunio (1950). *Economic organization of the Li tribes of Hainan Island.* Yale Southeast Asia Studies, Translation Series. New Haven: Yale University.

O'Gorman, D. A. (1893). Ginseng and its cultivation in Corea, *Journal of the Manchester Geographical Society.* **9**: 262-63.

Oiso, Toshio (1976). History of food and diet in Japan, *Progress in Food and Nutrition Science.* **2**: 35-48.

Oldt, Frank (1924-25). Night soil storage in relation to hookworm viability, *Lingnaam Agricultural Review.* **2**: 90-99.

_______ (1926). Night-soil as fertilizer, *China Medical Journal.* **40**: 1059-70.

Oliphant, L. (1860). *Narrative of the Earl of Elgin's mission to China and Japan in the years 1857, '58, '59.* New York: Harper and Brothers.

Olmo, H. P. (1976). Grapes: *Vitis, Muscadini* (Vitaceae), pp. 294-98 **In:** N. W. Simmonds (ed.), *Evolution of crop plants.* Longman: London and New York.

Orleans, Leo A. (1973). Population dynamics, pp. 193-213 **In:** Joseph R. Quinn (ed.), *Medicine and public health in the People's Republic of China.* DHEW Publication No. (N.I.H.) 73-67. Bethesda, Md.: Fogarty International Center, National Institutes of Health.

Orta, Garcia da (1913). *Colloquies on the simples and drugs of India.* New Edition (Lisbon, 1895), translated by Clements Markham. London: Henry Southeran and Company.

Osgood, Cornelius (1963). *Village life in old China: A community study of Kao Yao, Yünnan.* New York: Ronald Press Company.

Ossendowski, Ferdinand (1924). *Man and mystery in Asia.* New York: E. P. Dutton & Company.

Padmanaban, G. (1980). Lathyrogens, pp. 239-63 **In:** Irvin E. Liener (ed.), *Toxic constituents of plant foodstuffs.* Second Edition. New York: Academic Press.

Pal, Mohinder, and T. N. Khoshoo (1974). Grain amaranths, pp. 129-37 **In:** J. Hutchinson (ed.), *Evolutionary studies in world crops.* London: Cambridge University Press.

Palladius, Archimandrite (1876). Elucidations of Marco Polo's travels in North-China. Drawn from Chinese sources, *Journal of the North China Branch, Royal Asiatic Society (N.S.).* **10**: 1-53.

Palmieri, Richard Pietro (1976). *Domestication and exploitation of livestock in the Nepal Himalaya and Tibet: An ecological, functional, and culture historical study of yak and yak hybrids in society, economy, and culture.* Unpublished Ph.D. Dissertation in Geography, University of California, Davis.

Pálos, Stephen (1971). *The Chinese art of healing.* New York: Herder and Herder.

Parker, E. H. (1897). Diet and medicine in China, *Cornhill Magazine.* **2**: 175-83.

Parker, Edward Harper (1903). *China past and present.* London: Chapman and Hall.

Parkinson, S. (1984). The contribution of aroids in the nutrition of people in the South Pacific, pp. 215-24 **In:** S. Chandra (ed.), *Edible aroids.* Oxford: Clarendon Press.

Parry, John W. (1945). *The spice handbook.* Brooklyn, N.Y.: Chemical Publishing Company.

_______ (1969). *Spices.* New York: Chemical Publishing Company. 2 vols.

Parsons, James J. (1962). *The green turtle and man.* Gainesville: University of Florida Press.

Pasour, E. C., Jr. (1985). Kudzu: The government gift that keeps on growing, *Wall Street Journal (Western Edition),* Monday, September 30, p. 14.

Passmore, R., B. M. Nicol, M. Narayana Rao, G. H. Beaton, and E. M. DeMaeyer (1974). *Handbook on human nutritional requirements.* Geneva: World Health Organization.

Passwater, Richard A. (1980). *Selenium as food and medicine.* New Canaan, Conn.: Keats Publishing.

Pearson, Richard (1983). The Ch'ing-lien-kang culture and the Chinese Neolithic, pp. 119-45 **In:** David N. Keightley (ed.), *The origins of Chinese civilization.* Berkeley: University of California Press.

Peck, Richard E., Margaret Chuang, Gordon E. Robbins, and Milton Z. Nichaman (1981). Nutritional status of Southeast Asian refugee children, *American Journal of Public Health.* **71**: 1144-48.

Penyapol, A. (1958). A preliminary study of the sea turtles in the Gulf of Thailand, *Natural History Bulletin of the Siam Society.* **17**: 23-36.

Percival, William Spencer (1889). *Land of the dragon.* London: Hurst and Blackett, Limited.

Pereira, R. A. (1937). Chinese domesticated ducks, *Hong Kong Naturalist.* **8**: 159-65.

Perry, Lily M. (1980). *Medicinal plants of East and Southeast Asia: Attributed properties and uses.* Cambridge and London: MIT Press.

Peterson, Wendy (1973). *The hot-cold food concept: An introduction,* Seminar Report in Geography, University of California, Davis. 22 pp.

Phillimore, J. S. (trans.) (1912). *Philostratus: In honour of Appolonius of Tyana.* Oxford: Clarendon Press. 2 vols.

Phillips, Henry (1827). *History of cultivated vegetables.* Second Edition. London: Henry Colburn. 2 vols.

Phillips, John C. (1923). *A natural history of the ducks.* London: Longmans, Green and Company. 4 vols.

Phillipson, Beverley E., Douglas W. Rothrock, William E. Connor, William S. Harris, and D. Roger Illingworth (1985). Reduction of plasma lipids, lipoproteins, and apoproteins by dietary fish oils in patients with hypertriglyceridemia, *New England Journal of Medicine.* **312**: 1210-16.

Pillsbury, Barbara L. K. (1978). "Doing the month": Confinement and convalescence of Chinese women after childbirth, *Social Science and Medicine.* **12**: 11-22.

Pimental, David, Wes Jackson, Marty Bender, and Walter Pickett (1986). Perennial grains: An ecology of new crops, *Interdisciplinary Science Reviews.* **11**: 42-49.

Piper, Charles V., and William J. Morse (1923). *The soybean.* New York and London: McGraw-Hill Book Company.

Piper, Majorie Ray (1970). *Sunset oriental cook book.* Menlo Park, Cal.: Lane Books.

Pirazzoli-t'Serstevens, Michèle (1982). *The Han civilization of China.* Translated by Janet Seligman. Oxford: Phaidon.

Platt, B. S. (1936). An approach to the problems of infant nutrition in China, *Chinese Medical Journal.* **50**: 410-24.

_______ (1962). *Tables of representative values of foods commonly used in tropical countries.* Privy Council, Medical Research Council, Special Report Series, No. 302. London: Her Majesty's Stationery Office.

Platt, B. S., and S. Y. Gin (1938). Chinese methods of infant feeding and nursing, *Archives of Disease in Childhood.* **13**: 343-54.

Plucknett, D. L. (1976). Edible aroids: *Alocasia, Colocasia, Cyrtosperma, Xanthosoma* (Araceae), pp. 10-12 **In:** N. W. Simmonds (ed.), *Evolution of crop plants.* London and New York: Longman.

Pollard, S. (1921). *In unknown China.* Philadelphia: J. B. Lippincott Company.

Polunin, Nicholas V. C., and Njoman Sumertha Nuitja (1981). Sea turtle populations of Indonesia and Thailand, pp. 353-62 **In:** Karen A. Bjorndal (ed.), *Biology and conservation of sea turtles. Proceedings of the World Conference on Sea Turtle Conservation, Washington, D.C., 26-30 November 1979.* Washington, D.C.: Smithsonian Institution Press.

Pool, Robert (1987). Hot and cold as an explanatory model: The example of Bharuch District in Gujarat, India, *Social Science and Medicine.* **25**: 389-99.

Pope, Clifford H. (1935). *The reptiles of China.* Natural History of Central Asia, Vol. 10. New York: American Museum of Natural History.

Popenoe, Wilson (1920). *Manual of tropical and subtropical fruits.* New York: Macmillan.

Porkert, Manfred (1974). *The theoretical foundations of Chinese medicine: Systems of correspondence.* Cambridge, Mass., and London, England: MIT Press.

Portères, Roland (1976). African cereals: *Eleusine*, fonio, black fonio, Teff, *Brachiaria, Paspalum, Pennisetum,* and African rice, pp. 409-52 **In:** J. R. Harlan, J. M. J. de Wet, and A. B. L. Stemler (ed.), *Origins of African plant domestication.* The Hague and Paris: Mouton Publishers.

Porterfield, W. M. (1922). References to the algae in the Chinese classics, *Bulletin of the Torrey Botanical Club.* **49**: 297-300.

_______ (1933). Bamboo, the universal provider, *The Scientific Monthly.* **36**: 176-83.

_______ (1938a). Bamboo as it is grown for food in China, *Journal of the New York Botanical Garden.* **39**: 282-88.

_______ (1938b). Ko, the kudzu vine, provides food, shade, clothes, and medicine, *Journal of the New York Botanical Garden.* **39**: 203-5.

_______ (1938c). The Chinese water-chestnut, Scirpus tuberosus, *Journal of the New York Botanical Garden.* **39**: 32-38.

_______ (1938d). Trapa bicornis, a water-chestnut known to the Chinese as ling ko, *Journal of the New York Botanical Garden.* **39**: 134-38.

_______ (1939a). Chinese ginger, *Journal of the New York Botanical Garden.* **40**: 208-11.

_______ (1939b). The yam-bean as a source of food in China, *Journal of the New York Botanical Garden.* **40**: 107-8.

_______ (1940a). Seeds of the ginkgo, *Journal of the New York Botanical Garden.* **41**: 185-88.

_______ (1940b). The arrowhead as a food among the Chinese, *Journal of the New York Botanical Garden.* **41**: 45-47.

_______ (1941a). Lotus plants for ornament, food, and medicine, *Journal of the New York Botanical Garden.* **42**: 280-86.

_______ (1941b). The soybean, *Journal of the New York Botanical Garden.* **42**: 181-89.

_______ (1943). Luffas as they are used by the Chinese, *Journal of the New York Botanical Garden.* **44**: 134-38.

_______ (1944). Taro, ancient food plant of tropical regions, *Journal of the New York Botanical Garden.* **45**: 11-16.

_______ (1951). The principal Chinese vegetable foods and food plants of Chinatown markets, *Economic Botany.* **5**: 3-37.

Portugal, T. R., E. O. Ladines, S. S. Ardeña, L. Resurreccion, C. R. Medina, and P. M. Matibag (1983). Nutritive value of some Philippine seaweeds. Part II. Proximate, amino acid, and vitamin composition, *Philippine Journal of Nutrition.* **36**: 166-72.

Powell, John B. (1945). *My twenty-five years in China.* New York: Macmillan Company.

Prakash, Om (1961). *Food and drinks in ancient India.* Delhi: Munshi Ram Manohar Lal.

"The prevalence of infanticide in China" (1886). *Journal of the North China Branch, Royal Asiatic Society (N.S.).* **20**: 25-50.

Price, Frank W. (1948). *The rural church in China.* New York: Agricultural Missions.

Pritchard, Peter C. H., and Pedro Trebbau (1984). *The turtles of Venezuela.* [Oxford, Ohio]: Society for the Study of Amphibians and Reptiles.

"Production of fruit in Fukien Province" (1928). *Chinese Economic Journal.* **3**: 1022-35.

Pruthi, J. S. (1976). *Spices and condiments.* New Delhi: National Book Trust, India.

"P'u-êrh tea" (1889). *Bulletin of Miscellaneous Information, Royal Gardens, Kew.* 1889: 118-20, 139-42.

Pulleyblank, E. G. (1983). The Chinese and their neighbors in prehistoric and early historic times, pp. 411-66 **In:** David N. Keightley (ed.), *The origins of Chinese civilization.* Berkeley: University of California Press.

Purseglove, J. W. (1968). *Tropical crops: Dicotyledons.* New York: John Wiley & Sons. 2 vols.

_______ (1975). *Tropical crops: Monocotyledons.* London: Longman.

_______ (1976). Millets: *Eleusine coracana, Pennisetum americanum* (Gramineae), pp. 91-93 **In:** N. W. Simmonds (ed.), *Evolution of crop plants.* London and New York: Longman.

Raibaudi, M. (1931). La badiane, *Parfums de France.* **9**: 95-107.

Rajagopalan, R., B. R. Baliga, and J. V. Bhat (1960). Saffron as a possible source of riboflavin, *Proceedings of the National Institute of Sciences of India.* **26**, Part A, Supplement 1: 128-34.

Randhawa, M. S., and Prem Nath (1959). *Farmers of India. Vol. 1.* New Delhi: Indian Council of Agricultural Research.

Randoin, Lucie, and Jeanne Boisselot (1941). Détermination de la valeur antiscorbutique du fruit comestible d'*Actinidia chinensis* Planchon, source exceptionnellement riche en vitamine C, *Comptes Rendus des Séances et Mémoires de la Société de Biologie et de Ses Filiales (Paris).* **135**: 209-12.

Rao, G. N. S. (1965). *Use of seaweeds directly as human food.* Indo-Pacific Fisheries Council, Regional Studies, No. 2. Bangkok: FAO Regional Office for Asia and the Far East.

Rattenbury, Harold B. (1944). *China, my China.* London: Frederick Muller.

Raveret-Wattel, C. (1884). Extraits des procès-verbaux des séances de la société. Séance générale du 13 Juin 1884, *Bulletin de la Société Nationale d'Acclimatation de France.* **1 (Série 4)**: 599-607.

Rawski, Evelyn Sakakida (1972). *Agricultural change and the peasant economy of South China.* Harvard East Asian Series, No. 66. Cambridge: Harvard University Press.

Ray, Joges-Chandra (1933). I. Food and drink in ancient India, *Man in India.* **13**: 217-39.

_______ (1934). II. Food and drink in ancient India, *Man in India.* **14:** 15-38.

Ray, Priyadaranjan, and Hirenda Nath Gupta (1965). *Caraka Samhita* (A Scientific Synopsis). New Delhi: National Institute of Sciences of India.

Ray, Richard, and Lance Walheim (1980). *Citrus: How to select, grow and enjoy.* Tucson: HP Books.

Read, Bernard E. (1921). Some factors controlling the food supply in China, *China Medical Journal.* **35**: 1-7.

_______ (1929). A comparison of the materia medica of India and China, *Lingnan Science Journal.* **8**: 423-40.

_______ (1936). *Chinese medicinal plants from the Pen Ts'ao Kang Mu.* Third edition of a botanical, chemical and pharmacological reference list. Peking: Peking Natural History Bulletin.

_______ (1940). Insects used in Chinese medicine, *Journal of the North China Branch, Royal Asiatic Society (N.S.).* **71**: 22-32.

_______ (1976). *Chinese materia medica: Animal drugs.* Chinese Medicine Series, Vol. 4. Reprinted from the Peking Natural History Bulletin, 1931. Taipei: Southern Materials Center.

_______ (1977a). *Chinese materia medica: Avian drugs.* Chinese Medicine Series, Vol. 3. Reprinted from the Peking Natural History Bulletin, 1932. Taipei: Southern Materials Center.

_______ (1977b). *Chinese materia medica: Dragon and snake drugs.* Chinese Medicine Series, Vol. 2. Reprinted from the Peking Natural History Bulletin, 1934. Taipei: Southern Materials Center.

_______ (1977c). *Chinese materia medica: Fish drugs.* Chinese Medicine Series, Vol. 2. Reprinted from the Peking Natural History Bulletin, 1939. Taipei: Southern Materials Center.

_______ (1977d). *Chinese materia medica: Insect drugs.* Chinese Medicine Series, Vol. 2. Reprint of a monograph published by the Peking Natural History Bulletin, 1941. Taipei: Southern Materials Center.

_______ (1977e). *Chinese materia medica: Turtle and shellfish drugs.* Chinese Medicine Series, Vol. 3. Reprinted from the Peking Natural History Bulletin, 1937. Taipei: Southern Materials Center.

_______ (1977f). *Common food fishes of Shanghai.* Chinese Medicine Series, Vol. 6. Reprint of monograph published in Shanghai, 1939. Taipei: Southern Materials Center.

_______ (1977g). *Famine foods listed in the Chiu Huang Pen Ts'ao.* Chinese Medicine Series, Vol. 6. Reprint of monograph published in Shanghai, 1946. Taipei: Southern Materials Center.

Read, Bernard E., and G. K. How (1927). The iodine, arsenic, iron, calcium and sulphur content of Chinese medicinal algae, *Chinese Journal of Physiology.* **1**: 99-108.

Read, Bernard E., Lee Wei Yung, and Ch'eng Jih Kuang (1940). *Shanghai foods.* Second Edition. Shanghai: Chinese Medical Association.

Read, Bernard E., and W. Wagner (1940). Shanghai vegetables, *China Journal.* **33**: 206-20, 259-71.

Rebel, Thomas P. (1974). *Sea turtles and the turtle industry of the West Indies, Florida, and the Gulf of Mexico.* Revised Edition. Coral Gables, Fla.: University of Miami Press.

Recent archaeological discoveries in the People's Republic of China (1984). Paris: United Nations Educational, Scientific and Cultural Organization; Tokyo: Centre for East Asian Cultural Studies.

Redgrove, H. Stanley (1933). *Spices and condiments.* London: Sir Isaac Pitman and Sons.

Regmi, Puskal Prasad (1982). *An introduction to Nepalese food plants.* Kathmandu: Royal Nepal Academy.

"Regulations governing sale of dog-meat promulgated by Taipei police" (1960). *China Post,* January 27.

Rehder, Alfred (1914-15). Synopsis of the Chinese species of Pyrus, *Proceedings of the American Academy of Arts and Sciences.* **50**: 225-41.

_______ (1940). *Manual of cultivated trees and shrubs hardy in North America.* New York: Macmillan.

_______ (1949). *Bibliography of cultivated trees and shrubs hardy in the cooler temperate regions of the Northern Hemisphere.* Jamaica Plain, Mass.: Arnold Arboretum of Harvard University.

Reichelt, Karl Ludvig (1968). *Truth and tradition in Chinese Buddhism.* Reprint of the Second Edition, 1928. Translated by Kathrina van Wagenen Bugge. New York: Paragon Book Reprint Corp.

Reid, E. (1936). The fluorine content of some Chinese food materials, *Chinese Journal of Physiology.* **10**: 259-72.

Reiss, F. (1936). A contribution to the cutaneous manifestations of vitamin A deficiency, *Chinese Medical Journal.* **50**: 945-48.

Renfrew, Jane M. (1973). *Palaeoethnobotany: The prehistoric food plants of the Near East and Europe.* London: Methuen and Co.

"Report of the American schistosomiasis delegation to the People's Republic of China" (1977). *American Journal of Tropical Medicine and Hygiene.* **26**: 427-62.

Reynolds, Philip K., and C. Y. Fang (1940-41). The banana in Chinese literature, *Harvard Journal of Asiatic Studies.* **5**: 165-81.

Richard, C., and Nguyèn-Thi-Làu (1961). Les oeufs de tortue de mer (*Chelonia mydas*), aliment traditionnel vietnamien: Composition chimique et valeur alimentaire, *Revue d'Élevage et de Médicine vétérinaire des Pays tropicaux.* **14**: 329-35.

Rick, Charles M. (1976). Tomato: *Lycopersicon esculentum* (Solanaceae), pp. 268-73 **In:** N. W. Simmonds (ed.), *Evolution of crop plants.* London and New York: Longman.

Riedel, C. George (1932). Modern bee culture and the possibilities of its development in China, *Chinese Economic Journal.* **11**: 274-81.

Riedel, Dietmar (1961). World fisheries, pp. 41-75 **In:** Vol. 1 of Georg Borgstrom (ed.), *Fish as food.* New York and London: Academic Press.

Rinaldi, E., C. Costagliola, L. Albini, G. de Rosa, G. Auricchio, B. de Vizia, and S. Auricchio (1984). High frequency of lactose absorbers among adults with idiopathic senile and presenile cataract in a population with a high prevalence of primary adult lactose malabsorption, *Lancet.* **1**: 355-57.

Roach, F. A. (1985). *Cultivated fruits of Britain: Their origin and history.* Oxford: Basil Blackwell.

Roberts, Emmanuel (1931). *Vegetable materia medica of India and Ceylon.* Colombo: Plâté Limited.

Robertson, R. C. (1936). The transmission in China of helminths by vegetables, *Chinese Medical Journal.* **Supplement 1**: 418-22.

Robinson, Roy (1984). Cat, pp. 217-25 **In:** Ian L. Mason (ed.), *Evolution of domesticated animals.* London and New York: Longman.

Roche, Jean, and Serge Lissitzky (1960). Etiology of endemic goitre, pp. 351-68 **In:** *Endemic goitre.* WHO Monograph Series, No. 44. Geneva: World Health Organization.

Rockhill, W. Woodville (trans.) (1884). *The life of the Buddha and the early history of his order. Derived from Tibetan works in the Bkah-Hgyur and Bstan-Hgyur.* London: Trübner and Company.

Roi, Jacques (1955). *Traité des plantes médicinales chinoises.* Encyclopédie Biologique, No. 47. Paris: Paul Lechevalier.

Rondot, Natalis (1848). *Étude pratique du commerce d'exportation de la Chine.* Paris: Librairie du Commerce, Chez Renard.

Rosengarten, Frederic, Jr. (1969). *The book of spices.* Philadelphia: Livingston Publishing Company.

_______ (1984). *The book of edible nuts.* New York: Walker and Company.

Ross, Edward Alsworth (1911). *The changing Chinese.* New York: Century Company.

Rowe, David Nelson, and Willmoore Kendall (eds.) (1954). *China: An area manual.* Chevy Chase, Md.: Operations Research Office, Johns Hopkins University. 2 vols.

Roxburgh, William (1820-24). *Flora Indica, or descriptions of Indian plants.* Edited by William Carey. Serampore: Mission Press. 2 vols.

Roy, Pratap Chandra (trans.) (1919-30). *The Mahabharata of Krishna-Dwaipayana Vyasa.* Calcutta: Datta Bose and Co. 11 vols.

Roy, Sachin (1960). *Aspects of Padam-Minyong culture.* Shillong: North-East Frontier Agency.

Rozin, Elisabeth (1983). *Ethnic cuisine: The flavor principle cookbook.* Brattleboro, Vt., and Lexington, Mass.: Stephen Greene Press.

Rozin, Paul (1978). The use of characteristic flavorings in human culinary practice, pp. 101-27 **In:** Charles M. Apt (ed.), *Flavor: Its chemical, behavioral, and commercial aspects. Proceedings of the Arthur D. Little, Inc., Flavor Symposium, 1977.* Boulder, Colo.: Westview Press.

Rozin, Paul, and Deborah Schiller (1980). The nature and acquisition of a preference for chili pepper by humans, *Motivation and Emotion.* **4**: 77-101.

Rozin, Paul, and T. A. Vollmecke (1986). Food likes and dislikes, *Annual Review of Nutrition.* **6**: 433-56.

Rozin, Paul, Lori Ebert, and Jonathan Schull (1982). Some like it hot: A temporal analysis of hedonic responses to chili pepper, *Appetite: Journal for Intake Research.* **3**: 13-22.

Ruttle, Jack (1986). New fruit of the north, *Rodale's Organic Gardening.* **33**: 4 (April), 64-67.

Ryder, Edward J., and Thomas W. Whitaker (1976). Lettuce: *Lactuca sativa* (Compositae), pp. 39-41 **In:** N. W. Simmonds (ed.), *Evolution of crop plants.* London and New York: Longman.

Sabban, Françoise (1986a). Court cuisine in fourteenth-century Imperial China: Some culinary aspects of Hu Sihui's *Yinshan zhengyao, Food and Foodways.* **1**: 161-96.

_______ (1986b). Un savoir-faire oublié: le travail du lait en Chine ancienne, *Zinbun: Memoirs of the Research Institute for Humanistic Studies - Kyoto University.* **21**: 31-65.

Sacks, Frank M. (1977). A literature review of *Phaseolus angularis* - the adsuki bean, *Economic Botany.* **31**: 9-15.

Sakai, William S. (1983). Aroid root crops: *Alocasia, Cyrtosperma,* and *Amorphophallus,* pp. 29-83 **In:** Harvey T. Chan, Jr. (ed.), *Handbook of tropical foods.* New York and Basel: Marcel Dekker.

Sakamoto, Nobuko (1977). *The People's Republic of China cookbook.* New York: Random House.

Salaman, Redcliffe N. (1949). *The history and social influence of the potato.* Cambridge: Cambridge University Press.

Samagalski, Alan, and Michael Buckley (1984). *China: A travel survival kit.* South Yarra, Victoria, Australia; and Berkeley, Cal.: Lonely Planet Publications.

Sammons, Thomas (1918). Production of popped rice in China, *U.S. Department of Commerce, Bureau of Foreign and Domestic Commerce, Commerce Reports.* **1**: 182.

Sampson, Theos. (1869a). Grapes in China, *Notes and Queries on China and Japan.* **3**: 50-54.

_______ (1869b). The banyan or yung tree, *Notes and Queries on China and Japan.* **3**: 72-73.

_______ (1869c). The p'u-t'i tree, *Notes and Queries on China and Japan.* **3**: 100-105.

Sarkar, K. R., B. K. Mukherjee, D. Gupta, and H. K. Jain (1974). Maize, pp. 121-27 **In:** Joseph Hutchinson (ed.), *Evolutionary studies in world crops.* London: Cambridge University Press.

Sass, Lorna J. (1981). Religion, medicine, politics and spices, *Appetite.* **2**: 7-13.

Sauer, Carl O. (1952). *Agricultural origins and dispersals.* New York: American Geographical Society.

_______ (1963). *Plant and animal exchanges between the Old and the New Worlds.* Notes from a seminar presented by C. O. Sauer, edited and compiled by Robert M. Newcomb. Los Angeles: R. M. Newcomb, Los Angeles State College.

Sauer, Jonathan D. (1950). Amaranths as dye plants among the Pueblo peoples, *Southwestern Journal of Anthropology.* **6**: 412-15.

_______ (1957). Recent migration and evolution of the dioecious amaranths, *Evolution.* **11**: 11-31.

_______ (1967). The grain amaranths and their relatives: A revised taxonomic and geographic survey, *Annals of the Missouri Botanical Garden.* **54**: 103-37.

_______ (1971). A reevaluation of the coconut as an indicator of human dispersal, pp. 309-19 **In:** Carroll L. Riley, J. Charles Kelley, Campbell W. Pennington, and Robert L. Rands (eds.), *Man across the sea: Problems of pre-Columbian contacts.* Austin and London: University of Texas Press.

_______ (1976). Grain amaranths: *Amaranthus* spp. (Amaranthaceae), pp. 4-6 **In:** N. W. Simmonds (ed.), *Evolution of crop plants.* London and New York: Longman.

_______ (1977). The grain amaranths and their relatives: A revised taxonomic and geographic survey, pp. 13-24 **In:** *Amaranth round-up.* Emmaus, Pa.: Rodale Press.

_______ (1983). *Cocos nucifera* (Coco, Cocotero, Coconut), pp. 216-19 **In:** Daniel H. Janzen (ed.), *Costa Rican natural history.* Chicago and London: University of Chicago Press.

Schaaffhausen, Reimar v. (1952). Adlay or Job's tears—a cereal of potentially greater economic importance, *Economic Botany.* **6**: 216-27.

Schaafsma, G., and R. Visser (1980). Nutritional interrelationships between calcium, phosphorus and lactose in rats, *Journal of Nutrition.* **110**: 1101-11.

Schafer, Edward H. (1954). *The Empire of Min.* Rutland, Vt., and Tokyo, Japan: Charles E. Tuttle Company.

_______ (1959). Falconry in T'ang times, *T'oung Pao.* **46**: 293-338.

_______ (1962). Eating turtles in ancient China, *Journal of the American Oriental Society.* **82**: 73-74.

_______ (1963). *The golden peaches of Samarkand: A study of T'ang exotics.* Berkeley and Los Angeles: University of California Press.

_______ (1967). *The vermilion bird: T'ang images of the south.* Berkeley and Los Angeles: University of California Press.

_______ (1970). *Shore of pearls.* Berkeley and Los Angeles: University of California Press.

_______ (1977). T'ang, pp. 85-140 **In:** K. C. Chang (ed.), *Food in Chinese culture.* New Haven and London: Yale University Press.

Schafer, Edward H., and Benjamin E. Wallacker (1957-58). Local tribute products of the T'ang Dynasty, *Journal of Oriental Studies.* **4**: 213-48.

Schery, Robert W. (1972). *Plants for man.* Second Edition. Englewood Cliffs, N.J.: Prentice-Hall.

Schiffman, James R. (1985). Korea is cleaning up its act to get ready for the '88 Olympics, *Wall Street Journal (Western Edition),* Thursday, November 14, pp. 1, 24.

Schmid, O. J. (1967). Für die menschliche Ernährung wichtige Inhaltsstoffe mariner Algen, *Botanica Marina.* **10**: 41-46.

Schnell, R. (1957). *Plantes alimentaires et vie agricole de l'Afrique Noire.* Paris: Éditions Larose.

Schoff, Wilfred H. (trans.) (1912). *The periplus of the Erythraean Sea.* New York: Longmans, Green and Co.

Schram, Louis M. J. (1954). *The Monguors of the Kansu-Tibetan frontier: Their origin, history, and social organization.* Transactions of the American Philosophical Society, N.S., Vol. 44, Part 1. 138 pp.

Schroeder, C. A., and W. A. Fletcher (1967). The Chinese gooseberry (*Actinidia chinensis*) in New Zealand, *Economic Botany.* **21**: 81-92.

Schultze-Motel, Jürgen (1972). Die archäologischen Reste der Ackerbohne, *Vicia faba* L., und die Genese der Art, *Die Kulturpflanze.* **19**: 321-58.

Schumann, Carl (1883). *Kritische Untersuchungen über die Zimtländer,* Ergänzungsheft No. 73, Petermanns Mitteilungen. Gotha: Justus Perthes.

Schwabe, Calvin W. (1979). *Unmentionable cuisine.* Charlottesville: University Press of Virginia.

_______ (1984). *Veterinary medicine and human health.* Third Edition. Baltimore: Williams and Wilkins Company.

Schwimmer, Morton, and David Schwimmer (1955). *The role of algae and plankton in medicine.* New York and London: Grune and Stratton.

Sealy, J. Robert (1958). *A revision of the genus Camellia.* London: Royal Horticultural Society.

"Selenium in the heart of China" (1979). *Lancet.* **2**: 889-90.

Senft, Joseph P. (1980). Protein quality of amaranth grain, pp. 43-47 **In:** *Proceedings of the Second Amaranth Conference.* Emmaus, Pa.: Rodale Press.

Senft, Joseph P., C. S. Kauffman, and N. N. Bailey (1981). *The genus Amaranthus: A comprehensive bibliography.* Emmaus, Pa.: Rodale Press.

Seymour, Thomas Day (1963). *Life in the Homeric Age.* New York: Biblo and Tannen.

Shabad, Theodore (1972). *China's changing map: National and regional development, 1949-71.* Revised Edition. New York and Washington: Praeger.

Shamasastry, R. (ed. and trans.) (1951). *Kautilya's Arthasastra.* Fourth Edition. Mysore: Sri Raghuveer Printing Press.

Shamberger, Raymond J. (1981a). Animal and epidemiological studies relating selenium to heart disease, pp. 391-94 **In:** Julian E. Spallholz, John L. Martin, and Howard E. Ganther (eds.), *Selenium in biology and medicine.* Westport, Conn.: AVI Publishing Company.

_______ (1981b). Selenium in the environment, *Science of the Total Environment.* **17**: 59-74.

_______ (1983). *Biochemistry of selenium.* New York and London: Plenum Press.

Shamberger, Raymond J., and Charles E. Willis (1976). Epidemiological studies on selenium and heart disease, *Federation Proceedings.* **35**: 578.

Shapiro, H. L. (1932). The riverine life of Canton, *Natural History.* **32**: 540-51.

Shaw, Philip E. (1980). Loquat, pp. 479-491 **In:** Steven Nagy and Philip E. Shaw (eds.), *Tropical and subtropical fruits.* Westport, Conn.: AVI Publishing Company.

Shen, Jin-hua (1980). Rice-breeding in China, pp. 9-30 **In:** International Rice Research Institute and the Chinese Academy of Agricultural Sciences, *Rice improvement in China and other Asian countries.* Manila: International Rice Research Institute.

Shen, T. H. (1951). *Agricultural resources of China.* Ithaca: Cornell University Press.

Shiang, Elaine, and Frederick P. Li (1971). The yin-yang (cold-hot) theory of disease, *Journal of the American Medical Association.* **217**: 1108.

Shibasaki, K., and C. W. Hesseltine (1962). Miso fermentation, *Economic Botany.* **16**: 180-95.

Shibata, Shoji (1974). Some chemical studies on ginseng, pp. 69-76 **In:** *Proceedings of International Ginseng Symposium.* Seoul: Central Research Institute, Office of Monopoly, Republic of Korea.

Shih, Kuo-heng (1944). *China enters the machine age: A study of labor in Chinese war industry.* Cambridge: Harvard University Press.

Shih, Shêng-han (1962). *A preliminary survey of the book Ch'i Min Yao Shu, an agricultural encyclopaedia of the 6th century.* Second Edition. Peking: Science Press.

_______ (1982). *On Fan Shêng-Chih Shu, an agriculturalist book of China written by Fan Shêng-Chih in the first century B.C.* Peking: Science Press.

Shinozaki, H. (1978). Discovery of novel actions of kainic acid and related compounds, pp. 17-35 **In:** Edith G. McGeer, John W. Olney, and Patrick L. McGeer (eds.), *Kainic acid as a tool in neurobiology.* New York: Raven Press.

Shirokogoroff, S. M. (1924). *Social organization of the Manchus.* Royal Asiatic Society, North China Branch, Publications, Extra Vol. 3. Shanghai: North China Branch, Royal Asiatic Society.

_______ (1925). *Anthropology of eastern China and Kwangtung Province.* Royal Asiatic Society, North China Branch, Publications, Extra Vol. 4. Shanghai: North China Branch, Royal Asiatic Society.

Shirokogoroff, S. M., and V. B. Appleton (1924). Growth of Chinese, *China Medical Journal.* **38**: 400-413.

Shiu, Iu-nin (1933). Lemons of Kwangtung with a discussion concerning origin, *Lingnan Science Journal.* **12** (Supplement): 271-94.

Shryock, John K. (1966). *The origin and development of the state cult of Confucius.* Reprint of work published originally in 1932. New York: Paragon Book Reprint Corp.

Shurtleff, William, and Akiko Aoyagi (1975). *The book of tofu.* Brookline, Mass.: Autumn Press.

_______ (1976). *The book of miso.* Soquel, Cal.: Autumn Press.

_______ (1977). *The book of kudzu: A culinary and healing guide.* Brookline, Mass.: Autumn Press.

Siegel, Ronald K. (1979). Ginseng abuse syndrome: Problems with the panacea, *Journal of the American Medical Association.* **241**: 1614-15.

"The silk industry in Kwangtung Province" (1929). *Chinese Economic Journal.* **5**: 604-20.

Simmonds, N. W. (1962). *The evolution of the bananas.* London: Longmans, Green and Co.

_______ (1966). *Bananas.* Second Edition. London: Longmans.

_______ (1976a). Bananas: *Musa* (Musaceae), pp. 211-15 **In:** N. W. Simmonds (ed.), *Evolution of crop plants.* London and New York: Longman.

_______ (1976b). Sugarcanes: *Saccharum* (Gramineae-Andropogoneae), pp. 104-8 **In:** N. W. Simmonds (ed.), *Evolution of crop plants.* London and New York: Longman.

Simmonds, P. L. (1885). *The animal food resources of different nations.* London and New York: E. and F. N. Spon.

Simmons, James Stevens, Tom F. Whayne, Gaylord West Anderson, Harold Maclachlan Horack, and collaborators (1944). *Global epidemiology: A geography of disease and sanitation. Vol. 1: India and the Far East, the Pacific area.* Philadelphia: J. B. Lippincott Company.

Simon, James E., Alena F. Chadwick, and Lyle E. Craker (1984). *Herbs: An indexed bibliography 1971-1980.* Hamden, Conn.: Archon Books.

Simonds, Nina (1988). A scholar guides China's food policy, *New York Times (Western Edition)*, January 18, pp. 15-16.

Simoons, Frederick J. (1954). The non-milking area of Africa, *Anthropos.* **49**: 58-66.

_______ (1960). *Northwest Ethiopia: Peoples and economy.* Madison: University of Wisconsin Press.

_______ (1961). *Eat not this flesh.* Madison: University of Wisconsin Press.

_______ (1970). The traditional limits of milking and milk use in southern Asia, *Anthropos.* **65**: 547-93.

_______ (1971). The antiquity of dairying in Asia and Africa, *Geographical Review.* **61**: 431-39.

_______ (1981a). Dogs as human food in Northwest Africa, *Appetite: Journal for Intake Research.* **2**: 253-66.

_______ (1981b). Geographic patterns of primary adult lactose malabsorption: A further interpretation of evidence for the Old World, pp. 23-48 **In:** David M. Paige and Theodore M. Bayless (eds.), *Lactose digestion: Clinical and nutritional implications.* Baltimore and London: Johns Hopkins University Press.

_______ (1982). A geographic approach to senile cataracts: Possible links with milk consumption, lactase activity, and galactose metabolism, *Digestive Diseases and Sciences.* **27**: 257-64.

Simoons, Frederick J., and Elizabeth S. Simoons (1968). *A ceremonial ox of India: The mithan in nature, culture, and history.* Madison, Milwaukee, and London: University of Wisconsin Press.

Simoons, Frederick J., and James A. Baldwin (1982). Breast-feeding of animals by women: Its socio-cultural context and geographic occurrence, *Anthropos.* **77**: 421-48.

Sinclair, Walton B. (1972). *The grapefruit: Its composition, physiology, and products.* Berkeley: Division of Agricultural Sciences, University of California.

Singh, L. B. (1976). Mango: *Mangifera indica* (Anacardiaceae), pp. 7-9 **In:** N. W. Simmonds (ed.), *Evolution of crop plants.* London and New York: Longman.

Sinoda, Osamu (1977). The history of Chinese food and diet, *Progress in Food and Nutrition Science.* **2**: 483-97.

Sivin, Nathan (1980). Science in China's past, pp. 1-29 **In:** Leo Orleans (ed.) *Science in contemporary China.* Stanford: Stanford University Press.

Skinner, Charles M. (1911). *Myths and legends of flowers, trees, fruits, and plants, in all ages and in all climes.* Philadelphia and London: J. B. Lippincott Company.

Skvortzow, B. V. (1920). On some varieties of peanuts grown in China, *Journal of the North China Branch, Royal Asiatic Society (N.S.).* **51**: 142-45.

Smeds, Helmer (1955). The ensete planting culture of eastern Sidamo, Ethiopia, *Acta Geographica.* **13**: 4, 1-39.

Smith, A. K. (1949). *Oriental methods of using soybeans as food, with special attention to fermented products.* U.S. Department of Agriculture, Agricultural Research Administration, Bureau of Agricultural and Industrial Chemistry, AIC-234. Peoria, Illinois: Northern Regional Research Laboratory.

Smith, Allan K., and Sidney J. Circle (1972). *Soybeans: Chemistry and technology. Vol. 1: Proteins.* Westport, Conn.: AVI Publishing Company.

Smith, Arthur H. (1894). *Chinese characteristics.* New York: Fleming H. Revell Company.

Smith, D. G., and E. G. Young (1955). The combined amino acids in several species of marine algae, *Journal of Biological Chemistry.* **217**: 845-53.

Smith, D. Howard (1968). *Chinese religions.* London: Weidenfeld and Nicolson.

Smith, F. Porter (1969). *Chinese materia medica.* Second Revised Edition by Ph. Daven Wei. Taipei: Ku T'ing Book House.

Smith, John (1882). *A dictionary of popular names of the plants which furnish the natural and acquired wants of man, in all matters of domestic and general economy.* London: Macmillan and Company.

Smith, P. M. (1976). Minor crops, pp. 301-24 **In:** N. W. Simmonds (ed.), *Evolution of crop plants.* London and New York: Longman.

Smith, William (192-). *A dictionary of the Bible.* New York: Fleming H. Revell Company.

Smythies, Bertram E. (1981). *The birds of Borneo.* Third Edition. Sabah, Malaysia: The Sabah Society with the Malayan Nature Society.

Snapper, I. (1965). *Chinese lessons to Western medicine.* Second Edition. New York and London: Grune & Stratton.

Snogerup, Sven (1980). The wild forms of the *Brassica oleracea* group (2n=18) and their possible relations to the cultivated ones, pp. 121-32 **In:** S. Tsunoda, K. Hinata, and C. Gómez-Campo (eds.), *Brassica crops and wild allies: Biology and breeding.* Tokyo: Japan Scientific Societies Press.

Somadikarta, S., and R. Anggorodi (1962). Sea turtle eggs as food, *Communicationes Veterinariae.* **6**: 73-75.

Sood, Ranjana, and Charanjit M. Bhat (1974). Changes in ascorbic acid and carotene content of green leafy vegetables on cooking, *Journal of Food Science and Technology.* **11**: 131-33.

Soothill, W. E. (1923). *The three religions of China.* Second Edition. London and New York: Oxford University Press.

Sopher, David E. (1964). Indigenous uses of turmeric (*Curcuma domestica*) in Asia and Oceania, *Anthropos.* **59**: 93-127.

_______ (1965). *The sea nomads.* Memoirs of the National Museum, No. 5. Singapore: National Museum.

Soubeiran, J. Léon, and P. Dabry de Thiersant (1874). *La matière médicale chez les chinois.* Paris: G. Masson.

Sowerby, Arthur de C. (1925). *A naturalist's note-book in China.* Shanghai: North-China Daily News & Herald.

_______ (1931). The edible bird's nest swift, *China Journal.* **14**: 135-37.

_______ (1933). Cricket gourds and culture in China, *China Journal.* **18**: 157-62.

_______ (1940). *Nature in Chinese art.* New York: John Day Company.

Spallholz, Julian E., John L. Martin, and Howard E. Ganther (eds.) (1981). *Selenium in biology and medicine.* Westport, Conn.: AVI Publishing Company.

Spence, Jonathan (1977). Ch'ing, pp. 259-94 **In:** K. C. Chang (ed.), *Food in Chinese culture.* New Haven and London: Yale University Press.

Spencer, Joseph E. (1935). Salt in China, *Geographical Review.* **25**: 353-66.

_______ (1942). The Szechwan village tea house, *Journal of Geography.* **41**: 52-58.

Spices: A survey of the world market (1982). Geneva: International Trade Centre UCDTAD/GATT. 2 vols.

Srivastava, R. P. (1968). Saffron growing in India, *Indian Drugs.* **6**: 17-21.

Stafford, Allen E. (1983). Mango, pp. 399-431 **In:** Harvey T. Chan, Jr. (ed.), *Handbook of tropical foods.* New York and Basel: Marcel Dekker.

Stallard, Louise (1973a). *Cooking Hunan-style.* New York: Drake Publishers.

_______ (1973b). *Cooking Szechuan-style.* New York: Drake Publishers.

"Stanford morality" (1983). *Wall Street Journal (Western Edition),* Monday, July 25, p. 12.

Steele, W. M. (1976). Cowpeas: *Vigna unguiculata* (Leguminosae-Papilionatae), pp. 183-85 **In:** N. W. Simmonds (ed.), *Evolution of crop plants.* London and New York: Longman.

Stein, M. A. (trans.) (1900). *Kalhana's Rajatarangini: A chronicle of the kings of Kashmir.* London: A. Constable. 2 vols.

Steinke, W. E., J. K. Wang, J. R. Carpenter, and R. S. de la Pena (1984). The use of taro silage as animal feed in the Pacific, pp. 225-34 **In:** S. Chandra (ed.), *Edible aroids.* Oxford: Clarendon Press.

Stevenson, Paul H. (1925). Collected anthropometric data on the Chinese, *China Medical Journal.* **39**: 855-98.

Stobart, Tom (1970). *The International Wine and Food Society's guide to herbs, spices and flavourings.* [Newton Abbot] and London: David & Charles.

Stonor, C. R., and Edgar Anderson (1949). Maize among the hill peoples of Assam, *Annals of the Missouri Botanical Garden.* **36**: 355-404.

Stout, A. B. (1933). Gum-jum or gum-tsoy: A food from the flowers of daylilies, *Journal of the New York Botanical Garden.* **34**: 97-100.

_______ (1934). *Daylilies.* New York: Macmillan.

_______ (1935). New developments in daylilies, *Journal of the New York Botanical Garden.* **36**: 205-16.

Stübel, H. (1938). The Yao of the province of Kuangtung, *Monumenta Serica.* **3**: 345-84.

Stutley, Margaret, and James Stutley (1977). *A dictionary of Hinduism.* London and Henley: Routledge and Kegan Paul.

Su, Tsu-fei (1938). Studies on Hunan local food products. III. The vitamin C content of "vegetable water," *Chinese Medical Journal.* **54**: 351-57.

Su, Tsu-fei, and D. Y. Liu (1938). Studies on Hunan local food products. I. The chemical composition and vitamin C content of Hunan citrus fruits, *Chinese Medical Journal.* **54**: 40-50.

Summerfield, John (1984). *Fodor's Beijing (Peking), Guangzhou (Canton), and Shanghai.* New York: Fodor's Travel Guides.

Sun, Marjorie (1985). Stanford president upholds Mosher expulsion, *Science.* **230**: 298-99.

Sung, T. Y. (1940). History of gastro-intestinal diseases in China, *Chinese Medical Journal.* **58**: 324-38.

Sung, Ying-hsing (1966). *T'ien-kung k'ai-wu; Chinese technology in the seventeenth century.* University Park and London: Pennsylvania State University Press.

Suwelo, Ismu Sutanto, Njoman Sumertha Nuitja, and Iman Soetrisno (1981). Marine turtles in Indonesia, pp. 349-51 **In:** Karen A. Bjorndal (ed.), *Biology and conservation of sea turtles. Proceedings of the World Conference on Sea Turtle Conservation, Washington, D.C., 26-30 November 1979.* Washington, D.C.: Smithsonian Institution Press.

Swaminathan, M. S. (1984). Rice, *Scientific American.* **250**: 1, January, 80-93.

Swender, Phillip T., and Alan R. Sandler (1977). Vietnamese and Cambodian orphans 1975-the first look, *Clinical Pediatrics.* **16**: 264-69.

Swingle, Walter T., and Philip C. Reece (1967). The botany of Citrus and its wild relatives, pp. 190-430 **In:** Vol. 1 of Walter Reuther, Herbert John Webber, and Leon Dexter Batchelor (eds.), *The citrus industry.* Revised Edition. Berkeley: Division of Agricultural Sciences, University of California.

Swinhoe, Robert (1870). Catalogue of the mammals of China (south of the River Yang-tsze) and of the Island of Formosa, *Proceedings of the Zoological Society of London.* 1870: 615-53.

_______ (1871-72). Narrative of an exploring visit to Hainan, *Journal of the North China Branch, Royal Asiatic Society (N.S.).* **7**: 41-91.

Symons, Van Jay (1981). *Ch'ing ginseng management: Ch'ing monopolies in microcosm.* Occasional Paper No. 13. Tempe: Center for Asian Studies, Arizona State University.

Synge, Patrick M. (comp.) (1980). *Lilies: A revision of Elwes' Monograph of the Genus Lilium and its supplements.* London: B. T. Batsford Ltd.

Tachard, Guy (1981). *A relation of the voyage to Siam, performed by six Jesuits sent by the French king, to the Indies and China in the year 1685.* Bangkok: White Orchid Press.

Taft, Clarence E. (1965). *Water and algae: World problems.* Chicago: Educational Publishers.

Takakusu, J. (trans.) (1896). *A record of the Buddhist religion as practised in India and the Malay Archipelago (A.D. 671-695), by I-Tsing.* Oxford: Clarendon Press.

Takemoto, T. (1978). Isolation and structural identification of naturally occurring excitory amino acids, pp. 1-15 **In:** Edith G. McGeer, John W. Olney, and Patrick L. McGeer (eds.), *Kainic acid as a tool in neurobiology.* New York: Raven Press.

Tamiya, H. (1960). Role of algae as food, pp. 379-89 **In:** *Proceedings of the Symposium on Algology, 1959.* New Delhi: Indian Council of Agricultural Research.

Tamura, Tadashi (1961). Carp cultivation in Japan, pp. 103-20 **In:** Vol. 1 of Georg Borgstrom (ed.), *Fish as food.* New York and London: Academic Press.

Tan, Swee Poh, and Erica F. Wheeler (1980). Food intakes and growth of young Chinese children in London, *Community Medicine.* **2**: 20-24.

_______ (1983). Concepts relating to health and food held by Chinese women in London, *Ecology of Food and Nutrition.* **13**: 37-49.

Tanaka, Tyozaburo (1928-29). Citrus survey in the Orient region, *California Citrograph.* **14**: 122, 140-41.

_______ (1929a). Remarks on citrus and citrus relatives in China, *Lingnan Science Journal.* **7**: 337-48.

_______ (1929b). The best oranges of the Far East, *Journal of Heredity.* **20**: 37-45.

_______ (1954). *Species problem in Citrus.* Tokyo: Japanese Society for the Promotion of Science.

_______ (1976). *Tanaka's cyclopedia of edible plants of the world.* Tokyo: Keigaku Publishing Company.

Tang, P. S., and L. H. Chang (1939). A calculation of the Chinese rural dietary from crop reports, *Chinese Journal of Physiology.* **14**: 497-508.

Tang, T. H., F. C. Kou, and P. S. Tang (1936). Iodine content of some marine algae of the Shantung coast, *Chinese Journal of Physiology.* **10**: 377-78.

Tanner, Robert D., S. Shahid Hussain, Lindsey A. Hamilton, and Frederick T. Wolf (1979). Kudzu (*Pueraria lobata*): Potential agricultural and industrial resource, *Economic Botany.* **33**: 400-412.

Taylor, Ronald L. (1975). *Butterflies in my stomach.* Santa Barbara, Cal.: Woodbridge Press.

"Tea oil" (1925). *Chinese Economic Monthly.* **2**: No. 9 (June), 12-15.

Teas, J. (1981). The consumption of seaweed as a protective factor in the etiology of breast cancer, *Medical Hypotheses.* **7**: 601-13.

Temple, Richard Carnac (ed.) (1907-36). *The travels of Peter Mundy in Europe and Asia, 1608-1667.* Hakluyt Society Works, Second Series, No. 17, 35, 45-46, 55, 78. London: Hakluyt Society. 6 vols. in 5.

Thiselton-Dyer, T. F. (1889). *The folkore of plants.* New York: D. Appleton and Company.

Thiselton-Dyer, W. T. (1884). Note on the origin of *Cassia lignea, Journal of the Linnean Society.* **20**: 19-24.

Thomas, C. A. (1980). Jackfruit, *Artocarpus heterophyllus* (Moraceae), as source of food and income, *Economic Botany.* **34**: 154-59.

Thomas, P. (1980). *Epics, myths and legends of India.* Fourteenth Edition. Bombay: D. Taraporevala Sons & Co. Private Ltd.

Thompson, C. J. S. (1934). *The mystic mandrake.* London: Rider and Company.

Thompson, K. F. (1976). Cabbages, kales etc.: *Brassica oleracea* (Cruciferae), pp. 49-52 **In:** N. W. Simmonds (ed.), *Evolution of crop plants.* London and New York: Longman.

Thompson, Laurence G. (1975). *Chinese religion: An introduction.* Second Edition. Encino, Cal., and Belmont, Cal.: Dickenson Publishing Company.

Thomson, J. (1875). *The Straits of Malacca, Indo-China, and China.* London: Sampson Low, Marston, Low, & Searle.

Thomson, John Stuart (1909). *The Chinese.* Indianapolis: Bobbs-Merrill Company.

Thrower, L. B., and Yuk-Sim Chan (1980). Gau sun: A cultivated host-parasite combination from China, *Economic Botany*. **34**: 20-26.

Thrum, Thomas G. (1914). Pepeiao—*Hirneola polytricha, Hawaiian Almanac and Annual*. **40**: 201-2.

Thunberg, Carl Peter (1975). *Flora Japonica*. Facsimile of the First Edition of 1784. New York: Oriole Editions.

Tiêu, N.-C. (1928). Notes sur les insectes comestibles au Tonkin, *Bulletin Économique de l'Indochine*. **198**: 735-45.

Tiger, Lionel, and Reinhart Wolf (1985). *China's food: A photographic journey*. New York: Friendly Press.

Tikader, B. K. (1984). *Birds of the Andaman and Nicobar Islands*. Calcutta: Zoological Survey of India.

Ting, S. V. (1983). Citrus fruits, pp. 201-53 **In:** Harvey T. Chan, Jr. (ed.), *Handbook of tropical foods*. New York and Basel: Marcel Dekker.

Tolkowsky, S. (1938). *Hesperides: A history of the culture and use of citrus fruits*. London: John Bale, Sons & Curnow.

Topley, Marjorie (1954). Chinese women's vegetarian houses in Singapore, *Journal of the Malayan Branch, Royal Asiatic Society*. **27**: 51-67.

_______ (1970). Chinese traditional ideas and the treatment of disease: Two examples from Hong Kong, *Man*. **5**: 421-37.

_______ (1974). Cosmic antagonisms: A mother-child syndrome, pp. 233-49 **In:** Arthur P. Wolf (ed.), *Religion and ritual in Chinese society*. Stanford: Stanford University Press.

_______ (1976). Chinese traditional etiology and methods in Hong Kong, pp. 243-65 **In:** Charles Leslie (ed.), *Asian medical systems: A comparative study*. Berkeley: University of California Press.

Tow, Siow Kuan, and Edward O. Moll (1981). Status and conservation of estuarine and sea turtles in West Malaysian waters, pp. 339-47 **In:** Karen A. Bjorndal (ed.), *Biology and conservation of sea turtles. Proceedings of the World Conference on Sea Turtle Conservation, Washington, D.C., 26-30 November 1979*. Washington, D.C.: Smithsonian Institution Press.

Tregear, T. R. (1980). *China: A geographical survey*. New York: Halstead Press.

Trewartha, Glenn T. (1939). Field observations on the Canton Delta of South China, *Economic Geography*. **15**: 1-10.

Tropp, Barbara (1982). *The modern art of Chinese cooking*. New York: William Morrow and Company.

Trowell, H. C., J. N. P. Davies, and R. F. A. Dean (1954). *Kwashiorkor*. London: Edward Arnold.

Tseng, C. K. (1933). Gloiopeltis and other economic seaweeds of Amoy, China, *Lingnan Science Journal*. **12**: 43-63.

_______ (1935). Economic seaweeds of Kwangtung Province, S. China, *Lingnan Science Journal*. **14**: 93-104.

_______ (1944). Utilization of seaweeds, *Scientific Monthly*. **59**: 37-46.

_______ (1981). Marine phycoculture in China, pp. 123-52 **In:** *Proceedings of the Tenth International Seaweed Symposium*. Göteborg, Sweden, August 11-15, 1980. Berlin and New York: Walter de Gruyter.

_______ (ed.) (1983). *Common seaweeds of China*. Peking: Science Press.

Tsien, Tsuen-hsuin (1985). *Paper and printing*, Part 1 of Vol. 5 (Chemistry and chemical technology) of Joseph Needham, *Science and civilisation in China*. Cambridge: Cambridge University Press.

Tun, Li-ch'en (1965). *Annual customs and festivals in Peking*. Second Edition. Translated and annotated by Derk Bodde. Hong Kong: Hong Kong University Press.

Tung, Ch'eng-lang, Kuang-huei Wu, and Ch'uan-yang Wang (1958). The relative incidence of atherosclerotic heart disease in East China and its relationship to cholesterol, *Chinese Medical Journal*. **77**: 596-602.

Uchida, Itaru, and Masaharu Nishiwaki (1981). Sea turtles in the waters adjacent to Japan, pp. 317-19 **In:** Karen A. Bjorndal (ed.), *Biology and conservation of sea turtles. Proceedings of the World Conference of Sea Turtle Conservation, Washington, D.C., 26-30 November 1979*. Washington, D.C.: Smithsonian Institution Press.

Ukers, William H. (1935). *All about tea*. New York: The Tea and Coffee Trade Journal Company. 2 vols.

Underhill, M. M. (1921). *The Hindu religious year*. Calcutta: Association Press.

Underwood, Eric J. (1971). *Trace elements in human and animal nutrition*. Third Edition. New York and London: Academic Press.

_______ (1977). *Trace elements in human and animal nutrition*. Fourth Edition. New York: Academic Press.

United States Central Intelligence Agency (1975). *Peoples Republic of China: Administrative atlas*. Washington, D.C.: Central Intelligence Agency.

Unschuld, Paul U. (1985). *Medicine in China: A history of ideas*. Berkeley, Los Angeles, London: University of California Press.

Uphof, J. C. Th. (1959). *Dictionary of economic plants*. Weinheim (Bergstrasse): H. R. Engelmann (J. Cramer); New York: Hafner Publishing Co.; Codicote: Wheldon and Wesley, Ltd.

Valette, Paul (trans.) (1960). *Apulée: Apologie, Florides*. Paris: Société d'Édition Les belles lettres.

Vallaeys, G. (1948). Le "Coix Lacryma-Jobi," *Bulletin Agricole du Congo Belge*. **39**: 247-304.

Valtonen, M. H. (1984). Raccoon dog, pp. 215-17 **In:** Ian L. Mason (ed.), *Evolution of domesticated animals*. London and New York: Longman.

van Harten, A. M. (1970). Melegueta pepper, *Economic Botany*. **24**: 208-16.

van Veen, A. G. (1965). Fermented and dried seafood products in Southeast Asia, pp. 227-50 **In:** Vol. 3 of Georg Borgstrom (ed.), *Fish as food*. New York and London: Academic Press.

Vavilov, N. I. (1949-50). *The origin, variation, immunity and breeding of cultivated plants.* Translated from the Russian by K. S. Chester. Chronica Botanica, Vol. 13. Waltham, Mass.: Chronica Botanica Co.

Veith, Ilza (1949). *Huang ti nei ching su wên: The Yellow Emperor's classic of internal medicine.* Baltimore: Williams and Wilkins Company.

Verin, Pierre (1986). *The history of civilisation in North Madagascar.* Translated by David Smith. Rotterdam and Boston: A. A. Balkema.

Vermeer, Donald E. (1986). Geophagy in West Africa and the American South, *1986 AAAS Annual Meeting Program. Philadelphia, 25-30 May, 1986*, p. 82.

Vermeer, Donald E., and Ray E. Ferrell, Jr. (1985). Nigerian geophagical clay: A traditional antidiarrheal pharmaceutical, *Science.* **227**: 634-36.

Vickery, Kenton Frank (1980). *Food in early Greece.* Reprint of the 1936 edition. Chicago: Ares Publishers Inc.

Vickery, M. L., and B. Vickery (1979). *Plant products of tropical Africa.* London and Basingstoke: Macmillan.

Vidyârnava, Śrisá Chandra (1918). *Yajnavalkya Smriti, Book 1: The Âchâra adhyâya.* The sacred books of the Hindus, Vol. 21. Allahabad: Pàninî Office.

Vietmeyer, Noel D. (1987). The captivating kiwifruit, *National Geographic.* **171**: 682-88.

Vimokesant, S. L., S. Nakornchai, S. Dhanamitta, and D. M. Hilker (1974). Effect of tea consumption on thiamin status in man, *Nutrition Reports International.* **9**: 371-76.

Vishnu-Mittre (1966). Kaundinyapur plant economy in protohistoric and historic times, *Palaeobotanist.* **15**: 152-56.

_______ (1974). Palaeobotanical evidence in India, pp. 3-30 **In:** Joseph Hutchinson (ed.), *Evolutionary studies in world crops.* London: Cambridge University Press.

_______ (1977). Changing economy in ancient India, pp. 569-88 **In:** Charles A. Reed (ed.), *Origins of agriculture.* The Hague, Paris: Mouton Publishers.

Vishnu-Mittre, and H. P. Gupta (1966). Pollen morphological studies of some primitive varieties of maize (*Zea Mays* L.) with remarks on the history of maize in India, *Palaeobotanist.* **15**: 176-85.

Visser, T. (1969). Tea (*Camellia sinensis* (L.) O. Kuntze), pp. 459-93 **In:** F. P. Ferwerda and F. Wit (eds.), *Outlines of perennial crop breeding in the tropics.* Agricultural University Wageningen, Miscellaneous Papers, No. 4. Wageningen: H. Veenman and Zonen.

_______ (1976). Tea: *Camellia sinensis* (Camelliaceae), pp. 18-20 **In:** N. W. Simmonds (ed.), *Evolution of crop plants.* London and New York: Longman.

W. K. (1961). The state of nutrition in Communist China, *China Quarterly.* **7**: 121-27.

Waley, Arthur (trans.) (1962). *One hundred and seventy Chinese poems.* New Edition. London: Constable and Company.

Walker, Benjamin (1968). *The Hindu world.* New York: Frederick A. Praeger. 2 vols.

Walker, Egbert H. (1942-43). The plants of China and their usefulness to man, pp. 325-62 **In:** *Annual Report of the Board of Regents, Smithsonian Institution.* Washington, D.C.: Smithsonian Institution.

Walker, Egbert H. (1976). *Flora of Okinawa and the southern Ryukyu Islands.* Washington, D.C.: Smithsonian Institution Press.

Wallnöfer, Heinrich, and Anna von Rottauscher (1965). *Chinese folk medicine.* Translated by Marion Palmedo. New York: Bell Publishing Company.

Waln, Nora (1935). *The house of exile.* New York: Blue Ribbon Books.

Wang, C. S. (1934). Birds and Chinese tradition, *China Journal.* **21**: 308-11.

Wang, Cheng-fa (1942). Human nutrition of vitamin C in China. II. Effect of cooking on the total ascorbic acid content of vegetables, *Journal of the Chinese Chemical Society.* **9**: 141-50.

_______ (1943). Human nutrition of vitamin C in China. III. Plasma vitamin C content of Kweiyang residents, *Journal of the Chinese Chemical Society.* **10**: 80-82.

Wang, H., and F. W. Cheng (1935). A survey of iodine content in Chinese common salt, *Journal of the Chinese Chemical Society.* **3**: 343-54.

Wang, Hwa L., and Clifford W. Hesseltine (1970). Sufu and lao-chao, *Journal of Agricultural and Food Chemistry.* **18**: 572-75.

Wang, P. W. (1922). Chinese nuts—walnut, *Report of the Proceedings of the Northern Nut Growers Association.* **13**: 120-21.

Wang, Yongfa, Yan Yongshan, Xu Juijin, Du Ruofu, S. D. Flatz, W. Kühnau, and G. Flatz (1984). Prevalence of primary adult lactose malabsorption in three populations of northern China, *Human Genetics.* **67**: 103-6.

Wang, Zhongshu (1982). *Han civilization.* Translated by K.-C. Chang and collaborators. New Haven and London: Yale University Press.

Ware, James R. (trans.) (1966). *Alchemy, medicine, religion in the China of A.D. 320: The Nei P'ien of Ko Hung (Pao-p'u tzu).* Cambridge, Mass.: M. I. T. Press.

Warmington, E. H. (1974). *The commerce between the Roman Empire and India.* Second Edition. London: Curzon Press; New York: Octagon Books.

Wasson, Valentina Pavlovna, and R. Gordon Wasson (1957). *Mushrooms, Russia, and history.* New York: Pantheon Books. 2 vols.

Watkins, Ray (1976a). Apple and pear: *Malus* and *Pyrus* spp. (Rosaceae), pp. 247-50 **In:** N. W. Simmonds (ed.), *Evolution of crop plants.* London and New York: Longman.

_______ (1976b). Cherry, plum, peach, apricot and almond: *Prunus* spp. (Rosaceae), pp. 242-47 **In:** N. W. Simmonds (ed.), *Evolution of crop plants.* London and New York: Longman.

Watson, Ernest (1923). *The principal articles of Chinese commerce.* China. The Maritime Customs. II.-Special Series, No. 38. Shanghai: Statistical Department of the Inspectorate General of Customs.

Watson, James B. (1964). A previously unreported root crop from the New Guinea highlands, *Ethnology.* **3**: 1-5.

_______ (1965). From hunting to horticulture in the New Guinea Highlands, *Ethnology.* **4**: 295-309.

_______ (1968). Pueraria: Names and traditions of a lesser crop of the Central Highlands, New Guinea, *Ethnology.* **7**: 268-79.

Watson, William (1969). Early cereal cultivation in China, pp. 397-402 **In:** Peter J. Ucko and G. W. Dimbleby (eds.), *The domestication and exploitation of plants and animals.* Chicago: Aldine Publishing Company.

Watt, George (1889-96). *A dictionary of the economic products of India.* London: W. H. Allen and Co. 6 vols.

_______ (1904). *Coix* spp. (Job's tears): A review of all available information, *Agricultural Ledger.* **11**: 189-229.

_______ (1908). *The commercial products of India.* London: John Murray.

Watters, Thomas (1904-5). *On Yuan Chwang's travels in India, 629-645 A.D.* Oriental Translation Fund, New Series, Vol. 14. London: Royal Asiatic Society. 2 vols.

Webber, Herbert John (1943a). History and development of the citrus industry, pp. 1-40 **In:** Vol. 1 of Herbert John Webber and Leon Dexter Batchelor (eds.), *The citrus industry.* Berkeley and Los Angeles: University of California Press.

_______ (1943b). The commercial citrus regions of the world: Their physiographic, climatic, and economic characters, pp. 71-127 **In:** Vol. 1 of Herbert John Webber and Leon Dexter Batchelor (eds.), *The citrus industry.* Berkeley and Los Angeles: University of California Press.

Webber, Herbert John, Walter Reuther, and H. W. Lawton (1967). History and development of the citrus industry, pp. 1-39 **In:** Vol. 1 of Walter Reuther, Herbert John Webber, and Leon Dexter Batchelor (eds.), *The citrus industry.* Revised Edition. Berkeley: Division of Agricultural Sciences, University of California.

Wei, Ling Y. (1976). Theoretical foundation of Chinese medicine: A modern interpretation, *American Journal of Chinese Medicine.* **4**: 355-72.

Weinstein, Stanley (1987). *Buddhism under the T'ang.* Cambridge: Cambridge University Press.

Welch, Holmes (1958). *The parting of the way: Lao Tzu and the Taoist movement.* London: Methuen and Company.

_______ (1967). *The practice of Chinese Buddhism, 1900-1950.* Harvard East Asian Series, No. 26. Cambridge: Harvard University Press.

Wessells, Norman K. (1983). Peking's population policy and the return of infanticide, *Wall Street Journal (Western Edition)*, Monday, August 15, p. 13.

Wester, P. J. (1924). *The food plants of the Philippines.* Department of Agriculture and Natural Resources, Bureau of Agriculture, Bulletin No. 39. Manila: Bureau of Printing.

Wheatley, P. (1959). Geographical notes on some commodities involved in Sung maritime trade, *Journal of the Malayan Branch, Royal Asiatic Society.* **32**: 1-140.

Wheatley, Paul (1961). *The Golden Khersonese: Studies in the historical geography of the Malay Peninsula before A.D. 1500.* Kuala Lumpur: University of Malaya Press.

_______ (1965). A note on the extension of milking practices into Southeast Asia during the first millennium A.D., *Anthropos.* **60**: 577-90.

_______ (1975). Satyanrta in Suvarnadvipa: From reciprocity to redistribution in ancient Southeast Asia, pp. 227-83 **In:** Jeremy A. Sabloff and C. C. Lamberg-Karlovsky (eds.), *Ancient civilization and trade.* Albuquerque: University of New Mexico Press.

Wheeler, Erica, and Swee Poh Tan (1983). From concept to practice: Food behaviour of Chinese immigrants in London, *Ecology of Food and Nutrition.* **13**: 51-57.

White, Philip L., and Nancy Selvey (eds.) (1974). *Let's talk about food.* Acton, Mass.: Publishing Sciences Group, Inc.

Whyte, Robert Orr (1972). *Rural nutrition in China.* Hong Kong and New York: Oxford University Press.

_______ (1974). *Rural nutrition in monsoon Asia.* Kuala Lumpur, London, New York, Melbourne: Oxford University Press.

Wieger, L., and L. Davrout (1981). *Moral tenets and customs in China.* Reprint of 1913 edition. New York and London: Garland Publishing.

Wiens, Herold J. (1954). *China's march toward the tropics: A discussion of the southward penetration of China's culture, peoples, and political control in relation to the non-Han-Chinese peoples of South China and in the perspective of historical and cultural geography.* Hamden, Conn.: Shoe String Press.

Wilcox, A. N. (1958-62). The apple: 1. Systematics, pp. 637-45 **In:** Vol. 6 of H. Kappert and W. Rudorf (eds.), *Handbuch der Pflanzenzüchtung*. Second Edition. Berlin and Hamburg: Paul Parey. 6 vols.

Wilford, Edmund (1962). "Those great coops": Pigeon towers are a notable feature of the Persian landscape, *Architectural Review*. **132**: 443.

Wilkinson, P. B. (1944). Deficiency diseases in Hong Kong, *Lancet*. **2**: 655-58.

Williams, C. A. S. (1941). *Outlines of Chinese symbolism and art motives*. Third Revised Edition. Shanghai: Kelly and Walsh.

Williams, Louis O. (1957). Ginseng, *Economic Botany*. **11**: 344-48.

Williams, S. Wells (1895). *The Middle Kingdom*. New York: Charles Scribner's Sons. 2 vols.

Wilson, C. A. (1973). *Food and drink in Britain from the Stone Age to recent times*. London: Constable.

Wilson, Christine S. (1971). Food beliefs affect nutritional status of Malay fisherfolk, *Journal of Nutrition Education*. **2**: 96-98.

Wilson, Ernest Henry (1913). *A naturalist in western China*. Second Edition. London: Methuen & Co. 2 vols.

_______ (1929). *China: Mother of gardens*. Boston: Stratford Company.

Winfield, Gerald Freeman (1948). *China: The land and the people*. New York: Sloane.

Wing, Elizabeth S. (1983). Domestication and use of animals in the Americas, pp. 21-39 **In:** L. Peel and D. E. Tribe (eds.), *Domestication, conservation and use of animal resources*. Amsterdam: Elsevier.

Winslow, C. E. A., and K. W. Koh (1924). The mortality of the Chinese in the United States, Hawaii, and the Philippines, *American Journal of Hygiene*. **4**: 330-55.

Winters, Harold F. (1963). Ceylon spinach (*Basella rubra*), *Economic Botany*. **17**: 195-99.

Wit, F. (1969). Clove tree (*Eugenia caryophyllus* (Sprengel) Bullock et Harrison), pp. 163-74 **In:** F. P. Ferwerda and F. Wit (eds.), *Outlines of perennial crop breeding in the tropics*. Agricultural University Wageningen, Miscellaneous Papers, No. 4. Wageningen: H. Veenman and Zonen.

_______ (1976). Clove: *Eugenia caryophyllus* (Myrtaceae), pp. 216-18 **In:** N. W. Simmonds (ed.), *Evolution of crop plants*. London and New York: Longman.

Wittwer, Sylvan, Yu Youtai, Sun Han, and Wang Lianzheng (1987). *Feeding a billion: Frontiers of Chinese agriculture*. East Lansing: Michigan State University Press.

Witzell, W. N. (1983). *Synopsis of biological data on the hawksbill turtle*. FAO Fisheries Synopsis, No. 137. Rome: Food and Agriculture Organization of the United Nations.

Wolf, Arthur P. (1970). Chinese kinship and mourning dress, pp. 189-207 **In:** Maurice Freedman (ed.), *Family and kinship in Chinese society*. Stanford: Stanford University Press.

_______ (1974). Gods, ghosts, and ancestors, pp. 131-82 **In:** Arthur P. Wolf (ed.), *Religion and ritual in Chinese society*. Stanford: Stanford University Press.

Wolff, Peter H. (1972). Ethnic differences in alcohol sensitivity, *Science*. **175**: 449-50.

_______ (1973). Vasomotor sensitivity to alcohol in diverse Mongoloid populations, *American Journal of Human Genetics*. **25**: 193-99.

Wong, C. S. (1967). *A cycle of Chinese festivities*. Singapore: Malaysia Publishing House Limited.

Wong, K. Chimin, and Wu Lien-teh (1936). *History of Chinese medicine*. Second Edition. Shanghai: National Quarantine Service.

Wood, Clair G. (1974). Seaweed extracts: A unique ocean resource, *Journal of Chemical Education*. **51**: 449-52.

"World honey crop reports" (1984). *Bee World*. **65**: 58-61.

_______ (1985). *Bee World*. **66**: 131-33.

Wright, W. G. (1913). Edible flowers, *Pure Products*. **9**: 141-42.

Wu, Chentze Hsiang, and Shao-chang Soong (1941). The growth rate of Chinese, *American Journal of Physical Anthropology*. **28**: 357-73.

Xia, Bangmei, and Isabella A. Abbott (1987). Edible seaweeds of China and their place in the Chinese diet, *Economic Botany*. **41**: 341-53.

Xian, Y.-h., and H.-y. Zhang (1983). A new record of Chinese bird - edible-nest swiftlet, *Collocalia fuciphaga* (Thunberg), from Dazhou-Dao, Hainan Island, *Acta Zootaxonomica Sinica*. **8**: 125.

Xin, Wen (1984). The remains of a 7,000-year-old society at Hemudu Village, pp. 8-10 **In:** *Recent discoveries in Chinese archaeology*. Beijing: Foreign Languages Press.

Xing, Runchuan, and Yunming Tang (1984). Archaeological evidence for ancient wine making, pp. 56-58 **In:** *Recent discoveries in Chinese archaeology*. Beijing: Foreign Languages Press.

Xu, Yuan-zhong (trans.) (1982). *Su Dong-po—A new translation*. Hong Kong: Commercial Press, Ltd.

Yabuno, Tomosaburo (1962). Cytotaxonomic studies on the two cultivated species and the wild relatives in the genus *Echinochloa*, *Cytologia*. **27**: 296-305.

_______ (1966). Biosystematic study of the genus *Echinochloa*, *Japanese Journal of Botany*. **19**: 277-323.

Yamagata, N., and T. Yamagata (1972). Iodine content of thyroid glands of normal Japanese, *Journal of Radiation Research*. **13**: 81-89.

Yamamoto, I., T. Nagumo, K. Yagi, H. Tominaga, and M. Aoki (1974). Antitumor effect of seaweeds. I. Antitumor effect of extracts from Saragassum and Laminaria, *Japanese Journal of Experimental Medicine.* **44**: 543-45.

Yamamoto, I., T. Nagumo, M. Fujihara, M. Takahashi, Y. Ando, M. Okada, and K. Kawai (1977). Antitumor effect of seaweeds. II. Fractionation and partial characterization of the polysaccharide with antitumor activity from *Sargassum fulvellum*, *Japanese Journal of Experimental Medicine.* **47**: 133-40.

Yamamoto, T., and M. Ishibashi (1972). The content of trace elements in seaweeds, pp. 511-14 **In:** *Proceedings of the Seventh International Seaweed Symposium.* Sapporo, Japan, August 8-12, 1971. Tokyo: University of Tokyo Press.

Yang, C. K. (1959a). *A Chinese village in early Communist transition.* Cambridge, Mass: Technology Press, M.I.T.

_______ (1959b). *The Chinese family in the Communist revolution.* Cambridge, Mass: Technology Press, M.I.T.

_______ (1961). *Religion in Chinese society.* Berkeley, Los Angeles, London: University of California Press.

Yang, Chi-shih, and K. K. Huang (1934). An outbreak of pellagra in Nanking. A report of 30 cases, *Chinese Medical Journal.* **48**: 701-23.

Yang, E. F. (1938). Effect of method of preparation on the vitamin B1 content of rice, *Journal of the Chinese Chemical Society.* **6**: 55-65.

Yang, E. F., and B. E. Read (1940). Vitamin B1 content of Chinese plant beriberi remedies, *Chinese Journal of Physiology.* **15**: 9-18.

Yang, E. F., and M. Y. Dju (1939). The total and available iron in vegetable foods, *Chinese Journal of Physiology.* **14**: 479-88.

Yang, En-fu (1958). Nutritional evaluation of high cereal diets, pp. 83-92 **In:** Josué de Castro (ed.), *Hunger and food.* London: World Federation of Scientific Workers.

Yang, Guangqi, Junshi Chen, Zhimei Wen, Keyou Ge, Lianzhen Zhu, Xuecun Chen, and Xiaoshu Chen (1984). The role of selenium in Keshan disease, *Advances in Nutritional Research.* **6**: 203-31.

Yang, Martin C. (1945). *A Chinese village: Taitou, Shantung Province.* New York: Columbia University Press.

Yee, Rhoda (1975). *Chinese village cookbook.* San Francisco: Yerba Buena Press.

Yeh, Samuel D. J., and Bacon F. Chow (1973). Nutrition, pp. 215-39 **In:** Joseph R. Quinn (ed.), *Medicine and public health in the People's Republic of China.* DHEW Publication No. (N.I.H.) 73-67. Bethesda, Md.: Fogarty International Center, National Institutes of Health.

Yen, Douglas E. (1971). Construction of the hypothesis for distribution of the sweet potato, pp. 328-42 **In:** Carroll L. Riley, J. Charles Kelley, Campbell W. Pennington, and Robert L. Rands (eds.), *Man across the sea.* Austin and London: University of Texas Press.

_______ (1974). *The sweet potato and Oceania: An essay in ethnobotany.* Bernice P. Bishop Museum, Bulletin 236. Honolulu: Bishop Museum Press.

_______ (1976). Sweet potato: *Ipomoea batatas* (Convolvulaceae), pp. 42-45 **In:** N. W. Simmonds (ed.), *Evolution of crop plants.* London and New York: Longman.

Yetts, W. Perceval (1912). *Symbolism in Chinese art.* [Leiden: E. J. Brill].

Yeung, David L., Lilian W. Y. Cheung, and Jean H. Sabry (1973). The hot-cold food concept in Chinese culture and its application in a Canadian-Chinese community, *Journal of the Canadian Dietetic Association.* **34**: 197-203, 209.

Young, Robert A. (1954). Flavor qualities of some edible Oriental bamboos, *Economic Botany.* **8**: 377-86.

Yü, T. F. (1940). Effect of cooking on the vitamin C content of vegetables, *Chinese Medical Journal.* **57**: 523-34.

Yü, Te-tsun, and Chen-lung Yen (1956). Study on the Chinese species of genus Malus Miller, *Acta Phytotaxonomica Sinica.* **5**: 77-110 [in Chinese with English summary].

Yu, Wei-han (1982). A study of nutritional and bio-geochemical factors in the occurrence and development of Keshan disease, *Japanese Circulation Journal.* **46**: 1201-7.

Yü, Ying-shih (1967). *Trade and expansion in Han China.* Berkeley and Los Angeles: University of California Press.

_______ (1977). Han China, pp. 55-83 **In:** K. C. Chang (ed.), *Food in Chinese culture.* New Haven and London: Yale University Press.

Yule, Henry (trans. and ed.) (1866). *Cathay and the way thither; Being a collection of medieval notices of China.* Hakluyt Society Works, Nos. 36 and 37. London: Hakluyt Society. 2 vols.

_______ (1921). *The book of Ser Marco Polo.* Third Edition, revised by Henri Cordier. London: John Murray. 2 vols.

Yule, Henry, and A. C. Burnell (1903). *Hobson-Jobson; A glossary of colloquial Anglo-Indian words and phrases, and of kindred terms, etymological, historical, geographical and discursive.* New Edition edited by William Crooke. London: John Murray.

Zaneveld, J. S. (1959). The utilization of marine algae in tropical South and East Asia, *Economic Botany.* **13**: 89-131.

Zee, S. Y., and L. H. Hui (1981). *Hong Kong food plants.* Hong Kong: Urban Council.

Zeuner, Frederick E. (1963). *A history of domesticated animals.* London: Hutchinson.

Zeven, A. C., and J. M. J. de Wet (1982). *Dictionary of cultivated plants and their regions of diversity.* Wageningen: Centre for Agricultural Publishing and Documentation.

Zeven, A. C., and P. M. Zhukovsky (1975). *Dictionary of cultivated plants and their centres of diversity.* Wageningen: Centre for Agricultural Publishing and Documentation.

Zhang, Wen-cai (1981a). Development and outlook of citrus industry in China, pp. 987-90 **In:** Vol. 2 of the *Proceedings of the International Society of Citriculture.* Tokyo, Japan, November 9-12, 1981.

_______ (1981b). Thirty years achievements in citrus varietal improvement in China, pp. 51-55 **In:** Vol. 1 of the *Proceedings of the International Society of Citriculture.* Tokyo, Japan, November 9-12, 1981.

Zhang, Zhong-ge (1984). Cormorant, pp. 367-68 **In:** Ian L. Mason (ed.), *Evolution of domesticated animals.* London and New York: Longman.

Zheng, Guoqing (1987). Mao Tai Town, *China Pictorial:* 40-41.

Zohary, Daniel, and Maria Hopf (1988). *Domestication of plants in the Old World.* Oxford: Clarendon Press.

Zohary, Daniel, and Martin Hopf (1973). Domestication of pulses in the Old World, *Science.* **182**: 887-94.

Zohary, Daniel, and Pinhas Spiegel-Roy (1975). Beginnings of fruit growing in the Old World, *Science.* **187**: 319-27.

Zürcher, E. (1959). *The Buddhist conquest of China.* Sinica Leidensia, Vol. 11, Nos. 1 and 2. Leiden: E. J. Brill.

INDEX

A

abacá 234, 260
abortion 496
acorn 284, 290
adzuki bean 83, 84
agar-agar 181, 183, 187, 188, 189, 431
ahimsa 35, 41, 307, 466. *See also* Buddhism
alchemy 32, 41
alcoholic beverages 37, 82, 89, 91, 123, 128, 220, 257, 264, 277-78, 289, 448-54
  avoidance 36, 303, 363, 496
  in hot-cold classification 24, 454, 501
alcoholism 452-53, 465
alfalfa 149
algae, microscopic 182, 189
almond, Chinese 269-70, 288
amaranth 96, 141-43, 173, 286
American Chinese cooking, origins 57
Amorphophallus spp. 116, 119-21
An-yang 31, 39, 48
ancestors, food offered to 28, 29-30, 31, 40, 41, 113, 170, 362
Anderson, Eugene N. and Marja L. xxii, 457, 459
anise 387, 389-90
anise, star 47, 52, 370, 386-89, 418, 515
ant eggs 321
antelope horn 510
anthelmintics 148, 187, 246
aphrodisiacs 146, 157, 167, 173, 241, 261, 262, 308, 323, 324, 333-34, 359, 366, 408, 410, 420, 424, 436
apple 241, 243-44
apricot 217, 220, 258, 269-70, 288
arbor-vitae, Oriental 284
arenga palm 128, 131
arrowhead 101, 110-11, 129, 345
arrowroot 128
artichoke, Chinese 116, 119
asparagus 171-72, 177
asparagus lettuce 147-48, 173
aster 150

B

B vitamins. *See* niacin; riboflavin; thiamine
bacon 297
baking 24
bamboo and bamboo shoots 49-50, 53, 162-64, 176
banana and plantain 232-34, 260, 514
banyan tree 250, 263
barium poisoning 511
barley 49, 64, 73-74, 89-90, 97, 378, 465
barnyard millet (*Echinochloa* sp.) 80-81
basil, holy (*Ocimum sanctum*) 406-8, 422-23
basil, sweet 370, 406-8, 423
bean curd. *See* tofu
beans and peas 63-65, 68, 70-71, 74-75, 83-84. *See also* individual beans and peas
  preparation and consumption 86-88, 92
  toxicity 92-93
bear's gall 433, 434, 437
bear's paw 48, 433-34
bêche-de-mer 25, 358, 434-36, 437
beechnut 284, 290
beef 47, 301, 303-5, 331, 490
beer 89, 448, 452, 454
beet 116, 122, 417
beetles 322, 333
beriberi 445-46, 466, 479, 509
berries 192
betel nut 290, 418
    pepper 385, 418
bird's nest 25, 427-31, 436
birth procedures and risks 496-97
birth rates 506, 512
birth weights 497, 511
bitter cucumber (*Momordica charantia*) 151, 152, 153-54
black-eyed pea 83, 98
blood as food 50, 297, 300, 306
blueberry 255
bottle gourd 151, 152, 174, 285
box-thorn, Chinese 150
breads and their ingredients 15, 47, 89, 123, 158, 231, 234, 276, 291
brick tea 444, 446-47, 463, 464
broad bean 74-75, 92, 98
broccoli 135
Buck, John Lossing xxii
buckwheat 78, 91, 92, 95
Buddhism 23-24, 139, 160, 201, 202, 237, 238, 247, 273, 288, 307, 354, 358, 363, 389, 396, 399, 452. *See also* vegetarian dishes; vegetarianism
  animal liberation 345, 354, 358
  cuisine 37
  dairy product use 363, 456, 458, 459, 460, 466
  lotus 112-13
  onion and garlic avoidance 157-58, 175, 363
  sacred trees 249-50, 263
bull fighting 331
butter 455, 456, 459, 460, 461, 466, 478
butter tea 443, 446

C

cabbage 134-36, 172, 492
  dietary importance 135-36
  forms 47, 134
  nutritional value 136, 478, 479, 484

calamondin 204-5
calcium intake and deficiency 487-90, 510
calories 63, 475-76
camel 305
camel thorn 377-78
Cantonese cuisine 43, 44-45, 54-57, 70, 381, 513-15
dishes mentioned 55-56, 60, 84, 85, 86, 106, 108, 112, 123, 125, 130, 141, 148, 154, 166, 172, 177, 182, 183, 198, 210, 222, 224, 234, 254, 268, 270, 279, 282, 301, 304, 308, 312, 317, 318-20, 321, 323, 328, 308, 331, 358, 364, 367, 371, 376, 377, 406, 454, 461
carambola 235-37, 259, 261
cardamom 370, 400-402, 419, 420, 421, 515
carp 48, 338, 343-46, 349, 358
carrot 101, 108-9, 478
cashew nut 284
cassia. *See* cinnamon
castor oil 286
cat 305, 315-17, 333
caterpillars 322, 323-24
cattail 174
cattle, common 302, 303, 304. *See also* beef
cattle cult 304, 331
cattle sacrifice 30, 31, 34
cauliflower 135
cedar, Chinese 150
celery 166-67
cereals 15, 63-70, 72-74, 75-83, 84-86, 88-92, 476, 478, 482, 491. *See also* individual cereals
Ch'ang-an 252
Chang Ch'ien (Han traveller) 75, 252, 262, 279
Chang, Kwang-chih xxii
chayote 256
cheese 54, 457, 460, 461, 466, 478, 482
cheese, soybean 87, 88
Chen-chiang 49, 50
cherry 217, 222
chestnut 49, 267-69, 284, 288
ch'i concept 18-22
chicken 50, 53, 295, 297-99, 300, 301, 330, 375
chili oil 382, 386
chili pepper 24, 52-53, 56, 60, 385-86, 418, 478, 496
Ch'ing court dishes 58, 109
Ch'ing Ming 40
chive 47, 157, 158, 174
Chlorella spp. 183
cholera 148, 471, 472
cholesterol 349
chop suey 56
chow mein 56
Christianity 35, 245, 407-8, 452, 460, 499
cicada 334
cilantro 405, 406
cinnamon 52, 369, 370, 390-96, 419, 515
citron 200-203, 259
citrus ant 194, 255-56
citrus fruit 52, 194-206, 484
clam 344
clonorchiasis 471, 472, 473
clove 370, 402-3, 421
cockfighting 298
cocklebur 150
cocoa 440
coconut 276-78, 289, 290
coffee 439-40
common bean (*Phaseolus vulgaris*) 84, 92
common millet (*Panicum miliaceum*) 72-73, 91, 448, 450
confinement of mothers 500-503
Confucianism 26, 32-33, 34-35, 113, 273, 304, 452
Confucius 18, 28, 32, 33, 34, 303, 344, 372, 452
congee 24, 38, 82, 85, 88, 89, 109, 130, 288, 363, 445
cooking oil. *See* lard; oil seeds and cooking oil
Cordyceps fungus 176, 322, 323-24, 334
coriander 52, 370, 405-6, 422
cormorant, fishing with 340-42, 357
cottonseed oil 286
court dishes 123, 130, 245, 271, 278
crab 25, 56, 339, 350, 358
crickets and cricket fighting 325
cucumber 152
cucurbits 150-56
cuisines of China 43-60. *See also* Peking cuisine; Shanghai cuisine; Szechwan cuisine; Cantonese cuisine; etc.
curry 54, 414

D

date 194, 247-48
date-plum (*Diospyros lotus*) 216-17
death rates 506, 512
deep-frying 24
deer 34, 326
deer antler 25
deities, food offerings to 26-28, 30
delicacies 25, 112, 114, 273, 282, 333, 362, 364, 427-36
Digenia simplex 190
dill 370
dim sum 56-57, 60, 297
diphtheria 473
distilled liquors 264, 449-50, 451, 464, 465
divination 257, 298, 330, 352, 358, 361
dog and dogflesh 24, 25, 305, 309-15, 332, 333
donkey 305
Dragon Boat Festival 174, 254, 265
Dragon Gate 345
duck 50, 53, 60, 297-98, 299-301, 330, 331, 332, 345, 416
duckfighting 298
dumplings 47, 49, 88, 163, 183, 371, 381
dysentery 24, 216, 221, 246, 471

E

eel 106, 268, 338, 339, 344, 358
eggplant 169-70
eggs 375, 476
  avoidance 36
  fowl 361-65
  preparation 363-64, 489
  turtle 365-67
elm 150
ensete (*Ensete ventricosum*) 234-35
Euryale ferox 285-86
evodia 370, 383

F

fagara 52, 369, 370, 376, 382-83, 385
famine and famine food 45, 83, 102, 124, 129, 149, 173, 174, 290, 498, 507
fan-ts'ai principle 15-18, 43
favism 92, 98
feasts, festivals, and festival foods 15, 18, 25, 26, 30, 34, 35, 40, 49, 53, 85-86, 88-89, 113, 139, 146-47, 163, 174, 193, 200, 209, 218, 254, 260, 265, 270-71, 293-95, 301, 304, 308, 350, 361-62, 377, 380, 433, 436, 465. *See also* New Year
Fen-chiu 451
fenugreek 414-15
fermented foods 56, 335, 374, 446, 480. *See also* alcoholic beverages; vinegar
  dairy 459, 466
  fish 348-49
  soybean 88, 97
  tea 443, 446, 447, 448
fertility enhancement and symbols 106, 112, 115, 146-47, 245, 247, 249, 256, 307, 332, 361-62, 363, 372, 408
fertilizers 181, 183, 184, 189, 331, 332, 345, 507. *See also* night soil
fig 248-50, 262-63
finger millet 72, 80, 94
fish 337-50, 479, 489
  cultivation 343-46, 358
  iodine and goiter 491, 492-93
  methods of catching 339-43
  pastes and sauces 348-49
  perceptions as food 15, 25, 36
  preference for live fish 18, 339-40
  preparation and consumption 36, 357, 472, 508
  production and importance 47, 48, 49, 50, 53, 56, 337-38, 343, 344, 345, 357
  variety consumed 338-39
five, lists of 32, 41, 44, 58, 157-58, 310, 381, 406
five-spice powder 282, 382, 388, 392, 403
flax 286, 291
flesh, importance in diet 293-95, 329, 476
flesh avoidances 36
  beef 303-5, 331
  dog 310
  duck and chicken 301
  pigeon 307
  pork 303
  turtle 353-54, 357
floating gardens 172
flour 92, 121, 231, 276
  almond 270, 288
  amaranth 142
  barley 89
  buckwheat 91, 92
  chestnut 268
  Job's tears 82
  lotus seed 115
  maize 91, 97
  potato 125
  rice 446, 503
  sago 128, 129
  sorghum 91
  sweet potato 123
  wheat 88, 97
flowers in Chinese cuisine 37, 114, 196, 246, 263, 374, 423, 451, 452
fluke, liver 111, 276, 345
fluorine 349, 511
fly 322, 323
food avoidances 25-26, 31, 39. *See also* flesh avoidances; milk and dairy products; pregnancy and nursing; vegetarianism
food offerings,
  at time of burial 28-29, 40
  to dead ancestors 28, 29-31, 40, 41
  to deities 26-28, 30
fox 326
foxtail millet 72-73, 91, 100, 374, 448, 450
freshness of food 18, 301, 339-40
frog 326-28, 335, 344, 357
frugality in eating 18
fruit 24, 56, 60, 191-285, 374, 452, 476, 484, 501. *See also* individual fruits
fruit drinks 440
Fukien cuisine 50-51
  dishes mentioned 50, 59, 123, 162, 210
funerals 28-29, 34, 175-76, 218, 247, 453
fungi. *See* mushrooms and other fungi

G

galangal 370, 372-74
gambling 196
game 49, 54, 326
games, drinking 465
garland chrysanthemum 150
garlic 25, 47, 52, 157-59, 174-75, 374, 415
gau sun 165-66

gecko 335
geophagy 102, 129
ginger 24, 43, 47, 52, 56, 370-72, 373, 375, 415, 463
ginkgo nut 49, 271-74, 288, 289
ginseng 25, 370, 410-14, 424, 425
goat 34, 301-3
goiter 184, 491-93, 510-11
golden needles 52, 409-10, 423
goldfish 338, 343
good fortune, symbolism 39, 40, 196, 202, 218, 237, 345, 359, 361-62, 367, 407-8, 416
goose 305-6, 331, 332, 345
gooseberry, Chinese 228-29, 259
goosefoot (*Chenopodium album*) 150, 173-74
gourds. *See* cucurbits
grape 251-54, 264, 374, 449, 451-52, 454
grapefruit 199
grasshoppers and locusts 324-25
great burdock 116-17, 128
green turtle (*Chelonia mydas*) 352, 354, 355, 359
greens, vegetable 84, 105, 108, 109, 114, 122, 134, 136-50, 177, 479, 482, 484, 488, 490, 511
grilling 24
guava 254, 255, 264
guinea fowl 326

H

Hainan
- description of 7, 10-11
- cuisine mentioned 57

hair vegetable (*Nostoc commune*) 182
Hakka cuisine 57
ham 50, 54, 297
Hangchow 44, 49
hare 326
haw 227-28, 259
hawksbill turtle 354, 355, 358, 359
hazelnut 270-71
headcheese 54
hemp seed oil 286
hickory 283
Hinduism 31, 112, 201, 237, 384, 396, 399-400, 404
- spread of dairying 458
- food avoidances 157, 160, 174-75, 301, 359, 363
- holy basil of 406-8, 422-23
- sacred fig trees 249-50

Ho-mu-tu 3, 69, 94, 174, 274
Hokkien cuisine 57
Honan cuisine 48
- dishes mentioned 48, 59

honey 25, 234, 376-77, 416-17
honeybee 376, 416
hookworm and ancylostomiasis 470, 471, 472-73, 490, 508
Hopeh cuisine and dishes 45, 47, 58
horse bean. *See* broad bean
horsemeat 305
hot-cold foods 22-24, 38, 52, 157, 495-96, 500-501, 508. *See also* yin-yang
- foods classified as 24, 25, 84, 108, 109, 115, 134, 140, 145, 148, 149, 154, 158, 166, 167, 174, 186, 187, 193, 206, 209-10, 224, 234, 245, 254, 255, 276, 278, 297, 300, 304, 308, 312, 317, 333, 349-50, 372, 375, 385, 410, 439, 454, 463, 464

Hsi Wang Mu (Queen Mother of the West) 217
human head (*Dracontomelum dao*) 229, 260
Hunan cuisine and dishes 44, 54, 108, 177, 222, 234, 371, 386
Hupeh, dishes mentioned 86
hyacinth bean 83, 92

I

Ichang orange 205
immortality,
- search for 31-32, 41, 159, 174
- symbols of 217-18, 219, 334, 350

infanticide 497-500, 511
infants 496-506
- feeding 460, 462, 503, 512
- growth rates 504-6, 512
- illnesses and mortality 93, 490-91, 496-97, 504, 509
- weaning 502, 512

influenza 24
insects as human food 321-26, 333, 334-35
iodine, iodine deficiency, and goiter 184, 349, 491-93, 510-11
iron and iron deficiency 25, 490-91, 510

J

jackfruit 239-40, 261
Jainism 31, 175
Jerusalem artichoke 128
jicama 131
Job's tears 81-83, 95, 96
jujube 25, 39, 91, 223-25, 258, 374, 484, 501, 503
Jurchen 172, 256

K

K'ai-feng 44, 48
kale, Chinese 134, 136, 172, 479, 492
keratomalacia 478
Keshan disease 493-94
ketchup 56
Kitchen God 26-28, 40
kiwifruit 228-29, 259
kohlrabi 135, 172
konnyaku 121

koumiss 449, 451, 459, 461
kudzu 116, 117-19, 128, 130
kumquat 204, 374
Kwangtung, description 3-11
kwashiorkor 478

L

lactose malabsorption 457-58, 462, 466
lamb and mutton 26, 47, 58
Lao Tzu 33, 41
lard 286, 293, 297
lathyrism 93
leatherback turtle 355, 359
leek 47, 157, 158, 479
lemon and lime 203, 496
leppett tea 445-46
leprosy 473, 508
lettuce 146-48, 173
Li (people of Hainan) 257, 289, 465
licorice 226, 370, 414
lily and lily bulbs 162, 423, 501. *See also* golden needles
lima bean 84, 92
lime (compound) 91-92, 364, 418, 432, 492
lime (fruit). *See* lemon and lime
linden 263
ling nut 111, 274-76, 289
linseed oil 286
litchi 206-10, 256, 484, 515
liver 482, 490
locusts 324-25
longan 210-12, 484, 515
longevity, Taoist quest for 32
  foods contributing to 103, 115, 159, 334, 377, 429, 444, 447, 462-63
  symbols of 217-18, 221, 284, 350, 352, 354, 359, 361-62
loquat 212-13
lotus (*Nelumbium speciosum*) 49, 112-15, 130, 345
Lo-yang 48
luffa 151, 152, 153

M

Ma Po bean curd 53, 60
magic 25-26, 129, 196, 200, 217-18, 220, 247, 249, 297, 357, 403, 410, 413, 495, 503
maize 47, 64, 76-77
  consumption of, and pellagra 482-83
  nutritional value 73, 91-92, 97, 129
  preparation and use 86, 91-92, 97
Malabar nightshade 143-44
malaria 2-3, 473, 474
mallow 137-38, 172
maltose 378
Manchus 45, 47-48, 303
mandrake 410, 413-14, 423-24
mango 238-39
mangosteen 194
manioc 125-26, 128, 129
manna 377-78
Mao t'ai 451
Mao Tse-tung 31
Marx, Karl 24, 30
mat bean 83
mead 449
measles 25, 473
mei (*Prunus mume*) 258
melon (*Cucumis melo*) 49, 244-45, 285
Mencius 28, 32-33, 41, 427, 433
midwifery 496-97, 511
milk and dairy products 36, 41, 54, 363, 443, 454-63, 465, 466, 476, 478, 482, 512. *See also* soybean milk
  avoidance 454-60, 466.
milk, human 29, 462-63, 465, 512
milk, walnut 290
milkfish 344
millet 47, 72, 73, 91, 94, 402, 451, 479. *See also* individual millets
mint 415
Mioga ginger 370, 372
Mongols 45, 303, 459
monkey 329, 436
monkey head fungus 48, 59
moon cakes 88-89, 282
Moslems 35, 412, 490
  cuisine and restaurants 45, 47, 303
  food and drink 31, 54, 303-4, 307, 309, 310, 359, 366, 439-40, 452, 461
  plant introductions 389-90, 398, 439-40
mouse 333, 436
MSG 364, 415
mugwort 174
mulberry 230-32, 260
mule 305
mullet 344, 346, 358
mung bean 25, 83, 84, 92, 96, 482, 510
muscovy duck 305, 308-9
mushrooms and other fungi 48, 49, 53, 54, 56, 159-62, 165-66, 175-76, 501
mussel 339, 358
mustard
  greens 134, 479, 484
  seed 370
mutton and lamb 24, 58, 301, 303, 304-5
mutton wine 58
myrobalans 237-38, 261

N

Nan-Yüeh 3, 39, 54, 207, 208-9, 210, 352, 395, 419, 479
 contributions to China's food scene 82, 225, 233, 379, 383, 384, 402, 403, 404, 514-15
Needham, Joseph 450
New Year 196, 200, 218, 219, 284
 foods and gifts 27-28, 35, 86, 109, 115, 143, 146, 182, 202-3, 371, 382
New Zealand spinach (*Tetragonia expansa*) 150
niacin, deficiency and pellagra 482-83, 510
night blindness 509
night soil 159, 194, 296, 345, 470-72, 508
nine, lucky number 40
noodles 15, 18, 47, 49, 50, 56, 59, 88, 92, 97, 130, 139, 409, 445
 buckwheat 91, 92
 cellophane 92
 egg 363
 maize 91
 potato and sweet potato 123, 125
 rice 84, 96
 symbolism 362
 wheat 88, 89
Nosu (Black Lolo) 79
nuoc mam 348-49
nursing, of animals by women 465
nutmeg 370, 402, 404, 421-22, 514
nuts and seeds 24, 25, 49, 56, 267-91. *See also* individual nuts and seeds

O

oat 51, 79, 91, 95
octopus 25, 358
oil seeds and cooking oil 47, 88, 136, 138, 251, 283, 286-87, 290, 291, 297, 374, 464, 493
okra 170-71
olive (*Olea europea*) 225, 250-51, 263
olive, Chinese (*Canarium* spp.) 194, 225-26, 514
onions 26, 47, 50, 52, 156-59, 174, 370
 avoidance of 36, 174-75, 363
opium 508
oracle bones 282, 352, 358
oranges 18, 51, 194, 195-98, 205
osmanthus 263
osteomalacia 489, 490, 497
osteoporosis 510
otter, fishing with 342-43
owl 436
oyster and oyster farming 339, 344, 346-48, 358
oyster sauce 56, 304, 346, 347, 348, 358

P

padi-straw mushroom 515
pagoda tree 290
pancakes 47, 50-51, 58, 77, 88, 222, 301
pangolin 25, 329, 436
papaya 254, 255
paragonimiasis 472, 473
parasites, intestinal 158-59, 276, 296, 313, 345, 470-73, 484, 490, 507, 508, 510
parsley 150
parsnip 116, 122
partridge 326
pea 74-75, 94, 479
peach 217-20, 257, 258, 374
peacock 326
peanut 24, 280-82, 286, 290, 375, 492, 501
pear 241-43, 261, 262
pearl millet 72, 80, 492
Pearl River Delta 7-10
peepul tree 249-50, 263
Peking cuisine 44, 45-48, 303
 dishes mentioned 37, 47, 58, 59, 123, 125, 130, 139, 189, 244, 279, 282, 335, 371, 375, 409, 454
Peking duck 47, 58, 130, 222, 300, 301, 331, 381
Peking dust 268-69, 288
pellagra 482
pepper (*Piper* spp.) 52, 382, 383-85, 418. *See also* fagara; chili pepper
pepper, green 169, 478, 484
pepper, Melegueta 420
perilla 150, 286
persimmon 25, 213-16, 256-57, 479
phallic symbols 201, 249
pheasant 326, 335
phoenix tree 282
phosphorus 71, 107, 119, 348, 349, 489, 510
pickled foods 52, 114, 115, 119, 136, 148, 151, 157, 158, 167, 169, 221, 222, 236, 238, 244, 271, 321, 371, 375, 464, 484, 489
pig 34, 50, 124, 181, 295-97, 303, 329-30, 345
pig's feet 375, 489
pigeon 298, 305, 306-8, 331, 332
pili nut 258
pimpon 282-83
pine seeds 283-84
pineapple 254, 496
pistachio nut 279-80
plague 473, 474
plantain (*Musa paradisiaca*). *See* banana and plantain
plantain (*Plantago major*) 150, 174
plum 217, 221-22
pneumonia 473
Polo, Marco 49, 242, 252, 260, 291, 419
pomegranate 245-47, 262
popping of seeds 86, 96, 142
poppy oil 286
population and population increase 506, 512

pork 47, 50, 53, 54, 297, 303, 330, 490
porridge 15, 47, 82, 91, 92, 96, 97, 109, 130, 142, 270, 276, 490. *See also* congee
potato, white 124-25, 128, 131, 451, 484, 510
prawn 344
pregnancy and nursing 410, 478, 490-91, 494-96, 500-504, 510, 511
  foods prescribed and proscribed 24-26, 39, 40, 134, 154, 186, 193, 209-10, 224, 245, 297, 300, 304-5, 308, 350, 357, 372, 375, 410, 429, 454, 464, 489, 495-96, 500-502
protein and protein deficiency 63, 476-78
pu. *See* strengthening foods
p'u-erh tea 447
puma 333
pummelo 198-200
pumpkin seed 285
purslane 150

Q

quailfighting 298, 330
quince 226, 227, 258, 259

R

rabbit 34, 305, 331
raccoon dog (*Nyctereutes procyonoides*) 25, 332
radish 101, 107, 484
raisin tree (*Hovenia dulcis*) 194, 230, 260, 377
rapeseed oil 274, 286
rat 320-21, 333, 436
red bean (*Phaseolus calcaratus*) 83
red color, symbolism 25, 39, 109, 196, 209-10, 245, 274, 287, 361-62, 367, 402
red cooking and red-cooked dishes 50, 59, 111, 169, 183, 297, 308, 313, 320, 339, 363, 381, 389, 435
red-in-snow 376, 415
restaurants 15, 18, 36, 37, 44, 45, 47, 53, 54, 55-56, 57, 88, 155, 301, 303, 311-12, 314, 315, 316, 317, 318-20, 339, 346, 431, 436, 445, 459, 461. *See also* tea houses
rhinoceros horn 3, 25, 39, 510
riboflavin, intake and deficiency 481-82
rice 66-70. *See also* wild rice; wine, rice
  advantages over other crops 70
  dietary importance and production 47, 49-50, 51, 64, 70, 93
  domestication and early history 66-69, 94, 100-101, 104, 129
  forms and their characteristics 69-70, 94
  glutinous 69, 85-86, 94, 446, 448-49, 450
  in hot-cold classification 24
  in nutrition 70, 73, 490
  polished, and beriberi 480-81, 509
  preparation and consumption 18, 84-86, 96, 374, 451, 481
rice-paper plant (*Aralia papyrifera*) 284
rice-wine lees 453-54
rickets 489, 510
ritual use of plants, animals, foods, and drinks 41, 163, 196, 200-201, 202-3, 217-19, 249, 250, 298, 310, 361-62, 396, 397, 406-8, 444-45, 448, 453, 458, 465. *See also* sacrifice
roasting 24
role of food in Chinese life 13-15, 38
roots and tubers 99-131, 476, 482
  aquatic forms 99-100, 101, 109-15
round worm and ascariasis 471, 472, 473, 491, 508
Russian tea-trade 446-47, 464
rye 78-79, 95

S

sacred trees 272, 273
sacred turtles 353-54
sacrifice 28, 30, 40, 113, 202-3, 334, 448, 459
  animal 30, 31, 32, 34-35, 39, 41, 303, 310
safflower 286-87, 291, 398
saffron 397-99, 420
sago 128-29, 131
salad 105, 111, 112, 114, 130, 139, 145, 146-48, 153, 169, 171, 183, 508
salamander, giant 329, 436
salmon 339
salt 375-76, 415, 510
  iodine content of 492, 510-11
salt tea 443, 459, 463
salted foods 52, 163, 164, 216, 230, 268, 285, 297, 300, 357, 363-64, 365, 375-76, 416, 501
sapodilla 254
saté 97
sausage 297, 330
scallop 358
scarlet fever 473
schistosomiasis 471, 472, 508
scurvy 483-84
seaweed 179-90
  cultivation 179, 182-83
  importance in East Asia 179-80
  in hot-cold classification 24, 186, 187
  medicinal uses 180, 181, 182, 184, 186-88, 190, 491
  nutritional value 184-86, 189, 190
  per capita consumption 180, 188
  trade 180, 181, 182, 187, 189
  types eaten 180-81, 188, 189
  ways of preparing as food 183, 189
selenium deficiency and Keshan disease 493-94
senile cataracts 462
sesame 25, 47, 52, 53, 286, 291, 382, 489
Shang Ti (supreme deity) 30

Shanghai cuisine 44, 48-50
  dishes mentioned 49, 50, 59, 84, 125, 130, 139, 177, 212, 244, 268, 309, 328, 331, 406, 454
Shantung
  cuisine 45-47
  dishes mentioned 58, 130
  grapes and wine 451
shark's fin 25, 431-32
sheep 34, 301-3
shepherd's purse (*Capsella bursa-pastoris*) 150
Shou Hsing (God of Longevity) 218
shrimp 25, 338, 339, 349, 350, 357, 358
silkworm 305, 322, 326, 334-35
smallpox 25, 473, 474-75
smoked foods 52, 53, 60, 198, 215, 216, 221, 223, 225, 297, 434, 435, 445
snacks 38, 56-57, 60, 86, 87, 92, 115, 123, 127, 142, 157, 221, 268, 270, 276, 284, 297, 300, 322, 363, 364, 380, 381, 445
snail 328-29
snake 25, 317-20, 333, 339, 357, 436
snake gourd 152
social status and eating 26
sons, desire for 28, 29, 41, 115, 247, 269, 367, 410, 494, 498, 500, 511
Soochow 49, 59
sorghum 47, 73, 75-76, 95
  preparation and consumption 86, 90-91, 378, 417, 450, 451, 465
soup 50, 56, 60, 376, 406, 409, 441, 445, 484, 493
  bird's nest 429, 431
  ingredients and dishes mentioned 82, 87, 91, 92, 105, 108, 109, 112, 114, 115, 119, 125, 130, 136, 137, 139, 141, 143, 145, 146, 148, 153, 161, 162, 163, 167, 172, 181, 182, 183, 189, 221, 223, 224, 273, 282, 285, 297, 300, 308, 317, 323-24, 328, 331, 333, 338, 350, 363, 413, 435, 463
  mother's brew 501
  sweet 130, 218, 220, 270, 279, 282, 371
  shark's fin 431, 432
  snake 319-20
  turtle 352, 356-57, 367
  winter melon 154-56
sow thistle 149-50, 257
soybean 51, 70-71, 94. *See also* tofu
  flour 88
  in nutrition 71, 478, 479, 482, 490, 510
  milk 87, 440, 462, 502, 512
  oil 88, 286
  paste 47
  preparation and consumption 71, 86-88, 94, 97, 98
soy eggs 363
soy sauce 25, 43, 47, 56, 88, 369
Spanish fly 333-34
spices and flavorings 24, 369-425. *See also* individual spices and flavorings
spinach 139-40, 478, 488, 490
spinach beet 144-45
sprouts, bean and pea 92, 98, 484, 510
squash 151, 152, 285
squid 25
starch, types of 91, 111, 112, 114, 118, 119, 125, 126, 128, 129
steaming 24
stewing 24
stinkbug 322, 323
stir-frying 24, 56, 60, 286, 484
strawberry 229, 255, 260
strawberry, Chinese 229-30
strengthening foods 25, 38, 39, 86, 109, 115, 158, 161, 162, 167, 206, 212, 217, 224, 237, 254, 271, 297, 298, 300, 308, 312, 317, 320, 323-24, 329, 332, 333, 352, 357, 358, 372, 375, 385, 403, 410, 413, 414, 424, 429, 432, 433, 434, 435, 436, 495-96, 501
suckerfish or remora 355
sugar 56, 164, 285, 378, 379-81, 417, 476, 514
  barley 89
  beet 380, 417
  cane 51, 256, 378-80, 381, 417, 451, 514
  palm 128, 378
  sorghum 378, 417
sugar apple (*Annona squamosa*) 254
sugar pea 92
sunflower
  seed 285
  oil 286
sushi 472, 508
sweet flag 174
sweet potato 59, 102, 122-24, 129, 130, 131, 451, 465, 478, 482
sweet-and-sour dishes 45, 48, 58-60, 210, 254, 381, 489
sweet-and-sour sauce 47, 374
sweets 82, 85-86, 92, 114, 115, 125, 198, 256, 270, 276, 278, 279, 291, 377, 378, 379, 380, 381, 388, 417, 420, 436. *See also* sugar; soup, sweet
Swiss chard 144-45
sword bean 83, 92
symbolism of plants and animals 106, 221, 269, 284, 298, 334. *See also* red color, symbolism; yellow color, symbolism
  apple 243-44
  basil 407-8
  carp 345
  cherry 222
  citron 200-201, 202
  cow 304
  egg 361-62
  fig 248, 249-50
  goose, swallow, and mandarin duck 332
  lotus 112-13, 115
  orange 195, 196
  peach 217-18, 219
  persimmon 256
  pigeon 307
  pomegranate 245, 247, 262
  tea 444-45
  turtle 350, 352, 354, 358, 359

syphilis 475
Szechwan cuisine 44, 47, 51-54, 56, 59-60
dishes mentioned 53, 60, 86, 98, 107, 123, 125, 130, 161, 169, 177, 198, 270, 279, 282, 328, 331, 374, 375, 382, 386, 434
Szechwan duck 53, 60, 130, 301

T

tablet, ancestral 350, 361
Taiwan 234
Tan (people) 353-54
Taoism 18-22, 26-28, 31-32, 41, 113, 129, 174, 273, 459
foods to achieve longevity and immortality 218, 377, 462-63
lists of five forbidden foods 158, 406
possible invention of liquor distillation 450
vegetarianism and the struggle with Buddhism 31-35
views of alcohol consumption 452
tapeworm 313
taro 99, 100, 101, 104-7, 111, 128, 129
tea 364, 374, 440-48, 463, 464
as medicine 441
flavoring and scenting of 196, 263, 371, 377, 382, 388, 423
inhibits thiamine 481, 509
leaves consumed as a vegetable 441, 445-46, 464
role in reducing liquor consumption 453
teas, alternative
apple leaf 243, 262
apricot kernel 288
haw 227, 259
herbal 441, 464
litchi leaf 210
mint 415
myrobalan 261
orange 198
seaweed 181
tea eggs 363, 445
tea houses 60, 112, 445, 464
tea oil 286, 291
tea smoking 53, 60, 301, 445
Teochiu cooking 107
tetanus 473, 475, 496-97
thiamine, deficiency and beriberi 479-81, 509
Tibet and Tibetans 89, 91, 95, 363, 446, 461
tiger 25, 39, 436
tiger bones 510
tilapia 344, 358
toddy 278, 449
tofu 53, 60, 87-88, 96, 97
tomato 167-69
Torreya grandis nut 283
tortoise shell 3, 355-56, 359
toxicity of Chinese foods and spices 78, 111, 118, 121, 124, 167, 258, 264, 270, 283, 290, 389, 509
beans and peas 87, 92-93
ginkgo nut 274, 289
taro 106
yam 102, 105
trifoliate orange 205
tsamba 89, 97, 446
tuberculosis 473, 475
tuna 339
turkey 305, 309
turmeric 257, 370, 396-97
turnip 101, 109, 129, 492
turtle 350-59, 365-67
fresh water 352-54, 358
in divination 352, 358
preparation and consumption 356-57
sanctity 41, 353-54, 358
sea 352, 354-57, 359, 365-67
symbolism 350, 358, 359
typhoid fever 471
typhus 473

V

vegetables 15, 24, 476, 501. *See also* individual vegetables
vegetarian cuisine 37
vegetarian dishes 27, 37, 41, 45, 86, 89, 161, 182, 189, 268, 274, 409
vegetarianism 31-37, 41, 71, 301, 303-4, 310, 312, 363, 446
velvet bean 83
venereal diseases 475
vinegar 47, 50, 52, 56, 85, 128, 204, 220, 369, 374-75, 415, 489, 496
vitamin A and deficiency conditions 478-79, 509
vitamin C deficiency and scurvy 483-87
vitamin D, calcium, rickets, and osteomalacia 487-90

W

walnut 25, 39, 278-79, 284, 290
wampee 205-6
water, drinking 441, 463, 471, 492
water, in hot-cold classification of foods 24
water, sweetened 417
water buffalo 301-2, 303
waterbug, giant 322, 323
water chestnut (*Eleocharis dulcis*) 101, 111-12, 274, 345, 515
water cress 24, 145
water drop-wort 166, 167, 177
watermelon 244, 245, 284-85
water shield 148-49
water spinach (*Ipomoea aquatica*) 140-41, 172, 478
wax gourd 151
wax-producing insect (*Ericerus pe-la*) 334

weasel 436
weddings 34, 115, 219, 227, 237, 243, 247, 269, 274, 332, 361, 453
wet nursing 463, 502-3, 512
wheat 73-74, 374, 465, 479
dietary importance 47, 49, 51, 64-65
processing and consumption 88-89, 97
whiskey 501
wild rice (*Zizania* spp.) 165-66, 176, 177
wine 58, 158, 237, 246, 255, 277, 278, 320, 371, 382, 413, 423, 445, 448, 449, 450
grape 264, 449, 451-52, 454
millet 448-49, 450
rice 43, 47, 49, 50, 52, 56, 85, 308, 372, 375, 448, 449, 450-51, 501
terminology 37, 448
wine lees 50, 374
wine rice 86, 96
wine sediment paste 50, 59
winter melon (*Benincasa hispida*) 154-56
Worcestershire sauce 56
words, their sound as influencing food use 26, 182, 269
worms 321, 329

X

Xanthoceras sorbifolia 283
xerophthalmia 478, 509

Y

yam (*Dioscorea* spp.) 99, 100, 101, 102-4, 111, 128, 129
yam bean, American 131
Yangchow 49
yard-long bean 84
yellow color, symbolism 196, 377, 396, 399
yin-yang 22-24, 32, 298, 450, 473. *See also* hot-cold foods
symbol 37, 38
yogurt 54, 461
Yüeh people 3, 5, 265, 278
Yunnan 54, 243, 297, 461

Z

zedoary 370, 399-400